Geotechnical Engineering Congress 1991

Volume II

Proceedings of the Congress
sponsored by the
Geotechnical Engineering Division
of the
American Society of Civil Engineers

in cooperation with the
University of Colorado
U.S. Bureau of Reclamation

Hosted by the
Colorado Section, ASCE

Boulder, Colorado
June 10-12, 1991

Edited by Francis G. McLean
DeWayne A. Campbell and David W. Harris

Geotechnical Special Publication No. 27

Published by the
American Society of Civil Engineers
345 East 47th Street
New York, New York 10017-2398

ABSTRACT

This proceedings, *Geotechnical Engineering Congress 1991,* consists of papers presented at the Geotechnical Congress held in Boulder, Colorado from June 10-12, 1991. The congress provides a forum for examining current state-of-the-art in geotechnical engineering and forecasting future developments. Some of the topics covered are: 1) In situ testing; 2) computer applications; 3) developments in geotechnical instrumentation; 4) earth reinforcing systems; 5) hazardous waste materials; 6) numerical modeling; and 7) load and resistance factor design. This proceedings provides the engineer with the opportunity to assimilate new information and cope with the joint challenge and trauma that change inevitably brings.

Library of Congress Cataloging-in-Publication Data

Geotechnical Engineering Congress (1991: Boulder, Colo.)
Geotechnical Engineering Congress 1991: proceedings of the Congress/sponsored by the Geotechnical Engineering Division of the American Society of Civil Engineers in cooperation with the University of Colorado [and] U.S. Bureau of Reclamation: hosted by the Colorado Section, ASCE, Boulder, Colorado, June 10-12, 1991; edited by Francis G. McLean, DeWayne A. Campbell, David W. Harris.
p. cm.—(Geotechnical special publication; no. 27)
Includes indexes.
ISBN 0-87262-806-X
1. Engineering geology—Congresses. 2. Foundations—Congresses. 3. Soil mechanics—Congresses. I. McLean, Francis G. II. Campbell, DeWayne A. (DeWayne Allen) III. Harris, David W. (David William), 1951- . IV. American Society of Civil Engineers. Geotechnical Engineering Division. V. University of Colorado (Boulder campus) VI. United States. Bureau of Reclamation. VII. American Society of Civil Engineers. Colorado Section. VIII. Title. IX. Series.
TA703.5.G47 1991
624.1′51—dc20 91-17797
CIP

Library of Congress Catalog Card No: 91-17797
ISBN 0-87262-806-X
Manufactured in the United States of America.

CONTENTS

Session 1
Plenary Session I
Moderator: Stein Sture

Session 2A
IN SITU TESTING AND MEASUREMENT: 1
Session Organizer and Moderator: Peter K. Robertson
Co-Moderator: Ronald Y. S. Pak

Session 2B
COMPUTER APPLICATION: DATA BASES
Session Organizer and Moderator: Curt Lamprecht

Session 2C
GEOTECHNICAL INSTRUMENTATION—NEW DEVELOPMENTS
Session Organizer and Moderator: John Dunnicliff
Co-Moderator: Jay N. Stateler

Session 3A
IN SITU TESTING AND MEASUREMENT: 2
Session Organizer and Moderator: Peter K. Robertson
Co-Moderator: Harold W. Olsen

Session 3B
COMPUTER APPLICATION: EXPERT SYSTEMS
Session Organizer and Moderator: Kenneth Strzepek

Session 4C
RELIABILITY APPLICATIONS: 1
Session Organizers and Moderators: Herbert H. Einstein and Eric Vanmarcke

Session 5
Plenary Session II
Moderator: Charles C. Ladd

Session 6A
FOUNDATIONS: 2
Session Organizers and Moderators: Michael W. O'Neill and George G. Goble

Session 6B
COMPUTER APPLICATION: LABORATORY TEST CONTROL, DATA ACQUISITION, AND REPORT PREPARATION
Session Organizers and Moderators: Tom L. Brandon and Paul C. Knodel

Session 6C
REALIABILITY APPLICATIONS: 2
Session Organizers and Moderators: Herbert H. Einstein and Eric Vanmarcke

Session 7B
POSTER SESSION

Session 8A
EARTH REINFORCING SYSTEMS
Session Organizer and Moderator: Robert D. Holtz

Session 8B
COMPUTER APPLICATION: COMPUTER AIDED DESIGN AND DRAWING
Session Organizer and Moderator: Stephen G. Wright

Session 8C
LABORATORY TESTING AND EVALUATION OF RESULTS
Session Organizer and Moderator: David Elton

Session 9A
HAZARDOUS WASTES: 1
Session Organizer and Moderator: Ronald E. Smith

Session 9B
NUMERICAL MODELING: 1
Session Organizers and Moderators: John T. Christian and Stein Sture

Session 10A
HAZARDOUS WASTES: 2
Session Organizer and Moderator: Ronald E. Smith

Session 10B
NUMERICAL MODELING: 2
Session Organizers and Moderators: John T. Christian and Peter Bosscher

Session 10C
LOAD AND RESISTANCE FACTOR DESIGN
Session Organizer and Moderator: J. M. Duncan

A DATA ACQUISITION SYSTEM FOR DEFINING GEOLOGIC STRUCTURE AND SITE CHARACTERIZATION

T. Ted Miyake[1], A.M. ASCE, Edward A. Steiner[2], and Craig C. Lippus[3]

ABSTRACT: A data acquisition system for defining geologic structure and site characterization for use in stability analyses is described. Traditional investigation methods were combined with new technologies including borehole video camera and an improved natural gamma ray tool. Principles of natural gamma logging are discussed. The gamma ray tool provided valuable data with significant time and cost savings. Case studies from three southern California sites are presented. Use of the system at these sites allowed unequivocal lithologic correlations and more accurate landslide and slope stability analyses.

INTRODUCTION

The conclusions and recommendations which result from a geotechnical investigation depend a great deal upon the degree of accuracy with which a particular site can be characterized. This is especially true in slope stability studies in which the orientation of bedding planes (strata), continuity of structure, and previous mass movements must be adequately defined to make accurate engineering judgements. Not only must data be reliable and substantial, it must be obtained economically. Complex geologic forces and processes, however, can result in a heterogeneous mixture of earth materials making data acquisition and interpretation of geology difficult, time consuming, and costly. This is the case in many areas of tectonically active southern California.

This paper describes a system developed to obtain geotechnical data for analyzing local and regional geologic structure. The data was used to assess the gross stability of several sites along landslide prone areas of the southern California coast. A combination of factors made subsurface investigation and site characterization difficult. Near-surface bedrock was often hard and highly fractured. Exploration depths of up to 500 feet (152 m) were required. The bedrock consisted of rhythmically alternating beds of similar appearance. In some instances secondary mineralization and other processes had

---

1 - Project Engineer, Leighton and Associates, Inc., 2121 Alton Parkway, Irvine, California 92714

2 - Senior Project Geologist, Leighton and Associates, Inc., 2121 Alton Parkway, Irvine, California 92714

3 - Senior Staff Geophysicist, Leighton and Associates, Inc., 2121 Alton Parkway, Irvine, California 92714

significantly altered the appearance of the rock. These characteristics made unambiguous visual correlations between borings very difficult.

SYSTEM COMPONENTS

A synthesis of traditional investigation methods and improved technologies produced an effective data acquisition system which overcame the difficulties described. In addition to customary field mapping, aerial photograph analysis, and geologic trenching, the system includes: 1) large-diameter bucket-auger borings, 2) small-diameter diamond core borings, 3) borehole television camera, and 4) high-sensitivity downhole gamma-ray logging.

Large-Diameter Borings. Large-diameter bucket-auger borings, usually 2 foot (.6 m) in diameter, are a standard investigative tool in southern California, especially for landslide investigations. Because the ground water table is relatively deep and the bedrock is usually soft, depths up to 150 feet (46 m) can generally be achieved. These borings are logged by lowering a geologist down the hole. The geologic data obtained from these borings is highly reliable.

Conditions and project requirements limited the use of bucket-auger borings at the sites described. Difficult drilling in the hard rock, caving in fractured zones, and poor air quality inside the borings were some of problems encountered which limited the boring depths. Although drilled to less than optimal depths, these borings provided valuable supplemental data and verified the reliability of data gathered by other means.

Small-Diameter Core Borings. Wireline diamond core borings were used to investigate the site conditions below the limited depths of the bucket-auger borings. Although other drilling methods can be much faster and less expensive, the desire for good quality samples for testing and visual classification ruled out their use. Conventional core data such as rock quality designation (RQD) and percent of core recovery also proved useful. Core was 1 7/8 inch (47.6 mm) diameter.

Water with very small amounts of non-bentonitic polymer additives was used as the drilling fluid. This resulted in very clean boreholes with little or no mud caking. Other investigators in the region using air rotary methods had reported boreholes walls being coated with significant amounts of in-place bentonitic material below the water table (5).

Borehole Television Camera. The core borings were logged with a 2.15 inch (55 mm) diameter downhole television camera to obtain bedding and fracture orientations and observe ground water conditions. Equipped with a variable intensity light source, this camera can be operated underwater to depths of more than 10,000 feet (3050 m) and temperatures as high as 93° C without special thermal protection. It is gyrocompensated with continuous depth display and direction indicator.

Both black-and-white and color cameras were tested during these investigations. Although both cameras had the same high resolution picture, the borehole features appeared sharper with the black-and-white images. This was attributed to the lack of color contrast between different rock types. Images from the camera were recorded on VHS and large format (3/4 inch) video tape. Figure 1 is a typical image reproduced from the borehole video.

The strike of bedding and fractures were calculated by constructing a line perpendicular to a line connecting the high and low points on a bed as observed on the video display. Dip angles were estimated by trigonometric calculations using the difference in depth between the high side and low side of a bed. A professional grade video tape player with a frame by frame advance feature allowed interpolation of depths to one-hundredth of a foot (.25 mm). Since the borings for this project were continuously cored, dip angles were also measured directly on the core. Bedding plane and fracture orientations obtained with the camera corresponded well with nearby bucket-auger borings and surface

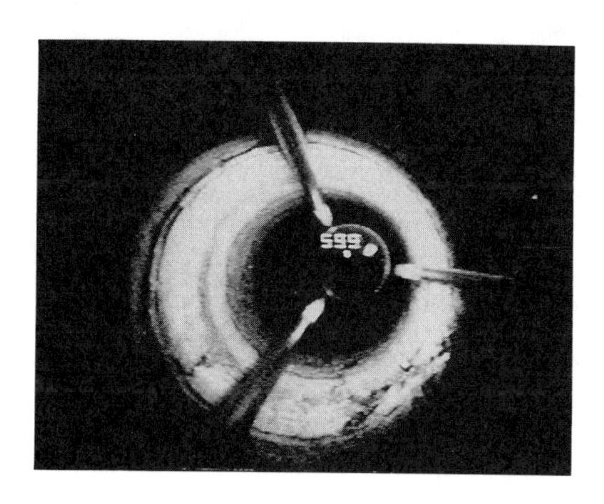

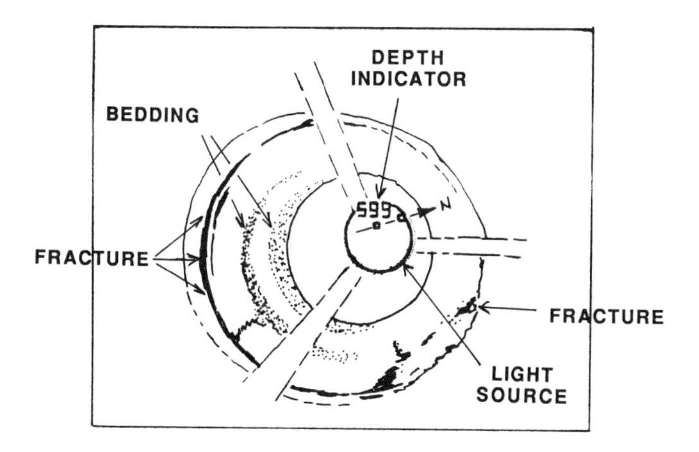

Figure 1. Typical Borehole Television Camera Image
(reproduced from video tape)

exposures. This data is believed to be as reliable (if not more) than those obtained with conventional oriented coring methods.

Borehole Geophysics. Because of the site characteristics that limited the use of more conventional exploration methods, a variety of borehole geophysical methods were considered. Most of these methods were developed in the petroleum industry, but have been used more frequently in recent years for geotechnical investigations of nuclear waste disposal sites and for ground water studies (3). Prior studies in the study area had reported the application of some of these techniques for site investigations (4). However, use of these tools for geotechnical engineering applications have generally been limited by unfamiliarity with the methods and their relatively high costs.

Geophysical tools initially considered for the investigations were those which could yield information on lithologic type, orientation of bedding, and allow correlation of identical strata between borings. They included: electrical resistivity, sonic log, natural gamma log, gamma-gamma log, neutron, and dipmeter. Nearly all of these tools required standing water in uncased borings to be effective. Unfortunately, the fractured nature of the bedrock resulted in most borings having complete loss of circulation and precluded retention of sufficient water for their use. Gamma-gamma and neutron logging tools with radioactive sources were considered inappropriate due to unstable hole conditions and concern about losing such hazardous material underground. (In fact, during the same investigations a sonic tool was lost in a boring due to caving.) By a process of elimination gamma-ray and the television camera were chosen to provide deep structural data, to correlate the deeper lithologic units and to supplement the bucket-auger and core data.

NATURAL GAMMA LOGGING

Gamma Radioactivity. Naturally occurring radioactivity is present in nearly all rocks and minerals due to the presence of uranium or thorium decay series, or potassium-40. Uranium and thorium have higher gamma radiation but are relatively rare compared to potassium-40. The relative gamma emissions between them, therefore, is roughly equal (4). These materials are generally present in trace quantities in many minerals, especially potassium rich feldspars and micas. Therefore, naturally occurring radioactivity is generally most abundant in igneous rocks of granitic composition or in sedimentary rocks derived from these types of rocks.

Uranium and thorium are concentrated by adsorption and ion exchange in clays. Fine-grained detrital rocks such as claystones, shales and siltstones generally have relatively high gamma ray radiation as do volcanic rocks such as tuff (ash rocks) of granitic composition. Volcanic rocks such as basalt, diabase, and tuffs of similar chemical composition have a low gamma ray signal.

Sandstones, limestones and other sedimentary rocks that generally do not contain minerals rich in potassium have low gamma ray signals. Uranium is water soluble, however, and can be transported and redeposited, thereby changing the gamma ray signal of the host sedimentary rock. Uranium deposits in the Four Corners region are often deposited in sandstones and these sandstones and often result in anomalously high gamma ray signals (5). There are exceptions and each individual site should be cored to correlate the gamma ray log to the rock in the core.

Gamma Ray Logging. The key premise to the value of gamma ray logging is that particular lithological units have unique gamma ray "signatures". These signatures can be used to correlate rock units from boring to boring. In our investigations, a conventional gamma tool was initially used but the resulting data lacked resolution and proved difficult to analyze. This prompted the geophysical contractor, Barbour Well Surveying, to rebuild the tool, increasing the volume of the thallium-doped sodium iodide crystal to four times that of the original scintillation crystal. The tool was fitted into a thin stainless steel housing rather than the thicker aluminum which is usually used. The stainless steel housing was strong, yet thin enough to permit the tool to be lowered down the center of the drill stem. Side walls of the housing were 1/8 inch (3.2 mm) thick.

The logging speed was also reduced to 5 feet (1.5 m) per minute which, when combined with a time constant of 1/2 second, resulted in a sampling interval of 1/2 inch (13 mm). Logging with the new, more sensitive tool yielded significantly better results as shown in Figure 2. It shows two gamma logs of the same test boring; one is the log from a conventional tool, the other is from the new tool.

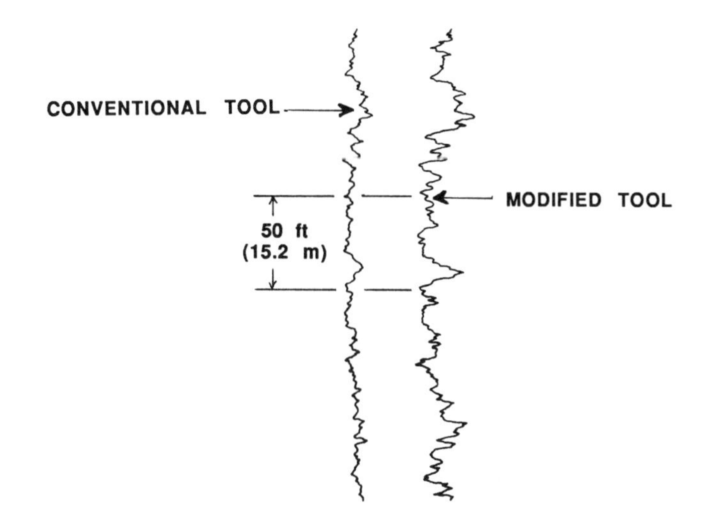

Figure 2. Comparison of Gamma Ray Logs Between Conventional and Modified Tool

Gamma ray logs with the new tool were highly repeatable and had better resolution. The quality of the logs allowed unequivocal correlations of gamma ray signatures between borings (Figures 3 and 4).

Specific gamma ray signatures were also correlated with specific lithologic units by comparing the gamma ray logs with the corresponding core. High gamma ray signals were found to indicate lithologies such as siltstone, organic-rich claystone, and shale. Low gamma ray returns indicated basalts, dolostones, sandstones and bentonitic material. Although many of the rocks were considerably altered by secondary mineralization, it did not significantly change the gamma radiation characteristics of the rock (6). Primarily, dolomite associated with regional volcanism has been emplaced in pore and fracture space. Although, dolomite has low gamma levels, it did not affect the primary gamma signature of many of the rock.

A major difference noted in our investigations was that the tuffs and bentonites had low gamma signals. Both tuff and clay are often expected to have high gamma ray signals. The volcanic rocks in this area, however, have the chemical composition of basalt. Local bentonitic clays are generally derived from the weathering of the tuffs. The bentonitic clays, such as the distinctive Portuguese Tuff unit, have a distinctively low gamma ray signature.

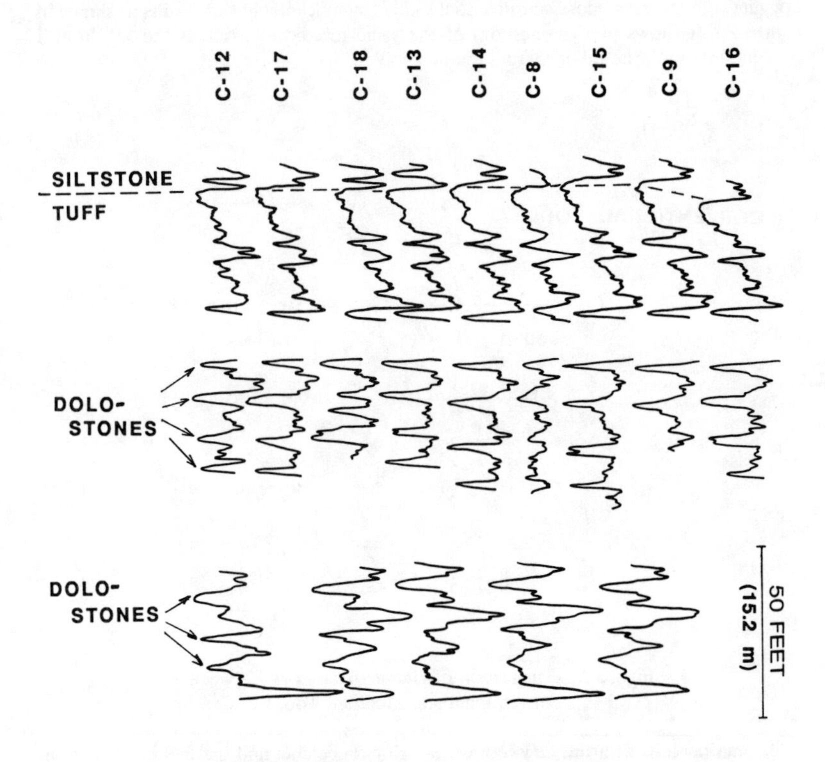

Figure 3. Detailed Gamma Logs Segments from Nine Borings
Illustrating Unequivocal Correlations of
Gamma Ray Signatures
(Adapted from Steiner and Lippus, 1990)

Potassium has been concentrated in fecal pellets of marine organisms (Ehlig; oral communication, 1989). The pellets are in turn concentrated in quiet water, low energy environments such as deep basins. The highest gamma ray levels in the areas of our studies occur in organic rich claystones which were deposited in this type of environment.

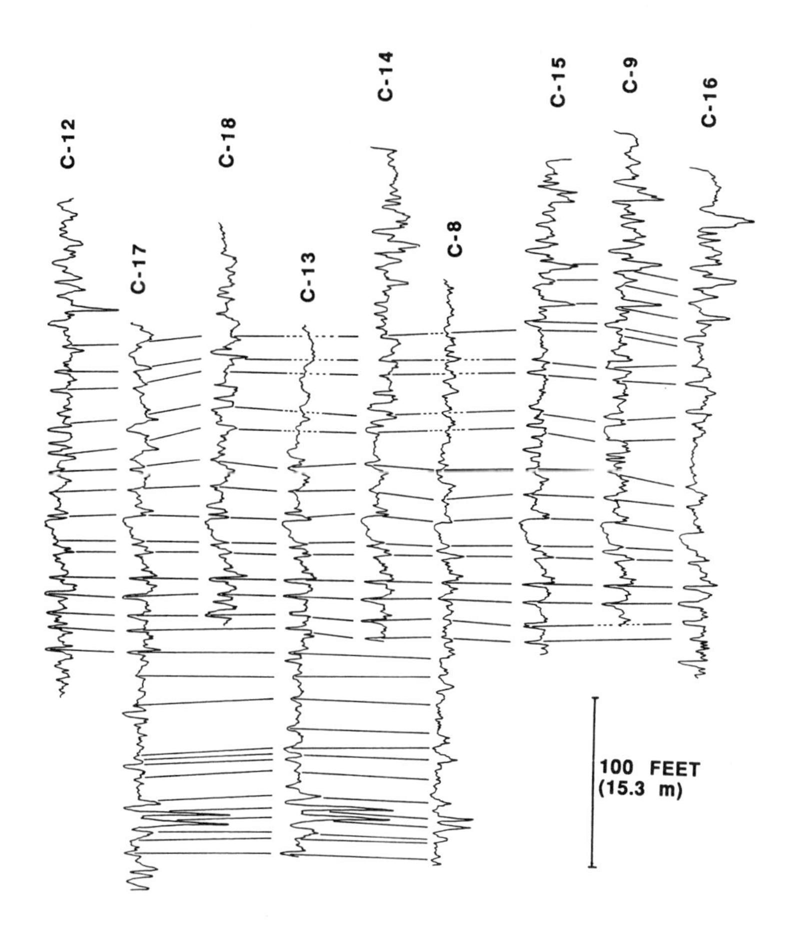

Figure 4. Gamma Ray Correlation Chart From South Shores Area
(Adapted from Steiner and Lippus, 1990)

CASE STUDIES

Natural gamma ray logging was an essential part of site investigations in three projects in southern California. In the first two cases the projects were located in the Palos Verdes Hills area in southwestern Los Angeles County, California. This area is well known for its large active landslides. The third example is located in southern Orange County, California. It also is located in a landslide prone environment in a coastal region.

Resort Site. This proposed resort site is located on the southern flank of the Palos Verdes Peninsula. The peninsula consists of a structural block which has been uplifted and folded into an elongated dome. Bedrock on the project site consists of the shale, siltstone, sandstone, tuffaceous sediments and volcanics of the Monterey Formation. The tuffaceous materials are frequently altered to bentonitic clays which control many of the large, recently active landslides on the Palos Verdes Peninsula.

The eastern portion of the site includes part of the South Shores landslide, a 145-acre (59 ha) landslide which extends along the coastline for a distance of approximately 3000 feet (915 m) and inland for approximately 4000 feet (1220 m). It is an ancient feature with well-developed topographic expression. Approximately 50 permanent residences were constructed on the eastern portion of the landslide in the late 1950's. A mobile home park was constructed near the center of the landslide in the 1970's, following extensive geotechnical study. The landslide is a structurally controlled failure contained in an open synclinal flexure that plunges toward the sea. Landsliding resulted from weak bedding and adverse structure which is daylighted at the beach. These conditions may have been aggravated by higher ground water levels during the Pleistocene epoch. Rupture occurred within a 2 to 5 foot (.6 to 1.5 m) thick zone of highly bentonitic tuff, with abundant slickensided surfaces. Radiometric age dating indicates that the latest movement of the landslide was at least 16,200 years ago (1). Younger landslides have been mapped along the coastline near the western portion of the resort site. The goals of the exploration program included determining the extent of the South Shores rupture surface, the geologic structure beneath the remainder of the resort site, and if other thick layers of weak bentonite were present on the resort site. It had been suggested that the rupture surfaces of these smaller landslides were part of the bentonite layer controlling the South Shores landslide.

Correlations of bucket-auger observations and of core from borings across the resort site were difficult because bedrock in the area consist of a repetitive sequence of rocks with similar appearance. Bedrock units also changed in character across the site, with gradational contacts. Secondary mineralization as described above and weathering had masked the appearance of the bedrock. Although easily differentiated sections of bedrock were exposed in the sea cliff as a result of differential weathering, unweathered bedrock sections often lacked sufficient visual lithologic characteristics for unambiguous correlations.

As shown in Figure 4 the gamma ray logs indicated a number of obvious correlations were possible. These correlations were checked against the core from each boring. Numerous core samples could be correlated between borings with the aid of the gamma logs. The gamma ray correlations were more difficult to interpret in the more highly altered area to the west of the landslide. Even here, however, it was possible to correlate a siltstone-tuff contact and an organic shale with a high, distinctive double peak signature with the rest of the resort site over a distance of about 1 mile (1.6 km).

Peacock Hill. Peacock Hill is located approximately 2 miles (3.2 km) north of the resort site in the Portuguese Bend landslide complex. This area consists of approximately 2 square miles (1035 ha) of land which has experienced repeated episodic landsliding during the last 600,000 years. Like the South Shores area, the Monterey Formation at the site is inclined at relatively shallow angles toward the ocean. Most of the rocks are slightly younger than those at the resort site, although our recent work suggests that some portions of the sites contain rocks of the same age. The rocks in the Portuguese Bend district include a distinctive seventy-foot (21 m) thick bentonitic tuff called the Portuguese Tuff which underlies much of the area.

Peacock Hill itself lies just north and uphill of the modern 260-acre (105 ha) Portuguese Bend Landslide. This landslide has continued to move since it was activated in 1956 with maximum displacements of up to 35 feet (10.7 m) per year. Total historic lateral offset is over 700 feet (213 m) (7). The rupture surface of the Portuguese Bend Landslide is along the Portuguese Tuff.

The exploration program at Peacock Hill was intended to determine the location and thickness of the Portuguese Tuff beneath the hill, as well as two marker beds below the tuff. Previous studies had suggested that the stability of Peacock Hill was due in part to an anticlinal fold. Subsurface data was used to determine the amplitude and wavelength of this structural feature for stability analysis.

Because the rock was less altered than at the resort site, visual correlations of lithologic units were much easier; the massive Portuguese Tuff was especially recognizable in the cores and bucket auger borings. The gamma ray logs verified the visual correlations, provided abundant additional data and allowed numerous additional correlations. Gamma ray logs were again correlated with rock core recovered from the borings. The gamma ray signatures were matched with rock types as successfully as in the South Shores area.

One discovery at Peacock Hill was particularly interesting and is significant in terms of understanding the geology of the Palos Verdes region. The large double peak signature present in the resort site logs also appears to be present at Peacock Hill (Figure 5). Geologic studies over several decades have defined the lithostratigraphy of the Palos Verdes Peninsula (1). However, the complexity of the regional geology makes it difficult to place particular sites within the stratigraphic column. Our investigations demonstrate the potential for correlation of geologic units across relatively large distances with the aid of gamma ray logging.

At South Shores and Peacock Hill we also investigated the use of gamma ray logs to determine the presence and location of landslide rupture surfaces. If bedrock above or within a rupture surface were sufficiently disturbed, one would assume that gamma signatures would not correlate well within these zones. Visual analysis of logs from within both ancient landslides showed the rupture surfaces to have no definitive signatures apart from the signature of the rock itself. This is apparently due to the landslides being block glide failures. Bedrock has moved parallel to bedding planes along rupture surfaces which are too thin to resolve with the gamma tool or because relatively little disruption of the bedrock within the slide has occurred.

Borings were drilled around the periphery of Peacock Hill in areas which appeared from aerial photographs to be subsidiary landslides. In these peripheral landslides, gamma ray correlations could be made below the but not above the rupture surfaces. Bucket auger borings adjacent to the core borings indicated these landslides were rotational failures, with backward rotation of the bedrock blocks and pervasive internal fracturing. As a result, differing types of bedrock were juxtaposed across the landslide boundary. Therefore, what was not possible with the block glide landslides, was possible in these rotational landslides; gamma ray correlations could be use to estimate the location of the rupture surface.

The Irvine Coast. Gamma ray logging was also successfully used at a site located 40 miles to the southeast of Palos Verdes, in Orange County, California. Bedrock of the Monterey Formation again consisted of repeated layers of sandstone, siltstone and claystone with remarkably similar appearance. As on the Palos Verdes Peninsula, the hardness of the bedrock limited the use of the standard bucket-auger borings. Visual correlations of rock units were at best tenuous but high-quality gamma logs provided distinct correlations between borings.

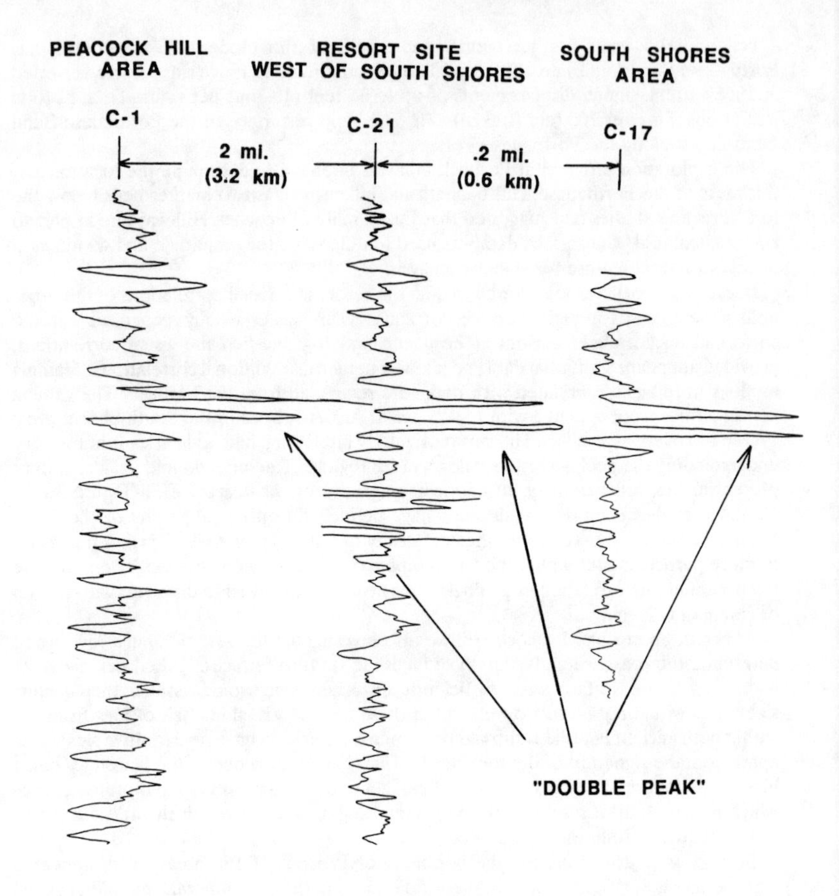

Figure 5. Distinct Double Peak Signature In Three Borings Along the Palos Verdes Coast

SUMMARY AND CONCLUSIONS

Defining local and regional geologic structure is one of the most critical aspects of assessing an area's slope stability conditions. The unique synthesis of traditional methods and the improved technologies described has proved to be an effective data acquisition system for geotechnical site investigations in a difficult region. This system provided the detailed information necessary to characterize the stratigraphy and structure of several relatively large sites and allowed more accurate stability analyses. Findings from these investigations showed that the methods may be useful for correlating stratigraphy over great distances.

Borehole geophysical logging with an improved natural gamma ray tool was a major component of the system. Apart from the high-quality data it provided, gamma ray logging proved advantageous in other respects. Gamma ray logging was possible in wet or dry holes. Retention of fluid in most of the borings was essentially impossible and economically infeasible. Logging was possible in cased or uncased holes with little signal attenuation. Considerable time and cost savings were also realized because the narrow diameter tool allowed logging of borings through the NQ drill stem (1 7/8 inch inside diameter). This logging technique reduced the amount of casing required for zones of severe caving or squeezing. It also virtually eliminated the risk of damage or loss of the gamma tool.

Certain technologies are often developed for and confined to a particular field of investigation. In many cases it does not take much imagination to realize that the data from such tools may be of immediate benefit in geotechnical investigations. What remains to be done is the transfer and application of these technologies to a particular geotechnical problem. A petroleum engineer or geologist may view the natural gamma tool as a relatively lowly member in the family of sophisticated oil exploration tools. However, the case studies presented in this paper underscore the continuing need to adapt such technologies to the more "near-surface" (and relatively low budget) world of geotechnical engineering. Although natural gamma logging has been used by geotechnical engineers for many years, relatively simple improvements made to the technology for our projects significantly increased its utility and enhanced the tool's geotechnical value.

ACKNOWLEDGEMENTS

The authors wish to thank the Hon Development Company for permission to use much of the data presented in this paper as well our employer for supporting our efforts in this undertaking.

REFERENCES

1. CL Conrad and PL Ehlig, The Monterey Formation of the Palos Verdes Peninsula, Calif. in Geology of the Palos Verdes Peninsula and San Pedro, Ed PJ Fischer & Mesa^2 Inc., SEPM/AAPG Guidebook, Los Angeles, 1987.
2. PL Ehlig, Portuguese Bend Landslide Complex, Southern California, Centennial Field Guide, Ed ML Hill, GSA, L.A., 1987, 179-174.
3. AB Esmilla, MB Phipps & JE Slosson, Application of Oil Well Technology and Continuous Coring to Landslide Investigation pp. 2.7-2.12 in Geology of the Palos Verdes Peninsula and San Pedro, Ed PJ Fischer & Mesa^2 Inc., SEPM/AAPG Guidebook, L. Angeles, 1987.
4. WS Keys, Borehole Geophysics Applied to Ground-Water Investigations, Nat. Well Water Assoc., Dublin, OH, 1989, 313 p.
5. CF Park Jr. & RA MacDiarmid, Ore Deposits, WH Freeman and Co., San Francisco, 1970, 522 p.
6. ME Ray, Geologic Investigation, Grading Stabilization Measures, and Development of the South Shores Landslide in Landslides and Landslide Abatement, Palos Verdes Peninsula, So. Calif., 78th Ann. Mtg. Cordilleran Sec., Geol. Soc. of America, 1982, 29-38.

7. EA Steiner & CL Lippus, Use of Downhole Geophysical Methods in a Geotechnical Investigation of the South Shores Landslide, Proc. 31st US Rock Mech. Symp. in Golden, CO, Balkema Pub., 1990.
8. WM Telford, LP Geldart, RW Sheriff & DA Keys, Applied Geophysics, Cambridge Univ. Press, Cambridge, 1976, 860 p.

A WORKBENCH FOR ENVIRONMENTAL DATA PROCESSING AND DISPLAY

Neno Duplancic, M. ASCE[1]

ABSTRACT: Two significant problems of hazardous waste site characterization process are (1) how to manage a large collection of information during exploration program and (2) how to interpret this amount of information. A state-of-the-art database management system has been developed that organizes this information and gives value to data. The system records, processes, and displays analytical and environmental data and helps interpreting the data. Easy-to-use interfaces are provided to standard off-the-shelf software packages. A unique feature of the system is a reference chemical database consisting of over 2,500 of the most commonly tested chemical substances. The system, which is entirely menu-driven, runs on MS-DOS microcomputers and UNIX-based workstations.

INTRODUCTION

Public demand for a cleaner environment, combined with a stricter government regulations, has spawned the development of a new environmental industry that focuses on the management of toxic wastes and remediation of contaminated sites. One of many challenging technical problems that engineers have to deal with as this new industry grows and matures is the management of huge amounts of data generated during the site characterization process. Generally, this information is not confined to the site geology and soil properties as is characteristic of the classical geotechnical site characterization, but also extends to the nature and extent of contamination. As a result, investigations have become more lengthy and complex. The standard observational method that served the geotechnical community so well for many years is no longer accepted. In order to properly characterize the site and to minimize uncertainty associated with the site characteri-

1. Director of site operations and engineering, International Technology Corporation, 4585 Pacheco Boulevard, Martinez, California 94553.

zation, engineers are forced to implement massive site investigation programs.

In addition to geotechnical testing, ground water, surface water, sediment, soil, and waste samples are chemically analyzed to assess the nature and extent of contamination. The number of parameters for which samples are tested varies from site to site and can be as high as several hundred. Environmental Protection Agency's (EPA) list of priority pollutants, for example, contains 127 parameters. As the pressure for a cleaner environment increases, it seems a forgone conclusion that the number of lists and the number of chemicals on each list will increase in the coming years.

This new approach to the site characterization process has led to an explosion in the amount of data that engineers and others must evaluate [1]. This paper describes a comprehensive database management system that has been developed to assist in the management and processing of these data.

STANDARD INFORMATION FLOW

The recovery of ground water and geological samples for identification and laboratory testing is a major objective of hazardous waste site remedial investigation programs. After samples are collected, a comprehensive laboratory testing program is carried out in the field and analytical laboratories to determine the engineering properties of the soil and the nature and extent of contamination. The traditional procedure of data collection during the remedial investigation process results in a relatively fixed flow of information. Samples are sent to the laboratory where they are prepared and analyzed. Typically, a report summarizing the results of the analyses is sent to the project office from several weeks to several months later. In some cases, the data must then undergo extensive validation. As a result of this process, some data are rejected outright or accepted with qualifications. If any analyses or graphical displays are to be generated, the data must be manually entered onto a computer. Several iterations of this process may occur due to demands to extend the scope of the investigation.

This process is tedious and time consuming. Errors and misunderstanding inevitably occur as data are passed from one person to the next. An intelligent database management system has been developed that is intended to assist engineers and scientists in overcoming some of the problems associated with the analysis and reporting of the enormous quantities of data generated during hazardous site investigations.

SOFTWARE ORGANIZATION

International Technology's Environmental Database Management System (ITEMS) is a computerized, menu-driven system for processing and storing environmental data. The system consists of standard forms for entering and editing geotechnical, hydrogeological, sampling, and chemical data; various tables, each consisting of a set of named columns, for storing the data; a secondary database containing information (Chemical Abstract Series (CAS) numbers, names, chemical/physical properties, and regulatory limits) on over 2,500 chemicals; a report generator; and variety of application programs that link the central database to graphics and contouring packages.

The central database, consisting of several levels of menus, input screens, and the report generator, was developed using the ORACLE Relational Data Base Management System (RDBMS) and the "C" programming language. For the generation of graphical displays, the system relies upon commercially available programs (e.g., AutoCAD, SURFER, and GRAPHER). The links that exist to these and other software packages operate in such a way that data are moved transparently without the need to exit from one program and load another.

Figure 1 shows the overall organization of ITEMS. Its various components, including the mechanisms for data entry and capture, the central and secondary databases, and the graphics and analysis module, are described in the sections below.

DATA ACQUISITION

ITEMS provides two different ways to enter data into the central database. Users can choose to (1) manually input field, analytical, and mapping data using standard data entry screens; or (2) import files containing these data types. Included with the system are over 40 data entry forms for entering data. These forms are used to record stratigraphic and lithologic information from soil and rock borings; monitoring well construction details; geotechnical data; well development records; hydraulic conductivity measurements; water level measurements; survey data; and ground water, surface water, sediment, soil and waste sampling information. When setting up a project database, users can select the forms they will need and specify the order in which the chosen forms will appear on the menus used to access these forms. By so doing, the system can be configured to meet the needs of particular projects.

Although they appear as simple screen displays for entering and modifying data, the data entry forms used in ITEMS are considerably

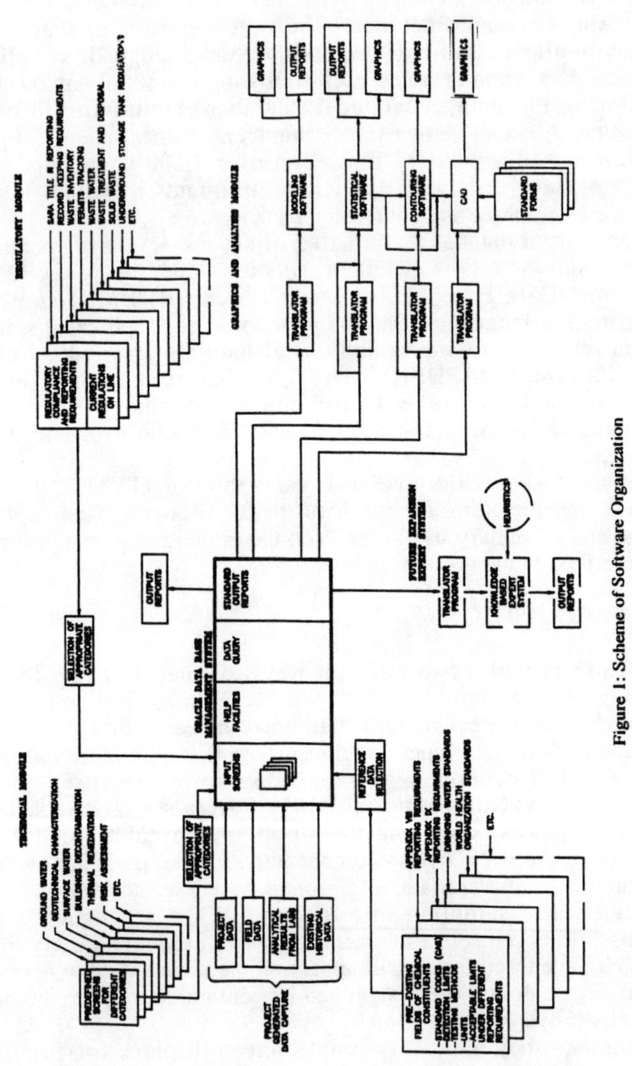

Figure 1: Scheme of Software Organization

more complex than this. Many of the forms have triggers that are executed when data are entered or the cursor is moved into a field. These are used to validate entries or display related data from another form. Data validation can result in some values being rejected. In other cases, where a value may not be incorrect but inconsistencies or missing data are detected, a warning message is displayed.

A second way of inserting records into the central database is to upload the data in electronic format. ITEMS provides a standard file format for the interchange of analytical data. Instead of reporting analytical results on paper, laboratories can transmit results electronically (via modem or floppy disk). The data in these files can then be directly imported into the database tables. This procedure not only reduces the amount of time involved in processing analytical data, it also reduces the inevitable errors that arise when data issued by a laboratory must be entered manually into a computerized system from printed reports.

CENTRAL DATABASE MANAGEMENT SYSTEM

ITEMS central database provides a means to enter, store, maintain, update, and retrieve data in support of the goals and objectives of a project. Aside from the data entry forms described above, the central database consists of several levels of menus, tables for storing the data, an on-line help facility, various mechanisms for querying the database, and a sophisticated report generator.

All data processing in ITEMS (aside from the entry of data) is done using menus. This includes gaining access to data entry forms, selecting what standard report is to be printed, specifying what information is to be included in the report, and linking to other application programs. As such, users do not have to be familiar with any computer language or database management system software to perform routine data processing. This not only makes the system easy to use but also easy to learn.

When questions arise concerning a particular data entry form, users can access ITEMS' on-line help facility. These help screens contain information on the purpose and structure of the form, the identity of not null fields, the type of information that is to be entered in specific fields, and special keystrokes that facilitate data entry and cursor movement within the from. Certain data fields also have their own help screens that provide the user with more detailed instructions. For example, the Description field in the Soil Boring form provides guidance on the use and proper ordering of terms to characterize soil layers.

The process of retrieving information from the central database is called querying the database. Queries must be performed in order to view, edit, update, or delete records from the database. They are also necessary to extract information for reporting purposes or to transfer data to other software packages. Within ITEMS, a variety of methods of querying the database are available. This flexibility allows users to select the method best suited to the task at hand.

One of the simplest ways that users can use to retrieve information is to enter queries on the same forms they used to initially enter the data. Queries can be constructed so that all or only those records that fit certain criteria are retrieved. These search criteria are expressed in terms of the values found in certain fields. Queries can also be initiated via ORACLE's database add-in for LOTUS 1-2-3, or via other ORACLE products such as Easy*SQL.

One of the major weaknesses of many database management systems and a source of continual frustration to their user populations is the inability to extract specific data sets for inclusion in a printed report. ITEMS offers a number of different standard reporting formats for presenting field and analytical data. As part of the report generator, users are provided with various selection criteria for specifying which information is to be included. For analytical data reports, for example, records can be selected based on sample number, sampling date, analytical parameter, and/or whether the result was above the detection limit. Users can also choose to print only those results that exceed established (e.g., Primary Drinking Water Parameter) or project-specific limits. Whereas project-specific limits must be entered manually, values of maximum concentration limits under different federal regulations and EPA guidelines can be retrieved from the secondary chemical database. Additional selection criteria are provided for ground water sampling and water level reports. For these, users can select which records are to be retrieved based on well number, measurement date, well screen elevation, well location, and/or geological monitoring zone.

An equally important component of any database management system is the capability to generate ad hoc reports. Such flexibility is necessary since it is not possible to anticipate all reporting requirements. ITEMS provides several different programs to generate ad hoc reports. All are members of the ORACLE RBDMS family (e.g., SQL*ReportWriter, Easy*SQL, and ORACLE for 1-2-3). Although all three packages are menu-driven, each requires some knowledge of the internal structure of the database.

REGULATORY AND CHEMISTRY DATABASE

A unique aspect of ITEMS is a secondary database of regulatory and chemistry information. Included in this database is information (names, CAS numbers, chemical/physical properties, and regulatory limits) on over 2,500 commonly tested chemical substances. When setting up a project database, users can choose which information they want to incorporate into their project tables. The first step in this process is the selection of chemicals that will be tested during the course of the project. Users are offered several different means of tagging the names of parameters to be tested. One set of menus allows selections to be made from lists of parameters cited in Federal regulations and guidelines. These lists range from the well-known (e.g, Appendix IX, Primary and Secondary Drinking Water parameters) to the obscure (Department of Energy Derived Concentration Guides for Members of the Public as cited in DOE 5400). Another set of menus groups chemicals based on method of analysis. Several hundred different methods are available, including EPA Methods for Chemical Analysis of Water and Waste (100-499 series), EPA Drinking Water Methods (500 series), EPA Test Methods for Evaluating Solid Waste (SW-846), and US Army Toxic and Hazardous Materials Agency (USATHAMA) Methods for Explosive Analysis. A third set of menus allows selections to be made from common chemical groupings (e.g., mctals, volatiles, radiological, and general chemistry parameters).

As parameters are selected, ITEMS copies their CAS numbers and names to the project-specific database. This eliminates the need to enter the names of substances manually, thus minimizing spelling errors. If the user prefers to use a synonym for one of the selected chemicals, a list of synonyms can be displayed and an alternative name chosen. Although compound names are printed in reports, CAS Numbers serve as the primary means by which results are identified in electronic files. They are used instead of names because they are generally easier to input, to sort by, and to query on. Furthermore, whereas substances are uniquely identified by their CAS number, each can have many synonyms.

Maximum concentration limits cited in Federal regulations and EPA publications can also be copied from the secondary database to the project-specific database. Within the report generator, concentrations of chemicals measured in samples can be compared to these established limits and a report prepared showing which samples exceeded these limits.

DATA PRESENTATION – REPORT QUALITY GRAPHICS

One of the biggest bottlenecks of environmental databases has been the lack of integration of the central database and graphics. A key feature of ITEMS is its capability to communicate with other software. All boring logs, well logs, well location maps, and geological cross sections can be generated automatically from the database using AutoCAD.

Interfaces also exist to GRAPHER (for generating XY graphs of changes in chemical concentrations or water levels over time) and SURFER or QuickSURF (for creating contour maps of water levels and contaminant concentrations in soil or water).

Selection of the data that is to be transferred to these graphics packages is made from menus within ITEMS. Translator programs have been developed in "C" that export the selected data to one of the aforementioned packages. Data transfers are accomplished in such a way that the user does not have to exit from one program, then load another. Instead, the application program is automatically loaded into the computer's memory. In some cases, the graphics display is generated automatically without further input. In other cases, the user must select from menu options within the application program in order to complete the generation of the display. Perhaps the most desirable aspect of this approach is that users have access to all of the features of each program. Thus, if they want to use a different interpolation function, alter the scale of a map or graph, or add additional notes to a boring log, they can do so.

The decision to use "off-the-shelf" programs for graphics has allowed to take advantage of the powerful capabilities of these packages, and in the future, reap the benefits as new features are added. It made no sense to develop capabilities that already existed in other readily available commercial packages.

HARDWARE REQUIREMENTS

ITEMS was initially developed for IBM-compatible 286- and 386-based microcomputers. However, because it is an ORACLE application, it can be ported to any of the over 100 computers and operating systems that ORACLE supports. The portability of Oracle and other selected software make it unlikely that a project will outgrow the capabilities of the software. As the data storage and processing requirements of a project expand, the system can simply be installed on a more powerful platform. Moving away from the DOS environment does cause compatibility problems for some of the application programs. However, this problem can be overcome by linking one or more microcomputers to a workstation, mini-, or mainframe computer. The more powerful computer can then be used for data storage and retrieval. Whenever graphic displays need to be generated, information can be downloaded to the PC-based system, then transferred to an application program for final processing.

CURRENT LIMITATIONS AND FUTURE EXPANSION

The current level of development can be used for automation of data collection, data management, decision making, and data presentation. The system can produce standard and ad-hoc reports, generate two-dimensional graphs and contour plots, and generate boring logs in graphical form. Despite the progress that has been made, much work remains to be done. High on the list of priorities is the enhancement of the interfaces to AutoCAD so users have even more flexibility concerning the types of figures that are produced. Future plans also call for the development of statistical routines that allow users to perform many of the tests cited in a recent EPA publication [2], and additional modules that address the data storage and reporting requirements of Section 302 and 313 of SARA, TSCA, CWA, and RCRA. Finally, at some point an expert-based system will be incorporated that would assist engineers and scientists in evaluating data and deciding among various courses of action.

Although not currently available, procedures that would allow two way communication between the various application programs and ITEMS will be developed. This will allow the central database to be updated concurrently with the graphics displays.

Other long-range plans include the development of a Geographical Information System within which users could access the ITEMS database. Because all Geographical Information systems must be built upon a solid database management system, the first priority has been to complete the central database. Eventually, the incorporation of GIS capabilities will allow users to bring up a map of a site and point to a location of interest. After specifying the selection criteria, all data records that satisfied the criteria would be retrieved and displayed on the screen.

SUMMARY AND CONCLUSION

The system described in this paper provides improved capabilities for the management and interpretation of environmental data. Data can be entered either electronically or manually into the system. Nearly all data processing activities are menu-driven. Output reports and graphics can be easily obtained at any stage of the project. The system relies on proven off-the-shelf software packages for graphic displays and translator programs written in the "C" programming language. These programs provide for a smooth transfer of data to the selected software packages without requiring the user to exit one program and load another.

REFERENCES

1. N. Duplancic and G. Buckle, "Hazardous Data Explosion," Civil Engineering, December 1989.
2. U.S. EPA Office of Solid Waste, Waste Management Division, "Statistical Analysis of Ground Water Monitoring Data at RCRA Facilities – Interim Final Guidance," February, 1989.

ARCHITECTURE OF AN EXPERT DATABASE SYSTEM FOR SOIL CLASSIFICATION USING CPT DATA

Adrian S. Chan[1] and Mehmet T. Tumay[2]

ABSTRACT: An expert database management system for soil classification using two types of data from electronic cone penetration testing (CPT) is introduced. These two types of data are defined as explicit and implicit CPT sounding data. Explicit CPT data are data from a direct comparison of laboratory and field results on the same sounding location. Implicit CPT data are generalized implied data that are presented in chart format from culmination of past experiences. The architecture of the system is primarily based on the priority given to the retrieval and storage of explicit data that are generated from CPT soundings. Secondarily, implicit data can be used in the decision making process under the option of the user or when explicit data are not available. Implicit data can also be updated and stored in the system for refinement of localized correlation for soil classification.

INTRODUCTION

The electronic quasi-static cone penetration test (QCPT) and piezocone penetration test (PCPT) have become increasingly acceptable as a subsurface investigation method for geotechnical engineering applications in the United States during the past fifteen years (Meigh, 1987). This is mainly attributed to the error free operation of the QCPT and the PCPT with an automatic data acquisition system, and the abundance of information and experience readily available on the QCPT and the PCPT method of interpretation worldwide. Archival documentation such as: ESOPT1, 1974; ASCE Specialty Conference, 1981; ESOPT2, 1982; In-situ '86 and ISOPT-1, 1988, made available multitude of correlations and interpretations of the QCPT and the PCPT data with important soil parameters for geotechnical engineering design and analysis.

[1] Project Engineer, International Technology Corporation, 1150 Leblanc Road, Port Allen, LA 70767.

[2] Professor, Department of Civil Engineering, Louisiana State University, Baton Rouge, LA 70803, currently Director, Geomechanics Program, National Science Foundation, Washington, DC 20550.

Since an exact theoretical solution is not currently available for the "direct" analysis of cone penetration mechanism, the interpretation of the QCPT and PCPT results are somewhat semi-empirical in nature. This implies that correlated equations and past experience are essential for the proper interpretation of cone data. Thus, side by side comparison of soil parameters of conventional geotechnical testing methods with QCPT and PCPT data will be beneficial for further improving and developing empirical correlations. Reviews of past QCPT and PCPT research confirmed the fact that most test data and information are reported in the form of goal specific ad hoc non computer-based record keeping system, such as in the form of charts and empirical equations (Schmertmann, 1978, Robertson and Campanella, 1984). This kind of information will serve well only when the data is static, however, this is normally not the case since empirical equations or chart specific information would need constant updating when more local QCPT and PCPT data become available.

In order to allow for more efficiency in future application of QCPT and PCPT data, a well organized computer-based data and information maintenance system that allows for storage and retrieval of both laboratory and insitu field data as well as existing chart format or empirical equations should be created.

DATA DESCRIPTION

Explicit Data. In the course of any insitu CPT research program, field performance of CPT always associates with either geotechnical laboratory testing or other form of insitu field testing equipment (i.e. pressuremeter test (PMT), dilatometer test (DMT), etc.). Field CPT data such as tip resistance, sleeve resistance from QCPT and excess pore pressure measurement from PCPT or laboratory testing data such as Atterberg Limits, triaxial compression strength, compressibility, grain size distribution etc. on disturbed/undisturbed samples are data that would be gathered in a CPT research program. These data are real data instances obtained explicitly by performing tangible testing from either insitu field or laboratory tests, and are thus termed the explicit test data.

Implicit Data. Abundance of knowledge and information involving the use of QCPT and PCPT sounding data for soil classification and estimation of strength/settlement parameters are reported in literature. These blocks of information are mostly implicitly presented by researchers for forecasting various situations. Implicit data, such as soil classification using QCPT and PCPT sounding data, are normally presented in different kinds of chart formats, such as presented by Douglas and Olsen, 1981, Jones and Rust, 1982, Tumay, 1985, Robertson and Campanella, 1984, Senneset et. al., 1989 and Robertson, 1990. Other forms of implicit data, such as the undrained shear strength of clay, are presented by researchers in a generalized model consisting of CPT sounding data and geotechnical laboratory data.

CONE PENETROMETER DATABASE SYSTEM (CPDS)

Database management systems (DBMS) are widely accepted as a tool for reducing the problem of managing a large collection of shared data for application programmers and end users. The users interact with the database of the DBMS by submitting requests for data selection (queries) or manipulation (updates). A data model is normally used to represent the actual database during implementation. Classic data models such as the relational model, the network model or the hiarchical model do not provide sufficient tool for representing engineering data, especially QCPT and PCPT data. In geotechnical testing, no two borings are identical, thus, the only linkage on two sets of test data from different test types for use in database implementation will be on samples collected from a close proximity three dimensional space. QCPT and PCPT data and geotechnical laboratory testing data from neighboring borings can be incorporated into a database using a database model, such as the geotechnical engineering database model (GEDM) (Chan, 1990).

The CPDS is an expert database management system that combines the explicit CPT information based on the GEDM and the implicit CPT information into a database management system. The notion of the system primarily stresses on the storage and retrieval of explicit data that is generated from CPT soundings program. That is, the most vital information from a CPT database should still be the direct data obtained from side by side borings. These data will be used for updating information contained in the implicit database. Secondarily implicit data is used for the decision making process under the option of the user when explicit data is not available or when the user deems necessary to check on previous experience. Implicit data can also be updated and retained in the system to ensue refinement of localized correlation for soil classification on the system.

SYSTEM ARCHITECTURE

The basic architecture of the CPDS exhibits a separation of data request on the explicit and the implicit database. It consists of three components: (1) an user-interface which consists of a query processing mechanism (engine) for accessing both the explicit and the implicit data, (2) an explicit database based on the GEDM, and (3) an implicit database based on previous backlog of research experience. The relationship between each component is illustrated in Figure 1. The separation of the database is essential, since it will allow for different databases to be loaded into the system at any one time. Each explicit or implicit database is stored under a database name which the users are allowed to select during the system initialization or any time during the session.

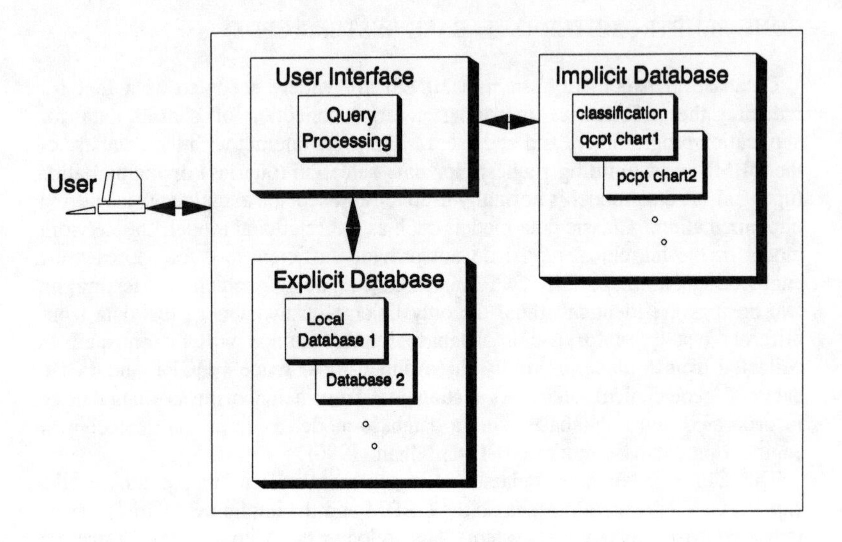

Figure 1. Architecture of the CPDS

User Interface. In order to efficiently utilize the database, the CPDS consists of a user friendly source providing engineering laboratory test data from in-situ tests data or vice versa to the users. The incorporation of a user-interface for the CPDS should be the most logical approach for users to communicate with the computer database.

Without a built-in user-interface, users who communicate with the database to retrieve information will need to learn the functions of whatever language or environment the DBMS is implemented, in this case, the Arity Prolog interpreter/compiler environment (Arity/Prolog, 1989). The user must be familiar with the concept of querying and be able to phrase the questions to the database in correct Prolog syntax. The user must know which predicates are available in the database to use and which of those predicates in particular will produce the answer to the question. This situation might be acceptable to Prolog programmers but is a burden to the casual user who wishes to make use of the database created in the Prolog environment. Some users may have no previous knowledge and experience of computers and computing terminology. Thus the existence of a user-interface will certainly enable the user to communicate with the database in a familiar format.

There are several forms of user-interface which could be chosen, for example: a dialogue system, a menu-driven system, a graphical system or a voice input system. A menu-driven user-interface is developed for this study since this is the most straight forward and the most common type of user interface currently employed by popular commercial software products. This menu-driven user-interface comprises a series of choices in the form of a "menu". Those choices are listed on the monitor screen where the user is requested to select one of the available options. This menu-driven system also comprises a hierarchy of other sub-menus. In this system, the user is initially presented with a main menu with options directing towards other more detailed sub-menus. After working through a succession of such sub-menus the user will eventually reach the actual required function. This menu-driven system thus reduces the chance of making input mistake.

With this user-interface, the users do not require to know where and how the data are organized and stored in the database during data query or update. The front-end user-interface system recognizes the meaning of the user's intention and apply the rules and facts concerning the relationship among the data to deduce a response. This response is then disclosed to the users in an easily understood format.

Data Model. The GEDM is an abstract multiple level object-oriented data model for modeling geotechnical engineering data (Chan, 1990). The GEDM is capable of modeling the semantic knowledge of the geotechnical engineering testing data by extending the Entity-Relationship Model (ERM) data analysis approach introduced by Chen (1976). In ERM, the database designer will describe the enterprise in terms of entities, relationships and attributes. Entities are distinct objects within a user enterprise whereas relationship are meaningful interactions between these objects. The entities and relationships are described by the attributes which are associated with a value set (domain) and can take a value from this value set. The GEDM extends the ERM by introducing a semi-abstract relationship type which allows data relationship between implied attributes in basic relationship set and entity set. This extension allows better organization of geotechnical engineering testing data in the database analysis stage and facilitate the eventual implementation of the GEDM into an underlying database system with the semantic knowledge of the data kept intact. The GEDM models the QCPT, the PCPT and the laboratory data as individual entity types, whereas relationship types are used to describe the abstraction level of the function relationship among data of the entities type. Figure 2 depicts a generalized model of the GEDM for soil classification using CPT and laboratory data.

Knowledge Base. The use of CPT sounding data to identify soil type have been proposed and used since the late sixties (Schmertmann, 1978, Sanglerat, 1972). Charts are created to allow soil engineers to select the appropriate soil type classification from QCPT/PCPT sounding data that falls within the limits of a defined zone. However, knowledge and information accumulated from these researches reflect generalized soil types and become stagnant once a chart is created and published, i.e. the boundary or zone inside these charts cannot be easily changed or modified to accomodate findings of local soil behavior type. In fact, these charts in no way are complete since usually not all data points and their respective soil types have actually been encountered by the researchers. Voids in the charts are filled and expanded by conjecture. However, the use of this knowledge at the time is an essential move in advancing and promoting the CPT technology when nothing else is available.

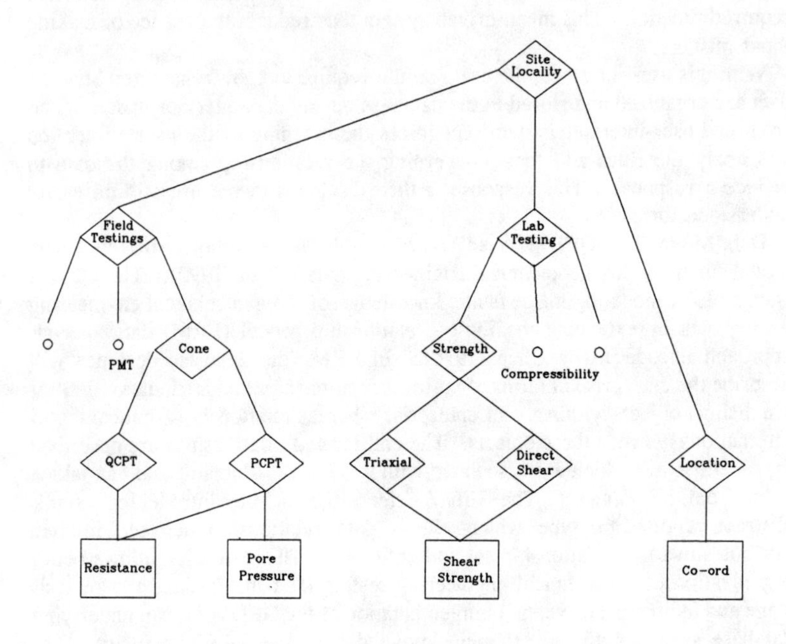

Figure 2. Simplified Representation of CPT in the GEDM

Soil classification and geotechnical soil parameter estimation using existing chart knowledge incorporated into the CPDS will provide one step further. The CPDS allows simultaneous changes along the boundaries in these charts to reflect local soil behavior type. The technique used in the implementation of soil classification chart information into the CPDS involved the layout of all the identifiable zone for each type of soil. These zones normally occupy a specific area prescribed by a polygon on the chart. Thus, each zone can be described by the corners of the polygon with its respective classification. These information will become the anchor basic knowledge of the charts.

IMPLEMENTATION

The CPDS is implemented using the language Prolog. The deductive reasoning feature of Prolog over database makes it ideal for a DBMS implementation. This language has a built-in database where rules and facts are stored. This built-in feature makes Prolog well suited for writing database and knowledge base applications.

A number of Prolog language interpreters and compilers are now commercially available running in various micro computer environment, such as the Turbo Prolog by Borland,. A Prolog language interpreter with compiling capability distributed by ARITY Corporation is chosen. This interpreter with compiling capability is a superset of the Edinberg's Prolog dialect which is considered as a complete Prolog language implementation. The CPDS is implemented into an IBM-compatible micro computer environment, which is the computing environment utilized by the Research Vehicle for Geotechnical Insitu Testing and Support (REVEGITS) at Louisian State University. REVEGITS is a fully computerized insitu testing vehicle that is capable of performing CPT, PMT and DMT. With the addition of the CPDS into the REVEGITS, data obtained from CPT sounding will be properly archived into local databases for future use.

Based on the architecture of the CPDS as depicted in Figure 1, the front-end user interface is implemented as two sub-systems. The first sub-system is the querry processing engine that comprises a series of facts and rules for submitting requests to the GEDM database. The rules and facts consist of the semantic and relationship among data of the GEDM. The second sub-system comprises facts and rules about the structure of the knowledge base where the implicit CPT information from past and present research are stored.

The explicit database is implemented into facts in the Prolog environment to describe each individual element of the GEDM. For example, the tip resistance and the sleeve resistance from QCPT for a particular location can be represented by a Prolog fact in the form of:

resistance(key,tip_resistance,sleeve_resistance). (1)

in which key is the linkage or pointer of the QCPT data to other available information around the vicinity. The real data instance for the resistance inside the database will look like:

resistance(3,55,5). (2)

in which the value 3 is the key that links this fact to other information in the database such as the location where this measurement is obtained, and the value 55 and 5 are the tip resistance and the sleeve resistance respectively in the unit of choice. All other elements of the GEDM are similarly implemented as Prolog facts in the database.

The implicit database is implemented as facts in the Prolog environment that represent the zone layout of a soil classification charts. For example, the soil classification type for using a particular chart can be represented by a predicate in the form of:

range(chart(corner_1, ..., corner_n,soil_type,soil_group). (3)

in which chart is the name of the soil classification chart in use and corner_1 to corner_n are the apexes of a zone bounded by a polygon in a classification chart. The real data instance inside the implicit database will look like:

range(chart((0,1),(1,1),(1.4,2),(0,1.5),very_soft_inorganic_clay,clay)). (4)

in which (0,1), (1,1), (1.4,2) and (0,1.5) are coordinate pairs of the corners of the polygon bounding the "very soft inorganic clay" zone.

These digitalized data are stored into individual data base file which is consulted into the CPDS when requested by user.

Updating of this information will be via another fact inside the implicit database in the form of:

area(center,radius,soil_type,soil_group). (5)

in which center and radius will define the area of the chart that needs to be updated with the soil_type and soil_group. For example, an area of the chart that needs to be updated is represented by the following fact as:

area((5.7,10),0.2,medium_inorganic_clay,clay). (6)

in which (5.7,10) and 0.2 is the center and the radius of the circle representing the area respectively.

This update feature allows users to change any classification charts with information obtained from local experience. Currently, two charts are incorporated into the CPDS. The Tumay (1985) chart is used with QCPT data while the Senneset (1989) chart is used with PCPT. Additional charts format can be implemented into the CPDS with identical protocol.

CPDS SESSION

A main menu will be presented to the user upon logging on to the CPDS. According to the archecture of the CPDS, either the explicit database or the implicit database can be selected. When the explicit database is selected, the default database will be loaded, however, the user is allowed to change to a specific database during the session. Querying into the database is simplified by the system with a set of sub-menu that consist of the most possible selective action of the user. User can either query individual test result from a single bore hole or query comparative result of field and laboratory result of adjacent bore holes. Conditions are set by picking the valid attributes allowed in the query.

Suppose a user requested to display all the available classification results (e.g. the Atterberg limits) performed on samples from all borings that have QCPT performed within the vicinity and that the tip resistance from QCPT is less than 100 MPa. First, the "adjacent boring" sub-menu of the CPDS should be picked, then, both the QCPT and the Atterberg limits test should be selected. A series of windows consisting all the allowable attributes will then be presented, in this case, tip resistance, friction resistance, liquid limit and plastic limit. The user can then select the tip resistance attribute for the restrictive match, in this case, the condition "less than 100 MPa" should be input at the system prompt. The system will then search for all the bore holes at each depth that has the required match.

The implicit database can be selected if the user request information directly from soil classification charts. The user can load a specific chart into the system that matches his requirement. Querying into the implicit database is very straight forward. Either QCPT or PCPT data will be needed to obtain soil classification information. The use of the implicit data is primarily for supplementing the decision making process when no explicit data are available.

CONCLUSION AND DISCUSSION

The concept of a CPDS to preserve and document CPT data and laboratory data from soil investigation into a database is presented. This CPDS has been implemented as a prototype model for the REVEGITS. By adhering to the architecture of the CPDS and the GEDM, regional CPT data and laboratory tests data for soil classification and strength parameters estimation can be implemented into individual explicit and implicit databases. In addition, by expanding other field instruments such as the PMT and the DMT into the GEDM, a comprehesive expert database system can be implemented that encapsulates most insitu and laboratory geotechnical engineering information.

ACKNOWLEDGEMENT

The research described in this paper was supported by the Louisiana Transportation Research Center (LTRC) under Contract No. 88-2GT (State Project No. 736-13-37, Louisiana HPR No. 0010(12)) and by the National Science Foundation Grant No. MSM-8705051.

REFERENCES

1 Arity/Prolog, Arity Corporation, Version 5.1, Massachusetts, 1989.

2 ASK Chan, An Expert Database System for Soil Classification Using Friction and Piezo Cone Penetration Test, Ph.D. Dissertation, Louisiana State University, Baton Rouge, 1990, in press.

3 PPS Chen, Entity Relationship Model, Toward a Unified View of Data, ACM TODS, Vol. 1, no. 1, March 1976, 9-35.

4 BJ Douglas & RS Olsen, Soil Classification Using Electric Cone Penetrometer, Ed GM Norris & RD Holtz, Proc. Symp. Cone Penetration Testing and Experience, Geot. Eng. Div., ASCE, St. Louis, Missouri, 1981, 209-227.

5 GA Jones & E Rust, Piezometer Penetration Testing, Proceedings of the Second European Symposium on Penetration Testing (ESOPT-2), Amsterdam, May 1982, 607-613.

6 AC Meigh, Cone Penetration Testing, CIRIA Ground Engineering Report, In-situ Testing, Butterworths, London, 1987, 141 p.

7 PK Robertson & PK Campanella, Guidelines for Use & Interpretation of the Electronic Cone Penetration Test, Hogentogler, MD, 1984, 175 p.

8 PK Robertson, RG Campanella, D Gillespie & J Greig, Use of Piezometer Cone Data, Proceedings of ASCE Specialty Conference, Blacksburg, Virginia, June 1986, 1263-1280.

9 PK Robertson, Soil Classification Using Cone Penetration Test, Canadian Geotechnical Journal, Vol. 27, No. 1, Feb. 1990, 151-158.

10 G Sanglerat, The Penetrometer and Soil Exploration, Elsevier Publishing Company, Amsterdam, Netherlands, 1972.

11 JH Schmertmann, Guidelines for Cone Penetration Test, Performance and Design, Federal Highway Administration, Report No. FHWA-TS-78-209, 1978, 145 p.

12 K Senneset, R Sandven & N Janbu, Evaluation of Soil Parameters from Piezocone Test, Transportation Research Record 1235, Transportation Research Board, Washington, DC, 1989, 24-33.

13 MT Tumay, Field Calibration of Electric Cone Penetrometers in Soft Soils, FHWA-LA, LSU-GE 85/02, 1985.

14 Turbo Prolog, Borland International, Inc., Version 1.1, California, 1986.

SHALLOW FOUNDATION DATA BASE

ASCE Shallow Foundation Committee

Jean-Louis Briaud[1] (chairman), G. Hossein Bahmanya, W. Michael Ballard, James R. Blacklock, Mark T. Bowers, James R. Carpenter, John R. Davie, Albert DiMillio, Guy Y. Felio, Adel Hanna, Thomas J. Kaderabek, Alan J. Lutenegger, William O. Martin, Joseph Ray Meyer, Michael W. O'Neill, Miguel Picornell, Fred Romani, Lawrence C. Rude, Robert L. Schiffman, Richard W. Stephenson, Harry Stewart, Charles E. Williams, Warren K. Wray

ABSTRACT: A shallow foundation data base is presented. The effort is a few years old and is described here so that feedback can be obtained from the membership. This article outlines the purpose, the logistics, the forms used to report each case history, the computer program, the current data base, and the plans for the future.

INTRODUCTION

In 1988, at a meeting of the ASCE Shallow Foundation Committee, Charles Williams who was then chairman of the committee proposed the idea of developing a shallow foundation data base. The idea was enthusiastically approved and the work began. Today, the leader of this effort in the committee is William Martin. The support for this data base development has come so far from the active work of the committee members, from the Geotechnical Engineering Division Executive Committee who provides funds for the mailing of a pamphlet to potential contributors, from the Federal Highway Administration who provides funds for the development of the electronic data base and from students at Texas A&M University who painstakingly place the data in paper files and in electronic files.

WHY IS A DATA BASE NEEDED?

The purpose of the data base project is to collect, organize, and disseminate case histories of shallow foundation behavior for the following reasons:

1 Professor, Department of Civil Engineering, Texas A&M University, College Station, TX 77843-3136, USA.

a. To reduce foundation costs. Shallow foundations are economical if total and differential settlements can be kept within tolerable limits. If this can be achieved a more costly deep foundation system is not necessary.

b. To provide data on soil compressibility characteristics. The hardest part of any settlement calculation is estimating the compressibility of the soil.

c. To preserve data on shallow foundation behavior. Many organizations have useful settlement data in their files, but don't have the time or money to find and organize these data. As a result, many case histories are effectively lost.

d. To organize the collective experience of the profession on this subject, and to make it available for future engineers. This data base could eventually lead to an expert system.

e. To identify types of soil for which compressibility data are not available, so that research needs can be identified.

f. To provide data to evaluate the degree of conservation (or lack thereof) in the design or construction of shallow foundations.

ORGANIZATION AND AVAILABILITY

The data are entered into a computerized data base, copies of which are available on a floppy disk. Each case history will be identified by a number and by key words, such as:

Geologic province
State, county, city/town
Geologic unit(s)
Soil type(s)
Consolidation state (normally consolidated/overconsolidated)
Structure or load type
Field investigation technique(s)
Laboratory test type(s)
Heave measurements
Total settlement measurement
Differential settlement measurement
Estimated soil compressiblity
Structural distress

The case histories are organized and stored at Texas A&M University. Hard copies of the published and unpublished case histories are maintained on file to provide backup for the data on disk. Should a user wish more specific information on an unpublished case history than contained on the disk, a copy of the full case history can be sent.

Contributors of unpublished case histories are asked to remove the name of their organization and that of their client, and to provide a statement to the effect that the client or sponsoring organization has permitted the inclusion of the case history in the data base.

To avoid copyright problems, copies of published case histories are not sent. The reference from which the case history was obtained is included on the disk.

COMPOSITION OF THE DATA BASE

In order to be as inclusive as possible, the data base contains case histories from the literature as well as from the files of consultants, public agencies and owners. Case histories selected for inclusion must be well documented and ideally should include the following:

Geologic history
Soil stratigraphy
Ground water information
Results of thorough subsurface investigations
Results of laboratory tests and in situ tests
Well defined structure geometry and loads
Description of settlement monitoring program
Load-settlement and time-settlement data
Any data on excess pore pressures versus time
Estimated soil compressibility characteristics

The specific information that would be useful under some of these headings is listed in the forms.

THE COMPUTER PROGRAM

The data base program is developed using DBASEIV. It is a menu driven interactive database. Several options exist in the main menu: input, viewing and analysis. The input option allows to enter new data, correct existing data or cancel data. The viewing option allows to look at any case history in the data base, and to create a subset of case histories satisfying a number of conditions. The analysis option allows to use a number of design methods to predict the response of a shallow foundation and to preform correlation studies. In the input option the forms come on the screen and are filled with the new data. In the viewing option, again the forms come on the screen and a number of graphs are displayed as visual aids. In the analysis option calculation results and correlation graphs are prompted on the screen.

The current data base has 46 footings. All these footings are on sand. Fourteen different methods to calculate the settlement of footings on sand have been incorporated and correlations can be performed between predicted and measured behavior.

THE PLANS FOR THE FUTURE

The plans include to continue to develop the program, to continue to input new data, and to seek a wider audience for the data base. In that respect a brochure has been prepared and will be distributed to a prepared list of potential contributors. The wish of the committee at this time is to receive as many quality case histories as possible. Later on, it is contemplated to develop this data base into an expert system.

SHALLOW FOUNDATION DATA BASE

Name: ____________________

Affiliation: ____________________

Address: ____________________

____

Phone Number: ____________________

Footing ID# ____________ (to be completed by Texas A&M University)

PROJECT DESCRIPTION: (project location, structure type, purpose)

____

____

FOOTING LAYOUT: (please sketch with plan and side view, indicate where settlement was measured)

Comments: ____________________

____

FOOTING DATA:

Elevation of natural ground surface ENGS (units) = ____________

Distance H_e (units) = ____________

Distance D_t (units) = ____________

Distance D_b (units) = ____________

Shape: ____________

Width or diameter (units): ____________

Length (units): ____________

Thickness (units): ____________

Embedment Depth (units): ____________

Comments: ____________

____

LOAD DATA: units: ____________

	Total Load	Static Dead Load	Static Live Load	Dynamic Load
Fx				
F_y				
F_z				
M_x				
M_y				
M_z				

Comments: ____________

____

SOIL PROFILE: depth to water table (units): ______________________

Layer No.	Depth to Bottom of layer (unit=)	Soil Class. (USC)	Soil Description

1. Distance D_i on diagram.

Comments: (geology) ______________________

____

____

SOIL DATA:

(ω, γ_t, LL, PI, SPT, CPT, P, p_c, OCR, c, ϕ, S_u, E, C_c, C_r, e, γ_{max}, γ_{min}, n, D_{50}, D_{10}, etc.)

Layer No.										

Comments: ____

____

____

FOOTING BEHAVIOR: please indicate the units inside the parentheses

Load ( )	Settlement ( )	Time ( )	Load ( )	Settlement ( )	Time ( )

Was the load measured or calculated? ______________________

How was the load measured? ______________________

How was the settlement measured? ______________________

Reliability and quality of data obtained? (10 = Excellent, 1 = Unreliable)

____

Adjacent footings? ______________________

GENERAL COMMENTS: (please attach any data sheet, boring, article or other hopeful documents)

____

____

____

____

PLEASE RETURN TO: DR. JEAN-LOUIS BRIAUD
CIVIL ENGR. DEPT.
TEXAS A&M UNIVERSITY
COLLEGE STATION, TX 77843-3136 USA
TEL: (409) 845-3736 FAX: (409) 845-6156

HIGH SPEED DATA ACQUISITION SYSTEM FOR FIELD TESTING

Harry E. Stewart[1], M.ASCE and Thomas D. O'Rourke[2], M.ASCE

ABSTRACT: Field testing of constructed facilities for live load and dynamic response requires data acquisition systems that are portable and rugged, and capable of collecting multi-channel data at a wide range of sampling frequencies. This paper describes the development of system hardware and control software for automated data acquisition of up to 48 channels of low and high level instrument signals. The system includes the necessary hardware and software for field calibration of instruments, sampling data at variable scan rates, and immediate playback of either the raw data or data reduced to engineering units. Also included is a discussion of the important characteristics necessary for collection of high speed data, with emphasis on the need for rapid field assessment of the data to assure that field instrumentation is performing properly. System features are identified which allow for changes in either field testing methods or data acquisition, based on feedback during monitoring.

INTRODUCTION

Field testing for live load or dynamic responses in civil engineering requires flexible methods for data acquisition that can be tailored to meet the specific test requirements. In addition, the acquisition system should be capable of rapid on-site data reduction, so that the quality of the measurements can be assessed, and changes in the experimental program can be made, if necessary, on a rational basis. Since field testing generally is expensive and often difficult to repeat, it is critical that useful data are obtained during the field test period.

This paper describes the development of system hardware and control software that has been used successfully on several geotechnical field experiments. The important components for high speed data collection are described. These components include not only hardware and software, but also personnel. The personnel must be trained in system usage, be able to configure the system for various applications, and be able to interpret the data readily so that field testing variations can be made on a rational basis.

1,2 - Assoc. Prof., Prof., School of Civil and Environmental Engineering, Cornell University, Ithaca, NY 14853-3501

DYNAMIC DATA SYSTEM

Dynamic data acquisition systems require specialized hardware capable of rapid sequential measurements, data analysis, and computer storage. The system can be broken down into three different components. First, there is signal conditioning, where the signal can be amplified. Next, the data acquisition system digitizes the signal, and then the computer controls, stores, and manipulates the readings. The dynamic data acquisition system described herein was developed in conjunction with several geotechnical research projects performed by Cornell University. Figure 1 shows a basic schematic of the dynamic data acquisition system.

Signal Conditioning. Instruments that can be used with the system consist of high output signal level instruments, such as accelerometers, pressure transducers, displacement transducers, and temperature sensors, as well as relatively low output signal level instruments, such as strain gages. The 48 channels used with the high-speed voltmeter and field effect transistor (FET) multiplexers can receive their input directly from the high signal level instruments, or from the amplifiers used with the low signal level instruments.

The signal conditioning and amplifier (SCA) units are Micro-Measurement (MM) Model 2120 components. Each unit contains two channels of independently controlled bridge excitation, output gain, and remote shunt calibration/verification circuitry. Thus, there are 24 dual channel SCAs providing complete, independent control for all of the instruments going into two 24-channel field effect transistor (FET) multiplexers. Five SCA units are mounted in each of five MM 2150 mounting racks. Each mounting rack contained a MM 2110 power supply. Shunt calibration/verification circuits are included in each SCA unit.

Shunt calibration can be used to calculate the calibration factor for each instrument, allowing the output voltage to be converted easily to engineering units, greatly simplifying data reduction.

In addition to the high speed components, the system can accommodate 20 channels of relay multiplexed instruments. The power for these channels is provided by two HP6200B power supplies, each supplying power to 10 channels.

The system components are mounted in a portable rack. The entire system was designed to provide versatility for use with almost any type of transducer, with a wide range of excitation voltages, amplified or unamplified signals, and scanned at either high or low sampling rates.

Data Acquisition Control System. The next main component of the acquisition system is a HP3852A data acquisition and control system. The system mainframe has a 5-1/2 digit integrating voltmeter, a 13-bit high-speed voltmeter, a 20-channel relay multiplexer, two 24-channel, high-speed field effect transistor (FET) multiplexers, a five-channel counter/totalizer, and a 16-channel digital input module.

The high-speed voltmeter is used with the high-speed FET multiplexers to make scanned voltage measurements at a maximum rate of 100 kHz (100,000 readings per second). This scanning rate is required for true dynamic measurements on a relatively large number of channels. The 13-bit (12-bit plus sign),

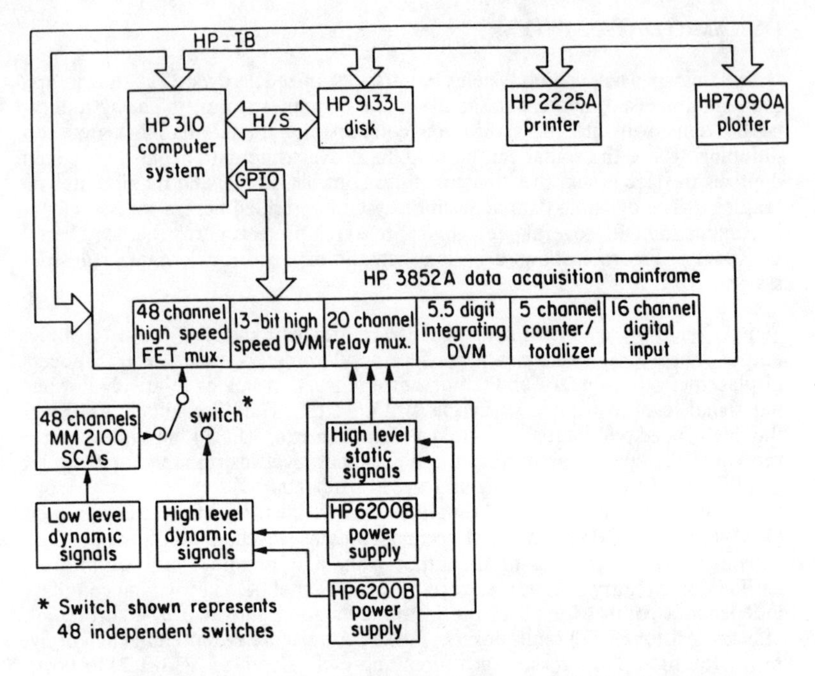

Figure 1. Schematic of High Speed Data Acquisition System

high-speed voltmeter has autoranging capabilities, with four ranges of ± 40 millivolts, ± 320 millivolts, ± 2.56 volts, and ± 10.24 volts. The autoranging feature causes no time delays in sensing variable signal levels from one channel to the next, and greatly enhances system speed. The four ranges allow the instrument signal output to be digitized at 9.8 microvolt, 78 microvolt, 0.63 millivolt, and 2.5 millivolt increments, respectively.

These increments provide for close following of the dynamic measurements, particularly when scanning at high speeds. The total number of instruments that may be used during any experiment is 48. Thus, when using all 48 channels, the maximum scanning rate per channel is 100,000 divided by 48, or approximately 2000 readings on each instrument each second. To ensure that the peak dynamic responses are captured, each gage response should be measured at an absolute minimum of two to four times the maximum anticipated response frequency, although oversampling by a factor of ten may be necessary for high frequency events such as blasting.

The 5-1/2 digit integrating voltmeter is used with the 20-channel relay multiplexer to make quasi-static readings that do not require high scanning rates. Quasi-static sampling rates typically would be taken at a rate of less than 1 Hz. Quasi-static measurements might be necessary, for example, to record events that do not vary much with time, or during static testing and system calibration.

The counter/totalizer is used as a system timer to record the initiation, duration, and termination of a data recording event. The digital input module is used to trigger the data collection upon a pre-selected signal. The measurement sequence is started when a trigger is activated.

Computer Controls. The last major component of the data acquisition system is the computer used for system control and reduction of the raw data to appropriate engineering units. The computer system consists of a HP310 processor with a HP9133L disk drive, containing a 40 Mbyte Winchester hard disk, and a 710 Kbyte floppy disk drive. Additional modules required for high-speed data transfer and floating-point calculations have been added to maximize data collection and enhance the subsequent reduction to engineering units. The HP310 system is enclosed in a field portable rack. A printer and plotter are used to complete the graphical display and hard copy capabilities.

The HP310 system is fully compatible with the HP3852A acquisition unit, and custom data acquisition software has been written. Data collected during the field experiments can be stored a) directly in the 13-bit high-speed voltmeter, b) transferred to the HP3852A mainframe memory, c) transferred to the HP310 computer memory, or d) transferred directly to the computer's 40 Mbyte hard disk. The maximum data scanning rate is 100 kHz when using modes a) and b), as given above. The data transfer rate when using modes c) and d) is approximately 70 kHz. All transfer rates are entirely suitable for both static and dynamic measurements.

Software. The software to control the measurement process, including data reduction and display, is as important as the hardware. The computer software to control the data acquisition has been developed within three main modules. Figure 2 shows the three modules schematically. Each module is a separate program, so they can be used at any time to monitor instruments, acquire data, or reduce data collected previously. An important feature of the software is that all modules were developed by the personnel who would be responsible for collecting and analyzing the field data. Although the programs were developed initially for a specific project, the approach taken was to develop program modules that were flexible and could be used in a variety of experimental situations.

The first software module is used to monitor the instruments prior to data collection. The output voltages from all transducers are displayed on the computer monitor, allowing the user to balance the instruments and determine which, if any, instruments are not functioning or are giving inconsistent readings. Identifying instruments that are not working properly allows the user to repair the instrument prior to the experiment, if possible, or substitute another instrument in its place prior to acquiring what otherwise would be worthless data. Shunt calibration circuits are provided for each instrument to provide remote transducer calibration and to verify proper response. While the sensor outputs are displayed, each instrument can be shunted, and the change in voltage is updated on the display. Since each channel has independent gain, the gains can be adjusted during this phase to achieve any desired calibration factor, using the same wiring circuitry that is used during the data collection phase. The calibration software calculates and stores the calibration factors for each instrument in a disk file

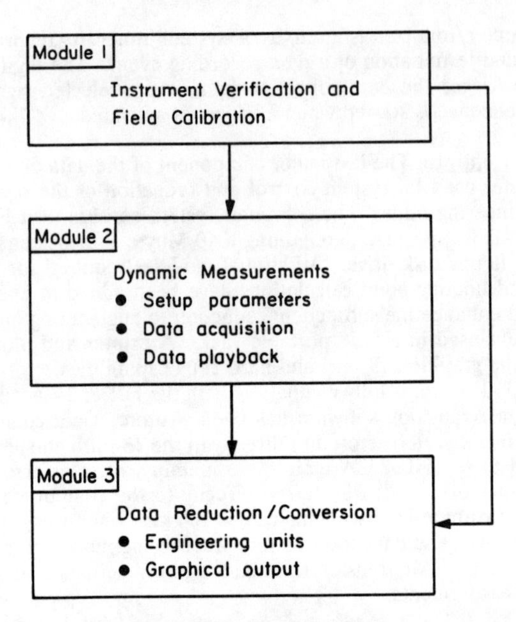

Figure 2. Software Modules

that can be used during data reduction.

The second software module is the main dynamic data acquisition program. This module is used to set the acquisition parameters such as the channels to scan, the sampling frequency, and the total duration of the acquisition phase. Checks are built into the program to ensure that hardware limits will not be exceeded; i.e., maximum total transfer rate, frequency, etc. All setup parameters are stored in an information file that becomes part of the data record. This information file is necessary to unpack the high speed data and assign the appropriate calibration factors. Once the setup parameters have been recorded, the system is ready to acquire data. Triggering options include keyboard initiation or triggering from a remote circuit. Remote circuit triggering, either from an instrument or using the counter/totalizer, allows for data pre-scan prior to a critical event.

All data collected (instrument output voltages) are transferred during data collection to both the hard disk and computer memory. Direct hard disk transfer is important, should there be a failure of an electronic component during acquisition. Partial recovery of data could be accomplished for an incomplete acquisition run. Generally, all data are copied to a floppy disk as well for permanent storage and subsequent data reduction.

The dynamic measurement module also has capabilities of playing back any of the instrument channels for immediate inspection of the raw data, prior to further processing or conversion to engineering units. By using the individual channel gains to set the instrument calibration factors to convenient numbers, the immediate playback feature can be used to determine very rapidly the instrument measurements. The playback portion of the program displays a figure of voltage versus time from any instrument selected. Convenient calibration factors (e.g., 1 microstrain per millivolt) allow the user to scan the entire record and visually determine the quality of the data. Calibration factors also can be used to display data in engineering units as well. This playback was used extensively in the dynamic system verification testing.

The third software module is used for final data presentation. This program module uses the calibration factors obtained from the first software module and the raw data from the second main data acquisition module. Data are reduced to engineering units and can be compensated for initial transducer imbalances by removing zero offsets from the data. Reduced data can be plotted versus time or plotted against another data set. The plots can be expanded to enlarge parts of the record and zoom in on important features. Minimum and maximum values are extracted from the data sets and displayed alongside the plots. The plots can be dumped to either the printer or plotter for further analysis.

Spectral distributions also can be displayed for frequency content analysis. An immediate analysis of the frequency content of any measurement is important. It allows the viewer to decide if the scanning rate was sufficient to capture the important signal characteristics and adjust the sampling frequency accordingly. This is particularly important when looking at high frequency content data, such as that generated by close-in blasting operations. Blasting vibrations near the source can contain significant energy at up to 300 - 400 Hz, and commercial blast monitoring equipment may not be capable of scan rates high enough to capture the true response. An immediate view of the spectral energy distribution lets the operator decide if the scan rate was sufficient, as well as identify fundamental field test responses.

APPLICATIONS

The data acquisition system described above has been used for several field testing programs. Data are presented from two of the projects to illustrate the system capabilities.

Pipeline Testing. Recent research projects with instrumented field test sites have focused on the development of design procedures for uncased gas pipelines (Ingraffea, et al., 1991; Stewart and Behn, 1991). These uncased pipelines must be designed to withstand the live load stresses imposed by vehicular traffic, as well as the stresses due to internal pressure and earth load. Two instrumented high-pressure steel pipelines were installed without casings, using auger boring methods, at the Transportation Test Center (TTC) in Pueblo, CO. Field experiments were conducted to measure pipeline response due to train loading. The effects of vehicle speed, internal pressure, and time since pipeline installation were investigated over a two-year period. Figure 3 shows profiles of the two pipelines. The 12.75-in. (324-mm) diameter pipeline has a wall thickness of 0.250 in. (6.4 mm) and a specified minimum yield strength of 42,000 psi (290 MPa). The 36-in. (914-mm) diameter pipeline has a wall thickness of 0.606 in. (15.4 mm) and specified minimum yield strength of 60,000 psi (414 MPa). The depth from the

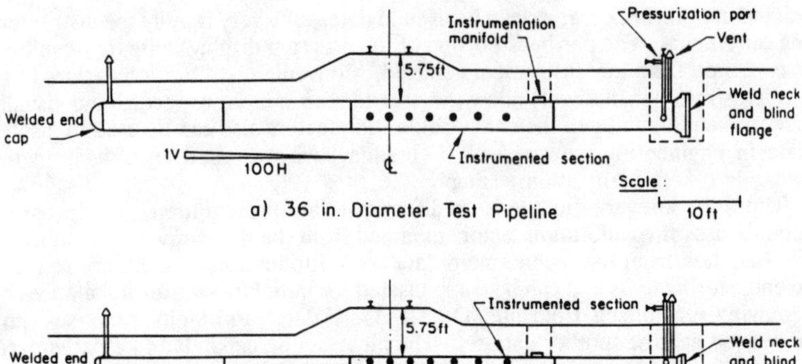

Figure 3. Profile Views of Test Pipelines

top of the railroad crossties to the crown of both pipes at the track centerline is 5.75 ft (1.75 m).

Both pipelines were instrumented prior to field installation (Stewart, et al., 1991). Pipe instrumentation consisted of strain gages, both internal and external, on the pipes, accelerometers, pressure transducers, and temperature sensors. Strain gages also were mounted on the rails directly above the pipes to measure the applied wheel loads. The strain gages on the pipes were oriented to measure both circumferential and longitudinal strains at the inside and outside crown, springlines, and invert. The locations of the instrument stations are shown on Figure 3 as solid circles. The gage locations correspond to locations on the pipelines directly beneath the outside rail, track centerline, inside rail, and other locations along the pipe's long axis sufficient to measure the distribution of strains along the pipeline.

Testing of the pipelines was initiated in July, 1988, with measurements made at four to six-month intervals through July, 1990. Baseline measurements of live load response were recorded from both the 12.75-in. (324-mm) and 36-in. (914-mm) diameter pipes, and special impact and impingement testing was conducted with the 36-in. (914-mm) pipeline.

Train loading was generated by loaded freight cars weighing 315,000 lb (1400 kN), producing 39.4-kip (175-kN) wheel loads. These freight cars are referred to as 125-ton cars, and are representative of the heavy loadings anticipated in the near future on U.S. revenue lines.

Figure 4 shows a typical data set. The sampling rate for the data shown in Figure 4 was 200 Hz. This sampling rate was more than adequate to capture the maximum pipeline strains at a depth of 5.75 ft (1.75 m), as well as the shape of the strain-time response of the pipe. However, the shear gage circuits on the rails needed to be sampled at a minimum of 200 Hz to capture the peak wheel-rail forces. A slower data sampling rate would not have allowed the peak rail

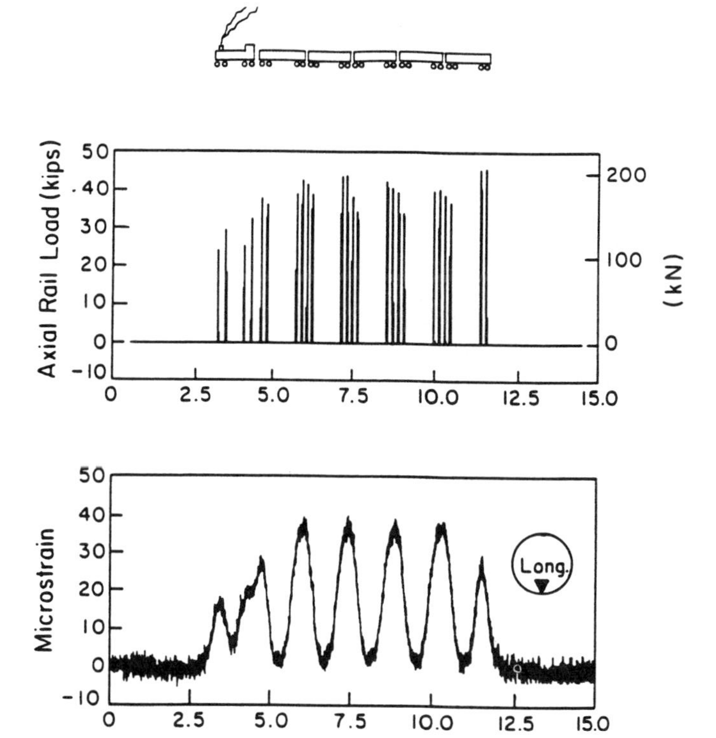

Figure 4. Dynamic Wheel Loads and Longitudinal Strains at Invert for 30 mph Train Speed

forces to be detected.

Blast Monitoring. Vibration frequencies from construction blasting often can contain significant energies at locations relatively close to the blast. New (1989) has reported significant explosive energies present at up to 400 Hz due to tunnel blasting at distances of 120 ft (36 m) from the blast source. Data acquisition rates below 1000 Hz could not guarantee that maximum responses were measured for frequency contents of 400 - 500 Hz. Even if the peak instrument responses were recorded, the shape of the wave form would be distorted greatly.

A blast monitoring program was completed recently using the dynamic data acquisition system described above. Five triaxial geophones were used to monitor particle velocities at locations ranging from 20 to 40 ft (6.1 to 12.2 m) from the blast face. Masonry structures between three and four stories high were within 20 to 100 ft (6.1 to 30.5 m) of the blast face, so vibration levels were of concern.

Charge weights of 3 to 5 lb (1.4 to 2.3 kg) per delay were used, with delay times ranging from 8 to 17 msec per detonation.

Figure 5 shows a typical record from a vertically oriented geophone located approximately 30 ft (9.1 m) from the blast face, and 10 ft (3.0 m) from a structure. The data sampling rate was 4 kHz per channel, and the scan duration was 5 sec. Thus, for the 15 channels, the total number of data recorded over the 5 sec duration was 300,000. Figure 5a shows the entire record from the blast and allows the maximum vertical particle velocity to be determined. However, the time scale needs to be expanded, as shown in Figure 5b, to see more detail. Figure 5c shows a further expansion of the time scale, isolating individual portions of the record. The shape of the particle velocity-time response also is preserved accurately when high data sampling rates are used. The plots shown in Figure 5 can be generated immediately after the blast, and can be used for establishing monitoring locations for subsequent blasts.

Figure 6 shows the results of a fast Fourier transform (FFT) of the entire record shown in Figure 5a. As described previously, the software can perform these analyses quickly to aid in planning data acquisition rates. The blast contained significant energies up to roughly 400 Hz, with two major peaks in the distribution near 200 and 300 Hz. Sampling at the Nyquist frequency of 400 - 600 Hz most likely would have missed the peak responses, and resulted in a distorted shape of the particle velocity-time curves.

The frequency distribution shown in Figure 6, and the relatively high vertical particle velocities shown in Figure 5, also point out clearly the need to anchor geophones firmly to the surfaces on which they are mounted. For example, the maximum vertical accelerations for a vertical particle velocity of $\pm$ 2 in./sec ($\pm$ 50 mm/sec) at 200 to 400 Hz would range from 6.5 to 13 gs. This presents no problem for upward ground accelerations, but the geophone would not remain in contact with the ground as the ground accelerated downward. The free-falling geophone then would strike the ground and result in a severe distortion of the signal. Merely adding sandbags over the geophone would not solve the problem and may worsen the situation, since the sandbags also would be falling near the geophone.

The data acquisition software can quickly display spectral response information, as shown in Figure 6, so that changes in data recording procedures can be made rapidly. It should be noted that typical commercial equipment used for blast monitoring often does not sample at a rate high enough to ensure that the maximum instrument responses will be recorded. Also, the electronic components in commercial systems may not have sufficient slew rates to capture rapidly fluctuating instrument voltages.

SUMMARY

The hardware and software components of a high speed data acquisition system were presented. The system is capable of monitoring simultaneously up to 48 channels of low or high level signals at a frequency of up to 2 kHz. Fewer channels can be scanned at higher rates. Computer control software initiates parameter setup, instrument verification and calibration, data collection, storage, and immediate playback of either raw data, data reduced to engineering units, or

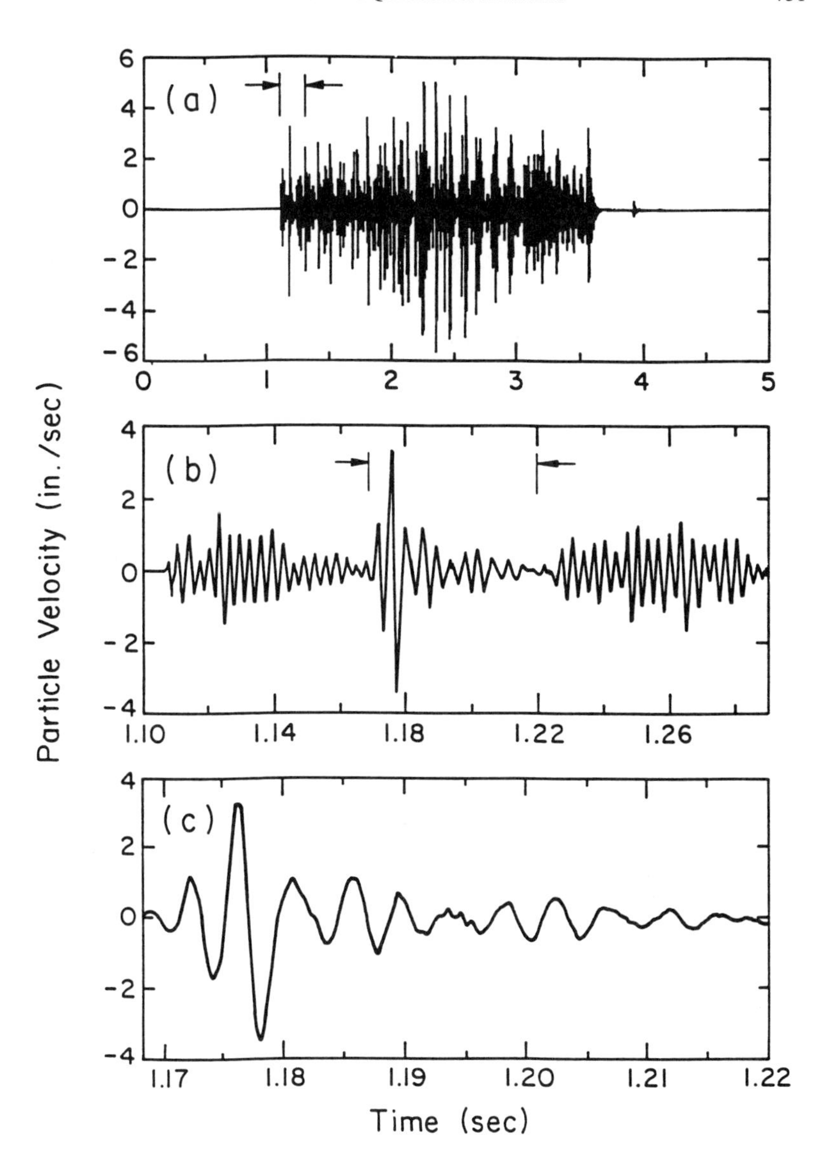

Figure 5. Measured Vertical Particle Velocity versus Time

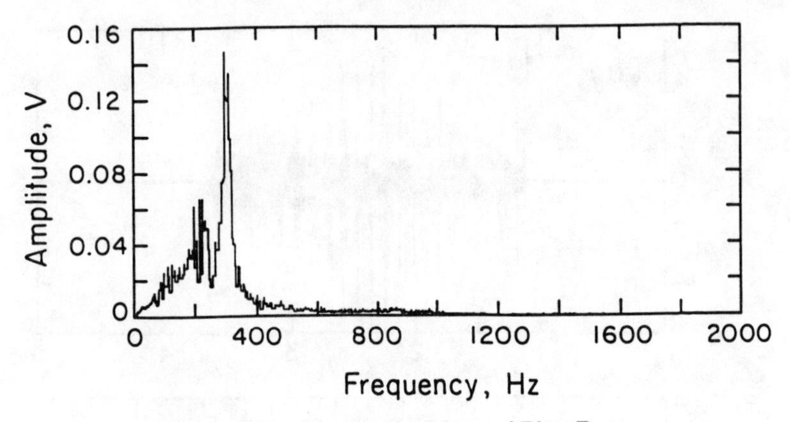

Figure 6. Fast Fourier Transform of Blast Data

frequency content analyses.

The flexibility of the system allows for rational changes in experimental procedures. For example, malfunctioning instruments can be identified quickly, and either replaced or eliminated from the scan list. Immediate inspection of the data from several instrument locations can identify whether additional measurement locations are necessary to capture an important field response.

Example applications were shown from two field experiments. The system has been in use for several years for ongoing pipeline research where a large number of instruments must be recorded at remote field sites. The ability to quickly inspect data when at remote locations is extremely important. The second application described the need for very high sampling rates for monitoring of vibrations caused by blasting at small distances from the blast source. Timely data reduction is important when making construction decisions and planning on-site monitoring locations.

ACKNOWLEDGMENTS

The equipment described in this paper was developed as part of research supported by the New York Gas Group (NYGAS) and the Gas Research Institute (GRI). K. J. Stewart prepared the manuscript and A. Avcisoy drafted the figures.

REFERENCES

1. AR Ingraffea, TD O'Rourke, HE Stewart, A Barry & CW Crossley, Guidelines for Uncased Crossings of Railroads and Highways, Proc. ASCE Spec. Conf. on Pipeline Crossings, Denver, Mar 1991.
2. BM New, Trial and Construction Induced Blasting Vibration at the Penmaenbach Road Tunnel, TRRL Res. Rpt 181, Department of Transport, Trans. & Road Res. Lab., Crowthorne, U.K., 1989, 11 p.

3. HE Stewart & MT Behn, Impact Effects on Pipelines Beneath Railroads, 70th Ann. Mtg, Transportation Res. Board, Washington, D.C., Jan 1991.
4. HE Stewart, MT Behn & TD O'Rourke, Instrumentation for High Pressure Pipelines Installed Using Auger Boring Methods, Proc. ASCE Spec. Conf. on Pipeline Crossings, Denver, Mar 1991.

A COMPUTER-ASSISTED AUTOMATED DATA ACQUISITION SYSTEM FOR REMOTE MONITORING

John T. Waggoner [1] and J.E. O'Rourke [2], F.ASCE

ABSTRACT: A computer-assisted automated data acquisition system (ADAS) was used to remotely monitor geotechnical instruments during construction of a cut-and-cover subway station in downtown Los Angeles. The on site (field) elements of the system consisted of more than 200 vibrating wire strain gages connected to a "data logger." Telephone service was provided to the data logger for remote access. Data logger programming, data collection from the data logger, data processing, and paper copy output were accomplished at office locations remote from the field site using a lap-top computer equipped with a modem. This relatively simple set up for instrument monitoring proved to be a powerful and reliable tool to track and analyze support system performance during construction. The system was cost-effective and provided information that could not be matched using manual methods.

INTRODUCTION

As with any new tool, ADAS's are becoming more widely used in engineering practice as workers become more familiar with available systems and become more comfortable with their capabilities. Available systems range widely in their capabilities, complexity and cost. The purpose of this paper is to describe a relatively simple ADAS that not only performed well during its approximately two years of use, but also provided useful capabilities beyond project requirements. This example shows that automated systems need not necessarily be complex and difficult to implement. Rather, they can be reliable, relatively simple to operate, cost effective and can provide information that cannot be matched through manual methods.

PROJECT BACKGROUND

Project Description. Initial sections of the Los Angeles Metro Subway project are under construction in downtown Los Angeles. One element of the system is the 5th and Hill Street Subway Station. The subway station was constructed by cut-and-cover methods; construction excavation support was provided by soldier piles, lagging, and pipe struts. The station is approximately 800 feet long, 65 feet wide and 65 feet deep. Project specifications required that the construction contractor install and monitor geotechnical instrumentation

1 - Associate, Woodward-Clyde Consultants, Santa Ana, CA, 92705
2 - Senior Associate, Woodward-Clyde Consultants, Oakland, CA, 94607

during construction. As is often the case for this type of project, the purpose of the instrumentation program was to monitor the performance of the temporary support system and verify its adequacy.

For this project, the required instruments included inclinometers, groundwater observation wells, settlement points, and strain gages. Although the ADAS could have been set up to also monitor other instruments in addition to strain gages, as discussed in the following paragraphs the ADAS was used only for the monitoring of strain gages.

Initial Instrument Requirements. Project specifications required that instruments be installed on the temporary support system, and that the instruments be monitored during excavation and construction until the support system was no longer needed. Because the construction contractor selected pipe struts for excavation support, strain gages were required. Further, the specifications required three "load zones" to be monitored: one near each end of the station and one near the center. Each of the load zones were to consist of three adjacent instrumented support members at each support level, resulting in nine instrumented struts per load zone and a total of 27 instrumented struts. The specifications also required that pipe strut bracing members be instrumented with three strain gages each, with the gages mounted circumferentially at 120 degrees. As a result, a total of 81 strain gages were required for the project.

Additional Instruments Requested. During the course of the project, the owner elected to expand the instrumentation program and requested that strain gages be installed on 42 additional pipe struts. Forty of these additional instrumented pipe struts received three strain gages each, similar to the original 27 pipe struts. Two of the additional pipe struts received six strain gages. As a result of this expansion, a total of 69 pipe struts were instrumented, with a total of 213 strain gages.

Monitoring Requirements. The required instrument monitoring schedule included daily readings of all gages where excavation was in progress within 50 feet of the instrumented support member and weekly readings if no excavation was completed within 50 feet. In addition, the data collected from the gages was to be submitted to the construction manager within 24 hours of obtaining readings. From this data, decisions regarding the adequacy of the temporary support system were based.

Daily strain gage readings were obtained during the excavation phase of the project that lasted approximately eight months. Following the excavation phase, weekly monitoring was performed for about 16 months, until the last of the instrumented struts were removed.

Approach to Instrument Monitoring. In considering an optimal approach to fulfill the strain gage monitoring requirements for the project, the authors decided that the monitoring requirements warranted the installation of an ADAS. It was estimated that the increased equipment cost of the ADAS (about $5,000 for the initial system configuration) would be easily recovered over the duration of the project through the greater efficiency of the ADAS and the resulting savings in labor cost for monitoring. Initially, the ADAS was installed and configured to automatically read up to 96 strain gages. Later, in response to the request to increase the number of strain gages, the system capacity was increased to handle up to 224 strain gages. The following sections describe the configuration and operation of the ADAS.

ELEMENTS AND CONFIGURATION OF FIELD EQUIPMENT

Vibrating Wire Strain Gages. According to project specifications, vibrating wire strain gages were required to monitor the change in strain at the designated locations on the pipe struts. To afford greater protection for the gages and lead wires, the project team installed the gages on the inside of the pipe struts during strut fabrication. The strain gages used were model VK-4100, manufactured by Geokon, Inc. of Lebanon, New Hampshire. With the gages previously installed inside the struts, final gage installation at the project site was reduced to connecting the gage lead wires to previously installed junction boxes at each load zone. Initial "no load" readings were obtained prior to strut pre-loading.

The vibrating wire strain gages also contained thermistors, which could have been used to monitor temperature variation. Although the data logger system was also capable of monitoring thermistors, temperature monitoring was not required.

Strain Gage Wiring. Two aspects of the strain gage wiring are worth noting. First, because the lead wires were run inside the pipe struts and along/inside steel members of the support system, the wires were relatively well protected. As a result, relatively little wire damage occurred. Second, rather than run a separate lead wire from each gage to the data logger, individual lead wires were connected to multi-conductor cables at junction boxes (see below). In using multi-conductor cables with a common ground, the wiring requirements (and thus installation costs) were significantly reduced. Wires were color-coded, labeled, and carefully spliced, and no significant problems were experienced.

Junction Boxes. Because load zones were required near each end of the station and near the center, a main junction box was installed at the top of each load zone, with subordinate junction boxes at the two lower levels. The junction boxes used were simple steel enclosures with terminal strips mounted inside for wire connections. These junction boxes were prefabricated and, where possible, were installed ahead of the struts to facilitate the wiring of gages to the system as instrumented struts were installed.

Multiplexers. A multiplexer is basically a circuit board on which a series of switches is mounted. By the successive opening and closing of these switches, the data logger addresses the system of sensors, one at a time. Through the use of multiplexers, the data logger's capacity for sensor connections can be expanded to accommodate several hundred sensors. For this project, two multiplexers were initially installed with the data logger to accommodate the 81 required strain gages. To accommodate the requested system expansion, two additional multiplexers were installed. Three of the multiplexers used had 64-channel capacity and one had 32-channel capacity, for a total capacity of 224 gages. The multiplexers were mounted in protective steel boxes and centrally located with respect to the load zones to reduce wiring requirements. The multiplexers were also provided by Geokon, Inc.

Data Logger. Several types of data logger systems are available from various instrument vendors. The data logger selected for this project was programmable and could automatically interrogate the sensors as well as process and store the acquired data. The actual data logger control module, model "CR-10", was manufactured by Campbell Scientific, Inc. of Logan, Utah. This module, together with other circuitry to enable the measurement of vibrating wire strain gages, was mounted in a steel enclosure and pre-wired for our use by Geokon, Inc. The data logger enclosure was mounted on the wall of the excavation, below street level, at about mid-station. This location was selected to reduce wiring requirements to the load zones.

For the purposes of this project, the data logger was programmed to interrogate all 213 of the strain gages every two hours, seven days per week. Although less frequent monitoring (daily or weekly) was required, the two-hour interrogation interval provided a means to check the validity of the readings. This capability is considered very important, and it is noted that as opposed to manual methods the more frequent monitoring could be performed at no additional cost.

During collection, the data gathered by the data logger is internally stored for later downloading and processing as desired by the user. Like any computer, the amount of data that can be stored in the data logger is a function of the amount of memory installed. At a data collection interval of two hours for the 213 strain gages, the "CR-10" data logger was capable of storing about five days of data, which amounted to approximately 13,000 individual values (including date, time, data logger temperature and battery voltage data with each set of readings).

A trickle charger was installed with the data logger to maintain a charge in the NiCad battery, which powered the data logger. On two occasions during the project, power to the trickle charger was interrupted. On these occasions, it was observed that with a fully charged battery, the data logger could function for about a week while interrogating over 200 gages every 2 hours.

Modem and Telephone Service. Key elements of the installation were the modem and telephone line. These elements allowed for remote programming of the data logger and remote data collection. The battery-operated modem was mounted inside the data logger's steel enclosure and plugged directly into the data logger. The telephone line was brought to the data logger from the basement of a building, which was located adjacent to the subway station excavation.

Field System Configuration. Figures 1 and 2 provide plan and profile views of the general locations of the load zones containing the 69 instrumented pipe struts and the locations of the various components of the data logger system.

ELEMENTS OF OFFICE EQUIPMENT

Lap-top Computer with Modem. A lap-top computer was used for programming and downloading of data from the ADAS as well as data reduction and generation of output. The computer was configured with 640k RAM, one 720k floppy disk drive, one 20 Mb hard disk and an internal 1200 baud modem. A lap-top computer was selected for use on the project to enable communication

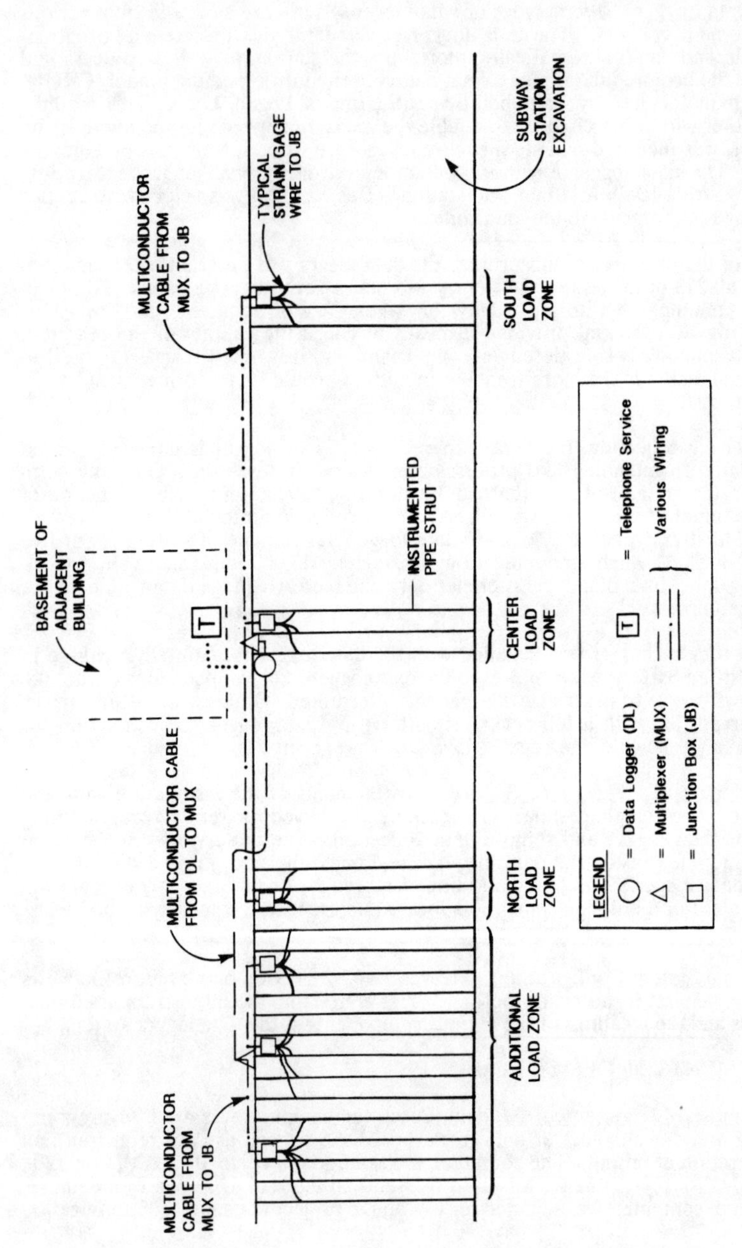

FIGURE 1. PLAN VIEW OF INSTRUMENT LAYOUT

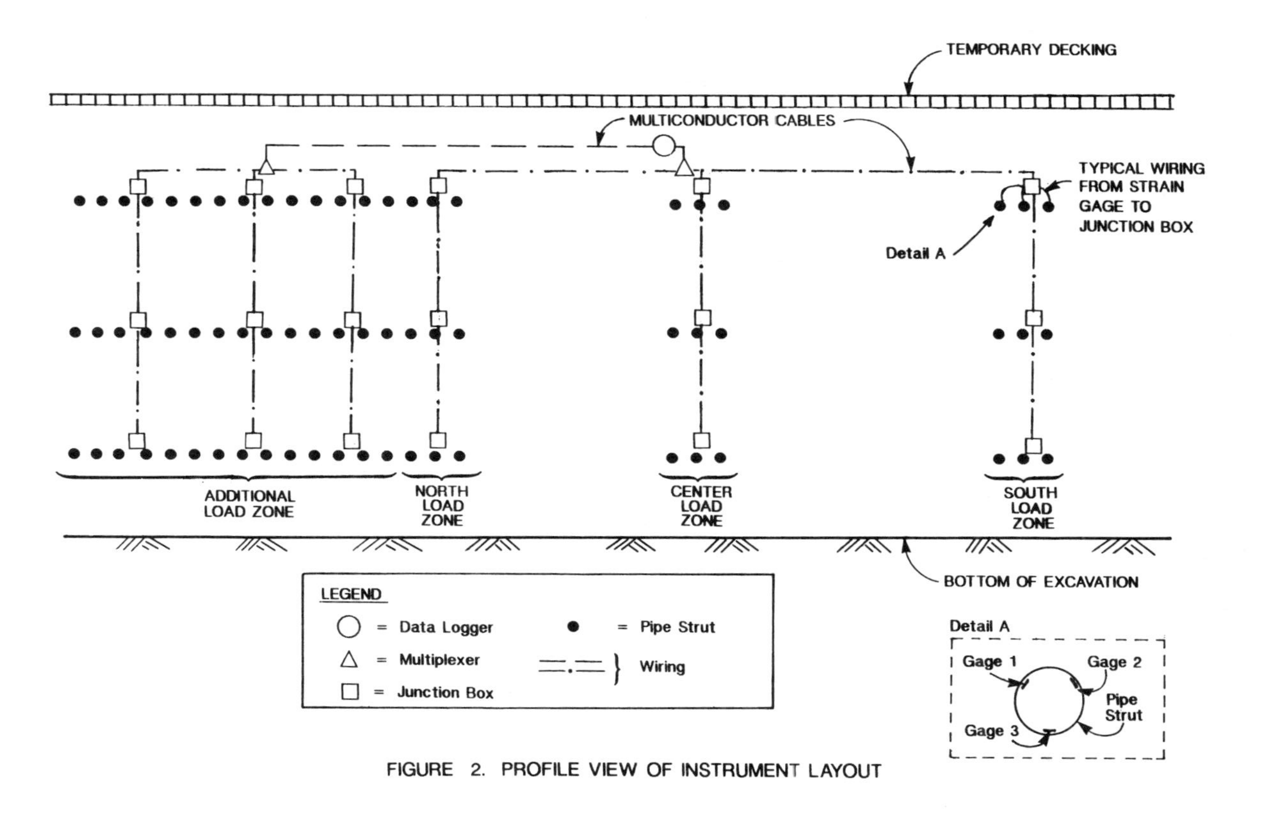

FIGURE 2. PROFILE VIEW OF INSTRUMENT LAYOUT

with the data logger from any location with telephone service. This mobility provided the opportunity to check the performance of newly installed system components before departing from the field site, and provided adequate computing power to process the large amount of data that was collected from our remote office locations. It is noted that a direct link could also be established with the data logger through an RS-232 port on both the computer and data logger enclosure.

Data Logger Software. The data logger was supplied with Campbell Scientific Inc.'s standard support software. The standard software included six separate programs that support various data logger operations. During the course of this project, only four of these programs were regularly used. The four programs provided for the following basic operations: 1) data logger program development, 2) telecommunication for program downloading to the data logger, 3) telecommunication for data retrieval from the data logger, and 4) raw data reformatting. Each of these four programs is described further below.

Data Logger Program Development. The data logger program was developed using a program editor. The program editor, "EDLOG" utilized an instruction set similar to a computer programming language. All of the data logger operations were controlled by the program. For example, instructions for the number and type of sensors were included in the program, as were the sensor interrogation interval, data processing instructions, and data storage. The data logger program can be simple or complex. For example, a simple program could include only interrogating a sensor and recording the raw reading. On the other hand, the data logger could be programmed to obtain and process data from many sensors, make comparative evaluations, and based on the results, activate relays that in turn can initiate actions such as sounding warning signals.

The programmable feature of the data logger provided great flexibility in setting up the monitoring system. Because remote access was possible, modifications to the data logger program could be made at any time. A program modification could be simply a change in the sensor interrogation interval, or perhaps the entry of a completely different program. In either case, the modification could be made remotely in a few minutes.

Telecommunications for Program Downloading. Once the data logger program was prepared, it was downloaded to the data logger using other manufacturer's software written for that purpose. Program downloading was accomplished using "TERM", a terminal emulator program that facilitated telecommunication with the data logger. Key options provided by the terminal emulator program included downloading and retrieving programs from the data logger, and real-time monitoring of data as it was being acquired by the data logger.

Telecommunications for Data Retrieval. Retrieval of data stored in the data logger was accomplished using "TELCOM" a program that enabled telecommunication with the data logger and facilitated the collection of data. For example, through the use of this program, the user could obtain all available data or only selected data. Selected data to be retrieved could be specified in terms of a number of sets of readings, or readings from a specific time interval. An example of raw data collected from the data logger is provided below in Table 1. It is noted that in this format, the data does not appear particularly useful. For this reason, a data reformatting program was used.

Table 1. Sample of Raw Strain Gage Data

118, 2055, 2578, 1743, 152, 602, 2052, 2573, 1740, 152, 802, 2049, 2570, 1740, 152, 1002, 2042, 2564, 1738, 152, 1202

Raw Data Reformatting. The "SPLIT" program was used to convert the raw data collected from the data logger, into a more useable form. This conversion program also provided the ability to make simple calculations. From the data in the format shown in Table 1, the reformatting program could produce output such as listed below in Table 2.

Table 2. Sample of Processed Data

1990 Day	Time	Gage 1	Gage 2	Gage 3
152	602	2055	2578	1743
152	802	2052	2573	1740
152	1002	2049	2570	1740
152	1202	2042	2564	1738

Spreadsheet Software. Although the data presented in Table 2 above is neatly formatted, additional calculations were necessary to obtain values such as load change per strain gage, average axial load per strut, and percent of maximum design load. Commercially-available spreadsheet software ("123" by Lotus Development Corporation) was used to make these calculations and to produce neatly formatted tabular output. A template spreadsheet was prepared that contained the basic information for all of the strain gages. The template spreadsheet contained initial strain values and formulas by which the other desired values were calculated. The spreadsheet was set up so that the user imported raw strain gage data into designated locations (cells). Spreadsheet cells into which the data was imported were referenced by formulas contained in other cells such that once the data were imported into the spreadsheet, the calculations were automatically completed. An "auto-execute macro" (standard feature of the spreadsheet program) was also implemented in the spreadsheet to automate the process of importing the raw strain gage data files, completing the calculations and producing output. By incorporating the auto-execute macro into the spreadsheet, the user had only to enter the spreadsheet program and load the template file to initiate processing. The process was further automated by assigning a special file name to the template spreadsheet, which caused it to be automatically loaded when the spreadsheet program was entered, and in turn initiating the auto-execute macro. An example spreadsheet print out is shown in Table 3.

Table 3. Example Final Strain Gage Data Printout

STRAIN GAGE DATA FILE - ADDITIONAL LOAD ZONE, LEVEL ONE
Contract A-145; 5th and Hill Station

Date: ______
Time: ______

Initial Data						Present Readings			
Gage	Reading (micro-strain)	Date	Strut Diameter (inches)	Approx. Strut Wall Thick. (inches)	Approx. X-sect. Area (sq.in.)	Reading (micro-strain)	Strain Change	Load Change (kips)	Average Axial Load (kips)
A1A1@	2550	06/08/89	30	0.75	68.9	2397	-153	306	
A1A2@	2487	06/08/89	30	0.75	68.9	2365	-122	244	
A1A3@	2477	06/08/89	30	0.75	68.9	2186	-291	582	377
A1B1	2700	06/08/89	30	0.75	68.9	2584	-116	232	
A1B2	2653	06/08/89	30	0.75	68.9	2560	-93	186	
A1B3##	2930	07/19/89	30	0.75	68.9	2688	-242	484	300
A1C1	2493	06/08/89	30	0.75	68.9	2481	-12	24	
A1C2	2675	06/08/89	30	0.75	68.9	2502	-173	346	
A1C3	2523	06/08/89	30	0.75	68.9	2240	-283	566	312
A1D1	2812	05/23/89	30	0.50	46.3	2575	-237	318	
A1D2	2380	05/23/89	30	0.50	46.3	2173	-207	278	
A1D3	2433	05/23/89	30	0.50	46.3	2201	-232	312	303
A1E1	1981	05/23/89	30	0.50	46.3	1839	-142	191	
A1E2	2296	05/23/89	30	0.50	46.3	2180	-116	156	
A1E3	2194	05/23/89	30	0.50	46.3	2046	-148	199	182

Batch Processing Files. Each element of the data retrieval, processing and output were individually automated through the respective software. To further automate the process from data collection through data output, simple batch processing files were prepared. A series of batch files were written so that depending on the desired output, a one word command could be typed on the lap-top computer to initiate the sequence of commands. For example, a three-line batch file called "WEEKLY.BAT" was written to collect and process data on a weekly basis. This batch file caused the computer to call the data logger, retrieve the desired readings, reformat the raw data, import the reformatted data into the template spreadsheet, make all necessary calculations, and print out summary tables of processed data. Similar batch files were written for acquiring data on a daily basis, and to process data for only certain pipe struts rather than for all the struts.

SUMMARY AND CONCLUSIONS

The above paragraphs describe the basic components of a relatively simple ADAS, which was used to monitor a large number of strain gages for a period of about two years. Although relatively simple, the ADAS performed well, was very reliable and provided capabilities beyond project requirements. Now having completed the project, the authors consider it to have been successful and consider the decision to use the ADAS a good one. As a result, this paper was prepared in an effort to share information on this positive experience.

Although the ADAS is described above as being relatively simple, and the overall experience on this project was positive, it is noted that several factors are considered important to the successful completion of this project and should be considered in similar applications. The items listed below are considered to have contributed substantially to the success of this project:

- Considerable planning went into this project. The authors tried to anticipate potential pitfalls associated with installing and operating an ADAS within a construction site, and carefully planned as many aspects of the project as possible to avoid problems.

- The authors enjoyed a good working relationship with the construction contractor. Personnel from Guy F. Atkinson Construction Company, the construction contractor for the project, provided valuable assistance and advice during the project, particularly with respect to locating equipment in "safe" places. In addition, it is appreciated that the construction personnel were generally cautious around the instrumentation.

- The authors were provided excellent support by the instrumentation equipment vendors.

- Back-up systems were included where practical. As mentioned above, although electric power was available on site, the ADAS was powered by battery and a trickle charger was installed to keep the battery fully charged. In addition, the multiconductor cables selected could accommodate many more gages than necessary, thus reducing the possibility of having to run new wiring in case of damage to some of the wires.

- The system as a whole was kept as simple as possible. From the gage wiring to the data reduction, simplicity was the aim in each step of the project. For example, all of the gage wires were color coded, and all gages were wired to the multiconductor cables and junction boxes in a similar manner, thus making wiring repairs easier. Also, the data reduction was accomplished using the data logger manufacturer's standard software, common spreadsheet software and batch files. In this way the data reduction could be more easily performed by other project staff, and changes in the data reduction did not require extensive re-programming.

In considering the conduct of similar projects, the authors would certainly consider using the same system because of its satisfactory performance on this project. However, invariably each project has its particular requirements and should be considered independently. In addition, although an automated system can provide excellent monitoring capabilities, some projects simply do not require such a system and manual readings are approprate. It is also noted that even with the significant capabilities of an automated system, regular site visits are still necessary to observe field conditions and confirm conclusions based on monitoring data.

ESTIMATION OF FOUNDATION SETTLEMENTS IN SAND FROM CPT

P.K. Robertson[1], M.ASCE

ABSTRACT: Many methods to estimate settlement from penetration test results rely on empirical correlations between penetration resistance and soil modulus. However, because of the non-linear behavior of soil the modulus varies with the degree of loading and this is generally not directly accounted for in most methods. Another important factor controlling the soil modulus is the stress history and age of the deposit. The current state of knowledge of correlations between soil modulus for sands and penetration resistance based predominantly on large calibration chamber studies is reviewed. This review incorporates recent findings regarding the influence of strain level, stress history and aging. A new correlation based on the CPT is proposed that directly incorporates the degree of loading from the structure and the stress history and age of the soil deposit. The proposed correlation is discussed and compared with other existing methods.

INTRODUCTION

Geotechnical engineers are often required to estimate the settlement of structures founded on sand. Because of the difficulty in obtaining undisturbed samples of granular soils many empirical methods have been developed to estimate settlements from in-situ tests. Most of these methods have been based on penetration tests, especially the Standard Penetration Test (SPT) (Terzaghi and Peck, 1967). However, due to the large variation in delivered energy to the SPT and the natural variability of most granular deposits, settlement estimates based on the SPT have generally been rather poor. Newer more reliable and repeatable in-situ tests, such as the Cone Penetration Test (CPT), offer the potential for improved methods to estimate settlement and some empirical methods based on the CPT have already been developed (De Beer, 1948; Meyerhof, 1965; Schmertmann, 1970).

Most methods to estimate settlement from penetration test results implicitly rely on empirical correlations between penetration resistance and soil modulus. However, interpretation of penetration test results suffer some limitations which make the assessment of soil modulus very difficult. This is further complicated by the difficult link to the relevant drainage conditions, stress paths and stress or strain level of the specific design project.

The deformation characteristics of a given soil depend on; stress and strain history of the deposit (Jamiolkowski et al., 1985), current level of mean effective stress, induced level of strain, effective stress path followed and a time factor, such as, aging

[1] - Professor, Dept. of Civil Engineering, University of Alberta, Edmonton, Alberta, Canada, T6G 2G7.

and creep. Therefore, the correct and safe use of correlations between penetration resistance and soil moduli are influenced qualitatively by the skill of the engineer to account for all these factors.

A brief review of current correlations between soil modulus for sands and penetration resistance based predominantly on large calibration chamber studies is presented and a new correlation is proposed that attempts to incorporate many of the important factors in a simple and logical manner.

EXISTING CORRELATIONS

There are many empirical methods to estimate settlement of foundations in sand based on penetration test results and observed performance. The work of Burland and Burbidge (1985) is one of the most recent and notable methods based on performance.

In the last decade there has been considerable improvement in our understanding of stress-strain behavior of sands combined with extensive study of penetration testing, mostly in calibration chambers. Results of these studies have shown that the stiffness of sands is very sensitive to stress history. However, stress history effects can be due to increases in horizontal stresses as well as strain hardening due to accumulated plastic strains. The increases in horizontal stresses are conventionally linked with mechanical overconsolidation, whereas strain hardening generally appears as a consequence of aging, dessication and low strain cyclic stress history.

Calibration chamber studies have shown that penetration resistance is strongly influenced by the current level of horizontal effective stress and is almost totally insensitive to the effects of plastic strain hardening (Lambrechts and Leonards, 1978, Baldi et al, 1986). This can be illustrated by the correlation shown in Figure 1 between CPT penetration resistance (q_c) and the secant Young's modulus at 25 % of the failure load (E'_{25}). The ratio of E'_{25}/q_c is significantly larger for overconsolidated sands (OC) than for normally consolidated (NC) sands (Jamiolkowski et al. 1988) and the correlation varies with relative density, as shown in Figure 1.

It is clear from Figure 1 that the correct application of empirical correlations between penetration resistance and moduli must take account of the strong influence of stress and strain history. Many existing methods that estimate settlements of shallow foundations on sands from penetration test results make a clear distinction between normally and overconsolidated sands. However, in practice, the stress history of a sand is usually unknown, which can be a major limitation of these correlations.

The correlation in Figure 1 is based on a secant modulus defined at 25 % of the failure load. This was thought to represent a reasonable average moduli for most shallow foundations with a factor of safety of around 3 to 4. However, recent numerical studies of shallow foundations have shown that the average strain level beneath well designed footings is around 0.1 % or less (Battaglio and Jamiolkowski, 1987; Jardine et al 1986; Burland, 1990).

In recent years there has also been accumulated evidence concerning the importance of aging in sands on correlations with penetration resistance (Skempton, 1986, Bellotti et al. 1989, Schmertmann, 1989, Mesri et al., 1990). An attempt to incorporate the important effect of aging in sands was made by Bellotti et al. (1989) and is shown in Figure 2. The moduli is defined as the secant Young's modulus at an axial strain level of 0.1% (E'_s). The use of relative density has been removed and replaced by a normalized penetration resistance and a new correlation has been added to represent aged, normally consolidated sands. The normalized penetration resistance to relate CPT q_c to sand relative density has been defined as follows;

$$q_{c1} = \frac{q_c}{P_a}\left(\frac{P_a}{\sigma'_{vo}}\right)^{0.5} \tag{1}$$

where: q_{c1} = normalized penetration resistance
q_c = measured penetration resistance, same units as P_a
σ'_{vo} = vertical effective stress at depth of CPT
P_a = atmospheric pressure in the same units as σ'_{vo}

The correlation shown in Figure 2 represents an improvement over previous correlations in that it includes the effect of aging. Aged sands are considered to represent most natural sands with an age > 1,000 years. However, the correlation in Figure 2 assumes that the average strain beneath well designed footings will always be 0.1%. It is likely that the average strain beneath footings on very dense or overconsolidated sands will be less than 0.1% and possibly slightly more for loose unaged sands, depending on the applied foundation stress. Hence, it would be useful to incorporate some form of degree of loading into the correlation.

Stroud (1988) has attempted to include a degree of loading into a correlation between modulii and SPT blow count, as shown in Figure 3. Stroud (1988) suggested that the ratio q_{net}/q_{ult} could be used as an indirect measure of average strain in the soil beneath a shallow footing, where q_{ult} is the ultimate bearing stress at the

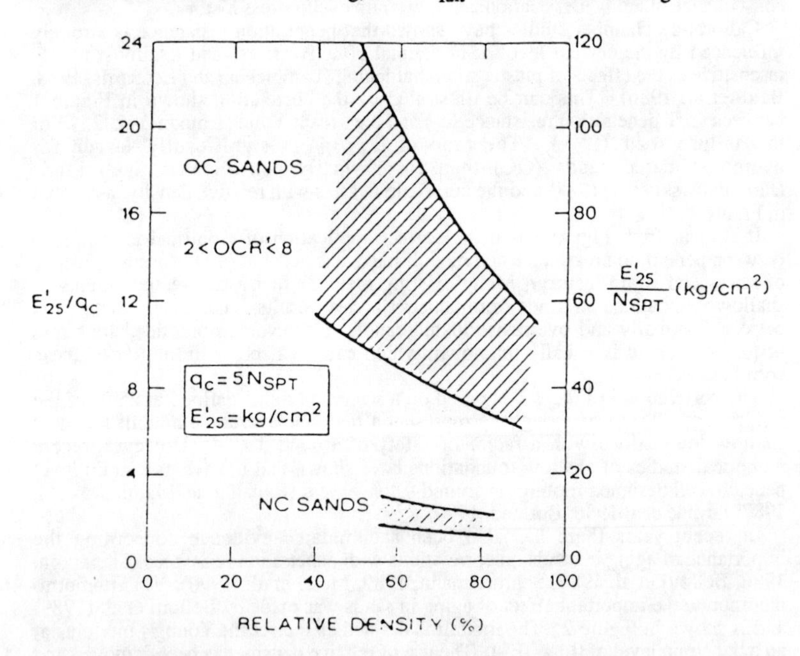

Figure 1. Relationship between CPT, q_c and SPT N_{60} and secant Young's Modulus at 25% failure load, E_{25} for uncemented, unaged silica sands. (After Ghionna and Robertson, 1987)

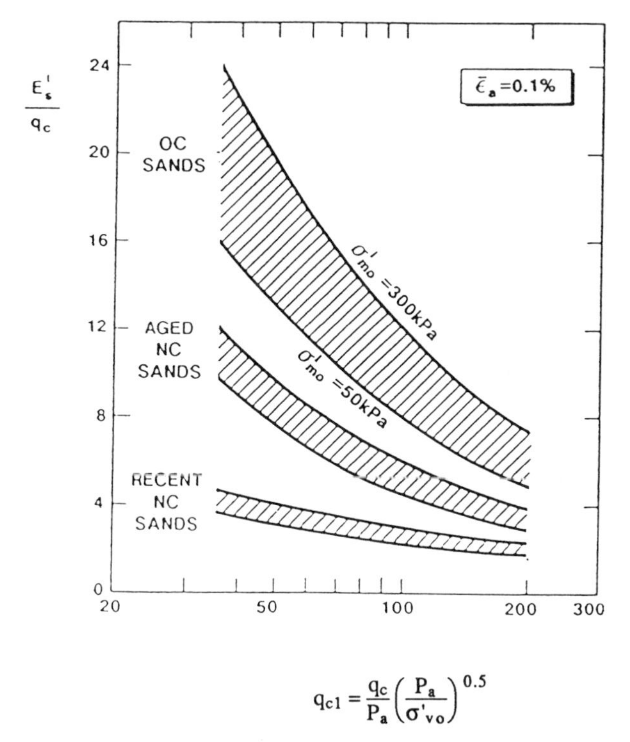

Figure 2. Relationship between CPT, q_c and secant Young's Modulus at 0.1% axial strain for uncemented, silica sands. (After Bellotti et al, 1989)

point of failure and q_{net} is the average net effective bearing pressure acting on the foundation. The value of the secant Young's modulus, E', shown in Figure 3 was estimated from the case histories referred to by Burland and Burbidge (1985) assuming a linear elastic theory and thus represents the average secant stiffness beneath the foundation under loading q_{net}. Stroud (1988) also showed that many of the case history foundations were found to have values of q_{net}/q_{ult} less than about 0.1. Stroud (1988) accounted for the variation of delivered energy to the SPT by 0correcting the data to an energy level of 60% by using N_{60}. It is also interesting to note that the correlation in Figure 3, which is based on observed foundation performance, shows a factor of about 2 or more difference in stiffness between overconsolidated and normally consolidated sands. This difference is consistent with that observed in laboratory studies as illustrated by the correlation in Figure 2. Most of the case histories involved sand deposits that were older that 1000 years in age.

The correlation proposed by Stroud (1988), and shown in Figure 3 represents an improvement in estimating the stiffness of sands from the SPT in that it includes the effects of stress history and the induced strain level in the soil. The concept of degree of loading will be applied to correlations based on the CPT in an effort to develop an improved relationship between q_c and E'.

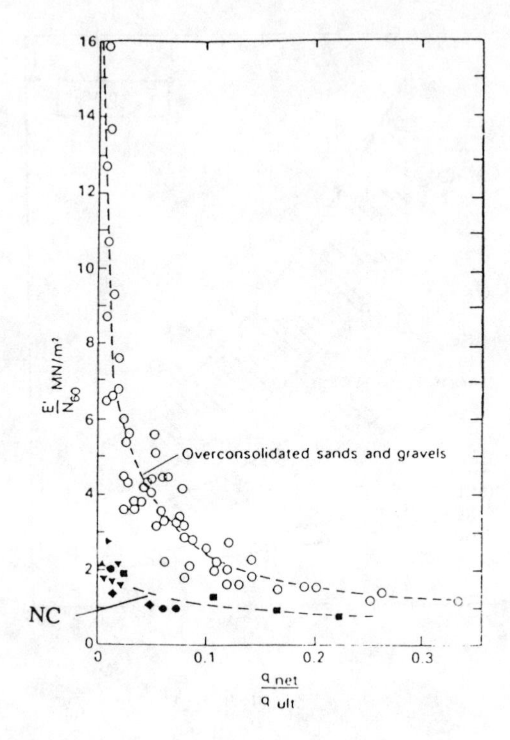

Figure 3. Variation of E'/N_{60} with degree of loading for overconsolidated and normally consolidated sands (Adapted from Stroud, 1988)

PROPOSED CORRELATION

Previous statements and comments can lead to a rather negative perception as far as the reliability of existing correlations between penetration resistance and the non-linear deformation moduli are concerned. However, a notable exception are correlations with the small strain shear modulus (G_o) measured at shear strain levels less than 10^{-3}% (Ohta and Goto, 1976; Sykora and Stokoe, 1983; Robertson 1982; Bellotti et al. 1986; Rix, 1984; Lo Presti, 1987).

A large amount of experimental data show that G_o in cohesionless soils is influenced very little by stress and strain history and is strongly controlled by void ratio and current level of effective stress (Lee and Stokoe, 1986). These same basic variables of void ratio and effective stress primarily control penetration resistance. An example of a CPT correlation between G_o and q_c based on extensive calibration chamber studies and laboratory resonant colunn test (RC) and verified by field results, is shown in Figure 4. By defining the modulus at a strain level of less than 10^{-3}% the ratio G_o/q_c appears to be almost independent of stress history (OCR).

Using the correlation shown in Figure 4 it is possible to develop a new relationship between q_c and the secant Young's modulus (E') defined at larger strain levels by

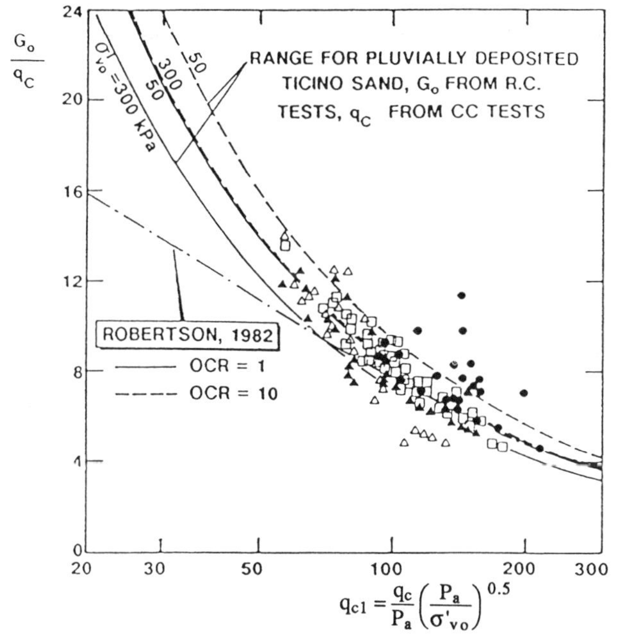

Figure 4. Relationship between CPT q_c and small strain shear modulus G_0 for uncemented silica sands (After Jamiolkowski and Robertson, 1988)

utilizing extensive laboratory data based on cyclic loading tests on a large variety of sands relating the decrease in modulus with increasing strain level, such as Seed and Idriss (1970), as shown in Figure 5. By applying the curves shown in Figure 5 an equivalent secant modulus for overconsolidated sands can be estimated at larger strains. The modulii apply only to overconsolidated sands since in both cases the loading is occurring below the yield surface. However, in order to relate the modulus reduction curves in Figure 5 to a degree of foundation loading (q_{net}/q_{ult}) it is necessary to establish a relationship between shear strain (γ) and q_{net}/q_{ult}. Stroud (1988) suggested using the studies of Eggestad (1963) to determine the average shear strain beneath model footings on normally consolidated sand for different degrees of loading, as shown on Figure 6. The shear strain was averaged over a depth of twice the footing width (2B).

The studies by Eggestad (1963) indicate that the relationship between degree of loading and the average shear strain is almost the same for a wide-range of relative densities. The work by Eggestad (1963) was performed using freshly deposited (i.e. recent), normally consolidated silica sand. Based on extensive laboratory studies and on observed field performance, Stroud (1988) suggested that the average shear strain for overconsolidated sands at the same degree of loading can be expected to be about half that of normally consolidated sand, as shown in Figure 6.

Combining the information in Figures 4, 5 and 6 it is now possible to develop a relationship between E'/q_c and q_{net}/q_{ult} for overconsolidated silica sands, as shown on Figure 7. The resulting correlation shows a high degree of variation with degree

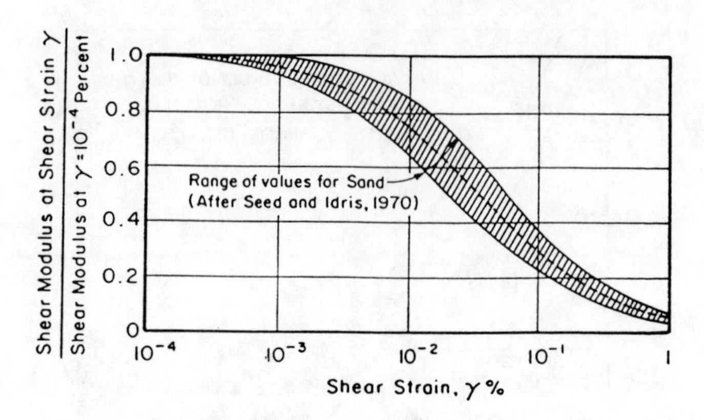

Figure 5. Variation of secant shear modulus with shear strain.

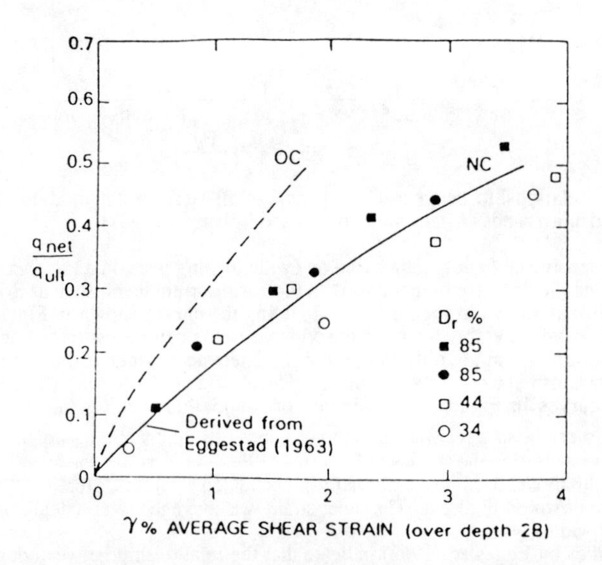

Figure 6. Approximate relationship between degree of loading and average shear strain beneath a footing on sand (After Stroud, 1988)

of loading and soil relative density, represented here by the normalized penetration resistance, q_{c1}. However, for the same applied foundation net bearing pressure (q_{net}) the degree of loading for a loose sand (i.e. q_{c1} = 30) will be much larger than for a dense sand (i.e. q_{c1} = 200) and the difference in the ratio E'/q_c will not be so large.

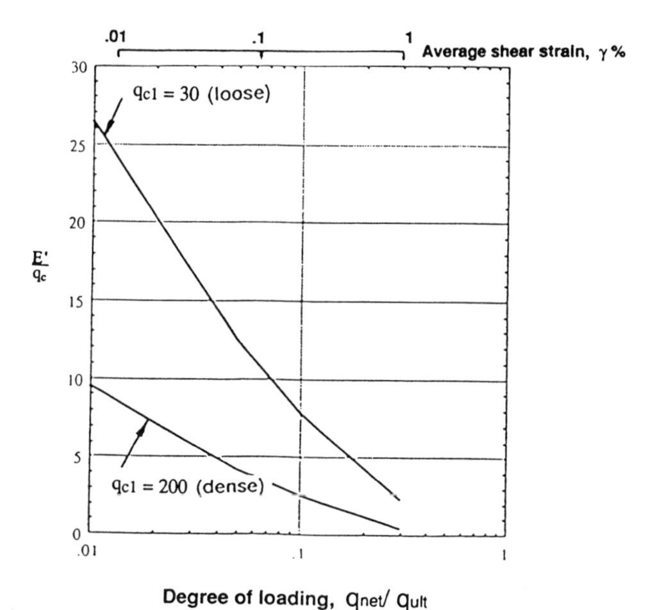

Figure 7. Proposed correlation between secant Young's Modulus E' and CPT q_c for overconsolidated silica sands based on degree of loading.

Based on previous observations, it would appear likely that the ratio E'/q_c derived from Figure 7 should be reduced by some factor when applied to normally consolidated sands. The following reduction factors are therefore recommended:

- Aged normally consolidated silica sands reduce E' by factor of 2,
- Recent (age < 1,000 years) normally consolidated silica sands reduce E' by factor of 3.

It is possible to compare the proposed correlation in Figure 7 with the SPT based method suggested by Stroud (1988) by assuming the following average relationship between the CPT and SPT penetration resistances for clean sand (Robertson et al. 1985):

$$\frac{q_c}{N_{60}} = 5 \qquad (2)$$

This comparison is shown in Figure 8. The CPT based correlation (Figure 7) was based on laboratory and field observations between q_c and the small strain shear modulus, Go, whereas the SPT based correlation (Stroud, 1988,) was based directly on field performance data (Burland and Burbidge, 1985). Although numerous approximate assumptions were made to develop the proposed correlation in Figure 7 the two correlations are remarkably similar for overconsolidated sands.

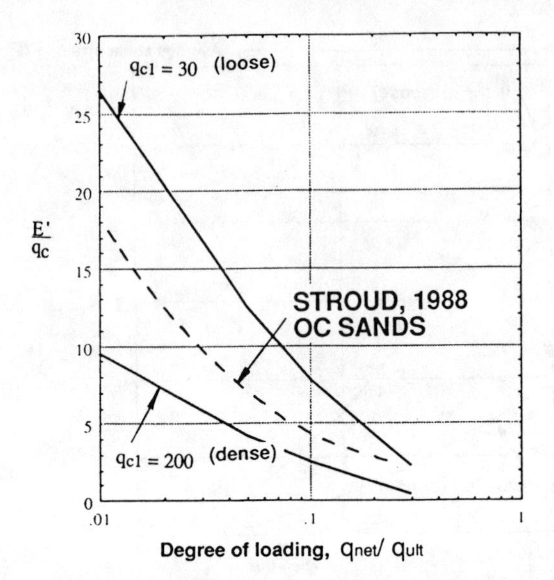

Figure 8. Comparison between proposed correlation based on CPT q_c and method based on SPT N_{60} by Stroud (1988) for overconsolidated sands.

Although the correlations in Figures 3 and 7 provide some improvement by including the degree of foundation loading to account for variations in induced strain levels, they still require some knowledge of stress history of the sand. Traditionally many natural sands have been considered to be normally consolidated. However, considerable evidence exits to suggest that most natural sands with an age greater than about 1,000 years behave as overconsolidated sands for most loading conditions. This is probably due to one or more of the following factors:

- aging
- cementation
- stress and strain history

These same factors have been recognized for some time to produce similar apparent preconsolidation in many clay soils.

WORKED EXAMPLE

To illustrate the application of the proposed correlation an example will be presented based on the following data:

Width of square foundation , B = 3m
Depth of foundation embedment, D = 1m
Unit weight of sand, γ =18 kN/m^2
Average CPT penetration resistance beneath proposed foundation, q_c = 15000 kN/m^2
Average in-situ vertical effective stress during CPT, σ'_{vo} = 100 kN/m^2
Proposed net foundation pressure, q_{net} = 400 kN/m^2.

The value of the foundation ultimate bearing stress, q_{ult} can be calculated using bearing capacity factors appropriate to local shear failure (Lambe and Whitman, 1969) estimated using direct correlations between penetration resistance and friction angle, ϕ', (Robertson and Campanella, 1983).

The normalized CPT penetration resistance is $q_{c1} = 150$. Using the correlation proposed by Robertson and Campanella (1983) the friction angle of the sand is $\phi' = 40^o$, and the ultimate bearing stress is $q_{ult} = 4000 kN/m^2$. Therefore, the foundation degree of loading (q_{net}/q_{ult}) will be approximately 0.10.

The settlement of the footing can be calculated using the following formulae derived from the theory of elasticity:

$$s = I_s \frac{q_{net} \cdot B}{E'} (1-\upsilon') \qquad (3)$$

where:
- s = settlement of foundation
- B = width of foundation
- q_{net} = net foundation pressure
- υ' = poisson's ratio
- E' = Drained Young's modulus
- I_s = influence factor depending on the shape and rigidity of the foundation and on the type of elasticity model.

Using the calculated degree of loading ($q_{net}/q_{ult} = 0.10$) and the normalized CPT penetration resistance ($q_{c1} = 150$) the following modulli values are estimated from Figure 7 and applying the elastic formulae the following settlements are calculated:

SAND TYPE	E'/q_c	Young's modulus, E' (Mpa)	Settlement, s, mm	Average strain, s/B, %
OC SAND	4	60	5	0.16
AGED NC SAND	2	30	10	0.33
RECENT NC SAND	1.34	20	14	0.47

The last column in the above table, (s/B), represents the approximate average vertical strain beneath the footing and shows that the average strain varies from 0.16% to 0.47% depending on the stress history of the sand.

If the previous correlation proposed by Belloti et al. (1989) and shown in Figure 2 had been applied the following values would have been calculated:

SAND TYPE	E'/q_c	Young's modulus, E' (Mpa)	Settlement, s, mm	Average strain, s/B, %
OC SAND	7	105	2	0.07
AGED NC SAND	4	60	5	0.16
RECENT NC SAND	2	30	10	0.33

The correlation in Figure 2 assumes that the average vertical strain is 0.1%. Unfortunately, if the net foundation pressure is kept constant ($q_{net} = 400 kN/m^2$) the average strain beneath the footing will vary, as illustrated by this example. The proposed correlation in Figure 7 appears to produce a slightly more conservative estimate of settlement than the previous correlation (Figure 2).

SUMMARY

Correlations between penetration resistance and deformation moduli for clean sands are a function of:

- stress history (OCR)
- aging
- strain level, and,
- relative density.

A correlation between CPT, q_c and an equivalent Young's moduli, E' has been proposed that incorporates all these factors. The influence of relative density can be accounted for by incorporation of normalized penetration resistance (q_{c1}). The strain level beneath most well designed shallow foundations is generally rather small (Jardine et al, 1986; Burland, 1990) and can be accounted for in a simple manner by using the degree of foundation loading, q_{net}/q_{ult}, as suggested by Stroud (1988).

The important effects of stress history (OCR) and aging have been incorporated but continued research is still required to quantify these factors. Further research is also required to better understand the influence of soil fabric on correlations between penetration resistance and stiffness. This is especially important for silty sands.

ACKNOWLEDGEMENTS

The financial support of the Natural Sciences and Engineering Research Council of Canada is gratefully acknowledged. The support, assistance and valuable discussions of Prof. M. Jamiolkowski and Prof. K.H. Stokoe II during the author's stay in Italy are also much appreciated.

REFERENCES

1. Baldi, G. et al. 1986. Interpretation of CPT's and CPTU's. II Part: Drained Penetration on Sands. Proc. IV Int. Geotech. Seminar on Field Instrumentation and In-Situ Measurements, Nanyang Tech. Inst., Singapore.
2. Battaglio M. & Jamiolkowski, M. 1987. Analisi Delle Defomazioni. XIII Ciclo Conferenze di Geotecnica di Torino, Italy.
3. Bellotti, R., Ghionna, V.N., Jamiokowski, M. and Robertson, P.K., 1989, Design Parameters of Cohesionless Soils from In-situ Tests, Transportation Research Boarch Conference, Washington, January.
4. Bellotti, R., et al., 1986. Deformation Characteristics of Cohesionless Soils from In-situ Tests. In-Situ '86 Proc. Spec. Conf. GED ASCE, Virginia Tech., Blocksburg.
5. Burland, J.B. and Burbidge, M.C., 1985, Settlement of Foundations on Sand and Gravel, Proceedings I.C.E., Part 1, 78, December, pp. 1325-1287.
6. Burland, J.B., 1990, Small is Beautiful- the Stiffness of Soils at Small Strains. Ninth Laurits Bjerrum Memorial Lecture, Canadian Geotechnical Journal, Vol. 26, No. 4, pp. 499-516.
7. De Beer, E.E., 1948, Settlement Records of Bridges Founded on Sand. Proc. II ICMFE, Rotterdam.
8. Eggestad, A., 1963, Deformation Measurements Below a Model Footing on the Surface of Dry Sand, Wiesbaden Settlement Conference. Volume 1, pp. 233-239.

9. Jamiolkowski, M., Ladd, C.C., Germaine, J.T., Lancellotta, R. 1985. New Developments in Field and Laboratory Testing of Soils. Theme Lecture, Proc. XI ICSMFE, San Francisco.
10. Jamiolkowski, M., Ghionna, V.N., Lancellotta, R. and Pasqualini E., 1988, New Correlations of Penetration Tests for Design Practice. Proc. Penetration Testing ISOPT-1, Florida.
11. Jardine, R.J., Potts, D.M., Fourie, A.B. & Burland, J.B. 1986. Studies of the Influence of Non-Linear Stress-Strain Characteristics in Soil Structure Interaction. Geotechnique, 36, 3.
12. Lambe, T.W. and Whitman, R.V. 1969, Soil Mechanics. John Wiley and Sons, Inc., New York.
13. Lambrechts, J.R. & Leonards, G.A. 1978. Effects of Stress History on Deformation of Sand. Journal of Geotechnical Engineering Division, ASCE, GT 11.
14. Lee, S.H.H. & Stokoe, K.H. 1986. Investigation of Low-amplitude Shear Wave Velocity in Anisotropic Material. Geotechnical Engineering Report GR86-6 Civil Eng. Dept. The University of Texas at Austin.
15. Lo Presti, D. 1987. Behavior of Ticino Sand During Resonant Column Test. Ph. D. Thesis, Technological University of Turin.
16. Mesri, G., Feng, T.W. and Benak, J.M., 1990. Postdensification Penetration Resistance of Clean Sands. Journal of Geot. Eng., ASCE, Vol. 116, No. 7, pp. 1095-1115.
17. Meyerhoff, G.G., 1965. Shallow Foundations. ASCE SM2 Vol. 91, March, p. 21.
18. Ohta, Y. & Goto, N. 1976. Estimation of S-Wave Velocity in Terms of Characteristic Indices of Soil. Bitsure-Tanko, 29(4), 34-41 (in Japanese).
19. Rix, G.J., 1984. Correlation of Elastic Moduli and Core Penetration Resistance. M.Sc. Thesis, Univ. of Texas at Austin.
20. Robertson, P.K. 1982. In-Situ Testing with Emphasis on its Application to Liquefaction Assessment. Ph.D. Thesis, University of British Columbia.
21. Robertson, P.K. & Campanella, R.G. 1983. Interpretation of Cone Penetration Tests. Canadian Geotechnical Journal Volume 20, 1983.
22. Robertson, P.K., Campanella, R.G. and Wightman, A., 1983. SPT-CPT Correlations. Journal of Geot. Eng. Div., ASCE Vol. 109, CT 11, pp. 1449-1459.
23. Schmertmann, J.H. 1970. Static Cone to Compute Static Settlement Over Sand. JSMF Div., ASCE, SM3.
24. Schmertmann, J.H., 1989. The Mechanical Aging of Soils. 25 Terzaghi Lecture, ASCE, New Orleans.
25. Seed, H.B. and Idriss, I.M., 1970. Soil Moduli and Damping Factors for Dynamic Response Analysis. Report No. UCB/EERC 70/10, Univ. of Cal., Berkeley.
26. Stroud, M.A., 1988. The Standard Penetration Test - Its application and Interpretation. Proceedings on Penetration Testing in the UK, Thomas Telford, London.
27. Skempton, A.W. 1986. Standard Penetration Test Procedures and the Effects in Sand of Overburden Pressure, Relative Density, Particle Size, Aging and Overconsolidation. Geotechnique, 36, No. 3, pp. 425-447.
28. Sykora, D.W. & Stokoe, K.H. 1983. Correlations of In-Situ Measurements in Sands of Shear Waves Velocity, Soil Characteristics and Site Conditions. Geotech. Eng. REP GR 83-33. Texas University, Austin.
29. Terzaghi, K. and Peck, R.B., 1967. Soil Mechanics in Engineering Practice. 2nd Ed., John Wiley and Sons, Inc., New York, N.Y.

CONE PENETRATION TESTING FOR IN-SITU EVALUATION OF LIQUEFACTION POTENTIAL OF SANDS

B. Mahmood-Zadegan[1], I. Juran[1], and M. T. Tumay[2]

ABSTRACT: Cone penetration tests have been increasingly used during the past two decades for soil stratification. Developments and engineering use of piezocone and seismic cone have enhanced the cone capabilities to identify soil strata and to provide data pertaining to the engineering properties of the in-situ soil.

Several investigators have proposed cone penetration based procedures to determine the in-situ liquefaction potential of sandy soil deposits. However, a comprehensive comparative analysis of these methods has not been reported in the literature.

In this study, friction and seismic cone penetration tests were conducted in two well documented sites in southern Imperial Valley, California. This paper is focused on the analysis of the test results using four widely used cone data interpretation methods for assessment of soil liquefaction potential. This comparative analysis demonstrates that although these methods rely on different empirical assumptions and involve different testing variables, they yield relatively consistent results. Results of this study are also compared with previously published data on the same sites.

INTRODUCTION

Conventionally, liquefaction potential of sandy soil deposits have been obtained through laboratory static (6,7,8,19) or Cyclic (9,26,28,37) tests. The laboratory procedures, however, present shortcomings with regard to simulation of the in-situ soil fabric and structure, the

---

1 - Assistant Professor, Professor and Head, respectively, Department of Civil & Environmental Engineering, Polytechnic University, Brooklyn, N.Y. 11201.

2 - Professor, Civil Engineering Department, Louisiana State University, Baton Rouge, LA 70803, currently Director, Geomechanics Program, National Science Foundation, Washington, D.C. 20550.

density and stress states, the induced stress path, and the depositional history (specially aging and cementation). Moreover, the stress/strain distribution within laboratory specimens is almost always non-uniform. Therefore, during the past two decades, new in-situ testing techniques have been developed
which resolve some of these shortcomings. The two most widely used techniques for in-situ evaluation of liquefaction potential of soils are standard penetration test, SPT, (18,29,30,38) and cone cenetration test, CPT, (21,24,30,40).

Seismic cone penetration tests have also been used to evaluate the liquefaction potential of sandy soil deposits in a number of investigations (4,14,17,22).

CONE PENETRATION TESTING FOR LIQUEFACTION PREDICTION: A COMPARATIVE ANALYSIS

Several CPT-based procedures have been proposed for the in-situ evaluation of the liquefaction potential of sandy soils. They can be mainly classified into two approaches:

The Seismic Cone Approach. A number of investigators (4,14,17,22) have reported evaluation of liquefaction susceptibility of sand deposits based on seismic cone test results. In this approach, the likelihood of liquefaction occurence, at a certain depth, is determined from the estimated values of the maximum ground acceleration and shear wave velocity (Fig. 1).

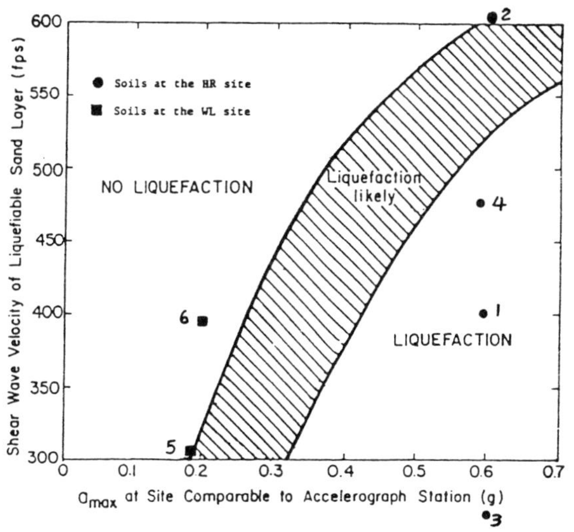

Figure 1. Liquefaction Evaluation Procedure Based on SW-Velocity (Adapted from Bierschwale and Stokoe, 1984)

Friction Cone Based Approaches. Several procedures for evaluation of liquefaction potential of sandy soils based on the friction cone penetration (CPT) test results have been proposed. These procedures can be classified into different categories:

CPT-SPT Correlations. In order to make use of the large data base of SPT results, several investigators (10, 24,27) have proposed semi-empirical and/or statistical correlations which would allow conversion of CPT tip resistance (q_c) values to equivalent SPT blow count (N) values. This in turn permits direct use of charts, such as that shown in Figure 2, which were originally developed for use with SPT results. In this figure τ/σ'_v represents the average induced field cyclic stress ratio. Its value is usually obtained using a semi-empirical approach (29).

Cyclic Stress Ratio-Tip Resistance Correlation. A number of investigators (24,30) have proposed procedures for evaluation of liquefaction potential of sandy soils by correlating the estimated average cyclic stress ratio induced in the field with the modified cone resistance (Q_c). Figure 3 presents one such procedure. A similar procedure has been proposed (40) based on a different size cone penetrometers.

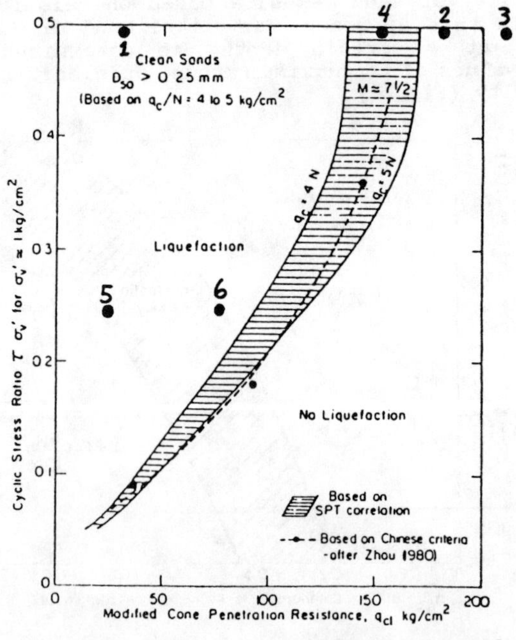

Figure 2. Liquefaction Evaluation Procedure Based on CPT-SPT Correlation (Adapted from Seed, 1983)

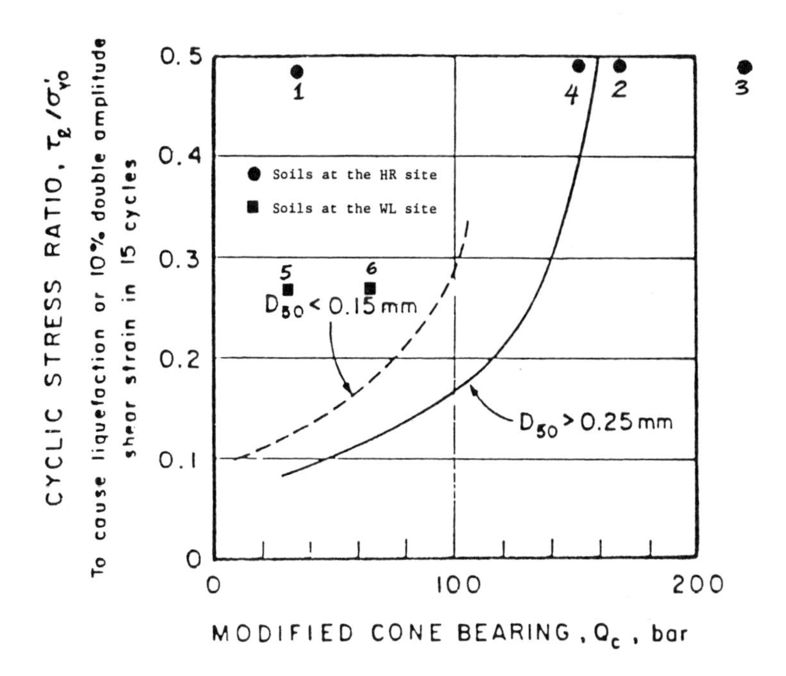

Figure 3. Liquefaction Evaluation Procedure Based on Tip Resistance (Adapted from Robertson and Campanella, 1985)

Cyclic Stress Ratio-Relative Density Correlation. CPT has been used extensively for determination of the in-situ soil density. Typically, a correlation is established between the tip resistance and soil relative density (2,25). Once established, such relationships can be used to directly correlate the average field cyclic stress ratio to soil relative density (Fig. 4).

Direct Observations. A liquefaction prediction chart has been proposed (24) based on observation of soil response during several earthquakes. This chart, shown in Figure 5, defines the envelope of most liquefiable soils, denoted as Zone A.

Comparative Evaluation of the Available Procedures. In order to assess the applicability of the procedures discussed above, the testing program, outlined in Table 1, including friction, seismic, and piezocone tests, was undertaken. Figures 6a and 6b illustrate the soil profiles at the two sites. The penetration test results indicated the presence of several sublayers of sandy deposit. A total of six sand

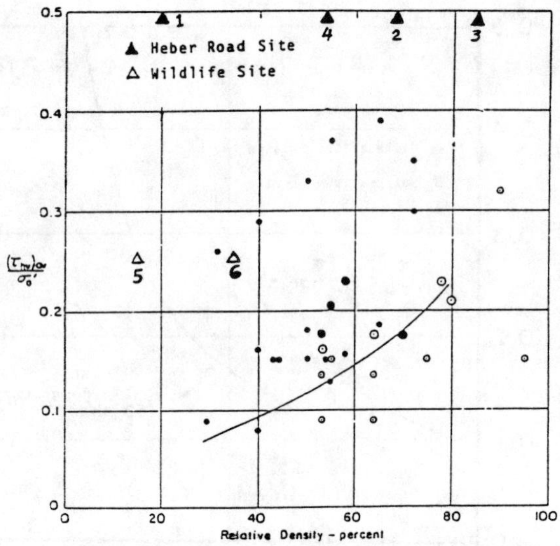

Figure 4. Liquefaction Evaluation Procedure Based on Relative Density Caculation (Adapted from Seed and Idriss, 1970)

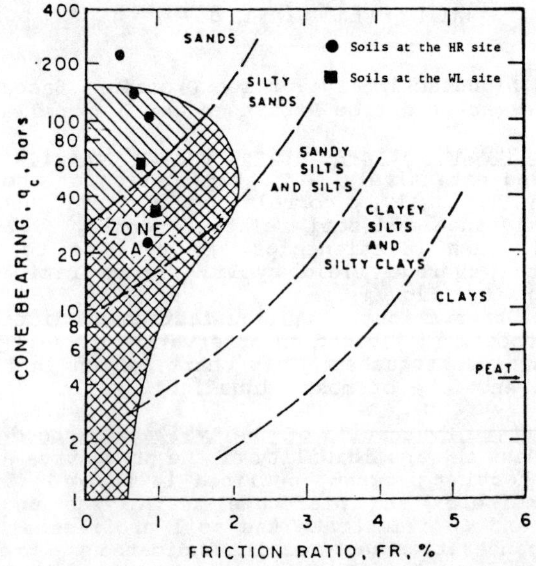

Figure 5. Liquefaction Evaluation Procedure Based on Direct Observations (Adapted from Robertson and Campanella, 1983)

SITE LOCATION	TEST TYPE	NO. OF TESTS	POROUS ELEMENT	PENETRATION RATE (cm/sec)
Heber Road	FC	4	N.A.	2.0
	SPC	3	Plastic & Ceramic	2.0 & 17.0
	DPC	4	Plastic & Ceramic	2.0 & 17.0
Wildlife	FC	3	N.A.	2.0
	SPC	1	Plastic	2.0
	DPC	3	Plastic & Ceramic	2.0 & 17.0

Table 1. The In-situ Testing Program

layers were selected and their liquefaction susceptibility was analysed using the four procedures discussed above. Table 2 summarizes the method predictions which seems to indicate that the four procedures yield relatively consistent evaluation of the liquefaction potential of the layers considered. The results of this study, presented in Figures 1 through 5 are in good agreement with those previously reported (39) for the two sites.

POTENTIAL USE OF DUAL PIEZOCONE DATA IN EVALUATION OF LIQUEFACTION POTENTIAL OF SANDY SOILS

Incorporation of piezometric element(s) in the electric cone has allowed the measurement of excess pore water pressures at different locations during penetration. Several investigators (12,23,31,35) have proposed charts for soil classification / stratification based on the piezocone penetration test (PCPT) data.

Previous experimental and theoretical studies (1,3,13, 14,15,16,19,21,23,34,35,41) have pointed out the individual merits of piezocone testing with pore pressure measurement capabilities on the shaft above the cone tip or the mid-section of the cone tip. The fundamental differences in strain paths at the cone tip and along the penetrometer shaft yields significantly different pore water pressure response.

The use of the new LSU / Fugro dual piezocone (DPCPT) have been reported in two recent studies (33,13) at various sites. Figure 7 illustrates the use of the LSU / Fugro DPCPT equipment in identifying loose (liquefiable) sand seams at the Heber Road site. As indicated in this figure, as the tip penetrates the loose, contractant sand seam, all three pore pressure records register an increase. However, the increase

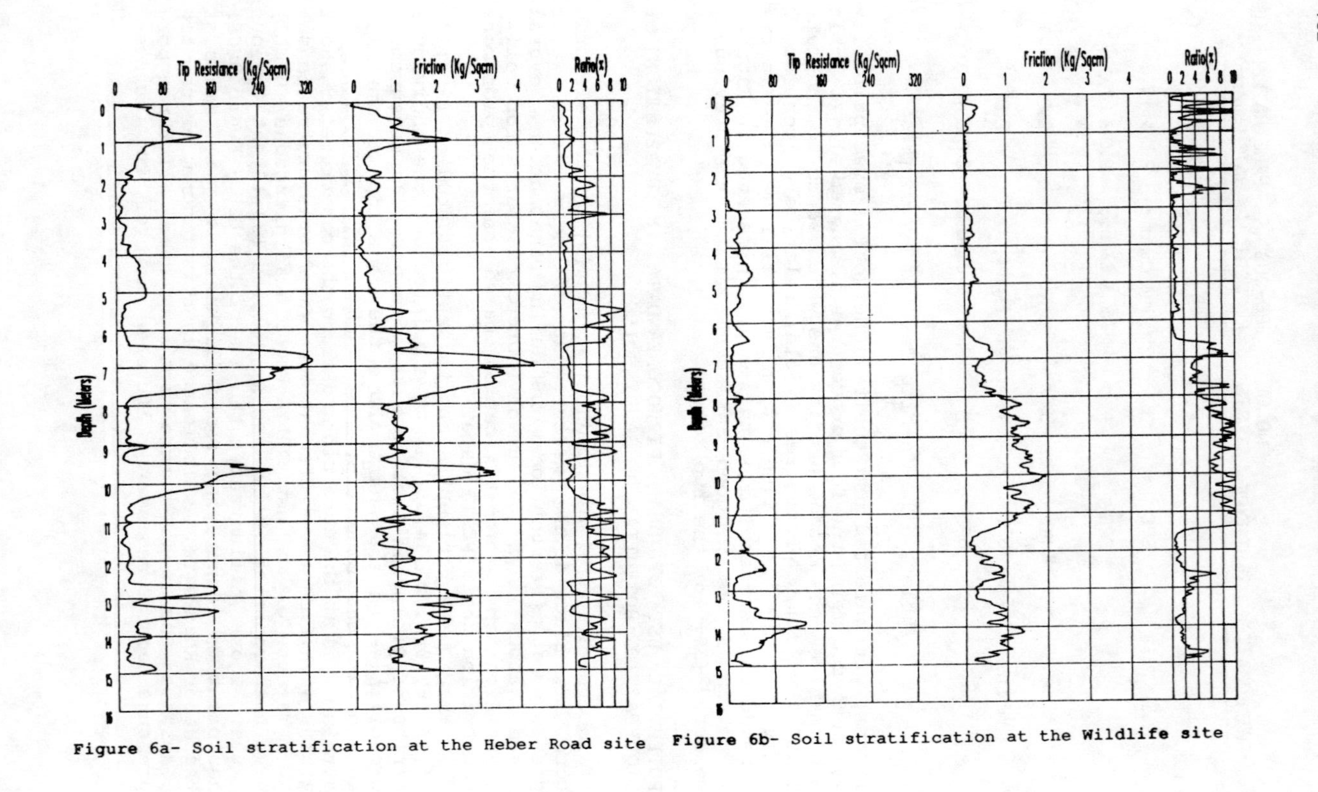

Figure 6a- Soil stratification at the Heber Road site

Figure 6b- Soil stratification at the Wildlife site

SITE LOCATION	DEPTH(m)	AVERAGE q_c(bar)	AVERAGE σ'_{vo} (psi)	AVERAGE D_r(%)	AVERAGE V_s(ft/sec)	EVALUATION PROCEDURE	LIQUEFACTION POTENTIAL
Heber Road (1)	2-5	30	10	20	400	I	Very Likely
						II	In Zone A, Very Likely
						III	Very Likely
						IV	Very Likely
Heber Road (2)	7.2-7.8	160	18	70	800	I	Not Very Likely
						II	Not in Zone A
						III	Not Very Likely
						IV	Not Very Likely
Heber Road (3)	9.2-10.5	240	22	85	150	I	Not Likely
						II	Not in Zone A
						III	Not Likely
						IV	Not Likely
Heber Road (4)	12.4-13.3	150	30	55	480	I	Likely
						II	In Zone A, Likely
						III	Likely
						IV	Likely
Wildlife (5)	2-3.5	30	7	15	300	I	Very Likely
						II	In Zone A, Likely
						III	Borderline Case, Likely
						IV	Very Likely
Wildlife (6)	4-6.5	60	12	35	400	I	Very Likely
						II	In Zone A, Likely
						III	Borderline Case, Likely
						IV	Very Likely

Table 2. Comparison of Liquefaction Evaluation Methods

in the excess pore pressure measured at the tip, shown in the figure as a solid line, is much larger, indicating a large increase in the octahedral normal stress around the tip. The tip resistance and the sleeve friction record further confirm the presence of a loose sandy layer at this depth.

At present, it seems that the DPCPT data can be best used for a qualitative enhancement of classification methodologies for sandy soils. The following classification guidline is proposed:

Classification	u_{tip}	u_{sleeve}
loose	positive, small	positive
medium dense	positive, medium	negligible
dense	positive, large	negative

CONCLUSIONS

A comparative analysis of four well known liquefaction prediction procedures is presented. The results of this study clearly indicate that the four procedures, although based on

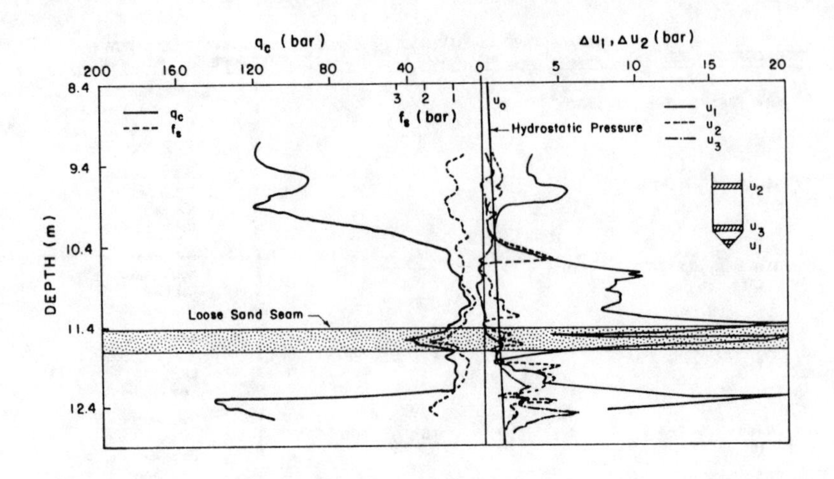

Figure 7. Use of DPCPT for Identification of Loose Sand Seams

different intrinsic assumptions and correlations, yield relatively consistent predictions.

The newly developed DPCPT seems to provide useful data for the evaluation of volume change tendencies of sandy soils subjected to hydrostatic or distortional stress paths. At this time, however, the data provided by the dual piezocone can only be used for qualitative predictions and further research and enhancement of the DPCPT data base are required for the development of quantitative liquefaction resistance assessment charts.

REFERENCES

1. A AlAwakati, On Problems of Soil Bearing Capacity at Depth," Ph.D. Diss., Duke University, Durham, 1975, 204 p.
2. G Baldi, R Bellotti, V Ghionna, M Jamiolkowski & E Pasqualini, Design Parameters for Sands from CPT, Proc. ESOPT II, Amsterdam, Vol. 2, 1982, 425-438.
3. MM Baligh, V Vivarant & CC Ladd, Cone Penetration in Soil Profiling,J.Geot.Eng.(ASCE), 106(4), April 1980, 447-461.
4. JG Bierschwale & KH Stokoe, Analytical Evaluation of Liquefaction Potential of Sands Subjected to the 1981 Westmoreland Earthquake, Geot. Eng. Report GR-84-15, Civ.

Eng. Dept., Univ. of Texas, Austin, 1985.
5. RG Campanella, & PK Robertson, Applied Cone Research, Proc. Symp. Cone Penetration Testing and Experience, GM Norris & RD Holtz (Eds.), Geot. Eng. Div. (ASCE), 1981, 343-362.
6. J Canou, Contribution a L'etude et a L'evaluation des Proprietes de Liquefaction D'en Sable, These de Doctorat, Ecole Nationale des Ponts et Chaussees, 20 mars 1989.
7. G Castro, Liquefaction of Sands, Ph.D. Diss., Harvard Univ., Cambridge, MA, 1969.
8. G Castro & SJ Poulos, Factors Affecting Liquefaction and Cyclic Mobility, Proc. ASCE National Convention, Liquefaction Problems in Geotechnical Engineering, Sept. 27-Oct. 1, 1977, 105-138.
9. R Dobry, RS Ladd, FY Yokel, RM Chung & D Powell, Prediction of Pore Pressure Buildup and Liquefaction of Sands During Earthquakes by the Cyclic Strain Methods, Building Science Series 138, National Bureau of Standards, U.S. Department of Comm., U.S. Government Printing Office, Washington, D.C., 1983.
10. BJ Douglas, RS Olsen & GR Martin, Evaluation of the Cone Penetrometer Test for SPT Liquefaction Asessment, Proc. Specialty Session No. 24: In Situ Testing to Investigate Liquefaction Susceptibility, ASCE National Convention, St. Louis, MS, Oct. 27, 1981.
11. WDL Finn, DJ Pickering & PD Bransby, Sand Liquefaction in Triaxial and Simple Shear Tests, J.SMFD(ASCE), 97(4), 639-659.
12. GA Jones & EA Rust, Piezometer Penetration Testing CUPT, Proc. ESOPT II, Amsterdam, Vol. 2, 1982, 607-613.
13. I Juran & MT Tumay, Soil Stratification Using Dual pore Pressure Piezocone Test (DPCPT), Transportation Research Board, 68th Annual Meeting, Jan. 22-26, Washington, D.C., Jan. 1989.
14. I Juran, B Mahmood-Zadegan & MT Tumay, Geotechnical Investigation and Piezocone Testing in Well Documented Liquefiable Sites in California," Research Report, Civ. Eng. Dept., Louisiana State Univ., December 1989.
15. PD Kiousis, GZ Voyiadjis & MT Tumay, A Large Strain Theory and Its Application in the Analysis of Cone Penetration, Intl. J. Num. Anal. Meth. Geomech., 12(1), Jan. 1988, 45-60. 16. T Lunne, TE Eidsmoen, D Gillespie & JD Howland, Laboratory and Field Evaluation of Cone Penetrometer, Proc. In-Situ 86, Specialty Conference, ASCE, Blacksburg, Virginia, 1986.
17. NCEE, Liquefaction of Soils During Earthquakes, National Committee on Earthquake Engineering, National Academy Press, Washington, D.C., 1985, 240 p.
18. Y Ohsaki, Y., Niigata Earthquakes, 1964 Building Damage and Soil Conditions," Soils and Foundations, 6(2), 1966, 14-37.
19. L Parez, M Bachelier & B Sechet, Pore Pressure Generated

During Penetration of the Cone, Proceedings, 6th ICSMFE (3), Vienna, 1976, 533-538.
20. SJ Poulos, G Castro & JW France, Liquefaction Evaluation Procedure, J.Geot.Eng.(ASCE), 111(6), June 1985, 772-791.
21. PK Robertson & RG Campanella, Interpretation of Cone Penetration Tests, Part I (Sand), Canadian Geot. J., 20(4), April 1983, pp. 718-733.
22. PK Robertson, RG Campanella, D Gillespie & A Rice, Seismic CPT to Measure In Situ Shear Wave Velocity, J.Geot.Eng. (ASCE), 112(8), Aug. 1986, 791-803.
23. PK Robertson, RG Campanella, D Gillespie & J Greig, Use of Piezometer Cone Data, Report No. 92, Soil Mechanics Series, Civ. Eng. Dept., Univ. of British Columbia, Vancouver, Canada, 1985.
24. PK Robertson & RG Campanella, Liquefaction Potential of Sands Using the CPT, J.Geot.Eng.(ASCE), 111(3), March 1985, 384-403.
25. JH Schmertmann, Guidelines for Cone Penetration Test, Performance and Design, U.S. Department of Transportation, FHWA Report No. TS-78-209, 1978.
26. HB Seed & KL Lee, Liquefaction of Saturated Sands During Cyclic Loading, J.SMFD(ASCE), 92(6), June 1966, 105-134.
27. HB Seed, Evaluation of the Dynamic Characteristics of Sands by In-situ Testing Techniques, Proc. Int. Conf. In-situ Testing and Investigation in Soil and Rock, May 18-20, Paris, 1983.
28. HB Seed & WH Peacock, Test Procedures for Measuring Soil Liquefaction Characteristics, J.SMFD(ASCE), 97(8), Aug. 1971, 1099-1119.
29. HB Seed, H.B. & IM Idriss, Simplified Procedure for Evaluating Soil Liquefaction Potential, J.SMFD(ASCE), 98(9), Sept. 1970.
30. HB Seed, IM Idriss & I Arango, Evaluation of Liquefaction Potential Using Field Performance Data, J.Geot.Eng.(ASCE), 109(3), March 1983, 458-482.
31. K Senneset, N Janbu & G Svano, Strength and Deformation Parameters from Cone Penetration Test, Proc., ESOPT II, Amsterdam, 1982, Vol. 2, 863-870.
32. BA Torstensson, The Pore Pressure Probe," Nordiske Geotekniske Mote, Oslo, Paper No. 34.1 - 34.15, 1977.
33. MT Tumay, Field Calibration of Electric Cone Penetrometers in Soft Soil, FHWA/LA/LSU-GE-85/02, Technical Information Service, Springfield, VA 22161, 1985.
34. MT Tumay, YB Acar & RL Bogges, Subsurface Investigations with Piezocone Penetrometer, Proc. Symp. Cone Penetration Testing and Experience, GM Norris & RD Holtz (Eds), ASCE, St. Louis, MS, 1981, 325-342.
35. MT Tumay, R Yilmaz, YB Acar & E Deseze, Soil Exploration in Soft Clays with the Electric Cone Penetrometer, Proc. ESOPT II, Vol. 2, 1982, 915-921.

36. MT Tumay, YB Acar, MH Cekirge & N Ramesh, Flow Fields Around Cones in Steady Penetration, J.Geot.Eng.(ASCE), 111(2), Feb. 1985, 193-204.
37. YP Vaid & JC Chern, Effect of Static Shear on Resistance to Liquefaction, Soils and Foundations, 23(1), Jan 1983, 47-60.
38. W Wang, Some Findings in Soil Liquefaction, Water Cons. and Hydro. Power Sci. Res. Inst., Beijing, China, August 1979.
39. TL Youd & MJ Bennett, Liquefaction Sites, Imperial Valley, California, J.Geot.Eng.(ASCE), 109(3), March 1983, 440-457.
40. SG Zhou, Evaluation of the Liquefaction of Sand by Static Cone Penetration Test, Proc. 7th World Conf. on Earthquake Engineering, Vol. 3, Istanbul, Turkey, 1980.
41. H Zuidberg, Piezocone Penetration Testing-Probe Development, ISOPT-1, Specialty Session No. 14, Orlando, March 1988, 21p.

PROGRESSIVE FAILURE AND PARTICLE SIZE EFFECT IN BEARING CAPACITY OF A FOOTING ON SAND

Fumio Tatsuoka[1], Michio Okahara[2], Tadatsugu Tanaka[3], Kazuo Tani[4], Tsutomu Morimoto[5], and Mohammed S. A. Siddiquee[6]

ABSTRACT: For air-pluviated Toyoura sand exhibiting post-peak strain-softening, the values of $N\gamma$ predicted by using the classic bearing capacity theories with the value of ϕ obtained from plane strain compression tests in which the σ_1 direction was normal to the bedding plane were found considerably larger than those observed in plane strain model tests. It is shown that this is because these theories ignore strength anisotropy of sand and progressive failure in the ground. Furthermore, $N\gamma$ values obtained by centrifuge tests were found larger than those obtained by the corresponding large scale 1g tests with a footing width of 23cm and 50cm. This is due to the particle size effect linked to shear banding since the so-called scale effect in 1g tests is understood to be the result of both the particle size and pressure-level effects while centrifuge tests only reflect the latter effect. Finally, the results of FEM analysis are presented which simulate very well the above-mentioned phenomena.

INTRODUCTION

Fig. 1 shows a range of the classic and widely used theoretical relations between the bearing capacity coefficient $N\gamma$ and the angle of internal friction $\phi = \arcsin\{(\sigma_1 - \sigma_3)/(\sigma_1 + \sigma_3)\}_{max}$ for a rough footing under plane strain conditions. Despite the various mathematical methods used in the classic methods, it is commonly assumed that ϕ is constant along the failure surface; i.e., ① ϕ is independent of pressure level, ② ϕ is isotropic, and ③ the material is perfectly plastic; thus the full strength is mobilized at all points simultaneously along a potential failure surface. While the effect of the pressure-level dependency of ϕ has been well recognized (e.g., 6), the use of the postulates ② and ③ seems still popular (e.g., 6, 7).

It is known, however, that sand beds made by raining air-dried particles through air are anisotropic (12) (see Fig.2). In Fig. 2, δ is

1- Prof., Institute of Industrial Science, Univ. of Tokyo, Roppongi, Minato-ku, Tokyo, 106, Japan
2- Head of Foundation Engineering Div., Public Works Research Institute, Ministry of Construction, Japan
3- Prof., Dept. of Agriculture, Meiji Univ.
4- Research Engineer, Central Research Institute of Electric Power Industry, Formerly graduate student of Univ. of Tokyo
5- Engineer, Ministry of Construction, ditto.
6- Graduate student, Univ. of Tokyo

the angle of the σ_1 direction relative to the bedding plane direction. It may be seen that ϕ is maximum and minimum when δ is 90° and around 30°. Besides, since sands, more significantly for denser sands, exhibit post-peak strain-softening, failure of sand mass is usually progressive (9). Thus, the use of ϕ obtained from plane strain compression (PSC) tests at δ = 90° with the classic theories over-estimates the average peak strength ϕ along a potential failure surface, which is still larger than the average mobilized friction angle ϕ_{mob} at the peak footing load.

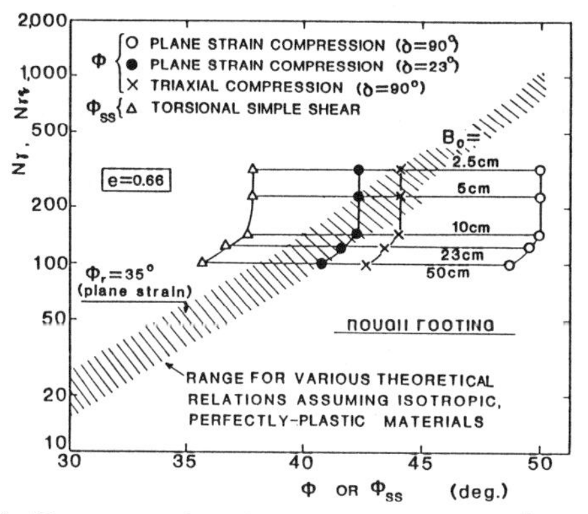

Fig. 1 Theoretical and experimental $N\gamma$-ϕ relations

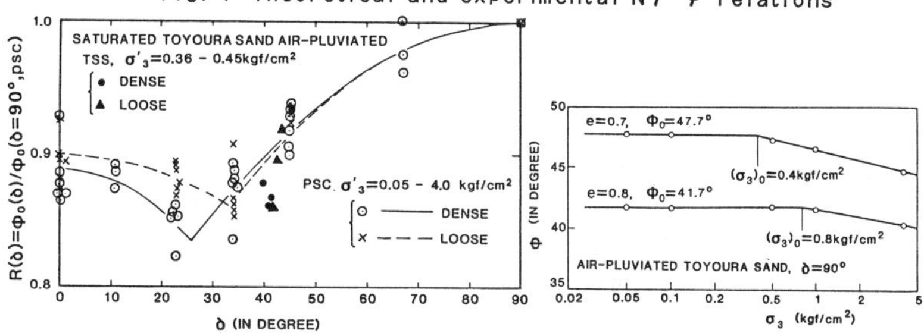

Fig. 2 Strength anisotropy of Toyoura sand (20, 21)

Fig. 3 Pressure-level dependency of Toyoura sand (20, 21)

STRENGTH OF AIR-PLUVIATED TOYOURA SAND

In the present study, specimens for element tests including PSC tests and sand beds for model bearing capacity tests were prepared by raining air-dried Toyoura sand, which is a uniform sand with a high content of quartz, sub-angular to angular particles, fines content= 0%, D_{50}= 0.16mm, U_c= 1.46, G_s= 2.64, e_{max}= 0.977 and e_{min}= 0.605. The strength and deformation properties of Toyoura sand used throughout the test program have been confirmed to be consistent.

Plane strain compression (PSC) tests were performed for void ratio

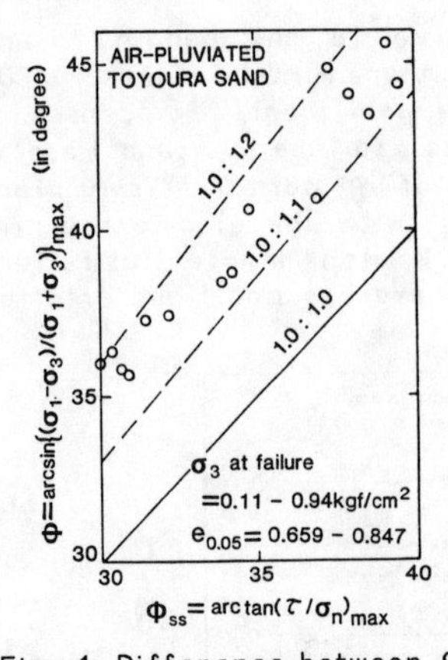

Fig. 4 Difference between ϕ_{ss} and ϕ in TSS tests

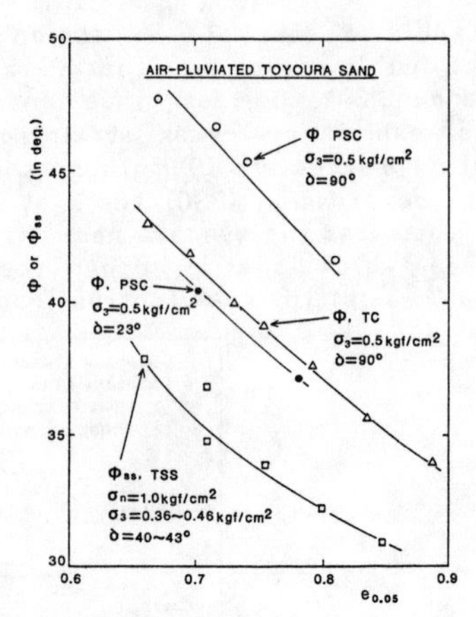

Fig. 5 $\phi(\phi_{ss})$ by different methods

e= 0.65-0.80 and σ_3= 0.05-4.0kgf/cm^2 (20, 21). It was found that the value of ϕ was rather independent of σ_3 below a certain level $(\sigma_3)_0$ (Fig. 3). This is one of the advantages of the use of this kind of sand in model tests, since ϕ at low pressure levels in models can be evaluated confidently. The empirical equations for $\phi = \arcsin\{(\sigma_1-\sigma_3)/(\sigma_1+\sigma_3)\}_{max}$ (in degrees) in PSC were obtained as:

$$\text{For } \sigma_3 \leqq (\sigma_3)_0 = 4.0\cdot(e-0.6);\quad \phi = \phi_0(\delta, e) = 59.5\cdot(1.5-e)\cdot R(\delta)$$
$$\text{For } \sigma_3 \geqq (\sigma_3)_0;\quad \phi(\delta, e, \sigma_3) = \{59.5\cdot(1.5-e) - 10\cdot(1.0-e)\cdot\log_{10}[\sigma_3/(\sigma_3)_0]\}\cdot R(\delta) \quad (1)$$

Here, $R(\delta)$ is the ratio of ϕ at a certain angle δ to that at δ = 90°.

A series of triaxial compression (TC) tests was also performed (4) and the empirical equations for ϕ at δ = 90° for e= 0.65-0.9 and σ_3= 0.029-4.0kgf/cm^2 were obtained as:

$$\text{For } \sigma_3 \leqq (\sigma_3)_0 = 2\cdot(e-0.55):\quad \phi = \phi_0(e) = 70\cdot e^2 - 151\cdot e + 113.6$$
$$\text{For } \sigma_3 \geqq (\sigma_3)_0;\quad \phi(\sigma_3, e) = \phi_0 - 4\cdot(1.4-e)\cdot\log_{10}[\sigma_3/(\sigma_3)_0] \quad (2)$$

Further, a series of torsional simple shear (TSS) tests, which are plane strain tests with continuous rotation of principal stress directions, were performed (14). It was found that when compared for the same values of δ, e and σ_3 at failure, the value of ϕ is very similar between PSC tests and TSS tests (14). For the TSS tests, similarly to conventional simple shear tests and shear box tests, the maximum angle of stress obliquity on the horizontal planes $\phi_{ss} = \arctan(\tau/\sigma_n)_{max}$ was obtained, which is given for e= 0.66 as:

$$\text{For } \sigma_3 \leqq (\sigma_3)_0 = 0.24;\quad \phi_{ss} = 37.9 \text{ (constant)}$$
$$\text{For } \sigma_3 \geqq (\sigma_3)_0;\quad \phi_{ss} = 37.9 - 3.4\cdot\log_{10}(\sigma_3/0.24) \quad (3)$$

As shown in Fig. 4 (23), ϕ_{ss} was 10-20% smaller than ϕ, which was in turn 10% smaller than ϕ in PSC at $\delta = 90°$ due to the strength anisotropy (see Fig. 2). Note that ϕ_{SS} by TSS tests may be different from $\phi_{DS} = \arctan(\tau/\sigma_n)_{max}$ obtained by conventional shear box tests, since ϕ_{DS} may be further affected by, at least, the box dimensions and the gap between the top and bottom boxes relative to particle size and the degree of the fixity of the top platen against rotation. At any rate, it is obvious that when using any $N\gamma$-ϕ relation, it should be taken into account that ϕ differs largely among different testing methods as shown in Fig. 5.

Fig. 6 View of the footing base, B_o= 50cm

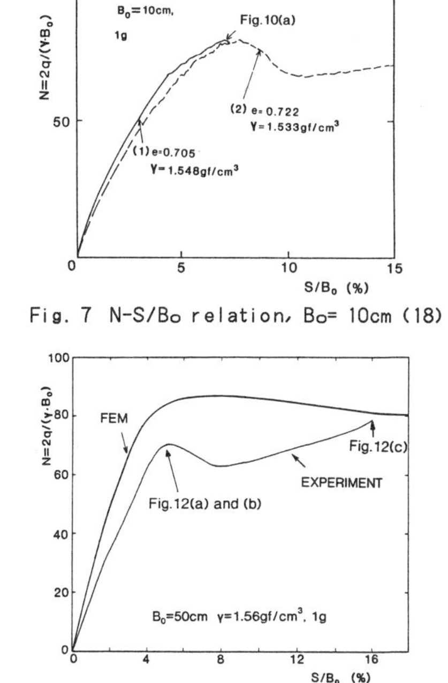

Fig. 7 N-S/B_o relation, B_o= 10cm (18)

Fig. 8 Experimental and analytical N-S/B_o relation, B_o= 50cm

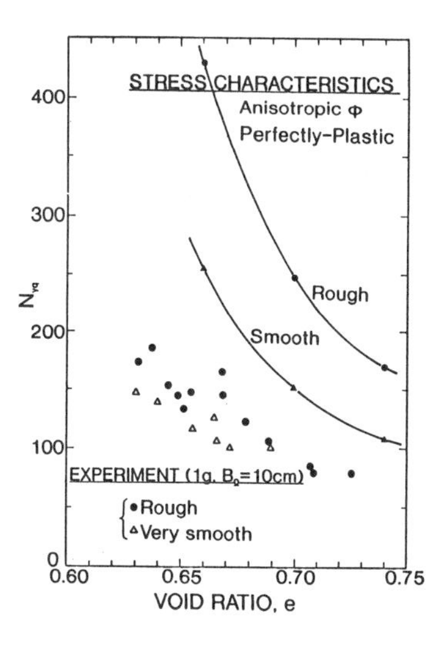

Fig. 9 Comparison of $N\gamma$ between experiments and perfectly plastic theory (18)

MODEL BEARING CAPACITY TESTS UNDER PLANE STRAIN CONDITIONS

Three types of model tests were performed with the use of a rigid footing which was not allowed to rotate and laterally translate. The base surface was made rough except the one having a lubricated base (see Fig. 9). Except the tests shown in Fig. 13(c), a footing was placed on the ground surface. The walls of the sand box and the pit were well lubricated with a 0.2mm-thick latex rubber membrane and a 0.1mm-thick silicone grease layer (this lubrication method is described in detail in 19). In the tests in the normal gravity (i.e., 1g), the footing load was measured at the central third of footing to minimize the side wall friction effect. Eleven load cells at the central third, as seen in Fig. 6, measured both shear and normal stresses. The $N\gamma q$ value is defined as the measured peak value of $N= 2 \cdot q/(\gamma \cdot B_0)$ (see Fig. 7) and its value for a certain density of sand was obtained from the averaged relation between $N\gamma q$ and the density for each value of B_0.

<u>Small-scale tests in 1g (18):</u> The footing width B_0 was 2.5, 5 and 10cm and the footing length was 40cm. The sand box had dimensions of 40cm in width, 183cm in length and 60cm in depth. Fig. 7 shows results of two typical tests. Here, the point of zero settlement was defined by extrapolating the initial linear portions of the measured relations to zero footing load conditions (11).

<u>Large-scale tests in 1g (22, 23):</u> For a footing on the ground surface, two tests with B_0= 23cm and four tests with B_0= 50cm were carried out in an underground pit of 200cm in width (equal to the footing length), 700cm in length and 400cm in depth at Public Works Research Institute (PWRI). Fig. 8 shows the result of one of the tests. Several tests for B_0= 23cm were also performed by using a footing with a depth. One of the side walls of the pit had a transparent window supported by steel stiffeners, through which the deformation of the model ground was observed. The lateral expansion of the window at the peak footing load was not greater than 0.04% of the pit width.

<u>Tests in a centrifuge:</u> Two series were performed at PWRI using a very stiff sand box of 10cm in width, 50cm in length and 30cm in depth with an about 25cm-thick sand model. The footing load was measured over the whole footing length. Densities of the models were measured before increasing the acceleration. The change in void ratio by the application of acceleration has been found less than 0.003. Herein reported are the results from the second series with B_0= 2 and 3cm.

COMPARISONS BETWEEN THEORETICAL AND EXPERIMENTAL RESULTS

The experimenatal data points shown in Fig. 1 are the ones for e= 0.66, which are the relationships between the values of $N\gamma q$ obtained from the 1g tests of surface footing and the values of ϕ from the PSC and TC tests and the values of ϕ_{ss} from the TSS tests. These values of ϕ and ϕ_{ss} were evaluated for the average pressure level (σ_3) in the model test, estimated by using Eqs. 1, 2 and 3 as σ_3= {peak average footing pressure q_u}/K_p, K_p= $(1+\sin\phi)/(1-\sin\phi)$. Note that the effect of the vertical footing displacement at the peak footing load, S_f, on the values of $N\gamma q$ can be considered very small, because the value of S_f was as small as $(0.05\text{-}0.08)\cdot B_0$, with a smaller value for a larger B_0. Note that when the effect of S_f on $N\gamma q$ is taken into account, the difference in $N\gamma$ between the theoretical and experimental values even increases. As may be seen from Fig. 1, the measured values of $N\gamma q$ were found significantly smaller than those predicted by the classic theories with the value of ϕ in PSC at $\delta = 90°$, particularly for a larger B_0. Such a tendency as above has already been reported for large-scale 1g model tests of footing on a slope (7).

It may be seen in Fig. 1 that the data points of solid circle with the minimum ϕ in PSC observed at $\delta = 23°$ are located at or near the theoretical relations. Thus, the data points with the value of ϕ averaged for $\delta = 0 - 90°$ should be located considerably below the theoretical

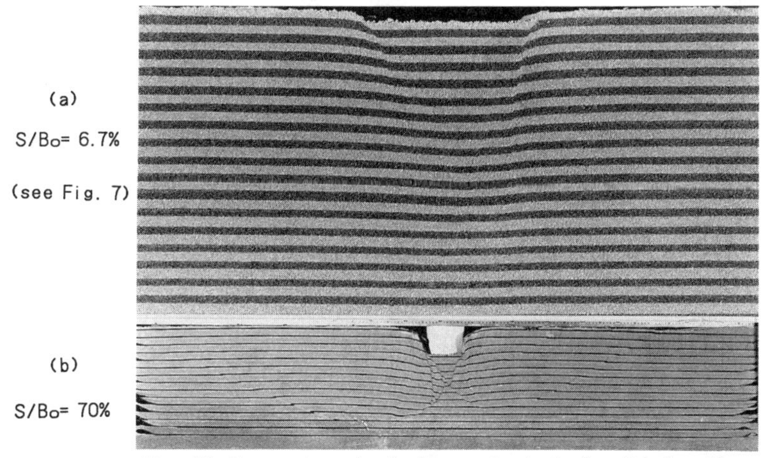

Fig. 10 Exposed central planes, tests in 1g with B_0= 10cm

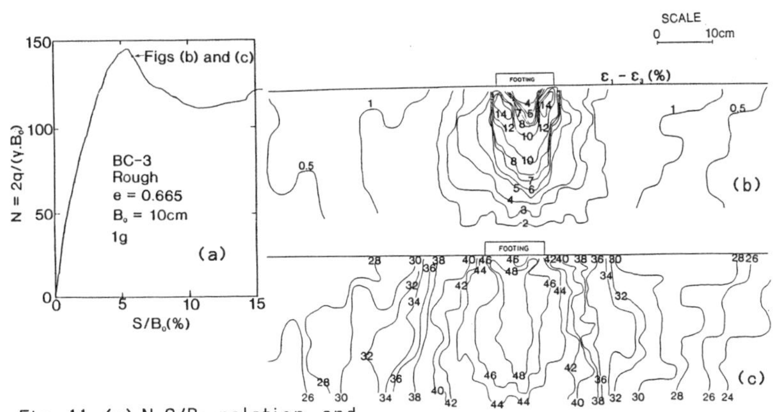

Fig. 11 (a) N-S/B_0 relation, and (b) strain field and (c) distribution of ϕ_{mob} at S/B_0= 6.7% (18)

relations. This means that the strength anisotropy alone is not sufficient to explain the discrepancy of the experimental results from the theoretical relations. As a result, the progressive failure should be taken into account in addition to the strength anisotropy. This point can be seen also from the comparison between theoretical values by the stress characteristics method and experimental results shown in Fig. 9 (18). Namely, for both rough and smooth footings with B_o= 10cm, the experimental values of $N\gamma q$ were found much smaller than the theoretical ones for which the sand was assumed to be perfectly-plastic, while the anisotropy (Fig. 2) and pressure-level dependency (Fig. 3) of ϕ (peak strength) as measured by the PSC tests were taken into account and the values of friction angle on the footing base measured at the peak footing load were used.

Some other researchers showed that experimental $N\gamma q$ values agreed well with those evaluated from the classic theory using the val-

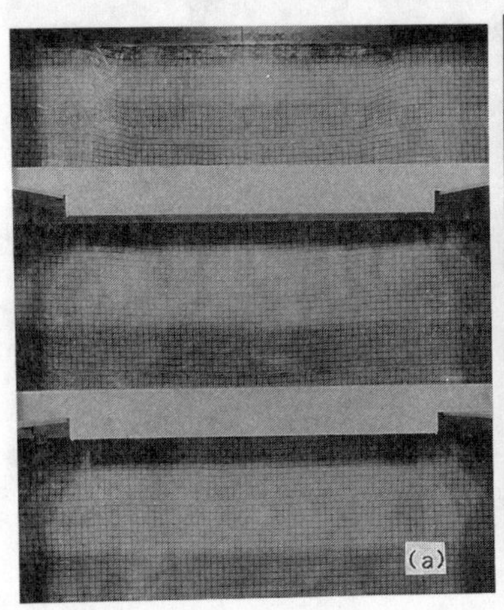

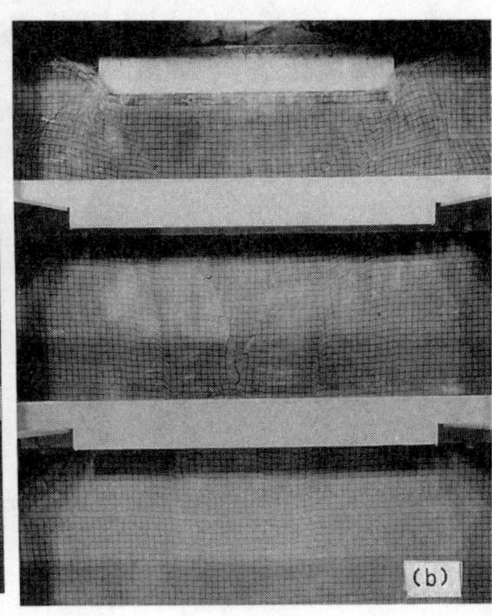

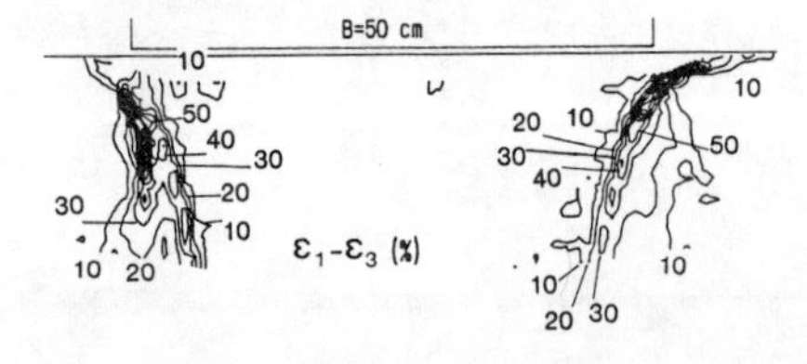

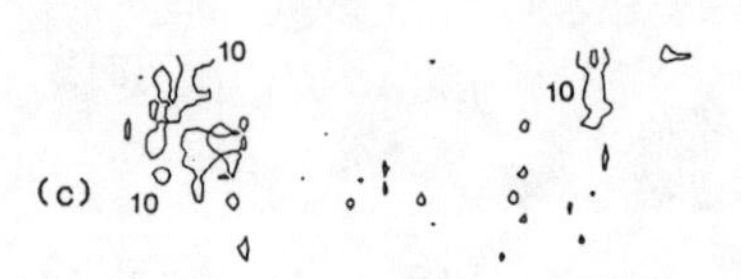

Fig. 12

Pictures at (a) S/B_o= 4.8% (see Fig. 8) and (b) S/B_o= 16%, and (c) strain field (mesh size 1cmx1cm), test in 1g with B_o= 50cm

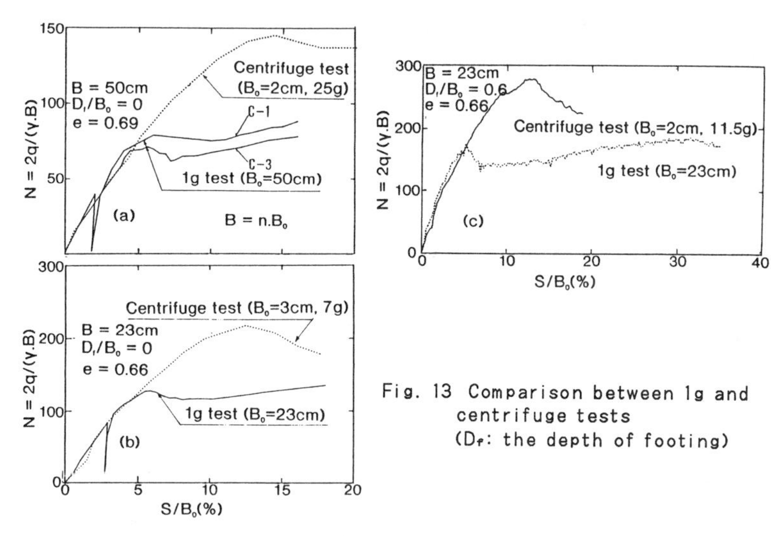

Fig. 13 Comparison between 1g and centrifuge tests (D_f: the depth of footing)

ue of ϕ_{DS} obtained from relevant shear box tests (e.g., 10). It may be interesting to note that also in Fig. 1, when the measured values of $N\gamma q$ are substituted into the theoretical relations, values of ϕ between ϕ in TC at $\delta = 90°$ and ϕ_{ss} are back-calculated. Consequently, such an agreement as above may be fortuitous and does not verify the isotropic perfectly-plastic theories. In addition, some other researchers (e.g., 5, 15) showed that experimental $N\gamma q$ values obtained from model tests using a strip or plane strain footing on sand ground were even larger than those evaluated from the classic theories using the value of ϕ obtained from the relevant PSC tests at $\delta = 90°$. However, such theoretical values could not be exceeded by experimental values. In such plane strain model tests, presumably the values of $N\gamma q$ may have been overestimated by some reason (e.g., due to the side wall friction).

In practice, the value of ϕ is often estimated from the standard penetration test N-values, and then, the value of $N\gamma$ is obtained by using one of the classic theories with the value of ϕ estimated as above. It is known, however, that such value of ϕ are usually much smaller than the value of ϕ in PSC at $\delta = 90°$, or even smaller than the value of ϕ in TC at $\delta = 90°$. The authors are understanding that common practice assumes that the result is balanced between the use of underestimated values of ϕ and the use of classic theories, or when not balanced, it is on the safe side.

On the other hand, one may consider that the use of the residual friction angle ϕ_r together with an isotropic and perfectly-plastic theory accounts for the effects of both strength anisotropy and progressive failure, and the result is on the safe side. In Fig. 1, the value of $N\gamma$ obtained from the theoretical relations for $\phi_r = 35°$ is around 50, which is much smaller than the measured values. This is generally the case especially for dense sands. Obviously, the use of ϕ_r cannot account for the effect of the density of sand, thus its use is not relevant for dense sands.

PROGRESSIVE FAILURE AND PARTICLE SIZE EFFECT

In two of the models in 1g with B_o =10cm, the sand grounds were moistened and their central sections were carefully exposed in order to take pictures (see Fig. 10) (18). The horizontal black bands are layers of black-dyed sand, placed only in a small width around the central section to minimize the effect of using them. It may be seen from Fig. 10(a) that at the peak load, two clear shear bands with a width of about 3mm (20 times the mean diameter of sand D_{50}) have already developed from both the sides of footing. The right-hand one has extended to a depth of around 6cm (= $0.6 \cdot B_o$). As may be seen from Fig. 10(b), it was not until S/B_o became as large as 0.7 that the outmost failure plane reached the soil surface. Fig. 11(b) shows a strain field at a moment just after the peak load in another test (see Fig. 11(a), constructed by a photogrametric method using two pictures taken before loading and at another moment from the outside of the sand box through the side wall. The strains were obtained from the displacement measured at the nodes of 1cm x 1cm grid printed on the surface of a rubber membrane (see Fig. 12a and b). It was confirmed that

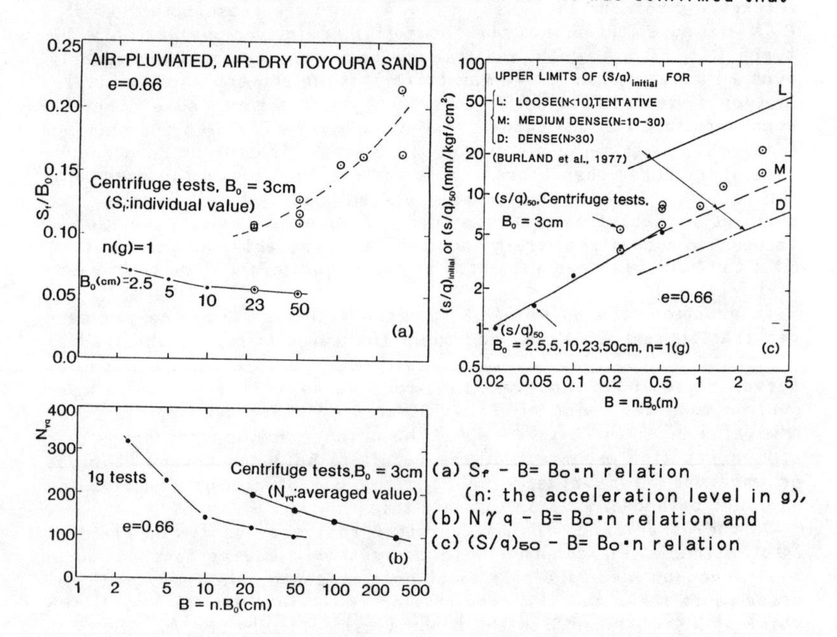

(a) S_f - B= $B_o \cdot n$ relation (n: the acceleration level in g),
(b) $N\gamma q$ - B= $B_o \cdot n$ relation and
(c) $(S/q)_{50}$ - B= $B_o \cdot n$ relation

Fig. 14 Comparison between 1g and centrifuge tests

the deformation of sand seen from the outside of the sand box was virtually the same with those seen on the exposed central plane (18). Since shear strains at the peak stress in the corresponding PSC tests were 4 - 7%, it can be learned from Fig. 11(b) that the failure was very progressive. In Fig. 11(c) is shown the distribution of the mobilized angle of friction ϕ_{mob} at the peak footing load obtained by substituting the measured local strains shown in Fig. 11(b) into such stress - 'local' strain relations in PSC as shown in Fig. 16 for the density of this model test. It may be seen that the values of ϕ_{mob} were close to the peak strength ϕ only in limited zones beneath the footing.

Fig. 12(a) shows the picture taken at the peak footing load for a 1g test with B_O= 50cm (see Fig. 8). Fig. 12(c) is the strain field constructed by using this picture. In this figure, zones without strain values are those occupied by the stiffeners used to achieve plane strain conditions (see Fig. 12a). It may be seen that shear strains are concentrated in two narrow bands. In Fig. 12(b) taken at S/B_O= 16% far after the peak footing load condition, a clear wedge has been formed, whereas the shear bands can be seen still only below the footing. All the results shown above, which are typical of all the tests, indicate that the failure in sand ground was always progressive.

Figs. 13 and 14 compare the results from the 1g tests with those from the centrifuge tests. It may be seen from Fig. 13 that the initial slope of load-settlement relation (as represented as $(S/q)_{50}$ at a half of the peak footing load in Fig. 14c) is very similar between the two types of tests. This result suggests that the centrifuge testing is most relevant for simulating a prototype footing in the case where the effect of shear band is not yet dominant. Further, it may be seen from Fig. 14 (a) and (b) that both the values of $N\gamma q$ and S_f/B_O are much smaller in the 1g test than in the corresponding centrifuge test. It is particularly interesting to note that in the 1g tests, the value of S_f/B_O decreased as B_O increased, while it was opposite in the centrifuge tests. This behavior may be explained as follows. First, for the same ratio of the footing settlement S to B_O, the value of S is larger for a larger B_O. Second, if the shear band width W_{SB} is independent of pressure-level (thus, independent of B_O), the ratio of W_{SB} to B_O decreases proportionally as B_O increases. Then, once shear bands have appeared, at the same value of S/B_O, the relative displacement across a shear band at corresponding points (e.g., at the same ratio of the depth from the ground surface to B_O) becomes larger for a larger B_O, particularly near the footing edges. This will lead to a larger ratio of shear band length relative to B_O for a larger B_O, which results in a less uniform distribution of ϕ_{mob} along the shear bands and in the ground, thus a larger degree of progressive failure. Third, shear bands start to develop before the peak footing load as seen from Figs. 10, 11 and 12. As a result, the value of $N\gamma q$ and S_f/B_O become smaller for a larger B_O, since both the values are controlled by the pattern of the distribution of ϕ_{mob} which is in turn controlled by the pattern of shear band. This effect may be called **the particle size effect**, as represented by **the ratio 'mean diameter D_{50}'/'footing width B_O'** as the first approximation (22).

So far, many researchers used the technique of modeling of models to demonstrate that the particle size effect is negligible in centrifuge tests (e.g., 13). Craig (2) pointed out, however, that "typically modelling has been carried out at scales varied by a factor of two or three and results within 5% are considered satisfactory, being close to the level of repeatability of experiments at a single scale."

The change of $N\gamma$ with the change in B_o in 1g has been called the scale effect. It may be seen from Fig. 15 that the tendency of the scale effect is similar for the 1g tests of the present study and the other 1g model tests collected by de Beer (3). Now, based on the results shown in Figs. 14 and 15, it can be concluded that the scale effect consists of **the particle size effect** and **the pressure-level effect**, the latter of which is typically observed in centrifuge tests with varying the g level using the same footing size and the same kind of sand. It is known that the pressure-level effect on $N\gamma$ is due to the effect of pressure level on both ϕ and the deformation properties of sand (e.g., 6).

PLANE STRAIN FEM ANALYSIS

A FEM code with an optimized dynamic relaxation (DR) solver developed by Tanaka (16, 17) was used. Namely, four-noded quadrilateral elements were used along with one point integration (reduced quadrature), which produced pseudo-equilibrium to improve the bounds of solution. The number of element for a half area was 203 (Fig. 17a), which was selected after a preliminary analysis of the effect of its number on the result.

An elasto-plastic analysis was performed using an isotropic strain--hardening-softening model. The yield function was Mohr-Coulomb type and the plastic potential was Drucker-Prager type. Stress - 'total strain (elastic plus plastic strains)' relations as a function of δ, as shown in Fig. 16, were used. The pre-peak parts were obtained from the corresponding PSC test results (20), after having been corrected for the bedding error. It was assumed that shear bands start to develop at the peak stress condition. Note that for the post-peak parts of the relationships shown in Fig. 16, strains are the ones defined locally within a band with a width of 1cm including a shear band obtained from another series of PSC tests at $\delta = 90°$ (21). The relations for the other angles δ were obtained from the one at $\delta = 90°$ by assuming that the rate of strain softening is independent of δ and the value of ϕ_{mob} becomes a constant residual value ϕ_r at the same relative lateral displacement across a shear band. These relations used were assumed independent of σ_3.

Elastic shear moduli were obtained as G (kgf/cm^2)= $900 \cdot (2.17-e)^2/(1+e) \cdot \sigma_m^{0.4}$ (σ_m; the mean principal stress in kgf/cm^2). An elastic Poisson's ratio of 0.3 was used. Further, the ratio of plastic strain increments $d\varepsilon_3/d\varepsilon_1$ was determined as follows. When the ratio of principal stress increments $d\sigma_1/d\sigma_3$ is larger than the current value of σ_1/σ_3, the volume change due to the increment of the mean principal stress σ_m was ignored and the Rowe's stress-dilatancy relation, $\sigma_1/\sigma_3 = -K \cdot d\varepsilon_3/d\varepsilon_1$ (K= 3.5) was used (thus, the flow rule was non-associated). When $d\sigma_1/d\sigma_3$ is smaller than the current value of σ_1/σ_3, the volume change due to dilatancy was ignored and the bulk modulus was obtained as K (kgf/cm^2)= 259·(2.17-e)2/(1+e)· $\sigma_m^{0.42}$, which was obtained from one dimensional compression tests. The footing was modeled by a very stiff material and no slip was allowed between the footing base and the ground surface.

An incremental integration technique by load controlling was used along with an equilibrium iteration tolerance of a force norm of 10^{-2}. Stresses were returned to the current yield surface by the substepping procedure. Probable hour glass modes were automatically suppressed by the nature of the DR solution (16, 17). Since geometric nonlinearity was not considered, the possible effects of the increase in the footing depth and continuous transition of the failing zone caused by the settlement of footing were ignored. Thus, the results at large footing settlements after the peak load are less reliable.

Fig. 17(b) shows the results of the following five cases of analysis which simulated the 1g tests (B_O= 50cm) shown in Figs. 8 and 17(b). Namely, the material property was perfectly plastic for the first three cases, (a), (b) and (c), while for the last two cases (d) and (e), the material exhibited post-peak strain-softening as shown in Fig. 16. Only in the case (e), the shear banding was considered.

(Case-**a**) ϕ is isotropic and independent of pressure-level; i.e., ϕ_o in PSC at δ = 90° with R(δ)= 1 for $\sigma_3 \leqq (\sigma_3)_O$ obtained from Eq. 1.

(Case-**b**) ϕ is isotropic and ϕ depends on pressure-level (Fig. 3).

(Case-**c**) ϕ is anisotropic (Fig. 2) and depends on pressure-level.

(Case-**d**) ϕ is anisotropic and depends on pressure-level. The effect of shear banding is neglected with the use of post-peak stress-strain relation for each FEM element equal to the corresponding relation in a shear band. That was obtained by increasing the strain rate after the peak by a factor of 1 cm/(shear band width= 0.3cm) for the relation for each δ shown in Fig. 16.

(Case-**e**) The properties of sand was modeled as close as to the actual ones; i.e., ϕ is anisotropic and depends on pressure-level, and the post-peak stress-strain relation for each FEM element is the one in the corresponding PSC specimen which has the same lateral area with the concerned FEM element. This stress-strain relation was obtained by changing the strain rate after the peak by a factor of 1 cm/$\sqrt{}$(area of element in cm^2) for the relation for each δ shown in Fig. 16. As a result, the post-peak stress-strain relation depends on the size of element and therefore, the result depends on the ratio of the shear band width to the footing size B_O, thus depends on the ratio D_{50}/B_O.

The result of Case-e is compared with the test result also in Fig. 8. The following points may be seen from Figs. 8 and 17(b).

(1) For Case-a, the peak footing load is similar to that obtained for ϕ_o in PSC at δ = 90° from the classic theories shown in Fig. 1. This result corresponds to that this analysis ignores the effects of the pressure dependency and anisotropy of ϕ and the progressive failure of ground as the classic theories shown in Fig. 1.

(2) In the perfectly-plastic analysis (Cases-a, b and c), the peak footing load is attained at a very large settlement of footing, because this is needed to mobilize the peak strength at all the points along the failure surfaces.

(3) The difference of the result between Case-c and Case-d indicates

that the effect of post-peak softening is significant. Note that the result of Case-d depends on the method of meshing, because the degree of strain localization depends on that. Namely, in the FEM analysis, the shear bands started to appear from the footing edges as in the tests. Thus, the analysis of Case-d is approximately equivalent to the analysis of Case-e for the case in which either the shear band size is

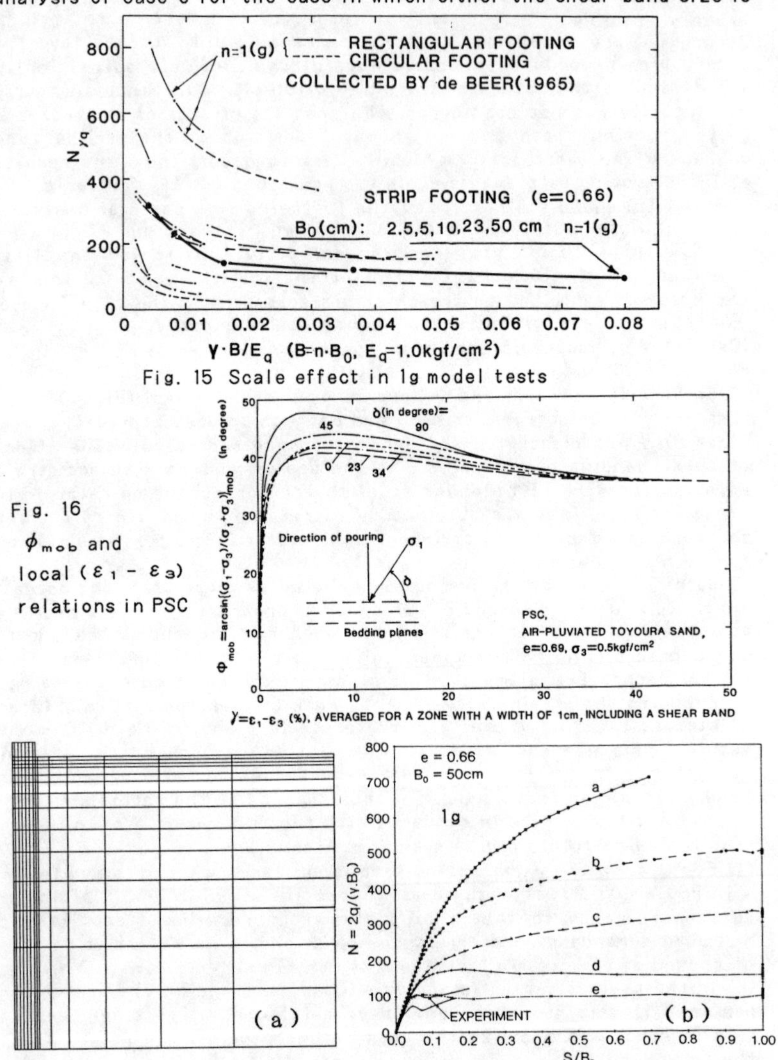

Fig. 15 Scale effect in 1g model tests

Fig. 16 ϕ_{mob} and local $(\varepsilon_1 - \varepsilon_3)$ relations in PSC

Fig. 17 (a) Meshing and (b) results of FEM analysis and test results

that of the meshes near the footing edges, 2.5cm, which is about 8 times the actual value, or the footing width B_0 is about 5cm with the actual shear band width (i.e., the simulation of the corresponding centrifuge test with B_0= 5cm at n~10(g)).
(4) Only the results of Case-e are close to the experimental results, particularly up to the moment of peak footing load.

CONCLUSIONS

(1) For air-pluviated Toyoura sand, the strength was anisotropic and the failure in the dense model ground was considerably progressive. Thus, when the value of ϕ from plane strain compression tests at δ = 90° was used, the isotropic perfectly-plastic theories largely overestimated the measured bearing capacity.
(2) Both the values of bearing capacity and footing settlement at the peak load in 1g tests with a footing width B_0 of 23cm and 50cm were larger than those in the corresponding centrifuge tests. Therefore, the particle size effect, or the effect of the ratio D_{50} (mean diameter of sand)/B_0, could not be ignored.
(3) The result of the FEM analysis was very close to the experimental results only when strength anisotropy, strain-softening and the effect of particle size associated with shear banding, in addition to the pressure-level dependency of the deformation and strength properties of sand, were taken into account.

ACKNOWLEDGEMENTS

The authors are deeply indebted to many of their colleagues who have joined this long-term research project. They also wish to express their gratitude to Dr. F. Molenkamp, Grondmechanica, the Netherlands, Dr. D. Leshchinsky, University of Delaware and Dr. J. Wu, University of Colorado, who kindly reviewed the manuscript.

REFERENCES

(1) Burland, J. B., Broms, B. B. and de Mello, V. F. B. (1977): Behaviour of foundations and structures, The SOA Report of Session 2, Proc. 9th ICSMFE, Tokyo, 2: 495-546
(2) Craig, W. H. (1988): On the use of a centrifuge, Proc. Centrifuge 88, Corte (ed.), Balkema, 1-6
(3) de Beer, E. E. (1965): Bearing capacity and settlement of shallow foundation on sand, Proc. Sympo. Bearing Capacity and Settlements of Foundations, Duke University, 15-33
(4) Fukushima, S. and Tatsuoka, F. (1984): Strength and deformation characteristics of saturated sand at extremely low pressures, Soils and Foundations, 24-4: 30-48
(5) Gemperline, M. (1988): Coupled effects of common variables on the behavior of shallow foundations in cohesionless soils, Centrifuge 88, Corte (ed.), Balkema, pp.285-292
(6) Graham, J. and Hovan, J. M. (1986): Stress characteristics for bearing capacity in sand using a critical state model, Canadian Geotechnical Jour., 11: 238-247

(7) Graham, J., Andrews, M. and Shields, D. H. (1988): Stress characteristics for shallow foundations in cohesionless slopes, Canadian Geotechnical Journal, 25: 238-249
(8) Hettler, A. and Gudehus, G. (1988): Influence of the foundation width on the bearing capacity factor, Soils and Foundations, 28-4: 81-92
(9) Kimura, T., Kusakabe, O. and Saitoh, K.(1985): Geotechnical model tests of bearing capacity problems in a centrifuge, Geotechnique, 35-1: 33-45
(10) Meyerhof, G. G. (1955): Influence of roughness of base and ground-water conditions on the ultimate bearing capacity of foundations, Geotechnique, 5: 227-242
(11) Molenkamp, F. and van Deventer, J. M. M. (1991): Scaling rule for the stiffness of foundations, The Journal of Engineering Mechanics, Vol. 116, No.12, Dec., pp. 2603-2624
(12) Oda, M. and Koishikawa, I. (1979): Effect of strength anisotropy in bearing capacity of shallow footing in a dense sand Soils and Foundations, 19-3: 15-28
(13) Ovesen, N. K. (1979): The use of physical models in design, Proc. 7th European Conf. on SMFE, Brighton, 4: 749-752
(14) Pradhan, T. B. S., Tatsuoka, F. and Horii, N. (1988): Strength and deformation characteristics of sand in torsional simple shear, Soils and Foundations, 28-2: 131-148
(15) Pu, J.-L. and Ko, H.-Y. (1988): Experimental determination of bearing capacity in sand by centrifuge footing tests, Centrifuge 88, Corte (ed.), Balkema, pp. 293-299
(16) Tanaka, T. and Kawamoto, O. (1988): Three-dimensional finite element collapse analysis for foundations and slopes using dynamic relaxation, Proc. Numerical Methods in Geomechanics, Insbruch, 1213-1218
(17) Tanaka, T. and Kawamoto, O. (1990): Numerical modelling for softening with localization of soil structures, Ingenieur-Archiv, submitted
(18) Tani, K. (1986): Mechanism of bearing capacity of shallow foundation on sand, Master of Engrg thesis, Univ. of Tokyo (in Japanese)
(19) Tatsuoka, F., Molemkamp, F., Torii, T. and Hino, T.(1984): Behaviour of lubrication lyers in element tests, Soils and Foundations, 24-1, 113-128.
(20) Tatsuoka, F., Sakamoto, M., Kawamura, T. and Fukushima, S. (1986): Strength and deformation characteristics of sand in plane strain compression at extremely low pressures, Soils and Foundations, 26-1: 65-85
(21) Tatsuoka, F., Nakamura, S., Huang, C. C. and Tani, K. (1989a): Strength anisotropy and shear band direction in plane strain tests of sand, Soils and Foundations, 30-1: 35-54
(22) Tatsuoka, F., Tani, K., Okahara, M., Morimoto, T. (1989b):Discussion on Hettler and Gudehus(1988), Soils and Foundations, 29-4: 146-154
(23) Tatsuoka, F., Huang, C.C., Morimoto, T. and Okahara, M.(1989c): Discussion on Graham et al. (1988), Canadian Geotechnical Jour., 26: 748-755

LIQUEFACTION ANALYSIS FOR NAVY STRUCTURES

J.M. Ferritto[1]

ABSTRACT: The Navy, forced by its mission to locate at the waterfront, often on loose, saturated cohesionless soils, faces a severe liquefaction threat. Research has been in progress to evaluate the Princeton University Effective Stress Soil Model. This paper will discuss that model and present validation studies comparing model predictions with centrifuge experiments and data recorded at a field site.

INTRODUCTION

The Navy has $25 billion worth of facilities in seismically active regions. Each year $200 million of new facilities are added in those seismically active areas. The Navy, because of its mission, must locate at the waterfront often on marginal land with a high water table. Seismically induced liquefaction is a major threat to the Navy. To understand the significance of liquefaction, it is important to note that the damage caused in recent earthquakes at the waterfront is largely attributed to liquefaction.

Present engineering practice does not consider the presence of the structure in determining the potential for liquefaction. Further, once liquefaction is evaluated, the resulting deformation state near a structure is not determined. Clearly this is inadequate to determine liquefaction induced consequences to a major structure. The finite element technique offers the potential for detailed analysis of soil structure problems and has been used extensively. Development of a constitutive model to predict pore pressure and characterize soils in terms of effective stresses (stress on soil grains causing deformation) rather than simple total stress offers the potential capability to analyze complex structures where liquefaction is possible.

The Naval Civil Engineering Laboratory (NCEL) undertook a review of available material models (1, 2, 3, 4). Comparison was made between test data and model predictions. Based on this study (1), it was concluded that the material model under development at Princeton University by Professor J.H. Prevost was able to predict the behavior of cohesive and cohesionless materials, and that its development was

1 - Project Engineer, Shore Facilities Department, Naval Civil Engineering Laboratory, Port Hueneme, CA 93043-5003.

implemented in a finite element code. A program of validating the soil model was undertaken using triaxial test data, centrifuge model tests and results of a field experiment.

PRINCETON UNIVERSITY EFFECTIVE STRESS MODEL

It was of interest to validate the material model developed at Princeton University by Professor J.H. Prevost to predict the behavior of cohesive and cohesionless materials (2).

The Princeton University Soil Model represents soil as a two-component system: the soil skeleton and the pore fluid. The hysteretic stress-strain behavior of the soil skeleton is modeled by using the effective-stress elastic-plastic model reported in Reference 3. The model is an extension of the simple multi-surface J_2-plasticity theory and uses conical yield surfaces. The model has been developed to retain the extreme versatility and accuracy of the multi-surface J_2-theory in describing observed shear nonlinear hysteretic behavior, shear stress-induced anisotropy effects, and to reflect the strong dependency of the shear dilatancy on the effective stress ratio in granular cohesionless soils. The model is applicable to general three-dimensional stress-strain conditions, but its parameters can be derived entirely from the results of conventional triaxial soil tests. The yield function is selected of the following form:

$$f = \frac{3}{2} (\underset{\sim}{s} - p \underset{\sim}{\alpha}) \cdot (\underset{\sim}{s} - p \underset{\sim}{\alpha}) - m^2 \ p^2 = 0 \tag{1}$$

where $\underset{\sim}{s} = \underset{\sim}{\sigma} - p\underset{\sim}{\delta}$ = deviatoric stress tensor; $p = 1/3 \text{ tr } \underset{\sim}{\delta}$ = effective mean normal stress; $\underset{\sim}{\delta}$ = effective stress tensor; $\underset{\sim}{\alpha}$ = kinematic deviatoric tensor defining the coordinates of the yield surface center in deviatoric stress subspace; m = material parameter. The yield function plots as a conical yield surface (Drucker-Prager type) in stress space, with its apex at the origin, as shown in Figure 1a. Unless $\underset{\sim}{\alpha} = \underset{\sim}{0}$, the axis of the cone does not coincide with the space diagonal. The cross section of the yield surface intersected by any deviatoric plane where p is constant is circular (3). Its center does not generally coincide with the origin, but is shifted by the amount $p \underset{\sim}{\alpha}$. This is illustrated in Figure 1a in the principal stress space. The plastic potential is selected such that the deviatoric plastic flow be associative. However, a nonassociative flow rule is used for its dilatational component (3).

In order to allow for the adjustment of the plastic hardening rule to any kind of experimental data, a collection of nested yield surfaces is used. The yield surfaces are all similar conical surfaces and a purely kinematic hardening rule is adopted. Upon contact, the yield surfaces are translated by the stress point and the direction of translation is selected such that no overlappings of the surfaces can take place. A plastic modulus is associated with each yield surface (3).

Complete specification of the constitutive model parameters requires the determination of 1) the initial position and sizes of the yield surfaces together with their associated plastic moduli;

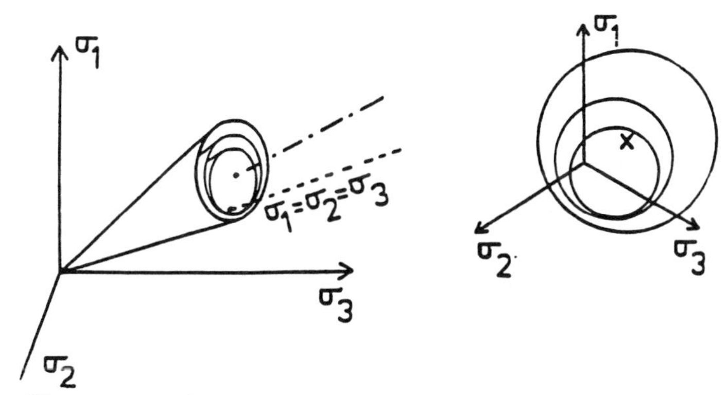

Figure 1a. Yield surfaces in principal stress space.

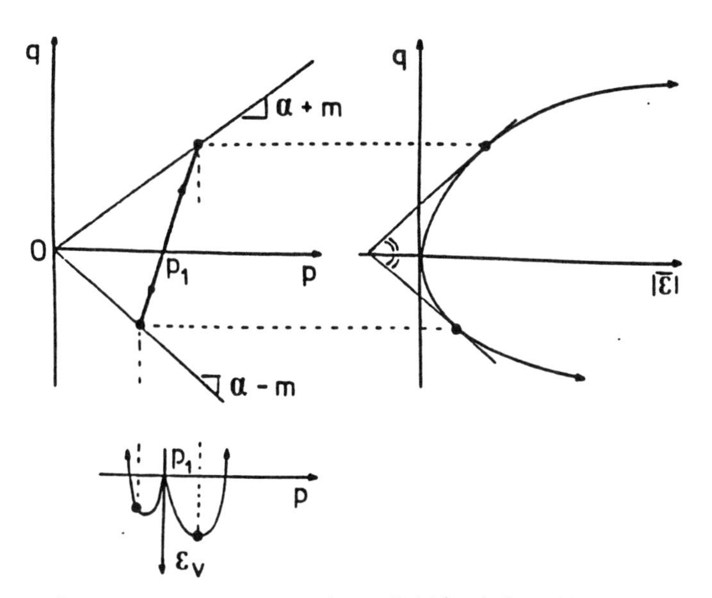

Figure 1b. Representation of triaxial soil test.

2) their size and/or plastic modulus changes as loading proceeds; and 3) the elastic shear and bulk moduli. The parameters used to define the constitutive soil model are obtained from monotonic drained triaxial compression and extension tests. Simple shear test data may be included as well.

Given the experimental stress-strain curves obtained in a triaxial shear soil test, identification of the model parameters associated with any given yield surface proceeds as follows (3). The smooth experimental shear stress-strain curves in compression and extension are approximated by linear segments along which the tangent (or secant) modulus is constant. The yield surface is identified by the condition that the slopes be the same in compression and extension, respectively, when the stress point has reached that yield surface. For a triaxial test (3) for which the two effective principal stresses are equal, $\sigma_2 = \sigma_3$. In order for the soil specimen to deform in an axisymmetric fashion ($\varepsilon_2 = \varepsilon_3$), the axes of loading must coincide with the principal axes of the anisotropic tensor α and $\alpha_2 = \alpha_3$. In the following, in order to follow common usage in soil mechanics, compressive stresses and strains are counted as positive and the discussion is presented in terms of the following stress and strain variables:

$$q = (\sigma_1 - \sigma_3) \quad p = (\sigma_1 + 2\sigma_3)/3 \tag{2}$$

$$\bar{\varepsilon} = (\varepsilon_1 - \varepsilon_3) \quad \varepsilon_\nu = \varepsilon_1 + 2\varepsilon_3$$

Eq.(1) then simplifies to:

$$f = (q - \alpha p)^2 - M^2 p^2 = 0 \tag{3}$$

where $\alpha = (\alpha_1 - \alpha_3) = 3\alpha_1/2)$. The trace of the yield surface onto the triaxial (q,p) stress plane consists of two straight lines of slopes $(\alpha + m)$ and $(\alpha - m)$, respectively (3), as illustrated in Figure 1b.

When upon loading in compression, the stress point reaches the yield surface f,

$$\eta_c = \left(\frac{q}{p}\right)_C = \alpha + m \tag{4}$$

whereas upon unloading in extension

$$\eta_E = \left(\frac{q}{p}\right)_E = \alpha - m \tag{5}$$

and

$$\frac{\dot{\varepsilon}}{\dot{q}} = \frac{1}{2G} + \frac{1}{H'} \frac{1 - \eta\dot{p}/\dot{q}}{1 + \frac{2}{9}\eta^2} \tag{6}$$

where G = Shear modules
H' = Plastic bulk modulus

$$\frac{\dot{\varepsilon}_\nu}{\dot{p}} = \frac{1}{B} \pm \frac{1}{H'} \frac{2}{\sqrt{6}} \frac{1-(\eta/\bar{\eta})^2}{1+(\eta/\bar{\eta})^2} \frac{\dot{q}/\dot{p} - \eta}{\left(1 + \frac{2}{9}\eta^2\right)^{1/2}} \tag{7}$$

where $\eta = \eta_C$; $\bar{\eta} = \bar{\eta}_C$ and $H' = H'_C$ in compression; $\eta = \eta_E$, $\bar{\eta} = \bar{\eta}_E$ and $H' = H'_E$ in extension.

The model parameters α, m, H'_C, H'_E, $\bar{\eta}_C$ and $\bar{\eta}_E$ are then simply computed from Eqs. (4) and (7), where $\dot{q}/\dot{p}$ = slope of the effective stress path (given by the specific stress path followed in the test). Therefore, once the elastic shear G and bulk B moduli are known, the identification procedure is straightforward and can easily be automated (2, 3).

The elastic shear modulus (low strain modulus) is determined by the steepest slope measured at the origin of the shear stress-strain curve, or better, through seismic-type measurements. The elastic B, and plastic H', bulk moduli associated with the stress point are determined by measuring the slopes $\dot{p}/\dot{\varepsilon}_\nu$ of small hydrostatic load-unload cycles at selected hydrostatic pressures, viz.

$$\frac{\dot{p}}{\dot{\varepsilon}_\nu} = B \frac{H'}{H'+B\sqrt{3}} \quad \text{(Load)}$$

$$\frac{\dot{p}}{\dot{\varepsilon}_\nu} = B \quad \text{(Unload)} \tag{8}$$

Typically, B = 2G/3.

Prevost (2) implemented the effective stress soil model into a finite element program DYNAFLO. It contains 2D 4-node plane elements with an axisymmetric option; 3D 8-node brick elements; 2D or 3D contact elements and slide line elements; and 2D or 3D truss and beam elements. The Navy sponsored a minor portion of the program development. The program is capable of solving static, consolidation and dynamic problems using implicit-explicit predictor-corrector time integration. Each node contains pore water velocity data, in addition to displacement data. As a 2-phase element, each element contains pore water pressure.

COMPARISON WITH LABORATORY TEST DATA

Five different sands ranging in relative density from 30 to 55 percent were studied (1). Test data required to develop model material parameters were obtained from a series of drained and undrained triaxial compression and extension tests conducted on a number of sands by universities and private geotechnical firms under NCEL sponsorship (4). Cyclic undrained tests and proportional loading tests were also conducted. With detailed sand test data available, NCEL began the task of validating the soil model by using the drained test data to determine model parameters and comparing predicted undrained monotonic and cyclic behavior with actual test data. In general the material model was capable of giving excellent representation of drained test data under a variety of loading conditions using parameters based on drained triaxial compression and extension. The model is capable of giving good agreement with undrained monotonic test data and can track the occurrence of liquefaction in cyclic tests in approximately the same number of cycles. Figures 2, 3, 4, and 5 give typical results for a fine silica sand at a relative density of 30 percent and a confining cell pressure of 2.0 kg/cm^2. The figures show stress difference, $(\sigma_1 - \sigma_2)$, as a function of vertical strain, ε, or mean stress, $(\sigma_1 + \sigma_2)/2$. The cyclic stress test, Figure 6, was a consolidated undrained slow cyclic triaxial test in which ± 4 kg/cm^2 stress difference was applied.

COMPARISON WITH CENTRIFUGE MODEL

The actual measurement of pore water pressure in the field during an earthquake would serve as the best source of data upon which to validate an effective stress soil model. Unfortunately these data are limited. The centrifuge has been used to study models of undrained soil deposits under seismic type excitation. This can serve as an approximate comparison given that the test data itself is only an indirect measure of field behavior and has errors associated with it.

The following sections will discuss two centrifuge tests which were modeled using DYNAFLO. In both cases the material model parameters were determined from triaxial test data on the same sand used in the centrifuge, compacted to the same relative density. The water table was established based on reported test conditions. Sinusoidal acceleration functions of the same magnitude were used in both cases. The computed results were obtained independently from the centrifuge test results and are not based on any post-centrifuge test data.

Soil Columns. The test procedures and test results are reported in Reference 5. The Monterey sand was pluviated in water in a stacked-ring apparatus. The model test was conducted on a centrifuge at a centrifugal acceleration of 100 gs, and was subjected to a sinusoidal base input acceleration. The corresponding prototype situation was analyzed. Triaxial tests were conducted on the same sand at the same relative density, 40 percent, as the model.

The centrifuge test modeled free-field conditions by using a stacked-ring device simulating a horizontally layered soil deposit. For the finite element analysis, the soil column was modeled with one vertical column of 20 two-dimensional 4-node plane elements, for a total height of 20 m. Each node was assigned four translational degrees of freedom: two for the soil skeleton and two for the fluid

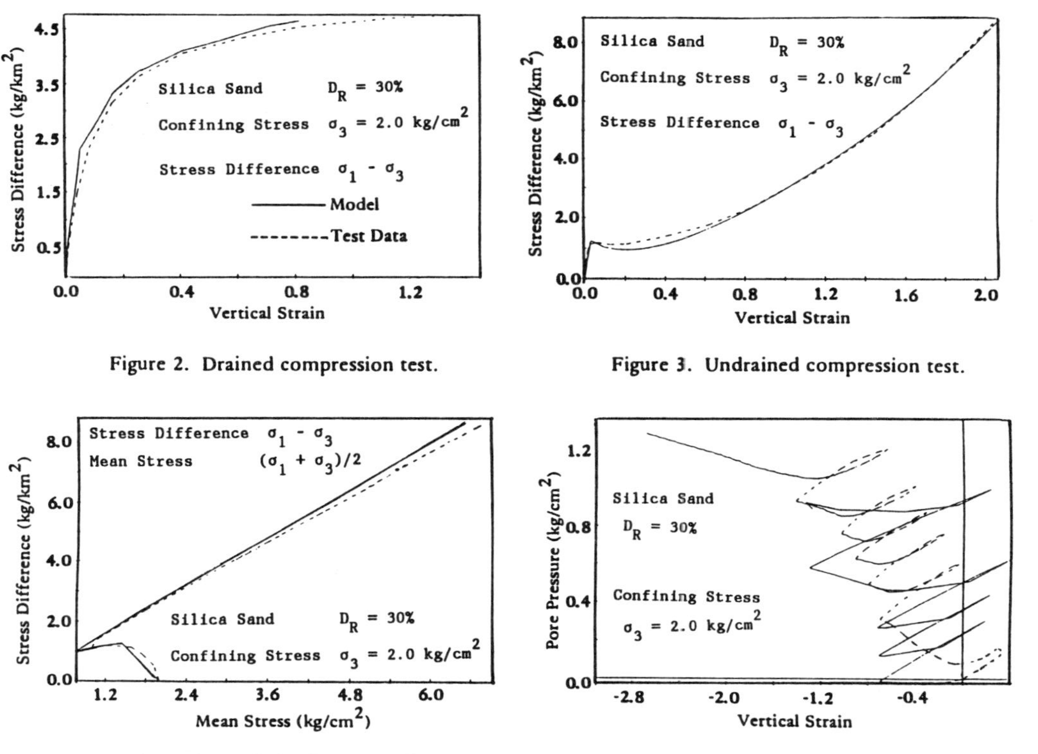

Figure 2. Drained compression test.

Figure 3. Undrained compression test.

Figure 4. Undrained stress path.

Figure 5. Cyclic test results.

phase (pore-water). The water table was located at the ground surface. Drainage of the pore-fluid was allowed to take place through the side and bottom boundaries and the ground acceleration was applied as a horizontal sinusoidal acceleration at the bottom boundary nodes. As a result of the shaking, excess pore-water pressures developed and partially dissipated in the soil column. The computed pore pressure ratios of 0.69 and 0.85 at two depths compare favorably with 0.74 and 0.86.

Footing. A model study from Reference 6 was selected for analysis. The soil used was Leighton-Buzzard 120/200 sand deposited in a stacked-ring apparatus by pluviating the sand in layers into water and then rodding to achieve the desired relative density of 55 percent. A brass weight simulating a footing was placed on top of the saturated sand deposit. The test was conducted in a centrifuge under centrifugal acceleration of 80g. The deposit was then subjected to sinusoidal base acceleration input motion. For the finite element analysis, the soil was discretized by using 240 elements and the brass footing by using two rows of 10 elements each (Figure 6). The brass footing is modeled as a one-phase elastic solid and a static pressure was applied to the top of the footing to produce the static bearing pressure. The water table was located at the ground surface. Drainage of the pore fluid was not allowed to take place through the bottom or side boundaries. A ground shaking was applied as a horizontal sinusoidal input acceleration at the bottom boundary nodes, with a maximum acceleration = 0.17g and a frequency of 1 Hertz for a duration of 10 seconds (10 cycles).

The stacked-ring apparatus controlled the side boundaries in the test. Test boundary conditions were simulated by DYNAFLO by assigning the same equation number to each nodal degree of freedom on the same horizontal plane for both side boundaries. The sand used in the centrifuge test was tested at the same relative density using monotonic compression and extension tests to determine the material model parameters.

An analysis of the computed acceleration time histories shows slight attenuation of base motion at the top of the footing. The computed acceleration time histories at the corners of the footings show rocking motions. Settlement beneath the footing increases continuously and almost linearly during the shaking. Additional significant settlements are not computed to occur after the shaking stops as noted in the test. As observed in the test, the pore-water pressure rises quickly in the "free-field" away from the structure. Directly under the structure, the pore-pressure increase is slower and always remains lower than the pore-pressure increase in the free-field at the same elevation. Immediately following shaking, the excess pore-pressures dissipate rapidly and reach their steady state conditions 5 seconds after the end of the shaking. This is further illustrated in Figure 6 which shows time histories for the vertical effective stress and excess pore-water pressures. Also shown in Figure 6 is a comparison of computed pore-pressure in comparison with the measured values. The computer simulation gives qualitatively good agreement showing liquefaction to occur in the same regions as observed in the test and also gives comparable pore-pressure ratios.

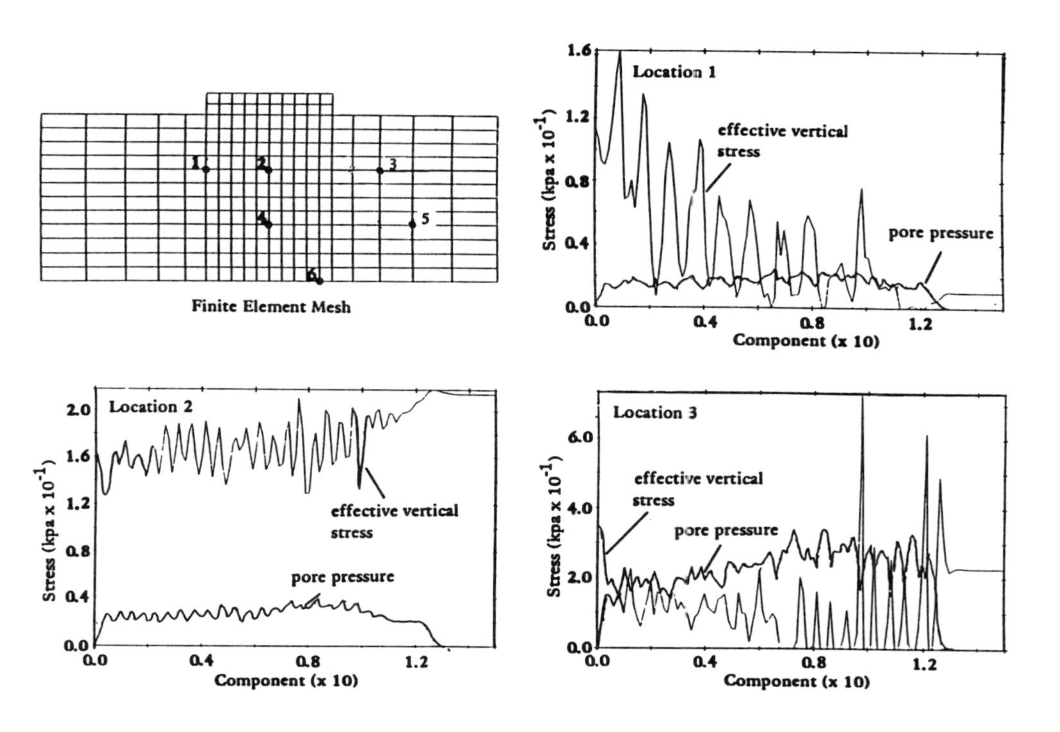

Figure 6. Footing analysis results.

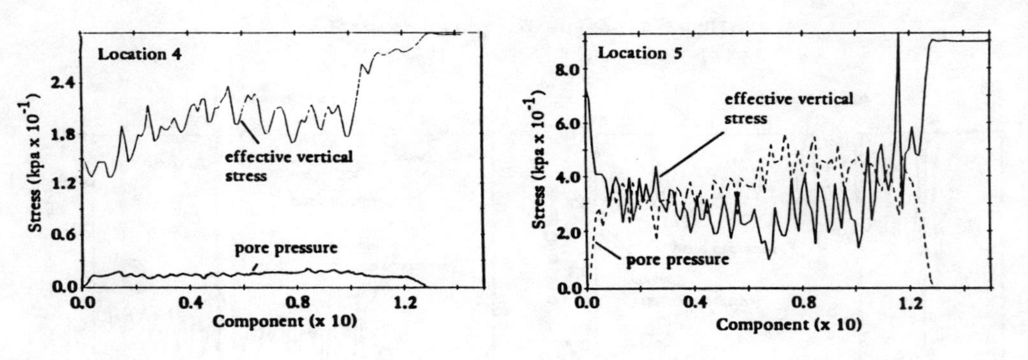

Peak Pore Pressure (kpa)

Location	Computer	Measured
1	20	20
2	30	30
3	30	25
4	15	12
5	50	65
6	40	60

Figure 6. Continued.

TAIWAN TEST SITE

The Electric Power Research Institute and the Taiwan Power Company designed and constructed a 1/4 scale and 1/12 scale reinforced concrete containment structure in Lotung, Taiwan. This site was selected for its high seismic activity. Under joint National Science Foundation and NCEL sponsorship, the University of California, Davis instrumented the site with piezometers to record pore pressure buildup during an earthquake.

The test site is located in the southern part of the Lanyang Plain of Northeastern Taiwan. The Lanyang River is located approximately 2 miles north of the test site. The plain is covered almost completely by alluvium. A detailed site exploration was completed including geophysical seismic up-hole and cross-hole tests, and a refraction survey. An Artesian pressure condition was noted at the site with a water table 2 to 3 feet below ground level. The potential for liquefaction was estimated to occur with acceleration levels as low as 0.06 to 0.20g.

A number of earthquakes in the magnitude range of 5 to 7 have been observed and recorded at the site. Two events at magnitude 6.2 in July 1986 and a magnitude 7.0 in November 1986 are being studied using DYNAFLO. The site contains silty sands and clayey silts. We have experienced some difficulty in computing the material parameters for the site based on triaxial test data. The model does not predict high enough compaction during the initial loading and dilates increasing strength more than the actual soil. The cause, perhaps lack of definition in the initial loading, is under study.

CONCLUSION

The material model developed by Prevost and implemented into the effective stress finite element code DYNAFLO has been evaluated. Results show the code is able to accurately predict undrained sand test results based on material parameters established from drained sand tests. The code has been used on several boundary value problems, two of which are reported herein, with satisfactory results. Further study is underway evaluating other materials besides sands.

REFERENCES

1. Prevost J.H., J. Ferritto, and R. Slyh (1986), "Evaluation and Validation of the Princeton University Stress Model," Naval Civil Engineering Laboratory, Technical Report R-919, Port Hueneme, CA.

2. Prevost, J.H. (1981), "DYNAFLOW: A Nonlinear Transient Finite Element Analysis Program," Princeton University, Department of Civil Engineering, Princeton, NJ.

3. Prevost, J.H. (1985), "A Simple Plasticity Theory for Frictional Cohesionless Soils," International Journal of Soil Dynamics and Earthquakes Engineering, vol 4 (1).

4. Ferritto, J.M. et al. (1983), "Utilization of the Prevost Soil Model for Soil-Structure Problems," Naval Civil Engineering Laboratory, Technical Memorandum M-51-83-09, Port Hueneme, CA.

5. Arulanandan, K., A. Anandarajah, and A. Abghari (1983), "Centrifugal Modeling of Soil Liquefaction Susceptibility," Journal Geotechnical Engineering, American Society of Civil Engineers, vol 109, no. 3, pp 281-300.

7. Whitman, R.V. and P.C. Lambe (1982), "Liquefaction: Consequences for a Structure," Proceedings of the Soil Dynamics and Earthquake Engineering Conference, Southhampton, U.K.

THE FUTURE OF GEOTECHNICAL CENTRIFUGES

William H. Craig[1], University of Manchester

ABSTRACT: The geotechnical centrifuge has become increasingly common as a tool in research and design studies, in the USA and elsewhere. New machines of varying capacity are being brought into use each year. The current status of the technique is reviewed and foreseeable developments which may foster increased emphasis on performance of structural components are reported.

INTRODUCTION

Although the principle on which the use of a centrifuge, for testing engineering models in which gravity induced body forces are significant was established as long ago as 1869 (1), the technique was first used in geotechnical studies only in the 1930s. It has been developed continuously in the USSR since then, but its use there was little known elsewhere and other usage was limited to the field of minor studies. There was work in geology and rock mechanics in the USA (2), which lapsed in the 1960s, around which time there was brief activity in South Africa, and photoelastic stress analysis of dams in the UK and China. In the broad field of geotechnical engineering an upsurge in activity can be dated back to 1969 and the simultaneous publication of papers from Japan, UK and USSR at the seventh International Conference on Soil Mechanics and Foundation Engineering (3, 4, 5). The technique may be considered to have 'come of age' and be relatively well established, with major hardware developments in several countries, Table 1. The first major, dedicated conference was held in Paris in 1988 (6) at about the same time as publication of a collection of state-of-the-art papers (7), including five from American authors, giving further insights to the historical development of centrifuge modelling in various countries and subject areas. The second dedicated conference takes place in 1991, in Boulder, Colorado.

The potential roles of the centrifuge in geotechnical engineering are many and the philosophies of usage vary widely. At this juncture it seems

1 - Lecturer, Dept. of Engineering, University of Manchester, Manchester, M13 9PL, UK.

appropriate to review the state of activities worldwide, with particular emphasis on developments in the USA, before venturing into the relatively mature years of this technology.

In a brief review of the 'Uses of a Centrifuge', the author has previously (8) highlighted four principal areas of activity where the centrifuge might make significant contributions, these being:-

a) Teaching.
b) Development of an understanding of the mechanics of geotechnical materials and structural forms.
c) Development of new methods of analysis (or modification of existing methods) for idealised but realistic geotechnical structures, by providing quantitative indications of the effects of parametric variations, once the overall mechanics are established.
d) Modelling site-specific situations to assist in design studies and project appraisal.

Ko (9) has used different sub-divisions which encompass essentially the same activities, with the exception of a contribution to education, namely:-

a) Modelling of prototypes.
b) Investigation of new phenomena.
c) Parametric studies.
d) Validation of numerical methods.

The role of the centrifuge in teaching is discussed briefly before a general discussion covering the remaining roles.

TEACHING

The potential of a centrifuge in teaching has been exploited since 1972 in Manchester University and has been recognised more recently in Japan and the USA. The principal benefit to an undergraduate or graduate student is seen to be the illustration of the effects of gravity induced self-weight stresses on structures. Such a demonstration aids the transition from the absorption of basic soil mechanics principles to the synthesis of the various components involved in the processes of design and analysis of geotechnical structures Simple models can be made of soil/structure interactions which demonstrate, for example, the effects of changes in structural flexibility. For modest capital cost a very powerful tool can be introduced to the teaching laboratory (10). At a time when overall pressure on the curriculum and the expansion of time spent on numerical computation has eaten into the time allocated for hands-on laboratory work, such a powerful demonstration of structural performance is time especially well spent.

CURRENT USAGE OF CENTRIFUGES WORLDWIDE

The role of the centrifuge is evident, but less clearly defined in research, development, engineering application and design. The trend in the last decade has been to develop a generation of highly sophisticated mechanical designs for major machines, typically with an acceleration capability of 200-300g. This follows an earlier generation of mechanically simpler, but possibly more durable, robust and adaptable, machines with capabilities of 100-200g. Today one major, European, manufacturer offers a range of large centrifuges designed specifically to meet the expanding demand from the geotechnical community.

It is of interest to review the contents of the Centrifuge 88 conference volume and assess the uses to which centrifuges, both new and old, are actually being put. This represents as objective a view as possible of the reality of world utilisation, which is preferred to a personal statement of opinion with in-built prejudices arising from standpoint and professional conditioning. It may be that this reality will be in conflict with the arguments sometimes presented to potential sponsors of new centrifuge building projects where budgets can run as high as $6,000,000 in the pursuit of machines with combinations of large model size and high acceleration capability.

In the conference proceedings there are a total of 78 papers of which 1 addresses the philosophy of usage, 22 are devoted exclusively to hardware and modelling technology and 3 are devoted to model studies of tectonic activity on a regional or global scale. The remaining 52 papers address essentially geotechnical engineering problems; of these, 14 papers are related to site-specific engineering situations and the remainder fall into two further broad categories:-

a) Assessment of the feasibility of centrifuge models. These either emanate from groups using new machines and acquainting themselves with the technology and its limits, or deal with possible applications to novel situations, especially in the area of dynamics (vibrations of structures, performance under quasiseismic or blast loading).
b) Generic and parametric studies of the performance of particular types of structure.

Perhaps surprisingly, not a single paper is devoted specifically to providing data for the verification of computer codes, though this has sometimes been cited in the past as a principal role for the centrifuge community (9, 11). It is pertinent to note that while many authors recognise the limitations of certain aspects of their models, none reports any major doubt about the use of the modelling concept.

Site-specific Models

As mentioned above, 14 of the Centrifuge 88 papers refer to site-specific studies. These emanate from groups working in 5 countries, Table 2, where it is seen that the specific site is sometimes a large scale field test rather than a truly independent primary structure. In several instances it is the performance of simulated structural components which is the primary subject of the work. The performance of centrifuge testing was in some instances in advance of (2), in parallel with (2), or after (1) the field test site work. In other instances, with true prototypes, model testing was carried out at the design or feasibility (4) stages of major projects and also at the time of re-evaluation of one structure at a critical stage in its working life and in another instance after a field failure. In only two of the studies was soil retrieved in an 'undisturbed' state from the field and subsequently built into the models, but in most cases true site soils were used in a reconsolidated condition or placed as fill materials in a manner similar to the field in order to avoid variations in the complex constitutive properties of replacement soils. In two cases the use of rockfill and crushed rock in the field necessitated modelling with substitute materials having scaled down grading and properties as near the prototype as possible.

Of the 14 projects, all but one involved testing at a single scale and a single acceleration level, Figure 1, - the exception being a wide-ranging study of pile performance at 17-92g with the modelling specific to a field test being carried out at 17g (data point included in the figure). A review of all the papers in the volume shows a similar pattern of acceleration/scale level usage, Figure 2. In certain studies a range of accelerations was used on different models and in this figure the highest level quoted has been plotted. In a small number of cases model performance was assessed by varying the acceleration, the so-called gravity turn-on approach, to induce failure or instability rather than by the application of external loading or perturbation at constant acceleration levels and the maximum levels are indicated in the same figure by cross hatching.

Figure 2 compares well with data plotted in Figure 3, abstracted from the papers to an earlier, smaller symposium on the Application of Centrifuge Modelling to Geotechnical Design held in Manchester in 1984.

Selection of Model Scale

Schofield (12) has argued that there is in general no particular requirement that a centrifuge be large, noting that the prime requirements are 'excellent instrumentation, skill in model preparation and keen observation of material behaviour in the model tests'. In practice there are severe limits recognised elsewhere (9, 13) dictated by the need to reproduce quite fine, but often critical, details within models. This results in the necessary selection of a modest linear scale reduction, which when coupled with a wish to avoid unnecessary boundary restraints, increases the desire to utilise large physical size of both model and centrifuge. The overall pattern of usage is clear - most testing is being performed at the lower end of the acceleration range although centrifuges with very high accelerations are available. That is to say that size limits are regularly reached while centrifuge acceleration limits rarely are. Operators are more comfortable with this situation than the reverse.

There will always be examples where there is a wish to test something a little larger than can be accommodated on any particular machine and recourse has to be made to either modelling part of the total structure or to accepting limited similarity for a complete model. An example of the first approach is the modelling of a single foundation pod for a proposed major concrete gravity tripod structure for offshore gas production. Testing took full advantage of both size and acceleration limits of the largest available centrifuge which was capable of subjecting models with a mass exceeding 3.5 tonnes to 120g. Problems encountered led to later, more successful, attempts to model the whole with reduced similarity - testing a 1:300 linear scale model under accelerations of 100g whilst utilising field soil at strengths either as in the prototype or reduced by a factor of 100/300.

Simulation of Construction Processes

The transition over twenty years from the application of the simplest gravity turn-on destruction of slope models to use of sophisticated loading systems and attempts to simulate complex field construction procedures has been rapid and promises to continue worldwide as electronic instrumentation and control systems are developed. While robotic systems are often massively inefficient (a severe handicap when rotating in a centrifuge) future developments in this area can be expected to lead to further advances in simulating construction processes and field operations. Reasonably successful devices for placing fills, by pluviation in flight (14, 15), for trafficking axle loads (16), for roller compaction (17) are among those which have already been used. Attempts to simulate excavation have been somewhat crude to date, draining fluid (18), removing soil en masse (19) or removing mechanical support in stages (20), but might now be tackled more realistically. Nor are the developments limited to construction. Recent studies (21) have focused on the problems of decommissioning offshore structures and the simulation of removal of individual jack-up spud cans from deep penetrations beneath soft seabeds.

Reporting and Use of Results

In discussing performance of structures subjected to quasi-static situations, deterministic loading patterns have usually been used and results are commonly presented in terms of specific loads, displacements, pressures etc. Most authors choose to present results at model scale, though some also quote numbers at prototype scale making use of the appropriate scaling relationships. Rarely are results presented at prototype scale alone. There is an implicit, widespread recognition that models are models and that the prototype which has been modelled and the true field structure are not identical. Idealised basic scaling relationships may be satisfied in terms of overall structure, loads, ground water conditions etc., but failure to include variations, and to follow identical stress paths during simulated construction or loading is inevitable to some degree, though this may be no more the case than with calculations on paper or in a computer. Thus the limitations of the centrifuge model approach in geotechnical engineering have to be recognised just as the limitations of the results from an in-situ SPT, laboratory triaxial shear strength test on soil, concrete cube or cylinder test, or the application of elastic or any other constitutive models used, have to be recognised in assessing the contribution to design decisions for real structures - the use of judgement is as essential as ever.

THE CENTRIFUGE AND THE STRUCTURAL ENGINEER

The 52 geotechnical engineering papers at Centrifuge 88, referred to above, can be classified in broad groups as shown in Table 3 with 35 being devoted to essentially static problems, of which 75% involved what might broadly be classified as soil/structure interactions in which the soil itself was not the primary structure under consideration as would be the case for a dam, embankment or consolidating landfill. While all of the foundation models (both shallow and deep) utilised external loading devices, without any attempt to simulate superstructure interaction, many of the other papers detail elaborate efforts to simulate as correctly as possible the properties of non-geotechnical components, their interactions with soil and rock and in particular the manner in which prototype loadings are applied. With larger machines becoming more available it is to be expected that future foundation studies will incorporate more realistic representations of linking between the several components of a total foundation system (single piles are rarely used in practice) and of superstructure stiffness, even if existing loading devices are used. For truly dynamic studies the mass distribution and other properties of the superstructure as well as the substructure will have to be incorporated and indeed there remains a potential role for structural model testing aboard centrifuges even under static loadings as envisaged by Phillips (1) but not yet explored by structural engineers.

SIMULATION OF EARTHQUAKE AND DYNAMIC LOADING

There has been a great deal of work already published in the field of dynamic modelling. Much of that in the area of impact and blast loading has been highly valued in areas as diverse as studies of impact crater formation on various bodies in the solar system (22), explosive catering (23) and weapons effects on buried structures (24). These effects are relatively easily simulated though there are limitations which have been openly recognised and acknowledged. For example soil container boundary reflections of waves from centrally detonated explosives are frequently observed, but do not detract from the critical primary wave passage effects, and Coriolis acceleration effects can lead to distortion of final crater dimensions as soils are ejected and subsequently fall back to ground.

Seismic simulation has been attempted in several laboratories, notably in Japan, UK, USA. There are several ways in which experimenters have generated quasi-seismic vibrations within rigid or flexible-walled containers hold soil and other structures, Table 4, and each has its merits and demerits which have been discussed elsewhere (25). As yet there is no consensus as to the 'best' system. One example is the Cambridge 'bumpy road' system which has operated for almost a decade and generated much valuable information. It provides a deterministic type of loading of ten nominally sinusoidal cycles of predetermined displacement at a single frequency, in one axis only, to the base boundary of a model container (26) . Electrical or servo-hydraulic shakers can provide more realistic representations of earthquake record spectra with wide range of frequency components. As control algorithms become more sophisticated (27), input can be provided to simulate specific documented acceleration or velocity records, e.g. from the El Centro earthquake, which can be used as a sort of proof test as is often used in numerical studies of prototype structures.

All seismic centrifuge studies so far have purported to model shaking in one horizontal dimension only and few attempts were made to measure other acceleration components in the early days. Although there is evidence that some, if not all, of the present shakers introduce substantial accelerations in other axes these are as yet unwanted, uncontrolled and often unquantified, but in a fast developing area of activity, mechanical refinements can be expected to reduce unwanted acceleration components to more reasonable levels. In practice of course real earthquakes generate multi-axis accelerations with large vertical components in some instances. In non-centrifuge studies much work has been done with multi-axis shaking tables. Arguably the future of centrifuge seismic studies lies in mounting a multi-axis shaking capability, but on the evidence of the existing single axis shakers, the step up in size and complexity for a useful servo-controlled shaker with multi-axis capability is some way off and the immediate future will see the development of single axis shakers capable of subjecting rather more massive models to more realistic spectra under sustained accelerations of 100g, or possibly 200g - though the arguments against going to smaller models and higher accelerations are as persuasive as ever when dynamic soil-structure interactions are to be assessed, given problems of instrumentation and the constraints of time scaling.

To date dynamic modelling in centrifuges has had a principal geotechnical focus. We are increasingly familiar with the ground conditions which led to high amplification of excitation over quite extensive areas of Mexico City in 1985, or with more limited extent as in the centre of Kirovakan in the Armenia quake of 1988, or the Marina area of San Francisco some 75km from the epicentre of of the Loma Prieta quake of 1989. Centrifuge studies have been helpful in gaining this appreciation, but it is clear that the move is now towards looking at dynamic soil/structure interactions to assess structural performance both above and below ground. More realistic modelling of above ground features will follow as centrifuge shaker capacities increase and structural engineers recognise the possibilities of assessing the effects of system non-linearities in realistic physical models.

THE CURRENT SITUATION IN THE USA

America now has three large centrifuges, at CU Boulder, UC Davis and RPI New York, which offer the geotechnical community a range of combinations of model size and acceleration capability. These machines are supported by others of smaller, but significant, size in these and other universities and in federal laboratories. Access to the engineering professional to centrifuges in defence industry establishments may be limited but the university machines have much to offer.

Graduate students have performed most of the work which has emerged from the smaller machines and an increased flow of interesting and useful results can be anticipated,which will contribute to the steady development of geotechnical endeavour. There are instances in the literature where engineers in commercial practice have gained access to machines (notably at Caltech), by sponsoring research or by buying machine time. The role of the big centrifuges is however still ill-defined. These machines are major national assets, largely funded out of the public purse. Within the US geotechnical community there is a view that the centrifuge groups have underperformed to date - concentrating on the protracted difficulties in getting the former NASA machine fully commissioned at Davis and ignoring the good work which has been done elsewhere and on the smaller machines at Davis. The perception will only change gradually, with the acceptance of results from a flow of significant projects on the major machines. In US academia it is difficult to find support for the infrastructure needed to sustain a steady throughput of tests of intelligently designed models and there remains a risk of continued under utilisation of these highly capitalised assets. Whilst this is not an uncommon situation in research laboratories, it will be particularly unfortunate if it pertains in a developing area of engineering. Those in everyday practice have much to gain in the short term on individual projects as well as in the longer term from investment in large scale experimental work which is still competitively priced when compared with field experiments. Those in universities with smaller centrifuges, and indeed with no machines, might also seek to utilise the machines of others. The sensitivities of college administrators seeking to contain academic activities within their own institutions should not be allowed to limit the opportunities to faculty and students - a measure of collaboration rather than excessive competition is the order of the day.

CONCLUSIONS

The concept of modelling aboard large centrifuges is now well established within the international geotechnical community and a wide range of studies has been conducted at the interface of soil-structure interaction, both static and dynamic. With increasing availability of large machines the possibilities are expanding. Increasing numbers of examples of modelling true field prototypes and large size field test structures are being documented - an indication of a sustained, broadly based increase in confidence in the technique which to an extent replaces an earlier, partly defensive, emphasis on modelling of models.

Most major centrifuges continue to operate in university laboratories and are controlled principally by teachers who have various, sometimes conflicting, commitments. The attitudes of the practising engineering profession vary from country to country, but contracting firms have provided backing for the second largest machine in Japan (28), albeit housed in a university, while the new machine in the Netherlands is operated in the private sector. The critical requirement for further uptake of centrifuge techniques by engineers in practice is ease of access to the necessary machinery within realistic limits of time. As the numbers of centrifuges and of engineers experienced in centrifuge work increases the restraints which may have inhibited potential users in the past will be reduced.

REFERENCES

1. Craig W.H. (1989). Edouard Phillips (1821-89) and the idea of centrifuge modelling. Geotechnique, Vol 39, pp697-700.
2. Cheney J.A. (1988). American literature on geotechnical centrifuge modelling 1931-1984. Ref 7, pp77-80.
3. Mikasa M., Takada N. & Yamada K. (1969). Centrifugal model test of a rockfill dam. Proc. 7th Int. Conf. Soil Mechanics & Foundation Engineering, Vol 2, pp325-333.
4. Avgherinos P.J. & Schofield A.N. (1969). Drawdown failures of centrifugal models. Proc. 7th Int. Conf. Soil Mechanics & Foundation Engineering, Vol 2, pp497-505.
5. Ter-stepanian G.I. & Goldstein M.N. (1969). Multi-storeyed landslides and strength of soft clay. Proc. 7th Int. Conf. Soil Mechanics & Foundation Engineering, Vol 2, pp693-700.
6. Carte J-F. (Editor) (1988). Proc. Int. Conf. on Geotechnical Centrifuge Modelling, Paris. A.A. Balkema, Rotterdam.
7. Craig W.H., James R.G. & Schofield A.N. (Editors) (1988). Centrifuges in Soil Mechanics A.A. Balkema, Rotterdam.

8. Craig W.H. (1988). On the uses of a centrifuge. Ref. 6, pp1-6.
9. Ko H-Y. (1988). Summary of the state-of-the-art in centrifuge model testing. Ref. 7, pp11-18.
10. Craig W.H. (1989). The use of a centrifuge in geotechnical engineering education. Geotechnical Testing Journal, Vol 12, pp288-291.
11. Scott R.F. (1988). Physical and numerical models. Ref. 7, pp103-118.
12. Schofield A.N. (1988). An introduction to centrifuge modelling. Ref. 7, pp1-9.
13. Rowe P.W. (1972). The relevance of soil fabric to site investigation practice. Geotechnique, Vol 22, pp195-300.
14. Beasley D.J. (1973). Centrifugal modelling of soft clay strata subjected to embankment loading. Ph.D. thesis, Cambridge University.
15. Horner J.N. (1982). Centrifugal modelling of multi-layer clay foundations subjected to embankment loading. Ph.D. thesis, King's College, London University.
16. Craig W.H. & Mokrani A. (1988). Effect of static line and rolling axle loads on flexible culverts buried in granular soil. Ref. 6, pp385-394.
17. McVay M. & Papadopoulos P. (1986). Long term behaviour of buried large span culverts. Journal of Geotechnical Engineering, Vol 112, pp424-441.
18. Lade P.J., Jessberger H.L., Makowski E. & Jordan P. (1981). Modelling of deep shafts in centrifuge tests. Proc. 10th Int. Conf. Soil Mechanics & Foundation Engineering, Vol 3, pp683-691.
19. Azevedo R.F. & Ko H-Y. (1988). In-flight excavation tests in sand. Ref. 6, pp119-124.
20. Craig W.H. & Yildirim S. (1976). Modelling excavations and excavation processes. Proc. 6th Euro. Conf. Soil Mechanics & Foundation Engineering, Vol 1, pp33-36.
21. Craig W.H. & Chua K. (1990). Extraction forces for offshore foundations under undrained loading. Journal of Geotechnical Engineering, Vol 116, No. 5, pp868-884.
22. Schmidt R.M. (1980). Meteor Crater: energy of formation - implications of centrifuge scaling. Proc. 11th Lunar & Planetary Sci. Conf., pp2099-2128.
23. Schmidt R.M. (1978). Centrifuge simulation of the Johnie Boy 500 ton cratering event. Proc. 9th Lunar & Planetary Sci. Conf., pp3877-3889.
24. Townsend F.C., Tabatabai H., McVay M., Bloomquist D. & Gill J.J. (1988). Centrifugal modelling of buried structures subjected to blast loadings. Ref. 6, pp473-479.
25. Whitman R.V. (1988). Experiments with earthquake ground motion simulation. Ref. 7, pp203-216.

26. Schofield A.N. (1981). Dynamical earthquake geotechnical centrifuge modelling. Proc. Int. Conf. on Recent Advances in Geotechnical Engineering & Soil Dynamics, University of Missouri, Rolla, Vol 3, pp1080-1100.
27. Ketcham S.A. (1989). Development of an earthquake motion simulator for centrifuge testing and the dynamic response of a model sand embankment. Ph.D. thesis, University of Colorado.
28. Fujii N., Kusakabe O., Keto H. & Maeda Y. (1988). Bearing capacity of a footing with an uneven base on a slope: direct comparison of prototype and centrifuge model behaviour. Ref. 6, pp301-306.

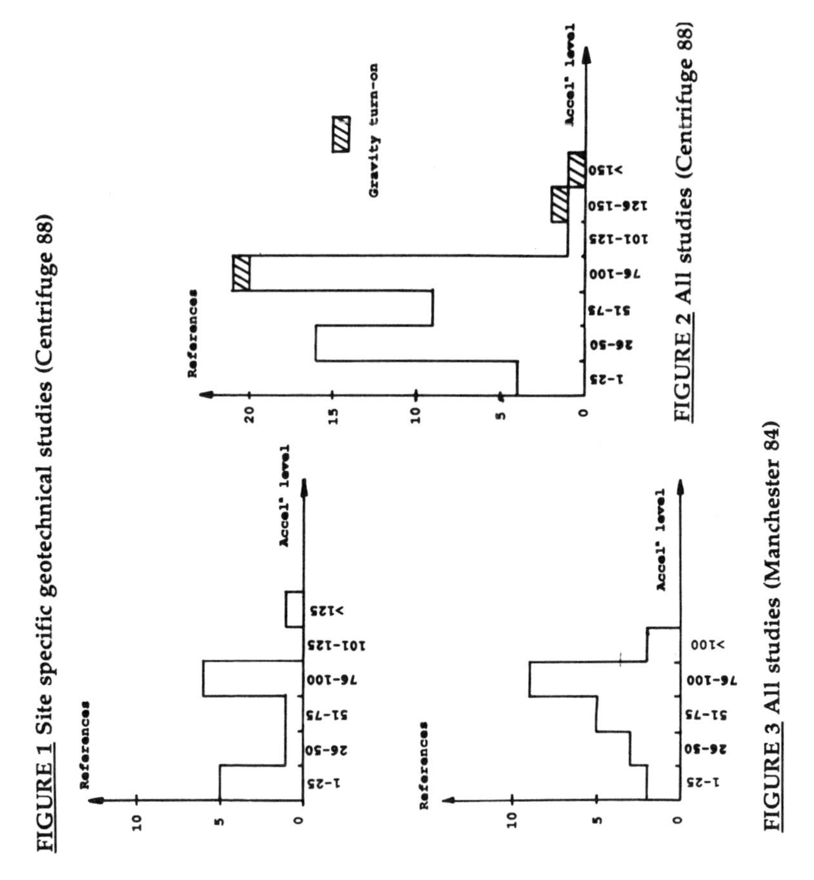

FIGURE 1 Site specific geotechnical studies (Centrifuge 88)

FIGURE 2 All studies (Centrifuge 88)

FIGURE 3 All studies (Manchester 84)

Canada	Denmark
France	Germany
Italy	Japan
Netherlands	P.R. of China
United Kingdom	United States of America
USSR	

TABLE 1 Countries with large centrifuges used for geotechnical engineering

Subject area	Number	Subtotal
Dams and embankments	4	35
Soil consolidation	5	
Anchors	2	
Soil reinforcement	6	
Surface foundation	8	
Deep foundations	3	
Undergound and buried structures	2	
Retaining structures	3	
Soil-structure interaction	2	
Dynamic models (impact, blasting, vibration, earthquakes)	13	
'New frontiers' (silos, heat transfer, faulting)	4	

Table 3 Subject areas of model studies - Centrifuge 88

Study	Scale	Dim.	Stage	Country
50m rockfill dam	90g	2D	Design	China
43m tailings dam	130g	2D	Working life reassessment	China
7m embankment on soft soil	100g	2D*	During construction	China
Consolidation of dredged fill	100g	1D	During construction	France
Bearing capacity of footing on slope	20g	2D*	Parallels field test	Japan
Bearing capacity of shallow foundation	14.5g	3D	Parallels field test	France
Foundation for offshore structure	100g	3D	Design•	France
Lateral loading of pile	17g	3D	In advance of field test	UK
Rolling axle loads above buried culvert	25g	3D	After field test	UK
Quay wall	70g	2D	After failure	China
Photoelastic analysis	40g	2D	Design	China
Earthfill protection against rockfall	100g	3D	Design	France
Dynamic earth pressure against bridge abutment 32m span	100g	2D	After construction	USA
Discharge from grain silo	10g	3D	In advance of field test	France

* Utilised undisturbed site specimens
• Project abandoned

TABLE 2 Site-specific centrifuge testing

Method	Comment
Trigger release, cocked springs	Free decay oscillations
Piezo electric	Appears limited in capacity
Explosive detonations	Limited control and repeatability
Mechanical excitation a) 'Bumpy Road' b) Rotating cams/link mechanisms	Generally repeated sinusiodal excitation at single frequency (possibly variable)
Servo controlled hydraulics	Mechanically complex and heavy, but capable of generating repeatable input motions with realistic spectra

TABLE 4 Methods of quasi-seismic excitation

LABORATORY SMALL SCALE MODELLING TESTS USING THE HYDRAULIC GRADIENT SIMILITUDE METHOD

Li Yan[1] and Peter M. Byrne[2]

ABSTRACT: A method of using hydraulic gradient to increase soil stresses in model tests is presented. The testing principle and procedure, and the factors affecting test results are discussed. Application to studying laterally loaded piles is presented. The corresponding scaling laws are evaluated using the "modelling of model" technique and found to follow the expected scaling relations as in centrifuge tests. The observed pile responses are found to be similar to centrifuge tests. The API (1987) laterally loaded pile design method is evaluated against the test results, and reasonably good agreement is obtained. It is shown that the hydraulic gradient similitude method is a simple and inexpensive way to test small scale models in the field stress condition.

INTRODUCTION

Model tests are used in almost all disciplines of engineering science research. In geotechnical engineering, small scale model tests are often used to study the complex nature of soil response and soil-structure interaction under a controlled condition. In the past, model tests were mostly performed under the normal stress condition of a one-gravity field. However, it is known that soil response depends upon the level of effective stress within the soil mass, especially for granular material. At different stress levels, granular material at a given density can behave in either a contractive or dilative manner. Thus, small scale model tests in a low stress field often fail to reveal some important phenomena which may exist at the prototype stress level.

In order to overcome this situation, it is desirable to perform a small scale test at field stress level conditions. Although the stress level in a model soil mass can be increased in conventional manner, such as by surcharge, the stress gradient within the soil mass, which is important to the soil structure response, cannot be correctly simulated. At present, one method of correctly simulating both field stress magnitude and distribution is the centrifuge technique, in which a small model scaled 1/n from the assumed prototype is tested under an 'n' times gravity field created by centripetal acceleration. With this escalated gravity field, the soil elements at homologous points of model and prototype have the same self-weight stresses, and a simulation of prototype behavior is then assumed from a set of scaling relations [11]. Although the principle is simple, the use of the centrifuge testing technique to simulate prototype behavior involves certain difficulties in practice. Very often, a complete similitude relation and various prototype soil conditions cannot be fully satisfied and simulated by the centripetal acceleration alone [3,13]. However, the main constraint to the use of centrifugal testing so far has

[1]Present Address: Klohn Leonoff Ltd., Vancouver, B.C.; formerly Ph.D. student, Dept. of Civil Engineering, University of British Columbia, Vancouver, B.C., Canada

[2]Prof., Dept. of Civil Engineering, University of British Columbia, Vancouver, B.C., Canada

been the high cost of equipment and testing, and the need for specially trained personnel to operate the system. For these reasons, the centrifuge technique is presently used only for some important structures and specialized research, and is not generally available in conventional soil mechanics laboratories. Many model tests reported in the geotechnical engineering literature are still performed under the low stress condition. This is not due to a lack of understanding of stress level dependency of soil behavior, but rather a lack of an easily accessible testing approach with which the field stress condition can be closely represented in the model.

This paper presents an alternative method, the hydraulic gradient similitude method (HGS) that was first used by Zelikson [18]. This method employs a high hydraulic gradient which causes a seepage force within the granular soil and creates a high body force. This body force results in a high stress level and distribution approaching field conditions. The principle of the testing method is described and the factors affecting the results are discussed. Application of the HGS method to the laterally loaded pile problem is presented.

THE HYDRAULIC GRADIENT SIMILITUDE METHOD

Testing Principle and Scaling Laws. Similar to the centrifuge modelling technique, the HGS method is just another way of increasing soil stresses in the model. The only difference is that the body force of model soils in HGS testing is effectively increased by the seepage force through the porous material rather than by centripetal acceleration.

For a model test subjected to a controlled downward hydraulic gradient, seepage force will increase the unit volume body force of a soil element by an amount of $i\gamma_w$. This is equivalent to increasing the unit weight of the material by $i\gamma_w$, then the effective unit weight, γ_m, of the model soil is:

$$\gamma_m = i\gamma_w + \gamma' \tag{1}$$

where i is the applied downward hydraulic gradient, γ_w is the unit weight of water if water is used in the test, and γ' is the submerged unit weight of soil. Thus, the vertical effective stress in the model soil has increased, and its distribution with depth is as shown in Fig.1, i.e.:

$$\sigma'_v = (i\gamma_w + \gamma') \cdot z \tag{2}$$

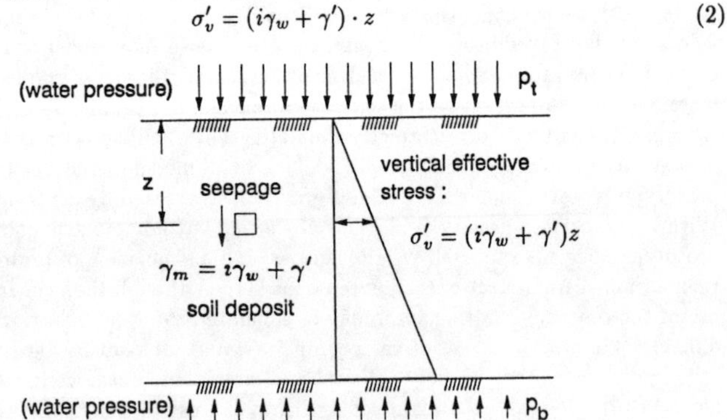

Figure 1 Soil Stress Condition in HGS Test

As compared to the assumed prototype condition, the model unit weight has been increased by a factor of N, i.e:

$$N = \frac{\gamma_m}{\gamma_p} = \frac{i\gamma_w + \gamma'}{\gamma_p} \tag{3}$$

where N is defined as the hydraulic gradient scale factor, and γ_p is the effective unit weight of the soil in the prototype, which could either be total or submerged unit weight depending upon the ground water conditions in the prototype soil. Thus, when the 1/n scaled model test is performed under a hydraulic gradient scale factor N=n, the stresses due to the self-weight of soils at homologous points of model and prototype will be the same, i.e., the scale factor for stress is unity. If the same soil is tested and the same stress path is followed in the model as in the prototype, the strains in the model and prototype will be the same [11], i.e., the scale factor for the strain is unity, while the displacements of the prototype will be larger than the model by the factor n=N. Thus, the scaling laws for the HGS tests are expected to be the same as in the centrifuge tests.

In the actual testing, however, the scaling laws related to the problems studied have to be verified experimentally, as many factors may not be scaled by the scaling laws due to technical limitations. In centrifuge tests, the "modelling of models" technique is often used, in which an assumed prototype behavior is simulated with different scaled models under different stress fields. In this paper, this same technique will be employed to examine the hydraulic gradient scaling laws for the laterally loaded pile problems.

Factors affecting HGS Tests in Sands. Since its development, HGS method has been successfully used in some model tests of anchor and pile problems [19,22], and footing foundation [15]. HGS tests have been compared with centrifuge tests [15,20], and similar results have been observed in the cases where comparison was possible. Recently, some researchers have combined the hydraulic gradient technique with centrifuge testing in which a large hydraulic gradient was used to consolidate large clay samples prior to centrifuge model testing [9,21]. However, possible factors affecting the HGS method have not yet been clearly identified. Herein, attempts are made to examine some factors affecting HGS test on sand.

In the HGS tests, a downward 1-D steady state flow occurs across the soil deposit. With this downward seepage flow, it is hoped that the fluid pressure decreases linearly with depth, giving a linearly distributed soil stress. This process can be described mathematically as follows based on the mass conservation principle:

$$\gamma_f V_z = C_1 \tag{4}$$

where V_z is the discharge velocity of the flow in z - direction, γ_f is the unit weight of the fluid, and C_1 is a constant.

Due to the large hydraulic gradient in the tests, a high seepage velocity may lead to non-Darcian flow in which the linear Darcy's law no longer exists. Criteria for which the Darcy's law becomes invalid have been suggested by many researchers based on theoretical and experimental results and show large discrepancies. Various empirical expressions were also proposed to fit the experimental data when the Darcian flow is invalid [12]. One such equation for sands is:

$$ki = V_z^m \quad \text{m} = 1 \text{ to } 2 \tag{5}$$

where i is the hydraulic gradient, and k represents some measure of permeability of the soil system. When m=1, Eq.(5) reduces to Darcy's law for the Darcian flow.

The hydraulic gradient, i, in the tests is:

$$i = \frac{\partial h}{\partial z} \quad \text{with} \quad h = h_e + h_p \tag{6}$$

$$i = 1 + \frac{1}{\gamma_f}\frac{\partial p}{\partial z} \tag{7}$$

where h is the total hydraulic head, h_e is the elevation head and h_p is the pressure head due to the pressure p. From Eqs. (4), (5) and (7), it follows:

$$\frac{\partial p}{\partial z} = \gamma_f \left[\frac{C_1^m}{k\gamma_f^m} - 1\right] \tag{8}$$

From Eq.(8), it can be seen that if γ_f is independent of the pressure, p, such as in an incompressible fluid flow, then the non-Darcian flow will also give a linearly distributed fluid pressure as in the case of Darcian flow. Provided the fluid pressure decreases linearly with depth, the soil effective stress will increase linearly with depth.

However, if the fluid is compressible due to either a compressible fluid or an unsaturated soil-fluid system, the values of γ_f will be a function of fluid pressure, p [5,6]. In that case, the fluid pressure will not linearly decrease with the depth, resulting in a nonlinear soil stress distribution.

In addition, k in the Eq.(8) is a function of soil void ratio and structure. Thus, theoretically, if the void ratio or structure changes during the tests, the artificial stress field created by the hydraulic gradient will be distorted. Its scaling effect can be evaluated experimentally using the "modelling of models" technique.

APPLICATION - LATERALLY LOADED PILES

In this paper, a model study of laterally loaded pile response is presented to illustrate the application of the HGS modelling test method. The corresponding scaling laws are evaluated experimentally. The test data are used to study the pile response to lateral load under different loading conditions and with this data base, methods of analysis can be evaluated.

Test Set-up and Procedure. A hydraulic gradient similitude testing device especially designed to examine the laterally loaded pile problem has been developed at the University of British Columbia [16]. A schematic of the device is shown in Fig.2. It consists of five major components: (1) soil container and air pressure chamber; (2) water supply and circulation system; (3) air pressure supply system; (4) pile loading system, and (5) data acquisition and control system.

The soil deposit was formed of uniform fine Ottawa sand using the "quick sand" sample preparation technique [15]. The soil deposit is 315.2 mm in height, and 404mm x 190mm in plan with the larger dimension in the pile loading direction. A centrifugal pump is used to supply the water to the top of the soil tank, and water table is maintained at 25 mm above the sand surface. The given hydraulic gradient is obtained by controlling air pressure in the air chamber and draining the water to a low pressure at the sand base. Thus, pore water pressure in the soil decreases with depth, giving escalated effective stresses that increase linearly with depth.

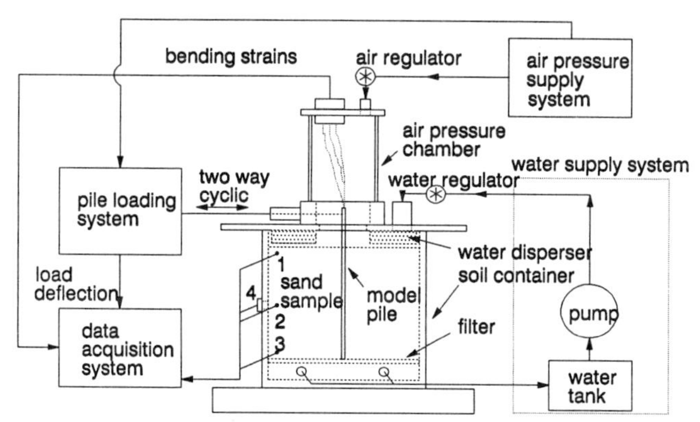

Figure 2 A Schematic of UBC HGST Device

Three model piles are made of 6.35, 9.53, 12.7 mm O.D alum. tubing. The 6.35 mm O.D. pile is instrumented with 8 pair of strain gauges along its length. The model piles are loaded laterally at the pile head, and the pile response is measured through the measurements of applied pile head load, pile head deflection and the pile bending moment along the pile length. In the tests, the model pile is hand pushed to the full depth into the sand deposit under the normal stress condition, as the pile driving device at high hydraulic gradient condition has not been developed. Limited test data from centrifuge tests have shown that the effects due to pile driving at $1g$ stress condition are small on the laterally loaded pile response [4].

Results and Discussion

Evaluation of Scaling Laws: The scaling laws implied in the HGS testing of laterally loaded piles are examined by using the "modelling of models" technique. A testing series is devised to produce a 0.3 m prototype pile at two loading eccentricities. The results are given in Fig.3 in terms of prototype pile head response which is obtained by multiplying the load and deflection measurements by N^2 and N respectively at the corresponding N-value. It can be seen from these results that the same prototype behavior is obtained from different model tests under scaled stress conditions. This verifies that the HGS modelling technique for laterally loaded pile tests is self-consistent and obeys the scaling principle as in the centrifuge tests. This also shows that the possible scale effects due to the distortion of soil permeability during the loading process are very small.

Pile Response and Predictions: With the HGS testing method, pile response to pile head lateral load has been studied under various conditions that include different soil stress levels, different loading eccentricity and pile head fixity, and different soil density [16]. Due to the space limitation, only the response of a 6.35mm O.D. free head pile embedded in dense sand under a loading eccentricity of 45mm is presented herein.

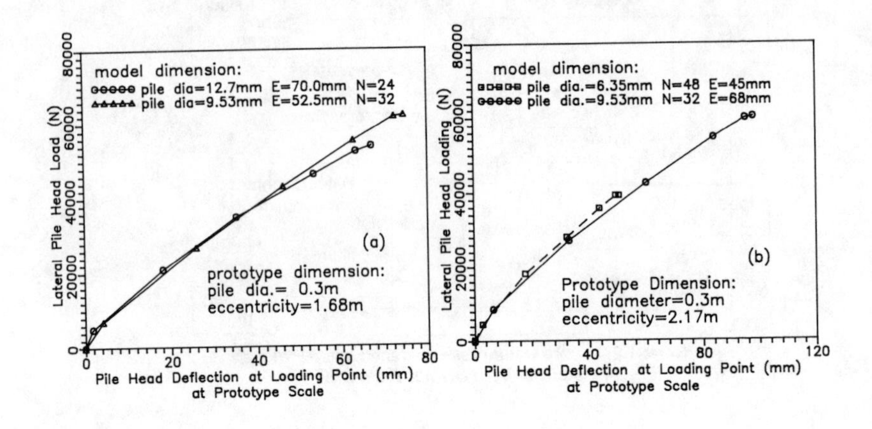

Figure 3 Evaluation of Scaling Laws for Laterally Loaded Piles; Prototype Dimension: (a) 0.3 m Pile with Eccen.=1.68m; (b) 0.3 m Pile with Eccen.=2.17m.

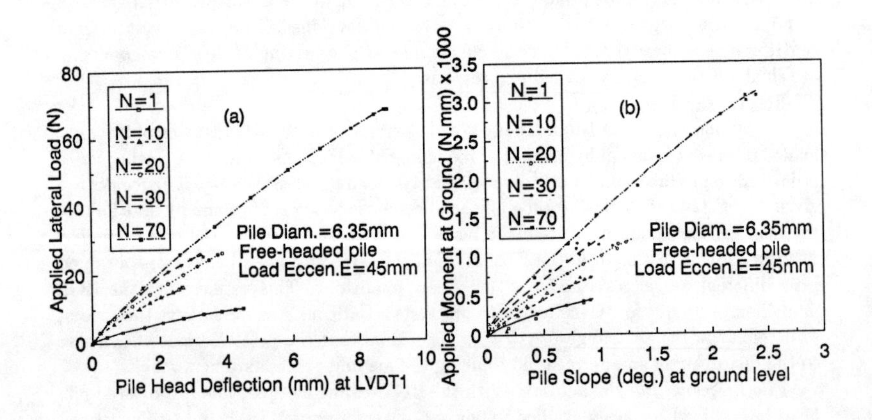

Figure 4 Pile Head Responses under Different Hydraulic Gradient

Fig.4 shows the pile head responses at different hydraulic gradients. The deflections are those measured at LVDT1 (20 mm above the loading point) and the slopes are those calculated at the ground level. It is seen from the figure that the pile head response is significantly influenced by the hydraulic gradient. As the hydraulic gradient increases, the pile response becomes stiffer. This influence is very marked as the hydraulic gradient scale factor N changes from one to ten, above ten the influence is less marked, especially for the pile head deflection. This indicates that the pile response under the lateral loading is strongly dependent upon the soil stress levels, or the relative soil-pile stiffness. These observation are similar in characteristics with those in centrifuge tests [2].

Fig.5 shows the corresponding bending moment distributions along the pile length. The measured bending moments have been normalized to the applied lateral load at the pile head so as to eliminate the bending moment difference due to the different loading magnitudes. It may be seen that the applied hydraulic gradient greatly changes both the shape and the magnitude of pile bending moment distribution. At $N=1$ (i.e. the low stress condition), in order to mobilize enough soil resistance to the applied load, the pile deflects down to almost its full length, resulting in significant bending moments along its full length. As the soil stress and soil stiffness increase due to the applied hydraulic gradient, the deflected pile length is greatly reduced. As a result, both the maximum bending moment and its depth decrease. At the hydraulic gradient scale factor, N, of 70, some negative bending moments are developed at depth. These observations indicate the importance of the correct field stress condition when studying soil-pile interaction using small scale model tests. Results from $1g$ model test may be misleading.

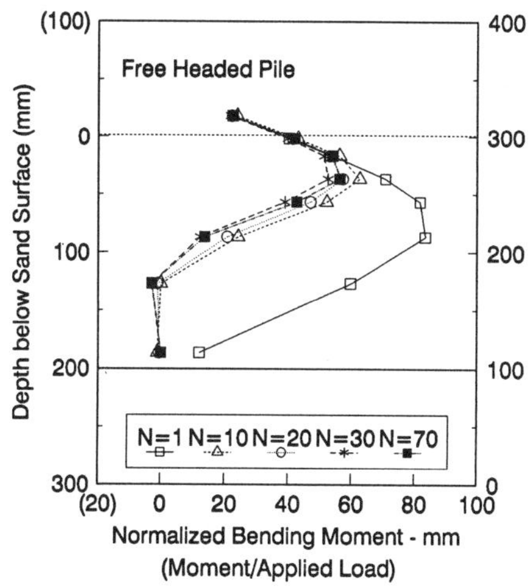

Figure 5 Pile Bending Moment Distribution

The soil-pile interaction in terms of P-y curves can be obtained from the bending moment distribution. A cubic spline is employed to fit the bending moment distribution, and the pile deflection, y, and soil reaction, P, are calculated from the numerical integration and differentiation of the cubic spline. Fig.6 shows the experimental P-y curves at four locations below the ground under different hydraulic gradient conditions. It is seen that at all depths the P-y curves at $N=1$ are very soft and reach some ultimate values at small pile deformation. However, as the soil stresses increase due to the applied hydraulic gradient, stiffer and less nonlinear soil-pile interaction curves are obtained. When these experimental P-y curves are plotted on a logarithmic scale, as shown in Fig.7, they are linear. This indicates that the derived P-y curves may be approximated by a power function. This is in agreement with other experimental P-y curves [8,17], and the theoretical results from finite element studies using a plane strain model [14].

Fig.8 compares the experimental P-y curves with those from API code (1987) [1] at the hydraulic gradient scale factor, N, of 48. It is seen that the P-y curves from API code do not resemble the experimental curves. The API curves are essentially bilinear while the experimental curves exhibit a parabolic shape. At the small pile deflection API code gives a stiffer P-y curve than observed, while at large deflection the P-y curve from API code reaches an ultimate pressure that is lower than the experimental data.

Nonlinear analyses of observed pile response were performed using the finite difference computer program LATPILE [10]. The soil-pile interaction P-y curves are prescribed along the pile using the API code. For the sake of comparison, the prediction using the experimental P-y curves is also made. Fig.9 shows the predictions for a free head pile response at the hydraulic gradient scale factor of 48. The predictions of pile head deflection at the loading point under different loading levels is shown in Fig.9(a). It is seen that API code gives slightly stiffer pile head response at deflections less than 1.4 mm. This is because of the stiffer API P-y curves at small deflection. At larger deflection, the prediction from API code becomes closer to the experimental data. Fig.9(b) shows the prediction of pile bending moments along the pile length at the loading level of 22.06(N). It is seen that the prediction from the API code slightly overestimates the maximum pile bending moment at this loading level. This may be caused by the low ultimate soil resistance specified by the API code at shallow depth. However, in general, despite the large difference in shape of the P-y curves between the API code and experimental result, reasonably good prediction of pile response is obtained based on the API design method.

SUMMARY AND CONCLUSION

The response of a soil-structure system to load is highly dependent on the stress level involved. This is because the stress-strain response of soil depends on stress level. Consequently, tests on small scale conventional models are unlikely to capture the response of large prototype structures. It is possible to overcome this problem with centrifuge tests in which the prototype stress level can be duplicated in a small model by applying very high centripetal accelerations, thus inducing large body forces. However, this is a very expensive testing procedure.

High stress can also be induced in small models by using very high hydraulic gradients to increase the body force as described in this paper. The testing principle and procedure, and the factors affecting test results were discussed. This technique was applied to the laterally loaded pile problems. The scaling laws were evaluated experimentally using the "modelling of models" technique in which the results of two different model tests simulating the same prototype condition were compared. The re-

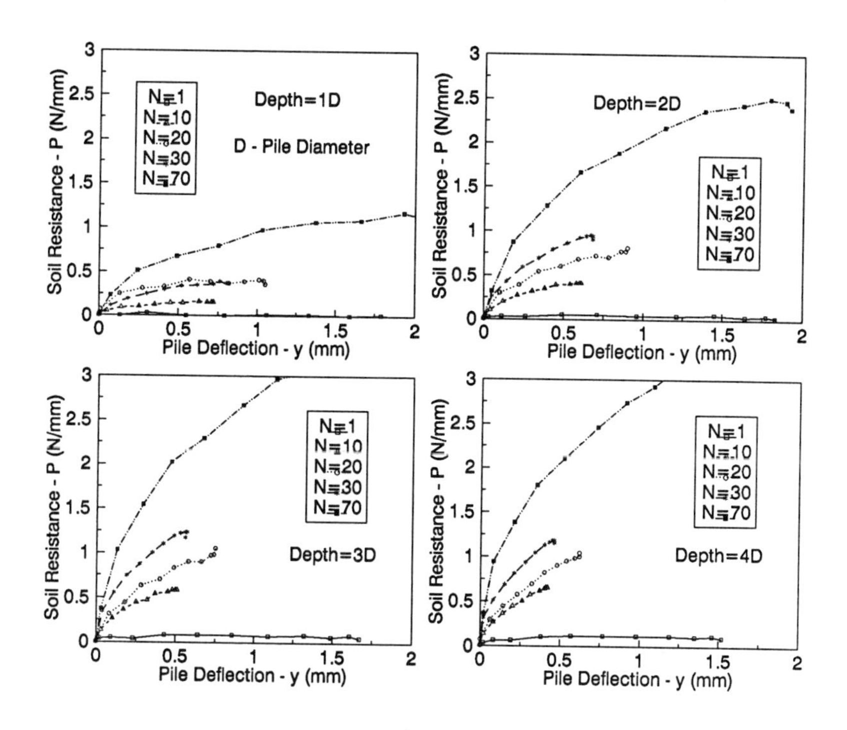

Figure 6 P-y curves at Different Hydraulic Gradients, Free Head

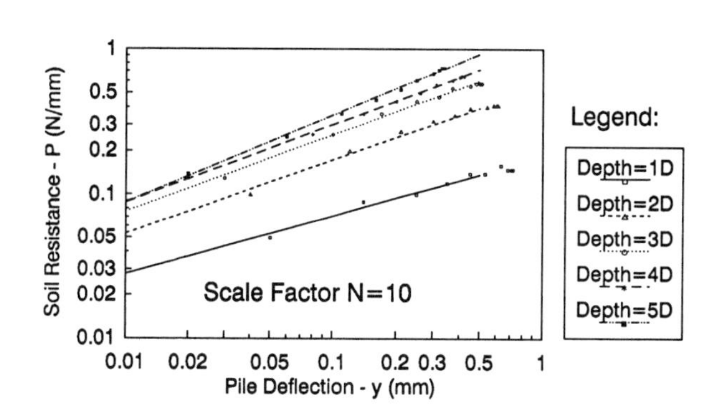

Figure 7 Experimental P-y curves in a Logarithmic Scale

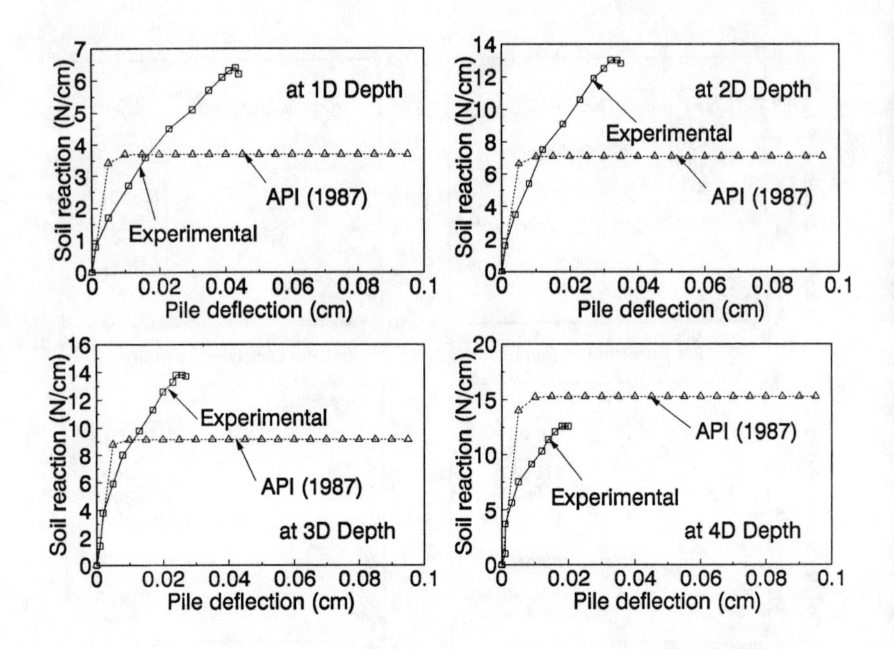

Figure 8 Comparison of Experimental P-y curves with API code

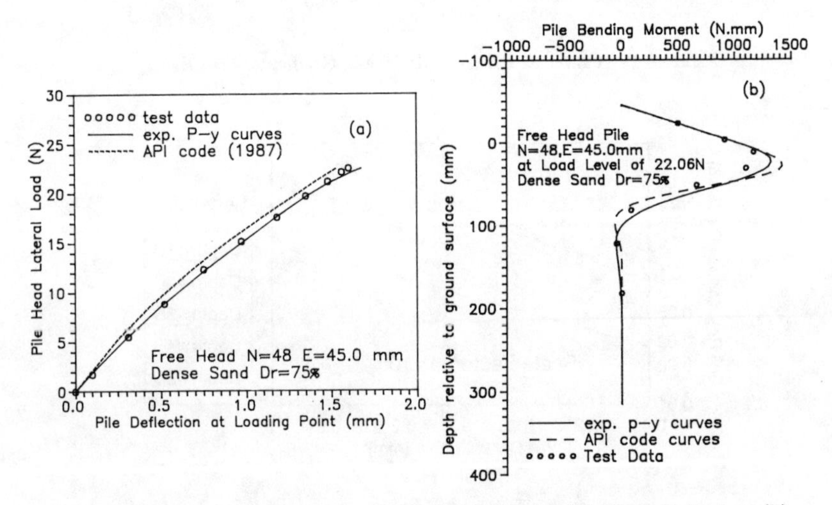

Figure 9 Prediction of Pile Responses using LATPILE; (a) Pile Head Deflection, (b) Bending Moment.

sults were found to be in very good agreement and verified that the HGS tests followed the expected scaling relations as in the centrifuge tests. This also confirmed that the distortion of soil permeability during the loading process has little effect on the test results.

Application to the study of laterally loaded pile response shows that the pile response to lateral pile head loadings exhibits strong soil stress level dependency and has similar characteristic behavior as that obtained from the centrifuge tests. Comparison between the P-y curves from API code and the experiment shows little resemblance in shape. At the small pile deflection, the API curves are stiffer than the experimental ones while at the large deflection API curves reach ultimate soil resistances that are not observed in the experimental curves. However, despite these differences, generally good agreement between API prediction and experimental data is obtained. The API code slightly underestimates the pile head deflection at a low loading range and overestimates the maximum pile bending moment at a large loading level.

This study demonstrates that the HGS modelling technique provides a simple and inexpensive means of loading soil and foundation at stress conditions corresponding to field conditions. This modelling technique is not as general as the centrifuge, but is much simpler to build and operate with a cost being about 10 to 100 times less expensive. For soil-structure interaction problems that involve horizontal and vertical boundaries such as footings and piles, HGS has the advantage that a test series can be carried out much faster than in the centrifuge.

ACKNOWLEDGEMENTS

Support from the Natural Science and Engineering Research Council of Canada and the University of British Columbia is gratefully acknowledged.

REFERENCES

1. API-RP2A, Recommended Practice for Planning, Designing and Constructing Fixed Offshore Platforms, American Petroleum Institute, Washington, D.C., 17th Ed., 1987
2. YO Barton, Laterally Loaded Model Piles in Sand; Centrifuge Tests and Finite Element Analyses, Ph.D. Thesis, Cambridge University, Engineering Department, 1982.
3. JA Cheney, Physical Modelling in Geotechnical Engineering, Proc. from a Centrifugal Workshop conducted during 12th Int. Conf. on S.M.F.E., San Francisco, 1985
4. WH Craig, Installation Studies for Model Piles, Proc. Symp. on the Application of Centrifuge Modelling to Geotechnical Design, ed. by W.H. Craig, Manchester, April 16-18, 1984
5. C Eckart, Properties of Water, Part II. The Equation of State of Water and Sea Water at Low Pressures and Temperatures, Amer. J. Sci. 256, 225, April, 1958.
6. DC Fredlund, Density and Compressibility Characteristics of Air-water Mixtures, Can. Geot. J., 13, 1976, 386-396
8. K Kubo, Experimental Study of the Behavior of Laterally Loaded Piles, Proc. 6th Int. Conf. SMEF., Montreal, Vol.II. 1965, 275-279.
9. IL Nunez & MF Randolph, Tension Pile Behavior in Clay - Centrifuge Modelling Techniques, Proc. Symp. on the Application of Centrifuge Modelling to Geotechnical Design, ed. by W.H. Craig, Manchester, April 16-18, 1984, 87-99
10. LC Reese, Laterally Loaded Piles: Program Documentation, Jnl. Geot. Engng. Div., ASCE, Vol. 103, NoGT4, 1977, 287-305.

11. KH Roscoe, Soils and Model Tests, J. of Strain Analysis, 3(1):1968, 57-64.
12. AE Scheidegger, Resistance to Flow through Porous Media, 3rd Ed., The MaC-cmillan Co.. 1974.
13. RV Whitman & K Arulanandan, Centrifuge Model Testing with Dynamic and Cyclic Loads, Proc. on Advances in the Art of Testing Soils under Cyclic Conditions, Ed. by Vijay Khosia, Detroit, Michigan, Oct.24, 1985, 255-285.
14. L Yan, Numerical Studies on Some Aspects with Pressuremeter and Laterally Loaded Piles, M.A.Sc These, Dept. of Civil Engng, Univ. of British Columbia, Vancouver, Canada, 1986.
15. L Yan & PM Byrne, Application of Hydraulic Gradient Similitude Method to Small-scale Footing Tests on Sand, Can. Geot. J., Vol.26, No.2, 1989, 246-259.
16. L Yan, Hydraulic Gradient Similitude Method for Geotechnical Modelling Tests with emphasis on Laterally Loaded Piles, Ph.D. Dissertation, Dept. of Civil Engng, Univ. of British Columbia, Vancouver, B.C., 1990.
17. I Yoshida & R Yoshinaka, A Method to Estimate Modulus of Horizontal Subgrade for a Pile, Soils and Found., JSCE, Vol.12, No.3, 1972.
18. A Zelikson, Geotechnical Models using the Hydraulic Gradient Similarity Methods, Geotechnique, Vol.19, 1969, 495-508.
19. A Zelikson, Rigid Piles in Sand under inclined Forces; Model Tests using the Hydraulic Gradient Similarity Method, Journal de Mécanique Appliquée, 2(2): 1978, 153-165.
20. A Zelikson & P Lequay, Some Basic Data on Piles under Static and Dynamic Loading From Stress Conserving Models, Proc. 3rd Int. Conf. on Numerical Methods in Offshore Piling, Nantes, May 21-22, 1986, 105-124.
21. A Zelikson, Simulation Methodology for Small Dynamic Models in Sea Bottom Clay, Symp. on Offshore and Arctic Operations, The American Society of Mechanical Engineers, New York, 1987, 209-215
22. A Zelikson, Hydraulic Gradient Simulation of Sequences of Pile Driving and Loading Tests, Preprint, 3rd Int. Conf. on Application of Stress-Wave Theory to Piles, Ottawa, Ontario. Ed. by B.H. Fellenius, 1988, 152-163.

MODELING OF MECHANICALLY STABILIZED EARTH SYSTEMS: A SEISMIC CENTRIFUGE STUDY

John Casey[1], David Soon[2], Bruce Kutter[3], and Karl Romstad[3]

ABSTRACT; Centrifuge model tests of mechanically stabilized earth retaining walls using different backfill materials were conducted. Yield acceleration levels within the reinforced mass during apparent sliding are deduced directly from the recorded accelerations and back calculated from the recorded displacements using the sliding block model. The experimentally deduced yield accelerations are found to be considerably lower than those determined from conventional limit analysis techniques.

INTRODUCTION

Very little field data exists on the earthquake behavior of reinforced soil retaining walls. Observations on the field performance of reinforced soil retaining walls during the Whittier Narrows earthquake (M 6.1) in 1987 and Loma Prieta earthquake (M 7.1) in 1989 are presented in Reference 1. Some damage was observed in both earthquakes, but no damage was observed in any instrumented walls. Early research (2,3,4,5) on the dynamic behavior of reinforced walls was done at UCLA and included static and dynamic testing of 1 g model walls and full scale testing of prototype walls subjected to blast and rotating mass vibrations. The model tests involved very lightly reinforced walls in which the failure usually involved breakage of the ties and failure surfaces were contained within the reinforced earth mass. Later testing by Nagel (6) used more realistically reinforced 1 g walls which demonstrated that the shape of the failure is strongly influenced by the density of the reinforcement. Nagel (6) used the theory of Bracegirdle (7) to explain the yield accelerations observed in his tests.

The current study by the authors was undertaken with the support of the California Department of Transportation to study the earthquake behavior of sound walls mounted on top of mechanically stabilized earth (MSE) walls and reinforced concrete cantilever walls using the centrifuge. Early results of this comparative study are presented by Casey (8). Some of the relevant centrifuge model scaling ratios for dynamic tests are summarized in Table 1 where N represents an arbitrary scaling factor. The centripetal acceleration provides a steady acceleration to simulate an increased acceleration due to gravity. Of particular importance is the scale factor of unity for stress. The stress dependency of friction angle and dilatancy is accounted for by testing the model in a centrifuge.

1 Engineer, CH2M Hill, 3840 Rosin Court, Sacramento, California
2 Graduate Student, Department of Civil Engineering, University of California, Davis
3 Prof., Department of Civil Engineering, University of California, Davis

EXPERIMENTAL PROCEDURES

The centrifuge tests were conducted using a servo-hydraulic shaking table mounted on the bucket of the 1 meter radius Schaevitz centrifuge at the University of California, Davis. A detailed description of the shaker system is given by Chang (9) and is not repeated here.

Table 1. Scale Factors for Dynamic Centrifuge Tests

Dimension	Scale Factor
Length, L	1/N
Velocity, L/T	1
Acceleration, L/T2	N
Time, T	1/N
Frequency, 1/T	N
Density, M/L3	1
Force, ML/T2	$1/N^2$
Stress, M/LT2	1

The model is constructed in a sample container with 11 inch x 22 inch plan area and a 7 inch height. The body of the hydraulic actuator is mounted to the swing bucket which functions as a reaction mass. The ram pushes the container from side to side to simulate the earthquake base motion. The maximum acceleration that can be imposed on the model is approximately 30 g, which corresponds to an 0.6 g prototype earthquake for a model scale factor, N = 50. Any desired acceleration time history can be integrated twice to obtain the displacement history and stored by the IBM AT computer. This computer controls the shaker to simulate the desired time history and acquires data from the experiment.

Model configuration and instrumentation are shown in Figure 1. Seven accelerometers and three to four displacement transducers measured horizontal motions and six strain gages measured strains in the top reinforcement grid. In this paper particular attention will be paid to accelerometers ACC1, ACC4 and ACC7 and displacement records LPT1 and LVDT2. A prototype twelve foot high sound wall, attached to a concrete slab extending eight feet back from the face and keyed into the backfill with 6 foot piles (16" diameter), was placed on top of the MSE wall . In this paper the response of sound wall will not be discussed.

The four models discussed in this paper represented 24 foot high retaining walls with three different backfills described in Table 2. The backfill material used in all models consisted of Nevada Sand, a fine uniform silica sand with a mean grain size of 0.15 mm and a coefficient of uniformity of 1.7. In the material designated Loose in Table 2, the relative density was only 32% and it had a friction angle (determined by triaxial tests) of only 30°. The material designated Cohesive was formed by combining 70% Nevada Sand with 30% Yolo Loam (a low plasticity silty clay) at a water content of 12%. The material designated Dense was simply the Nevada Sand compacted to a relative density of 95 to 100% with a resulting friction angle of 45°. Both the Cohesive and Dense backfills pass as Structural Backfills by Caltrans Standards.

Table 2. Backfill Characteristics

	Friction Angle (°)	Cohesion (psf)	Relative Compaction (pcf)	Relative Density (%)
Loose	30	0	93	32
Cohesive	34	500	92 @ 12% w/c	-
Dense	45	0	101	95-100

A coarse wire screen was used to simulate the MSE reinforcing mats which, in the prototype, consist of a mesh of 0.5 inch diameter rebar with the mats extending 16.8 feet (4.2 inches in model) back into the soil at a vertical spacing of 3 feet (0.75 inches model). In the prototype MSE design, each 14.5 foot by 3 foot face plate is supported by 4 mats, with each mat consisting of 5-16.8 foot bars (perpendicular to the wall face) spaced at 6 inches with 2 foot long bars spaced at 18 inches welded to the long bars. In the model, the wire screens were attached to aluminum face plates which were scaled to simulate standard reinforced concrete face plates of 6 inch thickness.

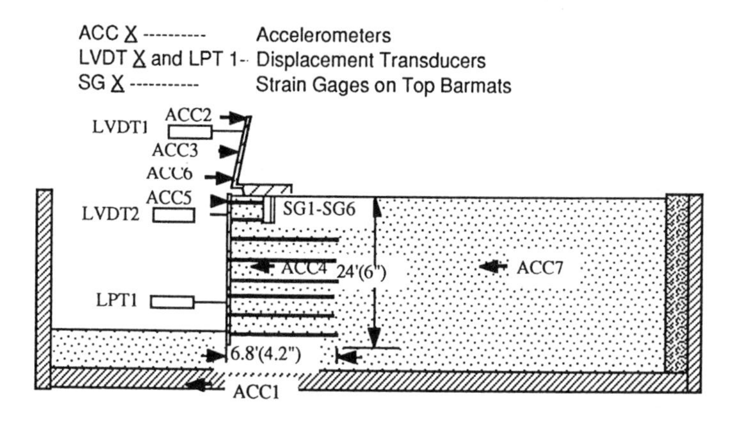

Figure 1, Instrumentation Types and Locations for the MSE Test Series

After the three different backfills were tested with the bar mats placed in a horizontal plane, it was decided to run another series of tests with the Loose backfill and with the bar mats inclined at 10° downward into the backfill. The purpose was to see if the failure surface within the reinforced earth mass could be altered to increase the horizontal acceleration levels required to induce permanent displacements and reduce the resulting outward movement of the wall. In the discussion that follows the four models will be referred to as Loose (model 5, earthquake d), Loose-10 degrees (model 8, earthquake d), Cohesive (model 7, earthquake d) and Dense (model 9, earthquake d). Each of the four models was subjected to a minimum of nine different earthquakes.

MODEL TEST RESULTS

Measured acceleration time histories from the El Centro (Imperial Valley Irrigation District, May 18, 1940, 8:37 pm PST, South 0°0' East) and San Fernando (8244 Orion Blvd., 1st Floor, Feb. 9, 1971, 6:00 am PST, N 0°0' West) were processed with appropriate filters and base line corrections to obtain displacement records. The displacement records could be magnified by any factor and used as a command signal to the shaker. After the data processing, and due to the idiosyncrasies of the shaker and centrifuge, the resulting actual input acceleration records will not precisely correspond to the original records in amplitude and frequency content. However, the resulting records are reasonable reproductions of the original events.

Figure 2 presents the acceleration records measured at the base of the MSE system during tests of the Dense backfill with the El Centro record magnified by factors of 2 and 5. Maximum prototype accelerations recorded during these events reached approximately 0.4 g and 0.6 g. Figure 3 presents the two percent damped acceleration response spectra computed using the measured accelerations at the base of the model for each of the four model types when subjected to the Filtered El Centro motion magnified by five. The shape of each of the spectra are very similar and the Loose-10 degree and Dense models appear to be receiving slightly more energy over the majority of the frequency range.

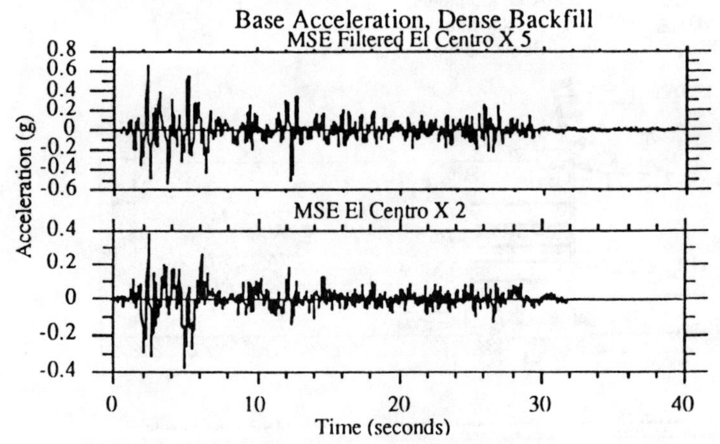

Figure 2. Two Base Acceleration Motions for the Dense Model

Figure 4 demonstrates displacement results at the top of the MSE wall throughout the earthquake time history for the Filtered El Centro x 5 base motion. Plotted below these motions is the time history of acceleration which produced the displacement of the Dense model. The base acceleration input motions to the other models are very similar as evidenced by the spectra in Figure 2. The first large acceleration pulse of approximately 0.6 g produces a significant permanent displacement at the top of the wall for each of the models. The Loose model continues to move out significantly during the first six seconds compared to the other three models. The sloping reinforcement at 10 degrees downward from the face plates is seen to reduce the total

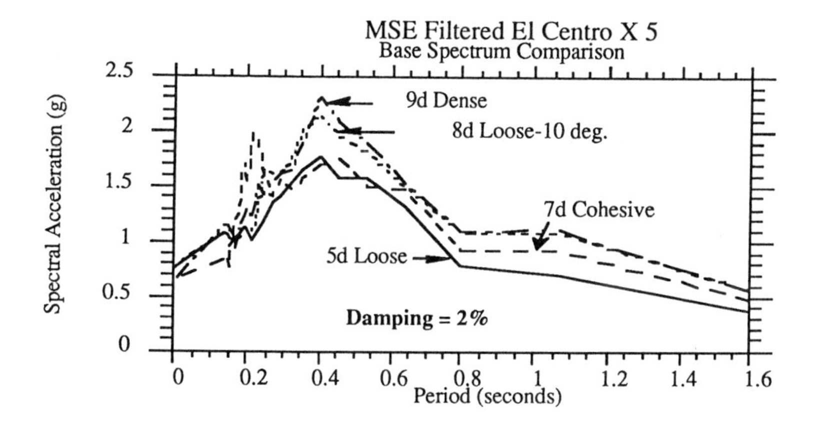

Figure 3. Base Acceleration Spectra for the Four Models w/ Filtered El Centro X 5

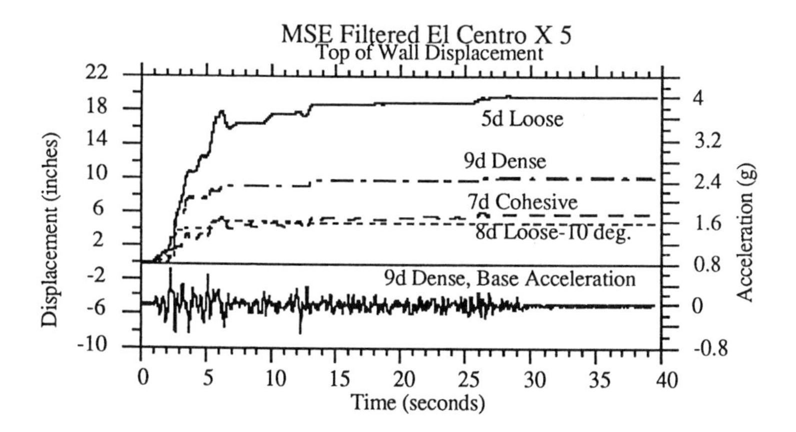

Figure 4. Top of Wall Displacement for Four Models with Filtered El Centro X 5

permanent displacement in this test by a factor of almost five compared to the horizontally placed grid. The total displacements of the Cohesive and Dense models are also significantly reduced compared to the Loose model.

Figure 5 shows the first six seconds of the base acceleration input motion (ACC1 in Figure 1), the corresponding acceleration two inches behind the facing within the reinforced earth mass at mid-height of the wall (ACC4 in Figure 2) and the displacement at the top of the wall for the Loose-10 degree model (LVDT2 in Figure 1). This model and test was chosen because all the permanent displacement for this model occurred within this time frame. The high frequency elastic oscillations of the measured permanent displacement have been filtered out to provide clarity. Positive accelerations correspond to outward movements of the wall. Four permanent displacements movements can be seen to occur. The first (less than 0.5 inches) occurs just before 2 seconds associated with a relatively small acceleration at both the base and behind the wall. The second permanent movement at about 2.5 seconds is about 3.5 inches at the top of the wall. This is due to the almost 0.8 g pulse at the base while the acceleration within the reinforced earth mass appears "clipped" at about 0.3 g. Later permanent movements at about 3.5 and 5.3 seconds are due to base accelerations in excess of 0.3 g while the acceleration within the reinforced mass is on the order of 0.25 g. Note that all the permanent displacements are associated with positive accelerations.

The value of peak acceleration within the reinforced earth mass 2 inches behind the wall face (ACC4) associated with permanent displacement appeared to vary with the time it occurred in a given event and how many earthquakes had preceded the event for a given model. To quantify the range of acceleration values associated with possible slippage it was decided to take each record and average up to the first six positive peak accelerations 2 inches behind the wall (ACC4) which were associated with permanent displacement. The total permanent displacement produced by these acceleration pulses was then divided by the number of corresponding pulses to provide one data point for each earthquake. This data, plotted in Figure 6, indicate that each model has a characteristic band of data points corresponding to limiting accelerations beyond which large permanent displacements occur. This is probably a conservative way of measuring the maximum acceleration within the sliding mass since low acceleration values associated with small permanent displacements are averaged in. The ranges of experimentally determined yield accelerations produced by this approach for all four models are presented in Table 3.

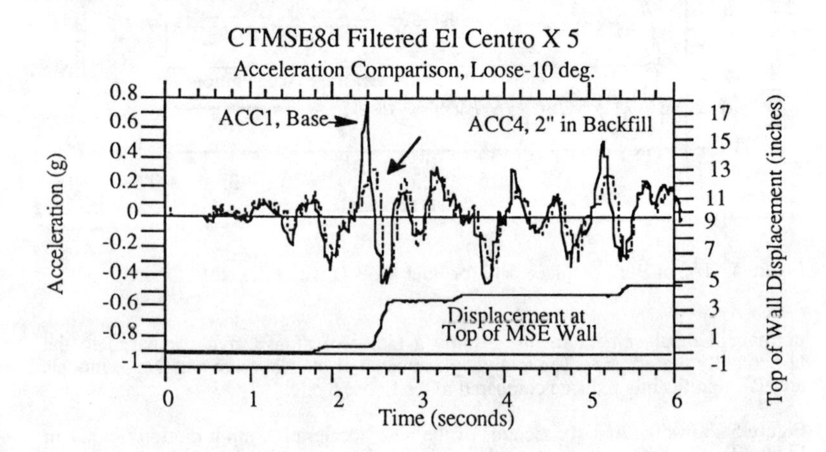

Figure 5. Cutoff Acceleration and Displacement for Loose-10 Degree Model

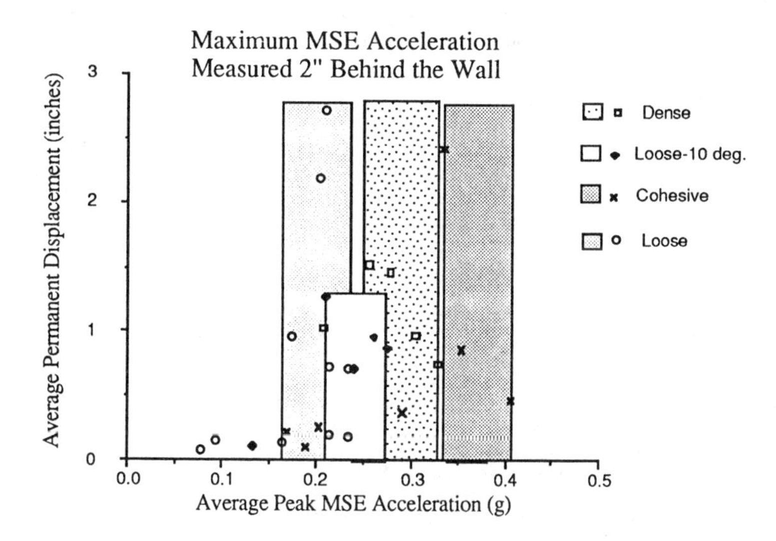

Figure 6. Interpreted Average Yield Accelerations from Measured Data

Table 3. "Yield" Accelerations for MSE Systems

Predicted Yield Accelerations for MSE

			Loose	Loose-10deg	Cohesive	Dense
Experimental	Observed		0.18-0.24	0.22-0.27	0.34-0.41	0.26-0.33
	Sliding Block		0.08-0.24	0.14-0.28	0.19-0.26	0.16-0.25
Theoretical	Snail (vers. 1.1)	OMS	0.16	0.20	0.55	0.39
		Wedge Method	0.22	0.30	0.67	0.53
	Coulomb		0.22	0.33	0.72	0.63

The failure mechanisms in the centrifuge tests were observed by placement of two grids of colored sand in the models. One of these grids was placed against the plastic window and another imbedded in a cross section along the centerline of the sample. By moistening the sand after the test, it was possible to dissect the sample to expose the central grid of colored sand as shown in Figure 7. The position of the ground surface and face plates on the photograph was highlighted by hand for clarity. An idealized interpretation of the observed failure mechanism is presented in Figure 8. Figure 8 also shows the deformed position of the soundwall on top of the MSE, but the soundwall is not the subject of the present paper. Primarily lateral shear is observed to occur within the reinforced block of soil. A graben appears to form behind the reinforced soil mass (see Figure 8). The most apparent failure plane occurs along a vertical plane along the back of the reinforcing coupled with a rear failure plane at 48-54°. The failure surfaces that pass between the barmats and at an angle through the backfill are not well defined and they are dotted in Figure 8. It is possible that there is a family of such failure planes, all of which undergo sliding, which smear out the failure zone into a broad band rather than a distinct surface. A similar phenomenon was discussed by Kutter and James (10) regarding observed deformation patterns in clay embankments during earthquakes.

The yield acceleration (k_y g), defined as the minimum lateral acceleration that will cause permanent displacements to occur for a rigid-plastic system, was calculated for the model retaining structures by assuming two distinct blocks developed as shown in Figure 9. On each of the sliding surfaces (including the vertical interface between the blocks), it was assumed that the Mohr Coulomb shear strength was mobilized. The first block is bounded by a failure plane, inclined at an angle q1 from the horizontal which passes from the toe of the wall through the reinforced soil mass. At the back of the reinforced soil mass (along the ends of the reinforcing mats) a vertical failure plane is assumed to develop. The second wedge in the backfill, behind the reinforced soil mass, was assumed to be triangular in shape, with the base of this wedge inclined at an angle, θ_2 from the horizontal. The loads carried by the mats were not included in the analysis since the failure planes did not intersect the reinforcing. Thus the angle θ_1 was taken to be 8°, the maximum angle between layers which can be developed.

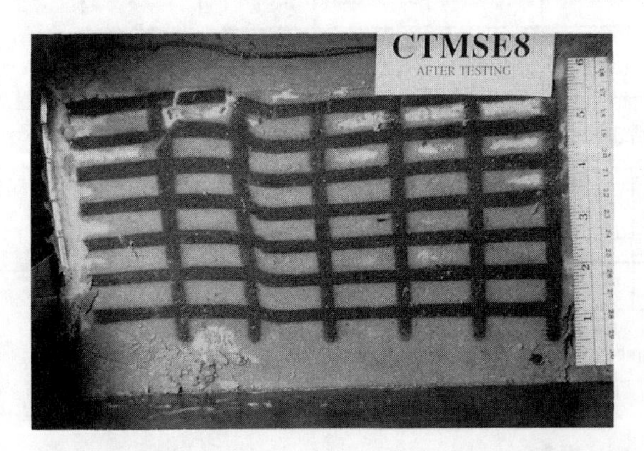

Figure 7. Photo of Observed Failure Modes for Loose-10 Degree Model

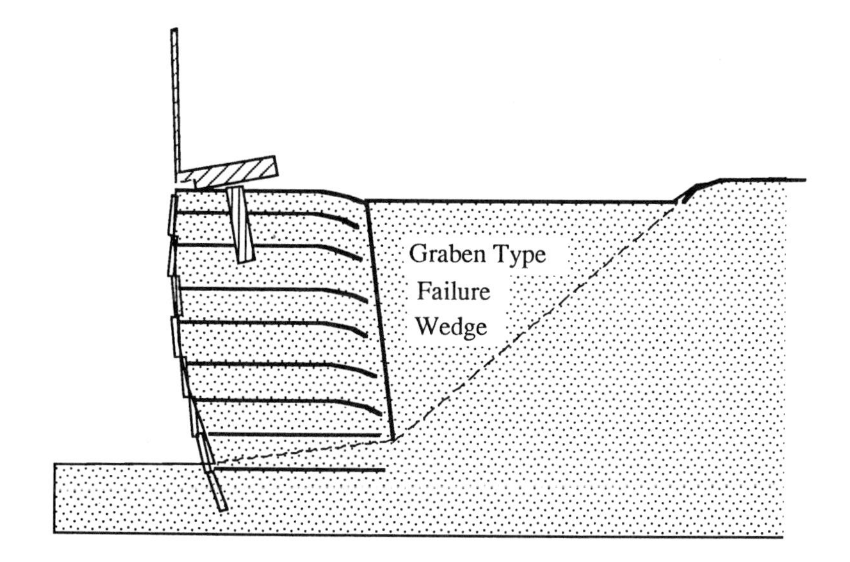

Figure 8. Idealized Mode of Failure Based on Observed Behavior

From observations of the centrifuge tests (see Figures 7 and 8), this assumption appears to be reasonable. The validity of this assumption probably depends on the density and stiffness of the reinforcement. The MSE design tested here may be classified as "over reinforced" since the amount of reinforcing appears to be sufficient to force the failure plane to avoid the reinforcement. This assumption may not be valid for the design of geotextile or other polymer geogrids that have less rigidity. The resulting yield acceleration coefficients calculated by this method are summarized in Table 3. Also included in Table 3 are results predicted by an unpublished computer program under development at the California Transportation Laboratory (11). The program calculates yield acceleration by both the Ordinary Method of Slices (OMS) and the Wedge Methods.

After calculation of the yield acceleration using the two block model, the displacements of the models could be predicted using a simple sliding block analysis, as depicted in Figure 9b. A parametric study indicated that the angle of the failure plane had very little effect on the predicted displacements for the sliding block analysis. The angle θ for the sliding block was assumed to be 20°. Figure 10 shows a series of predicted displacements for a test on a wall with Loose backfill. The results were calculated by numerical double integration of the difference between the input acceleration and the yield acceleration as suggested by Newmark (12). Using the calculated yield acceleration coefficient for the Loose backfill, $k_y = 0.22$, and using the measured base acceleration as the input motion, the bottom prediction for displacement of the sliding block was obtained. The predicted block displacements are about one third of the permanent displacements measured at the third face plate (LPT1 in Figure 1).

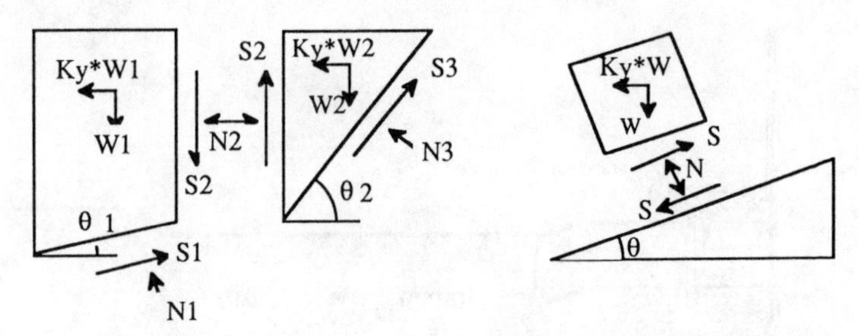

Figure 9a. Coulomb Model

Figure 9b. Sliding Block Model

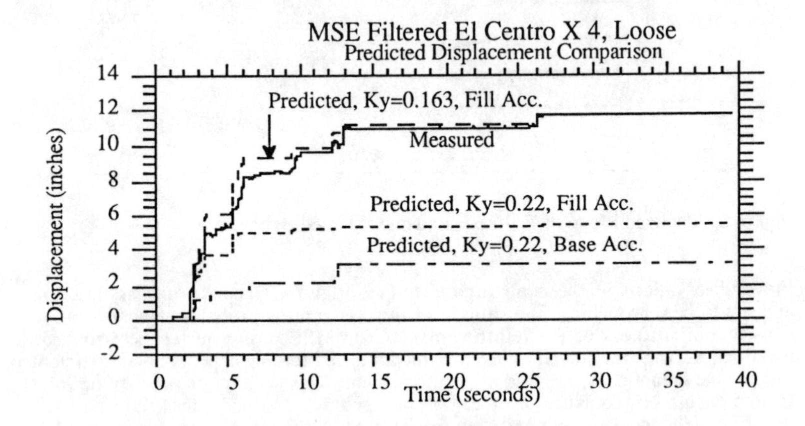

Figure 10. Measured Displacement and Sliding Block Predictions of Displacement

It was reasoned that amplification of the acceleration within the backfill is ignored if the base motion is used as input. In an approximate way, amplification may be accounted for by using the measured accelerations in the backfill, 10 inches behind the wall as an input motion for the sliding block. When this was done, larger permanent displacements were predicted as shown by the middle curve in Figure 10, but predictions still underestimated measured displacements by a factor of about two.

Finally, it was decided to use a trial and error back-analysis to determine what value of yield acceleration would have to be used in order to match measurements with predictions. It was found that a yield acceleration of 0.163 would provide a good match of the ultimate permanent displacement measured by LPT1 if the fill acceleration was used as the input motion. It also appears that the history of the development of permanent displacements is reasonably predicted by this method. The discrepancy between the calculated yield acceleration and the yield acceleration determined by back-analysis may be due to a combination of the following:

* A significant amount of permanent deformation may occur before the yield acceleration is reached (i.e. before the shear strength is fully developed),
* The rigid-plastic sliding block is not able to capture the elastic flexibility and resonant behavior of the structure, and
* A single block on a slope does not adequately model the actual failure mechanism.

Nevertheless, the results from the first five or six earthquakes imposed on all four models were back-analyzed to determine what yield acceleration would be required in order to match the permanent displacements measured by LPT1. This exercise provides quantitative comparisons of yield accelerations for the four different types of backfill conditions, approximately accounting for the differences (see Figure 3) in the input motions that were obtained from the shaker. The results of the back analysis are shown in Table 3. It is seen in this table that the wall with the Loose backfill had significantly lower yield accelerations than the other models. The Cohesive backfill appeared to have the greatest yield acceleration. It is also apparent that the sloping barmats provided a significant increase in the yield acceleration.

Figure 11 indicates that there is a general tendency for the yield acceleration to increase, perhaps due to densification of the backfill, as successive earthquakes are imposed on the models. The sequence of earthquakes was very similar for models 7, 8 and 9, with earthquake numbers 4 and 5 being significantly larger than earthquakes 1, 2, 3 and 6. There appears to be a general trend that the back-analyzed yield acceleration is greater for high levels of shaking (earthquake numbers 4 and 5) than for lower levels of shaking. During high level shaking, most of the displacement probably develops after the yield acceleration is reached (i.e. the shear strength is fully mobilized). During low level shaking, a major portion of the displacement may be the result of permanent deformations that can occur before the full Mohr-Coulomb shear strength is developed on the failure planes.

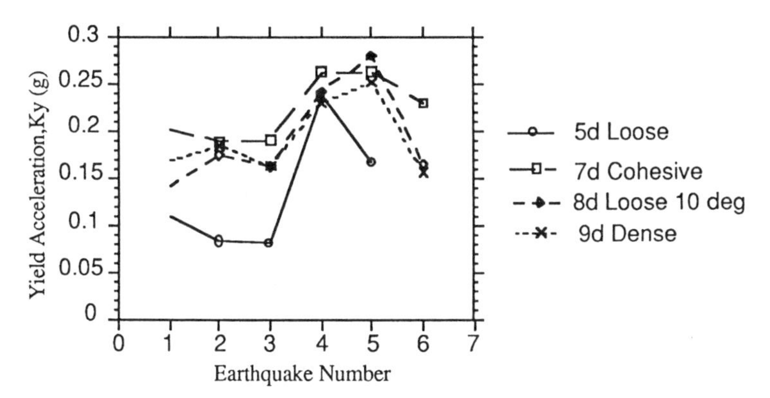

Figure 11. Yield Acceleration Predicted by Sliding Block vs. Earthquake Sequence

CONCLUSIONS

Centrifuge model testing is useful for providing a repeatable, realistic input motion that enables direct comparisons to be made regarding the seismic performance of different types of backfill materials.

Permanent displacements predicted by the rigid-plastic sliding block model significantly underestimate the measured displacements if the yield acceleration is calculated by conventional limit analysis. To a lesser extent, measured accelerations in the sliding mass also appear to be lower than calculated yield accelerations. The amplification of the input motion is not accounted for by the sliding block model. By using the backfill acceleration instead of the measured base acceleration as the input motion, the sliding block predictions were improved.

Inclining the mat reinforcement downward at 10° with respect to the horizontal in the very Loose material significantly raised the yield acceleration and lowered the permanent displacements.

In all models, yield surfaces within the reinforced soil mass appeared to be contained between the layers of mat reinforcement and the most apparent failure plane occurs along a vertical plane at the back of the reinforcing.

UNIT CONVERSIONS:
1 inch = 2.54 cm
1 lb = 4.448 N

ACKNOWLEDGEMENTS
This work was carried with funding from the Federal Highway Administration and the California Department of Transportation, Technical Agreement No. 54H320. The continued review and comments of Ken Jackura are gratefully acknowledged.

REFERENCES
1. Kutter, BL, JA Casey & KM Romstad, Centrifuge Modeling and Field Observations of Dynamic Behavior of Reinforced Soil and Concrete Retaining Walls, 4th U.S. Nat. Conf. on Earthq. Eng., EERI, Palm Springs, Ca, Vol. 3, May, 1990.
2. Richardson, GN & KL Lee, Response of Model Reinforced Earth Retaining Walls, Report to the National Science Foundation, Project GI 38983, UCLA, Feb. 1974.
3. Richardson, GN , D Feger, A Fong & KL Lee, Seismic Testing of Reinforced Earth Walls, J. Geot. Eng. (ASCE), Vol. 103, Jan. 1977.
4. Richardson, GN, Earthquake Resistant Reinforced Earth Walls, Symposium on Earth Reinforcement, ASCE, Pittsburg, Pa., 1978.
5. Wolfe, WE & D Rea, Earthquake Induced Deformations in Reinforced Walls, School of Engineering and Applied Science, UCLA, May, 1980.
6. Nagel, RB, Seismic Behavior of Reinforced Earth Walls, Department of Civil Engineering, University of Canterbury, Christchurch, N.Z., Research Report 85-4, March, 1985.
7. Bracegirdle, A, Seismic Stability of Reinforced Earth Retaining Walls, Bull. N.Z. National Soc. for Earthquake Eng., Vol. 13, No. 4, Dec. 1980.
8. Casey, JA, A Comparison of Two Earth Retaining Systems: A Dynamic Centrifuge Study, M.S. Thesis, University of California, Davis, 1989.
9. Chang, GS, Centrifugal and Numerical Modeling of Soil-Pile Building Interaction During Earthquake, PhD. Thesis, University of California, Davis, 1990
10. Kutter, BL & RG James, Dynamic Centrifuge Model Tests on Clay Embankments, Geotechnique 39 No.1, pp 91-106, 1989.
11. Jackura, K, Personal Communication, California Transportation Laboratory, Sacramento, California, 1990.
12. Newmark, NM, Effects of Earthquakes on Dams and Embankments, Geotechnique 15, No. 2, pp.141-156, 1965.

A PLANE STRESS DIRECT SHEAR APPARATUS FOR TESTING CLAYS

Luis E. Vallejo[1], M ASCE

ABSTRACT: The evolution of the shear zone structure plays an important role in the failure of clays subjected to direct shear. The shear zone evolution leads to a better understanding of how clays fail and mobilize their shear strength. However, the structure and evolution of the shear zone can not be directly visualized and recorded with current available procedures and equipment. The present study uses an open shear box direct shear apparatus, allowing continuous recording of the evolution of the shear zone structure. Using the new apparatus, the shear zone structure developed within uniform samples of kaolinite clay was recorded.

INTRODUCTION

Relationships between the shear stress and the displacements induced in clay samples can be obtained from laboratory and field shear testing of clays using the direct shear apparatus. From the shear stress-displacement relationships, the drained peak and residual shear strengths mobilized by the clay in the direct shear apparatus can also be evaluated (Skempton, 1964). Direct shear testing as it is currently conducted does not allow for the visualization and recording of the shear zone that develops during the loading of the samples (Bowles, 1978).

A knowledge of the evolution of the shear zone structure in clays as a function of the applied shear and normal stresses in the direct shear apparatus is very important in order to understand how clays fail and mobilize their drained peak and residual shear strengths (Morgenstern and Tchalenko, 1967; Skempton, 1964; Palmer and Rice, 1973).

The problem of directly recording the structure of the failure zone in a clay placed in a closed-box direct shear apparatus has been partially solved by Morgenstern and Tchalenko (1967) in the following way. They prepared a homogeneous block sample of kaoli-

---

1 - Associate Professor, Department of Civil Engineering, 949 Benedum Hall, University of Pittsburgh, Pittsburgh, PA 15261

nite clay by consolidating a clay-water slurry. From the block of clay, samples were cut for direct shear testing. The samples were all subjected to a normal pressure of 215.28 kPa in the direct shear apparatus and they were sheared up to different shearing stresses (Fig. 1). When the samples reached the selected shearing stresses (V.1 to V.4 in Fig. 1), the tests were stopped, the samples unloaded and removed from the shear box, and the discontinuities recorded. Figure 2(a) and 2(b) show the discontinuities that formed the shear zone at two different levels of shear stress (Fig. 1)

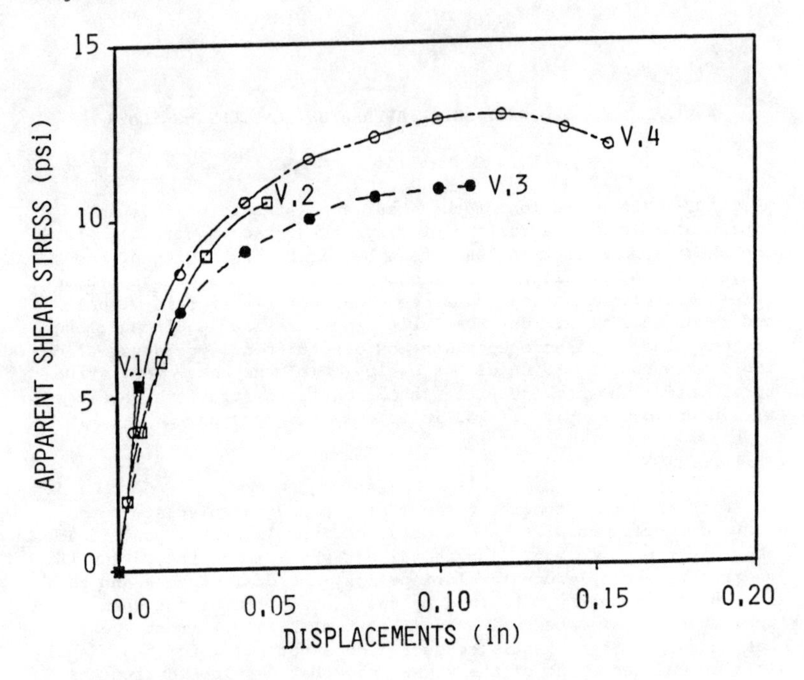

Figure 1. Stress-Displacements Relationships From Closed Box Direct Shear Tests On Four Samples of Clay. (Adapted from Morgenstern and Tchalenko, 1967, p. 313)

However, the above method could not determine the shear zone structure <u>while</u> the sample is subjected to a prescribed stress state. In addition, the direct shear tests carried out by Moregenstern and Tchalenko used only one normal stress (215.28 kPa). Thus the effect that the normal stress could have in the structure developed by the shear zone in clays was not investigated. In the present study, a direct shear apparatus that allows the continuous recording of the evolution of the shear zone structure in intact

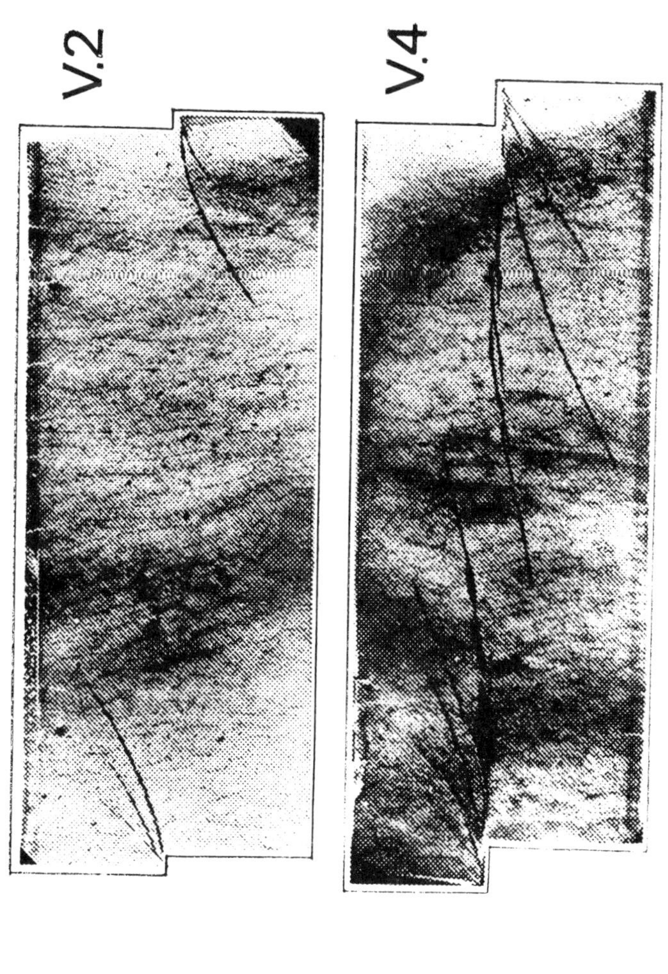

Figure 2. Discontinuities In A Closed Box Direct Shear Apparatus At Different Levels Of Stress (Adapted from Morgenstern and Tchalenko, 1967, p. 314)

clays as a function of the applied normal and shear stresses is presented. Application of the new apparatus is also introduced.

THE PLANE STRESS DIRECT SHEAR APPARATUS (PSDSA)

In order to be able to record the structure of the failure zone developed by clays when subjected to direct shear testing conditions, a new direct shear apparatus with an open shear box was developed. A description of the parts forming the apparatus is presented in Fig. 3(a). A photograph of the apparatus is shown in Fig. 3(b).

The Open Shear Box System. The open shear box consists of two main U sections that enclose a prismatic sample of clay. The top section is a moving section that transmits the normal and shearing loads to the soil sample. The movable top section rolls on a large base plate. The lower main U section of the open shear box is fixed to the base plate.

The two main U sections are separated a distance of 0.635 (Fig. 3(a)). Also, two smaller metallic U sections can be fitted inside the two main U sections to allow the testing of smaller samples. Table 1 lists the normal surface dimensions as well as the dimensions of the open (viewing) face. The evolution of the failure zone during the testing can be recorded by using a still camera or a video camera.

Table 1. Dimension Of Samples That Can Be tested in the PSDSA

Sample Size	Normal Loading Surface		Viewing Surface	
	Length (cm)	Width (cm)	Length (cm)	Width (cm)
(1)	(2)	(3)	(4)	(5)
Large	12.70	3.17	12.70	11.40
Small	7.62	3.17	7.62	7.62

The Loading Displacement Mechanisms. The normal and shearing loads are applied to the top movable section of the open shear box. The loads are generated by the rotation of a hand screw jack system that uses a reaction platten attached to the base plate (Fig. 3(a)). The magnitude of the normal and shearing loads transmitted to the soil sample are measured with a proving ring-dial gage system.

The proving ring for the normal load is attached to a horizontal rectangular bar which then transmits load to the top U section via roller bearings.

Displacements during the shearing process are measured by the use of dial gages attached to the movable section of the direct shear apparatus.

Since in the shear box of the new direct shear apparatus the soil samples are loaded in two directions only, the apparatus is

(a)

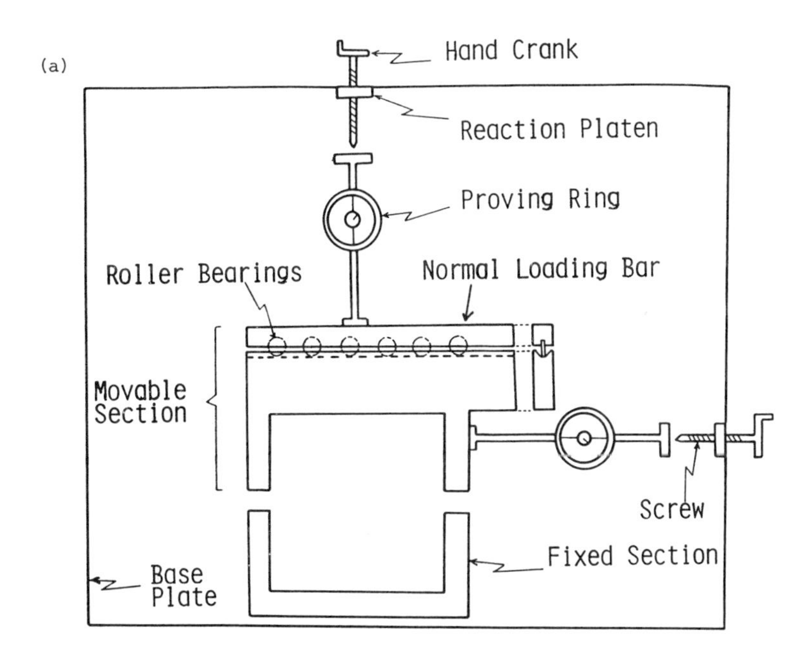

(b)

Figure 3. Description Of Plane Stress Direct Shear Apparatus

called the Plane Stress Direct Shear Apparatus (PSDSA for short).

APPLICATION OF THE PSDSA TO INTACT CLAY SAMPLES

Sample Preparation. For the experimental investigation, laboratory-prepared samples of kaolinite clay were used. Kaolinite clay was chosen because of its homogeneity. This is an important factor in that it facilitates the study of the growth and interaction of the discontinuities that form the shear zone without the adverse effect of microscale heterogeneities.

The kaolinite used had a liquid limit equal to 58% and a plastic limit equal to 28%. Dry kaolinite was mixed with destilled water to form a soft soil mass with a water content of about 40%. After the mixing was done, the clay water mixture was placed in a consolidometer 30 cm in diameter and consolidated under a normal pressure of 25.7 kPa for a period of 5 days. After unloading the consolidometer, specimens measuring 7.62 cm by 7.62 cm and 2.54 cm were cut from the clay block. After being cut, the samples were allowed to air dry for one day.

The dimensions of the samples, their water content and the normal pressures used for the testing in the PSDSA are shown in Table 2. The evolution of the failure surface was recorded using a 35 mm camera. The rate of shearing used in the PSDSA was about 1 mm/min.

Table 2. Sample Characteristics And Normal Pressures

Sample Number	Dimensions (cm)			Water Content %	Normal Pressures (kPa)
	Width	Height	Thickness		
(1)	(2)	(3)	(4)	(5)	(6)
1	7.62	7.50	2.27	26	0
2	7.62	7.50	2.46	28	118.6
3	7.60	7.62	2.44	24	362.4

Test Results And Analysis. The shear stress-displacement relationships obtained for the three samples tested in the PSDSA is shown in Fig. 4(a). Fig. 4(b) shows the relationships between the vertical and horizontal displacements that the samples experienced in the PSDSA.

An analysis of the shear stress-displacement relationships indicates that as the normal stress was increased in the PSDSA from zero to 362.4 kPa, the samples changed their response to shearing from that of a normally consolidated clay to that of an overconsilidated clay. That is, at high normal pressures the samples developed a peak shear resistance (point B_1 and C_1 in Fig. 4(a)) as well as a residual shear resistance (points B_2 and C_2 in Fig. 4(a)) which is characteristic of overconsolidated clays (Skempton, 1964).

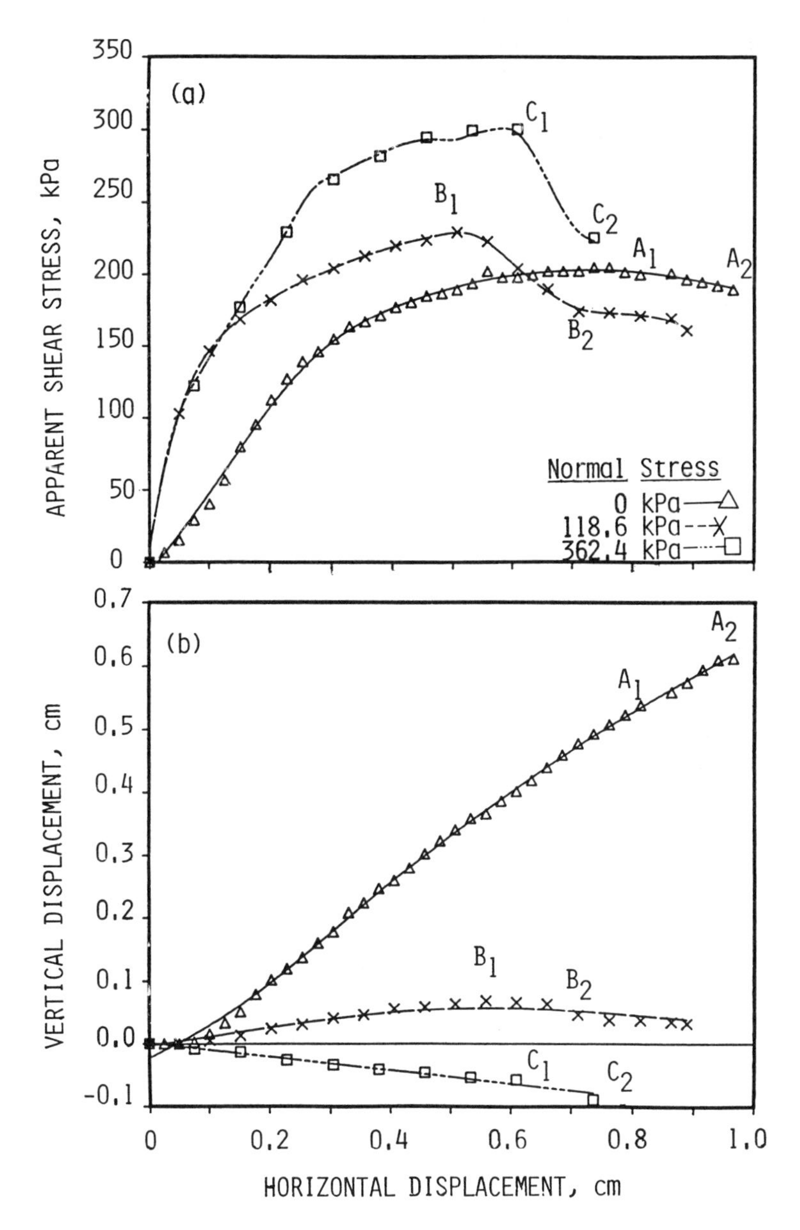

Figure 4. Stress-Displacements Relationships Obtained In The PSDSA

However, under a zero normal pressure, the samples of clay developed at peak shear resistance (points A_1 or A_2 in Fig. 4(a)) and no residual resistance to failure. This later response is typical of normally consolidated clays.

A look at the failure surfaces developed by the clay samples during direct shear (Fig. 5 and 6) shows that there is a marked difference in the shapes of the failure surfaces that developed. This is caused by the presence or absence of normal pressure during testing. At peak shear stress (point A_1 in Fig. 4(a)), the sample with zero normal pressure developed a straight, relatively smooth failure parallel to the general direction of shearing. At peak shear stress conditions (point A_1 in Fig. 4(a)), the straight failure surface was almost fully developed, needing only a small displacement of 0.1 cm (from point A_1 to A_2, in Fig. 4(a)) for its completion (Fig. 5(b)).

In contrast, the failure surfaces developed by the samples with normal pressures of 118.6 and 362.4 kPa developed failure surfaces that were not straight but had a zig-zag configuration (Fig. 6 shows the failure surface for the sample with normal pressure equal to 118.6 kPa which failed in a way similar to the sample under a normal pressure equal to 362.4 kPa). This zig-zag configuration did not resemble the straight failure plane that samples of intact clay are expected to develop when subjected to direct shear (Bowles, 1978).

Evolution Of Failure Surfaces Under Normal Pressure. Under a constant normal pressure of 118.6 kPa and at peak shear stress conditions (point B_1 in Fig. 4(a)) sample No. 2 (Table 2) first developed two discontinuities that were inclined with respect to the general direction of shearing (Fig. 6(a)). These two inclined discontinuities originated at the same time from the two edges of the samples. Morgenstern and Tchalenko (1967) also found similar discontinuities in their closed box direct shear testing of kaolinite clay (Fig. 2(a)). They called these discontinuities "edge discontinuities" and cited the high stress concentrations induced by the loading sides of the shear box as the reason for their formation.

As the stressing progressed to level B_2 (Fig. 4(a)), the clay located between the two edge discontinuities failed and developed a single discontinuity (Fig. 6(b)). This latter discontinuity formed in the direction opposite to that of the edge discontinuities. Tchalenko (1968) also found this discontinuity when testing kaolinite clay in a closed box direct shear apparatus and called it a "thrust discontinuity." The final failure surface in the sample formed when the edge discontinuity and the thrust discontinuity interacted and joined, producing a continuous failure surface with a zig-zag configuration (Fig. 6(b).

Comparison Of Results Between Closed and Open Shear Box Tests. A comparison of the test results obtained by Morgenstern and Tchalenko (1967) using a closed box direct shear apparatus and a constant normal pressure of 215.28 (Fig. 2) and those obtained in this study (Fig. 6) shows that the samples of kaolinite clay developed somewhat similar failure surfaces under direct shear.

(a)

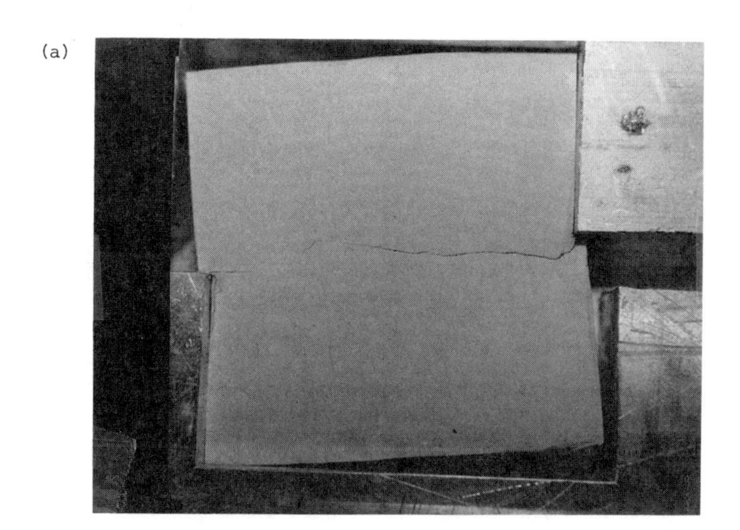

(b)

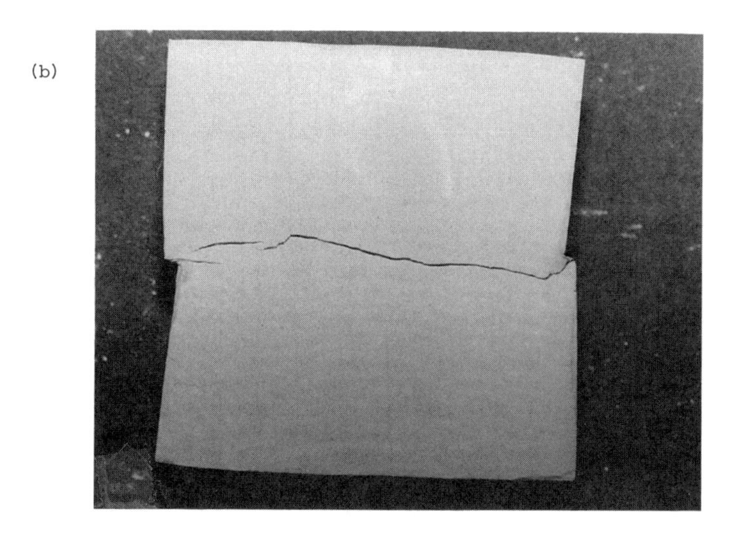

Figure 5. (a) Discontinuities Formed In The Clay In The PSDSA Under Zero Normal Pressure (Point A_1 in Fig. 4), (b) Discontinuities Under Zero Normal Pressure corresponding to Point A_2 in Fig. 4.

(a)

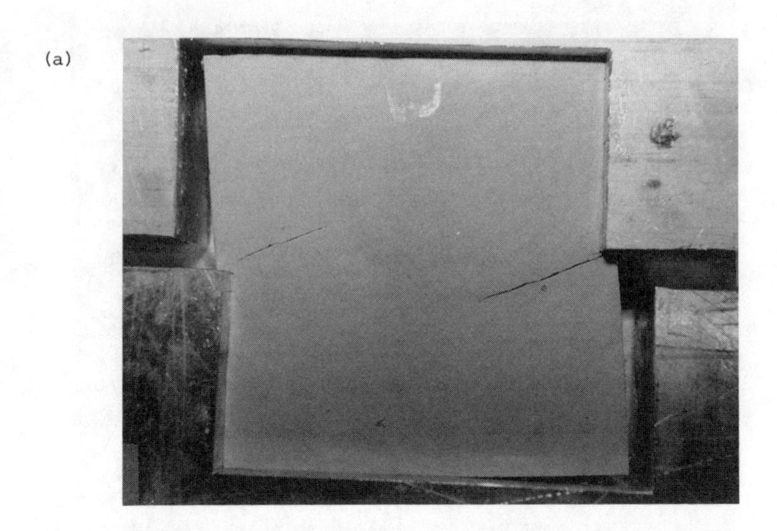

(b)

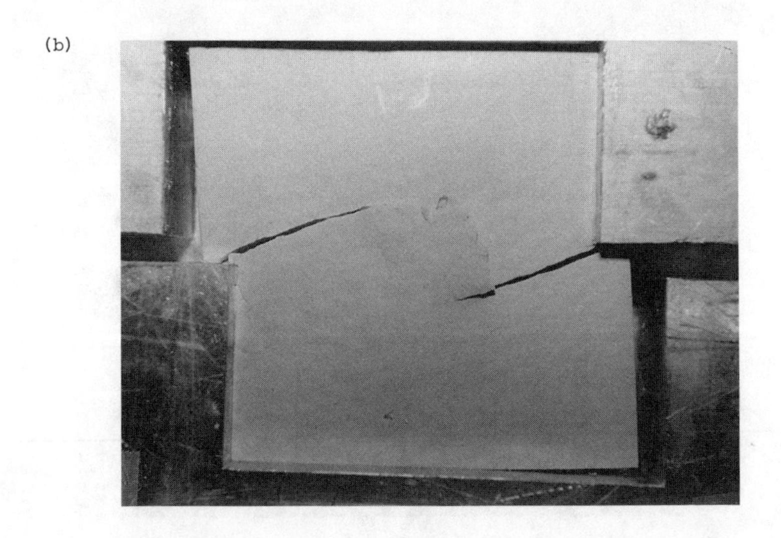

Figure 6. (a) Discontinuities Formed In The Clay In The PSDSA Under 118.6 kPa (Point B_1 in Fig. 4), (b) Discontinuities Under 118.6 kPa Corresponding to Point B_2 in Fig. 4.

The samples in the closed box direct shear apparatus developed multiple edge discontinuities that propagated toward the center of the samples (Fig. 2(b)). The number of these edge discontinuities seems related to the levels of shear stress applied to the samples (Fig. 2(a) and 2(b)). In the open shear box, however, only two edge discontinuities developed (Fig. (6).

The test results obtained by Morgenstern and Tchalenko (1967), however, were obtained by stopping the direct shear tests before their completion, and unloading the samples for the recording of the discontinuities. The changes that occur in the shear zone structure during unloading and preparation for recording are not known. Thus, there is some uncertainty as to whether what Morgenstern and Tchalenko recorded was actually the same shear structure that was present *while* the samples were under load in the direct shear box. The results recorded using the PSDSA do not have this uncertainty because there is no removal and unloading of the samples for the recording of the discontinuities.

In the closed box direct shear tests conducted by Moregenstern and Tchalenko, the effect of the normal pressure on the shear zone structure developed in clays was not investigated. In the present study it was found that the normal pressure had a marked influence on the shape of the failure surface developed by clays under direct shear stress conditions.

Failure Envelopes. The failure envelope corresponding to peak shear stress conditions (points A_1, B_1 and C_1 in Fig. 4(a)) has been plotted in Fig. 7. Also, the failure envelope for the residual shear stress conditions (points B_2 and C_2 in Fig. 4(a)) has been plotted in Fig. 7. The values of the corresponding peak (c_p, ϕ_p) and residual (c_r, ϕ_r) shear strength parameters are also shown.

CONCLUSIONS

A direct shear apparatus that uses an open shear box that allows the continuous recording of the evolution of the shear zone structure in clays has been introduced. From its application to intact samples of kaolinite clay the following conclusions have been drawn:

1) For samples under normal pressures other than zero and at peak shear stress conditions, two edge discontinuities were recorded to form at the two extreme sections of the clay samples. These discontinuities were inclined with respect to the general direction of shearing. When the samples reached residual shear stress conditions, a thrust discontinuity developed in the area between the two edge discontinuities. The thrust discontinuity formed in the opposite direction of that of the edge discontinuities.

2) The failure surface in the samples with a normal pressure greater than zero had a zig-zag configuration and did not resemble the smooth failure plane expected to develop in intact clay samples.

3) For the clay sample with zero normal pressure in the PSDSA, the failure surface induced by the applied shear stresses was a straight plane parallel to the direction of shearing.

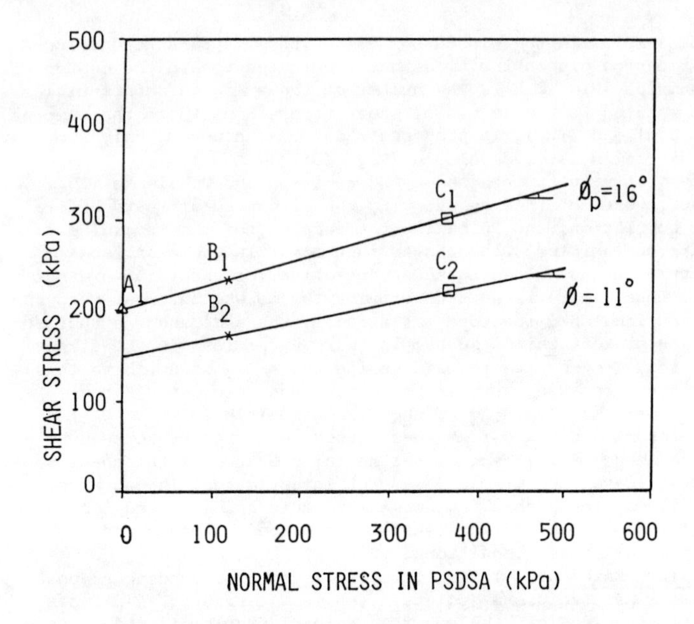

Figure 7. Failure Envelopes in the PSDSA

ACKNOWLEDGEMENTS

The work described herein was supported by Grant No. ECE-8414931 to the University of Pittsburgh from the National Science Foundation, Washington, D.C. This support is gratefully acknowledged.

REFERENCES

1. Bowles, J.E. (1978). Engineering Properties of Soils and Their Measurement, McGraw-Hill Book Co., New York.
2. Morgenstern, N.R. and Tchalenko, J.S. (1967). "Microscopic Structures In Kaolin Subjected to Direct Shear," Geotechnique, 17 (3), 309-328.
3. Palmer, A.C., and Rice, J.R., (1973). "The Growth Of Slip Surfaces In The Progressive Failure Of Overconsolidated Clay," Proceedings Royal Society of London, A332, 527-548.
4. Skempton, A.W., (1964). "Long-Term Stability of Clay Slopes," Geotechnique, 14 (2), 77-101.
5. Tchalenko, J.S. (1968). "The Evolution of Kink Bands And The Development of Compression Textures in Sheared Clays," Techtonophysics, 6 (2), pp 159-174.

STIFFNESS OF SANDS IN MONOTONIC AND CYCLIC TORSIONAL SIMPLE SHEAR

Supot TEACHAVORASINSKUN [1], Satoru SHIBUYA [2], Fumio TATSUOKA [3]

ABSTRACT: A laboratory investigation was carried to evaluate the stiffness of two kinds of Japanese sands for strains from 10^{-6} to those at peak strength. A hollow cylindrical specimen was anisotropically consolidated and subjected to drained simple shear in a fully automated torsion shear apparatus. With an aid of small strain measurements, the linear elastic response has been identified for the limiting shear strain of about 5 x 10^{-6} (0.0005%) below which the secant shear moduli from the monotonic and cyclic loading tests were virtualy identical. The elastic shear modulus, G_{max}, was scarcely influenced by the overconsolidation. A hyperbolic fitting using the measured G_{max} and the strength, τ_{max}, was found appropriate for the range of shear strain less than 10^{-3} (0.1%) to represent the stress-strain relationship as cyclically sheared, but it overestimated the stiffness when monotonically sheared.

INTRODUCTION

In geotechnical engineering practice, the simple shear mode of deformation is encountered in many situations, including level ground subjected to seismic loadings and soil elements along displacement piles. In these cases, both the pre-peak stress-strain behaviour and shear strength must be evaluated when analysing deformation of the ground or predicting the displacement of piles. In doing so, monotonic(=static) and cyclic(=dynamic) loadings should be distinguished according to the in-situ loading conditions.

The stiffness of isotropically consolidated sands subjected to cyclic loadings has been intensively investigated in the laboratory using a resonant-column apparatus or a torsion shear apparatus (e.g., Hardin and Drnevich, 1972; Iwasaki et al.,1978 among others). In these studies, the soil specimens were sheared under quasi simple shear conditions. In general, the strain levels which may be investigated in a resonant-column apparatus range from 10^{-6} to 10^{-4}, whereas those

---

1) Graduate student, Institute of Industrial Science, University of Tokyo, Roppongi 7-22-1, Minato-ku, Tokyo, Japan.
2) Research Assistant, ditto. 3) Professor, ditto.

investigated in a conventional torsional shear apparatus are usually larger than 10^{-4}. Yet a better understanding of effects of the loading conditions on soil stiffness is highly desirable so as to properly assess elastic properties of soils.

The overall nonlinear stress-strain relationship of soils is often modelled using a hyperbolic function which incorporates only two soil parameters; i.e., the maximum shear modulus, G_{max} and the maximum shear stress, τ_{max} (Kondner, 1963). Hardin and Drnevich (1972) showed that the hyperbolic stress-strain relationship for sands subjected to cyclic loadings underestimated the stiffness for the range of shear strains from 10^{-5} to the peak and vice versa for clays. However, the applicability of the hyperbolic model for monotonic stress-strain relationships has not been strictly examined over the full range of strains.

This paper presents the results of monotonic and cyclic torsion shear tests performed on two kinds of clean sands. In these tests, the shear strains were measured with an accuracy of the order of 10^{-6} (0.0001%). The results were interpreted with a particular attention paid on the effect of the different patterns of loadings on the secant shear modulus.

TESTS PERFORMED

The materials tested were fine quartz-rich Toyoura sand and fine-to-medium coarse Hamaoka sand (Fig.1). The test results are summarized in Table 1.

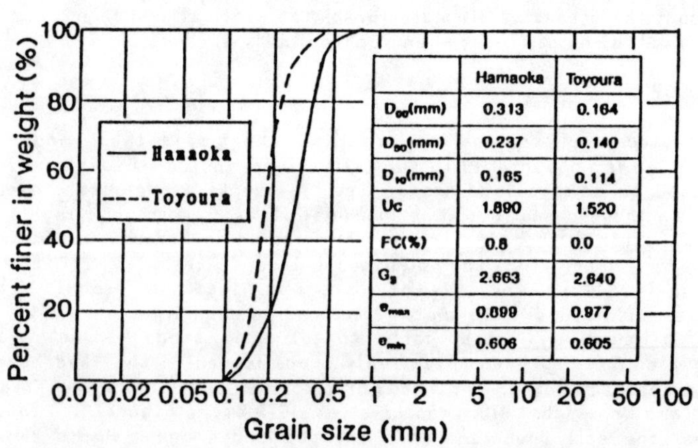

	Hamaoka	Toyoura
D_{60}(mm)	0.313	0.164
D_{50}(mm)	0.237	0.140
D_{10}(mm)	0.165	0.114
Uc	1.890	1.520
FC(%)	0.8	0.0
G_s	2.663	2.640
e_{max}	0.899	0.977
e_{min}	0.606	0.605

Fig.1 Grain size distribution of sands tested

Table 1 Summary of tests

TEST No.	$e_{0.05}$	σ_a'	K[1]	p'[2]	τ_{max}[2]	G_{max}[3]	OCR[4]
Toyoura Sand							
MTS06	0.694	1.00	0.36	0.58	0.75	930	1
MTS07	0.690	1.67	0.36	0.97	1.22	1240	1
MTS27	0.791	1.67	0.41	1.01	1.06	970	1
CTS01[5]	0.788	1.67	0.41	0.95	1.15	1050	1
CTS02[5]	0.662	1.67	0.36	0.96	1.26	1250	1
OMS03	0.688	1.00	0.58	0.72	0.78	1145	2
OMS05	0.682	0.50	0.99	0.48	0.41	875	4
OMS08	0.799	1.00	0.58	0.77	0.68	880	2
OMS09	0.782	0.50	0.98	0.51	0.38	730	4
Hamaoka Sand							
HDS04	0.633	0.80	0.50	0.53	0.75	900	1
CHS04[5]	0.623	0.80	0.50	0.53	-	930	1

1) $K = (\sigma_r'/\sigma_a')_{initial}$
2) in kgf/cm^2
3) $G_{max} = (\tau_{at}/\gamma_{at})$ at $\gamma_{at} \to 0$
4) OCR = $[\sigma_a'/(\sigma_a')_{max}]$
5) Cyclic test

Drained simple shear tests involving no radial and circumferential strains were performed on hollow cylindrical specimens of saturated Toyoura sand having a dimension of 6cm i.d. 10cm o.d. and 20cm high. The detailed descriptions of the torsion shear apparatus (TSA) have been given by Pradhan et al. (1988). The outline of the current servo-system is sketched in Fig.2. In this TSA, the torsional shear strain, γ_{at}, was applied in a strain controlled manner.

Specimens of Hamaoka sand were sheared under conditions of drained torsion shear involving no control of the lateral deformation of the specimen. Figure 3 shows a loading system used to generate the cyclic application of shear stress, τ_{at}. It incorporated a double-acting pneumatic Bellofram cylinder.

Figure 4 shows the newly developed instrumentation to measure γ_{at}. The 'Relay' system incorporates two proximity transducers, with the capacities of a 4 mm and a 8 mm, together with a rotational transducer (potensiometer). As shear strain increases, the measurement is relayed in sequence of a 4 mm proximeter, a 8 mm proximeter and the potensiometer for the ranges of γ_{at} corresponding to less than 0.3%, less than about 1% and more than 1%, respectively. The resolution of a proximity transducer with a 4 mm capacity is 0.4 μm in the current logging system. This enables the shear strain to be measured with an accuracy of about 1×10^{-6} (0.0001%).

The specimens of Toyoura sand were prepared by pluviating dry grains through air (i.e. air-pluviation method). The density was controlled by having different, but fixed, free fall heights during the preparation. The specimens were subjected to a suction of 0.05kgf/cm^2 and the initial void ratio, $e_{0.05}$, was measured (Table 1).

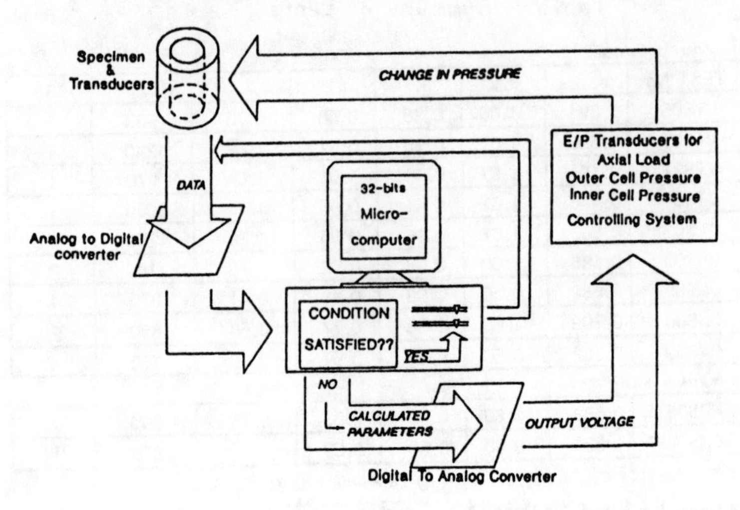

Fig.2 A servo-control used to simulate simple shear

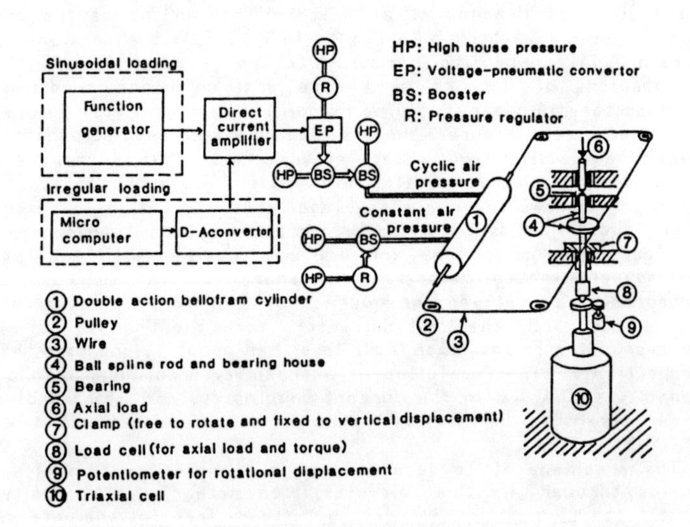

Fig.3 System to generate cyclic torsional shear for a test on Hamaoka sand.

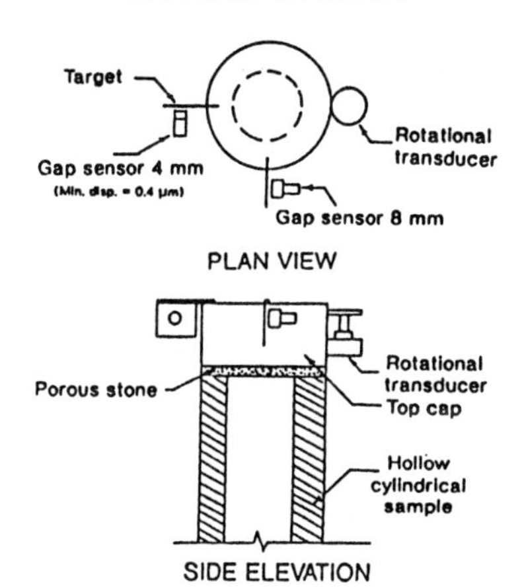

Fig.4 Instrumentation for measurements of shear strain

Each specimen was anisotropically consolidated, against a constant back pressure of 2kgf/cm², to reach a prescribed value of axial stress, σ_a. As shown in Fig.5, the value of K ($=\sigma_r/\sigma_a$) (c.f., σ_r: radial stress) was fixed in each test as 0.52 x $e_{0.05}$, which gives rise to approximately K_o conditions for normally consolidated sand (Okochi and Tatsuoka, 1984). The overconsolidated specimens were prepared by reducing both σ_r and σ_a, from a common stress point at σ_a=2 kgf/cm², along the K_{ou}-line shown in Fig.5. The Skempton's B-value measured at the end of consolidation was in excess of 0.96 for all the tests. The drained simple shear was performed using a constant shear straining of 0.01 percent per minute.

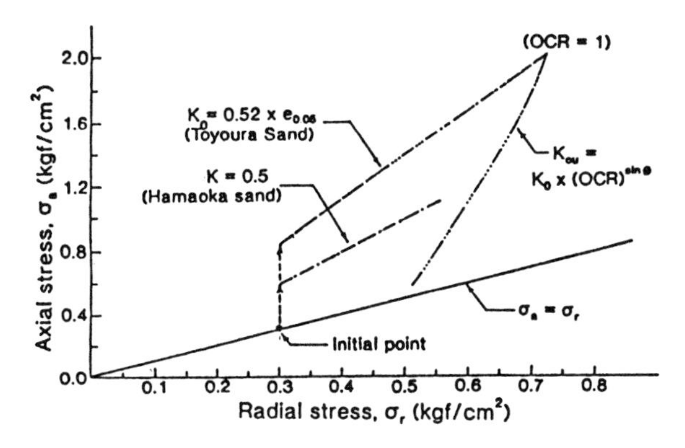

Fig.5 Consolidation paths applied

The specimens of Hamaoka sand were prepared by means of tamping method in which the air dried grains having a water content of about 0.19% were compacted in the hollow cylindrical moulds for total of eight layers. After $e_{0.05}$ was measured, the specimen was consolidated anisotropically by increasing both the suction (σ_r) and σ_a with K-value kept constant at 0.5 (Fig.5). The rate of shear straining in the drained monotonic loading test (HDS04) was 0.01% per minute. In the cyclic test CHS04, the drained stage-cyclic loadings were applied in a manner that the cyclic shear strain amplitude was increased in steps, each using a sinusoidal cyclic shear stress with a frequency of 0.1 Hz. The axial stress during shear was common for these tests at 0.8 kgf/cm^2. Although the volume change of the specimens was not directly measured in these tests, the current radii of the specimens were estimated based on the relationship between the stress ratio, τ_{at}/σ_a, and the volumetric strain, which was obtained from the other corresponding monotonic loading test performed on a saturated specimen with the similar void ratio of 0.630.

PRESENTATION OF TEST RESULTS

A variation of mean effective stress, $t=(\sigma'_1+\sigma'_3)/2$, during a drained monotonic simple shear on a loose Toyoura sand can be seen in Fig.6, in which the shear strain levels are indicated along the stress path. Note that the mean effective stress, $p' = (\sigma'_1+\sigma'_2+\sigma'_3)/3$, remained more or less constant up to a shear strain level of approximately 10^{-2} (1%); then it gradually increased as shear progressed. This strain level corresponded to the shear stress of about 50% of τ_{max}. This means that the shear modulus beyond the level of shear strain of about 1% was affected, to a certain extent, by the change in p'. The discussion of the test results therefore refers chiefly to the region with shear strains less than 1%.

Figure 7 shows the relationships between τ_{at} and γ_{at} for specimens of a loose and a dense Toyoura sand (MTS27 and MTS07). When the small strain behaviour was enlarged (Fig.7 b and c), the initial linear portion can be seen for a range of γ_{at} less than about 5×10^{-6}. In test MTS27, an unloading-reloading cycle was applied to examine an elastic property of the sand (Fig.7b). It can be seen that the response was not purely elastic, however the equivalent shear modulus, G_{eq}, corresponding to a double amplitude of shear strain of approximately 2×10^{-5} was more or less equal to the G_{max}-value that obtained from the virgin loading.

The stress-strain relationships of loose specimens with OCR=1 (MTS27) and OCR=4 (OMS09) are shown in Figs.8 and 9, in which both of τ_{at} and γ_{at} are plotted on a full logarithmic scale. Despite different values of p', it can be seen that i) the specimens exhibited a linear stress-strain relation at small strains, and ii) the OC sand had a slightly larger linearity-limit than the NC sand.

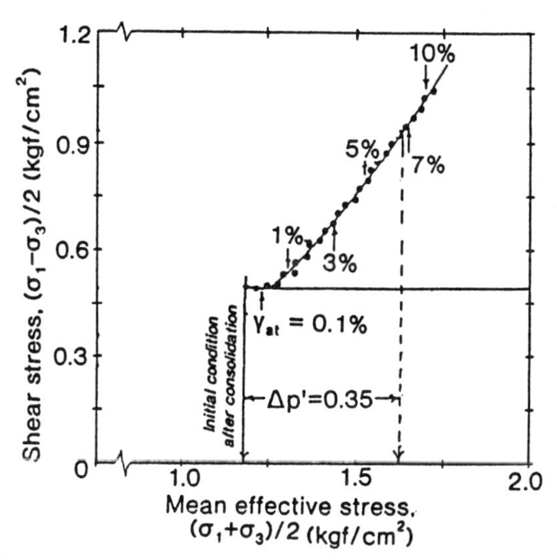

Fig.6 Typical of stress path during drained simple shear

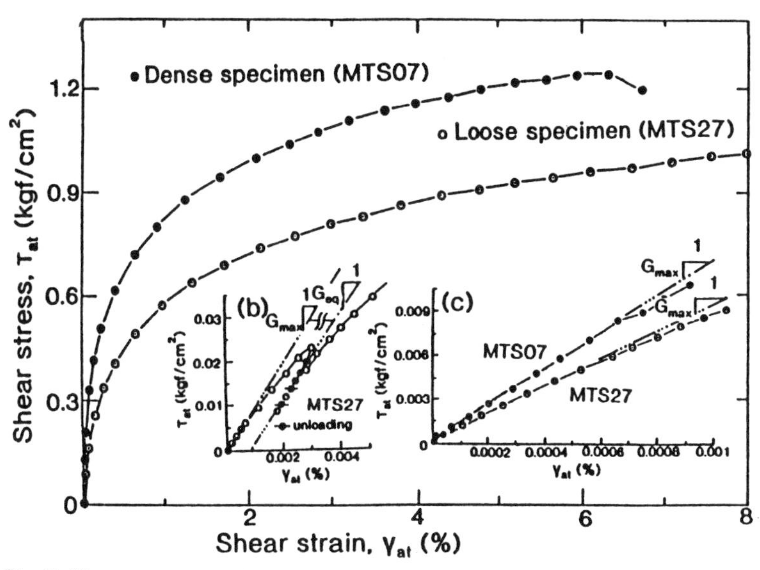

Fig.7 Stress-strain relationship for normally consolidated specimens of Toyoura sand

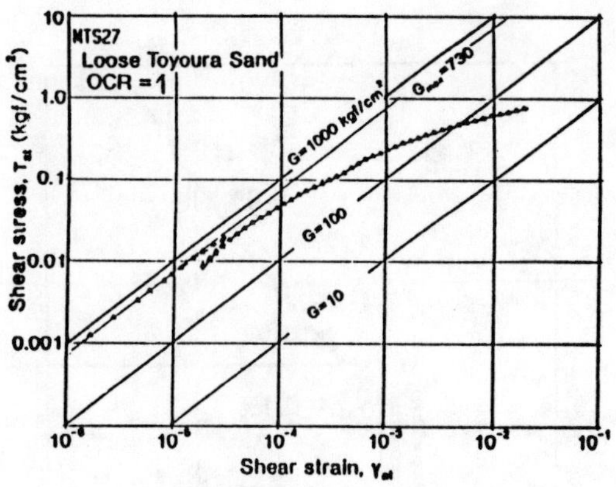

Fig.8 Stress-strain relationship (Toyoura sand, OCR=1)

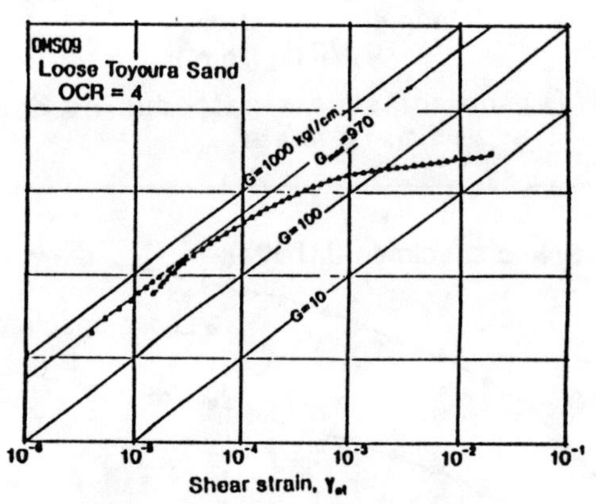

Fig.9 Stress-strain relationship (Toyoura sand, OCR=4)

DISCUSSIONS

Figure 10 shows a comparison of stress-strain relations of Hamaoka sand; i.e., τ_{at} versus γ_{at} in a monotonic test and $(\tau_{at})_{SA}$ versus $(\gamma_{at})_{SA}$ in a cyclic test (c.f., $(\gamma_{at})_{SA}$: single amplitude cyclic shear strain). Note that the secant shear moduli of these two tests practically coincided to each other as the shear strain level was less than about 10^{-5} (Fig.10a). The stress-strain response of the cyclic became stiffer than that of the monotonic test as the shear strain level increased. A similar comparison was made in Fig.11 (a) and (b), in which the hyperbolic stress-strain relation using the measured G_{max} and $(\tau_{at})_{max}$ is also drawn for comparison. Figure 12 shows the variations of secant shear moduli, G_{seo} and G_{eq}, as plotted against the corresponding shear strains. In the case of Toyoura sand (Fig.12a), the results of cyclic loading tests, obtained using a resonant-column device and a torsion shear apparatus for γ_{at} less than and larger than 4 x 10^{-4} (0.04%) (Iwasaki et al., 1978), are also plotted for comparison. The followings should be noted.

(i) In the range of small strain less than 5 x 10^{-6} (0.0005%), the secant shear moduli remained constant, and scarcely affected by the different types of tests and loadings.

(ii) The values of G_{eq} in the static cyclic test increased as the number of cycles increased.

(iii) The reduction of the secant shear modulus with the increasing shear strain level was faster in the monotonic than in the cyclic tests.

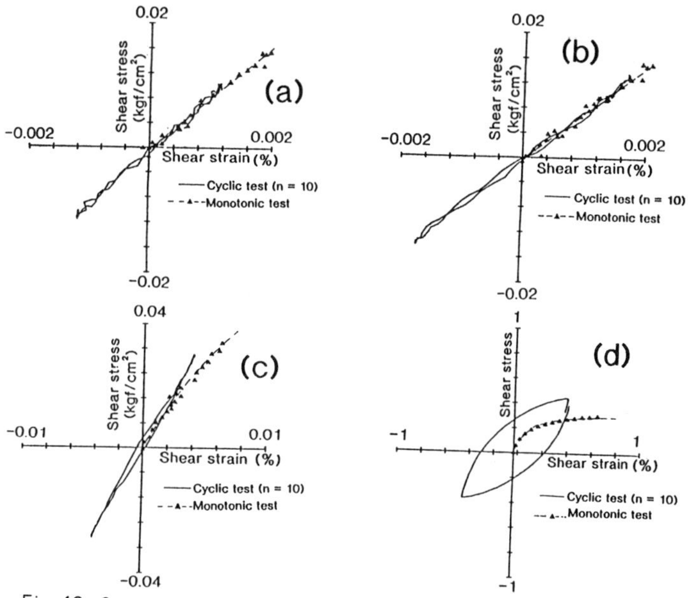

Fig.10 Comparison of stress-strain relations of Hamaoka sand subjected to monotonic and cyclic loadings

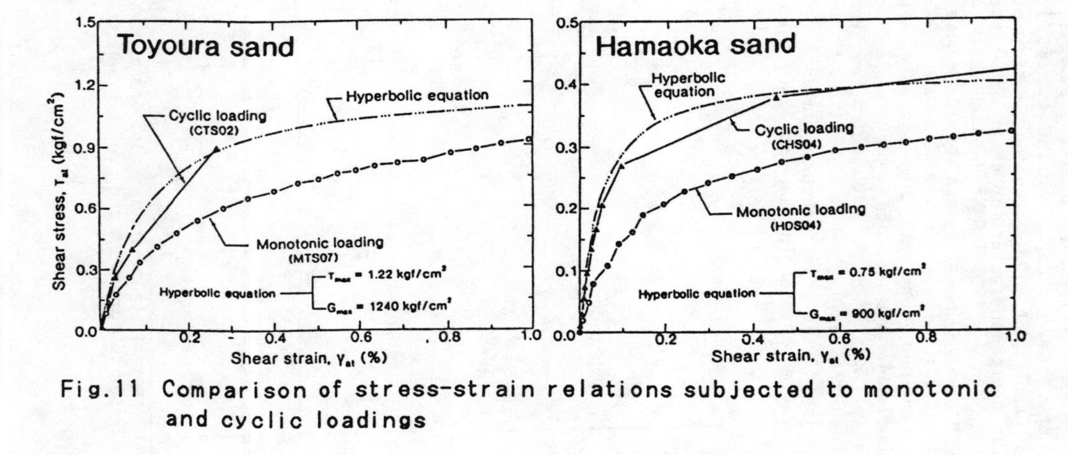

Fig.11 Comparison of stress-strain relations subjected to monotonic and cyclic loadings

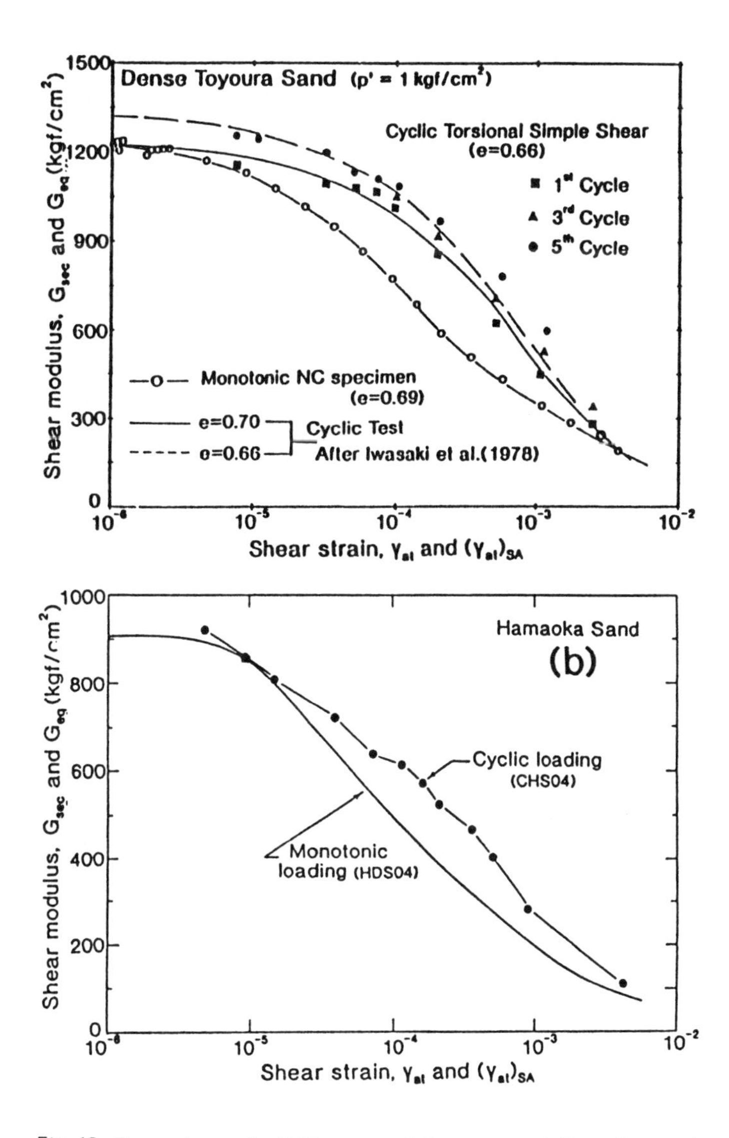

Fig.12 Comparison of stiffnesses of Toyoura and Hamaoka sands subjected to monotonic and cyclic loadings

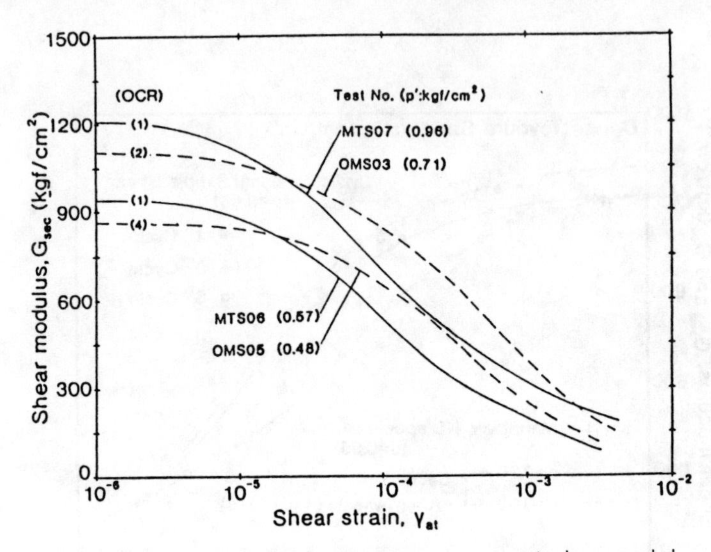

Fig.13 Effect of overconsolidation on secant shear modulus for dense specimens of Toyoura sand.

The coincidence of G_{eq} and G_{sec}, together with the linearity in the stress-strain relationship (see Fig.7 through Fig.12) means that the response of the sands was linear elastic for shear strains less than about 5 x 10^{-6} (0.0005%). Note also that the Masing's second rule, which assumes the stress-strain curve of the monotonic test to be the backbone curve of the cyclic test, did not hold for shear strains larger than the elastic limit.

The effect of overconsolidation on the stiffness can be seen in Figs.13 and 14. The G_{max} value was scarcely influenced by the OCR up to four if the different values of p' among the tests were properly considered (Fig.13). Note also that, in a plot of G_{sec}/G_{max} versus $e_{0.05}$, the OC specimens were obviously stiffer than the NC specimens for the shear strain levels beyond about 10^{-5} (0.001%).

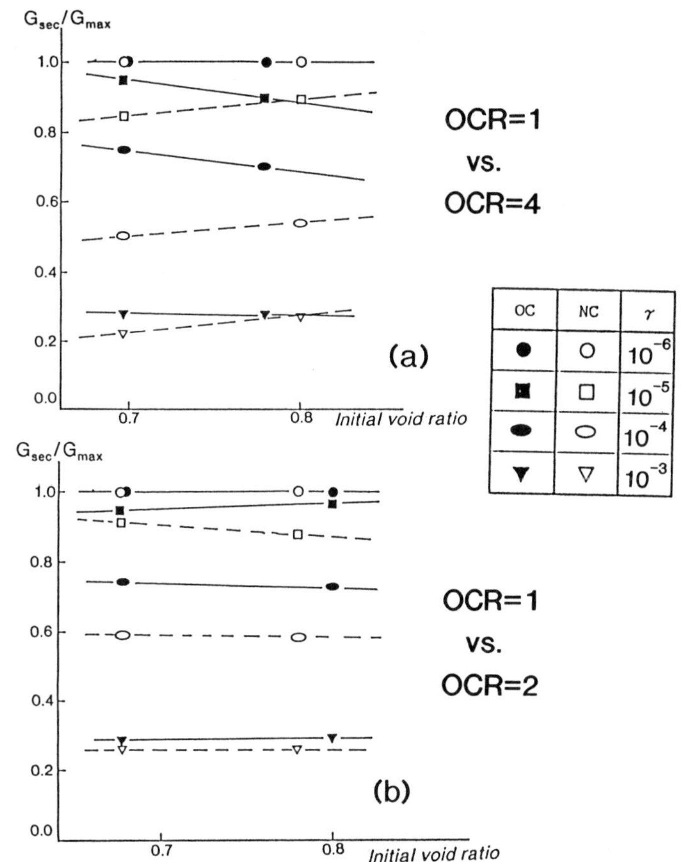

Fig.14 Effect of overconsolidation on secant shear modulus;
(a) p'= 0.5 - 0.6 kgf/cm^2 (b) p'= 0.7 - 1.2 kgf/cm^2

A hyperbolic equation (Kondner, 1963) is expressed as;

$$\tau = \frac{\gamma}{\dfrac{1}{G_{max}} + \dfrac{\gamma}{\tau_{max}}} \qquad (1)$$

For this hyperbolic stress-strain relationship, the secant shear moduli are given as;

$$G_{sec}/G_{max} \text{ or } G_{eq}/G_{max} = 1 - \tau/\tau_{max} \qquad (2)$$

The relationship between the secant shear modulus and the corresponding shear stress is shown in Fig.15, in which the examinations are made using the normalized forms as shown in Eq.(2). In each graph, the comparison was made for specimens having similar void ratio and p'. The hyperbolic relationship represented by Eq.(2) is also drawn as a yardstick. In the cyclic tests on Toyoura sand (CTS01 and CTS02), the specimens were eventually brought to failure by monotonically increasing the shear stress. In the case of Hamaoka sand (test CHS04), the strength was not directly measured; however, it was assumed that τ_{max} was not affected by the prestraining applied during the cyclic loadings. This hypothesis seems to be appropriate in the case of Toyoura sand; for instance, the values of τ_{max} of tests MTS07 and CTS02 (see Table 1) were similar to each other. It is suggested in Figs.11 and 15 that the relationship between G_{eq} and $(\gamma_{at})_{SA}$ in the cyclic tests may be approximated using a hyperbolic function shown in Eq.(1). On the other hand, the hyperbolic model obviously overestimated the stiffness of sands subjected to monotonic loadings for the ranges of density and p' examined. The degree of overestimation was remarkable; for example, at τ / τ_{max} equal to 0.5, the values of G_{sec}/G_{max} were around 0.2 or even less against the value of 0.5 predicted by the hyperbolic model. Note that this trend was more significant for the NC specimens (Figs.15 a and b).

Theoretically, no elastic stress-strain relation exists for an assembly of elastic-perfectly plastic particles without any bonding. Thus, one may consider that the G_{max}-value obtained from the monotonic loading tests was neither an elastic modulus nor that of the genuine 'virgin' specimen. For instance, Jardine et al. (1991) are skeptical of the existence of elastic region for clays by considering that the apparently linear G_{max} plateaux seen in resonant column test may involve the rate effects, which in turn mask the underlying non-linearity. It should be pointed out that any soil encountered in the field or prepared in the laboratory could have inevitably been subjected to prestraining. Note also that similar coincidence of G_{sec} and G_{eq} at extremely small strains was observed in triaxial tests and in plane strain compression tests using various types of soils (clays, sands and gravels) and soft rocks, wherein strains were measured locally over the central part of the specimens (Shibuya et al., 1990 and 1991, Tatsuoka et al., 1990 and 1991).

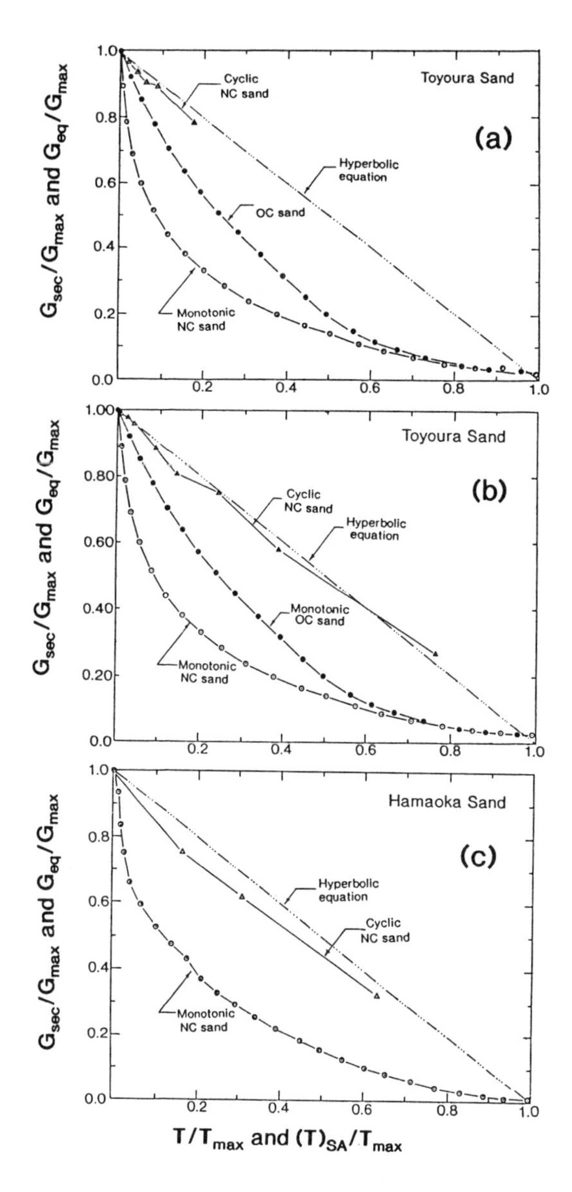

Fig.15 Secant shear moduli versus the corresponding shear stress; (a) Loose Toyoura sand, (b) dense Toyoura sand and (c) Hamaoka sand

CONCLUSIONS

(1) The linear elastic response was observed for shear strains less than about 1×10^{-5} (0.001%). In this region, the secant shear moduli from the monotonic and cyclic tests were practically identical, and scarcely affected by the overconsolidation ratio.
(2) The reduction of stiffness with the increasing shear strain was larger in monotonic loading tests than in cyclic loading tests.
(3) The relationship between the secant shear modulus and the shear strain may be approximated by a hyperbolic function for the cyclic loading. However, this relationship overestimated the stiffness of sands for monotonic loading, especially for normally consolidated specimens.

References

1) Hardin, B.O. and Drnevich, V.P., Shear modulus and dumping in soils: measurement and parameter effects, ASCE, 98-6, 1972, 603-624.
2) Iwasaki, T., Tatsuoka, F. and Takagi, Y., Shear moduli of sands under cyclic torsional shear loading, Soils and Foundations, 18-1, 1978, 39-56.
3) Jardine, R.J., St. John, H.D., Hight, D.W. and Potts, D.M., Some practical applications of a non-linear ground model, Proc. of 10th European Regional Conference on SMFE, Florence, 1991 (to be published).
4) Kondner, R.L., Hyperbolic stress-strain response: cohesive soil, Journal of ASCE, 89-1, 1963, 115-143.
5) Okochi, Y. and Tatsuoka, F., Some factors affecting K_o-values of sand measured in triaxial cell, Soils and Foundations, 24-3, 1984, 52-68.
6) Pradhan, T.B.S., Tatsuoka, F. and Horii, N., Simple shear testing on sand in a torsional shear apparatus, Soils and Foundations, 28-2, 1988, 95-112.
7) Shibuya, S., Kong, X.J. and Tatsuoka,F, Deformation characteristics of gravels subjected to monotonic and cyclic loadings - with particular reference to their small strain behaviour, Proc. of 8th Japan Earthquake Engineering Sympo., Tokyo, 1990, 771-776.
8) Shibuya, S., Tatsuoka,F, Abe, F. Teachavorasinskun, S. and Park C-S., Elastic properties of granular materials measured in the laboratory, Proc. of 10th European Regional Conference on SMFE, Florence, 1991 (to be published).
9) Tatsuoka, F., Shibuya, S., Goto, S., Sato, T. and Kong, X.J., Discussion on the Paper by Clayton et al., Geotechnical Testing Journal, 13-1, 1990, 63-67.
10) Tatsuoka, F., Shibuya, S., Teachavorasinskun, S. and Park C-S., Discussion on the Paper by Bolton and Wilson, Geotechnique, 1991, 659-662.

PERMEABILITY OF DISTURBED ZONE AROUND VERTICAL DRAINS

Atsuo Onoue[1], Nai-Hsin Ting[2], Student M.ASCE, John T. Germaine[3],M.ASCE and Robert V. Whitman[4], F.ASCE

ABSTRACT:

Homogeneous cylinders of clay, about 305 mm in diameter and 89 mm thick were prepared by sedimentation and consolidation. A rigid cylindrical drain was pushed into the center of a specimen. After trimming and reconsolidation three specimens were subjected to a radial gradient, with measurement of pore pressures at various locations. With one specimen, the time-rate of consolidation under increments of additional vertical stress was monitored. All three specimens were then trimmed to progressively smaller diameters, with horizontal permeability measurements at each stage. Water content distributions were recorded. The result is a remarkably consistent pattern for the radial variation of permeability for one clay.

INTRODUCTION

Numerous types of vertical drains are used to speed consolidation of clays following placement of embankments or fills. It has long been recognized that disturbance, induced by the installation of a mandrel to form a drain, may cause the permeability of the soil adjacent to the drain to be reduced, thus impeding the ability of a drain to accelerate consolidation (Barron, 1948).

There is intense shearing immediately adjacent to the mandrel in a region called a smear zone In addition, disturbance of the clay may extend to a considerable distance. The pattern, type and magnitude of induced straining has been analyzed by Baligh (1986). Excess pore pressures develop and subsequently dissipate, and clay undergoes some reconsolidation. Even in a homogenous clay, remolding and reconsolidation are expected to lead to a reduction in permeability. If the clay is stratified (i.e. varved) with alternating bands of smaller and larger permeability, remolding may mix together the layers causing a final permeability less than the horizontal permeability of the undisturbed soil.

---

1 - Shimizu Corporation, Tokyo, Japan; Visiting Engineer, Massachusetts Institute of Technology, Cambridge, MA.

2 - Garduate Research Assistant, Massachusetts Institute of Technology.

3 - Principal Research Associate, Massachusetts Institute of Technology.

4 - Professor of Civil Engineering, Massachusetts Institute of Technology.

Several theoretical analyses have been developed to predict the effect of a smear zone upon the rate of consolidation (Barron 1948, Hansbo 1981, Onoue 1988). In all such analyses, it is assumed that the soil around a drain can be represented by two zones, each having uniform permeability as shown in Fig. 1: an undisturbed region having the permeability of the natural soil, and a region near the drain with reduced permeability. Thus there are two key parameters: S, indicating the size of the smear zone, and η , giving the reduction of permeability within the zone

$$S \equiv r_s/r_w, \quad \eta \equiv k_{ho}/k_s \tag{1}$$

where r_s is the outer radius of the smear zone, r_w is the radius of the well, k_s is the horizontal permeability within the smear zone, and k_{ho} is the horizontal permeability of the undisturbed soil (see Fig. 1). Values of S and η have been backfigured from observations of rate of settlement and excess pore pressure dissipation during field tests (e.g. Aboshi and Inoue, 1986). There are essentially no direct data for the radial variation of permeability around a drain. Moreover, the predictions of the theories have on occasion been inconsistent with results from field tests (Ladd, 1990).

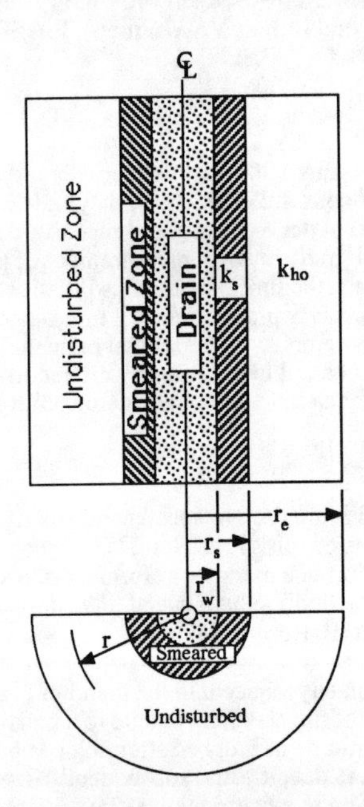

Figure 1: Schematic picture of soil cylinder with a vertical drain - two zone model.

In order to study directly the nature and extent of the disturbance caused by mandrel insertion, and the resulting radial variation of permeability, a series of small-scale experiments have been undertaken in the laboratory (Ting, 1990).

EXPERIMENTAL APPARATUS AND PROCEDURES

Initial Procedures

Specimen Preparation. All tests were performed using resedimented samples of Boston Blue Clay (BBC), a moderately sensitive illitic marine clay. The equipment and procedures for resedimentation and consolidation have been perfected during the past decade and are described by Germaine (1982) and Seah (1990). Samples were first consolidated to a large stress and then rebounded to a "final" state with an OCR=4. At this "final" state, horizontal and vertical effective stresses were approximately equal. The consolidometer had an inside diameter of 305 mm. A quantity of clay was used so as to obtain a specimen with a total thickness of about 140 mm. Three specimens were prepared, using the consolidation stresses given in Table 1. Samples prepared in this manner will be referred to as undisturbed samples. All specimens were backpressured to 400 kPa during consolidation and the subsequent tests.

Table 1. Characteristics of Test Specimens

Specimen	Max -$\bar{\sigma}_v$-Pa	Final -$\bar{\sigma}_v$-Pa	Final void ratio
B211	100	25	1.092
B212	200	50	0.978
B213	100	25	1.092

Insertion of Vertical Drain. The vertical drain consisted of a filter sleeve, of sintered bronze with very small pore openings. It had an external diameter of 32 mm. The bottom cap was rounded to facilitate penetration.

Before penetration, the loading plate for consolidation was replaced by a donut-shaped rubber bag, topped by wooden blocks, and the consolidometer was moved to a loading frame. The bottom end of the drain was placed in contact with the clay through the hole in the rubber bag, and the bag was pressured to the final vertical stress achieved at the end of consolidation. Then the drain was pushed into the clay by hand-cranking, at a rate of approximately 75 mm/min. The clay was allowed to reconsolidate under the applied pressure before undertaking further operations.

General Features of Tests

Three types of tests were conducted. Horizontal flow tests, from which permeability variations were inferred from the observed distribution of pore pressure with radius; a radial consolidation test, the results of which were compared to theoretical predictions; and horizontal permeability tests upon samples of progressively smaller diameter, giving data from which the radial variation of permeability could be inferred.

Horizontal Flow Tests. The general arrangement of these tests is shown in Fig. 2. The specimen was 264 mm in diameter by 89 mm high, confined between plexiglass plates (to avoid corrosion problems) at the top and bottom and a flexible membrane around the exterior circumference. Wax and vaseline were used between the clay and the top and bottom plates to obtain good contact and a good seal. Layers of filter cloth were placed between the membrane and the clay. The entire assembly was placed within a large chamber, and subjected to a confining pressure. Inward horizontal flow was induced by imposing a pressure difference of 20 kPa between the

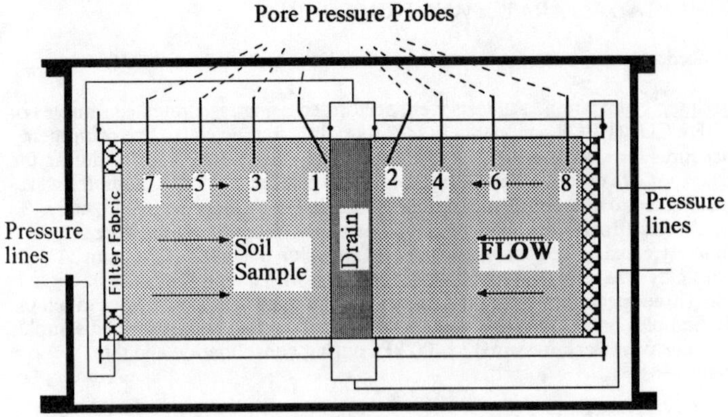

Figure 2: Apparatus for horizontal flow test

filter cloth on the outer circumference and the drain at the center. Eight pore pressure probes were inserted through holes in the upper cover plate, at various radii. The pore pressure probes were hypodermic needles, with outside and inside diameters of 1.27 and 0.64 mm, respectively inserted through openings in the top plate. They extended 10 mm into the clay. Two of the openings were inclined, so as to place the tips of the probes as close to the drain as possible.

To prepare the consolidated cake for these tests, upper and lower portions were trimmed away, leaving that part of the original specimen where the disturbance by penetration of the drain was most uniform. The upper and lower plates were placed; the assembly was then extruded from the consolidometer, the diameter trimmed, the outside filter cloth and membrane affixed and the "final" effective stress restored.

Radial Consolidation Test. One specimen (B213) was left in the original consolidometer and rested on an impermeable base plate, as shown in Fig. 3. It was loaded at the top through a rigid plate, through which pore pressure probes were inserted into the clay. After reconsolidation to the "final" stress, vertical stress was increased in increments with full consolidation permitted for each increment. Drainage of the clay occurred only through the central vertical drain. The principal data were those for the effective stress increment from 120 kPa to 200 kPa; that is, an increment of virgin consolidation. Following completion of the radial consolidation test, the specimen was trimmed further and a horizontal flow test was performed, with an effective stress of 50 kPa.

Horizontal Permeability Tests. Following each of the tests described above, the thickness was cut down to 76 mm, removing a short length of the central vertical drain. The diameters were trimmed successively to 114, 72 and 47 mm. At each diameter, a radial flow test was performed in a large triaxial cell using a small gradient, to evaluate the average horizontal permeability.

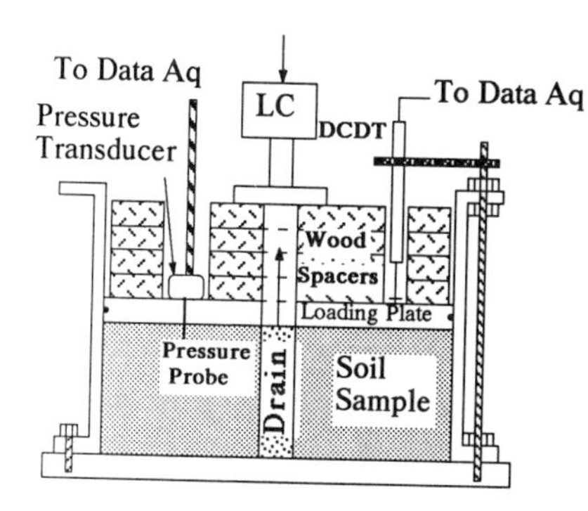

Figure 3: Radial consolidation apparatus

PROPERTIES OF THE CLAY

The batch of BBC used to prepare the specimens tested in this program had the following characteristics:

Liquid limit: 45.20 ± 0.44%
Plastic limit: 21.74 ± 0.44%
Plasticity index: 23.46 ± 0.42%

When consolidated $\bar{\sigma}_V$ = 100 kPa, the water content was 40.59 ± 0.57%.

An extensive series of permeability evaluations for resedimented BBC has been reported by Seah (1990) and Ting (1990), using controlled-rate-of-strain consolidation tests. Results are summarized in Fig. 4. Note that there is a factor of approximately 3 between the permeability of undisturbed and remolded clay, at the same void ratio. Permeability was measured during the consolidation of specimens B211-B213; the results fell within the scatter shown in Fig. 4. Direct permeability measurements were made on samples cut from the outermost (undisturbed) portions of one of the test specimens, so as to evaluate both horizontal and vertical permeability. The results are superimposed on Fig. 4, and show that this undisturbed clay is isotropic as regards permeability.

RESULTS

Effect of Drain Insertion on Void Ratio

The data points in Fig. 5 are void ratios, calculated from measured water contents, of samples of clay cut from the bottom of specimen B213 during trimming. Very similar data came from trimmings at the top of the specimen. Thus insertion of the drain and subsequent reconsolidation resulted in a void ratio decrease that extended to a considerable distance from the drain. The average void ratio for the undisturbed sample as a whole (see Table 1) was reached at r/r_W of about 6 to 7. Referring to Fig. 4, the maximum decrease in void ratio - about 0.1 or somewhat larger - corresponds to about a 30% to 40% decrease in permeability.

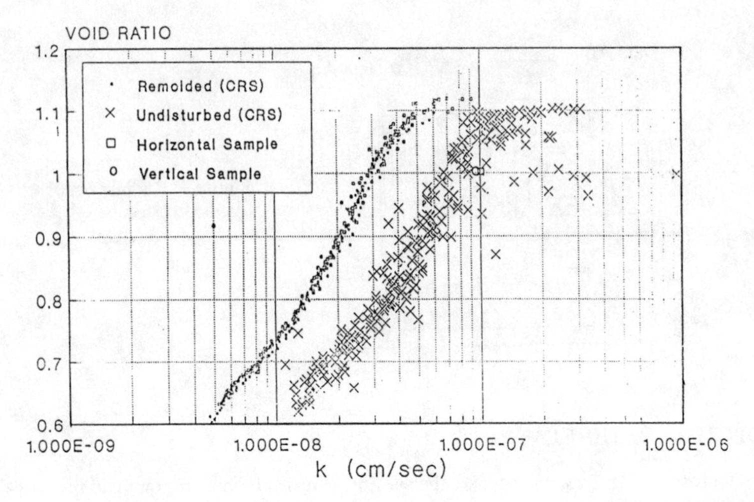

Figure 4: Coefficient of permeability versus void ratio

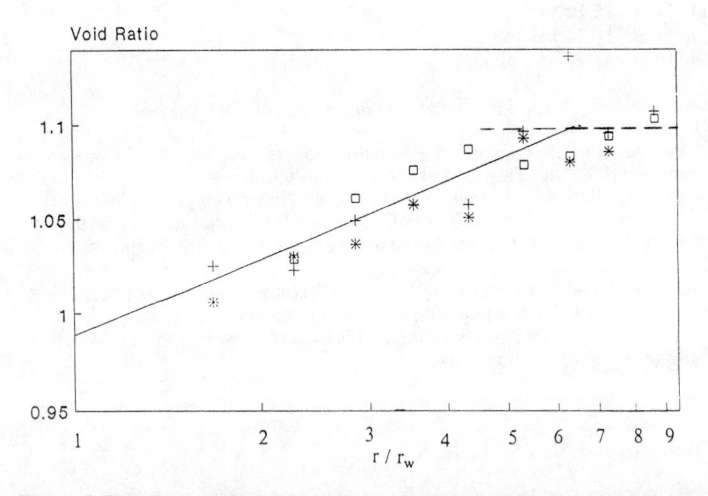

Figure 5: Distribution of void ratio after consolidation following insertion of drain

Average Horizontal Permeability

Average permeability refers to the value of permeability computed using the observed rate of seepage and the applied head difference during the horizontal permeability tests, using an equation for radial flow that assumes uniform permeability. Fig. 6 shows results for normalized average permeability from the three sets of tests, plotted against the normalized outside radius of the cylinder of soil being tested. The normalizing permeabilities are the values for undisturbed specimens at the appropriate void ratios.

To deduce the variation of permeability implied by these results, trends computed for different assumed distributions of permeability may be compared with the results. For example, Fig. 6 shows theoretical curves for normalized average permeability, computed using a two-zone model (Fig. 1) assuming a normalized smear zone radius S = 1.6 and several values for the permeability reduction ratio that provide reasonable fits to the data. The first and second columns of Table 2 give combinations of parameters providing reasonable fits.

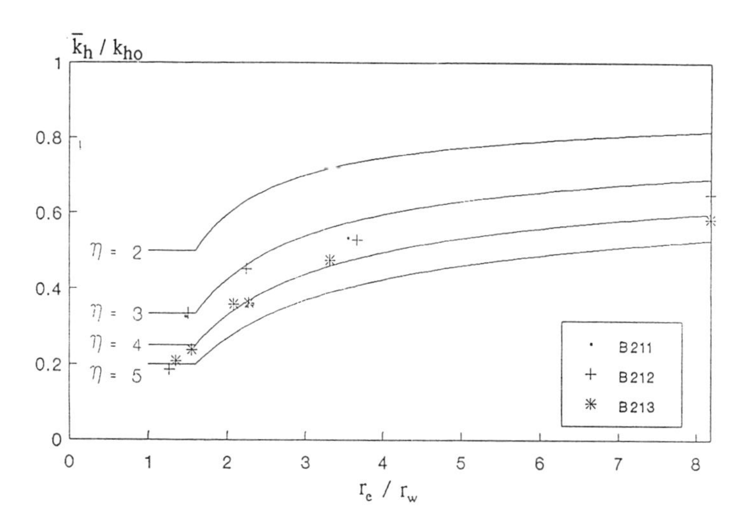

Figure 6: Comparison of theoretial curves for average permeability, assuming S =1.6, with data from horizontal permeabilty tests

Table 2. Parameters for Two-Zone Model Providing Reasonable Fits to Data

	Permeability reduction ratio η	
S	Hor. permeability test	Horizontal flow test
1.3	5	4
1.4	5	3.5
1.5	4.5	3
1.6	4	3
1.7	4	2.5

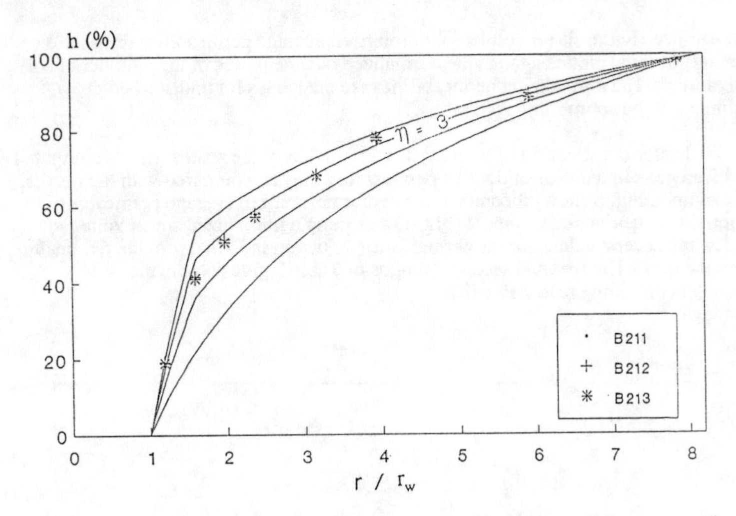

Figure 7: Comparison of pore pressures measured in horizontal flow test with predicted curves for 2-zone model with S = 1.6

Horizontal Flow Tests With Measured Pore Pressures.

Fig. 7 shows heads, normalized by the difference in head between the outer circumference of the sample and the drain, plotted against the normalized radius. The results are remarkably consistent. The figure also illustrates the fitting of the data by theoretical curves computed using a two-zone model. Parameters providing reasonable fits are listed in the first and third columns of Table 2. The writers believe these results to be more reliable than those from horizontal permeability tests.

Radial Consolidation Tests

Data from this test are curves of change-in-thickness for the specimen as a whole and pore pressure at several points, all as a function of time. Fig. 8 gives the observed results for vertical strain (change in thickness divided by original thickness), for the increment of effective vertical stress from 120 to 200 kPa. During this increment the soil is in the range of virgin compression, so that the process of consolidation is most consistent with that assumed in theories.

Fig. 9 illustrates the expected effect of the smear zone, using the two-zone theory (e.g. Hansbo, 1981) together with the values of parameters deduced from horizontal flow tests with pore pressure measurements. Here the time factor T is:

$$T = C_h t/d_e^2 \tag{2}$$

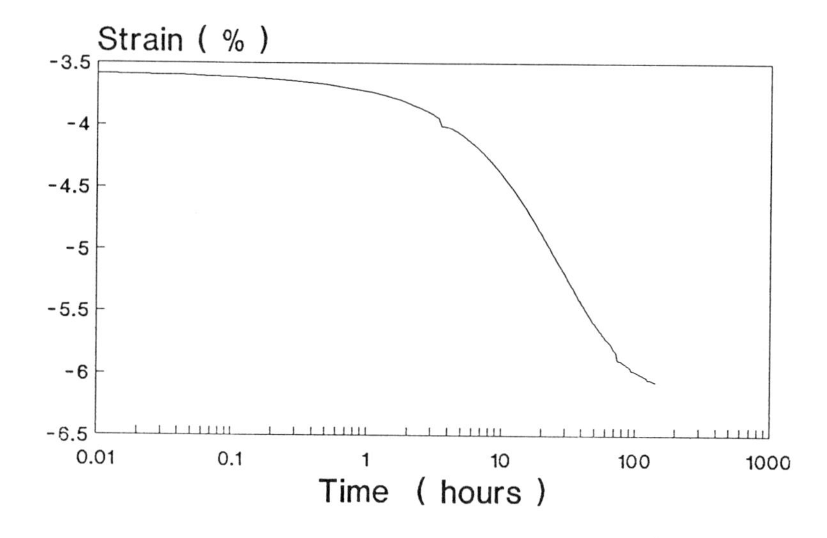

Figure 8: Vertical strain versus time, during radial consolidation test. The increment of vertical effective stress is from 120 to 200 kPa

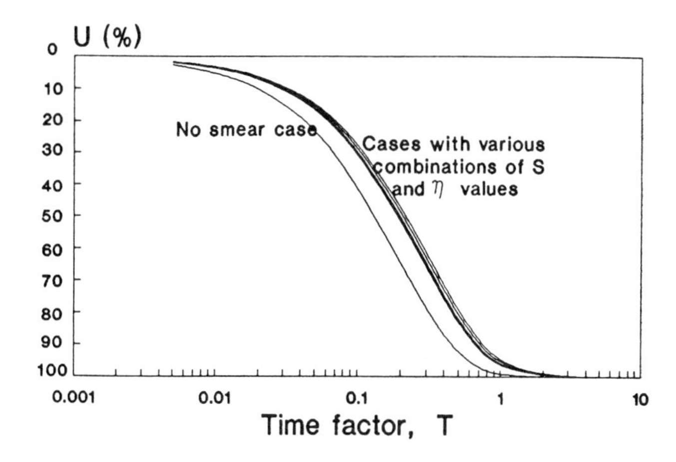

Figure 9: Theoretical curves for average consolidation ratio versus time, assuming 2-zone model of Hansbo

where C_h is the coefficient of horizontal consolidation for undisturbed clay, t is time and d_e is the effective diameter for the drain (which in our case is the diameter of the specimen). It may be seen that the choice of combinations of parameters, all of which provided reasonable fits to the results from the horizontal flow tests, makes little difference in the predicted result. At 50% consolidation, the T for the case of no smear is about 65% of that considering the expected effect of the smear zone.

Since the consolidation curves for the case with a smear zone are similar in shape to those for a uniform soil, it can be expected that the usual curve-fitting methods can be used to evaluate the parameters for the consolidation process. This procedure will yield an estimate for C_{ha} - the apparent average coefficient of consolidation for radial consolidation. Several fitting techniques - square root or log time, Aboshi and Monden (1963), and Asaoka (1978) - were applied to the change-of-thickness and pore pressure dissipation data. C_{ha} obtained from the pore pressure data by all methods was 1.55×10^{-3} cm^2/sec, with very little scatter. Similar values, but with more scatter, were calculated from the data for change of thickness.

Dividing C_{ha} by 0.65 gives, according to Fig. 9, C_h for undisturbed soil as 2.4×10^{-3} cm^2/sec. The compressibility m_v is evaluated from the strain vs. effective stress curve for specimen B213, for the stress increment here being considered. Using the definition for C_h, $k_h = 7.2 \times 10^{-8}$ cm/sec. Referring to the curve for undisturbed clay in Fig. 4, it is seen that this is a very reasonable value for k_{ho}.

Incidentally, m_v for specimen B213, after drain installation and reconsolidation to an effective stress greater than the maximum past stress, was substantially the same as that for specimen B212 in the same range of effective stress. Thus drain installation had little effect upon the compressiblity of the sample as a whole. This does not necessarily mean, however, that there was no effect upon m_v in the small, highly disturbed soil nearest the drain.

DISCUSSION

The foregoing results demonstrate that the two-zone theories of Barron, Hansho, Onoue and others can provide a satisfactory prediction for the process of radial consolidation and provide directly values for the key parameters for resedimented Boston Blue Clay. At the same time, it seems clear that the actual distribution of permeability varies more smoothly with radius, and the disturbance extends to a greater distance than assumed in these theories. For example, it appears more logical to divide the soil into three zones:

Zone I - the undisturbed zone. From the data for void ratio in Fig. 5, it is assumed that this zone begins at about $r = 6.5r_w$.

Zone II - a zone where drain installation causes a decrease in void ratio and hence in permeability.

Zone III - the remolded zone. Here it is anticipated that there is an additional reduction of k_h below that implied by the decrease in void ratio. The maximum decrease at the wall of the drain is that for full remolding.

Equations for the variation of permeability in Zones II and III have been developed from the data (Ting, 1990). Fig. 10 shows schematically the variation of permeability with radius according to these relations. Excellent agreement was obtained between

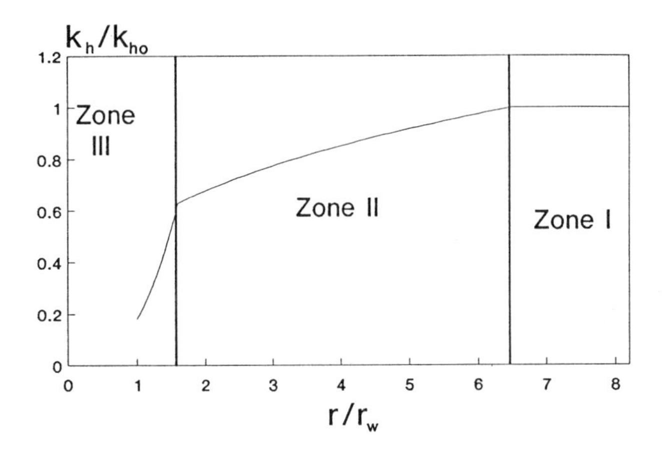

Figure 10: Suggested variation of permeability with radius, assuming S = 1.6

predicted and measured distributions of pore pressure during horizontal flow tests with the actual data, using S = 1.6. Analyses indicate that the decrease in void ratio and the further decrease in permeability by remolding are almost equally important.

FINAL COMMENTS

This study has examined one aspect of the affect of drain installation upon the permeability of clay around the drain: i.e. remolding and reconsolidation of a homogeneous clay. Only one clay for which $k_h \approx k_v$, has been tested. Two major conclusions have emerged:

* Solutions based upon the two-zone approximation - with η equal to 3 (the reduction in permeability by remolding at constant water content) and S=1.5 - 1.6 can provide satisfactory predictions for the rate of radial consolidation.

* The actual radial variation of permeability is more complex, resulting from a combination of void ratio decrease during reconsolidation after drain installation, and remolding (complete and partial) near the drain.

While these results come from small-scale tests, it is believed that behavior at full-scale will be essentially the same. A major remaining question is what happens in the remolded zone when a drain is inserted into a clay with $k_h > k_v$: e.g. a varved clay.

ACKNOWLEDGEMENTS

Support for this study was provided by the Shimizu Corporation. Professor C. C. Ladd provided invaluable advice.

LIST OF SYMBOLS

C_h = radial coefficient of consolidation = $k_{ho}/\gamma_w m_v$
C_{ha} = average coefficient for case with smear zone;
d_e = diameter of sample;
H = head between a radius and the drain, as percentage to total head lost;
k_h = coefficient of horizontal permeability
 $\bar{k}_h$ = averaged over radius r_e;
 k_{ho} = in the undisturbed zone;'
 k_s = in smear zone of 2-zone model;
k_v = coefficient of vertical permeability;
m_v = coefficient of compressibility (vertical deformation);
r = radius:
 r_e = of the sample;
 r_s = of smear zone;
 r_w = of drain well (mandrel) ;
S = normalized radius of smear zone, = r_s/r_w;
T = time factor of consolidation, = $C_h t/d_e^2$;
t = time of consolidation ;
U = average consolidation ratio;
γ_w = unit weight of water;
η = ratio of the coefficient of permeability in the undisturbed zone to the coefficient of permeability in the smear zone.

REFERENCES

Abashi, H. and Monden, H. (1963), "Determination of the Horizontal Coefficient of Consolidation of An Alluvial Clay," Proc. 4th Australia New Zealand Conf. SMFE, pp. 159-164.

Abashi, H and Inoue, T (1986), "Prediction of Consolidation of Clay Layers Especially in the Case of Soil Stabilization by Vertical Drains," Proc. IEM-JSSMFE Joint Symposium on Geotechnical Problems , Kuala Lumpur, pp.31-40.

Asaoka, A. (1978), "Observational Procedure of Settlement Prediction, "Soils and Foundations, p. 18, No. 4.

Baligh, M. M. (1986), "Undrained Deep Penetration, I: Shear Stresses", Geotechnique, Vol. 36, No. 4, pp. 471-485.

Barron, R. A. (1948), "Consolidation of Fine Grained Soils by Drain Wells," Trans. ASCE, Vol. 113. No. 2346, pp. 718-742.

Germaine, J.T. (1982), "Development of The Directional Shear Cell for Measuring Cross Anisotropic Clay Properties," Sc.D. Thesis, Department of Civil Engineering, MIT., Cambridge, MA.

Hansbo, S. (1981), "Consolidation of Fine-Grained Soils by Prefabricated Drains," Proc. of the 10th ICSMFE, Vol. 3, pp. 677-682.

Ladd, C. C. (1990), Personal communication.

Onoue, A. (1988), "Consolidation by Vertical Drains Taking Well Resistance and Smear into Consideration", Soils and Foundations, Vol. 28, No.3, pp. 165-174.

Seah, T. H. (1990), "Anisotropy of Resedimented Boston Blue Clay", Sc.D. Thesis, Massachusetts Institute of Technology, Cambridge, Mass.

Ting, N. H. (1990), "Effects of Disturbance on Soil Consolidation with Vertical Drains", S.M.Thesis, Massachusetts Institute of Technology, Cambridge,

INFORMATION SYSTEMS FOR ENGINEERING ORGANIZATIONS

C. D. Lamprecht[1]

ABSTRACT: The Bureau of Reclamation, Denver Office, Management Information System (MIS) is presented. The concepts and philosophies used to design and implemented the MIS are given and discussed. These include budget formulation, resource and workload scheduling, design resource controls, automated time and attendance, and design and management reports.

INTRODUCTION

The Bureau of Reclamation, within the Department of the Interior, is one of the largest water resource planning and engineering organizations in the Federal Government. Its geographic territory is the 17 Western States west of the Mississippi River. In order to better manage its large planning and design programs, Reclamation designed and implemented MIS. The following model presents the MIS used by Reclamation engineers and managers. It provides Reclamation managers and engineers with state-of-the-art tools that have increased productivity and proficiency in managing and controlling engineering design costs and scheduling.

Reclamation's MIS was custom designed for a large government engineering organization, however, the design concepts and principles easily adapt to private industry.

PHILOSOPHIES

Policy. MISs are not adaptable to all engineering organizations unless the existing engineering data systems are or can be automated to be accessed. Without this vital linkage to data, the MIS will fail for lack of current and accurate data and high overhead costs of maintaining individual data sources as opposed to one comprehensive system. Therefore, it is important to understand the commitment that must be made by management to provide the resources within the organization to initiate policy, methods, and procedures for designing, implementing and maintaining an MIS. Management must realize that an MIS is intrinsically valueless. The payoff comes in 'value added' associated with its design products during the design process and construction. Do not expect your organization's culture to change

[1] – Manager, Management Systems Office, Department of the Interior, Bureau of Reclamation, Assistant Commissioner–Engineering and Research, Mail Code D-3220, Denver Federal Center, PO Box 25007, Denver CO 80225

overnight, because earned creditability and trust in an MIS will be slow in the beginning; however, once established it will be of inestimable value.

Planning. The computer age has impacted all businesses and commerce. Improved efficiencies are a never ending endeavor. MIS has proved to be an effective tool to limit staff days and associated project costs by presenting the hard numbers (bottom line) that managers and supervisors require for analysis. To attain these means, success will depend on how well the MIS planning process is accomplished.

In business, profit and loss is the measurement of success. Governments and other institutional organizations measure profit by the success or failure of a given budget or program. Regardless of the measurement philosophy, the bottom line is to satisfy the customer's needs in terms of budget, schedule, and product quality. Reclamation recognized the fact that it could improve its engineering products by providing managers and engineers with better design and scheduling data by effectuating an MIS.

Control. Management must initiate the controls for managing the system's operation and for resolving the data input/output used within the MIS. Often, these will conflict with one another and may result in loss of credibility and trust, of the system data, by the users. A variety of controls for managing the MIS has been experienced.

An administrative control was implemented through the adoption of policy memorandums on the access and use of the MIS. These memorandums clearly state access and cost charge validations down to individual employees working on projects. The Management Systems Office (in Denver) acts as a control center for cost controls using an automated daily time and attendance system (ATA).

Another management control is limiting access by specific users with password and user validation. All MIS users have a unique identification number and password which sets their users access privileges to the data in the database. The ATA system has a validation control that the engineering design team managers establish; it limits user labor charges to specific projects.

DESIGN CONCEPT

Reclamation's MIS was custom designed to prepare, maintain, measure, and track the engineering staff days (budget) for the organization. The system serves two elements of the organization; i.e., administration and design engineers. Data used by the MIS is managed by a data collection system and data reporting system.

Architecture. Simplicity to the user in design and maintaining the relational data is the formula to success for Reclamation's MIS. It was achieved by use of relational design concepts now offered by state-of-the-art software and hardware platforms.

The Reclamation MIS consist of five relational modules as shown in figure 1. These modules are: (1) Program Information Tracking System (PITS), (2) ATA, (3) Resource, (4) Schedule, and (5) Total Quality Management (TQM). Each of the these modules and their relationships are explained in the following paragraphs.

State-of-the-art Relational Database Management Systems (RDBMS) are available that provide a high level of sophistication of data management power and flexibility for both administrative and engineering applications not found in traditional third generation languages (3GLs) such as Cobol and Fortran nor in other types of databases. Fourth generation language (4GL) tools have been developed which provide the ability to model and implement major applications much faster than with 3GL. As a result of power and flexibility, relational databases

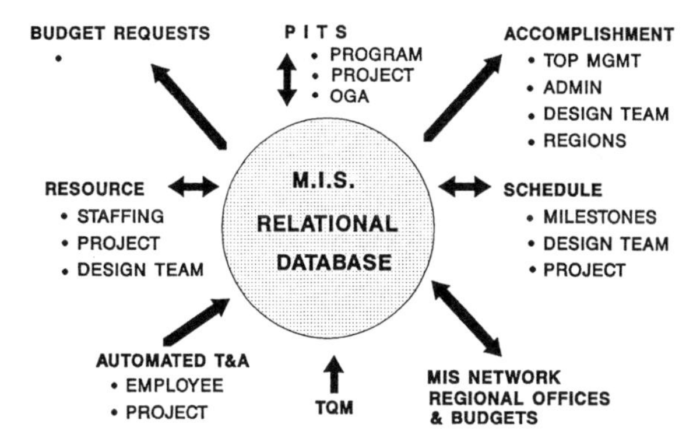

Figure 1. The MIS

have become an integral part of the design process at Reclamation—primarily in design scheduling, workforce resource allocation, time and attendance reporting, and program tracking. Reclamation developed its MIS application using Ingres, a state-of-the-art RDBMS, by first using VMS and now on UNIX operating system platforms. The database is easily queried and by using User's Data Management Systems (UDMS) report generator.

The budget formulation process is the foundation of the data collection system. It begins with the dynamic budget window concept. Figure 2 illustrates this concept.

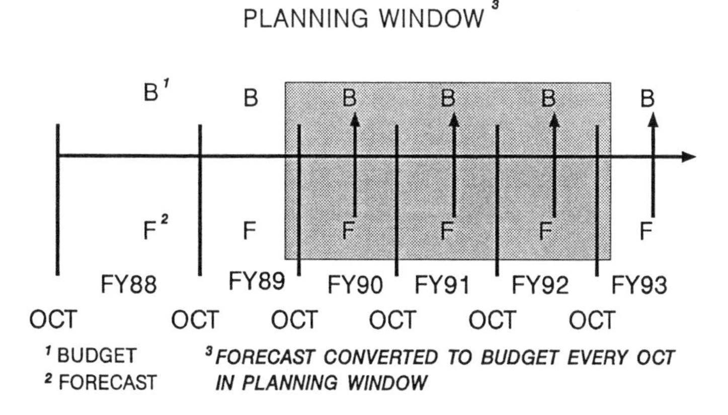

Figure 2. Budget window

The horizontal line is a time line through FY (fiscal years) where F (forecast) estimates are entered to accomplish a particular design task. As time moves along this horizontal axis, forecast estimates are revised as necessary. When the FY resource requirements are requested by the fiscal managers the forecast data shown in the budget window, shown in the shaded area, are redefined to B (budget) status and the FY forecast estimates ahead of the budget window continue the dynamic cycle. This concept has eliminated the annual panic budget preparation by providing timely and accurate resource needs on a real time basis.

This concept can be further defined in detail by the data cycle shown in figure 3. This cycle takes place everyday beginning with the individual cost center entering yesterday's employee time. These same data provide design engineers and managers with 1-day-old data on resource utilization and accomplishment on their projects as well as preparing a data file used to generate the employee payroll for the organization.

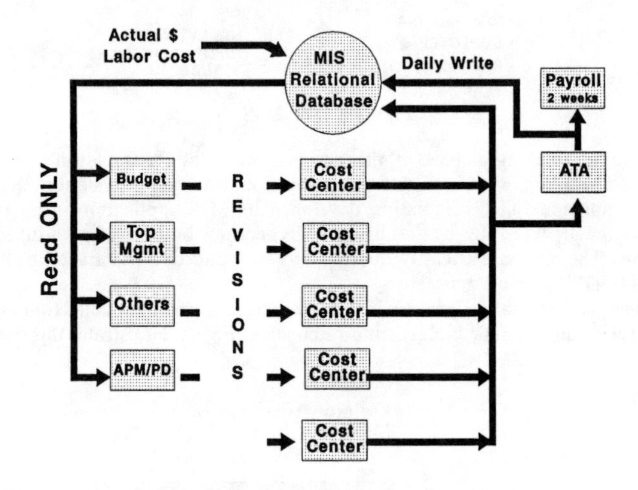

Figure 3. Data Cycle

Once the work request is entered into the database by the design manager, the involved design organizations are identified and validated, and cost centers make resource estimates to accomplish their portion of the design work. When all the involved offices agree to a project plan which includes these resource, cost estimates and a design scheduled work begins. As the work progresses, the MIS tracks actual staff days, labor and nonlabor costs, and the design schedule by using the modules described in the following paragraphs.

(1) PITS–This module contains three types of data: (a) budget request, (b) forecasted estimates, and (c) actual charges. It is used by managers to develop budget requests and workload estimates for fiscal year planning. It is used by design engineers to estimate and track required resources for and during project design. Figure 4 shows these relationships.

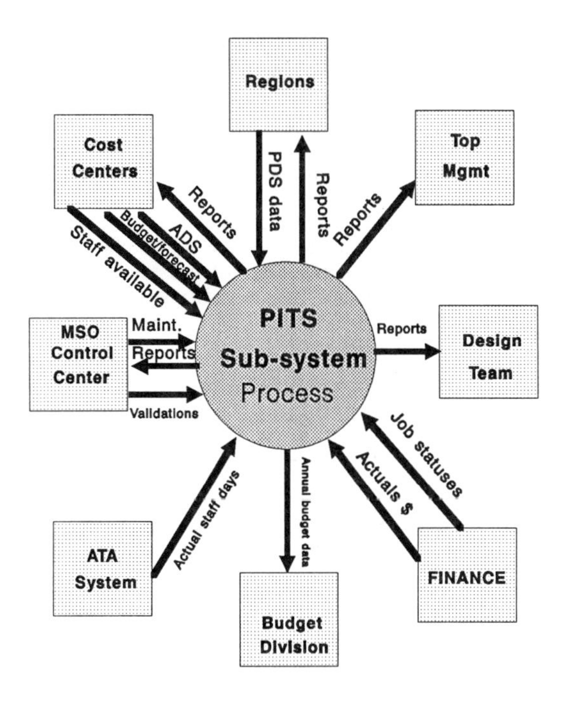

Figure 4. PITS Module

Simplicity was the major consideration in the MIS applications design meaning user friendly by menu driven operations. Reclamation's design program has two separate management environments; i.e., administrative activities and design formulation. Figure 5 shows these relationships.

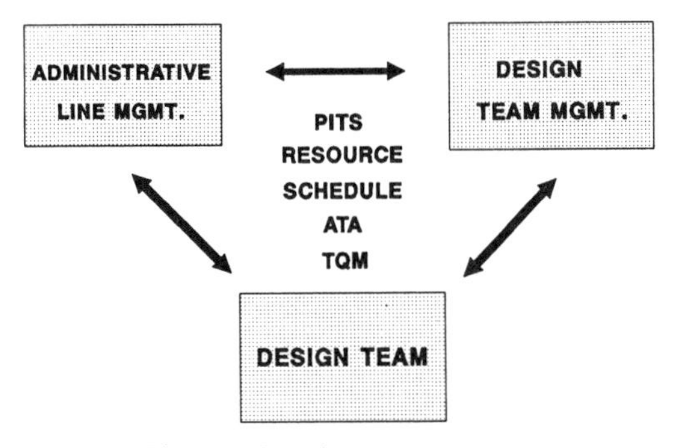

Figure 5. Matrix Management

The module requires only three data sets from the organization initially making an estimate of work. These are (a) by month, (b) estimate of average daily salary (ADS) per staff day, and (c) the number of staff days estimated by month to accomplish the design task. The average staff day cost can be adjusted depending on the type of work and the discipline that is assigned to the estimated staffdays. Figure 6 shows a typical organization workscreen for cost center (code) 3220 in the PITS module. It also shows the six types of work included in the workload and their calculated staff day cost (SDC) associated with each work type. The work screen provides the user with Reclamation's formula for calculating direct labor cost (DLC) as well as all SDS which includes additives and benefits. Each design organization maintains its own data sets for the required elements.

CODE: 3220

DLC = ADS x (1.0 + Leave + Benefits + ODE)
SDC = DLC x (1.0 + ICR)

Type	Leave	Benefits	ODE	ICR	ADS	DLC	SDC
APP	0.213	0.156	0.000	0.920	150.00	205.35	394.27
FOR	0.213	0.156	0.000	0.920	150.00	205.35	394.27
GAE	0.213	0.156	0.000	0.920	125.00	171.13	328.56
NPJ	0.213	0.156	0.000	0.000	125.00	171.13	171.13
ODE	0.000	0.000	0.000	0.000	0.000	0.000	0.000
OGA	0.213	0.156	0.000	0.920	150.00	205.35	394.27
PRJ	0.213	0.156	0.000	0.920	150.00	205.35	394.27

Figure 6. Organization Workscreen

(2) ATA–The ATA module is used to report daily time and attendance for approximately 1,100 employees. The data entered into this module is used to develop organization payroll and to track design accomplishment. This module is capable of providing managers and design supervisors with yesterday's data on staff day usage and labor costs. Figure 3 shows the daily data cycle of ATA data and its relationship to workload tracking.

(3) Resource–This module is used by the line manager and supervisor to assess the labor force required to accomplish the estimated workload. It allows manager and supervisor to schedule employees time to project design teams and to forecast optimum staff in the future. It can provide data by project and all resources charging to it, or by employee(s) and all their resultant charged time.

(4) Schedule–This module is used to maintain and track project design information and the schedule identified in the project design plan. It maintains the target flag dates, revisions to target flag dates and the met dates for each flag date as well as accomplishment data. Three categories of data are shown on a typical report. The top section provides the administrative text associated with the project; the second section provides the scheduled milestones and flag dates, revised dates, and met date; and the bottom section provides the actual, budget, and forecast data and notes.

(5) TQM–This module is used by the organization to manage employee concepts and team ideas, to aid management's insight and projections, and to improve the

organization's design products. It tracks ideas through TQM assessment process from entry through adoption or rejection.

SUMMARY

MIS is intrinsically valueless within an organization. Its value must be recovered in the 'value added' to the quality of the organizations design products in cost savings during the design process and construction of the project. Reclamation has proven this philosophy to be effective by lowering its design and construction costs. Reclamation management has committed to improving design productivity and performance and will continue to use and enhance its MIS toward accomplishing these goals.

GLOSSARY

APP Appropriated funding–All jobs that have an appropriated fund source
ADS Average Daily Salary (base salary)
APM ACER Project Manager
ACER Assistant Commissioner–Engineering and Research
CODE Same as Cost Center
DLC Direct Labor Cost
FOR Foreign–These jobs being per for foreign governments
GAE General Administrative Expense — These are overhead jobs which have a limited allocation for each year
LEV Leave related jobs — All jobs that are related to leave (vacation, sick, etc.)
MIS Management Information System
MSO Management Systems Office
NPJ Nonproject
ODE Organization Distributive Expense — These jobs are charged to supervisory type personnel and recovered by an additive on all labor costs.
OGA Other Government Agencies — This is work performed for other Government agencies such as Environmental Protection Agency, Bureau of Land Management, and National Park Service
PITS Program Information System
PRJ Project work — These jobs are performed on cost reimbursement basis, generally for regional offices.
PD Principal Designer (design team manager)
PDS Program Data Sheet (request for work)
RDBMS Relational Data Base Management System
SDC Staff Day Cost
WCF Working Capital Fund — The cost associated with these jobs is recovered in overhead additives

FINANCIAL MANAGEMENT IN THE ENGINEERING FIRM

Charles A. Parkhill, C.P.A.[1]

ABSTRACT: This paper discusses the concepts and procedures involved in the financial management of an engineering practice. First, the importance of financial management within an engineering firm is discussed. Definitions of the various aspects of financial management are provided and suggestions are made regarding the appropriate data to be used for analysis. Following this discussion the paper will briefly outline the interface of computerized systems with this management responsibility.

INTRODUCTION

Financial Management is both a necessary and critical process for any business whether it is a geotechnical engineering firm or an auto manufacturer or real estate brokerage. While entrepreneurs and owners may be well versed in the technology and specific work tasks involved in providing a given service, all of this knowledge is useless unless the firm is able to sustain financially. Without proper cash flow and profit management, the finest engineers cannot continue to practice. While it is commendable that many firm's owners put quality and client satisfaction at the top of their priority list, it is condemnable if financial management is not given equal weight.

PROJECT AND FIRM FINANCIAL MANAGEMENT

To begin, it is important to point out that there are two types of financial management which need to take place in an engineering practice, project financial management and firm financial management, hereinafter referred to as project management and firm management since we are only dealing with the financial aspects of management in this paper.

Project management is the hour to hour, day to day monitoring

1 - C.P.A., President, Parkhill & Company, 2900 W. Maple, Suite 107, Troy, MI. 48084

of the specific projects within a firm. This topic holds the key to a firm's financial survival. For it is the sum of all of the individual projects' profit or loss that equals the firm's overall profit or loss. The task of managing individual projects should be performed by project managers and reviewed by principals.

Firm management, the job of the firm's principals , is the review, understanding and interpretation of the sum of the projects' profits or losses. In combined form, these numerical results make up the financial statements of a firm.

PROJECT MANAGEMENT

For project management data to be useful it must be 1) accurate and 2) timely (within three business days of the close of the timesheet period, in this author's opinion).

Data Compilation. The financial data that must be analyzed by project managers is the result of an organized process involving the following key elements.

TIMESHEET PREPARATION The timesheet is the starting point for all financial data collection within any engineering firm. The importance of the accuracy and timeliness of this document cannot be overstated. Errors here will cause invalid final reports and require time consuming adjustment to correct. Late timesheets cannot be tolerated. A single late timesheet may hold up the reporting process for the entire firm. Staff members must be educated by the firm and coached by the project manager in the proper techniques.

TIMESHEET EDIT STAGE This procedure, performed by the project manager is an attempt to find errors in timesheet entry prior to the final posting of project cost reports. At a minimum this involves project managers reviewing timesheets prior to posting. At its best this involves the sorting of transactions by project, for a more encompassing review by the project manager.

TIMESHEET POSTING The process of recording each transaction from the timesheet to the proper project cost report. This posting must include the staff members name, number of hours spent (categorized by the service provided), the date of the transaction, the hourly cost rate for the individual, the hourly billing rate for the individual (if invoicing is to be part of this system), and the extensions of cost and billing totals.

EXPENSE POSTING Project expenses such as printing, travel and consultants must also be recorded in the project cost report. The information is obtained from outside vendor invoices or an employee expense report. The posted information should include the description of the expenses, the vendor or employee name, the date of the

transaction, the cost of the item and (for invoicing purposes) the status (reimbursable or not) and the amount to be invoiced to the client.

Project Reports and Preparation. While there are literally dozens of project reports that can be generated, there are three key reports that this paper will address. Each firm, publication or system you encounter may use different terminology, but the information contained in each report is the important issue.

PROJECT DETAIL REPORT (FIGURE A)
This report contains the greatest level of detail regarding a project's financial status. It contains entry level detail for all labor and expense transactions. This report is actually the one created through the posting routines enumerated in the section Data Compilation.

PROJECT PROGRESS REPORT (FIGURE B)
The most important project report. It summarizes the detailed transactions by levels representing the work performed on the project (commonly referred to as phases or tasks) and compares the actual hours and costs to budgeted hours and costs. Further it totals the labor costs, adds firm overhead plus project expenses to arrive at a total cost (or spent) for the project; to compare to income, to yield a profit or loss for that specific project. In addition, this report will compare the percent complete (by phase or task) reported by the project manager to the percentage actually spent (of budget). The first section relates to labor. Totals by phase and task are accumulated from the project detail report and posted to this report. Following the total labor cost, the firm's overhead expenses are allocated to this project. The rate is more fully discussed later but essentially it is the rate used to allocate indirect expenses such as health insurance and rent to the project cost. An overhead rate of 1.5 means that for every $1.00 of direct, project labor, this firm spends $1.50 for overhead items. After allocating over head, project expenses are added to this report. A total cost for the project can then be calculated. This cost is compared to income to yield a profit or loss for this project. Further enhancements to this report are made by adding budgeted hours and costs (by phase or task) to the report and finally by recording the percentage spent of budget compared to the estimated percent complete provided by the project manager.

OFFICE EARNINGS REPORT (FIGURE C) A single line summary by project. This report contains the least level of detail, showing key financial information for each project.

Compiled from information contained in the project progress report.

Report Analysis. This is of course the substance of project management. Even if all of the above steps are followed to produce accurate and timely reports, they are useless unless used to properly analyze the financial status of the project(s). Project managers and principals must be instructed in the proper use of the basic reports; how to read them, where to look for problems and what steps to take upon recognizing a problem. We must remember that the readers of these reports are typically not trained in the financial area and thus this education must be ongoing. It is also important to caution against over analyzing. With experience, the project manager and principal will learn which areas need detailed attention and which require only a cursory review. It is the author's opinion that the review of project financial status should occur no less frequently than twice a month. In larger firms, once per month may suffice for the principal's review of the office earnings report but in all cases at least twice per month for the project progress report. Following are some detailed analysis points by project report.

PROJECT DETAIL REPORT This report should be analyzed only if this level of detail is required following the analysis of the project progress report. If a problem or inconsistency is uncovered during that review it may be necessary to review the detail to pinpoint the problem.

PROJECT PROGRESS REPORT This report deserves the closest scrutiny, on a regular basis. Profit or loss may be the first financial indicator to look at but, it is not enough to stop there. Beyond overall project profitability, each phase of a project needs to be examined for compliance with budget. A project may have an overall profit but the current phase of work is in trouble but a profitable previous phase is covering the current problem. In this case examination of the overall profit, with a conclusion of financial health would have precluded the reader from being alerted to a current problem which could be addressed. Each phase should be examined in relation to budget along with review of the percent completes. If the report says that you have spent 72% of a phase's budget and your project manager has reported a 50% completion of the phase, you may have a problem. This is where examination project detail report may be required. If this kind of analysis lead to the conclusion that this discrepancy in percent complete is due to performing work outside of the scope of your services and should thusly be invoiced to your client as an additional service, herein lies the value of project management. However most cases are not so advantageously resolved. More often than not, a budget overrun results from internal inefficiencies that must be dealt with internally. This type of conclusion could recommend several strategies including tighter con

trol for the remainder of the project, using lower cost personnel on the project, modifying future tasks or simply recognizing that there are fewer hours left to complete the project than originally anticipated if the project is to produce the projected profit.

OFFICE EARNINGS REPORT This is an overview of all projects. Profit/loss,percent complete and unbilled services are the key items for review here.

The project progress report should be reviewed by the project manager, with reference to the project detail report where necessary. The office earnings report should be reviewed by principals, with reference to the project manager where necessary.

Managing the financial status is the key to the financial health of a firm. It is only the sum of all project's profit and loss that make up the financial condition of the firm as a whole.

FIRM MANAGEMENT

Financial Reports. While CPA's may be confident that they can list the specific documents that should be included in a financial statement package for all businesses, an engineering firm should be concerned about having the documents in the package that allow the principals to understand and evaluate the financial performance of the firm. A firm with a financially savvy principal may require one single report, while a typical engineering firm may have 6-10 documents making up the package. Following is a list of the more important parts of the financial statement package for an engineering firm.

RATIO REPORT (FIGURE D) A summary of key firmwide financial ratios. At a minimum these should include:

Overhead Rate - total indirect expenses divided by direct labor.

Net Multiplier - Net fees (total fees less reimbursables and consultants) divided by direct labor.

Productivity - Direct labor divided by total labor.

In addition to these basic ratios a firm should include any others that are relevant to the firm's principals.

INCOME STATEMENT (FIGURE E)

A line item statement of income and expenses culminating in profit or loss for the firm. At a minimum an income statement for an engineering firm should include sections for income, direct expenses, indirect expenses and profit/loss.

BALANCE SHEET
A statement of a firm's assets, liabilities and equity (What you own, what you owe and what's left - what you're worth).

SUBSIDIARY DETAIL
Detailed back up of line items on the income statement and balance sheet, if necessary.

CASH FLOW SUMMARY
A summary on a daily and monthly basis of cash receipts, cash disbursements and cash balance.

ACCOUNTS RECEIVABLE ANALYSIS
Detail of your accounts receivable, including client and project identification, typically aged from your invoice date.

ACCOUNTS PAYABLE ANALYSIS
Detail of your accounts payable, including vendor and due date.

Report Preparation. The proper thought process of the personnel preparing the financial statement package should be: keep the information clear and concise and show overall summarizations vs. detail documentation. The theoretical goal would be to put all of the critical data on a single 8 1/2" x 11" sheet of paper. Principals have neither the time nor the desire to examine a 1" thick printout with details down to the transaction level. Graphics should be used if the principals relate to that medium better than to rows and columns of numbers. All of the aforementioned reports are derived from source documents such as invoices and cash receipts or are accumulations from the detailed general ledger of the firm.

Report Analysis. The key documents for overall firm financial analysis are the ratio report and the income statement. The other reports should be reviewed for other management purposes or for substantiation to the two reports identified. Once again, education is critical. Most principals are not trained financial people.

Once understood, the ratio report can hold the key to analyzing the firm's financial condition. The overhead rate is the single most important ratio. Knowing your firm's overhead rate means you know what your breakeven point is and allows you intelligent decision making when negotiating fees. If for example your firm's overhead rate is 1.5 then your breakeven point is 2.5 times direct labor cost (1.00 for direct labor + 1.50 for overhead expenses). Therefore, when setting fees you know you must negotiate a fee of 2.5 times the projected labor cost of doing the work plus a profit, meaning you may set your fees at 3.0 times labor cost. This leads into the second ratio, net multiplier. This represents the multiplier you actually achieve on your labor. While you may target a 3.0 multiple, you may run over budget on fixed fee projects bringing your effective multiplier down to 2.6. With a net multiplier of 2.6 and an overhead rate of 1.5 (2.5 breakeven) your profit margin is only 4% (.1 divided by 2.6). It is the margin between the net multiplier and the breakeven point (overhead + 1.00) that is critical. The third component is productivity. As productivity goes up, the overhead rate comes down. In times of economic slowdown for a firm, productivity will drop, overhead will rise, thus leaving a smaller profit margin. Productivity is a key indicator of workload and staffing requirements. Properly used these three ratios and their relationship to each other lend a great deal of insight into a firm's financial condition. The income statement, from which the key ratios are derived, offers a detailed look at where dollars are earned and spent. One can analyze the relationship of items to total income, expressed as percentages, or to a preconceived financial budget or to prior years income statements or ideally all of the above.

COMPUTERIZED SYSTEMS

While all of the reports described above can be prepared manually, computerized systems make the process much more efficient and can aid in the timeliness factor. The proper computerized project management/financial software, tailored specifically for the (geotechnical) engineering firm should provide reports ranging from full detail (source transactions) to overall firm financial analysis (ratio report). The key to a good computerized system is integration. That means that by entering "source documents" one time, the data from these transactions is sent to several locations within the system to help perform multiple tasks. For example, a timesheet is a source document. The entering of a timesheet's data into a good computerized system will generate:

1 - A Paycheck

2 - A Record of How the Employees Time was Spent

3 - A Posting of Hours and Cost to a Project Management Report

4 - An Invoice to the Client (if hourly billing is used)

5 - The Distribution of the Employees Salary between Direct and Indirect Labor in the Companies Financial Records

Item #5 is an example of how source documents affect the financial statements of a firm. Therefore, a good system will use transactions from project management, accounts receivable, accounts payable and payroll to build the financial data necessary to create the financial statements. The system should then be capable of producing all of the required financial reports, from detailed general ledgers, for accountants use, to the ratio report, a key summary of the firm's financial positions.

CONCLUSION

Financial management is vital to any engineering firm. Properly practiced it encompasses both project and firm financial management, each being unique but inexorably tied together. The procedures and resulting reports, whether prepared manually or with the assistance of project management/financial software packages, are simply tools. It is the responsibility of the principals of the firm to use the tools to properly direct the firm.

PROJECT NAME: Raskel Residence
PROJECT CODE: 84001
PROJECT PRINCIPAL: Sam Polunsky
PROJECT MANAGER: Jack Jones
PROJECT START DATE: 03/01/84
LAST TIMESHEET: 10/19/90

Alexander and Allison
PROJECT MANAGEMENT SYSTEM
PROJECT PROGRESS REPORT
FOR THE PERIOD 12/01/90 - 12/31/90

FIGURE B

	SPENT-THIS-PER		SPENT-TO-DATE		---BUDGET----		-%COMPL-		---RATES---		------- BALANCE -------		
	HOURS	DOLLARS	HOURS	DOLLARS	HOURS	DOLLARS	EXP	RPT	BUDGT	EFFEC	DOLLARS	HOURS @BUDG	HOURS @EFFEC
Schematic design													
Project Administration	0	0	30	215						7.17			
Design	0	0	52	1042						20.03			
Engineering	0	0	30	967						32.25			
TOTAL	0	0	112	2224	423	3874	57	50	9.16	19.86	1649	180	83
LABOR TOTAL	285	2887	414	5368	2822	25794	21	63	9.14	12.97	20426	2235	1575
DPE @ 0.0000		0		0		0					0		
OVERHEAD @ 1.5000		4331		8052		38691					30639		
LABOR, DPE & OVERHEAD	285	7218	414	13421	2822	64485	21	63			51064		
REIMBURSABLE EXPENSES													
Consultants-Civil		0		500									
OTHER DIRECT EXPENSE TOTAL		0		1267									
PROJECT TOTALS		7218		14687		67785	22	65			53098		

------------------------------FINANCIAL ANALYSIS ------------------------------

	TOTAL COMP	BILLED	A/R	EARNED INCOME	SPENT	PROFIT (LOSS)	PERCENT PROFIT
JOB-TO-DATE	70000	16618	7663	23480	14687	8792	37.4

FIGURE C

Alexander and Allison
PROJECT MANAGEMENT SYSTEM
OFFICE EARNINGS REPORT - PROJECT TO DATE
FOR THE PERIOD 12/01/90 - 12/31/90

PROJECT CODE	PROJECT NAME		TOTAL COMP	PCT COMP	EARNED INCOME	BILLED	UNBILLED SERVICES	RECEIVED	A/R	SPENT	PROFIT (LOSS)	OVERHEAD MULT	NET MULT
83021	Euphoria Towers	A	1200000	6	43968	42067	1901	0	42377	34949	9019	1.5000	3.310
84001	Raskel Residence	P	70000	65	23480	16618	6862	9055	7663	14687	8792	1.5000	4.135
85001	NBT branch Office	P	0	0	5625	0	5625	0	0	8438	-2813	1.5000	1.667
87001	Clark County High School	P	90000	32	36202	23310	12892	0	23310	45858	-9656	1.5000	2.264
88000	Kroger Store #66	P	120000	93	112015	82800	29215	61000	21800	107534	4481	1.5000	2.645
	FINAL TOTALS	P	1480000	18	221290	164795	56494	70055	95150	211466	9824	1.5000	2.769

(Columns TOTAL COMP through NET MULT are headed PROJECT TO DATE.)

FIGURE A

Alexander and Allison
PROJECT MANAGEMENT SYSTEM
PROJECT DETAIL REPORT
FOR THE PERIOD 12/01/90 - 12/31/90
Status conditions printed: ALL

PROJECT NAME: Raskel Residence
PROJECT CODE: 84001
PROJECT PRINCIPAL: Sam Polunsky
PROJECT MANAGER: Jack Jones
PROJECT START DATE: 03/01/84

EMPLOYEE CODE	EMPLOYEE NAME	DESCRIPTION	STT	DATE	HOURS TYPE	HOURS	LABOR RATE	LABOR AMOUNT	DC	BILLING RATE	BILLING AMOUNT	INVOICE NUMBER
	S - Schematic design											
	2 - Project Administration											
AB	Bill Alexander		B	06/19/89	REG	10.000	10.00	100.00	0	24.80	248.00	856
FJ	Jeanine Fromholz		S	01/13/84	REG	20.000	5.75	115.00	1	30.00	600.00	856
	TOTAL Project Administration					30.000	7.17	215.00		28.27	848.00	

XYZ ENGINEERS — FIGURE E
INCOME STATEMENT

	THIS MONTH	RATIO	YEAR TO DATE	RATIO
INCOME				
PROF FEES	1,281,000	86.8	3,123,600	87.2
REIMB INCOME	183,000	12.4	441,000	12.3
OTHER INCOME	12,000	0.8	19,100	0.5
TOTAL INCOME	1,476,000	100.0	3,583,700	100.0
DIR EXPENSES				
DIR SALARIES	294,000	19.9	823,000	23.0
CONSULTANTS	208,000	14.1	490,000	13.7
OTHER DIR EXP	8,500	0.6	27,500	0.8
REIMB EXPENSES	140,500	9.5	484,000	13.5
TOT DIR EXPENS	651,000	44.1	1,824,500	50.9
GROSS PROFIT	825,000	55.9	1,759,200	49.1
INDIR EXPENSES				
INDIR SALARIES	57,000	3.9	225,000	6.3
PAYROLL BNFTS	85,000	5.8	263,000	7.3
OFFICE EXPENSE	100,500	6.8	299,000	8.3
LEGAL & FINC EXP	68,000	4.6	202,000	5.6
AUTO & TRVL EXP	39,500	2.7	96,500	2.7
DEPREC & AMORT	31,500	2.1	131,000	3.7
MRKTNG/PR/BUS DV	48,000	3.3	117,000	3.3
TOT INDIR EXPENS	429,500	29.1	1,333,500	37.2
INCOME BFR TAXES	395,500	26.8	425,700	11.9

RATIO REPORT — FIGURE D

	THIS MONTH	YEAR TO DATE
OVERHEAD RATE	1.46	1.62
NET MULTIPLIER	3.65	3.20
CONSULTANTS (% OF	16.2%	15.7%
DIRECT LABOR (% O	23.0%	26.3%
PRODUCTIVITY:		
DOLLARS	83.8%	78.5%
HOURS	78.0%	73.0%

PERFORMANCE OF A WELDED WIRE WALL WITH POOR QUALITY BACKFILLS ON SOFT CLAY

D. T. Bergado[1], A.M.ASCE, C. L. Sampaco[2], R. Shivashankar[3], M. C. Alfaro[2], L.R. Anderson[4], M.ASCE and A.S. Balasubramaniam[5], F.ASCE

ABSTRACT: An experimental and a full-scale welded-wire wall of 5.7 m height was built on soft Bangkok clay utilizing three different poor-quality, cohesive-frictional backfill soils in each of the 3 sections. The subsoil consists of about 6 m thick soft clay overlain by an uppermost 2 m thick weathered crust and underlain by a stiff clay layer. An extensive instrumentation program was employed to monitor the behavior of the subsoil, the lateral movements of the face of the wall, and the strains in the reinforcements. It was observed that excessive settlements and lateral movements of the soft clay subsoil influenced very much the variations in the vertical stresses at the base of the wall and the tensile stresses of the reinforcements wherein the wire mesh facing also played an important role in the arching effects. Pullout resistances of dummy mats from field pullout tests were also found to be affected by the deformation of the weak subsoil conditions. An overall assessment of the wall behavior suggests a significant deviation from mechanically stabilized earth walls resting on a comparatively good foundation subsoils. Laboratory pullout tests on the three backfills with normal pressures up to 130 kN/m^2 proved that even with such poor-quality backfill materials, the pullout resistances increased with increasing confining normal stresses as observed for good quality granular backfill materials, but were much affected by the moisture content and degree of compaction of the soil. It can be concluded that the welded-wire wall system can be effectively used to reinforce poor-quality and marginal-quality backfill materials on soft foundations.

1 - Assoc. Prof., Geotech. and Transp. Engineering Div., Asian Inst. of Technology, G.P.O. Box 2754, Bangkok 10501, Thailand.
2 - Research Associates, Geotech. and Transp. Engineering Div., Asian Inst. of Technology, G.P.O. Box 2754, Bangkok 10501, Thailand.
3 - Doctoral Candidate, Geotech. and Transp. Engineering Div., Asian Inst. of Technology, G.P.O. Box 2754, Bangkok 10501, Thailand.
4 - Prof., Dept. of Civil and Environmental Engineering, Utah State Univ., Logan, Utah 84322-4110, U.S.A.
5 - Prof., Geotech. and Transp. Engineering Div., Asian Institute of Technology, G.P.O. Box 2754, Bangkok 10501, Thailand.

INTRODUCTION

In the Southeast Asian region, most of the coastal plains are characterized by a recent deposit of marine sediments consisting of a topmost layer of compressible clay underlain by alternating layers of stiff clay and dense sand with gravel. Typical of these features is the flat, deltaic-marine deposit of the Chao Phraya Plain in Central Thailand, wherein Bangkok metropolis is located, which covers a width of about 200 km and a north-south dimension of about 300 km. The typical near surface subsoil profile at the campus of the Asian Institute of Technology, located about 45 km north of Bangkok is shown in Fig. 1 with the corresponding geotechnical properties. The presence of a thick layer of compressible clay can pose considerable problems to infrastructure constructions within the coastal plain. Because of the weak subsoil conditions, earth structures such as road embankments are subject to height restrictions of about 3.4 m with a gentle slope of 3H to 1V to avoid stability failures. But even with this height, highway embankments can undergo excessive settlements of about 2 m in about 10 years time, sinking below their maximum flood level, and requiring costly reconstruction and maintenance works (3). On the other hand, rapid expansion and high cost of land often compels for steep and high embankments without the normally wide and flat slopes. To alleviate such problems, mechanical stabilization by earth reinforcement using either steel or polymer grids can be a viable alternative. However, high quality granular backfill materials suitable for such structures are not readily available and thus, are expensive due to high transportation cost. Hence, the use of locally-available backfill materials is indeed practical and imperative for economic reasons. This paper is a partial result of the on-going research project at the Asian Institute of Technology, concerning the

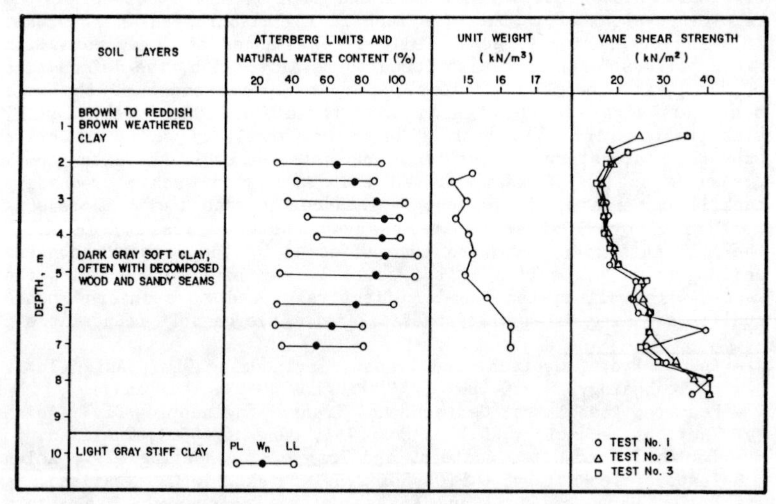

Figure 1. Typical Subsoil Profile and Geotechnical Properties

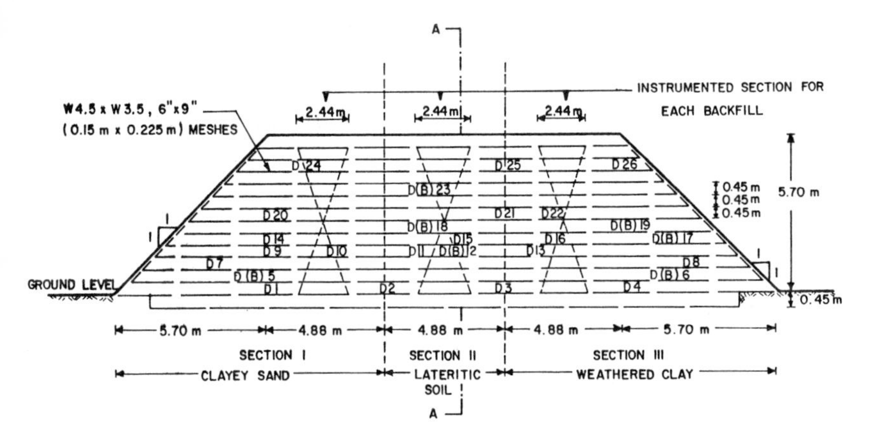

Figure 2. Front View of the Welded Wire Wall

investigation of the potential use of widely available, but poorer in quality, backfill materials in walls and embankments stabilized mechanically by welded wire mesh reinforcements. The research project involves the construction of a full-scale welded wire wall on compressible foundation and the study of soil-reinforcement interaction by laboratory and field pullout tests. Results of laboratory pullout tests have been published earlier (2,4,5). Detailed description and analysis of the results will be published later.

THE WELDED WIRE WALL

The welded wire wall is divided into 3 sections along its length consisting of about 14.64 m long at the top, and composed of 3 different backfill materials (Fig. 2). It has a vertical welded-wire

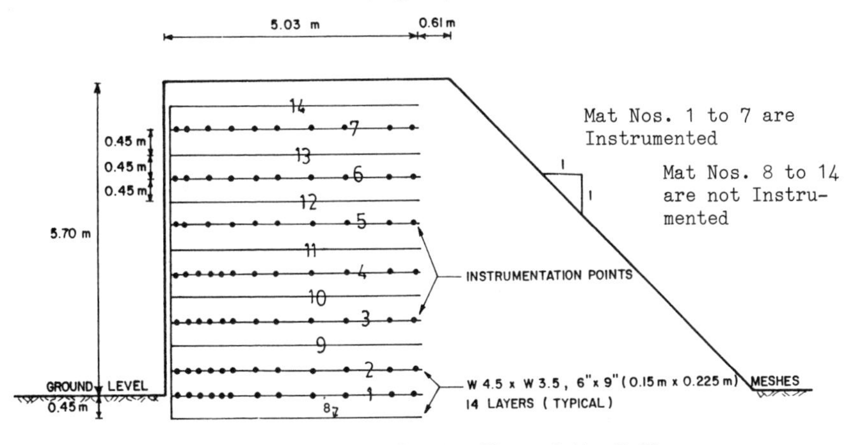

Figure 3. Typical Section View of the Wall

TABLE I SUMMARY OF BACKFILL SOIL PROPERTIES

SOIL	G_s	W_L (%)	W_p (%)	% PASSING SIEVE No. 200 (0.075 mm)	W_{opt} (%)	$\gamma_{d max}$ (kN/m³)	DIRECT SHEAR		UU	
							C (kN/m²)	ϕ (DEGREES)	C (kN/m²)	ϕ (DEGREES)
CLAYEY SAND	2.55	—	—	44.28	14.0	17.0	38.0	26.8	42.0	23.5
LATERITIC RESIDUAL SOIL	2.61	23.20	39.19	17.91	11.5	19.3	88.0	40.2	80.0	32.5
WEATHERED CLAY	2.67	21.04	45.00	82.94	22.0	16.3	129.0	30.7	118.0	31.5

reinforced wall with a wire mesh facing unit on one side, and a sloping, unreinforced embankment along the opposite side as shown in the section view in Fig. 3. Table 1 summarizes the relevant geotechnical properties of the backfill materials used in the wall. The reinforcing mats used were 2.44 m wide by 5.0 m long galvanized, welded wire mesh of W4.5 x W3.5 (diameter of 6.0 mm and 5.40 mm, respectively) wires on a 6 x 9 inches (0.15 x 0.225 m) grid openings. Seven of these mats were instrumented with strain gages for each section as shown in Fig. 3. The fill layers between reinforcing mats

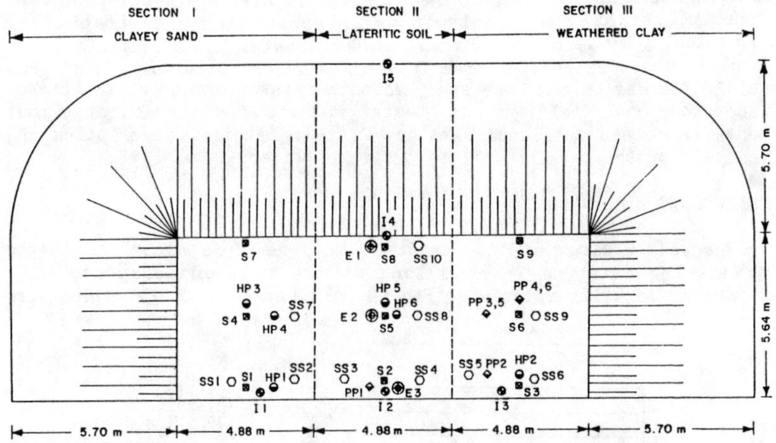

DEPTH OF INSTRUMENTATION IN METERS (BELOW GROUND SURFACE)

LEGEND	No. OF INSTRUMENTATIONS									
	1	2	3	4	5	6	7	8	9	10
▣ SURFACE SETTLEMENT PLATES (S)	0.45	0.45	0.45	0.45	0.45	0.45	0.45	0.45	0.45	—
○ SUBSURFACE SETTLEMENT PLATES (SS)	5.30	3.00	6.00	3.00	6.00	3.00	3.00	3.00	6.00	3.00
INCLINOMETER CASINGS (I)×	10.85	10.85	10.25	8.50	8.50	—	—	—	—	—
HYDRAULIC PIEZOMETERS (HP)	5.00	5.00	3.00	6.00	7.00	4.00	—	—	—	—
◆ PNEUMATIC PIEZOMETERS (PP)	5.00	2.00	3.00	6.00	(2.00)+	(4.00)+	—	—	—	—
⊕ EARTH-PRESSURE CELLS (E)	0.45	0.45	0.45	—	—	—	—	—	—	—

× UP TO TOP OF EMBANKMENT + 5 & 6 IN EMBANKMENT

Figure 4. Schematic Plan View Layout of Field Instrumentation

were placed and compacted in 3 equal compaction lifts up to a total thickness of about 0.45 m corresponding to the vertical spacing of the reinforcing grids. Each lift was compacted by a combination of a hand tamper and a roller to the specified density of about 95% standard Proctor compaction. The placement moisture content was maintained within 1-2% on the dry side of optimum as verified by the Troxler nuclear densitometer. An instrumentation program, primarily consisting of strain measurements on the seven instrumented layers in each section, was developed to evaluate the performance of the welded wire wall. In addition, dummy mats as shown in Fig. 2 were also instrumented and embedded at different levels of each section for field pullout tests. Additional mats for corrosion observation were embedded at different locations for later retrieval. The schematic layout of the field instrumentation is given in Fig. 4.

LATERAL DISPLACEMENTS, SETTLEMENTS, AND EXCESS PORE PRESSURE

The typical plots of the lateral movements for inclinometer I2 are shown in Fig. 5(a) which is similar to that of inclinometers I1 and I3. After 228 days from the end of construction period, the maximum outward lateral movement measured at the top of the wall face was about 300 mm. The maximum lateral movement in the subsoil at 3 m depth was about 120 to 150 mm which indicated the potential location of shear failure surface corresponding to the weakest soft clay layer (Fig.5(b)). The rate of lateral movement in the subsoil was, however, observed to be decreasing with time. The direction of the subsoil lateral movements in I4 and I5 are opposite and of smaller magnitudes compared to that near the face which seems to indicate that the soil is being squeezed out from the front and from the back, but predominantly from the front corresponding to the heavier load. The surface settlement-time relationships at different sections of the embankment are shown in Fig. 6. The surface settlements at the front along the longitudinal section beneath the interconnected wire mesh facing of the wall (S1,S2,S3) have been almost identical with a magnitude of 0.90 m. Similarly, the subsurface settlements at 6 m

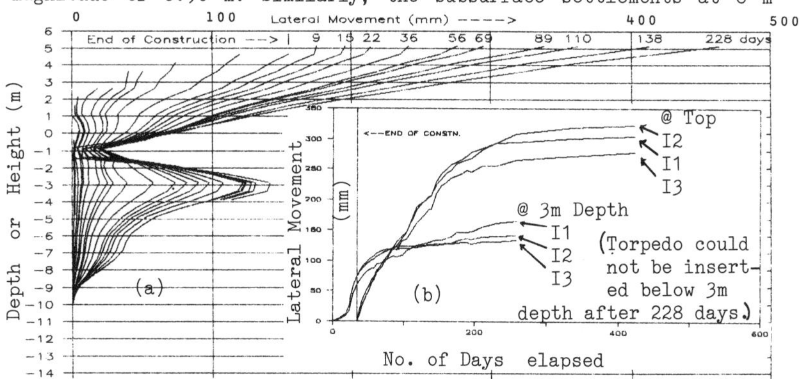

Figure 5a Typical Plots of the Lateral Movement of Wall & subsoil
Figure 5b Lateral Movements at Top and at 3 m Depth with Time

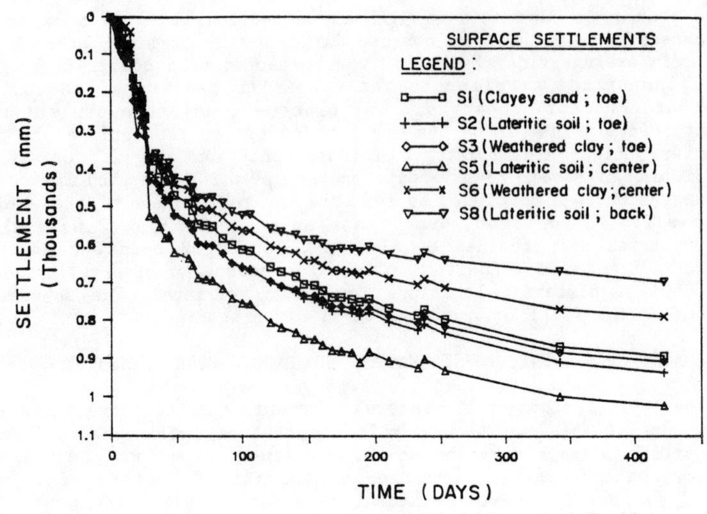

Figure 6. Surface Settlement-Time Relationships at 3 Sections

depth were also found to be nearly identical at 0.25 m. However, at 3 m depth below the same longitudinal section, settlement plate SS4 at the middle section settled an amount of 0.51 m which is comparably lower than its adjacent settlement plates SS2 and SS6 which settled at 0.67 m and 0.75 m, respectively. The maximum surface settlement occurred at the center plate S5 below the middle section (lateritic soil) such that the overall settlement pattern at the surface indicates a dish-like configuration. The vertical (d) and lateral (δ) deformations were plotted in d-(δ/d) coordinates, similar to the diagram for construction control of embankments on soft ground (10). As shown in Fig. 7, the plots are way below the critical boundary curves of p_j/p_f = 0.90 which was the suggested failure criterion for assessing the safety of embankments. Piezometer readings taken at different locations beneath the wall indicated that the porewater pressures continued to dissipate, though at a very slow rate.

EARTH PRESSURES AT THE BASE OF THE WALL

To determine the vertical pressure distribution at the bottom of the wall, SINCO pneumatic total pressure cells were used (Fig.4). During the construction of the first 4 layers, the base pressure at 0.50 m behind the face (E1) was increasing (from 1 to 29 kN/m^2) and higher than E2 and E3 (each recorded a constant pressure of 1 kN/m^2), implying that the center of pressure is located near the face. After 8 layers, the center of pressure tends to have been shifted backwards as was observed from the base pressure readings, probably due to the increase in embankment weight, and with it, the increase in the surface settlements at the center. By the end of the twelfth layer, the base pressures recorded in all the three cells were nearly the same at 55 kN/m^2. The surface settlements near these points were also

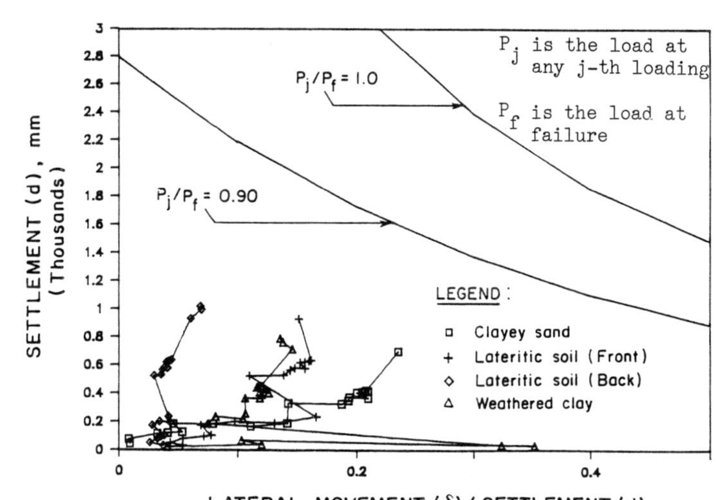

Figure 7. Plots of Settlement Against the Ratio of Lateral Movement to Wall Settlement

about the same at 260 mm. However, towards the end of construction, E2 recorded a base pressure of 70 kN/m^2 which is greater than 63 kN/m^2 at E1, and much greater than the pressure of 50 kN/m^2 at E3, resulting to the drastic increase of settlement at S5 (Fig.8). It has been observed throughout the post-construction phase that any abrupt increase in E2 is followed by a release of pressure at E1. When E1 starts increasing from its lowest value, there will be at first a slight release in the value of E2, and thereafter E2 again starts increasing gradually at first for a while and then at some stage an abrupt increase. This whole process seems to develop a cyclic variation in the base pressures caused by arching effects due to interconnection of reinforcements at the facing units of the wall. This process is expected to continue until consolidation of the subsoil is completed. Any abrupt increase in the value of E2 coupled with large settlement therein at the center, was also reflected by the sharp increase in porewater pressures at the center and the variable mode of strain generation in the reinforcing mats as discussed in the next section.

TENSION IN THE REINFORCING WIRES

The 21X datalogger with a multiplexer and storage module was used to store the data from the temperature compensating electrical-resistant strain gages which were mounted on the longitudinal wires in pairs that were diametrically opposite each other. The reinforcement tensions immediately after construction and four other periods after construction, are depicted in Fig. 8 for the middle section of the wall, including the settlement profiles and base pressures. After 22 days from the end of construction, it is seen that some mats displayed a sudden release of stresses in all sections corresponding to the abrupt decrease in earth pressure near the face (E1) to almost a zero value, with E2 and E3, retaining almost the same magnitude. After 89 days, the earth pressures at 3 locations were all found to increase drastically at almost the same rate which subsequently increased the strains in the reinforcement for all layers. The maximum lateral pressures immediately after construction

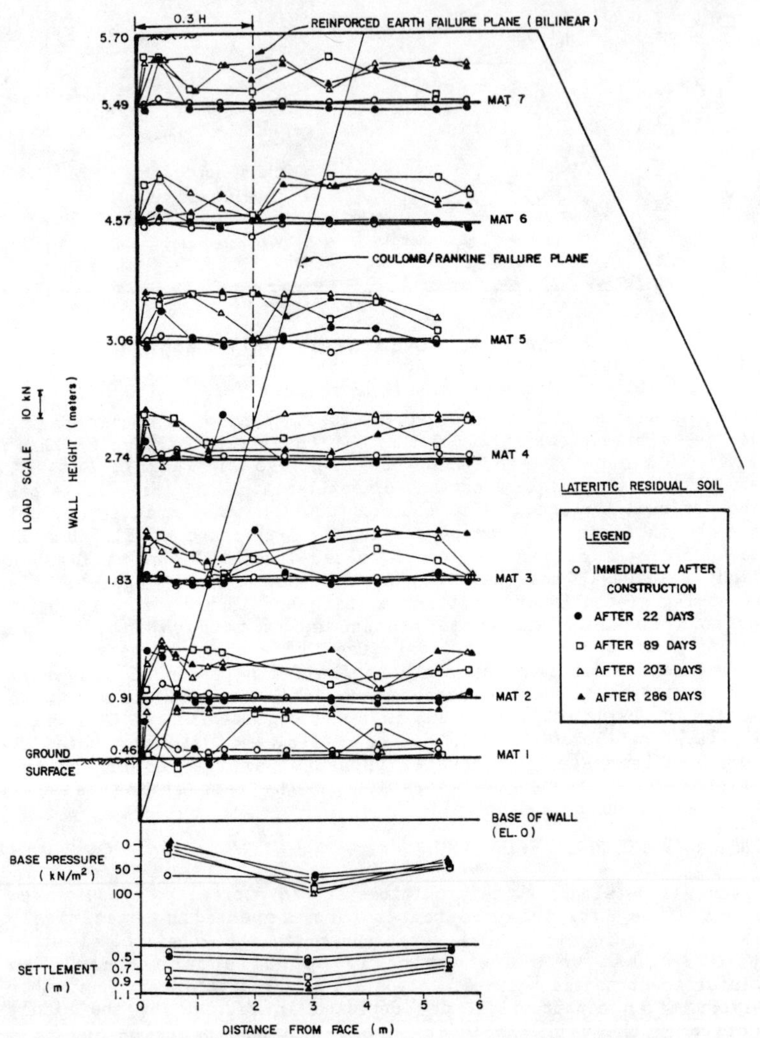

Figure 8. Reinforcement Tensions Immediately After Construction and Four Periods After Construction (Middle Section)

Table 2. FIELD AND LABORATORY PULLOUT TEST RESULTS

Details	DUMMY MAT NOS. (REFER TO FIG.2)				
	23	21	15	12	11
Mat Size (LXT)	W12XW5	W4.5XW3.5	W7XW4.5	W12XW5	W4.5XW3.5
Mesh Size	6"X9"	6"X9"	6"X9"	6"X9"	6"X9"
M X N	4 X 5	4 X 5	4 X 6	4 X 5	4 X 0
Overburden (m)	1.50	2.40	3.33	3.80	3.80
Pullout (mms)	136.7	147.5	144.3	126.0	126.7
Embedment (m)	2.117	2.035	2.037	2.046	2.045
Pt-field (KN)	108.83	57.13	40.55	54.82	25.37
Pt-Lab (KN)	95.60	81.50	102.20	133.30	56.20

(LXT)=Longitudinal X Transverse; Pt is Total Pullout Force;
M is Number of Longitudinal Bars & N is Number of Transverse bars

are plotted with depth in Fig. 9 for the middle section, and were compared to existing earth pressure theories on reinforced soil structures. The measured values immediately after construction were higher than the coherent gravity and tie-back structure hypotheses but seems to be closely predicted by the compaction theory proposed by Ingold (9). It is interesting to note how arching effect has altered the lateral pressures in the lateritic soil section, which is supposed to be the strongest backfill used in the wall, by yielding lower measured lateral pressures than the other 2 sections. This effect was verified from field pullout tests on dummy mats located at the center section which indicated contrasting results with theoretical expectations in terms of lower pullout resistances than both weathered clay and clayey sand, and strikingly, decreasing pull-out resistances with increasing overburden pressures (Table 2). On the other hand, the pullout resistances from laboratory pullout tests show increasing pullout capacity with increasing normal pressures (2,4,5).

Lateral Earth Pressure Coefficient, K. Typical variations of reinforcement tension during construction as each lift of the backfill was placed above the mat are given in Fig. 10 for the middle section. The graphs were replotted to show the variation of tension with distance from the face of the wall as shown in Fig. 11. The maximum value of K was obtained corresponding to the height of the backfill above each mat and plotted as shown in Fig. 12 for the lateritic residual soil at the middle section of the wall. The plots of K indicate an increasing trend as we approach the top of the wall, with most of these values way above the limiting active value (K_a). K_a was determined from the expression:

$$K_a = \frac{1 - \sin\phi}{1 + \sin\phi}$$

ϕ was determined from UU triaxial shear tests on partially saturated and compacted backfill soil samples (compacted to 95% of standard Proctor's density on the dry side of optimum).

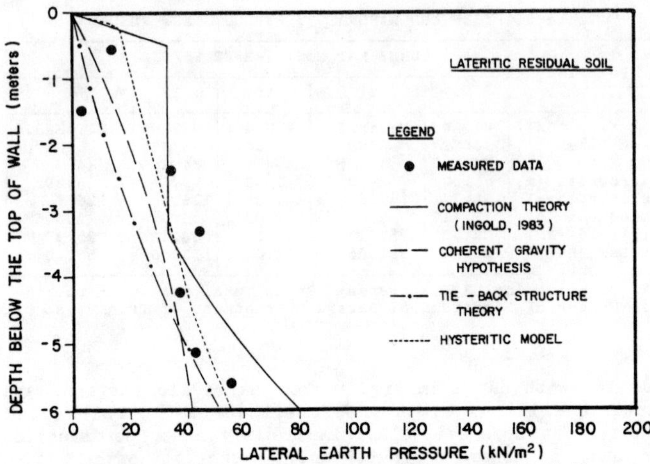

Figure 9. Observed and Predicted Maximum Lateral Earth Pressures at the Middle Section Immediately After Construction

Similar trend was obtained for the other 2 sections of the wall. This variation is significantly different from those reported for welded wire walls with high quality backfill on comparatively hard foundations (1) as well as for reinforced earth walls (11). These deviations maybe attributed to the flexibility of the foundation subsoil and to the residual pressures induced by compaction.

Foundation compressibility can enhance lateral displacement of the wall face as the wall is constructed. The lateral movement necessary to develop the fully active case (K_a) had been reported to be only a minimal fraction (1/1000) of the wall height (14). Any further movement of the face will increase the lateral pressure (7,8,14). In this study, the maximum lateral movement measured for the wall, immediately after construction, amounted to about 0.15 m which is much higher than the required displacement of 0.0057 m to develop the fully active case. For a grid-reinforced soil wall, this continuous outward wall movement may cause the full mobilization of the passive resistance in front of the transverse members, thereby inducing larger strains in the longitudinal members. On the other hand, experimental evidence using compacted sand and silty clay backfills have shown that the maximum lateral earth pressures throughout the height of the wall are significantly higher than were calculated from the at-rest (K_o) values, especially at the top meter of the wall (7,8,12). These high values were attributed to the stresses induced by compaction. Although magnitudes of these stresses depend on the type and size of the compaction plant employed, the maximum stresses at the very top of the wall cannot exceed the lateral pressure calculated using the coefficient of earth pressure for unloading, K_o' (6,13). The value of K_o' was assumed to have a value of $1/K_o$, where K_o was calculated from Jaky's equation ($K_o = 1 - \sin\phi'$) (7,8). Figure 12 typically shows the envelopes of K for the middle section which appeared to have a similar trend as the other 2 sections of the wall. The K-envelope varying

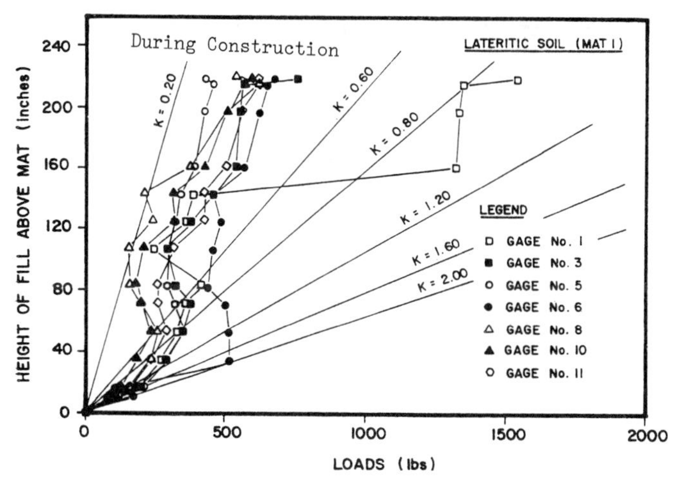

Figure 10. Typical Variation of Reinforcement Tension with Backfill Height Above the Mat (1 lb = 0.00445 kN, 1 in = 0.025 m)

from K_a to K_o' may be taken as the upperbound. The other envelope which varies from K_a to K_o, where K_o was calculated assuming that the compacted backfill is overconsolidated with an OCR of 8, seems to be more appropriate. These recommended envelopes for K may be applicable to walls constructed up to 6 m height on soft and compressible foundation, and utilizing poor quality backfill materials having both friction and cohesion. This result may also be overconservative for walls which will undergo only very small lateral movements. In addition, the values of K reported herein were calculated from construction data of the wall and did not take into account the increase in vertical pressures due to overturning moment. If such a factor is considered, lower backfigured values of K will result.

Maximum Tension Line. The response generated by the wall due to foundation compressibility creates a unique situation wherein existing theories on earth pressures may not be directly applicable. Current design methods use either classical earth pressure theories such as those of Coulomb and Rankine, or an empirical design method which usually involves the assumption of an equivalent fluid pressure distribution. While this situation is seldom, if not encountered in practice, it can be said that the present design method presupposes the very specific combinations of a rigid wall rotating actively about its toe with a backfill acted on by gravitational forces only. For reinforced soil walls, the maximum tension line was reported to define a failure surface or wedge of a Coulomb/Rankine type failure plane, bilinear failure plane, or the log spiral failure plane. Any of these conditions may not be satisfied if foundation flexibility gets into the picture as in the case of the base pressure distributions cited earlier (Fig. 8). It was, however, found that even with this variable mode of earth pressure at the base of the wall, the maximum tension line for this study seems to closely

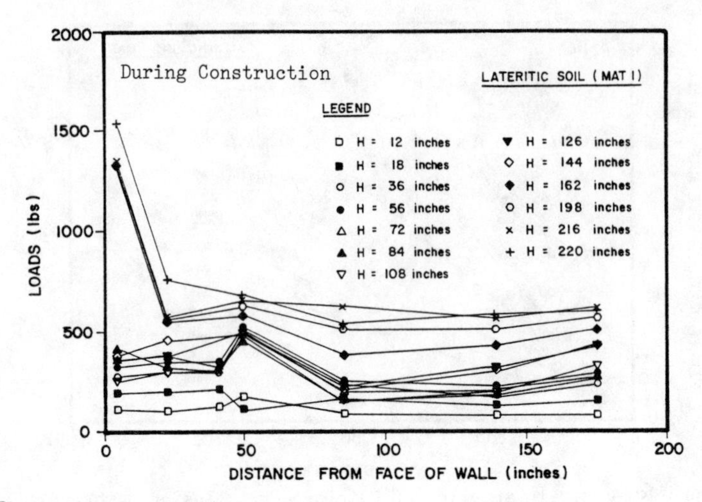

Figure 11. Typical Variation of Reinforcement Tension with Distance From Face of Wall (1 lb = 0.00445 kN, 1 in = 0.025 m)

conform to the log spiral failure plane (farther from the face than the Coulomb/Rankine failure plane) at the lower half of the wall, and conform to the reinforced earth failure plane (closer to the face than the Coulomb/Rankine failure plane) for the upper half of the wall, especially at the two outer sections (I and III).

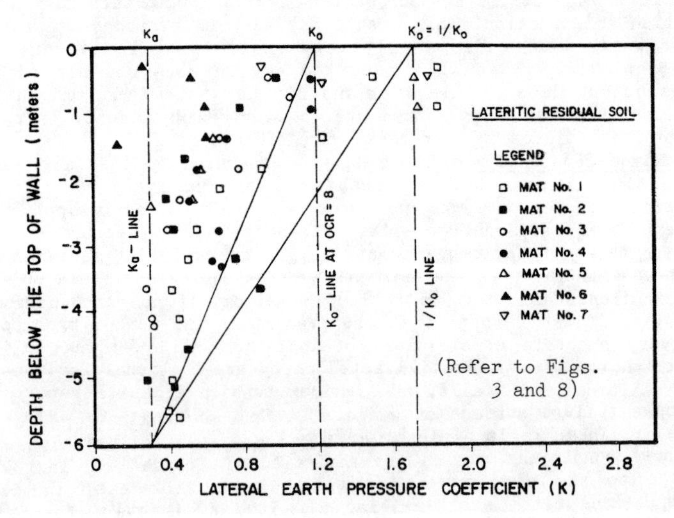

Figure 12. Envelopes of Lateral Earth Pressure Coefficient, K at the Middle Section During Construction

CONCLUSIONS

The 5.7 m high welded wire wall with cohesive-frictional backfills on soft Bangkok clay, its stability being never endangered both during construction and post-construction phases inspite of the excessive settlements and lateral movements, is an ample proof that mechanically stabilized system can be effectively used to reinforce poor quality and marginal quality backfill materials on soft foundations as was also confirmed from laboratory and field pullout tests conducted in this study. Furthermore, if the subsoil was also stabilized by some effective method, it would have drastically reduced the total and differential settlements and lateral movements, resulting in improved stability of the embankment system. It was also found that the variation of earth pressure at the base of the wall, the strains in the reinforcing members, the lateral earth pressure coefficient K, and the location of maximum tension line are strongly affected by foundation compressibility and the effects of backfill compaction which subsequently caused the overall behavior to deviate from those observed on reinforced walls with granular backfill and constructed on a relatively rigid foundation.

REFERENCES

1. LR Anderson, KD Sharp & OT Harding, Performance of a 50-ft. High Welded Wire Wall, Geotech. Spec. Publication No. 12 (ASCE), Ed JP Welsh, 1987, 280-308.
2. DT Bergado, CB Cisneros, R Shivashankar, MC Alfaro & CL Sampaco, Pullout Resistance of Steel Geogrids with Weathered Clay Backfill, Proc. Symp. App. Geosynthetic and Geofibre in SE Asia, Petaling Jaya, Selangor Darul Ehsan, Malaysia, 1989, 1-26-1-34.
3. DT Bergado, S Ahmed, CL Sampaco & AS Balasubramaniam, Settlements of Bangna-Bangpakong Highway on Soft Bangkok Clay, J. Geotech. Eng. Div. (ASCE), 116(1), January 1990, 136-155.
4. DT Bergado, CL Sampaco, R Shivashankar, MC Alfaro, LR Anderson & AS Balasubramaniam, Interaction of Welded Wire Reinforcement with Poor Quality Backfill, Proc. 10th SE Asian Geotech. Conf. (1), Taipei, April 1990, 29-34.
5. DT Bergado, CB Cisneros, CL Sampaco, MC Alfaro & R Shivashankar, Effect of Compaction Moisture Contents on Pullout of Steel Geogrids with Weathered Clay Backfill, Proc. 4th Intl. Conf. on Geotextiles, The Hague, Netherlands, May-June 1990.
6. BB Broms, Lateral Earth Pressures due to Compaction of Cohesionless Soils, Proc. 4th Budapest Conf. on SMFE, 1974, 373-384.
7. DR Carder, RG Pocock & RT Murray, Experimental Retaining Wall Facility-Lateral Stress Measurements with Sand Backfill, Rep. No. LR 766, Trans. and Road Res. Lab., Crowthorne, 1977, 19 pp.
8. DR Carder, RT Murray & JV Krawczyk, Earth Pressures Against an Experimental Retaining Wall Backfilled with Silty Clay, Rep. No. LR 946, Trans. and Road Res. Lab., Crowthorne, 1980, 21 pp.
9. TS Ingold, The Design of Reinforced Soil Walls by Compaction Theory, The Structural Engineer, 61A(7), July 1983, 205-211.

10. M Matsuo & K Kawamura, Diagram for Construction Control of Embankment on Soft Ground, Soils and Foundations, 17(3), September 1977, 37-52.
11. DP McKittrick, Reinforced Earth: Application of Theory and Research to Practice, Proc. Symp. on Soil Reinforcing and Stabilizing Techniques, Sydney, 1978.
12. RT Murray & JB Boden, Reinforced Earth Wall Constructed with Cohesive Fill, Proc. Intl. Conf. Soil Reinforcement (2), Paris, 1979, 569-577.
13. PW Rowe, A Stress-Strain Theory for Cohesionless Soil with Application to Earth Pressure At-Rest and Moving Walls, Geotechnique, 4(2), 1954, 70-88.
14. K Terzaghi, Large Retaining Wall Tests (I): Pressure of Dry Sand, Engineering News Record, 112, 1934, 136-140.

COMPARATIVE STUDY OF A GEOGRID AND A GEOTEXTILE REINFORCED EMBANKMENTS

Tarik Hadj-Hamou[1], A.M.ASCE and Reda M. Bakeer[2], A.M.ASCE

ABSTRACT: The predicted and actual field performance of a geogrid and a geotextile reinforced embankments built in the Lower Mississippi Valley are compared. The two embankments under consideration are full scale models instrumented with strain gages, inclinometers, settlement plates and piezometers. The results of the study showed that the field deformations were within the predicted design values. However, deep seated movements in the soft clay foundation soils were observed at both sites. The effect of the reinforcements on controlling movements of the embankments are compared by normalizing vertical settlements and horizontal movements of the embankments with respect to their geometry.

INTRODUCTION

Hurricane and flood protection levees are built in Louisiana according to the guidelines established by the US Army Corps of Engineers (CE). These guidelines stipulate the height of the levee and its minimum factor of safety against failure. When a new levee was under consideration in Marrero, Louisiana, the CE recommended a final crown elevation of 10 feet (3.05 m) and a factor of safety of at least 1.3. Similarly, when an existing levee was slated for retrofitting according to the new estimate of a 100 year storm, the CE recommended a height of 14.5 feet (4.42 m) with safety factor of at least 1.3. These requirements combined with the poor soil conditions and the site constraints posed potential problems for both levees which resulted in similar but not identical solutions. In general the soil conditions at both sites consist of thick deposits of very soft and sometimes organic alluvial clays. Each levee had to be "squeezed in" between a marsh and an existing drainage canal that could not be relocated. Conventional design and construction would result in embankments with rather flat slopes of one vertical to 3 or 4 horizontal (1V to 3H or 4H) and large stabilizing berms covering a very wide area of reclaimed marshland. This solution was

[1] Assoc. Prof., Dept. of Civil Eng., Tulane Univ., New Orleans, LA 70118.

[2] Asst. Prof., Dept. of Civil Eng., Tulane Univ., New Orleans, LA 70118.

deemed unacceptable because of its ecological consequences and its high cost. The alternative solution was to reinforce the embankments with geosynthetics to allow for narrower bases and smaller berms thereby reducing the encroachment on the marsh and the amount of borrowed material. The remainder of this paper details the predicted and measured behaviors of the two embankments, and a comparison of the effectiveness of the two types of reinforcement on controlling their movements.

BACKGROUND

Geology. The two construction sites are located in the southeastern part of the state of Louisiana on the Central Gulf Coastal plain of the modern delta of the Mississippi River which projects into the Gulf of Mexico. The depositional history for general engineering applications ranges from the Pleistocene Epoch to present time. The sediments are typically divided into natural levee, point bar, and backswamp deposits. Natural levees are the slightly elevated ridges consisting of sands and silts deposited along both sides of the river during flood stages. Point bar deposits are the direct results of the lateral migrations of the river. Over the past 5000 years, seven major deltas were formed in the coastal area of Louisiana (8). During the delta migration process, erosion occurs from the banks and the coarsest materials are redeposited immediately downstream as point bars on the convex bank of the river. Backswamp deposits are formed by the deposition of sediments in the shallow ponded areas of over bank flows. They consist primarily of thinly laminated clays and silts which sometimes have a high organic content. Initial urbanization in the New Orleans region developed on the natural levee and point bar deposits. However, as the city expanded it has become more and more common to utilize other marginal sites such as swampy and marshy deposits.

Stability Analysis. The stability of the two levees was assessed using the Corps of Engineer-Lower Mississippi Valley Division Method of Planes Analysis, also commonly known as the Wedge Method (2). The levees were designed for the end of construction case utilizing a factor of safety of 1.3 for the levees themselves and a factor of safety of 1.5 for analyses extending to the adjacent canals. The method is based on limit equilibrium analysis and it expresses the factor of safety as the ratio of the sum of the resisting forces R to the sum of the driving forces D. The effect of the geosynthetic reinforcement is considered as a tensile force, T, added to the resisting forces. Initial analyses were performed on unreinforced sections to determine the required tensile strength of the geosynthetics where the tensile strength, T, needed to maintain a factor of safety of 1.3 is calculated as follows;

$$T = 1.3\ D - R \qquad (1)$$

The length of development of the geosynthetics was computed to ensure adequate resistance against pullout.

GEOGRID TEST SECTION

Description. The geogrid test section denoted as the Westminster Levee is located on the Westbank of the Mississippi River in the town of Marrero in Jefferson Parish. It is now part of the North/South reach of the finished one mile (1.6 km) long levee protecting the Westminster residential subdivision. The test section is 350 feet long (106.7 m), 10 feet high (3.05 m), 10 feet (3.05 m) wide at the crown, and 136 feet (41.45 m) wide at the base including two stabilizing berms. The centerline of the section is 165 feet (50.29 m) east of the edge of an existing drainage canal. The levee is constructed with a central core of semi-compacted fill placed on a working pad of sand fill. The stabilizing berms are constructed of uncompacted clay fill placed over the sand pad. The slopes of the main levee are 1V on 4H, and those of the berms are 1V on 3H, as shown on Figure 1. The reinforcement consists of a bottom layer of Tensar SS1 installed only to facilitate construction operations, and two reinforcing layers of high density polyethylene Tensar SR2 (UX1500) geogrids located at elevation +0.5 and +2.0 feet (0.15 and 0.61 m) National Geographic Vertical Datum (NGVD).

The soil conditions along the entire reach were inferred from 21 soil borings. From natural ground elevation (Elev. 0.0 NGVD) to a depth of 15 ft (4.57 m) the soil consists of extremely soft to very soft brown and gray clay, organic clay, and humus. This deposit is underlain by a stratum of very soft to soft gray clay intermixed with silt pockets to a depth of approximately 55 feet (16.76 m).

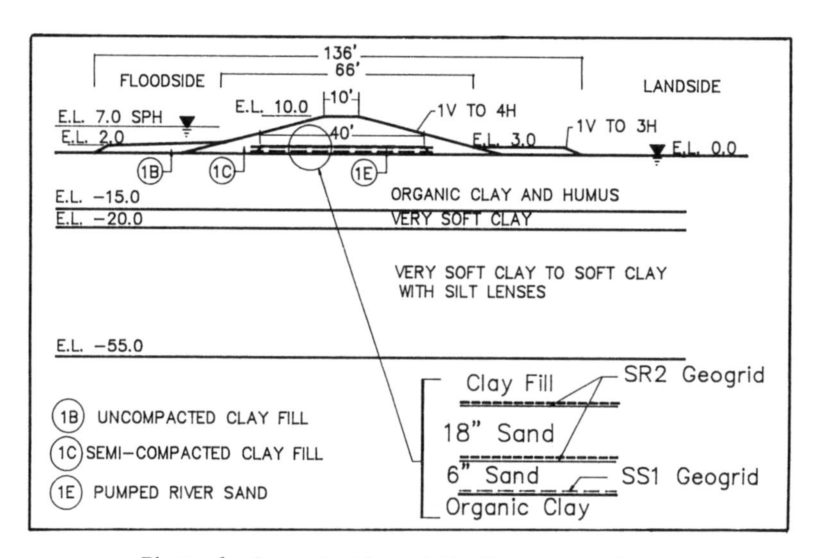

Figure 1: Cross Section of the Westminster Levee

From that depth and continuing to a depth of approximately 76 feet (23.17 m) is a stratum of medium soft gray clay containing sand pockets and shell fragments. Pleistocene deposits of stiff to very stiff overconsolidated gray and tan clays intermixed with sand extend from 76 to 87 feet (23.17 to 26.52 m).

The soil testing program consisted of water content determination, Atterberg Limits, unconfined compression test, consolidated undrained (R) and unconsolidated undrained (Q) triaxial tests, and consolidation tests. The natural water content varied from 40% to 350% in the organic clay and from 40% to 80% in the soft clay. The design shear strength profile was selected based on the results from field vane, unconfined compression, and triaxial tests. For design purposes, the undrained shear strength was assumed constant and equal to 150 psf (7.18 kN/m^2) from depth 0 to 20 feet (0 to 6.1 m) and to increase with depth at a rate of 7.15 psf/ft (1.12 kN/m^2/m) thereafter.

The field monitoring package consisted of three inclinometers, two settlement plates, four piezometers, and 34 strain gages attached to the geogrids, as shown on Figure 2. The installation and monitoring was a collaborative effort of several agencies; the CE, the Louisiana Department of Transportation and Development (LDOTaD), the Louisiana Transportation Research Center (LTRC), and the geotechnical consulting firm Eustis Engineering (EE).

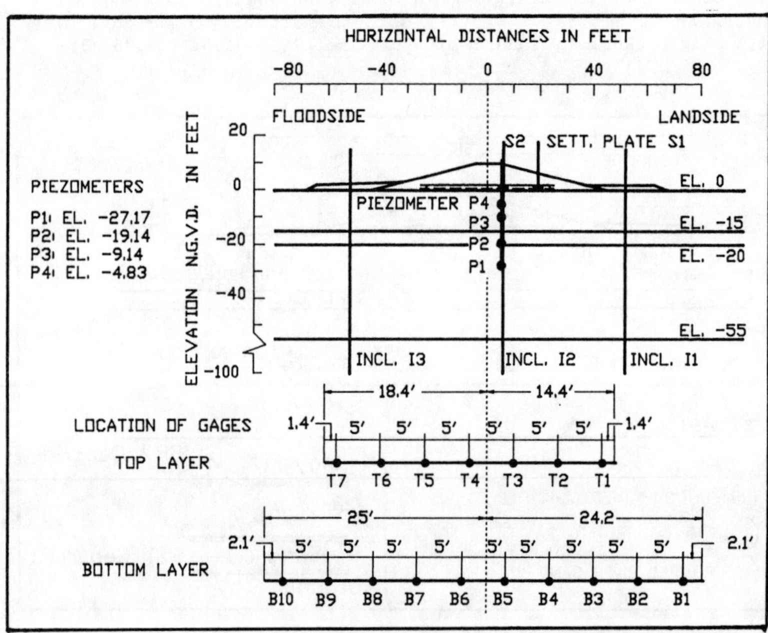

Figure 2: Layout of the Instrumentation at the Westminster Levee

Predicted Performance. Slope stability analyses indicated possible failure wedge mechanisms at elevations -20, -30 and -42 feet NGVD (6.1, 9.14 and 12.8 m) with factors of safety ranging from 1.09 to 1.24 for unreinforced section and from 1.26 to 1.54 for a reinforced section of the same geometry. Settlement of the levee was estimated using Terzaghi's one-dimensional consolidation theory modified for multi-layer systems (4) utilizing the results of four consolidation tests performed by EE and CE. Due to the extremely poor soil conditions at the site, the maximum ultimate settlements were expected to reach 6 feet (1.83 m) under the centerline of the section at point A, 4.6 feet (1.4 m) at point B, and 1.5 feet (0.46 m) point E at the toe of the berms, (refer to Figure 2). About 75% of these settlements was expected to occur in the organic layer and up to 2 feet (0.61 m) of settlement was expected to develop within the construction period of about 60 days. The coefficient of consolidation of the medium soft gray clay was computed as 10 ft^2/yr (0.93 m^2/yr). Strains in the geogrids were estimated using the computed deformation profile and the expected loads from the potential failure mass to be in the order of 2% at the centerline of the bottom layer and 1.8% in the top layer.

Observed Performance. A complete description of the field performance of the test section was reported by Hadj-Hamou et al (3). The instruments were installed on varying dates during the construction period spanning from October 26, 1987 to December 2, 1987. Consequently, the first readings were taken after some fill had already been hauled in place. Maximum pore pressures were recorded when the section reached its final elevation of 10 feet (3.05 m). The pore pressure dissipation data taken over 140 days was used to compute a field coefficient of consolidation of 263 ft^2/yr (24.43 m^2/yr) in the top organic clay layer and 58 ft^2/yr (5.39 m^2/yr) in the medium soft gray clay or about 6 times faster than that computed in the laboratory. These ratios do confirm, however, that most of the settlement measured after 140 days took place in the top organic layer. Maximum settlements at that time reached 3.1 feet (0.94 m) under the centerline of the embankment. Maximum horizontal movements in the embankment cross section (normal to centerline), were recorded after 60 days of the completion of construction as reported in Table 1. No measurable increase was observed in the horizontal movements of the embankment after 60 days. Maximum strains of 0.57% and 1.7% was recorded in the top and bottom layers of geogrid, respectively.

Table 1. Maximum Horizontal Movement - Geogrid Section

Inclinometer Number	Maximum Movement (in ; m)	Elevation NGVD (ft ; m)	Direction* of Movement
I1	1.0 ; 0.0254	-18.0 ; 5.49	West
I2	2.2 ; 0.0559	-20.0 ; 6.10	West
I3	1.9 ; 0.0483	-15.0 ; 4.57	East

* Flood side on the East.

GEOTEXTILE TEST SECTION

Description. The geotextile test section denoted as reach A shown on Figure 3 is a prototype of a proposed enlargement of an existing 13 mile (20.8 km) reach of levee located near the Gulf of Mexico in lower Plaquemines Parish. The crown of the existing levee was at approximately elevation 7.5 NGVD and it was to be raised to elevation 14.5 NGVD on one side and 14.3 on the other side. The center line of the new embankment is 160 feet (48.8 m) off that of a drainage canal and is 20 feet (6.1 m) off the center line of the existing levee on the canal side. The crown of the new levee is 8 feet (2.13 m) wide and its base is 122 feet (37.19 m) wide. The new levee has side slopes of 1V on 4H on the Gulf side and 1V on 3H on the canal side with no stabilizing berms. The 500 feet (152.4 m) long section is reinforced with polyester geotextile of high strength of 1700 lbs/in (96.3 kN/m) at 5% strain. Both ends of the fabric were folded back to provide additional resistance against pullout, as shown on Figure 3. The geotextile was placed over the existing ground and marsh grass after the top of the existing levee was excavated to elevation +5.0 NGVD. Poorly graded river sand was placed on the fabric to a depth of 4 feet (1.2 m) and hauled silty clay was used for the impermeable cover.

Soil samples, laboratory tests and the soil investigation were all conducted by the New Orleans District of the U.S. Army Corps of Engineer (NOD). Soil samples were collected from two soil borings on the Gulf side of the new levee and one soil boring on its canal side. Additional soil information was obtained from the earlier investigation conducted for the existing levee. The simplified soil profile of the site shows that the top 11 feet (3.36 m) are predominantly high organic clay (CH). The next 13 feet of the deposit consist of four thin layers of silt of low plasticity (ML),

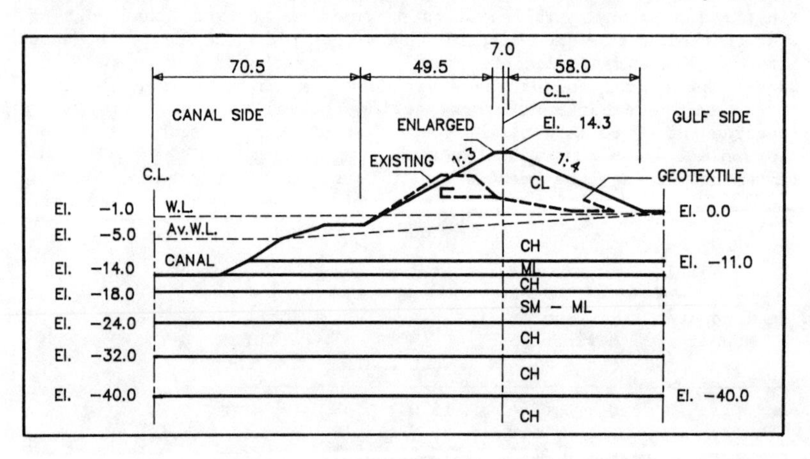

Figure 3: Cross Section of the Reach A Section

a CH clay layer, a silty-sand (SM), and a second ML layer. The soil deposit below elevation -24.0 feet (-7.3 m) consists of layers of highly organic clay (CH). The water table is at elevation -1.0 NGVD, whereas the average water level in the canal is at elevation -5.0.

Properties of the soil layers were obtained from the results of R and Q type triaxial tests at different confining pressures. The natural water content of the top organic clay varies from 76% to 135% and averages about 54% in the other deposits. The liquid limit of the top organic clay has extremely high values of 98 to 113 and it ranges from 43 to 75 in the remainder of the deposits. The plastic limit varies slightly throughout the clay deposits with an average value of 21. At the toe of the levee on the canal side, the undrained shear strength was assumed constant and equal to 150 psf (7.18 $kN/^2$) in the top 11 feet (0 to 3.36 m) and to increase gradually with depth at a rate of 10.0 psf/ft (1.57 $kN/m^2/m$) thereafter to a maximum of 550 psf (26.3 kN/m^2) at elevation -40 NGVD. The top clay layer is slightly weaker on the Gulf side but the strength of the deeper layers is essentially identical on both sides of the levee. Examination of the consolidation curves indicated that the CH layers are all normally consolidated except for the top layer, specifically on the canal side, which is slightly overconsolidated under the weight of the existing levee. Variation of the wet density of the subsoils with depth shows that the soils near the surface have wet density between 86 and 95 pcf (137.4 and 151.9 kN/m^3) then it increases below elevation -10 NGVD to between 100 to 105 pcf (159.9 and 167.9 kN/m^3).

Two profiles across the test levee were instrumented each with 3 inclinometers, 2 settlement plates, and 4 piezometers, as shown in Figure 4. The geotextile was instrumented with two types of strain gages; mechanical transducers manufactured by the CE and foil gages along three profiles. Construction, installation and monitoring of the instrumentation was carried out by NOD and Plaquemines Parish authorities.

Predicted Performance. The stability analysis of the new levee was performed using the NOD version of the wedge method (2) and the Spencer and Simplified Bishop options of the UTEX-2 slope stability program (6). The existing levee, outlined by the dashed lines on Figure 3, had a factor of safety of about 1.1. Based on the wedge analysis, raising its crown to elevation 14.5 NGVD while maintaining the same slopes would result in a factor of safety of 1.02 for a failure surface at elevation -32 NGVD. On the other hand, the reinforced section shown in Figure 3 has a factor of safety of at least 1.33 at the same failure surface. The most critical circular failure was found to be toward the canal and passing through elevation -11 NGVD tangent to the top ML layer. At that particular elevation, the factor of safety of an unreinforced section with the same geometry is only 0.67. Circular arc analyses showed that the reinforced section has a factor of safety of 1.36 at elevation -40 NGVD and that a minimum fabric tensile strength of 270 lbs/in (14.8 kN/m) is required to maintain a safety factor of 1.00 at elevation

-11 NGVD. Meanwhile, the stronger fabric used in the construction of the levee with a tensile strength of 1700 lbs/in would provide the target factor of safety of 1.3.

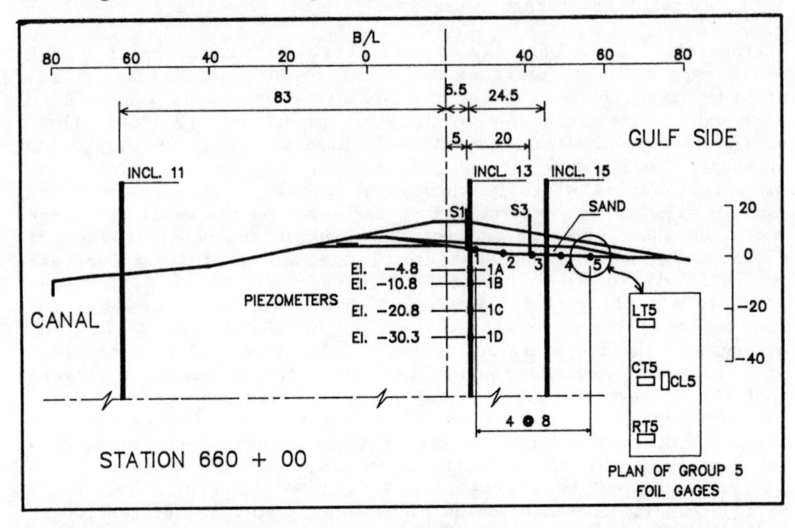

Figure 4: Layout of the Instrumentation at the Reach A Section

Settlement calculations were based on the results of consolidation tests performed by the CE on CH clay samples at different depths. The consolidation tests indicated that the clay in the top layer is slightly overconsolidated to normally consolidated, whereas the clay at the deeper layers is normally consolidated. The calculated coefficient of compressibility ranged between 0.14 and 0.22. The results of the field measurements from two settlement plates in the first 6 months following construction were used by the NOD (8) to develop two empirical equations to predict time-rate settlement for two locations 5 ft Gulf side of centerline and 25 ft Gulf side of centerline:

$$S = 0.5875 \, Ln(t) - 1.4036 \quad (2)$$

$$S = 0.6387 \, Ln(t) - 1.0644 \quad (3)$$

where S is the settlement and t is the time elapsed since 22 October, 1986. A long-term settlement of 3.34 feet (1.02 m) was computed from the one-dimensional consolidation analysis at the location of the settlement plate under the crown, whereas a settlement of 4.11 feet (1.25 m) was calculated at the location of the second plate under the slope. The time-settlement plot showed that most of the settlement would take place in the first 8 years after construction (8). The values predicted by the empirical equations are listed in Table 2 below.

Table 2. Predicted Settlements - Geotextile Section (8).

Elapsed Time (yrs)	Date	Settlement Center Line (inch ; mm)	Settlement Under Slope (inch ; mm)
0.5	22 Apr 87	1.66 ; 6.5	2.26 ; 8.9
1.0	22 Oct 87	2.07 ; 8.2	2.70 ; 10.6
2.0	22 Oct 88	2.47 ; 9.7	3.15 ; 12.4
3.0	22 Oct 89	2.71 ; 10.7	3.40 ; 13.4
4.0	22 Oct 90	2.88 ; 11.3	3.59 ; 14.1
5.0	22 Oct 91	3.00 ; 11.8	3.73 ; 14.7
10.0	22 Oct 96	3.41 ; 13.4	4.17 ; 16.4

Observed Performance. Bakeer et al (1) and Duarte and Satterlee (5) present a full description of the results of the instrumentation program. Construction began on October 17, 1986 with excavating the existing levee to elevation +5.0 NGVD and was completed on November 24, 1986. The instruments were installed on October 22, 1986 after laying the fabric and adding about 3 feet (0.91 m) of sand.

Consistent patterns of progressive lateral movement towards the Gulf were observed in all inclinometers and corroborated by the strains measured in the geotextiles. An average movement of 10.75 inches (0.273 m) was measured at the crown toward the Gulf in January 27, 1988. The horizontal movement indicated by the four inclinometers at the fabric level was also toward the Gulf and averaged 10.6 inches (0.269 m) which translates into an average strain of about 1.26% in the 70 feet (24.4 m) wide fabric. This value, even though approximate, is consistent with the average reading of all foil gages on the Gulf side of 1.4%. A plane of relatively large horizontal movement was detected at elevation -11 NGVD indicating a potential failure surface as predicted in the original design. However, measured movements were all toward the Gulf in contrast with the pattern detected in design. The shear strength of the foundation soil was higher on the canal side because it was consolidated under the weight of the existing embankment. Accordingly, this situation forced the observed pattern of movements toward the Gulf. This particular condition was not accounted for in the slope stability analysis. Movement along the other anticipated failure surface at elevation -40 and -32 NGVD were barely measurable, (on the order of an inch).

Maximum settlement of 1.99 feet (0.61 m) and 2.51 feet (0.77 m) was recorded below the crown and at the slope on the Gulf side on October 23, 1987. These values are in agreement with the values predicted during the design phase based on one-dimensional consolidation and the empirical equations of 2.07 and 2.7 feet (0.63 and 0.82 m), respectively. The settlement recorded in the field started at a rate of 0.13 ft/wk (40 mm/wk) then gradually decreased to about 0.005 ft/wk (1.5 mm/wk) after one year of construction. The empirical equations represent an initial rate of 0.15 ft/wk and 0.023 ft/wk (46 and 7 mm/wk) after one year.

The difference between the readings of the two inclinometers located at the crown and slope was 5.14 inches (0.13 m) on August 24, 1987. The horizontal distance between these two inclinometers is 24 feet (7.32 m) which correspond to an appropriate strain of 1.39% in the fabric. On the same day, the average net readings of the four strain gages on the fabric within that region indicated a net strain of 1.51%. Both values of strain are below the 5% strain assumed in the design.

The net deflected area between the two inclinometers was computed as 7.84 ft^2 (0.73 m^2) by subtracting the deflected areas under the two inclinometer. Area of the settlement bowl between the two settlement plates was computed as 49 ft^2 (4.56 m^2). Consequently, the lateral spread of the embankment is about 16% of its vertical settlement. It should be noted that lateral deformation of the section may not be symmetrical due to the weaker soils on the Gulf side. However, due to the lack of instrumentation on the canal side, this stipulation can not be confirmed.

COMPARATIVE STUDY

The use of reinforcement in both sections produced stable designs and narrower embankments than a traditional design. Although the two embankments have different geometries, a comparison can be made with respect to the effectiveness of the geosynthetics. The saving in marsh land was estimated at about 2500 ft^2/ft (762.5 m^2/ft) at Reach A and at about 500 ft^2/ft (152.5 m^2/ft) at the Westminster levee. In addition to the additional height of 4.5 ft (1.37 m) required for the reach A, a conventional design would have required the construction of a dike in the marsh to dry the construction area, the excavation of a 10 ft deep trench (3.05 m), and hydraulic dredging of river sand to fill the trench. In both sections, the geosynthetics were not used to their full capacites after more than one year of construction. At Reach A, the maximum measured strain was around 2.6% after 2 years compared to an allowable design value of 5%. The maximum strain in the geogrid at the Westminster levee was about 1.7% converting to 1715 lbs/ft (25 kN/m) or about 73% of the allowable maximum of 2350 lbs/ft (34.3 kN/m). Horizontal and vertical movements at both sites were significant but within acceptable limits considering the poor subsoil conditions. Both embankments showed similar patterns of deformation, a large vertical downward movement accompanied with deep seated horizontal movement away from the center line of the embankment. The relative magnitude of these deformation can be compared by computing vertical and horizontal strains. Maximum settlements have been normalized by the height of fill at the location of each settlement plate and are reported in Table 3. At both sites, the loss of height is on the order of 28 to 30% except at the crown of reach A where it is about 14.3%. The lesser settlement in the reach A levee is attributed to the possible rebound caused by the removal of the top 2.5 feet (0.76 m) of the existing levee and to the pre-consolidation of the soft foundation clay under the weight of the existing levee. These conditions may have also resulted in the lesser settlement under the Gulf slope of the reach A levee situated on relatively virgin soil than at its crown situated above the existing levee.

Maximum elongations at different elevations have been normalized by the width of the embankment at those elevations. The elongation is defined as the difference between the readings of two inclinometers. It was observed that most of the movement at the Westminster section occurred at a depth of about 18 feet (5.49 m) with a total elongation of near 5 inches (0.13 m) measured by inclinometers I1 and I3 located 104 feet (31.7 m) apart. Table 4 lists the maximum elongation (L) recorded at both sections, the width (W) of the embankment and the resulting ratios. At Reach A, most of the movement took place along the geotextile with a maximum elongation of 5.1 inches (0.13 m).

Table 3. Measured and Normalized Vertical Movements

Section	Crown			Slope		
	D (ft) (m)	H (ft) (m)	D/H (%) (%)	D (ft) (m)	H (ft) (m)	D/H (%) (%)
Geotextile	2.04	14.3	14.3	2.56	9.3	27.5
	0.62	4.36	14.3	0.78	2.83	27.5
Geogrid	3.10	10.5	29.5	2.17	7.5	29.0
	0.94	3.20	29.5	0.66	2.29	29.0

Table 4: Maximum Normalized Horizontal Movement

Section	L (in ; m)	W (ft ; m)	L/W (%)
Reach A	5.1 ; 0.13	24.0 ; 7.3	1.8
Westminster	5.0 ; 0.13	104.0 ; 31.7	0.4

CONCLUSIONS

The use of geosynthetics allowed for the design of much narrower embankments, considerable conservation of precious marsh land, and reduction of the required borrowed material which more than offset the price of the reinforcement. The instrumentation program was essential in verifying the stability of the proposed design and in revealing the behavior of embankments founded on weak soils. At both embankment, due primarily to the high compressibility of the soils , there is a loss of up to 30% of their height. Horizontal spreading amounted to at most 1.8% of the initial width of the section.

ACKNOWLEDGMENTS

The authors would like to acknowledge the assistance of Mr. F. Duarte and Mr. G. Satterlee of the US Corps of Engineers, NOD and Mr. W. Gwyn of Eustis Engineering for providing part of the data used in this paper. The authors are also grateful to Mr. W. Sherman for his review of the manuscript and to J.N.C. Hoskins for use of his computing facility.

REFERENCES

1. RM Bakeer, TA Hadj-Hamou, FM Duarte, and GS Saterlee, Field Test of a Geotextile Reinforced Levee, Meeting of the portation Research Board, Paper No. 890239, Washington D.C., 1990.
2. WW Caver, Slope Stability in Soft Layered Soil Systems, MS Thesis, Oklahoma State Univ., 1973.
3. TA Hadj-Hamou, RM Bakeer, and WW Gwyn, Field Performance of Geogrid Reinforced Embankment, Meeting of the Transportation Research Board, Paper No. 890240, Washington D.C., 1990.
4. Department of the Navy, Soil Mechanics - Design Manual 7.1 NAVFAC DM-7.1, Naval Facility Engineering Command, Alexandria, VA, 1982.
5. FM Duarte, and GS Satterlee, Case Study of a Geotextile Reinforced Levee on a Soft Clay Foundation, Proc., Geosynthetics '89, (1), San Diego, Ca, 1989, 160-171.
6. EV Jr Edris, UTEXAS-2 Slope Stability Package, User Manual, Department of the Army, Waterways Experiment Station, Corps of Engineers, I.R. GL-87-1 (1), Aug. 1987.
7. CR Kolb, and JR Van Lopic, "Geology of the Mississippi River Deltaic Plain, Southern Louisiana", Department of the Army, Waterways Experiment Station, Corps of Engineers, T.R. 3-483, 1958.
8. New Orleans District, US Army Corps of Engineer, Geotextile Reinforced Levee Test Section, Reach A Hurricane Levee, Tropical Bend, Louisiana, Draft Report, Nov. 1987.

A REINFORCING METHOD FOR EARTH RETAINING WALLS USING SHORT REINFORCING MEMBERS AND A CONTINUOUS RIGID FACING

Osamu Murata[1], Masaru Tateyama[1], Fumio Tatsuoka[2], Kazuyuki Nakamura[3] and Yukihiko Tamura[3]

ABSTRACT: To develop a reinforcing method suitable for most on-site soils, six full-scale test embankments having near vertical slopes were constructed, five using near-saturated clay and one using sand. This method uses relatively short sheets of a grid for sandy soils and a stiff woven/non-woven composite having a function of drainage for cohesive soils. Further, a continuous rigid facing is placed on the wrapped surface of the wall which has first been constructed with the aid of gabions. The long-term behavior and the loading test results of the last two embankments are described. The reconstruction of about 2.5km-long railway embankments has started in 1990.

INTRODUCTION

A study has been undertaken for the last decade to develop a reinforcing method which satisfies the following requirements:
(A) On-site soils as obtained from an excavation work, including sands with a high fines content and even near-saturated cohesive soils, can be used as the backfill soil. Their use will be a large cost saving, compared with the cost for the use of cohesionless soil transported from a remote place and the treatment of excavated soil.
(B) It can be used for the reconstruction of existing embankment making a gentle slope steep to produce a wider crest area, without a large amount of excavation work. Fig. 1 shows its example.
(C) Deformation of slope, especially the settlement at the crest, should be very small, so as to be used for railway embankments.
(D) It should be reasonably inexpensive so that it could be used for large lengths as, for example, railway or highway embankments.

PROPOSED REINFORCING METHOD

The first four 4-5.5m high full-scale test embankments having a slope between 0.2-0.5 in H to 1.0 in V with various facing types were constructed using a very problematic volcanic ash clay called Kanto loam, which had at the time of filling a high water content of 100-120%, a high degree of saturation of 83-90% and a low dry density of 0.55-0.65 g/cm^3 (7-9, 14). They were reinforced with one kind of spun-

1- Research Engineers, Railway Technical Research Institute, Hikari-cho, Kokubunji City, Tokyo, 185, Japan
2- Prof., Institute of Industrial Science, Univ. of Tokyo
3- Research Engineer, Tokyu Construction Co., Ltd.

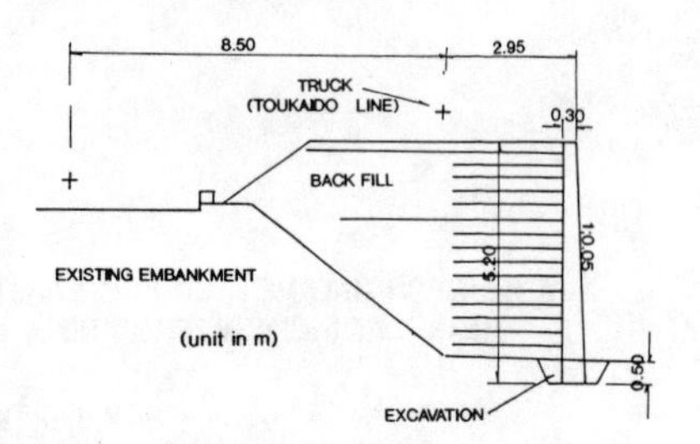

Fig. 1
Reconstruction of railway embankment at Amagasaki City (the height of wall is 5.2m)

-bond non-woven geotextile made of 100% polypropylene. Based on their long-term behavior together with laboratory model test results (10), the following method has been proposed (6, 10, 11).

(A) Planar geotextile sheets are used so as to reduce the required anchorage length by increasing the contact area with the backfill, compared to metal strips. While a stiff and high-strength grid is suitable for cohesionless soils, a composite of non-woven and woven geotextiles, among others, is suitable for cohesive soils to facilitate drainage and also to ensure high tensile rigidity and strength. The function of drainage is essential for the effective compaction of near-saturated cohesive soils and for both the high soil shear strength and high bond strength. The latter is achieved by maintaining suction (negative pore water pressure) in the backfill during heavy rainfall (3, 8) and by maintaining the backfill under drained conditions for load application. The importance of the drained condition was demonstrated by two drained triaxial compression tests performed on undisturbed Kanto loam taken from No. 2 embankment, either unreinforced or reinforced with the geotextile used for that embankment (Fig. 2a). The drained strength increased by reinforcing, while the initial rigidity decreased by large compression of the geotextile (Fig. 2a). On the other hand, in the undrained test (Fig. 2b, c), the maximum effective principal stress ratio increased by reinforcing as in the drained test, but the maximum deviator stress did not increase noticeably because of a larger positive excess pore water pressure developed due to large compression of the geotextile.

(B) Relatively short reinforcement members with a length of, say, 30% of the wall height are used so that it can be used for the reconstruction of embankment (Fig. 1). It has been confirmed that the reduction in the stability of the slope by using shorter reinforcements is compensated by using a planar geotextile (Item A) and using a continuous rigid facing (Item C) (7, 10, 11).

(C) A continuous rigid facing is placed directly on the wrapped wall surface for increasing the stability of wall and for reducing the lateral deformation and the settlement of the backfill, by enhancing the reinforced zone and the facing together to behave like a monolith. Its use also increases the resistance against mechanical damage, fire and the deterioration of geotextile when exposed to sun light.

The local and overall rigidity of facing can contribute to the stability of wall in various ways (7, 10), see Fig. 3. In fact, three walls of the test embankments made by wrapping around geotextile sheets (Type A in Fig. 3) deformed largely during natural heavy rainfalls and

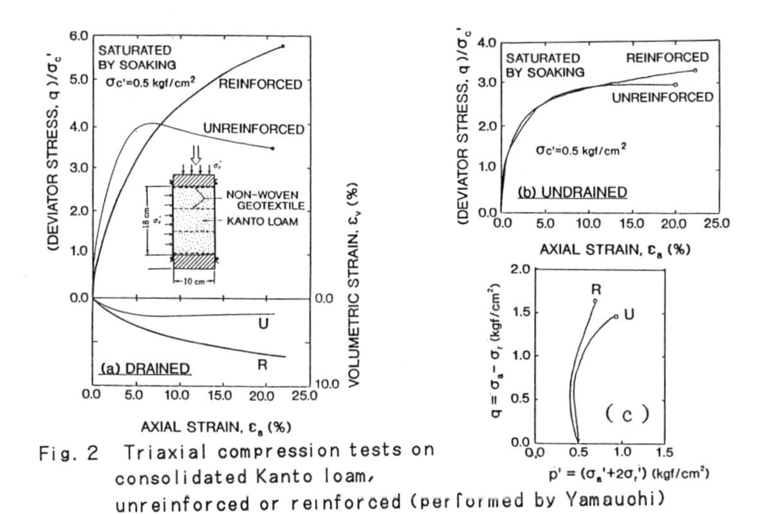

Fig. 2 Triaxial compression tests on consolidated Kanto loam, unreinforced or reinforced (performed by Yamauchi)

earthquakes. This was due to the local compressional failure in the soil layers behind the wall face, because of the lack of the confinement to the soil layers of the wrapping geotextile sheets. Other three walls was constructed with the aid of gabions (Type B2 in Fig. 3), and other two slopes with the aid of discrete concrete panels (Type C). On the wrapped surface of the other wall, an about 8cm-thick shotcrete layer was placed (Type D). Despite their steep slope of 0.2-0.3 in H to 1.0 in V, all these slopes with facings of Types B2, C and D were very stable against natural heavy rainfalls and earthquakes, due mainly to a high degree of compaction of the backfill near the slope face and the confinement by the facing. Further, in the laboratory, the model walls were loaded with a footing (see Fig. 14a). They were more stable in the order of facing types D, C, B and A. With Type D, the failure surface always passed through the toe of wall, while for the other facing types, particularly for Type A, it passed through the intermediate height of the slope face. Thamm et al. (13) also showed that for a 3.6 m-high wrapped reinforced wall (Type B1) with a near vertical wall face, a local failure occurred by concentrated loading on the crest only in the upper part of wall.

(D) A stage construction method (Fig. 4) is employed. First the wall is constructed with the aid of gabions, wrapped around them with geotextile sheets. After major part of the compression of the backfill and/or the supporting ground has occurred, a continuous rigid facing is placed by either of the following methods or another; (a) A lightly steel-reinforced concrete slab is placed directly on the wrapped slope face so that the slab does not separate from the slope face. (b) A lightly steel-reinforced precast concrete plate is erected, with leaving some space between its back face and the slope face. Or discrete modules (e.g., concrete blocks) are piled up and later tightened up by penetrating reinforcement bars through them. The space is sub-

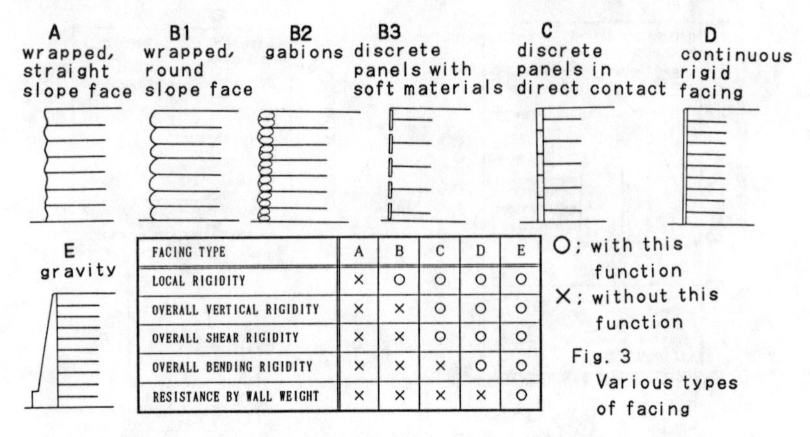

FACING TYPE	A	B	C	D	E
LOCAL RIGIDITY	×	○	○	○	○
OVERALL VERTICAL RIGIDITY	×	×	○	○	○
OVERALL SHEAR RIGIDITY	×	×	○	○	○
OVERALL BENDING RIGIDITY	×	×	×	○	○
RESISTANCE BY WALL WEIGHT	×	×	×	×	○

○; with this function
×; without this function

Fig. 3 Various types of facing

sequently filled with fresh concrete. In this method, the final height of the wall is controlled by the stable height of the wrapped wall without a continuous rigid facing.

Several other methods were proposed to alleviate problems associated with the use of a continuous rigid facing (Fig. 5):

(a) In Fixed type (Fig. 5a), when the backfill is compacted and filled up with reinforcements connected to the back face of facing, relative settlement may occur between the facing and the backfill due to the compression of the backfill and/or the supporting ground and this may damage the connection between the reinforcements and the facing. Therefore, sometimes heavy compaction immediately behind the facing is avoided, but it would not be a final solution and also on the expense of reducing the stability of wall. On the other hand, in Separated type (Fig. 5b) (1), the facing is not in contact with the wrapped wall. In Sliding type (Fig. 5c) (4), the reinforcements are permitted to slide relative to the facing by means of slideable attachments. In Compressional layer type (2, 5, 15), the reinforcements are not connected to the facing and a compressive layer is placed on the back face of facing. The stage construction method is free from this problem. Note also that in this method, if large relative settlement occurs after the wall is completed, the gabions are expected to elimi-

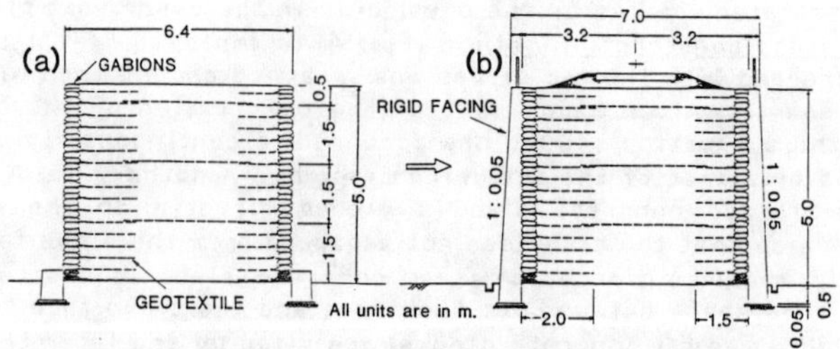

Fig. 4 Stage construction method used for JR No.1 embankment (sand)

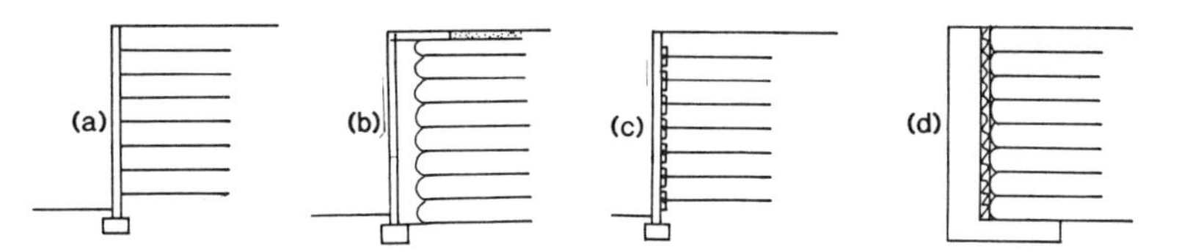

Fig. 5 Reinforcing methods using a continuous rigid facing
(a) Fixed type with a facing fully propped at filling,
(b) Separated type (1),
(c) Sliding type with a facing fully propped at filling (4), and
(d) Type using a compressional layer on the rear of facing fully propped or fixed at filling (2, 5, 15)

nate a large degree of stress concentration to the connection between the facing and the reinforcements.

(b) By compacting heavily the backfill near the facing, the facing and the reinforced zone together can behave like a monolith, thus the wall becomes more stable and less deformable. However, if the facing is not well supported during filling the backfill, the compaction immediately behind the facing should be avoided so as not to induce excessive lateral deformation of the facing (15). In Separated type, heavy compaction can be achieved by using a temporary inflation system placed between the facing and the wrapped wall face (1). In the stage construction method, the soil near the crest of slope can be compacted heavily with a compaction machine.

(c) The development of large tensile stress in the reinforcements is essential for their efficient use. In Fixed and Sliding types, when the facing is well fixed during filling the backfill, only small tensile strain may be developed in the reinforcing members with large earth pressure activated on the back face of facing. Some amount of tensile strain is mobilized only by removing the support of the facing after the full height of wall is completed. Thus, when a relatively extensible geotextile is used, relatively large displacement of facing may occur, inducing some unpreferable large relative settlement between the backfill and the facing. In Separated type, a sufficient amount of tensile strain may be mobilized during the heavy compaction of backfill. In Compressional layer type, by collapsing the compressional layer during compacting the backfill, some amount of tensile strain may be mobilized in the reinforcing members with smaller earth pressure activated on the facing. In the stage construction method, sufficient amount of the tensile strain can be developed during filling the backfill and very small earth pressure would be activated on the facing placed subsequently.

(d) Fixed type facing and the facing made by the stage construction method have all the kinds of facing rigidity shown in Fig. 3, while the facings of Sliding type and Compressional layer type have only some of them. While it is on the safe side to ignore in the design their contribution to the stability of wall, at the same time it is less economical. Thus, this contribution is to be duly relied on.

TEST EMBANKMENTS JR NOS. 1 AND 2

JR (Japan Railway Co.) embankments Nos. 1 and 2 were constructed by the stage construction method from the end of 1987 and after the observation of their behavior by the end of 1989 (6), loading tests were carried out in 1990 to bring them to failure. JR No. 1 (Fig. 6) used a sand having a mean diameter of 0.2mm and a fines content of 16%. The reinforcement was a grid consisting of members made of polyester, covered with PVC to increase its durability, having a rectangular crosssection of 0.9mm x 3mm with an aperture of 20mm. The grid had a tensile rupture strength of 2.8 tonf/m and an initial tensile modulus of 1.0 tonf/m at an elongation of 5% at a strain rate of 5 %/min. JR No. 1 had six test segments; five had a continuous rigid facing of unreinforced concrete slab with two horizontal lightly reinforced construction joints in each, with some amount of the gravity resistance (Type E in Fig. 3). Only one had a facing of discrete panels (Type C in

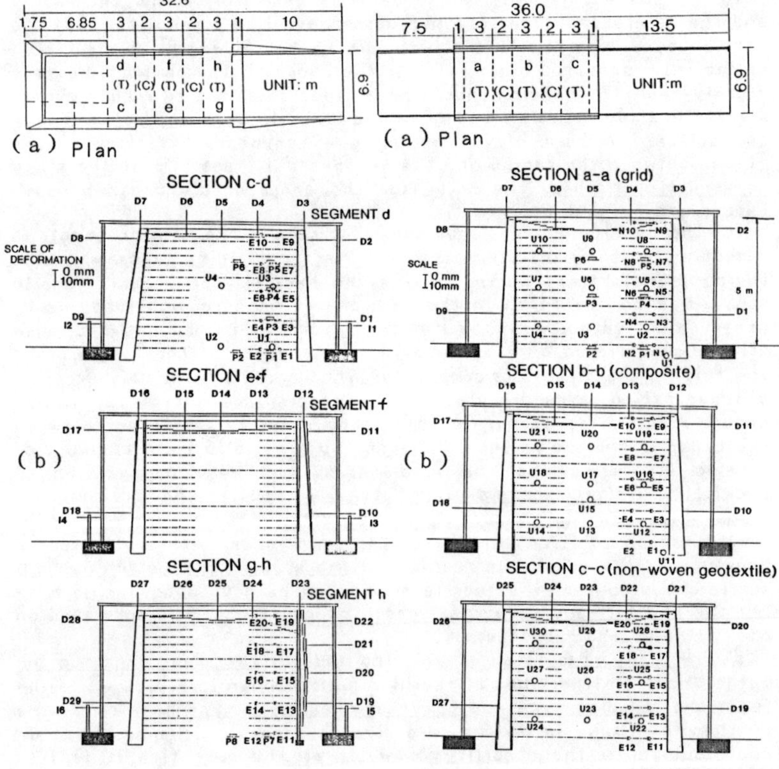

Fig. 6 JR No.1 embankment (sand) Fig. 7 JR No.2 embankment (clay)

(T; test section, and C; control section)

Fig. 3). Each panel had an area of 0.6m x 0.6m and a weight of 40 kgf.

JR No. 2 (Fig. 7) used Kanto loam with a water content of 120-130% and a degree of saturation of about 90% and the dry density of 0.55-0.60 g/cm^3. All the six test segments had a continuous rigid facing. Two test segments in Section a-a were reinforced with a grid as used for JR No. 1, sandwiched between gravel drainage layers. Other two in Section b-b used a composite of high tensile-rigidity woven geotextile sandwiched between non-woven geotextile sheets with a rupture strength of 1.8 tonf/m and a initial tensile modulus of 1.4 tonf/m at a 5% strain. The other two in Section c-c used a non-woven geotextile with a rupture strength of 0.7 tonf/m as used for the first four clay test embankments.

Figs. 6b and 7b show the cross-sections before and after deformation (until Feb. 1989) (note that the scales for embankment and deformation differ 60 times). All the test segments having a continuous rigid facing exhibited very good performance with a very small settlement of 1cm or less at the center of the crest over one and a half years after the construction (see also Fig. 8). Only Segment **h** of JR No.1 with a discrete-panel facing deformed relatively largely. Correspondingly, tensile strain in the reinforcements in Segment **h** was larger than that in Segment **d** (Fig. 8a). It was also found that while using the same kind of soil, Kanto loam, the deformation of JR No. 2 was much smaller than that of the walls with a facing of Type B2 of the fist four test embankments (7-9), primarily because of the additional use of a continuous rigid facing. The effect of facing type shown above is well in accordance with the laboratory model test results (10).

Fig. 9 shows the vertical distribution of hydraulic potential h (= the pore water pressure u - the gravity potential $\gamma_w \cdot z$) at the time when the antecedent precipitation index (API) was maximum. A larger API means a more wet condition (8). It was found that suction was maintained in the clay fill JR No. 2 with larger value in the reinforced zones, while almost no suction in the sand fill JR No. 1. A positive value of dh/dz means downwards percolation of pore water and vise versa. It can be deduced that in the reinforced zones, water percolated from both the crest and bottom of fill, which seems to indicate that the reinforced zones were always kept "dry". It is important that this clay test embankment JR No. 2 also has been very stable despite several times of natural heavy rainfalls.

Three sections of JR No.1 were loaded at their crests using a footing with a 2m x 3m base (Figs. 10 and 11). Each 3m-wide test section was separated from each other by a 2m-wide control section in between. A layer of two plywood sheets with a layer of grease in between had been placed between two sections to achieve plane strain conditions. The length of geotextile was 2m, except in Segment **f** (1.5m). The footing was located at a setback of 2m from the edge of the backfill of Segments **d**, **f** and **h** with the center of footing off the center of the crest by 15cm. Of the two segments in each section, the one which was considered weaker in advance failed (i.e., Segments **d**, **f** and **h**). The effect of the different length of geotextile between Segments **d** and **f** and that of the different facing types between Segments **d** and **h** may be seen in Fig. 11. Note that the horizontal displacement of the wall was largest at the top of wall for Segments **d** and **f**, but not for Segment **h** (see Fig. 10). Therefore, if the largest horizontal displacement along he wall height is plotted in Fig. 11, the difference between these segments becomes larger. Segments **d** and **f** yielded when a crack appeared in the upper construction joint (see Figs. 10 and 11). Therefore, it seems that if the bending rigidity of the construction joints

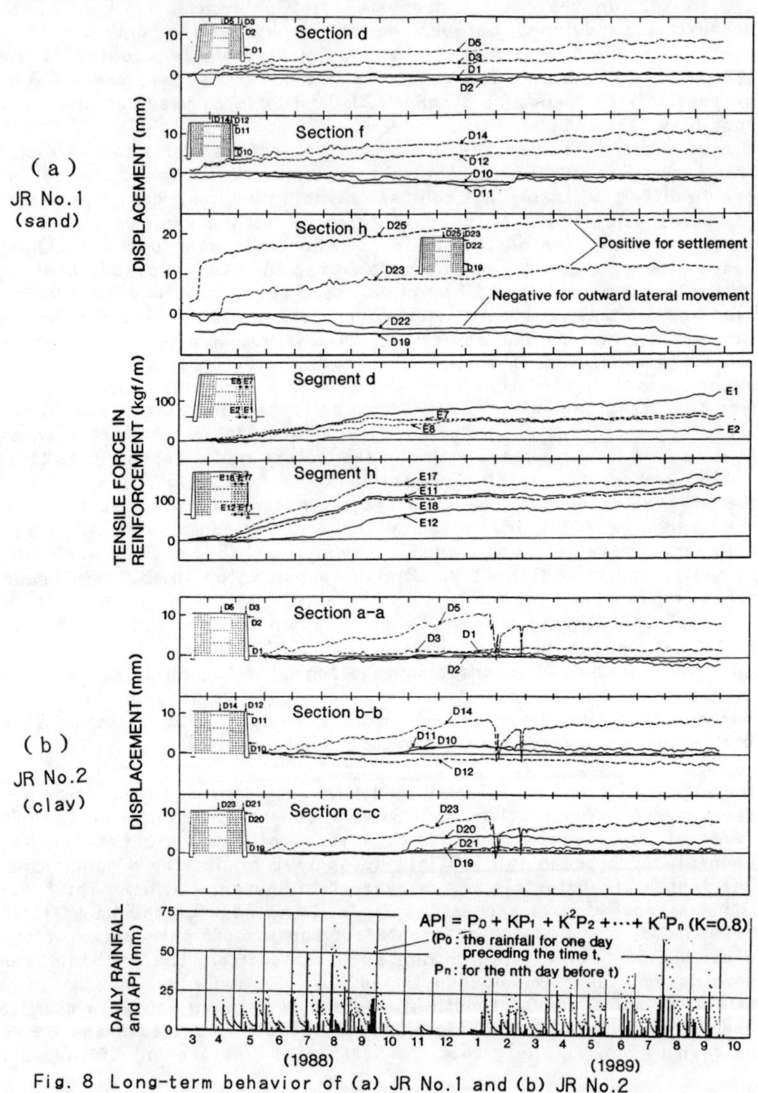

Fig. 8 Long-term behavior of (a) JR No.1 and (b) JR No.2

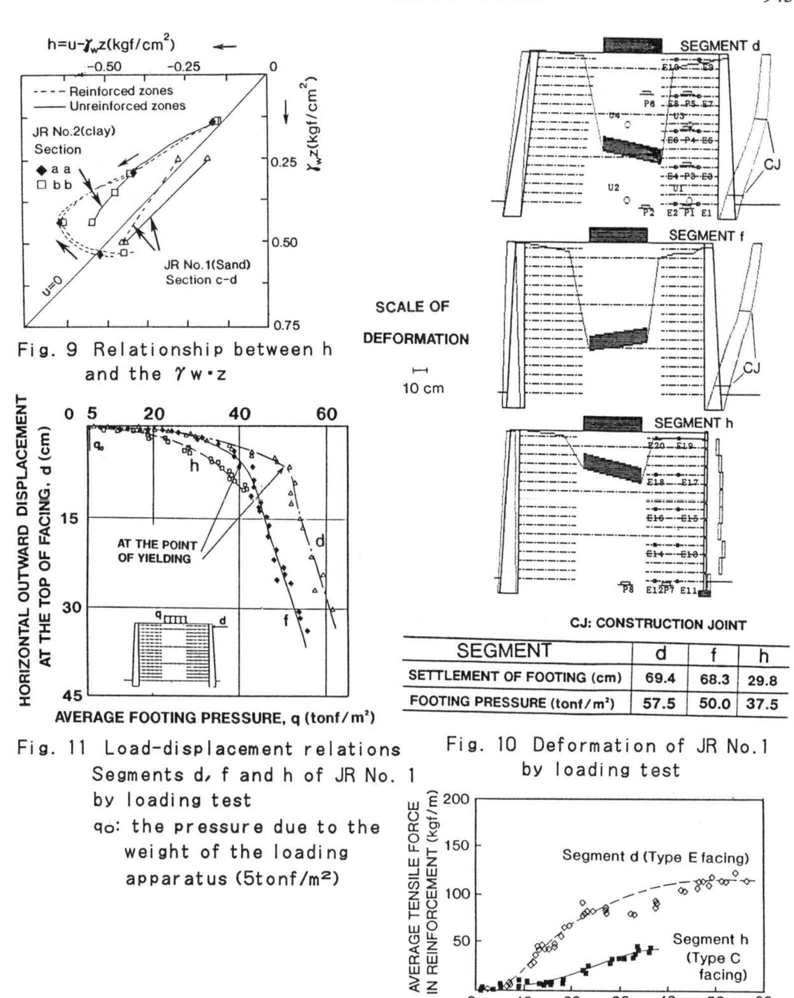

SEGMENT	d	f	h
SETTLEMENT OF FOOTING (cm)	69.4	68.3	29.8
FOOTING PRESSURE (tonf/m²)	57.5	50.0	37.5

Fig. 9 Relationship between h and the $\gamma w \cdot z$

Fig. 11 Load-displacement relations Segments d, f and h of JR No. 1 by loading test
q_o: the pressure due to the weight of the loading apparatus (5tonf/m²)

Fig. 10 Deformation of JR No.1 by loading test

Fig. 12 Average tensile strain developed due to loading, JR No. 1

had been larger, the strength of the walls would have been larger. Indeed, this behavior shows the important role of the overall rigidity of facing for the stability of reinforced walls. The pattern of deformation of each facing seen in Fig. 10 was found very similar to that observed in the corresponding laboratory model test (10). The reinforcement functioned more effectively in Segment **d** having a Type E facing than in Segment **h** having a Type C facing (Fig. 12). This was due to that for such type of facing (Types D and E), larger earth pressure can be activated on the facing, as observed in the laboratory model tests reported in (10).

A simplified stability analysis by the two-wedge limit equilibrium method was performed for the laboratory model walls with a Type D facing (Fig. 14a) (10) and Segment **d** with a Type E facing of JR No. 1 by using the following assumptions (see Fig. 13). **(a)** Failure occurs by over-turning, since for the dimensions of this case, both experimentally and analytically, the safety factor was smaller for over-turning than for the failure by sliding out. **(b)** The safety factor is defined as $SF = \sum Ti \cdot Xi / \Delta M$. Ti is the tensile force in the reinforcement working at the height Xi on the failure surface **abd**. ΔM is the resisting moment additionally required to stabilize the corresponding unreinforced wall. To obtain ΔM, the friction angles δ_w and δ_f between the two wedges and on the back face of facing, respectively, are equal to the soil friction angle ϕ, while their actual values may be smaller than ϕ (note $\delta_f = 0$ for the facing types A and B in Fig. 3). ΔM is also a function of either the location and magnitude of the reaction R on the failure surface **ab** (Fig. 14a), or the distribution pattern of earth pressure on the back face of facing (these two conditions are not independent of each other). This time, the latter was given (Fig. 13b and c). **(c)** The effect of reinforcing is reflected directly only in the change of failure plane from the Coulomb failure plane as for unreinforced slopes to a deeper one. Namely, the increase in the shear

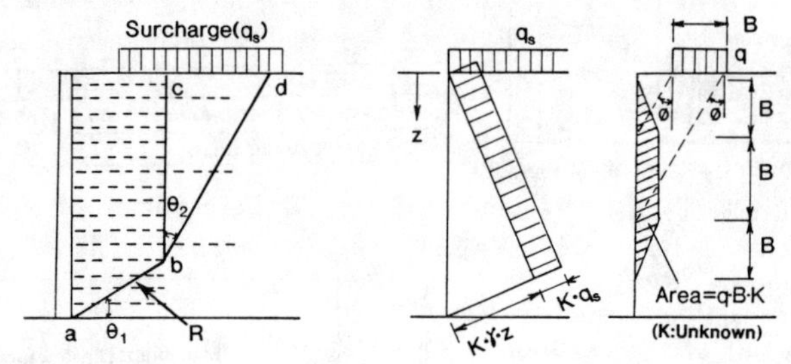

Fig. 13 Assumptions used in the limit equilibrium method

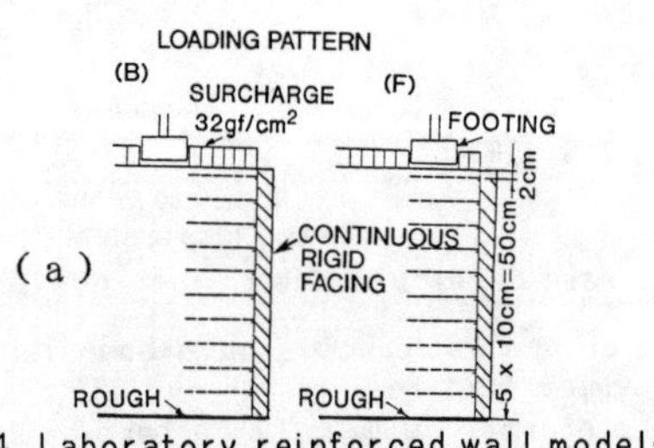

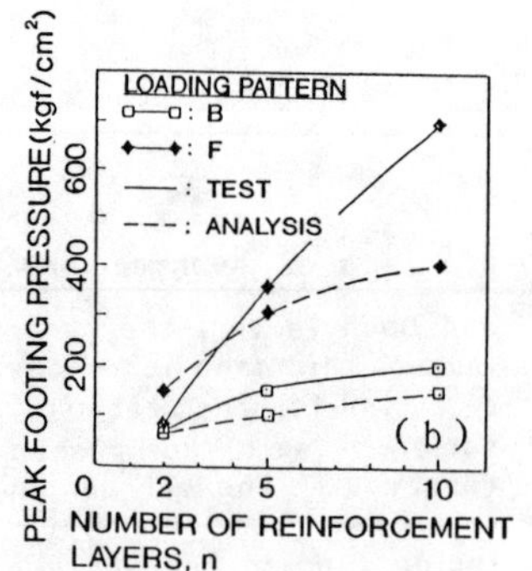

Fig. 14 Laboratory reinforced wall models of Toyoura sand (10):
(a) dimensions and loading method (the case of n= 10 shown)
(b) results of stability analysis

strength of soil due to the increase in the normal force caused by the reinforcement tensile force is not taken into account. This assumption may lead to a safe-side result. **(d)** Ti is obtained by integrating the bond stress for the anchorage length; $\int 2 \cdot \sigma_v \cdot \tan\phi \cdot dl$. σ_v is the normal stress acting on the plane of reinforcement (= the self weight of soil plus the uniform surface pressure q_a plus the pressure by the footing pressure distributed within the backfill at an angle of ϕ relative to the vertical). For the laboratory model tests, Ti was always less than its rupture strength, while for JR No.1, this was upper-bounded by the rupture strength (i.e., 2.8 tonf/m). **(e)** The failure plane **ab** always passes through the toe of wall, because of its sufficiently large overall rigidity of the facing types D and E. The smallest value safety factor SFmin was seeked by changing the location of the point **b** and the angles θ_1 and θ_2. The footing load for SFmin= 1 was then obtained. **(f)** ϕ of Toyoura sand for the laboratory model tests was obtained from the plane strain compression tests (12). The strength of sand of JR No. 1 was assumed as ϕ= 30° and 35°.

For the laboratory model tests, experimentally and analytically, the footing pressure which bring the wall to failure increased as the number of reinforcement layers, n, increased (Fig. 14b). In the case of n= 10, despite the use of short reinforcement, the wall without the footing pressure was very stable, and it failed only by at a very large footing pressure. The discrepancy of the measured and calculated values is due partly to the assumption concerning the pressure distribution pattern of the earth pressure on the back face of facing. Namely, the gravity center of the measured earth pressure was lower than the assumed ones (Figs. 13b and c). The maximum footing pressure computed for Segment **d** of JR No. 1 are 48 and 70 tonf/m² for ϕ= 30° and 35°, which are not very different from the measured value. A more refined analysis using measured values of ϕ is now underway.

A series of shaking table tests of the scaled models (1/10 and 1/2) of Jr. No. 1 also were performed for the aseimic design. Further, JR No.2 clay embankment was loaded in late 1990. In July 1990, to support railway trucks, the reconstruction of 5-7m high existing embankments with a total length of about 2.5km has started at three sites by using the reinforcing method described in this paper.

CONCLUSIONS

A reinforcing method for walls has been described, which is characterized by the use of on-site soils, the use of relatively short geotextile sheets (having a drainage function for cohesive soils) and the stage construction of continuous rigid facing placed on the wrappped surface of the wall constructed with the aid of gabions.

ACKNOWLEGEMENTS

The authors gratefully acknowledge Dr. H. Tarumi, Railway Technical Research Institute, Dr. H. Yamauchi, Penta-Ocean Construction Co., Prof. J. T. H. Wu, University of Colorado at Denver, Mr. K. Iwasaki, Mitsui Petrochemical Industry Ltd and Mr. H. I. Ling, graduate student of Univ. of Tokyo for their support and cooperation.

REFERENCES

(1) Delmas, Ph., Grourc, J. P., Bilvet, J. C. and Maticard, Y. (1988): Geotextile-reinforced retaining structures: A few instrumented examples, Proc. Int. Geotextile Sympo. on Theory and Practice of Earth Reinforcement (IS Kyushu), Japan, 511-516

(2) Edgar, T. V., Puckett, J. A. and D'Spain, R. B. (1989): Effects of geotextiles on lateral pressure and deformation in highway embankments, Geotextiles and Geomembranes (G&G), 8: 275-292

(3) Fabian, K.(1990): Time dependent behaviour of geotextile reinforced clay walls: Proc. 4th Int. Conf. on Geotextiles, Geomembranes and Related Materials (IC on GG and RM), 1: 33-38

(4) Jones, C. J. F. P. (1985): Earth reinforcement and soil structures, Butterworths.

(5) McGown, A., Andrews, Z. Z. and Murray, R. T. (1988): Controlling yielding of the lateral boundaries of soil retaining structures, Proc. of Sympo. on Geosynthetics for Soil Improvement, GE Div., ASCE, 193-210.

(6) Murata, O., Tateyama, M, and Tatsuoka, F.(1990): Steep slpoes reinforced with a planar geotextile having a rigid facing, Proc. 4th IC on GG and RM, 1: 122

(7) Nakamura, K., Tatsuoka, F., Tamura, Y., Iwasaki, K. and Yamauchi, H. Roles of facing in reinforcing steep clay sopes with a non-woven geotextile, Proc. of IS-Kyushu, 1988, 553-558.

(8) Tatsuoka, F. and Yamauchi, H. (1986): A reinforcing method for steep clay slopes with a non-woven geotextile, G&G, 4: 241-268

(9) Tatsuoka, F., Tamura, Y. Nakamura, K., Iwasaki, K. and Yamauchi, H. (1987): Behaviour of steep clay embankments reinforced with a non-woven geotextile having various face structures, Proc. the Post Vienna Conf. on Geotextiles, Singapore, 387-403

(10) Tatsuoka, F., Tateyama, M. and Murata, O.(1988): Earth retaining wall with a short geotextile and a rigid facing, Proc. 12th ICSMFE, 1989, 2: 1311-1314

(11) Tatsuoka, F., Murata, O., Tateyama, M., Nakamura, K., Tamura, Y., Ling, H.I., Iwasaki, K. and Yamauchi, H. (1990): Reinforcing steep clay slopes with a non-woven geotextile, Proc. International Reinforced Soil Conference, Glasgow

(12) Tatsuoka, F., Okahara, M., Tanaka, T., Tani, K., Morimoto, T. and Siddiquee, M. S. A. (1991): Progressive failure and particle size effect in bearing capacity of footing on sand, Proc. ASCE Geotech. Engineering Congress 1991, Denver, ASCE (this conference)

(13) Thamm, B. R., Krieger, B. and Krieger, J. (1990): Full-scale test on a geotextile reinforced retaining structure, Proc. 4th IC on GG and RM, 1: 3-8

(14) Yamauchi, H., Tatsuoka, F., Nakamura, K., Tamura,Y. and Iwasaki, K. (1987): Stability of steep clay embankments reinforced with a non-woven geotextile, Proc. Post Vien Conf. on Geotextiles,370-386

(16) Wu, J. T. H. and Helwany, H. B. (1990): Alleviating bridge approach settlement with geosynthetic reinforcement, Proc. 4th IC on GG and RM, 1: 107-111

CONTRIBUTION OF VEGETATION ROOTS TO SHEAR STRENGTH OF SOIL

Lakshmi N. Reddi[1], A.M.ASCE

ABSTRACT: A theoretical model is developed for the shearing resistance of a root-reinforced soil block. The shear strength contribution of a root system is formulated in terms of axial forces developed in the root segments. The model is applied to a hypothetical soil block containing a Hemlock root system. The results show that the root system contributes significantly to shearing resistance of the soil block within a certain range of shear distortion. The amount of contribution and the range of shear distortion are found to be dependent upon tortuosity of the root system, mechanical properties, and orientation of the roots.

INTRODUCTION

Vegetation is often a significant component of the soil environment. It has been well recognized in the recent past that vegetation plays an important role in the stability of slopes by offering good protection against surface erosion and mass-movement. Study of landslides on several hillslopes in south-eastern Alaska and westside of the Cascade Range show the cause-and-effect relationship between vegetation removal and mass-movement (Bishop and Stevens, 1964; Gray, 1982). Croft and Adams (1950) attributed the occurrences of landslides in the Wasatch Mountains of Utah to loss of mechanical support from root systems of trees due to burning of timber and excessive grazing by livestock. Similar field observations on the role of vegetation in stabilizing slopes have prompted several investigators (Endo and Tsuruta, 1969; Kassif and Kopelovitz, 1968; Waldron, 1977; Ziemer, 1981; Gray, 1982) to conduct experimental studies on the shear strength of soil root systems. These experimental studies conclude that the shear strength of soils increases linearly with the bulk weight of roots per unit volume of soil.

---

1 - Asst. Prof, Dept. of Civil, Environmental & Coastal Engineering, Stevens Institute of Technology, Hoboken, NJ 07030

The experimental investigations on soil-root systems have led to a series of theoretical models (Wu, 1976; Waldron, 1977; Gray, 1982; McOmber, 1981) which attempted to develop an understanding of the mechanism of soil reinforcement by roots. All the theoretical models to date consider only single root systems anchored in a stiff layer that lies below a potential slip surface. The roots' contribution to shear strength was formulated based on tensile stress mobilized in the roots due to distortion of the shear zone.

In light of the above, the present study is initiated to develop a formulation of the shearing resistance of a soil block containing a typical root configuration. The model is developed such that the geometrical characteristics needed of the root system are minimal. The significance of the root-reinforcement is assessed by applying the model to a hypothetical Hemlock tree root system.

SHEAR STRENGTH MODEL OF A SOIL-ROOT SYSTEM

The model will be developed using the root configuration shown in Figure 1. The following notation is adopted throughout the analysis to identify the roots in the system (Fig. 1).

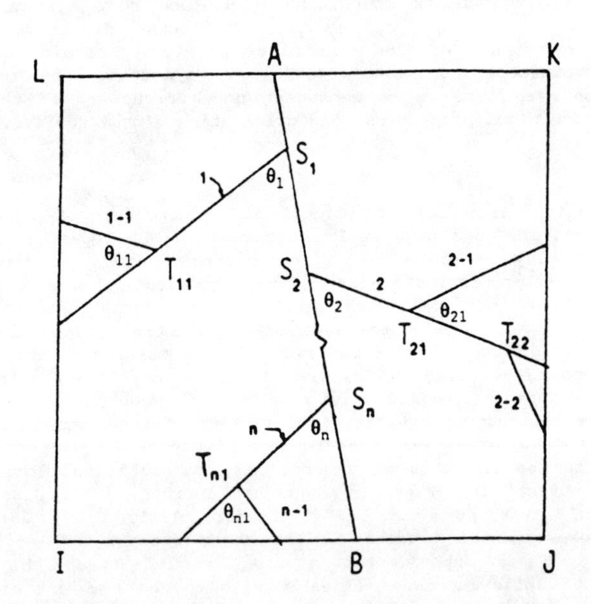

Figure 1: Hypothesized root configuration

- A single numeral 'i' denotes the number of the tertiary root branching from the lateral root AB.
- A double numeral 'i-j' denotes the jth fourth-order root branching from the ith tertiary root.
- S_i denotes the point at which the ith tertiary root meets the lateral root.
- T_{ij} denotes the point at which the jth fourth-order root meets the ith tertiary root.
- The prime denotes the displaced position (for example, T'_{ij} denotes the displaced position of the point T_{ij}).
- L_i and L_{ij} denote lengths of i and i-j roots within the soil block.
- F_i and F_{ij} denote axial forces in i and i-j roots.
- Δ_i and Δ_{ij} denote deformations of i and i-j roots.
- θ_i denotes the angle between the lateral and the ith tertiary roots and θ_{ij} denotes the angle between i^{th} and $(i-j)^{th}$ roots.

The analysis of the soil-root system has been carried out assuming the roots to be straight. However, as will be shown in the next section, tortuosity of the roots can be taken into account with no loss of generality of the theory. The following assumptions are made in the present model.

1. Strains are uniform in the root-reinforced soil block.
2. The entire root system in the soil block lies in one horizontal plane thus limiting the model to be two-dimensional.
3. The roots are flexible, of uniform diameter, and linearly elastic.
4. The roots take only axial stresses and no bending stresses.
5. Roots fail only in tension. No slippage between root and soil can occur.

The assumption of the elastic nature of vegetation roots was verified by Wu (1976). The validity of the other assumptions needs to be checked using field investigations. The Mohr-Coulomb equation for the soil-root system can be written as

$$S = C + \Delta S + \sigma \tan\phi \tag{1}$$

where S = shear strength of the root-reinforced soil, C = cohesion of the soil, ΔS = shearing resistance of the soil due to the effect of reinforcement, and σ = normal force acting on the soil.

When the root-reinforced soil block is given a certain shear deformation (Fig. 2), axial forces are developed in the branches of the root system. The axial force in the branch AS_1 (F_{AS1}) of the lateral root can be resolved into components perpendicular and tangential to the soil block. The perpendicular component acts as normal force on the block and the tangential component directly resists shear. These components contribute to the terms σ and ΔS in Eq. 1.

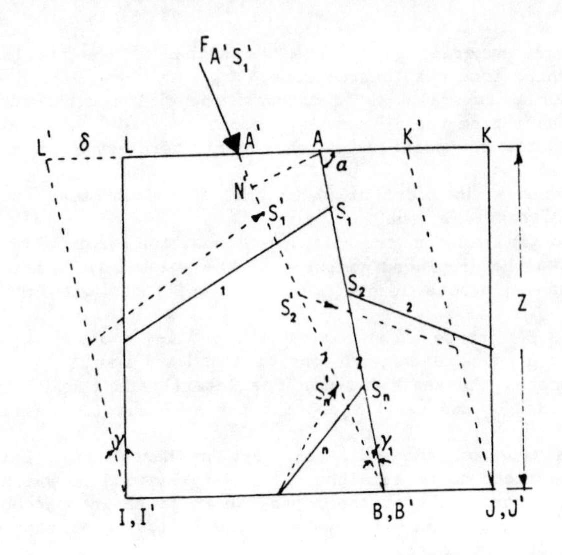

Figure 2: Deformation of root-reinforced soil block

It can be seen from Fig. 2 that tertiary roots mainly act as restraints to the movement of the lateral root AB. Therefore, they can be conveniently represented as inclined forces acting on the displaced root AB. The force in AB is thus a function of the deformation of root AB and the inclined forces from the tertiary roots. Superimposing these two effects gives

$$F_{AB} = F'_{AB} + F''_{AB} \tag{2}$$

where F'_{AB} = Force due to deformation of the root AB, and F''_{AB} = Force due to tertiary roots. Since the end A in Fig. 2 is fixed to the upper face of the block,

$$\overline{AA'} = \delta \cong \gamma Z$$

The elongation of the root AB due to the deformation of the soil block is

$$\Delta_{AB} = \overline{AN'} = \delta\cos(\alpha - \gamma_1) \tag{3}$$

where $\gamma_1 \cong \delta/L_{AB}$, and L_{AB} is the length of the lateral root AB. Therefore, the tensile force in AB, due to shear deformation of the soil block is,

$$F'_{AB} = (\Delta_{AB}/L_{AB})\ EA \tag{4}$$

where E = Elastic Modulus of the root, and A = Area of cross-section of the root.

To calculate F''_{AB} , axial forces in tertiary roots should be evaluated first. Determination of axial forces in the tertiary roots requires a knowledge of the deformations of the roots.

Suppose, in Fig. 3, ξ is the angle made by a root with positive x-axis in the anti-clockwise direction. When the block is deformed, it is evident that roots with ξ in the first and third quadrants (0°-90° or 180°-270°) will be subjected to compression and roots with ξ in the second and fourth quadrants (90°-180° or 180°-360°) will be subjected to tension. A general expression for the deformation of the ith root in compression and tension can be obtained from the geometry of Fig. 4.

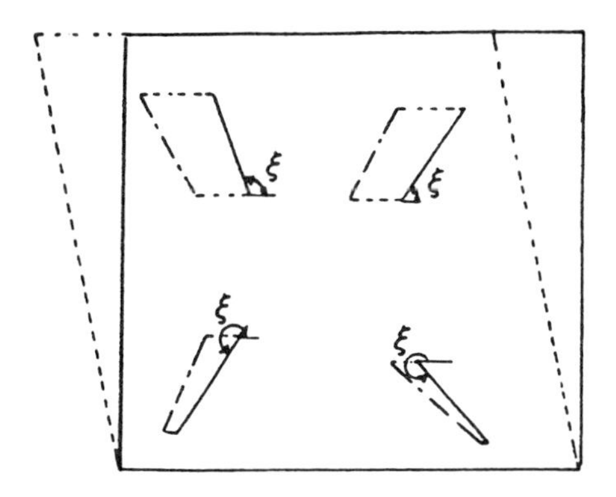

Figure 3: Identification of tensile and compressive roots.

Applying the cosine rule to the triangle S_iPR in the case of compression root and to the triangle S_iPM in the case of tension root,

$$\Delta_i(\text{compression}) = L_i - \sqrt{p_i^2 + L_i^2 - 2p_iL_i\cos(180-\theta_i-\alpha)} \tag{5}$$

$$\Delta_i(\text{tension}) = \sqrt{p_i^2 + L_i^2 - 2p_iL_i\cos(180-\theta_i-\alpha)} - L_i \tag{6}$$

Having obtained the deformations of the points S_i, the axial forces (F_i) in the tertiary roots can be calculated as

$$F_i = (\Delta_i/L_i)EA \tag{7}$$

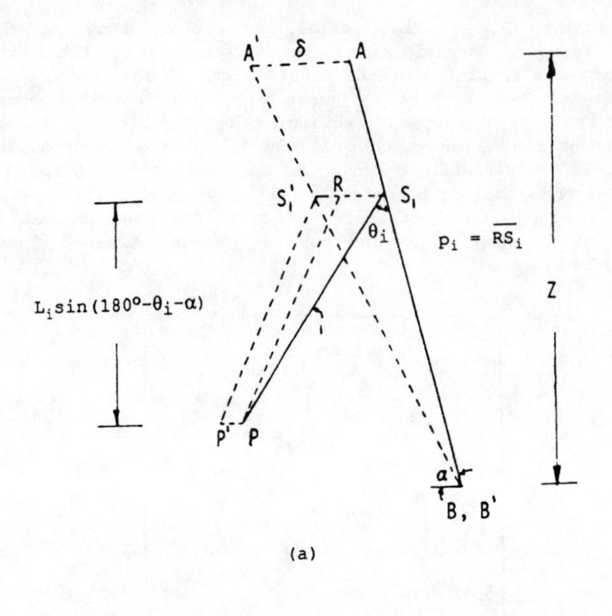

(a)

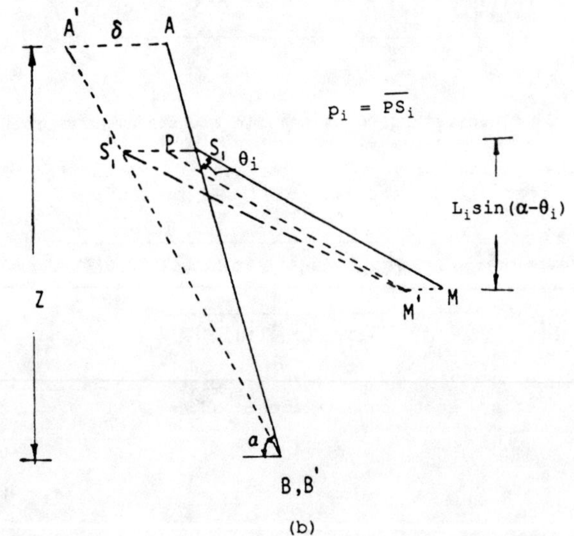

(b)

Figure 4: Deformation of a) compression root and b) tension root

The axial forces (F'') in various branches of the root due to F_i can now be calculated from the equilibrium conditions of AB in Fig. 5.

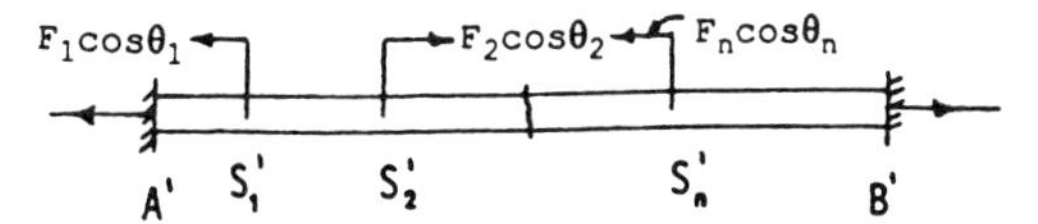

Figure 5: Axial forces on the lateral root

Equilibrium in the horizontal direction gives

$$F''_{A'S'} + F''_{B'S'n} = \sum_{i=1}^{n} F_i \cos_i \tag{8}$$

Since the positions of A and B are determined by δ alone, there would be no change in the length of AB due to the action of $\Sigma F_i \cos_i$. Therefore,

$$(F''_{A'S'1}) L_{A'S'1} + (F''_{A'S'1} - F_1 \cos_1) L_{S'1S'2} + \ldots + (F''_{S'nB'}) L_{S'nB'} = 0 \tag{9}$$

Solving Eqs. 8 and 9, F''_{AS1} can be obtained.

The total axial force (F_{AS1}) in AS_1 can be evaluated by superimposing $F''_{A'S'1}$ and F'_{AB} (Eq. 2). Now, in Eq. 1,

$$\Delta S = F'_{A'S'1} \cos(\alpha - \gamma_1) \text{ and } \sigma = F'_{A'S'1} \sin(\alpha - \gamma_1)$$

Therefore, the shear strength of the soil-root system is

$$S = C + F'_{A'S'1} \cos(\alpha - \gamma_1) + F'_{A'S'1} \sin(\alpha - \gamma_1) \tan\phi \tag{10}$$

For improved accuracy, the effect of fourth and higher order roots that emanate from the tertiary roots should be incorporated in the above analysis. The effect of fourth order roots on the axial force (F_i) of the ith tertiary root is similar to the effect of tertiary roots on the axial force of the lateral root. In other words, equations similar to (8) and (9) will be developed for tertiary roots and F_i calculated similar to F_{AS_1}. Similarly, the effect of fifth and higher order roots can also be included in the analysis.

APPLICATION OF THE MODEL

The analytical model developed in the previous section is applied to a soil block containing a Hemlock tree root system. The parameters of the root system were obtained from south-eastern Alaska by the Ohio State University teams. The estimated mean parameters are as follows:

Distance (l) between successive branch points = 0.30 m
Angle (θ) between main root and branch root = 45^0

Tortuosity = 1.20
Diameter of the roots = 0.003 m
Elastic Modulus = $1.9*10^5$ kN/m^2
Tensile Strength = 200 N

The tortuosity of the root system was obtained as the average ratio of tortuous distance to straight line distance between any two points on the root segments. Supposing that the lateral root is oriented at an angle of 80° to the horizontal in the soil block, a 3-branch root system is constructed as shown in Fig. 6. For the sake of simplicity, only lateral and tertiary roots are considered and higher order roots are neglected. From an examination of Hemlock tree root samples, it appears that there is an equal opportunity for the branch roots to lie on either side of the lateral root.

When the soil block in Figure 6 is deformed, tensile forces are developed in roots 1 and 3 and compressive force is developed in root 2. However, due to initial crookedness, root 2 buckles thereby causing negligible compressive force. The buckling load in the compression root is estimated (using strength of materials approach for column buckling under lateral loads) to be 9 N, which is negligible when compared to the tensile strength of roots.

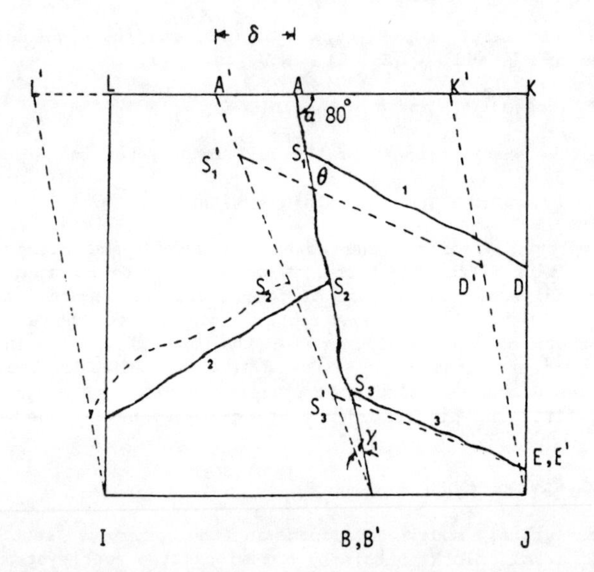

Figure 6: Soil block containing Hemlock root system

In the process of deformation of the soil block, the lateral root AB will be stretched and will become straight at some stage of deformation. Tensile force is developed in AB only after the deformation exceeds this stage. Depending on the mechanical properties of tertiary roots, it is likely that the tertiary roots in tension break before the tortuosity of the lateral root is overcome. In the present problem however, the tertiary roots are found to contribute to the axial forces in the lateral root.

Figure 7 shows shear distortion versus shearing resistance curves. Curve I shows the contribution of the tertiary roots and the lateral root, and curve II is based on the contributions from the lateral root alone. The figure illustrates that the tortuosity of the root system is overcome at a shear distortion of about 0.38. Beyond this stage, the root system contributes significantly to shear strength of the soil block until the lateral root fails in tension. Thus, the shear distortion at which roots become effective is governed by tortuosity of the root system, and the shear strength contribution of roots is governed by mechanical properties of the root system. It can be seen that the effect of tertiary roots is to accelerate the failure of the lateral root and to narrow the range of shear distortion within which the roots contribute to shearing resistance of the soil block. The peak shear strength contribution remains unchanged since it is limited by the tensile strength of the lateral root alone. Fig. 8 shows the stress strain curves for three angles of orientation of the lateral root (α). The curves are obtained assuming the roots to be infinitely elastic. As the figure shows, the stress-strain curves are found to be steeper for greater angles of α. Also, for a given tortuosity, mobilization of tensile force in the roots and contribution to shear strength of the soil starts at smaller δ for smaller angles of α.

SUMMARY AND CONCLUSIONS

A theoretical model is developed for the shearing resistance of a root-reinforced soil block. The geometrical characteristics used to construct the root system are tortuosity, distance between two successive branch points and angle between main and branch roots. The contribution of the root system to shearing resistance of soil is formulated in terms of axial forces developed in the root segments. The model is applied to a soil block containing a Hemlock root system to assess the significance of root reinforcement. Three factors controlling the shear strength contribution of roots are identified to be: i) tortuosity of the root system, ii) mechanical properties of roots, and iii) orientation of roots in the soil.

The conclusions drawn from this study must be viewed with respect to the limitations of the model. The geometrical characteristics of root system used in this study are admittedly simple. In general, certain root systems may not be so well-behaved as to clearly identify the lateral and tertiary roots. More accurate models of root geometry including random-walk models are currently being researched.

Considering the strain to be uniform in the soil block has its obvious limitations. Non-uniform strain in the soil block together with adhesive forces between roots and soil cause non-uniform axial forces in roots. Composite materials approaches are currently being researched to account for these factors.

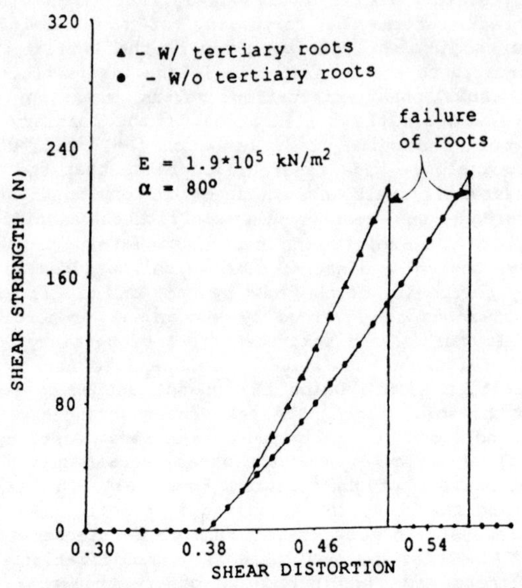

Figure 7: Shear distortion vs. shear strength for soil with Hemlock root system.

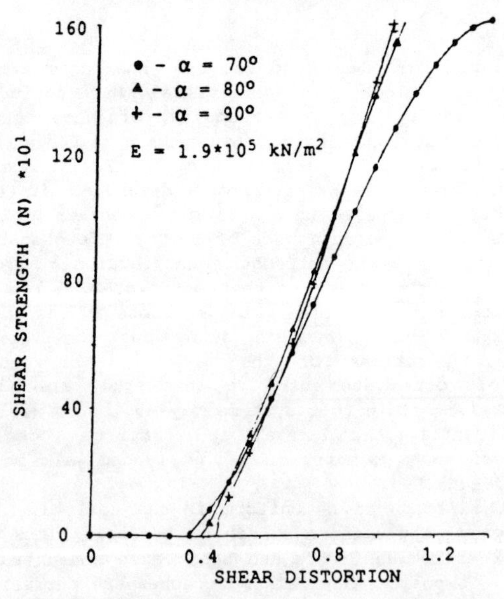

Figure 8: Shear distortion vs. shear strength curves for soil with infinitely elastic root system.

REFERENCES

1. DM Bishop & ME Stevens, Landslides on Logged Areas in Southeast Alaska, USDA Forest Service Research Paper, NOR-1, 1964, 18 p.
2. AR Croft & JA Adams, Landslides and Sedimentation in the North Fork of the Ogden River, U.S. Forest Service Research Paper, INT-21, 1950.
3. T Endo & T Tsuruta, Effects of Tree Roots Upon the Shearing Strength of Soil, Annual Report of the Hokkaido Branch, No. 18, Tokyo Forest Experiment Station, Tokyo, 1969.
4. DH Gray & O AT Leiser, Biotechnical Slope Protection and Erosion Control, Van Nostrand Reinhold Company, New York, 1982.
5. G Kassif & A Kopelovitz, Strength Properties of Soil-Root Systems, Department of Technion Research and Development Foundation, Ltd., Technion, Israel Institute of Technology, Haifa, Israel, 1968.
6. RM McOmber, Pullout Resistance of Roots Embedded in Cohesive Soils (M.S. Thesis), The Ohio State University, Columbus, Ohio, 1981.
7. LJ Waldron, The Shear Resistance of Root Permeated Homogeneous and Stratified Soil, Soil Science of America Journal, 41(5), 1977, 843-849.
8. TH Wu, Investigation of Landslides on Prince of Wales Island, Alaska, Geotechnical Engr. Report, No. 5, Dept. of Civil Engr., Ohio State University, Columbus, Ohio, 1976, 94 p.
9. RR Ziemer, Roots and the Stability of Forested Slopes, International Association of Hydrological Sciences, No. 132, 1981, 343-361.

SAND-ANCHOR INTERACTION IN ANCHORED GEOSYNTHETIC SYSTEMS

S.J. Vitton[1], M.ASCE and R.D. Hryciw[2], M.ASCE

ABSTRACT: A series of ribbed anchor pullout tests was conducted in sand to study the behavior of anchors used in anchored geosynthetic systems (AGS). A large triaxial testing tank was constructed and used to investigate the effects of confining stress. Two dissimilar sands were tested in loose and dense conditions. Large displacement two-way cyclic loading tests were conducted to simulate the loading history of the anchor. The initial peak load resistance for dense sands corresponded to pullout results observed in earth reinforcement. However, significant loss in pullout resistance occurs upon load reversal. Continued cycling resulted in large degradation of both the driving and pullout resistance. Possible mechanisms that may contribute to the load loss are presented.

INTRODUCTION

Anchored geosynthetic systems (AGS) were developed by Koerner (6,7) for in-situ stabilization of soil slopes that are at or near failure. These systems combine a surface deployed geosynthetic with an anchoring system of driven reinforcing rods. The anchors are driven through reinforced openings in the geosynthetic to a depth sufficient to achieve anchorage. The geosynthetic is then fastened to the anchor and the anchor is driven an additional distance, thereby tensioning the geosynthetic and creating a curved geosynthetic-soil interface. This tensioning and curvature imparts compressive stress to the soil and a pullout load to the anchor. The stress transferred to the soil increases the shear resistance along potential failure surfaces, thereby increasing stability. Soil consolidation and stress relaxation in the geosynthetic may require anchor redriving after the initial installation.

A major factor in the success or failure of an AGS is the ability of the anchors to resist pullout. Therefore, an investigation was conducted to develop an understanding of soil-anchor interaction during pullout as well as during driving and redriving of the anchors. The present paper presents the results of this investigation and discusses possible load transfer mechanisms between sands and anchors in AGS.

---

[1]Assistant Professor, Dept. of Civil Engineering, The University of Alabama, Tuscaloosa, AL 35487-0205

[2]Assistant Professor, Dept. of Civil Engineering, The University of Michigan, Ann Arbor, MI 48109-2125

Background The procedure for installing an AGS is outlined by Koerner (5). Small diameter, ribbed steel rods (rebar) typically 1/2 in. (1.3 cm) in diameter or larger are used as anchors. The anchors are driven into the soil using a vibropercussion, pneumatic or similar type of hammer. The anchors are driven to approximately 75 to 90% of their designed depth, connected to the geosynthetic and then driven the remaining distance, thereby tensioning the fabric and exerting a pullout load on the anchor.

For smooth anchors the main load transfer mechanism is interface skin friction. The coefficient of interface friction (μ) typically ranges from (0.5 to 0.8)tanϕ', where ϕ' is the angle of internal friction (8). The value of μ increases with surface roughness. However, beyond a critical roughness the failure develops through the sand. Uesugi, et al. (14,15,16) have shown that the height of the shear zone is about five times D_{50}, where D_{50} is the diameter corresponding to 50% finer in the particle-size distribution curve. Further increase in the surface roughness beyond the critical roughness does not increase interface friction. Therefore, an upper limit for μ is tanϕ'. Since ϕ' itself, is a function of the test boundary conditions and other effects including induced and inherent anisotropy, some questions remain as to which ϕ' value to use for estimating μ. A lower limit would be the residual or constant volume friction angle of the sand, ϕ'_{cv}. Since, large relative motion between anchors and sand will occur in AGS, the use of ϕ'_{cv} in the present study is particularly appropriate.

For ribbed anchors the load transfer mechanisms are far more complicated than in smooth anchors and may include passive resistance of the soil against the ribs. For the ribbed reinforcement used in reinforced soil systems, Mitchell and Villet (8) state that "proven theoretical means for computing the relative contributions (of friction and passive resistance) are not available, and actual data are very limited ... accordingly, the most reliable values of friction coefficient are obtained by direct measurement." As a consequence, in situations where both side friction and passive soil resistance occur an apparent friction coefficient μ^*, is used. Schlosser and Elias (12) indicate that the values of μ^* for a dense sand vary from 0.5 for smooth reinforcements to over 6.0 for ribbed reinforcements. It should also be noted that for dense sands the greater the confining stress is the more restricted dilation of the sand becomes. The result of this is a decrease in the apparent coefficient of friction (12).

Hryciw and Irsyam (3) have studied the pullout resistance of plane ribbed inclusions in sand. They found that a very distinct grain structure develops between the ribs during shearing. The front face of each rib compress the soil skeleton thereby developing a zone of passive resistance while leaving a loose zone at the back face of each rib. Hryciw (2) has shown that for 0.1 in. x 0.1 in. (2.5 mm x 2.5 mm) square ribs, a spacing of approximately 1.3 in. (3.3 cm) maximizes the pullout resistance of ribbed inclusions in dense sand. As the spacing increases beyond 1.3 in. (3.3 cm), the amount of slippage along the sand-inclusion interface increases. For spacing less than 1.3 in. (3.3 cm) a full passive zone does not develop and the shear surface is entirely through the sand mass rather than along the soil-anchor interface (4).

EXPERIMENTAL INVESTIGATION

Testing System To study the load transfer mechanisms between a rebar-anchor and sand under large displacement cyclic loading, a large triaxial testing tank was designed and constructed. The inside diameter is 16.7 in. (42.4 cm) and the height is 36 in. (91.4 cm). Horizontal and vertical confining stresses can be applied independently to the soil through latex rubber membranes built into the sides and the top platen. The triaxial testing tank is illustrated in Figure 1.

An MTS 22 Kip (98 kN) closed loop servo-controlled hydraulic actuator was used for controlled compressive and tensile loading of the anchor. All tests were displacement controlled. An IBM-PC, with a digital to analog card, provided program control. A 5000 lbf (22 kN) load cell measured anchor load. Displacement of the anchor was measured with a ± 4 in. (±10.2 cm) LVDT.

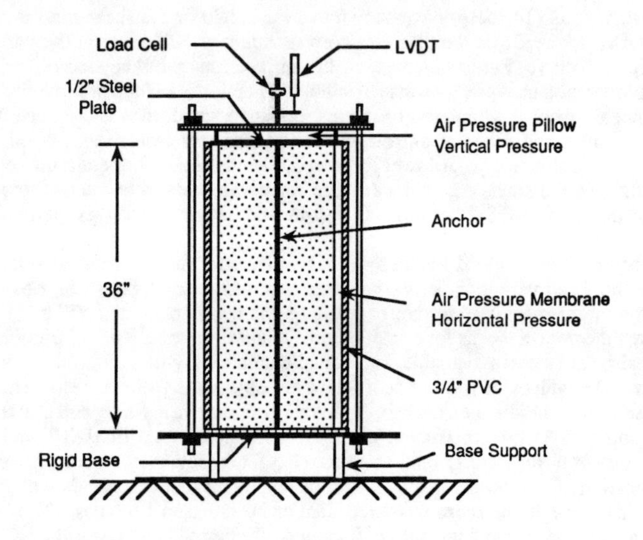

Figure 1. Triaxial test tank for anchor pullout tests.

Soil Description Two sands were tested: an Ottawa 20-30 sand and a glacial outwash sand termed Glazier Way Sand. The Ottawa 20-30 sand is a coarse, poorly graded (SP), subrounded quartz sand, while the Glazier Way sand is a fine to medium, poorly graded (SP) brown subangular sand containing quartz, feldspar, limestone and mafic minerals. The following parameters characterize the two sands:

Ottawa 20-30 Sand	Glazier Way Sand
D_{10} = 0.60 mm	D_{10} = 0.13 mm
D_{50} = 0.72 mm	D_{50} = 0.32 mm
C_u = 1.1	C_u = 2.9
e_{min} = 0.51	e_{min} = 0.40
e_{max} = 0.70	e_{max} = 0.76

Model Anchors Number 3, grade 60, Laclede steel rebar was used for the model anchors. A cross-section of the anchor is shown in Figure 2. The outside diameter of the rebar's ribs is 0.40 in. (10.2 mm), the diameter of the shaft is 0.35 in. (9.0 mm), while the spacing of the anchor ribs is 0.25 in. (6.4 mm). The nominal dimension of 3/8 in. (9.5 mm) was used for computation of interface friction. As shown in Figure 1, the test tank is designed to allow the anchor to extend through the bottom of the tank. A smooth 3/8 in. rod was drilled and pressed onto the anchor base so that a rib-free section of the anchor could pass through a teflon sleeve built into the base of the platform, thus preventing the release of sand from the bottom of the test tank.

Boundary Effects The top and sides of the tank are controlled stress boundaries and the applied confining stress remained constant throughout each test. The tank to anchor diameter ratio was greater than 40 to insure simulation of free field conditions. The base was rigid. A stress boundary at the base would have been more representative of in-situ conditions and it is believed that the rigid base did affect the test results to some degree as will be subsequently discussed.

Soil Placement In actual field installation the anchors are driven into the soil. However, to gain a basic understanding of rib behavior in sand, including development of peak resistance, test samples were prepared by first placing an anchor in the triaxial tank and then preparing the sand around it. Testing was performed at loose and dense soil conditions. Loose Ottawa 20-30 sand and Glazier Way sand were prepared with a 2 in. (5.1 cm) PVC pipe. The pipe was filled with sand and slowly raised allowing the sand to flow out in a loose condition. Relative densities were measured to be below 10% for the Ottawa 20-30 sand and between 10 and 15% for the Glazier Way sand. Two methods had to be used to prepare dense samples. An air pluviating system was used for the Ottawa 20-30 sand and relative densities were measured to be between 95 and 100%. Dense Glazier Way sand was prepared in 1.5 in. (3.8 cm) lifts with vibratory compaction. Relative densities for both dense Ottawa 20-30 and dense Glazier Way sand for the anchor pullout tests were measured to be approximately 95%.

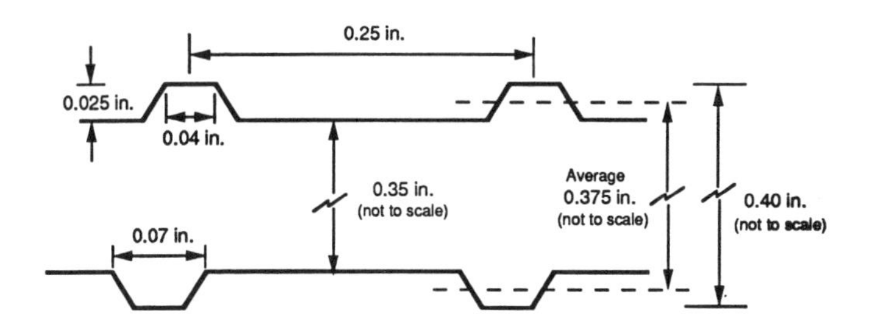

Figure 2. Cross-section of the rebar-anchor.

Soil Loading and Testing After the anchor and soil had been placed into the test tank an effective isotropic confining pressure, σ_o' of 5, 10 or 15 psi (34, 69, 103 kN/m^2) was applied. Tests were conducted in displacement control using a constant anchor displacement rate of 0.0023 in./s (0.06 mm/s). The first loading cycle was a "push" or compressive loading of the anchor to simulate driving conditions. In all of the pullout tests, the anchor was displaced downward 1.75 in. (44 mm) and then 1.75 in. back to its starting position. Hereafter, the subscript *peak* is used for the highest or peak load reached on the first cycle of loading, *res* is used for the post peak residual load and *pull* is used for the residual load upon anchor pullout.

RESULTS

1. The load versus displacement curves for tests conducted at $\sigma_o' = 5$ psi are shown in Figure 3. Corresponding curves for 10 and 15 psi showed the same features, although as would be expected, the loads increased with increasing confining stress. Due to laboratory constraints, samples were not all cycled the same number of times. A summary of all test results is given in Table 1. The initial loading in all tests resulted in a relatively high peak load, P_{peak}, as the anchor was displaced from its at-rest position. The apparent coefficient of friction, μ^* corresponding to the P_{peak} values are plotted in Figure 4 against μ^* values at similar normal stresses σ_n reported by Schlosser and Elias (12) for ribbed strips.

2. After achieving P_{peak}, continued displacement results in establishment of a residual load, P_{res}. In dense sand P_{peak} was followed by a fairly high P_{res}. It is interesting to note that the uniform subrounded Ottawa 20-30 sand quickly dropped from P_{peak} to P_{res}, while the less uniform subangular Glazier Way sand had a gradual reduction of load from P_{peak} to P_{res}. Several tests were conducted in which the first direction of anchor movement was upward. These tests revealed identical behavior, therefore the high residual strength in dense sand is not attributable to the rigid base of the triaxial tank. In loose sand P_{peak} was followed by a significant load reduction before a constant P_{res} was established.

3. In both dense and loose tests, a constant residual load was allowed to develop prior to reversing the loading to simulate anchor pullout. On pullout no distinct peak load was observed and P_{pull} remained relatively constant throughout the pullout range as is seen in Figure 3. On the first cycle, significant load loss occurred from P_{res} to P_{pull} for the tests in dense sand. The ratio P_{pull}/P_{res} for dense sand tests was 0.55 ($\pm$0.05). For loose sands the loss was considerably less with an average P_{pull}/P_{res} of 0.86 ($\pm$0.15).

4. Upon load reversals, a zone of negligible resistance to anchor movement was observed; the length of this zone was approximately 0.015 in. (0.38 mm) for Glazier Way sand and 0.03 in. (0.76 mm) for Ottawa 20-30 sand. This zone of negligible resistance developed only after several cycles in loose sands. It is interesting to note that the lengths of these zones were approximately equal to D_{50} for each sand. However, it is reasonable to presume that the rib dimensions would also affect the length of these zones.

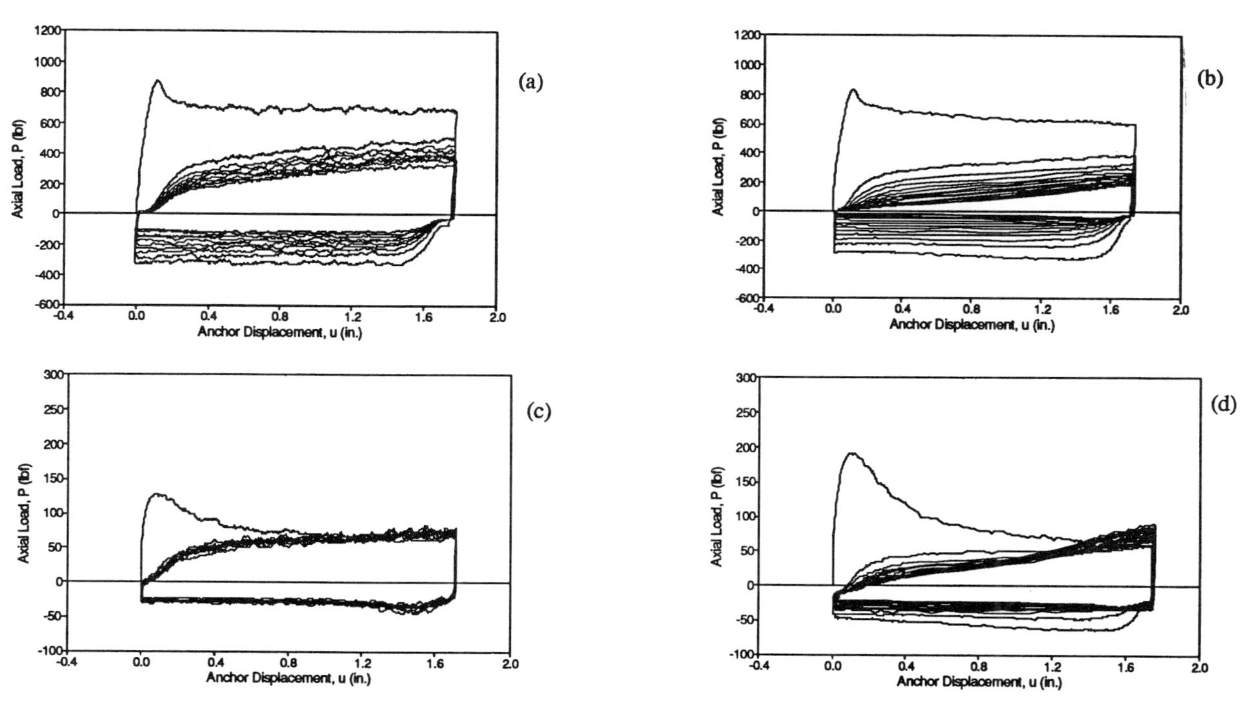

Figure 3. Two-way cyclic load-displacement test at 5 psi, (a) dense Ottawa 20-30 sand, (b) dense Glazier Way sand, (c) loose Ottawa 20-30 sand, dense Glazier Way sand.

5. After each load reversal, the displacement to reach full strength mobilization in both dense and loose sand was approximately 0.25 in. (6.4 mm) or one rib spacing. Smooth anchors required considerably less displacement, about 0.12 in. (3 mm) to mobilize the full interface friction.

6. In dense sand, severe degradation of the interface friction occurred with cycling as shown in Figure 5 for Glazier Way sand. The degradation was greater for Glazier Way than for Ottawa 20-30 sand. In fact, after ten cycles the average μ^* for Glazier Way sand was 0.25 or $0.32\tan\phi'_{cv}$, where ϕ'_{cv}=38°. For Ottawa 20-30 sand, after ten cycles the average μ^* was 0.64 or $1.06\tan\phi'_{cv}$, where ϕ'_{cv}=31°.

Table 1 Summary of Anchor Pullout Tests

Sand Type	σ_o (psi)	Total Cycles N	First Cycle						Last Cycle	
			P_{peak} (lbf)	P_{res} (lbf)	P_{pull} (lbf)	μ^* peak	μ^* res	μ^* pull	P_{pull} (lbf)	μ^* pull
Ott-L	5	8	128	70	44	0.60	0.33	0.21	31	0.15
Ott-L	10	10	277	115	103	0.65	0.27	0.24	56	0.13
Ott-L	15	6	462	235	224	0.73	0.37	0.35	170	0.27
GW-L	5	22	190	60	64	0.90	0.28	0.30	30	0.14
GW-L	10	18	393	172	136	0.93	0.41	0.32	30	0.07
GW-L	15	18	585	217	175	0.92	0.34	0.28	25	0.04
Ott-D	5	12	880	700	325	4.15	3.30	1.53	113	0.53
Ott-D	10	5	1770	1358	820	4.17	3.20	1.93	575	1.36
Ott-D	15	12	2500	1950	1137	3.93	3.07	1.79	472	0.74
GW-D	5	15	826	600	320	3.90	2.83	1.51	43	0.20
GW-D	10	14	1423	934	500	3.36	2.20	1.18	51	0.12
GW-D	15	7	1925	1355	767	3.03	2.13	1.21	250	0.39

Ott-L = Ottawa 20-30 Loose Sand GW-L = Glazier Way Loose Sand
Ott-D = Ottawa 20-30 Dense Sand GW-D = Glazier Way Dense Sand

7. Loose Ottawa 20-30 sand maintained an average μ^*_{pull} of 0.18 or $0.30\tan\phi'_{cv}$ with cycling. Degradation of the side friction did not occur and in fact a slight increase in interface friction resulted with cycling. The loose Glazier Way sand, on the other hand, exhibited significant degradation of side friction with cycling resulting in an average μ^*_{pull} of 0.08 or $0.10\tan\phi'_{cv}$ after an average 19 cycles.

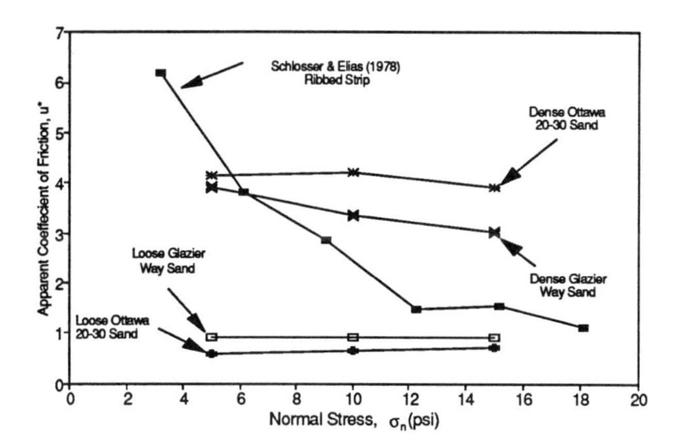

Figure 4. Apparent coefficient of friction versus normal stress including data published by Schlosser & Elias (12).

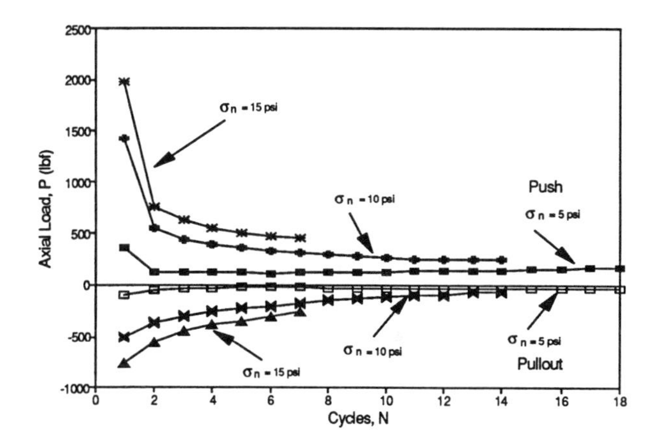

Figure 5. Load degradation with cycling for dense Glazier Way sand at 10 psi.

8. A visual inspection of the sand around the anchor was made after testing. For the Ottawa 20-30 sand, a 0.6 in. (15 mm) diameter zone of black sand grains surrounded the anchor. The sand particles were blackened by the abrasion of the iron oxide surface of the rebar-anchor. Consequently, a polishing of the rebar anchor was occurring while the iron oxide was coating the Ottawa sand particles.
   Polishing of the rebar-anchor occurred in the Glazier Way sand as well. This resulted in the light brown sand turning light gray. The extent of this gray zone was difficult to determine. A major difference between the Glazier Way and Ottawa sand in the zone around the anchor was that the Glazier Way sand was noticeably finer and more uniform in particle size than the surrounding sand.

9. On all downward loading cycles, the load gradually increased with displacement until load reversal. It is believed that the rigid base of the testing tank was responsible for this behavior.

DISCUSSION

1. It is not possible to study the micro-mechanics of soil-rib interaction in an axisymmetric test configuration. However, in a related study, Irsyam (4) traced the movements of sand grains and monitored the development of failure surfaces around plane ribbed inclusions. While a distinct passive zone was observed for 0.1 in. x 0.1 in. square ribs spaced 1.3 in. apart, no such passive zone was observed for the rib shape and spacing shown in Figure 2 and used in the present study. Therefore, the rebar-anchor behaves merely as a rough surface.

2. Preliminary tests were performed on smooth rods of 3/8 in. (9.5 mm) diameter. The computed values of μ^*_{peak} were fairly consistent with results presented by Schlosser and Elias (12) for smooth strips and Uesugi et al. (14) for smooth plates. When ribbed anchors were tested, μ^*_{peak} for dense sand was similar to values observed by Schlosser and Elias for ribbed strips. Although passive resistance has been cited as a possible mechanism for high μ^*_{peak} values for ribbed inclusions (8), this mechanism was not a factor for the anchors in the present study as discussed above. Two alternative explanations are therefore offered for the high computed μ^*_{peak} for dense sand. First, the tank to anchor diameter was sufficiently large to simulate free field conditions and therefore allow stress concentrations to develop normal to the anchors due to sand dilation. Second, unlike for plane strips, the surface area used for computation of μ^*_{peak} for cylindrical anchors does not correspond to the actual area of the shearing surface. Irsyam (4) has shown that the development of μ^*_{peak} is accompanied by grain movement as far away as 0.75 in. (19 mm) from the interface. For plane inclusions, the area of a potential "shearing surface" does not change with distance from the interface. For cylindrical inclusions, the area of a shearing surface could be far larger than the interface area computed from the nominal anchor diameter.

3. After μ_{res} has been reached a distinct shear zone, approximately five grain diameters in height develops during post peak shear as observed by Irsyam (4). Therefore, the reduction of μ^*_{peak} to μ_{res} is probably due to a reduction in the total area of the shear surface. However, the normal stresses due to dilation may continue to increase since a majority of the volumetric increase occurs between μ^*_{peak} to μ_{res}.

4. The pullout load, P_{pull} on the first cycle was consistently lower than P_{res}. This probably resulted from the loss of the dilational induced normal stress increases that developed during the initial push. This is confirmed by Irsyam's (4) visual observation of sand grain movement towards the ribbed surface upon load reversal. The inward movement of grain is also likely to facilitate the development of circumfential arching thereby further reducing the normal stresses.

   A loss of interface friction upon load reversal was also observed by Rao and Venkatesh (11) in their study of uplift behavior of piles in sand. They observed decreases in skin friction of up to 80% from the initial driving skin resistance and found that the decrease in pullout resistance was associated with increasing surface roughness.

5. Although degradation of interface friction with cycling has been reported by numerous researchers (1,9,13), the degradation seen in the test results for ribbed anchors is considerably more rapid and severe, especially for Glazier Way sand. It is believed that load degradation for dense sands occurs due to the reduction of the shear zone surrounding the anchor with cycling. Additional mechanisms causing significant loss of load in the non-uniform subangular Glazier Way sand are particle segregation and reorientation. As the anchor is cycled the larger grains are pushed away from the shear surface by the ribs, while smaller grains move towards it. The agglomeration of fine particles around the ribs results in poor interlocking of grains with the ribs and the reorientation of particles results in decreased interlocking between particles. Neither mechanism would occur to any significant measure in the uniform subrounded Ottawa 20-30.

   Another mechanism that could account for the loss of interface friction is the coating of sand grains with iron oxide from the abrasion of the anchor. Oda et al. (10) have reported that sand grains coated with talcum powder suffered a 50% reduction in overall friction angle.

CONCLUSIONS

A study of the interaction of sand with ribbed anchors was performed. The tests were conducted in a large triaxial tank. To simulate driving conditions, the anchors were driven past the peak load until a constant residual load developed. For dense sand the residual load was relatively close to the peak load, while for loose sand the residual load value was considerably lower.

The most significant observation, however, was the loss of pullout resistance with cycling. On the first load reversal in dense sand a decrease in pullout resistance of approximately 45% occurred. For both dense and loose sand severe interface friction degradation then followed with continued cycling. The only exception to the degradation of interface friction was for loose Ottawa 20-30 sand which did not degrade but actually increased to some degree. For all load reversals in both dense and loose sand the displacement to mobilize full shear resistance was 0.25 in. or one rib spacing. Proposed mechanisms responsible for side friction degradation are the loss of the increased normal stress on the anchor due to initial dilation, development of circumferential arching, particle segregation, and particle coating.

ACKNOWLEDGMENTS

This study was part of a research program supported and funded by Air Force Office of Scientific Research under Grant No. AFOSR-88-0166 to the University of Michigan. The authors thank Mr. Masyhur Irsyam for contributing the results of his study of grain movements to the discussion of the ribbed anchor pullout tests.

REFERENCES

1. DV Holmquist and H Matlock, Resistance-Displacement Relationships for Axially-Loaded Piles in Soft Clay, Proc. Offshore Tech. Conf., Houston, Texas, 1976, OTC2474.
2. RD Hryciw, Load Transfer Mechanisms in Anchored Geosynthetic, University of Michigan Report to AFOSR on Grant No. 88-0166, 1990.
3. RD Hryciw and M Irsyam, Shear Zone Characterization in Sands by Carbowax Impregnation, ASTM, Geotechnical Testing Journal, March, 1990. pp. 49-52.
4. M Irsyam, Load Transfer Mechanisms in Anchored Geosynthetic Systems, thesis to be submitted to the University of Michigan, Ann Arbor, MI, in partial fulfillment of the requirements for the degree of Doctor of Philosophy.
5. RM Koerner, Designing with Geosynthetics, Prentice-Hall, Englewood-Cliffs, NJ, 1986, pp 118-123.
6. RM Koerner, Slope Stabilization Using Anchored Geotextiles: Anchored Spider Netting, Proceedings Spec. Engineering for Roads and Bridges Conference, PennDot, 1984, Harrisburgh, PA.
7. RM Koerner and JC Robbins, In-situ Stabilization of Slopes Using Nailed Geosynthetics, Proceedings of the Third International Conference on Geotextiles, Vienna, Austria, April 7-11, 1986, PP. 395-400.
8. JK Mitchell and WCB Villet, Reinforcement of Earth Slopes and Embankments, National Cooperative Highway Research Program Report 290, Transportation Research Board, national Research Council, Washington DC, June, 1987, pp. 22-31.
9. HG Poulos, Cyclic Axial Response of Single Pile, J. Geotech. Engrg. Div., ASCE, 107(GT1), Jan. 1981, pp. 41-58.
10. M Oda, K Junichi, and S. Nemat-Nasser, Experimental Micromechanical Evaluation of the Strength of Granular Materials: Effects of Particle Rolling, Mechanics of Granular Materials: New Models and Constitutive Relations, Elsevier, Amsterdam, 1983, pp. 21-29.
11. KSS Rao and KH Venkatesh, Uplift Behavior of Short Piles in Uniform

Sand, Soils and Foundation, Vol. 25, No. 4, Dec. 1985, pp. 1-7.
12. F Schlosser and V Elias, Friction in Reinforced Earth, Proc. ASCE Symposium on Earth Reinforcement, Pittsburgh, PA, 1978, pp. 735-761.
13. JP Turner and FH Kulhawy, Drained Uplift Capacity of Drilled Shafts Under Repeated Axial Loading, J. Geotech. Engrg., ASCE, Vol 116 No. 3, March 1990, pp. 470-491.
14. M Uesugi, and H Kishida, Frictional Resistance at Yield Between Dry Sand and Mild Steel, Vol. 26, No. 4, Dec. 1986, pp. 139-149.
15. M Uesugi, H Kishida and Y Tsubakihara, Behavior of Sand Particles in Sand-Steel Friction, Soils and Foundations, Vol. 28, No. 1, Mar. 1988, pp. 107-118.
16. M Uesugi, H Kishida and Y Tsubakihara, Friction Between Sand and Steel Under Repeated Loading, Soils and Foundations, Vol. 29, No. 3, Sept. 1989, pp. 127-137.

DESIGN OF RETAINING WALLS REINFORCED WITH GEOSYNTHETICS

Tony M. Allen[1], M.ASCE and Robert D. Holtz[2], F.ASCE

ABSTRACT: This paper is a state-of-the-art review of current design practices for earth retaining walls reinforced with geosynthetics. After a brief history and a listing of appropriate applications, the behavior of these structures is discussed, and current design methods for internal and external stability are presented and evaluated. The properties required for design are also described, with appropriate consideration of backfill, soil-geosynthetic interaction properties, and modifications because of installation damage, creep, and chemical aging. Facing and seismic design are also briefly discussed, and recommendations for future research are given. Based on this review, it appears that geosynthetic walls provide a cost effective adaptable solution to retaining wall needs, and they are relatively straightforward to design. The main problem hindering their widespread use is the difficulty in assessing long-term strength of the geosynthetic. Once this problem is solved, geosynthetic walls are likely to become the standard retaining wall in geotechnical engineering practice.

INTRODUCTION

Retaining walls are used in places where slopes are uneconomical or not technically feasible. Retaining walls with reinforced backfills are very cost effective, especially for higher walls, and their flexibility makes them very suitable for soft foundations and to survive earthquakes.

The concept of reinforcing the backfill behind retaining walls was developed by H. Vidal in France in the mid-1960's. Since then, several thousand walls reinforced with steel strips have been successfully built around the world. The use of geosynthetics as reinforcing elements started in the 1970's because of concerns about the possible corrosion of steel strips in reinforced earth. Corrosion rates of metals in soils are very difficult to predict, even when galvanizing or other corrosion protection is used. Furthermore, geosynthetic walls have been found to be 30 to 50 percent less expensive than many other reinforced soil systems.

Reinforced soil retaining walls using geosynthetics are shown in concept in Fig. 1. Considerable economy can be achieved when the geosynthetic is also used as the facing (Fig. 1b); the reinforcement sheets are wrapped over the

---

[1]Asst. Found. Engr., Materials Lab., QM-21, Washington State Dept. of Transportation, P.O. Box 167, Olympia, WA 98504

[2]Prof., Dept. of Civil Engr., FX-10, Univ. of Washington, Seattle, WA 98195

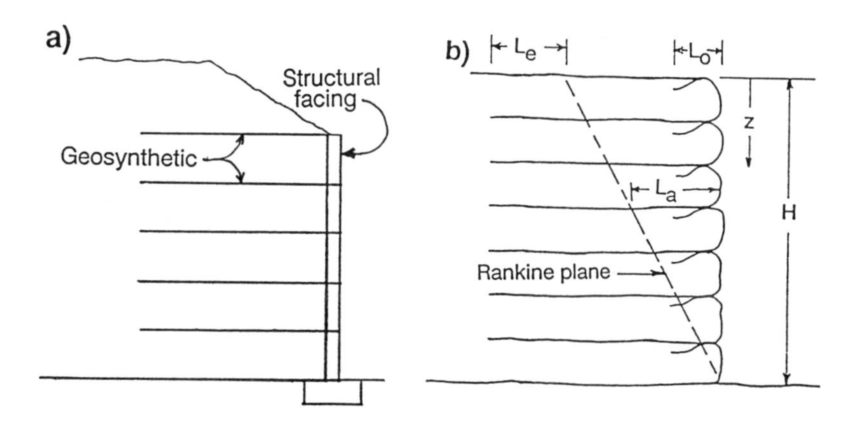

Figure 1. Concept of geosynthetically reinforced walls: (a) with precast concrete face panels; (b) wrapped geosynthetic facing.

backfill and tucked in to make a kind of sandbag, and successive layers are backfilled and stacked to form the wall. However, such facings may be less attractive and less durable than precast concrete or other structural face elements. Full height as well as segmental structural facings are also possible.

Historical Development. Bell and Steward (1977) described the first full scale geotextile reinforced soil wall constructed in the U.S. in 1974 in Oregon. Additional walls were built in Washington and New York (Douglas, 1982). In Colorado, ten different geotextiles and designs were tested full scale (Bell, et al., 1983). These tests indicated that geotextile reinforced walls are economical and practical, and wall performance showed that the design procedures were very conservative.

Other examples of geotextile-reinforced walls were given by Puig, Blivet, and Pasquet (1977). Geotextiles were also used to reinforce the vertical downstream face of an earth dam spillway in France (Kern, 1977). In Sweden, experiments on geotextile reinforced walls were carried out by Holtz and Broms (1977, 1978). Excellent case histories of geosynthetic reinforced walls can be found in the proceedings of the 2nd, 3rd, and 4th International Conferences on Geotextiles as well as the 1987, 1989, and 1991 Geosynthetics conferences.

Other excellent references on geosynthetic wall design are also available. For example, in spite of a misleading title, Mitchell and Villet (1987) contains a wealth of information on reinforced walls. Jarrett and McGown (1988) also have a number of useful papers on the subject, including Yako and Christopher's (1988) detailed compilation of North American case histories. Finally, the FHWA has recently completed a major research project on reinforced soil structures, and the project report of that research (Christopher, et al., 1989) is particularly recommended.

This paper reviews the applications and behavior of geosynthetic reinforced retaining structures. A number of design procedures and the properties required for design are described and evaluated. Also discussed are the design of the wall facing and seismic design. Recommended design procedures and research needs are also given.

GEOSYNTHETIC WALL APPLICATIONS

Conceptually, geosynthetic reinforced walls can be used in any fill wall situation. They can be especially useful in settlement and earthquake prone areas due to their flexibility. The relatively wide wall base width required for geosynthetic walls generally precludes their use in cut situations, similar to other reinforced soil walls. The majority of geosynthetic walls built to date have been for temporary applications such as preload fills and support of detour roadways, or for relatively short design life permanent applications, such as low volume road construction, slide stabilization, and construction in remote areas. The use of geosynthetic walls for long design life permanent applications has been minimal but will likely grow as designers gain confidence in the durability of geosynthetic reinforcement.

The level of confidence needed for the design of a geosynthetic wall depends on the critical or noncritical nature of the application (Carroll and Richardson, 1986). The criticality of an application depends on the design life, maximum wall height, the soil environment, and risk of loss of life and impact to the public and to other structures if failure occurs. Assessment of criticality is rather subjective, and sound engineering judgement is required. The engineer or regulatory authority should determine the critical nature of a given application, and guidelines for assessing the critical nature of a structure are presented in Table 1. These guidelines can be adjusted somewhat with the implementation of a comprehensive performance monitoring program which will give an early warning of impending failure.

Note that guidelines are not given for noncritical temporary applications regarding risk of loss of life, loss of services, and/or economic loss if failure occurs. Provided appropriate design procedures are used, the main source of risk regarding geosynthetic wall performance is in the prediction of long-term geosynthetic strength. Such prediction is not an issue for temporary walls, except possibly in the most extreme environments. Therefore, risk is always low for temporary walls regarding loss of life, services, and economic loss.

If any of the requirements in Table 1 for noncritical applications are not met, the wall should be considered critical. Geosynthetics may be used in critical

Table 1. Definition of Criticality of Geosynthetic Reinforced Walls

Structures	Noncritical and Temporary	Noncritical and Permanent
Structure design life	≤ 5 years	≤ 50 years
Maximum wall height	15 m	5 m
Soil chemistry	3 < pH < 10	chemically neutral; 6 < pH < 8 and low concentrations of deleterious chemicals
Average soil temperature	≤ 20° C	≤ 20° C
Risk of loss of life if failure occurs	----	low
Loss of services and/or economic loss if failure occurs	----	minimal
Reinforced structure supports other structures	no, unless other structures are temporary and noncritical	no

applications if the geosynthetic manufacturer can provide adequate scientific data about the long-term design strength of the geosynthetic.

GEOSYNTHETIC WALL BEHAVIOR

Loads. Soils reinforced with geosynthetics are a composite material, and their behavior depends on the interaction between soil and geosynthetic. Tensile stress in the geosynthetic is transmitted to the soil by friction. At failure, a well defined surface approximating a Rankine failure surface develops, as shown by model studies (Bell, et al., 1975; Holtz and Broms, 1977). The failure plane separates an active zone behind the wall face from a resistant zone. The soil within and behind the reinforced wall is subject to lateral earth pressures, as conceptually shown in Fig. 2. Note that the backfill materials and external loading conditions may be different within and behind the reinforced zone, and thus the horizontal stress conditions may also be different in these zones.

Other loads which may act on a reinforced wall include surcharges, live loads, water loads, and environmental loads such as frost action and swelling soils. Seismic loads may also need to be considered; they include both inertial forces on the wall mass and an increase in earth pressure behind the reinforced fill.

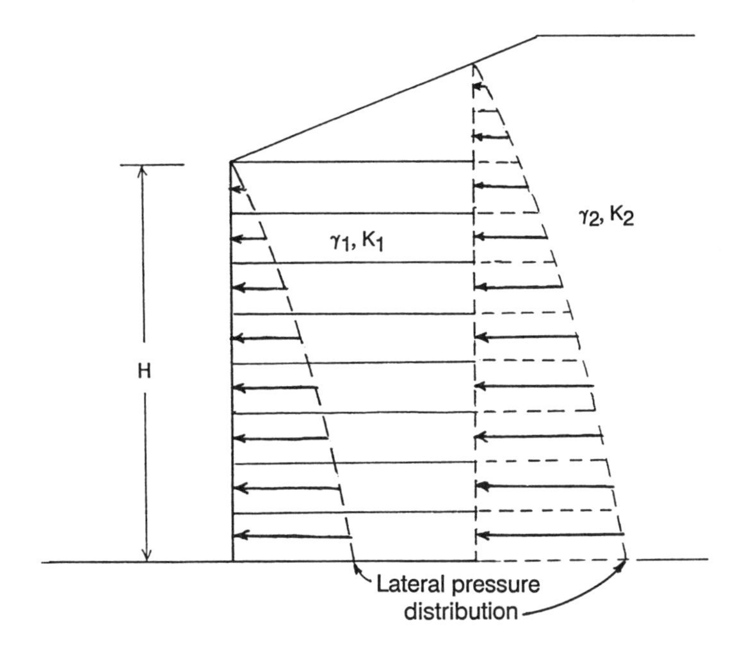

Figure 2. Lateral earth pressures acting within and behind reinforced section.

Performance Criteria and Modes of Failure. The relevant performance criteria for geosynthetic walls are stability and acceptable deformations. Adequate stability is of course necessary but not sufficient to insure acceptable performance. To investigate stability, all possible modes of failure must be considered. Failure can occur either internally or externally, i.e., within the reinforced soil mass or outside it. The factors of safety for each failure mode are numerically different and often applied differently (Bonaparte, et al., 1987). For internal stability, for example, the factor of safety is commonly applied to the soil and reinforcement strength. For external stability, as in conventional geotechnical stability calculations, the factor of safety is applied to the resistive forces or moments.

The other performance criterion, acceptable wall deformations, is determined by the movements required to mobilize the shear resistance in the reinforcement at the working stress level. As noted by Bonaparte, et al. (1987), allowable wall deformations depend on the tolerance of the facing to movements, the visual impact of a deformed wall face, and on serviceability criteria for supported structures such as pavements, bridges, cranes, etc.

External stability is evaluated by assuming that the reinforced soil mass acts as a rigid body, although in reality the wall system is really quite flexible. It must resist the earth pressure imposed by the backfill retained by the reinforced mass and any surcharge loads. Potential external modes of failure to be considered are:

- Sliding on the wall base
- Overturning of the wall
- Bearing capacity of the foundation
- Overall slope stability (sliding or rotation on a deep seated failure surface)

Seismic stability should also be checked in earthquake prone areas.

Internal stability basically considers two failure possibilities: (1) rupture or excessive elongation of the reinforcement, and (2) pullout of the reinforcement.

CURRENT DESIGN METHODS

Reviews of geosynthetic wall design methods have been made by Christopher and Holtz (1985), Mitchell and Villet (1987), and Claybourn (1989), among others. Our summary will focus on how the various methods treat certain key aspects of wall design.

Internal Stability. The proposed design methods are:

1. Forest Service-tieback wedge (Steward, et al., 1977)
2. Broms (1978)
3. Leshchinsky (Leshchinsky and Volk, 1985; Leshchinsky and Perry, 1987; Leshchinsky and Boedeker, 1989)
4. FHWA (Christopher, et al., 1989)
5. Others (Jewell, 1988; Gourc, et al., 1986 and 1988; Rowe and Ho, 1988).
6. Working stress methods

Design will be discussed in terms of (a) assumed stress distribution, (b) how reinforcement tensile forces are determined, and (c) how reinforcement pullout forces are determined. Additional comments will be made about the methods in Items 5 and 6 above.

(a) Stress distribution

Possibilities (Fig. 3) include the classical Rankine distribution, which is triangular (Steward, et al., 1977), rectilinear (Broms, 1978), bilinear, and the

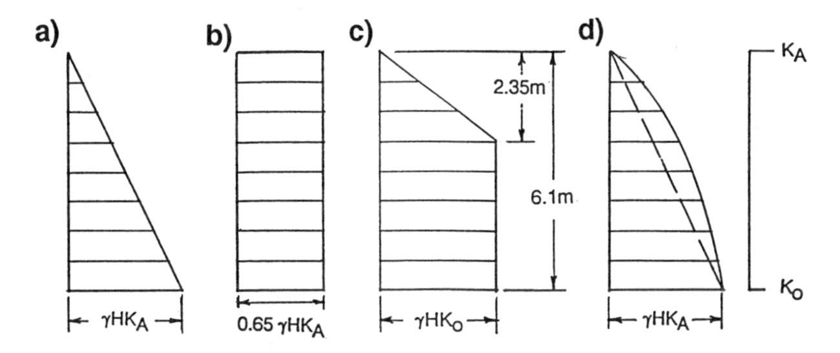

Figure 3. Lateral stress distributions used for design: (a) Rankine, (b) uniform, (c) bilinear, and (d) coherent gravity (after Simac, et al., 1990).

coherent gravity distribution, with K_o at the top and gradually decreasing to K_A at 6 m (20 ft) from the top (Christopher, et al., 1989). Both the uniform and coherent gravity distributions are semiempirical. Leshchinsky and co-workers assume failure occurs on a logarithmic spiral failure surface through the reinforced soil mass, but for the distribution of horizontal force to the reinforcement, they assume that a triangular stress distribution applies.

The FHWA method (Christopher, et al., 1989) uses the classical Rankine tieback wedge stress distribution for geotextiles, but for geogrids they assume an earth pressure coefficient of 1.5 K_A at the top of the wall reducing proportionally to K_A at $H \geq 6$ m (20 ft).

(b) Reinforcement tensile forces

How the tensile forces in the reinforcement are determined follows from the assumed stress distributions discussed above (Fig. 3). The Forest Service tieback wedge method simply determines the reinforcement force by

$$T_{applied} = S_v K (\gamma z + q)\, FS \tag{1}$$

where

S_v = vertical spacing of the geosynthetic sheets
K = coefficient of lateral earth pressure
γ = unit weight of backfill soil
z = depth below top of wall of layer under consideration
q = added vertical stress at depth z due the surcharge load
FS = factor of safety

Originally, the at-rest lateral earth pressure coefficient was used for design (Steward, et al., 1977). This method was later modified to use the Rankine lateral earth pressure coefficient (Mitchell and Villet, 1987). The recommended factor of safety is 1.2 to 1.5.

Broms (1978) proposed the same lateral earth pressure distribution (Fig. 3b) as is assumed for the design of tieback walls (Terzaghi and Peck, 1967), or

$$\sigma_{reinf} = S_v(0.65\gamma HK_A) \tag{2}$$

Broms uses the allowable long term geosynthetic tensile strength to determine the required reinforcement spacing. No other factor of safety is applied.

The Leshchinsky method uses

$$\sigma_{reinf} = \frac{z\gamma HT_m (FS)}{n} \tag{3}$$

where T_m = nondimensional mobilized equivalent tensile resistance
n = number of reinforcement layers

and the other terms were defined previously. Leshchinsky and Boedeker (1989) provide a chart for determination of T_m if a mobilized value of ϕ is known; i.e., $\phi_m = \tan^{-1}(\tan\phi/FS)$.

Note that Eq. 3 has the same basic form as Eq. 1, where H/n is equivalent to S_v for uniform spacing; thus, T_m is a lateral earth pressure coefficient. Though the method used by Leshchinsky and co-workers to obtain T_m is substantially different than that used to obtain K_A, T_m is equal to K_A if the reinforcement layers are assumed to remain horizontal.

FHWA recommends the reinforcement stress be determined by the tieback wedge method (Eq. 1), except that K is based empirically on the type of geosynthetic (geogrids or geotextile) and its stiffness factor S_r. S_r is determined from the secant modulus at 5% strain divided by the average vertical spacing of the geosynthetic. No specific recommendations on safety factors are given.

(c) Reinforcement pullout forces

Reinforcement pullout forces are determined in a similar manner for all methods except Broms and Leshchinsky, who combine the pullout calculations with an overall sliding resistance determination. The required embedment length L_e (Fig. 1b) extending beyond the assumed failure surface is calculated from force equilibrium of the reinforcement at the elevation of the layer under consideration and the friction developed on both surfaces of the geosynthetic, or

$$L_e = \frac{\sigma_{reinf} (FS)}{2\gamma z \tan\phi_{SG}} \tag{4}$$

where ϕ_{SG} = soil-geosynthetic friction angle.

FHWA (Christopher, et al., 1989) have a similar but more complicated approach to soil-geosynthetic friction. Their denominator of Eq. 4 is $2\gamma z f^* \alpha$, where f^* is a modified friction coefficient (e.g., Schlosser and Elias, 1978) and α is a "scale factor" to be obtained from pullout tests. For geotextiles, $f^* = \tan\phi_{SG}$, although they recommend ϕ_{SG} be determined from a direct shear test (it is a lower bound friction angle and thus is conservative). For geogrids, a complex interaction formula is given, which considers the percent open area of the grid, and the thickness and longitudinal spacing of the transverse members (see Christopher, et al., 1989, for details).

Recommended factors of safety for pullout vary from zero (Leshchinsky), 1.3 (Broms), 1.5 (FHWA), to 1.75 (Forest Service). The Forest Service also recommends a minimum length L_e of 0.9 m (3 ft) to account for uncertainties in the location of the potential failure surfaces. This minimum almost always governs the pullout design with sheet geosynthetics and vertical spacings less than 0.5 m.

(d) Other approaches

Gourc, et al. (1986; 1988) present a design procedure which considers the reinforcement to be an anchored membrane that can locally displace at the failure surface. They then utilize the "perturbation" slope stability method commonly used in France. Jewell (1988) presents a modified tieback wedge design procedure. Variable reinforcement lengths and spacings are proposed that may be more efficient but would not ordinarily be practical to construct. Rowe and Ho (1988) suggest the use of the finite element method for wall design. Although probably not a viable design tool except for the most complex projects, such procedures are necessary if deformations are to be properly considered.

(e) Working stress methods

All of the procedures discussed thus far are limiting equilibrium methods, or at least they utilize limiting equilibrium assumptions with empirical adjustment to account for conditions at working stresses. Strain compatibility of the soil and reinforcement is not taken into account. A good discussion of the issues regarding strain compatibility are given by Beech (1987). Christopher, et al. (1989) have found that the stiffness of the reinforcement does have a significant impact on the reinforcing stresses, as discussed previously.

Researchers have recently begun to develop methods which attempt to account for strain compatibility and stiffness of the reinforcement (i.e., working stress methods). The strain compatibility approach proposed by Juran, et al., (1990) shows promise; however experimental validation and a more practical calculation method are required before it is likely to be accepted by most wall designers. Additional research on working stress methods is warranted.

External Stability. For external stability, the reinforced soil mass is considered to act as a rigid body, and ordinary procedures are used to compute the factors of safety for sliding, overturning, bearing capacity, and overall slope stability. For sliding resistance, FHWA (Christopher, et al., 1989) recommend that no wall friction be assumed between the back of the reinforced section and the retained backfill when geosynthetic reinforcement is used. Because of the flexibility of geosynthetic reinforced sections, it seems unlikely that the reinforced mass would overturn as a rigid block, although there is some evidence this mode of failure can occur with cohesive backfills (Goodings, 1989 and 1990). However, overturning should be checked. For foundation bearing capacity calculations, the resultant force on the foundation is inclined and eccentric, and thus a nonuniform and reduced area vertical stress distribution is applicable. Conventional bearing capacity theory assumes a uniform (i.e., flexible) loading, which is appropriate for geosynthetic walls.

PROPERTIES OF GEOSYNTHETIC WALL COMPONENTS

Material properties needed for the design of geosynthetic walls include backfill and subgrade soil properties, soil-geosynthetic friction, and short and long-term geosynthetic strength and deformation. Soil properties, which can be obtained from standard geotechnical site and laboratory investigations, include (1) shear strength, density, and compressibility of the backfill and foundation soils; (2) backfill permeability, if there is a concern about the ability of the backfill to drain; (3) backfill chemical characteristics, if there is concern about geosynthetic durability; and (4) soil modulus, if a sophisticated analysis is used. Values for shear strength and density may be conservatively assumed based on previous experience, especially if the construction specifications are well defined and quality inspection of the backfill materials and their placement are insured.

Geosynthetic properties required for internal stability include tensile strength and soil-geosynthetic friction. Bonaparte, et al. (1987) present a detailed discussion of how these properties should be obtained for design, and they include a discussion of the effects of time (creep) and installation on the design tensile strength. FHWA (Christopher, et al., 1989) also give recommendations for obtaining the required design properties of geosynthetics. A comprehensive review of long term tensile strength determination methods including the effect of environmental factors such as ultraviolet radiation, chemical aging, and biological degradation, has been prepared by Allen (1991). Only a brief summary will be given herein.

The roughness or adhesion of the surface of the geotextile structure is necessary for the development of soil-geosynthetic friction, whereas geogrids may additionally rely on the passive bearing resistance of their cross ribs (Mitchell and Villet, 1987). It has generally been assumed for wall design (Bell, 1980) that the soil-geosynthetic friction angle ϕ_{SG} for granular soils is equal to $2/3\phi$, where ϕ is the angle of internal friction of the backfill material. Studies by various researchers have shown this to be a conservative assumption, at least for granular soil types where soil creep is not a significant problem (Holtz, 1977; Ingold, 1982; Mitchell and Villet, 1987; Bonczkiewicz, et al., 1988). It is sufficient to use an assumed ϕ_{SG} of $2/3\phi$ for noncritical applications with limit equilibrium design methods. Pullout tests are desirable in critical applications, especially if widely spaced reinforcement layers are anticipated, if strips of geosynthetic rather than continuous sheets are utilized, or if fine-grained backfill soils which may tend to creep are anticipated. However, care should be taken when interpreting the results of pullout tests on geogrids because of the very complex relationship of junction strength to pullout load.

Junction strength cannot be equated directly to pullout resistance since each junction is only subjected to a portion of the total pullout load. The spacing of the junctions transverse and parallel to the direction of stressing will influence the magnitude of the portion of the total pullout load applied to each junction. Considering the nonuniform distribution of stress along the geogrid in the direction of stressing, the relationship of junction strength to pullout load is complex indeed, and is a subject of some controversy (Carroll, 1988; Simac, 1988). Current AASHTO-AGC-ARTBA Task Force No. 27 guidelines for geosynthetic walls require that the summation of the junction strengths within a 300 mm length of the grid be equal to or greater than the ultimate strength of the grid element to which they are attached (Task Force 27, 1989). If this criteria is met, then the ultimate strength of the grid does not need to be reduced to account for junction strength. This appears to be a reasonable minimum criteria for junction strength until more research is performed to evaluate this property and its effect on geosynthetic structure performance.

Even if the junctions in a geogrid do fail prematurely, the consequences of a premature junction failure are not clear at this time. Though it is obvious that loss of the crossribs could significantly reduce the soil-geogrid interaction coefficient, and recent research appears to bear this out (Berg and Swan, 1990), this loss may not be all that important regarding pullout design of geosynthetic walls. Current practice for pullout requires a minimum pullout length, L_e, of 0.9 m (3 ft), which is considerably greater than required by pullout calculations for typical design geometries and granular backfills. It is possible, however, that a reduced interaction coefficient could allow horizontal deformation in the wall to increase, as the grid could become less restrained by the soil. The amount of additional deformation which could result from this unknown.

Cohesive backfills, a problem in conventional retaining structures, are also a problem in geosynthetic walls. Poor drainage and potential creep are difficult problems to deal with in any case. Clean granular backfill should be used if at all possible, but if not, soil-geosynthetic interaction tests are advisable. Both short term tensile tests and creep tests should be performed. Some research has been conducted on cohesive backfills (Goodings, 1989 and 1990) and more is underway in Colorado (Chou, 1990).

Long-Term Geosynthetic Tensile Strength. The geosynthetic strength at any time during the design life of the wall must be greater than the tensile stresses applied to the geosynthetic. The lack of information regarding this issue is the main stumbling block preventing more widespread use of geosynthetic walls. Part of the problem is the large variation in the molecular chemistry and structure of the three most common geosynthetic polymers, polypropylene (PP), polyethylene (PE), or polyester (PETP). Different kinds of PP, PE, and PETP have different abilities to resist environmental degradation.

Geosynthetics lose tensile strength due to the following reasons: ultraviolet (UV) light degradation, installation damage, creep, biological degradation, and chemical aging, and Allen (1991) provides a detailed discussion of each factor. Unless the wall face is left uncovered, strength loss due to UV degradation occurs only at the beginning of the wall design life. Strength loss due to installation damage also occurs only at the beginning of the wall design life. Strength losses due to creep, biological degradation, and chemical aging can occur throughout the life of the wall. Chemical aging includes degradation mechanisms such as oxidation, environmental stress cracking, and hydrolysis. The effect of all these losses on the geosynthetic design strength is considered as follows (Task Force 27, 1989; Koerner, 1990):

$$S_a = \frac{S_{ult}}{FS_{UV} \times FS_{ID} \times FS_{CR} \times FS_{CA} \times FS_{BD} \times FS_{SM} \times FS_{JCT}} > T_{applied} \quad (5)$$

where,

S_a = allowable long-term geosynthetic strength;
S_{ult} = ultimate wide width strength of geosynthetic;
$T_{applied}$ = tension applied to geosynthetic;
FS_{UV} = partial factor of safety for ultraviolet light degradation;
FS_{ID} = partial factor of safety for installation damage
FS_{CR} = partial factor of safety for creep;
FS_{CA} = partial factor of safety for chemical aging;
FS_{BD} = partial factor of safety for biological degradation;
FS_{SM} = partial factor of safety for strength reductions due to seams and connections;
FS_{JCT} = partial factor of safety to account for effect of junction strength of grid structures.

Because the partial factors of safety are multiplied, Eq. (5) assumes the worst case regarding the synergism of the various degradation mechanisms. Considering how little is known about possible synergism, this worst case assumption is probably appropriate at this time.

If the geosynthetic is left exposed at the face of the wall, UV degradation must be considered. A minimum strength retention of 70 percent after 500 hours in a weatherometer (ASTM D-4355) for geosynthetics is recommended where extended exposure at the wall face is expected. Although soil burial does stop UV degradation, there is some evidence that extended exposure to UV light before soil burial may accelerate the degradation processes after burial (Jailloux and Segrestin, 1988). Therefore, UV exposure before soil burial should be minimized. FS_{UV} can be taken as 1.0 for the portion of the geosynthetic which is buried provided that the manufacturer's recommendations for storage and handling are followed.

Tables 2, 3, and 4 are recommended ranges of partial factors of safety for strength losses due to construction damage, creep, and chemical aging in chemically neutral soils and at average soil temperatures of 20°C or less. These factors of safety should only be used for preliminary design or for walls in noncritical applications. If the application is considered critical, product specific scientific data appropriate for the expected soil environment should be obtained, as discussed previously.

The partial safety factors in Table 2 for high survivability should be used if the expected installation conditions are not known during design. Higher safety factors than shown in Table 2 may be needed if installation conditions are expected to be severe, such as with highly angular gap-graded gravels and initial soil lift thicknesses of less than 15 cm (6 in.) In that case on-site installation damage tests should be conducted. The high end of the safety factor ranges shown in Tables 3 and 4 should be used if little is known about the specific geosynthetic and if the proposed wall is at the limits of the guidelines set in Table 1 for noncritical applications. A value of 1.0 is recommended for FS_{BD} for noncritical applications.

Table 2. Partial Factor of Safety to Account for Installation Damage (after Allen, 1991).

Geosynthetic Polymer	Geosynthetic Type	Geosynthetic Weight (g/m²)	Range Of Safety Factor: High Survivability	Range Of Safety Factor: Low Survivability
PP and HDPE	Nonwoven	< 270	2.0	1.15
		> 270	1.8	1.05
	Woven	< 270	2.5	1.2
		> 270	1.4	1.1
	Grid	all weights	1.4	1.0
PETP	Nonwoven	< 270	3.2	1.25
		> 270	1.8	1.1
	Woven	< 270	?	?
		> 270	2.2	1.4
	Grid	all weights	?	?

Table 3. Partial Factors of Safety to Account for Creep (after Allen, 1991).

Application	Polymer Type	Range of FS
Noncritical, temporary	PP	2.5 to 3.0
	PE	2.3 to 3.0
	PETP	1.7 to 2.0
Noncritical, permanent	PP	3.3 to 5.0
	PE	2.7 to 5.0
	PETP	2.0 to 2.5

Table 4. Partial Factors of Safety to Account for Chemical Aging (after Allen, 1991).

Application	Polymer Type	Range of FS
Noncritical, temporary	PP	1.0 to 1.25[a]
	PE	1.0 to 1.25[a]
	PETP	1.0 to 1.25[a]
Noncritical, permanent	PP	1.25 to 2.0[b]
	PE	1.1 to 1.5 ?
	PETP	1.25 to 2.0[c]

[a] Range depends on severity of soil chemistry.

[b] High end of range reflects potential for antioxidants to leach out. If antioxidants in proposed geosynthetic are known to be highly leach resistant, use low end of range.

[c] Does not apply to determination of strength loss of PETP where in contact with a concrete facing. Higher safety factors may be needed.

RECOMMENDED DESIGN PROCEDURES

Internal Stability. Reinforcement stresses should be calculated using the tieback wedge method, as shown in Eq. 1. A factor of safety FS of 1.2 for temporary walls and 1.5 for permanent walls is recommended.

The active earth pressure coefficient with a triangular pressure distribution should be used for flexible faced geotextile walls. As discussed previously, there is some evidence from the recent FHWA study (Christopher, et al., 1989) that higher K coefficients may be needed for geogrid and possibly high modulus geotextiles walls, probably due to higher compaction stresses as well as the additional stiffness that geogrids and high modulus geotextiles possess. Compaction stresses can develop more readily against the relatively stiff concrete facing systems which are often used with geogrid walls. However, it is not clear from this research just how much influence the stiffness of the facing system had on reinforcement stress levels. If relatively stiff facing systems are utilized to construct geotextile walls, higher compaction stresses may develop. Until more is known, a K coefficient between the active and at rest condition, as is shown in Fig. 26, page 103 of Christopher, et al. (1989) for geogrids, is also recommended for geotextiles. (This means in effect that a coherent gravity pressure distribution will be used.) Additional research is needed to verify the effect of different facing systems on the lateral earth pressure coefficient. The calculated reinforcement stresses should be equated to the long-term design strength S_a using Eq. 5 to determine the required reinforcement tensile strength.

The total width of the wall required to insure that it is stable against pullout is equal to $L_a + L_e$, as shown in Fig. 1b. If the wall is noncritical, the vertical spacing of layers is 0.5 m (18 in.) or less, and if continuous sheet reinforcement, granular backfill, and a minimum L_e of 0.9 m (3 ft) are used, no calculations are necessary. All methods will show that L_e as calculated is less than 0.9 m (3 ft). If the wall application is critical, if the reinforcement is not continuous and/or widely spaced, or if cohesive backfill is used, L_e should be calculated using the method by Christopher, et al. (1989, p. 107), or pullout tests should be conducted. A minimum factor of safety for pullout of 1.5 should be used. The dimensions of the active zone, which will define L_a, should be determined assuming a Rankine failure surface.

The presence of heavy or unusual surcharges are not specifically addressed by the above recommendations. The FHWA manual (Christopher, et al., 1989) provides some guidance and should be consulted for these special cases.

Though several methods have been proposed to deal with lateral wall deflection, none appear to be really adequate in our opinion and much research is needed in this area. The most widely used method of dealing with deflection (Task Force 27, 1989) consists of limiting the design load applied to the reinforcement to the load at the 5% strain level. This procedure appears to be unwarranted based on observations by Bell and Steward, (1977), Berg, et al., 1986, Simac, et al., (1990), and Christopher, et al., (1990), in which measured reinforcement strains have been less than 1%. The use of the tieback wedge method appears to provide sufficiently low design load levels to insure that reinforcement strains and the resulting face deflections will be small.

External Stability. The external stability of geosynthetic walls should be checked for all the possible modes of failure discussed previously. Conventional methods of analysis are used to evaluate each of these modes of failure, as detailed by Mitchell and Villet (1987) and Christopher, et al. (1989). Minimum recommended factors of safety for each of these modes of failure are as follows:

Sliding: 1.5
Overturning: 2.0
Bearing capacity: 2.0
Overall slope stability: 1.5

Generally, overturning will not control the stability requirements for the wall, but it should not be ignored. There is some evidence that wall deflection at the face will increase significantly if the wall width is too short, especially if surcharges are present (Christopher, et al., 1989) or cohesive backfills are used (Goodings, 1989 and 1990). FHWA (Christopher, et al., 1989) suggest an empirical relationship to estimate relative lateral displacements on walls with granular backfills during construction which was developed from finite element analyses, small scale model tests, and very limited field evidence from 6 m (20 ft) high test walls. The procedure predicted wall face movements of a 12.6 (40 ft) high geotextile wall that were slightly greater than observed (Holtz, et al., 1991).

WALL FACE DESIGN

A variety of facing systems have been used for geosynthetic walls. They can be installed (1) as the wall is constructed or (2) after the wall is constructed. Facings installed as the wall is constructed include segmental and full height precast concrete panels, interlocking precast concrete blocks, welded wire panels, gabion baskets, treated timber facings, and geosynthetic face wraps. In these cases, the geosynthetic reinforcement is attached directly to the facing element. Shotcrete, cast-in-place concrete facia, as well as precast concrete or timber panels, can be attached to steel bars placed or driven between the layers of the geosynthetic wrapped wall face at the end of wall construction after wall movements are complete.

The success of a geosynthetic wall is highly dependent on the type of facing system used and the care with which it is designed and constructed. Aesthetic requirements often determine the type of facing systems. Anticipated deflection of the wall face, both laterally and downward, may place further limitations on the type of facing system selected. Tight construction specifications and quality inspection are needed to insure that the wall face is constructed properly; otherwise an unattractive wall face, or a wall face failure, could result.

The one geotextile wall "failure" reported in the literature was in fact the failure of the wall facing system (Richardson and Behr, 1988). The wall was 4 m (13 ft) in height and utilized 150 × 150 mm (6 × 6 in.) timbers to form the face. The design of the face did not consider any potential deformation of the geotextile during construction. Vertical spacing between reinforcement layers which was too great, as well as lack of compaction at the face also contributed to the problem.

Facings constructed as the wall is constructed must either allow the geosynthetic to deform freely during construction without any buildup of stress on the face, or the facing connection must be designed to take the stress. Although most wall design methods assume that the stress at the face is equal to the maximum horizontal stress in the reinforced backfill, recent research suggests that some stress reduction does occur near the face. The FHWA (Christopher, et al., 1989) recommends that the stress at the face at any depth below the top of the wall be varied linearly from approximately 75% of the maximum geosynthetic stress at the top of the wall to 100% of this stress at the base of the wall. This guideline appears reasonable and is therefore recommended.

Higher face connection stresses are possible due to bending of the connection caused by vertical settlement of the wall. This can be a problem especially if full height precast panels are used, or if compaction at the wall face is poor. Heavy compaction at the wall face can also create high connection stresses. It is best to use lightweight hand-compactors within 1 m of the wall face. In any event, the level of compaction influences facing stresses.

The long-term strength of the facing connections must also be considered in facing design, as it is at the face where the wall environment (moisture, pH, temperature, and abrasion) is often most severe. If metallic connections are utilized, corrosion protection for the connections should be considered. If geosynthetic connections are utilized, installation damage, creep and chemical durability must be considered. A case in point is the wall at Poitiers, France, where polyester reinforcement strips, which were connected directly to the concrete facing, lost 45% of their strength in 17 years (Leflaive, 1988). The concrete created a highly alkaline environment and hydrolysis of the polyester resulted. Had this been allowed to continue, the wall facing system would have surely failed.

The use of a geosynthetic wrap for the wall facing should, in general, be limited to short design life applications (i.e., less than 5 yr). Geosynthetic face wraps can degrade with time due to ultraviolet exposure, and they are subject to fire and vandalism. Regarding ultraviolet light degradation, some of the heavier geogrids may provide acceptable strength at the wall face for design lives of 50 years or more (Wrigley, 1987).

Methods used to calculate required pullout length may also be used to calculate the required face embedment length, Lo, as shown in Fig. 1b. A minimum embedment length of 0.9 m or 3 ft should be specified due to constructability requirements. Constructability requirements would also dictate that the vertical spacing of reinforcement be limited to a maximum of 0.4 m, if a single layer forming system is used to construct walls with a geosynthetic face wrap. Greater spacings could be used if a multi-layer external forming and bracing system is utilized.

Connections for facing systems installed on geosynthetic face wrapped walls after all wall settlement and lateral movement are complete need only be designed to resist overturning, seismic, and gravity forces on the facing itself.

SEISMIC DESIGN

In seismically active areas, an analysis of the stability of the geosynthetic reinforced wall under seismic conditions should be carried out. For temporary structures, a formal analysis is probably not necessary. For permanent structures, seismic analyses can range from a simple pseudo-static analysis to a complete dynamic soil-structure interaction analysis such as might be performed on earth dams and other critical structures.

Seismic analysis procedures for reinforced earth walls with metallic reinforcement and concrete facings are quite well established, and Vrymoed (1990) presents a good review of these procedures. For geosynthetic reinforced walls, Christopher, et al. (1990) recommend the generally conservative pseudo-static Mononabe-Okabe analysis. Their analysis correctly includes the horizontal inertial forces for internal seismic resistance, as well as the pseudo-static thrust imposed by the retained fill on the reinforced section. This approach is also described by Mitchell and Villet (1987).

Because of their inherent flexibility, properly designed and constructed geosynthetic walls are probably able to more than adequately resist seismic loadings, but high walls in earthquake prone regions should be checked. The facing connections must also be able to resist the inertial force of the wall facia which could occur during the design seismic event, as well as stress build up behind the face resulting from strain incompatibility between a relatively rigid facing system and the relatively extensible geosynthetic reinforcement. Much research is needed to evaluate the effect of seismic forces on geosynthetic walls with "rigid" facings.

RECOMMENDATIONS FOR FUTURE RESEARCH

Future research on geosynthetic walls should be concentrated in the following areas: 1) long-term geosynthetic design strength, 2) calculations of geosynthetic stress levels, especially with regard to working stress conditions and surcharges, 3) determination of wall deflections, 4) assumption of a rigid body in external stability calculations, 5) seismic design, especially of the facings and connections, and 6) effect of cohesive soil backfills on design methodology and material properties.

CONCLUDING REMARKS

Retaining wall backfills reinforced with geosynthetics have proven to be a very cost effective alternative to both conventional retaining structures and backfills reinforced with steel strips or bar mats. Geosynthetic walls are easy to build and they are adaptable to variable height, grade, and alignment requirements. They are very flexible and can tolerate large foundation settlements without distress. Once the long term strength and durability problems are solved, geosynthetic walls are likely to become the standard retaining wall in geotechnical engineering practice.

ACKNOWLEDGEMENTS

Prof. J.R. Bell reviewed the manuscript and made numerous technical and editorial comments which significantly improved the paper. The manuscript was prepared by Kim Hartsoch and Kristin Mingus.

REFERENCES

Allen, TM (1991) Determination of Long-Term Tensile Strength of Geosynthetics: A State-of-the-Art Review, Proc. Geosynthetics '91 Conference, Atlanta (submitted for publication).

Beech, JF (1987) Importance of Stress-Strain Relationships in Reinforced Soil System Design, Proc. Geosynthetics '87, New Orleans, (1), 133-144.

Bell, JR (1980) Design Criteria for Selected Geotextile Installations, Proc. 1st Canadian Symposium on Geotextiles, Calgary, 35-37.

Bell, JR, Barrett, RK and Ruckman, AC (1983) Geotextile Earth-Reinforced Retaining Wall Tests: Glenwood Canyon, Colorado, Transportation Research Record 916, 59-69.

Bell, JR and Steward, JE (1977) Construction and Observations of Fabric Retained Soil Walls, Proc. Int. Conf. on Use of Fabrics in Geotechniques, Paris, (1), 123-128.

Bell, JR, Stilley, A.M., and Vandre, B. (1975), Fabric Retained Earth Walls, Proc. 13th Annual Eng. Geology and Soils Eng. Symposium, Boise, 271-287.

Berg, RR, Bonaparte, R, Anderson, RP, and Chouery, VE (1986) Design, Construction, and Performance of Two Geogrid Reinforced Soil Retaining Walls, Proc. Third Int. Conf. on Geotextiles, Vienna, (II), 401-408.

Berg, RR and Swan, DB (1990) Evaluation of Pullout Mechanisms, Proc. Int. Reinforced Soil Conf., Glasgow preprint No. 4/9.

Bonaparte, R, Holtz, RD and Giroud JP, (1985) Soil Reinforcement Design Using Geotextiles and Geogrids, Geotextile Testing and the Design Engineer, J.E. Fluet, Ed., ASTM STP 952, 69-116.

Bonczkiewicz, C, Christopher, BR, and Atmatzidis, DK (1988) Evaluation of Soil-Reinforcement Interaction by Large-Scale Pull-Out Tests, Transportation Research Record 1188, 1-18.

Broms, B.B. (1978) Design of Fabric Reinforced Retaining Structures, Proc. ASCE Symposium on Earth Reinforcement, Pittsburgh, 282-303.

Carroll, R Jr. (1988) Specifying Geogrids, Geotechnical Fabrics Report, (6), 2, 40-43.

Carroll, RG and Richardson, GN (1986) Geosynthetic Reinforced Retaining Walls, Proc. of the Third Int. Conf. on Geotextiles, Vienna, (II), 389-394.

Chou, N. (1990) Preliminary Design of CDOH Geofabric Test Wall with Cohesive Backfills, Colorado Division of Highways, Denver, 9.

Christopher, BR, Gill, SA, Giroud, J-P, Juran, I, Mitchell, JK, Schlosser, F, and Dunnicliff, J (1989) Reinforced Soil Structures, Vol. I., Design and Construction Guidelines, 287 ; Vol. II, Summary of Res. and Systems Info., 158; FHWA, Federal Highway Administration Report FHWA-RD-89-043.

Christopher, BR and Holtz, RD (1985), Geotextile Engineering Manual, Federal Highway Administration, Report FHWA-TS-86/203, 1044.

Christopher, BR, Holtz, RD, and Allen, TM (1990) Instrumentation for a 12.6m High Geotextile-Reinforced Wall, Proc. International Reinforced Soil Conference, Glasgow, Preprint No. 1/10.

Claybourn, AF (1989) A Comparison of Design Methods for Geosynthetic-Reinforced Earth Walls, M.S. Thesis, Univ. of Colorado at Denver, 90.

Douglas, GE (1982) Design and Construction of Fabric Reinforced Retaining Walls by New York State, Presented to Annual Meeting, Transportation Research Board.

Goodings, DJ (1989) Effects of Poorly Drained Backfill on Geotextiles for Earth Reinforcement of Vertical Soil Slopes, Univ. of Maryland, Report AW089-322-046, 51.

Goodings, DJ (1990) Research on Geosynthetics in Reinforced Cohesive Retaining Walls at the Univ. of Maryland, Geotechnical News, (8), 2, 23-25.

Gourc, JP, Ratel, A, and Delmas, P (1986) Design of Fabric Retaining Walls: the Displacements Method, Proc. Third Int. Conf. on Geotextiles, Vienna, (II), 289-294.

Gourc, JP, Ratel, A, and Gottelaud, P (1988) Design of Reinforced Soil Retaining Walls: Analysis and Comparison of Existing Design Methods and Proposal for a New Approach, The Application of Polymeric Reinforcement in Soil Reinforcement, Proc. of NATO Advanced Research Workshop, Kingston, Ont., Jarrett and McGown, Eds., Kluwer Academic Publ., 459-505.

Holtz, RD (1977) Laboratory Studies of Reinforced Earth Using a Woven Polyester Fabric, Proc. Int. Conf. on the Use of Fabrics in Geotechniques, Paris, (3), 149-154.

Holtz, RD, Allen, TM, and Christopher, BR (1991) Displacements of a 12.6 m High Geotextile-Reinforced Wall, Proc. X European Conf. on Soil Mechanics and Foundation Engineering, Florence (submitted for publication).

Holtz, RD and Broms, BB (1977) Walls Reinforced by Fabrics--Results on Model Test, Proc. Int. Conf. on the Use of Fabrics in Geotechniques, Paris, (1), 113-117.

Holtz, RD and Broms, BB (1978) Small Scale Model Tests of Fabric-Reinforced Retaining Walls, Proc. Symposium on Soil Reinforcing and Stabilizing Techniques in Engineering Practice, Sydney, 151-158.

Ingold, TS, (1982) Some Observations on the Laboratory Measurement of Soil-Geotextile Bond, Geotechnical Testing Journal, ASTM, (5), 3/4, 57-67.

Jailloux, JM and Segrestin, P (1988) Present State of Knowledge of Long Term Behavior of Materials Used as Soil Reinforcements, Proc. of the Int. Geotechnical Symposium on Theory and Practice of Earth Reinforcement, Kyushu, Japan, 105-110.

Jarrett, PM and McGown, A, Ed., (1988), The Application of Polymeric Reinforcement in Soil Retaining Structures, Proceedings of the NATO Advanced Research Workshop, Kingston, Ont., 637.

Jewell, RA (1988) Reinforced Soil Wall Analysis and Behaviour, The Application of Polymeric Reinforcement in Soil Retaining Structures, PM Jarrett and A McGown, Eds., Kluwer Academic Press, 365-408.

Juran, I, Ider, HM, and Farrag, K (1990) Strain Compatibility Analysis for Geosynthetics Reinforced Soil Walls, J. of Geot. Eng., ASCE, (116), 2, 312-329.

Kern, F (1977) An Earth Dam with a Vertical Downstream Face Constructed Using Fabrics (in French), Proc. of the Int. Conf. on the Use of Fabrics in Geotechnique, Paris, (1), 113-117.

Koerner, RM (1990) Determination of the Long-Term Design Strength of Stiff Geogrids, GRI Standard Practice GG4, Geosynthetic Research Institute, Philadelphia.

Leflaive, E (1988) Durability of Geotextiles: the French Experience, Geotextiles and Geomembranes, (7), 1, 23-31.

Leshchinsky, D and Boedeker, RH (1989) Geosynthetic Reinforced Soil Structures, J. of Geot. Eng., ASCE, (115), 10, 1459-1478.
Leshchinsky, D and Perry, EB (1987) A Design Procedure for Geotextile Reinforced Walls, Geotechnical Fabrics Report, (5), 4, 21-27.
Leshchinsky, D and Volk, JC (1985) Stability Charts for Geotextile Reinforced Walls, Transportation Research Record 1031, 5-16.
Mitchell, JK and WCB Villet (1987) Reinforcement of Earth Slopes and Embankments, NCHRP Report 290, Transportation Research Board, 323.
Puig, J, Blivet, JC, and Pasquet, P (1977) Earth Fill Reinforced with Synthetic Fabric (in French), Proc. Int. Conf. on the Use of Fabrics in Geotechniques, Paris, (1), 85-90.
Richardson, GN and Behr, LH (1988) Geotextile-Reinforced Wall: Failure and Remedy, Geotechnical Fabrics Report, (6), 4, 14-18.
Rowe, RK, and Ho, SK (1988) Application of Finite Element Techniques to the Analysis of Reinforced Soil Walls, The Application of Polymeric Reinforcement in Soil Retaining Structures, PW Jarrett and A McGown, Eds., Kluwer Academic Press, 541-553.
Schlosser, F and Elias V (1978) Friction in Reinforced Earth, Proc. of the Symposium on Earth Reinforcement, Pittsburgh, ASCE, 735-763.
Simac, MR (1988) Geogrid Article Prompts Response, Geotechnical Fabrics Report, (6), 4, 6-8.
Simac, MR, Christopher, BR, and Bonczkiewicz, C (1990) Instrumented Field Performance of a 6m Geogrid Soil Wall, Proc. Fourth Int. Conf. on Geotextiles, Geomembranes, and Related Products, The Hague, (1), 53-59.
Steward, J, Williamson, R and Mohney, J (1977) Guidelines for Use of Fabrics in Construction and Maintenance of Low-Volume Roads, USDA, Forest Service, Portland, OR. Also published as Report No. FHWA-TS-78-205.
Task Force 27 (1989) Design Guidelines for the Use of Extensible Reinforcements (Geosynthetic) for Mechanically Stabilized Earth Walls in Permanent Applications, AASHTO-AGC-ARTBA Committee on Materials, 10.
Terzaghi, K and Peck, RB (1967) Soil Mechanics in Engineering Practice, Wiley, (2nd Ed.) 729.
Vrymoed, J (1990) Dynamic Stability of Soil-Reinforced Walls, Transportation Research Record 1242, 29-38.
Wrigley, NE (1987) Durability and Longterm Performance of Tensar Polymer Grids for Soil Reinforcement, Materials and Science Technology, (3), 161-170.
Yako, MA and Christopher, BR (1988) Polymerically Reinforced Retaining Walls and Slopes in North America, The Application of Polymeric Reinforcement in Soil Retaining Structures, P.M. Jarrett and A. McGown, Eds., Kluwer Academic Press, 239-283.

DESIGN OF GEOSYNTHETICALLY REINFORCED SLOPES

Barry R. Christopher[1], and Dov Leshchinsky[2]

ABSTRACT: Virtually all design methods of geosynthetically reinforced slopes are based on limit equilibrium analyses. Presented is a brief review of some analyses with an emphasis on assumptions and concepts associated with the inclusion of geosynthetics. Although numerous reinforced slopes were constructed based on limit equilibrium analysis, very little field tests were conducted to verify this analysis predictions. The results of a field program, specifically performed to ascertain the limit equilibrium analysis, is briefly presented. Generally, it appears that current design methods are adequately conservative. Subsequently, a design procedure is presented.

INTRODUCTION

There are two main purposes for using reinforcement in engineered slopes:

1. to provide improved compaction of the edge of a slope, thus decreasing the tendency for surface sloughing, and
2. to increase the stability of the slope, particularly after a failure has occurred or if a steeper than "safe" unreinforced embankment slope is desirable.

For the first application, geosynthetic reinforcement placed at the edges of the embankment slope have been found to provide lateral resistance during compaction (Iwasaki and Watanabe, 1978 and Christopher and Holtz, 1985). The increased lateral resistance allows for an increase in compacted soil density over that normally achieved. Edge reinforcement also allows compaction equipment to more safely operate near the edge of the slope. Even modest amounts of reinforcement in compacted slopes have been found to reduce sloughing and slope erosion. For this application, the design is simple: place a geotextile, geogrid, or wire mesh reinforcement that will survive construction at every lift or every other lift along the slope. Only narrow strips about 1.2 to 2 m in width are required and have to be placed in a continuous plane along the edge of the slope. This application may be combined with reinforcement in a steep slope, with the short strips installed in between the reinforcing strips, thus stabilizing the slope face.

The design of reinforced, steep slopes requires an analysis which is more extensive then that needed for designing unreinforced slopes (Christopher, et al. 1990). The basic objective then is to specify the layout and required tensile strength of the geosynthetic. Most reinforced soil analyses are modified versions of classical limit equilibrium slope

[1]Technical Manager, Polyfelt, Inc., Atlanta, GA 30328
[2]Assoc. Prof., Dept. of Civil Engineering, University of Delaware, Newark, DE 19716

stability methods. The modifications must include additional assumptions on top of those already introduced in the unreinforced analysis. Kinematically, the potential failure surface in a reinforced *homogeneous* slope is assumed typically to be defined by the same idealized geometry (not location, though) as in the unreinforced case (e.g., circular, log spiral, bi-linear wedge). This greatly simplifies the computation and agrees well with the available experimental results. Statically, the inclination (e.g., horizontal, tangential) and distribution (e.g., linear, constant, etc. with depth) of the reinforcement tensile force along the selected failure surface must be postulated. The tensile capacity of a reinforcement layer is taken as the minimum of its allowable pullout resistance behind the potential failure surface and its allowable design strength. The practice assumes that the target factor of safety for the reinforced steep case is equally "safe" as a flatter unreinforced slope with the same factor of safety.

CONCEPTS OF ANALYSIS: BRIEF REVIEW

A detailed summary of several analyses is contained in Christopher and Holtz (1985). The basic approach using a circular arc (e.g., Fellenius' or Bishop's method of slices) has been modified by, for example, Christie and El-Hadi (1977), Phan, et al. (1979), Fowler (1982), Ingold (1982) Koerner (1986), Bangratz and Gigan (1984), Christopher and Holtz (1985,1989), Gourc, et al. (1986), Studer and Meier (1986), Berg, et al. (1989). Logarithmic spiral procedures have been modified by Juran and Schlosser (1978) and Leshchinsky and Reinschmidt (1985). The plane wedge approach has been proposed by Schlosser and Vidal (1969), Lee, et al. (1972), Segrestin (1979) and Jones (1985). An approach based on a bilinear wedge failure surface was presented by Romstad, et al. (1978), Stocker, et al. (1979), Jewell (1982), Murray (1982, 1985), Jewell, et al. (1984), Schneider and Holtz (1986), and Bonaparte and Schmertmann (1988). Edris and Wright (1987) have modified Spencer's method using a general shape slip surface. A method for evaluating complex soil profiles with a non-regular layout of reinforcement is presented in the section describing the design procedure. Several of these analyses have been used to develop design charts to determine the reinforcement requirements for simple slopes. Following is a brief review of three such methods. The critique, rationale, and concepts presented in this review, however, are typical to all methods.

Leshchinsky and his colleagues have extended the log spiral failure mechanism to deal with geosynthetics reinforcement. The statical formulation is rigorous in the sense that all three limiting equilibrium equations are explicitly satisfied for the minimum factor of safety. A unique feature of their formulation is that the distribution of stress normal to the slip surface is part of the solution thus making the problem statically determinate. The inclination of the reinforcement force at the slip surface was taken as orthogonal to the radius defining their intersection with the log spiral. This implies maximum contribution of reinforcement to resist failure (Leshchinsky and Reinschmidt, 1985, and Leshchinsky and Perry, 1989). Such a mechanism is identical to rotational failure of a rigid body where the movement vector is opposing the geosynthetic tensile force. Leshchinsky and Volk (1986) and Leshchinsky and Boedeker (1989) showed that this inclination has very little effect on the required tensile resistance as compared to a horizontal inclination (i.e., typically, the as-installed position) when *granular* backfill is considered; however, the effect is significant for cohesive backfill. Furthermore, the trace of the critical slip surface is deeper for the orthogonal case.

To estimate the distribution of tensile resistance with depth, Leshchinsky and Volk (1985,1986) and Leshchinsky et al. (1986) considered virtual rotation, δ, of a rigid body defined by the log spiral–see Fig. 1. The geotextile at the lowest elevation, denoted by R_{g1} in Fig. 1, will elongate the most (i.e., $R_{g1} \cdot \delta$) and therefore, will reach its allowable tensile strength t_1 first. Once t_1 is fully developed, and assuming only breakage failure mode is possible, a collapse resembling a row of dominoes falling down will occur; that is, all other sheets will fail one after the other in an upward orderly manner.

This mechanism implies that while sheet No. 1 is approaching its strength, all other sheets are not necessarily near failure (i.e., not at t_1) and therefore, their tensile forces need to be specified at that instance. By assuming that up to the elongation that corresponds to t_1, the load-elongation relationship is linear and by utilizing the mechanism shown in Fig. 1, the adjusted tensile force distribution relative to t_1 at the limit state was written as

$$t_j = t_1 \exp[-\tan\phi(\beta g_j - \beta g_1)/F_s] \tag{1}$$

where t_j is the tensile force developed in sheet No. j. Subsequently, the problem was fully defined. Fig. 2 shows the distribution of a typical tensile resistance for frictional soil.

Note that the mobilized strength of the top sheet is about 70% of that at the bottom. Also note that the vector describing the tensile resistance is plotted in its orthogonal orientation. Fig. 3 shows the effects of reinforcement on the critical slip surface and the associated normal stress distribution. Note that larger reinforcement strength deepens the potential slip surface and increases the compressive normal stress over it. Fig. 4 is a typical design chart relating the slope inclination, ϕ, c, γ, H, number of reinforcing sheets and F_s. Although Leshchinsky and Volk (1985) and Leshchinsky et al. (1986) provide information regarding the trace of the surface along which maximum reinforcement strength is required, they do not provide an explicit guide for specifying the required total embedment length.

Leshchinsky and Boedeker (1989) provide a limited discussion about the layout; however, they utilized a different distribution than the one expressed by Eq. (1) resulting with larger required tensile strength. An increase in the required tensile resistance that is directly proportional to the overburden pressure was employed. Although not given explicitly in their paper, the following rationale was used to select this tensile distribution. Refer to Fig. 5. For a given slope m and a free-draining soil defined by (γ, ϕ), the required tensile resistance of the uppermost reinforcement sheet can be calculated rigorously for a failure surface that passes through this sheet only; i.e., $t_a = \gamma d^2 T$, where T is a nondimensional parameter resulting from the limiting equilibrium analysis. The critical log spiral within the uppermost soil layer emerges at point $\underline{A}$. The force t_a acts at the "toe"-point $\underline{A}$–physically counterbalancing the earth pressure against the slope face. Next, stability analysis limited to the two upper layers, will result with critical surface No. 2 emerging through $\underline{B}$ and a tensile resistance $t_b = 2\gamma d^2 T$. Since t_a is taken as a known quantity, t_b can be calculated without resorting to assumptions regarding the tensile resistance distribution. Physically, t_b counterbalances the earth pressure against segment $\underline{AB}$. Similar to the assumption in tieback analysis, t_a is assumed to act behind surface No. 2; i.e., it is fully transferred from the slope surface. Thus, the reinforcement acts as a tieback carrying load from the active zone into the stable

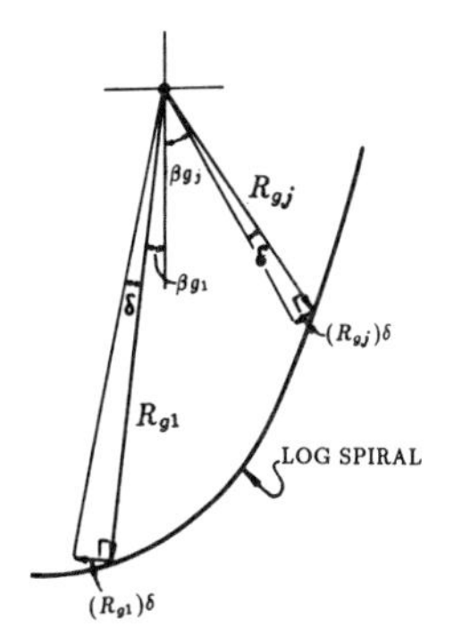

Fig. 1. Rigid Body Rotation of a Log Spiral.

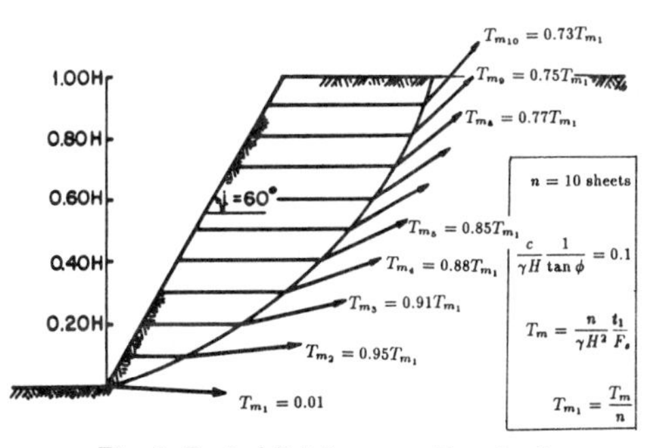

Fig. 2. Typical Reinforcement Reaction in a Non-Cohesive Slope.

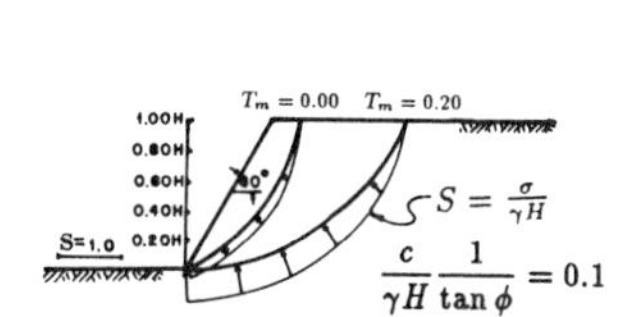

Fig. 3. Typical Effects of Reinforcement on Slip Surface and the Associated Normal Stress Distribution.

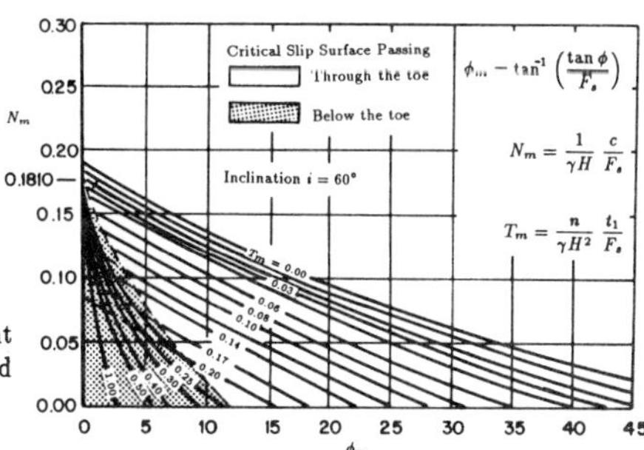

Fig. 4. Typical Design Chart for a Rigid Body Rotational Mechanism.

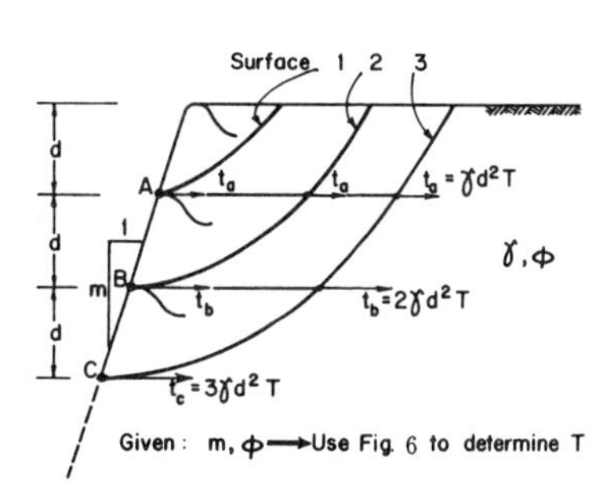

Fig. 5. Rationale for Non-Rigid Body Rotational Mechanism.

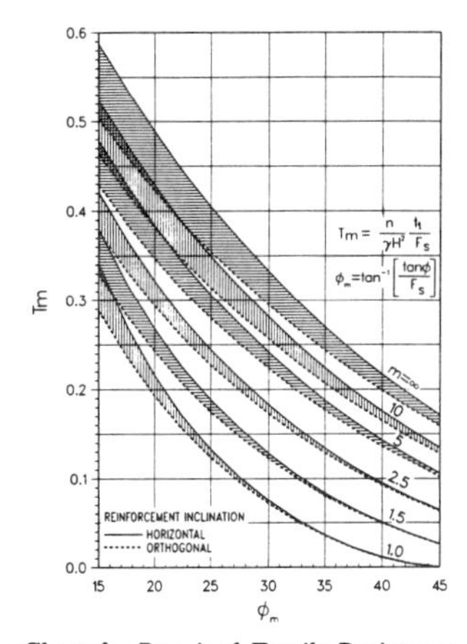

Fig. 6. Design Chart for Required Tensile Resistance.

zone. Continuing with the above rationale for additional layers one concludes that this tieback mechanism results with t_j which is directly proportional to the overburden pressure and that it acts along a critical surface extending between the crest and toe and passing through all the sheets. Fig. 6 shows the corresponding design chart for the required tensile resistance. Note that it contains a range of results considering two possible extreme inclinations of reinforcement force: horizontal (as installed) and orthogonal to the log spiral radius (maximum contribution of reinforcement). The force in reinforcement layer j above the lowest sheet (i.e., $j = 1$) is

$$t_j = t_1 \left(1 - \frac{y_j}{H}\right) \tag{2}$$

where y_j is the elevation of sheet j above toe. Since the critical slip surface corresponding to the orthogonal inclination makes a slightly deeper cut as compared to the horizontal tensile force direction, Leshchinsky and Boedeker (1989) presented design charts only for this inclination; e.g., Fig. 7. Once t_j has been determined from Fig. 6 and Eq. (2), one can estimate the required anchorage length to resist pullout behind the trace of the slip surface given in Fig. 7 using the following equation:

$$\ell_{e_j} = \frac{t_j}{2k(\tan\phi)\bar{\sigma}_j} \tag{3}$$

where k = a parameter relating the coefficient of friction at the soil-geosynthetic interface and $\tan\phi$; $\bar{\sigma}_j$ = the average overburden pressure above ℓ_{e_j}.

Taking the internally stable reinforced zone as a monolithic block, Leshchinsky and Boedeker (1989) assessed its external stability using a two-part wedge mechanism. They conservatively assumed the interwedge force to be horizontal. Fig. 8 provides the required length of reinforcement for various values of k and is based on external stability considerations.

Because of space limitation, Leshchinsky and Boedeker (1989) did not elaborate on practical layout specifications, nor on an examination of stability from a conventional slope stability concept as compared to the tieback approach. In practice, all reinforcing sheets are typically specified to have the same length. Alternative layouts are presented by Schmertmann, et al. (1987) and Leshchinsky and Perry (1987). The tieback mechanism approach makes Leshchinsky and Boedeker (1989) analysis compatible with the one typically used for reinforced soil walls. However, using a conventional slope stability approach there might be a log spiral surface extending outside the reinforced zone for which a lower factor of safety exists. To check this, a study was conducted for two extreme design values of ϕ: 20° and 40°. The log spiral analysis for the orthogonal reinforcement inclination was modified assuming the tensile resistance outside the sliding mass, but within the reinforced zone, to be as before (i.e., linear, as for the tieback approach–Fig. 6). Fig. 9 shows the results of this study. Note that ℓ_s equals the total embedment length, ℓ, minus the anchorage length, ℓ_e (see Fig. 7). Since ℓ_e is typically small relative to ℓ, it is clear from Fig. 9 that the layout prescribed by Leshchinsky and Boedeker (1989) may be somewhat short using the slope stability approach. However, steep slopes are no different from reinforced soil walls in the sense of stability, and for walls the tieback analysis has been proven to be safe. Limited field experience with a 63° slope (Fannin and Hermann, 1990) indicates that even when the reinforcement is shorter than that predicted by the conventional slope stability analysis, satisfactory performance is obtained. In fact, it can be verified that the shorter reinforcement in this field case history approximately corresponds to Leshchinsky and Boedeker (1989) tieback analysis.

It is interesting to point out a seemingly unreasonable pattern in Fig. 9 for the required ℓ_s when $\phi = 20°$; i.e., for slopes flatter than 65°, longer embedment length is needed as the slope flattens to ensure internal stability. Intuitively, one would expect less reinforcement (strength and length) as a slope flattens. This and the inability to rationally specify the tensile resistance distribution, especially for slip surfaces which are only partially within the reinforced zone, demonstrate difficulties associated with conventional slope stability type of analysis in conjunction with reinforcement. Conversely, the tieback approach is consistent within the framework of its assumptions and, equally important, with reinforced soil wall design for which extensive experience has been gained. Furthermore, it allows for a rational design of the required geosynthetic's re-embedment length at the slope face so as to resist t_j.

Schmertmann, et al. (1987) developed design charts that are based on a synthesis of different analysis methods. The conceptual layout of the reinforcement, which is especially suited for repair of failed slopes, is shown in Fig. 10. The assumptions employed in their analysis are noted on this figure. The coefficient of friction between backfill soil and reinforcement sheets was taken as $0.9 \tan\phi$, an appropriate value for geogrids. A uniform surcharge q is replaced by a modified slope height

$$H' = H + \frac{q}{\gamma} \tag{4}$$

Although the conversion of load to extra soil layer expressed in Eq. (4) is attractive for usage in design charts, it may be unconservative. That is, a distributed load q has no defined shear strength, whereas an equivalent soil layer, $(H' - H)$ thick and exerting pressure q, has internal shear strength which adds to stability if used as such in a design chart.

Three wedge models were used by Schmertmann, et al. (1987)–Fig. 11. The horizontal and vertical force equilibrium for each wedge were assembled and solved in nondimensional form. Interwedge friction angles, δ, were assumed to be equal to the factored soil friction angle, ϕ_f, defined as $\phi_f = \tan^{-1}(\tan\phi/F_s)$. This assumption shortens the required length at the bottom by about 30-40% as compared to $\delta = 0$. The two-part wedge model shown in Fig. 11.a, modified after Jewell, et al. (1984), was used to calculate the total required reinforcement tensile resistance, T_t. All tensile forces were assumed to be horizontal and their distribution directly proportional to depth below the slope crest. A search was performed for the critical combination of nodal point location and wedge angle, θ, producing $\max(T_t)$. In their design chart, T_t is presented by the nondimensional coefficient K:

$$K = 2\frac{T_t}{\gamma(H')^2} \tag{5}$$

Fig. 12.a is a design chart providing the value of K. In essence, K is equivalent to T_m given in Leshchinsky and Boedeker's (1989) chart - Fig. 6.

Schmertmann, et al. (1987) assumed that the reinforcement length at the bottom of the slope, L_B, would be controlled by external stability requirements and that the reinforcement length at the top of the slope, L_T, would be controlled by internal stability requirement. The single wedge model shown in Fig. 11.b was used to predict L_T. A search was performed to find θ producing the maximum horizontal reinforcement

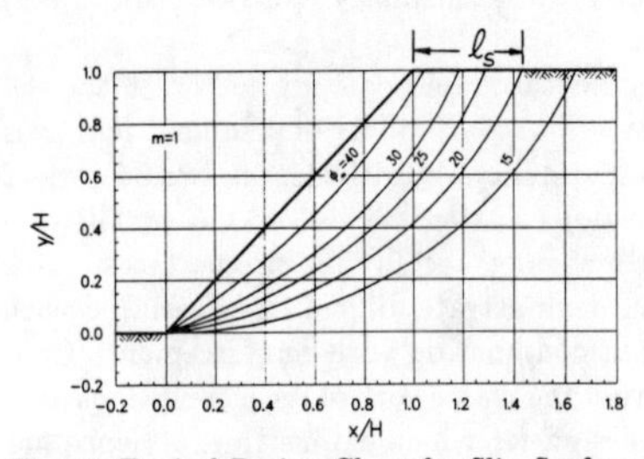

Fig. 7. Typical Design Chart for Slip Surface Trace and Embedment Length.

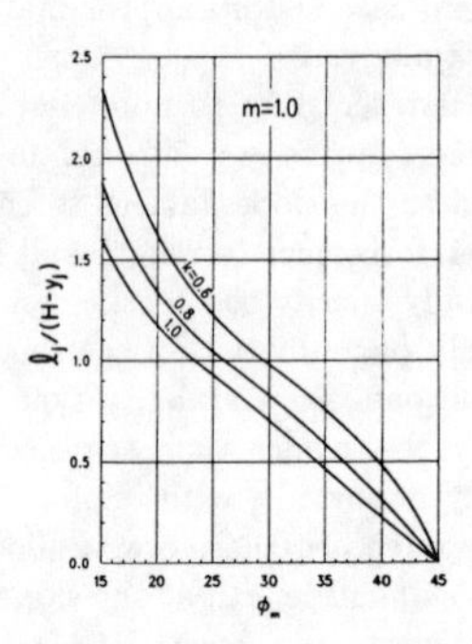

Fig. 8. Typical Design Chart for External Stability and Embedment Length.

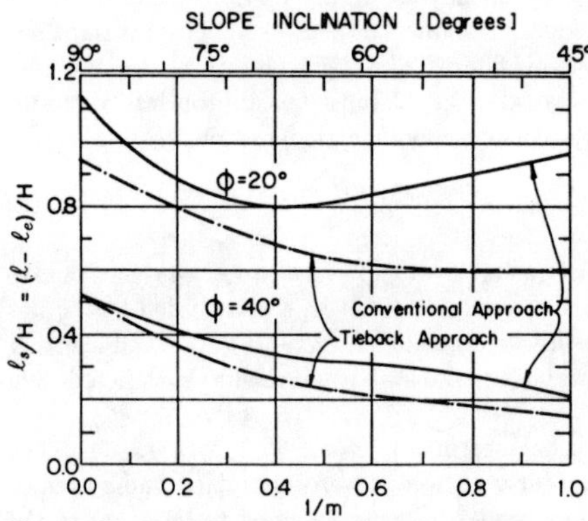

Fig. 9. Required Embedment Length: Conventional Slope Stability Analysis vs. Tieback Approach (both using log spiral surface).

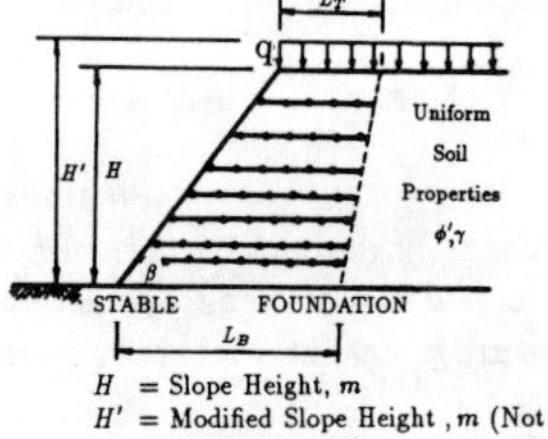

H = Slope Height, m
H' = Modified Slope Height , m (Not to exceed 1.2 H)
q = Uniform Surcharge on Top of Slope, kPa
β = Slope Angle from Horizontal, deg.
ϕ' = Effective Soil Friction Angle, deg.
γ = Moist Unit Weight of Soil, kn/m^3
L_B = Reinforcement Length at Base of Slope, m
L_T = Reinforcement Length at Top of Slope, m

ASSUMPTIONS:

1. ϕ' only ($c' = 0$) analysis is appropriate.
2. Uniform soil properties throughout slope.
3. Stable slope foundation.
4. Flat slope face and horizontal slope crest.
5. No pore pressures within the slope.
6. No seismic loading.
7. Uniform surcharge load on top of slope.
8. Horizontal reinforcement layers with coefficient of interaction equal to 0.9.

Fig. 10. Slope Geometry Parameters and Assumptions.

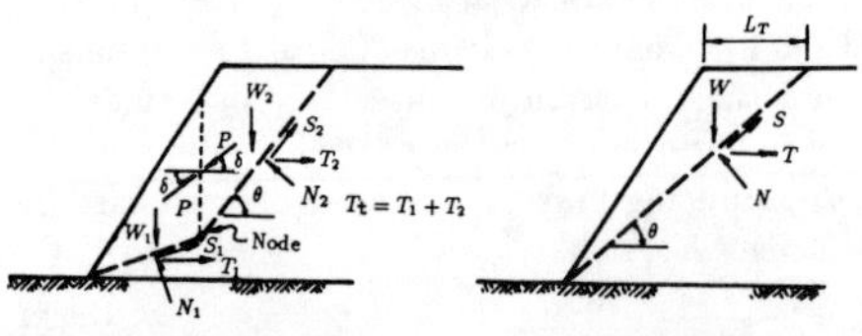

c) MODEL FOR L_B

DEFINITIONS:

W = Soil Weight
N = Normal Force on Shear Plane
S = Shear Resistance on Shear Plane
P = Interwedge Force
δ = Interwedge Friction = ϕ'_1
T = Horizontal Reinforcement Tensile Force
θ = Angle of Shear Plane
μ = Coefficient of Interaction

Fig. 11. Simple Wedge Models.

force, T. The intersection of this critical wedge with the crest was assumed to define a minimum value of L_T. This model was found to predict greater values of L_T than the two-part wedge used to calculate T_t. Such an approach seems to be inconsistent since two different mechanisms are used to assess the same aspect of stability. Furthermore, in both cases a linear distribution of tensile force with depth was assumed. It is difficult to visualize a situation where such identical distribution exists simultaneously within the same structure for two different mechanisms. It should be pointed out that Schmertmann, et al. (1987) used also Bishop's modified method to evaluate their K and L_T by considering, this time, circular surfaces passing through and possibly behind the reinforced soil zone. In this method they assumed T to be inclined at $(0.25\ \alpha)$ to the horizontal where α is the slope of failure circle at the reinforcement intersection. Further, it appears that a linear distribution of tensile force with depth was assumed again for all circles. While K predicted by the circular and wedge mechanisms was close, L_T determined by Bishop's was up to 40% larger than the simple wedge results. Subsequently, K and L_T predicted by the simple wedge models were used as a baseline for the chart shown in Fig. 12.b. These baseline values were then increased, apparently only *in some cases*, in order to satisfy or exceed Bishop's analysis results.

Analysis based on two-part wedge shown in Fig. 11.c was used to predict L_B. This model represents a sliding failure mode in which no reinforcement tension is mobilized across the two wedges. Spencer's method was used to evaluate L_B obtained from the two-part wedge. Not surprisingly, both methods yielded close results. Recommended values for L_B are shown in Fig. 12.b.

Jewell (1990) presented a comprehensive approach to geosynthetically reinforced earth structures. Using an argument similar to the one presented in Fig. 5 (termed "similarity" of potential slip surfaces), Jewell concludes that the reinforcement tensile force distribution must be directly proportional to depth, that is, it must react to lateral soil stresses which increase linearly with depth. Unlike Leshchinsky and Boedeker (1989), however, Jewell considers an infinite number of similar failure surfaces developing parallel to each other between the deepest surface (passing between toe and crest) and the slope surface. This, in a sense, is analogous to Rankine's analysis for vertical walls. Consequently, Jewell concludes that the reinforcement tensile force must remain horizontal since any "kinking" requires width and therefore cannot occur simultaneously through all equally critical infinite surfaces. Using the log spiral failure mechanism, Jewell assembled the moment equilibrium for a surface passing through all the sheets. Minimization with respect to the factor of safety resulted in the required tensile resistance. The results were presented in a nondimensional chart, similar in organization to Fig. 12.a, and, not surprisingly, identical in values to Fig. 6 for horizontal inclination only.

Line OF in Fig. 13 represents the critical log spiral for which maximum reinforcement resistance, F_{Req}, is required. To prevent failures such as defined by surface OG, reinforcement with constant length L_R was selected. This L_R is analog to L_T in the Schmertmann, et al. (1987) presentation. The resulting L_R/H is presented in Fig. 14. Evidently, Jewell assumed the same linear distribution of tensile force, truncated in case of failure behind the reinforced zone (Fig. 13.a), although the argument of similarity of surfaces is not valid for surfaces like OG (Fig. 13.b). It is not clear that non-similar surfaces, especially those outside the reinforced zone, should generate the same reaction as the critical ones.

Jewell suggests to compensate for a slight reduction of pullout resistance immediately below surface OG by decreasing the reinforcement spacing and thus increasing the available stress for equilibrium. An empirical relationship is suggested to increase K_{Req} to a design value K_d to compensate for that slight loss of bonding (Fig. 13.c)

$$K_d = \frac{K_{Req}}{1 - L_B/L_R} \tag{6}$$

where L_B is the bond length at the base of the slope needed to develop the available force in the reinforcement–Fig. 13.b. In essence, L_B is analogous to ℓ_{e_1} in Eq. (3).

External stability (or "direct sliding") was analyzed by Jewell (1990) using the mechanism shown in Fig. 15. Force equilibrium for this two-part wedge results with the second chart is shown in Fig. 14. If critical, the direct sliding governs the required reinforcement length at the base. In this case, the layout can have the same configuration as Schmertmann, et al. (1987). The wedge in Fig. 15 slides through two distinct regions (unreinforced: friction = $\tan\phi$, and along a sheet: friction = $k\tan\phi$). The value of k used in Fig. 14 (i.e., the interaction coefficient) was taken as 0.8.

It should be noted that the practical layout shown in Fig. 10 can be equally applicable to Leshchinsky and Boedeker (1989) design charts. Subsequently, it is interesting to compare predictions based on Jewell (1990), Schmertmann, et al. (1987) and Leshchinsky and Boedeker (1989) charts using an example problem. Given a 1(H):1(V) slope, 10 m high to be reinforced with 10 equally spaced sheets. For design, the backfill soil has: $\gamma = 20kN/m^3$ and $\phi = 25°$. Fig. 16 illustrates the results of this comparison. Note that because of the conceptual difference in the internal stability analysis, Leshchinsky and Boedeker's (1989) layout is significantly shorter at the crest. However, because of conservative assumption in their external stability analysis, the reinforcement is longer at the bottom. The required unfactored tensile strength is nearly the same for the three analyses.

VERIFICATION OF LIMIT EQUILIBRIUM METHODS

The equilibrium approaches reviewed in the previous section have been successfully utilized to design numerous reinforced fill slopes. A study of polymeric reinforced soil structures in North America by Yako and Christopher (1987) identified approximately 100 reinforced slope structures. Although the procedures have a competent theoretical and empirical basis, until recently surprisingly little research had been performed to ascertain the accuracy of these methods. The review by Yako and Christopher could not identify any instrumented field projects.

Recognizing the need for verification of the limit equilibrium design approach, a field evaluation program was developed as part of a Federal Highway Administration study on the "Behavior of Reinforced Soil" (Christopher, et al., 1990). In that program, four, 7.6 m high, steep sloped (two at lV:lH and two at 2V:1H) structures, reinforced with both geotextiles and geogrids were constructed and instrumented. Each slope was constructed to a width of 15 m using a silt type soil. The silt had 70 to 90% particles finer than 0.07 mm and was placed at 95% of Standard Proctor density (moist unit weight = 130 lb/ft^3). Triaxial test results indicated $\phi = 35°$, $c = 0.05\,kPa$. Details concerning the reinforcement used for construction of each slope are listed in Table 1.

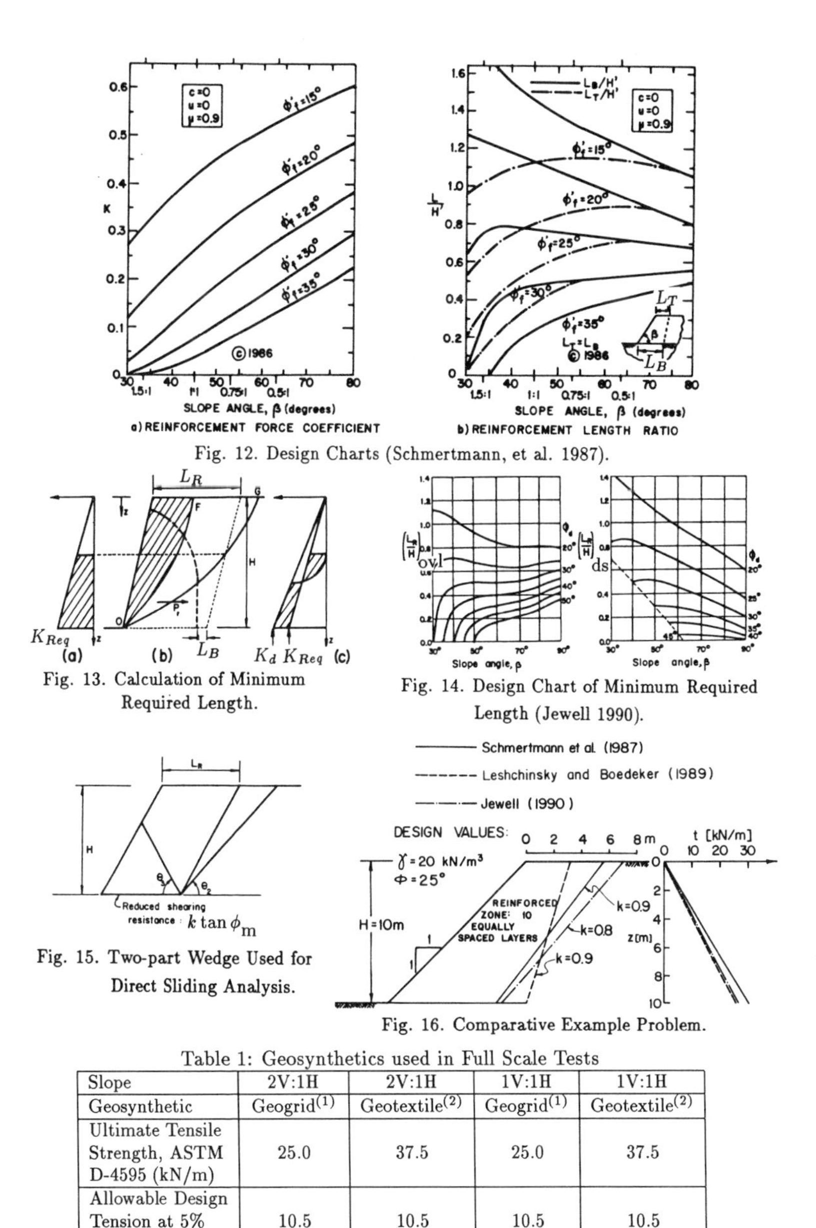

Fig. 12. Design Charts (Schmertmann, et al. 1987).

Fig. 13. Calculation of Minimum Required Length.

Fig. 14. Design Chart of Minimum Required Length (Jewell 1990).

Fig. 15. Two-part Wedge Used for Direct Sliding Analysis.

Fig. 16. Comparative Example Problem.

Table 1: Geosynthetics used in Full Scale Tests

Slope	2V:1H	2V:1H	1V:1H	1V:1H
Geosynthetic	Geogrid(1)	Geotextile(2)	Geogrid(1)	Geotextile(2)
Ultimate Tensile Strength, ASTM D-4595 (kN/m)	25.0	37.5	25.0	37.5
Allowable Design Tension at 5% Strain (kN/m)	10.5	10.5	10.5	10.5

(1) Polypropylene welded strip; (2) Polyprolylene woven

The slopes were designed using a rotational failure limit equilibrium analysis approach as it was found to provide the most critical reinforcement requirement. The results were compared and agreed well with the design charts developed using Jewell's (1981,1982) approach. For all structures, eight layers of reinforcement with a uniform spacing of 0.75 m were used. A total length of 4.3 meters was used for each layer which was beyond the design pullout length required for either material: this allowed a comparative evaluation of internal stress without considering length as a factor. Based on the allowable tension of 10.5 kN/m assumed of both materials, the reinforcement arrangement provided a design factor of safety, SF, of 1.3 for the 1V:1H slopes and an SF approaching 1.0 for the 2V:1H slopes.

Slope behavior was evaluated by instrumenting and monitoring external and internal horizontal movement and reinforcement strains during and after construction. The instrumentation layout is shown in Fig. 17. Construction of the slopes was completed in November, 1987. Periodic monitoring has been performed since completion. Fig. 18 shows a representative plot of the distribution of tension measured in the reinforcement. The results of a finite element model (Adib, 1988 and Adib, et al. 1990) prediction of the tension are also shown on the plot. Superimposed on the plot is the predicted location of maximum tension in the reinforcement from the circular arc limit equilibrium approach which is in excellent agreement with the actual maximum values observed in the field and from the numerical model.

Figs. 19 and 20 show the maximum tension measured in each reinforcement layer from the field and the computer model as compared to the computed maximum tension from the circular limit equilibrium method. As can be seen, the limit equilibrium methods are somewhat conservative for estimating the maximum tension in the lower layers of reinforcement which is usually the most critical zone of reinforcement. The conservatism in the results has been attributed to the confining effect of the reinforcement on the soil and of the soil on the reinforcement which tends to increase the composite soil/geotextile strength in the area of the reinforcement over the assumed for the individual components (i.e., the inherent stress-strain characteristics of the geosynthetic may be improved by the confining effect of the soil and/or the strength for the soil may be improved due to the increased lateral confinement provided by the reinforcement). The geosynthetic/soil interaction group effect is ignored in the limit equilibrium analysis.

Based on both the field and numerical model results, it would appear that the limit equilibrium approach provides a suitable, albeit somewhat conservative, design approach. This conclusion and the question of the equally "safe" assumption with regard to the reinforced and unreinforced factors of safety were further substantiated in a recent study of the uncertainty in the reinforced slope design methods by Cheng and Christopher (1991). Using probabilistic methods to evaluate the results of the FHWA project, they found that the reinforced slope design was actually more reliable than a flatter unreinforced design at the same factor of safety since the variability of the properties of the soil are greater than the variability of the properties of the reinforcement. They also found that the reliability of the design increases as the steepness of the slope increases.

As a result of their studies, the Federal Highway Administration has selected a circular arc limit equilibrium analysis as an appropriate design tool. However, since the predictions of reinforcement tension by the three chart methods presented in the previous section (Figs. 6 and 12) provide results similar to the circular arc, either approach would thus appear applicable. A step by step design procedure incorporating the circular arc approach is included in the following section.

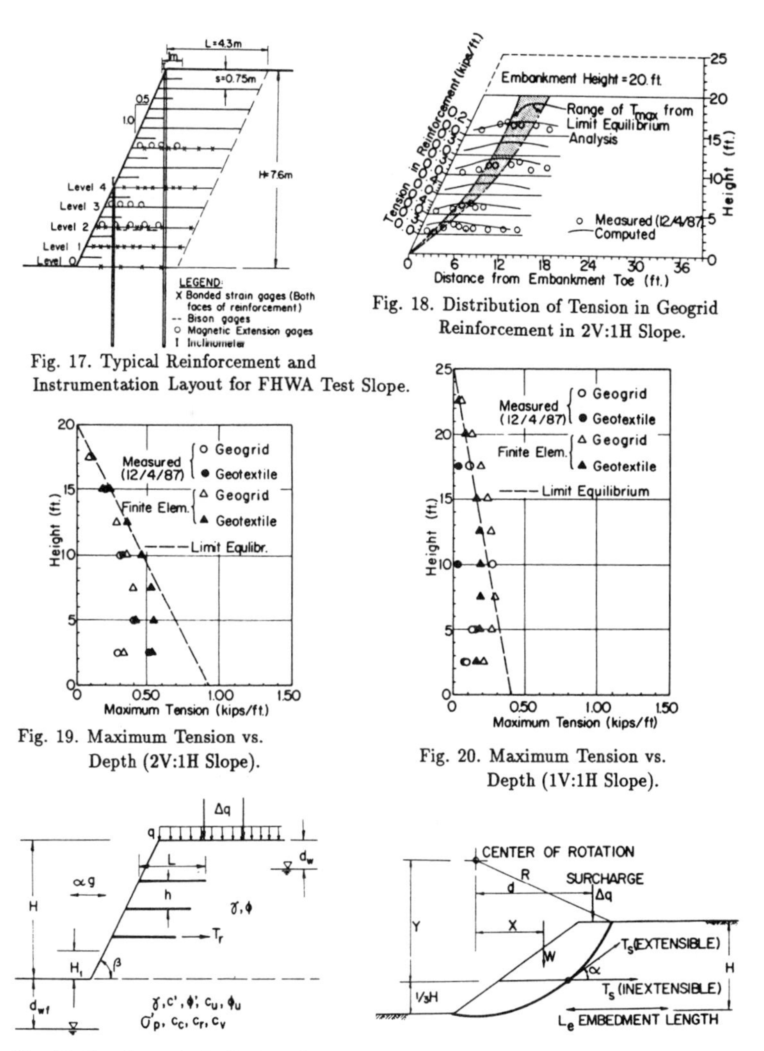

Fig. 17. Typical Reinforcement and Instrumentation Layout for FHWA Test Slope.

Fig. 18. Distribution of Tension in Geogrid Reinforcement in 2V:1H Slope.

Fig. 19. Maximum Tension vs. Depth (2V:1H Slope).

Fig. 20. Maximum Tension vs. Depth (1V:1H Slope).

Fig. 21. Requirements for Design of Reinforced Slopes.

Fig. 22. Notation for Circular Arc Failure Mechanism.

RECOMMENDED DESIGN PROCEDURE

The steps that should be followed in the design of a reinforced soil slope as outlined in this section are recommended regardless of which internal stability approach is selected. The procedure assumes that the slope is to be constructed on a stable foundation and includes the possibility of a deep seated failure.

For slope repair applications, it is also very important to identify the cause of the original failure to ensure that the new reinforced soil slope will not have the same problems. If a high water table exist, particular attention has to be paid to drainage. In natural soils, it is also necessary to identify any weak seams that might affect stability.

Step 1. Establish the Geometric and Loading Requirements for Design-Fig. 21.

Step 2. Determine the Engineering Properties of the Natural Soils in the Slope: strength and consolidation parameters.

Step 3. Determine Properties of Available Fill

- i. Gradation and plasticity index.
- ii. Compaction characteristics
- iii. Recommended lift thickness for backfill material
- iv. Peak shear strength parameters, c_u, ϕ_u, and c', ϕ'.
- v. Chemical composition of soil that may affect durability of reinforcement, (i.e., pH, chloride content, oxidation agents). Do not use soil with pH > 12 or pH < 3.

Step 4. Establish Performance Requirements

- i. External stability: sliding, deep seated, compound failure, dynamic loading and settlement.
- ii. Internal stability
  - · Slope stability: F.S. 1.3 or greater.
  - · Allowable tensile force per unit width of reinforcement T_a
  - · For geosynthetics: $T_a = \frac{T_{ult}(CRF)}{(FD \cdot FC \cdot FS)}$

    where: CRF = Creep reduction factor
    FD = Durability factor of safety
    FC = Construction damage factor of safety
    FS = Overall factor of safety for uncertainty

    Pullout Resistance F.S. = 1.5 for granular soil with a 3-foot minimum length in the resisting zone.
    Use F.S.= 2 for cohesive soils.

Step 5. Evaluate Internal Stability Requirements

- The chart methods discussed previously (Figs. 6 or 12) may be used for determining the reinforcement requirements for simple structures. That is structures with a homogeneous granular soil mass supported by a competent, firm foundation, having a piezometric water level below the base of the slope and supporting a relatively low surcharge (i.e., uniform surcharge less than 0.1 times the height of the slope). More than one method should be used to provide a check of the results.
- Alternatively, as a check and for complex structures (e.g., weak foundation and anticipated seepage problems are anticipated), use a modified conventional rotational slip surface methods commonly used for unreinforced slopes . Fig. 22

illustrates the modifications required to incorporate reinforcement into conventional rotational slip surface methods. This approach can accommodate fairly complex conditions depending on the method used (e.g., Bishop, Janbu, etc.).

The following design steps and calculations are necessary for the rotational slip surface method:

a. Check Unreinforced Stability

To determine the size of the critical zone to be reinforced, analyze the slope without reinforcement using conventional stability methods. Use both circular and wedge-type surface shapes to consider failure through the toe, through the face (at many elevations) and in deep seated surfaces below the toe. Examine the full range of potential failure surfaces found to be less than or equal to the target safety factor for the slope FS. Plot all of these surfaces on the cross-section of the slope. The surface closest to the face of the slope that just meets the target factor of safety roughly defines the limits of the critical zone to be reinforced.

Critical failure surfaces extending below the toe of the slope are indications of deep foundation and edge bearing capacity problems that must be addressed prior to completing the design. For such cases, a more extensive foundation analysis is warranted and foundation improvement measures should be considered.

b. Calculate the total reinforcement tension T_s required to obtain the required factor of safety FS_R for each potential failure circle inside the critical zone in step a that extends through or below the toe of the slope using the following equation:

$$T_S = (FS_R - FS_U)\frac{M_D}{D} \tag{7}$$

where

FS_U = factor of safety for unreinforced slope.

T_S = sum of allowable tensile force (considering rupture and pullout) in all reinforcement layers intersecting the failure circle.

M_D = driving moment about the center of the failure circle.

D = radius of circle R for extensible reinforcement (i.e., assumed to act tangentially to the circle at $H/3$ above the slope base).

The largest T_S calculated establishes the required design tension, T_{max}.

c. Determine the strength of reinforcement, T_{max} using the charts in Figs. 6 or 12 and compare with step b. If substantially different, recheck steps a and b.

d. Determine the distribution of reinforcement:

- For low slopes ($H \leq 6m$) assume a uniform distribution of reinforcement and use T_{max} to determine spacing or reinforcement requirements in step e.
- For high slopes (H > 6m), divide the slope into 2 (top and bottom) or 3 (top, middle and bottom) reinforcement zones of equal height and use a factored T_{max} in each zone for spacing or reinforcement requirements in step e. Assuming a triangular distribution, the total tension in each zone are found from:

  For 2 zones: $T_{top} = 1/4T_{max}$; $T_{bottom} = 3/4T_{max}$

  For 3 zones: $T_{top} = 1/6T_{max}$; $T_{middle} = 1/3T_{max}$; $T_{bottom} = 1/2T_{max}$

e. Determine reinforcement vertical spacing S_v:

- For each zone, calculate the design tension T_d requirements for each reinforcing layer in that zone based on an assumed S_v or, if the allowable reinforcement strength is known, calculate the minimum vertical spacing and number of reinforcing layers N required for each zone based on:

$$T_d = T_a R_c = \frac{T_{zone} S_v}{H} = \frac{T_{zone}}{N}$$

where

S_v = multiples of compaction layer thickness for ease of construction

T_{zone} = maximum reinforcement tension required for each zone:

= T_{max} for low slopes (H < 6 m)

= T_{top}, T_{middle}, and T_{bottom} for high slopes (H > 6 m)

- Check to minimize reinforcement requirements, use short [1.2 to 1.5 m] lengths of intermediate reinforcement layers to maintain a maximum vertical spacing of 60 cm or less for face stability and compaction quality. Intermediate reinforcement should be placed in continuous layers and need not be as strong as the primary reinforcement.

f. For critical or complex structures, and when checking a complex design, step b should be repeated for a potential failure above each layer of primary reinforcement to make sure distribution is adequate.

g. Determine the reinforcement lengths required:

The embedment length L_e beyond the most critical sliding surface found in step b (i.e., circle found form T_{max}) must be sufficient to provide adequate pullout resistance. For the method illustrated in Fig. 22, use:

$$L_e = \frac{T_a FS}{F* \cdot \alpha \cdot \sigma'_v \cdot 2b}$$

where

$F*$ = the pull-out resistance (or friction-bearing-interaction) factor

α = a scale effect correction factor

σ'_v = the effective vertical stress at the soil-reinforcement interfaces

$2b$ = effective unit perimeter in which b = width of strip, grid, or sheet.

The pull-out resistance factor $F*$ can most accurately be obtained from pull-out tests performed in the specific backfill to be used on the project. Minimum value recommended for L_e is 1 m. For cohesive soils, check L_e for both short and long-term pullout conditions. For long-term evaluation, conservatively use ϕ' with $c = 0$. For short-term evaluation, conservatively use ϕ with $c = 0$ or run pullout tests.

Plot the reinforcement lengths obtained from the pullout evaluation on a slope cross section containing the rough limits of the critical zone determined in step a. The length of the lower layers must extend to or beyond the limits of the critical zone. The length required for sliding stability at the base (i.e., external stability) will generally control the length of the lower reinforcement levels. Upper levels of reinforcement may not be required to extend to the limits of the critical zone provided sufficient reinforcement exists in the lower levels to provide the FS_R for all circles within the critical zone (e.g., see steps a and b). Make sure that the sum of the reinforcement passing through each failure surface is greater than T_S, from step b, required for that surface. Only count reinforcement that extend a meter or more beyond the surface to account for pull-out resistance. If the available reinforcement is not sufficient, increase the length of reinforcement not passing through the surface or increase the strength of lower level reinforcement. Simplify the layout by lengthening some reinforcement layers to create two or three sections of equal reinforcement length. Reinforcement layers do not generally need to extend to the limits of the critical zone, except for the lowest levels of each reinforcement section.

h. Checking design lengths of complex designs.

When checking a design that has zones of different reinforcement length, lower zones may be over reinforced to provide reduced lengths of upper reinforcement levels. In evaluating the length requirements for such cases, the pullout stability for the reinforcement must be carefully checked in each zone for the critical surfaces exiting at the base of each length zone.

Step 6. Check External Stability

The external stability of a reinforced soil mass depends on the ability of the mass to act as a stable block and withstand all external loads without failure. Failure possibilities include sliding and deep seated overall instability as well as compound failures initiating internally and externally through the reinforced zone. If applicable, foundation settlement and seismic stability (Christopher, et al. 1990) should be evaluated.

The external stability must be checked for both short and long-term conditions and can be performed using the limit equilibrium approach detailed in Step 5.a.

CONCLUSIONS AND FUTURE DEVELOPMENT

The state of the practice in the form of current design procedures, along with field verification of design performance, has been briefly reviewed. The main conclusion is that most limit equilibrium design approachs, commonly used in practice, will provide appropriate results for evaluating reinforcement requirements.

The current focus in reinforced soil development is in the areas of reinforcement dimension optimization, low cost facing systems and lower quality fill material. Cohesive soils are currently analyzed using undrained parameters. To allow utilization of effective parameters, research on products that allow inplane drainage is needed. Product development is leading to the potential of constructing reinforced soil slopes at a lower cost than conventional flatter unreinforced slopes, making such systems the standard of practice for slope construction.

REFERENCES

Adib, M.E. (1988). "Internal earth pressure in earth walls," Ph.D. Dissertation in Civil Engineering, University of California, Berkeley.

Adib, M.E., Mitchell, J.K. and Christopher, B.R. (1990). "Finite element modeling of reinforced soil walls and embankments," Proc. Conf. on Design and Performance of Earth Retaining Structures, ASCE, Geotechnical Special Publication No. 25, 409-423.

Bangratz, J.L. and Gigan, J.P. (1984). "Methode rapide de calcul des massifs cloues." Proc. Int. Conf. Insitu Soil and Rock Reinforcement, Paris, 293-299.

Berg, R.R., Chouery-Curtis, V.E. and Watson, C.H. (1989). "Critical failure planes in analysis of reinforced slopes." Proc. Geosynthetics '89, San Diego, Vol. 1, 269-278.

Bonaparte, R. and Schmertmann, G.R. (1988). "Reinforcement extensibility in reinforced soil wall design." *The Application of Polymeric Reinforcement in Soil Retaining Structures.* Proc. NATO Adv. Research Workshop, Kingston. Kluwer Academic Publishers, The Netherlands, 409-57.

Cheng, S. and Christopher, B.R. (1991). "A probabilistic review of geotextile reinforced slope design," Proc. Geosynthetics '91, Atlanta, GA.

Christie, I.F. and El-Hadi, K.M. (1977). "Some aspects of the design of earth dams reinforced with fabric." Proc. Int. Conf. on Use of Fabrics in Geotechnics, Paris, Vol. 1, 99-103.

Christopher, B.R., Gill, S.A., Giroud, J.P., Juran, I., Mitchell, J.K., Schlosser, F., and Dunnicliff, J. (1990). *Reinforced Soil Structures Volume 1: Design and Construction*, FHWA Report No. FHWA-RO-89- 043, Washington, D.C.

Christopher, B.R. and Holtz, R.D. (1985). "Geotextiles Engineering Manual," National Highway Institute, FHWA, Washington, D.C., Contract No. DTFH61-80-C-00094.

Christopher, B.R. and Holtz, R.D. (1989). "Geotextile Design and Construction Guidelines Manual," National Highway Institute, FHWA, Under Contract to GeoServices, Inc., Contract No. DTFH61-86-R- 00102.

Edris, E.V. and Wright, S.G. (1987). "User's guide: UTEXAS2 slope-stability package." *Instruction Report GL-87-1*, U.S. Army COE, Waterways Experiment Station, Vicksburg, MS.

Fannin, R.J. and Hermann, S. (1990). "Performance data for a sloped reinforced soil wall." *Canadian Geotechnical Journal*, Vol. 27, No. 5, 676-686.

Fowler, J. (1982). "Theoretical design considerations for farbic-reinforced embankments." Proc. 2nd Int. Conf. on Geotextiles, Las Vegas, Vol. 3, 665-670.

Gourc, J.P., Ratel, A. and Delmas, P. (1986). "Design of fabric retaining walls: the displacement method." Proc. Third Int. Conf. on Geotextiles, Vienna, Vol. 2, 289-294.

Ingold, T.S. (1982). "An analytical study of geotextiles reinforced embankments." Proc. 2nd Int. Conf. on Geotextiles, Las Vegas, Vol. 3, 683- 688.

Jewell, R.A. (1981). "A computer design method for equilibrium soil structures using limit equilibrium analysis," Report prepared by Binnie & Partners, London, for Netlon, Ltd., 6pp.

Jewell, R.A. (1982). "A limit equilibrium design method for reinforced embankments on soft foundations." Proc. 2nd Int. Conf. on Geotextiles, Las Vegas, Vol. 3, 671-76.

Jewell, R.A., Paine, N., and Woods, R.I. (1984). "Design methods for steep reinforced embankments," Proc. Sumposium on Polymer Grid Reinforcement, Institute of Civil Engineering, London, pp. 18-30.

Jewell, R.A. (1990). "Strength and deformation in reinforced soil design," Keynote paper on Reinforced Soil, 4th International Conference on Geotextiles, Geomembranes and Related Products, The Hague, The Netherlands, May 29.

Jones, C.J.F.P. (1985). *Earth Reinforcement and Soil Structures*, Butterworths, England.

Juran, I. and Schlosser, F. (1978). "Theoretical analysis of failure in reinforced earth structures." Proc. Symp. Earth Reinforcement, ASCE, Pittsburgh, 528-555.

Koerner, R.M. (1986). *Designing with Geosynthetics*, Prentice-Hall, Inc., Englewood Cliffs, NJ.

Lee, K.L., Adams, B.D., and Vagneron, J.J. (1973). "Reinforced earth retaining walls." *Journal of Soil Mechanics and Foundations Division* , ASCE, Vol. 99, No. 10, 745-64.

Leshchinsky, D. and Reinschmidt, A.J. (1985). "Stability of membrane reinforced slopes," *Journal of Geotechnical Engineering*, ASCE, Vol. III, No. 11, 1285-1300.

Leshchinsky, D. and Volk, J.C. (1985). "Stability charts for geotextile reinforced walls," *Transportation Research Record 1031*, 5-16.

Leshchinsky, D. and Volk, J.C. (1986). "Predictive equation for the stability of geotextile reinforced earth structure," Proc. of the 3rd Int. Conf. on Geotextiles, Vienna, Vol. 2, 383-388.

Leshchinsky, D., Volk, J.C. and Reinschmidt, A.J. (1986). "Stability of geotextile-retained earth railroad embankment," *Geotextiles and Geomembranes*, Vol. 3, Nos. 2 & 3, 105-128.

Leshchinsky, D. and Perry, E.B. (1987). "A design procedure for geotextile reinforced walls," *Geotechnical Fabrics Report*, Vol. 5, No. 4, 21-27.

Leshchinsky, D. and Boedeker, R.H. (1989). "Geosynthetic reinforced soil structures," *Journal of Geotechnical Engineering*, ASCE, Vol. 115, No. 10, October, 1459-1478.

Leshchinsky, D. and Perry, E.B. (1989). "On the design of geosynthetic-reinforced walls," *Geotextiles and Geomembranes*, Vol. 8, No. 4, 311-323.

Murray, R.T. (1982). "Fabric reinforcement of embankments and cuttings," Las Vegas, Nevada, Vol. 3, 707-713.

Murray, R.T. (1985). "Reinforcement techniques in repairing slope failures." Polymer Reinforcement in Civil Engineering, Thomas Telford, London, 47-53.

Phan, T.L., Segrestin, P., Schlosser, F. and Long, N.T. (1979). "Etude de la stabilite interne et externe des ouvrages en terre armee par deux methodes de cercles de rupture." Proc. Int. Conf. Soil Reinforcement, Vol. I, Paris, 119-23.

Romstad, K.M., Al-Yassin, A., Hermann, L.R. and Shen, C.K. (1978). "Stability analysis of reinforced earth retaining structures." Proc. Symp. Earth Reinforcement, ASCE, Pittsburgh, 685-713.

Schlosser, F. and Vidal, H. (1969). "La terre armee." Bull. Liason du Laboratorie Central des Ponts et Chaussees, 41, 101-44.

Schmertmann, Chouery-Curtis, Johnson and Bonaparte (1987). "Design charts for geogrid-reinforced soil slopes," Proc. of Geosynthetics '87 Conf., New Orleans, LA., Vol. 1, 108-120.

Schneider, H.R. and Holtz, R.D. (1986). "Design of slopes reinforced with geotextiles and geogrids." *Geotextiles and Geomembranes*, Vol. 3, No. 1, 29-51.

Segrestin, P. (1979). "Design of reinforced earth structures assuming failure wedges." Proc. Int. Conf. Soil Reinforcement, Paris, Vol. I, 163-8.

Stocker, M.F., Korber, G.W., Gassler, G. and Gudehus, G. (1979). "Soil nailing." Proc. Int. Conf. Soil Reinforcement, Paris, Vol. 2, 469-74.

Studer, J.A. and Meier, P. (1986). "Earth reinforcement with nonwoven fabrics: Problems and computational possibilities." Proc. Third Int. Conf. on Geotextiles, Vienna, Vol. 2, 361-5.

Yako, M.A. and Christopher, B.R. (1987), "Polymerically reinforced retaining walls and slopes in North America." *The Application of Polymeric Reinforcement in Soil Retaining Structures*, Proc. NATO Adv. Research Workshop, Kingston. Kluwer Academic Publishers, The Netherlands, 239-282.

DESIGN OF REINFORCED EMBANKMENTS
RECENT DEVELOPMENTS IN THE STATE-OF-THE-ART

by D. N. Humphrey[1] and R. K. Rowe[2], Members, ASCE

ABSTRACT: Reinforced embankments constructed on soft soils are generally designed using simplified procedures based on limiting equilibrium. One commonly used set of procedures was proposed by Haliburton, et al. (1978). It requires that the following failure modes be checked: bearing capacity, global stability, deformation, pullout, and lateral spreading. While this set of procedures generally produces satisfactory designs there is considerable uncertainty as to the details of the analysis and validity of underlying assumptions. This paper presents recent developments in the state-of-the-art within the context of this set of procedures. It is shown that bearing capacity should take into account the strength increase with depth and limited thickness of foundation soil. Typical values for allowable strain in the reinforcement may be unconservative. Allowable strain should be evaluated on a case-by-case basis.

INTRODUCTION

Geosynthetic reinforcement has proved to be a valuable technique for constructing embankments on soft soils. Reinforced embankments are generally designed using simplified procedures based on limiting equilibrium. One commonly used set of procedures was proposed by Haliburton, et al. (1978) and refined by others such as Fowler and Koerner (1987). It requires that the following failure modes be checked: bearing capacity, global stability, deformation, pullout, and lateral spreading. These failure modes are illustrated in Fig. 1. While this set of procedures generally produces satisfactory designs there is considerable uncertainty as to the details of the analysis and validity of underlying assumptions.

This paper will present recent developments in the state-of-the-art within the context of the set of procedures proposed by Haliburton, et al. (1978). Results from plasticity theory, limiting equilibrium, and finite element analyses will be used to show the limitations of the procedures and how they can be improved. However, analysis of pullout and lateral spreading failure modes is straight forward and will not be discussed here. Reference is made

---

1 - Assist. Prof, Dept. of Civil Engineering, Univ. of Maine, Orono, Maine 04469, USA.

2 - Prof, Geotechnical Research Centre, Faculty of Engineering Science, The University of Western Ontario, London, Ontario N6A 5B9, Canada.

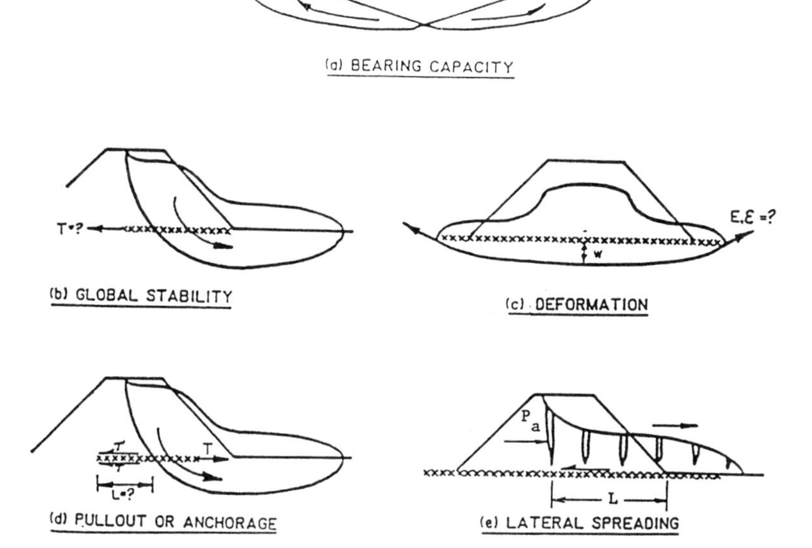

Figure 1. Failure Modes for Reinforced Embankment Design (Modified from Fowler and Koerner, 1987)

to Fowler and Koerner (1987), Christopher and Holtz (1985) and Holtz (1989) for a discussion of these failure modes.

Much of the following discussion makes use of finite element (FE) analyses to examine the validity of Halibruton's method. It would be preferable to use actual case histories but regrettably there are too few well documented case histories to allow this to be done. The FE analyses were done using one or more of the following programs: LEPSSIA (Rowe, 1982); AFENA (Carter and Balaam, 1982); or PS-NFAP (McCarron, 1985; Humphrey, 1986; Humphrey, et al., 1986). These programs properly model the large deformations which can occur in reinforced embankments on very compressible foundations such as peat and the layer-by-layer construction of the embankment itself. Space does not permit a more detailed discussion of these programs.

BEARING CAPACITY

Bearing capacity gives an upper bound solution to the allowable vertical stress that can be applied by a highly reinforced embankment. It can be easily calculated so it is generally the first check to be made when investigating the feasibility of reinforced embankments. Haliburton, et al. (1978) proposed that the bearing capacity q_u be calculated using

$$q_u = c_u N_c \tag{1}$$

where c_u is the undrained shear strength of the foundation soil and N_c is the bearing capacity factor. Haliburton, et al. (1978) recommended that N_c be taken as 5.14 which is the value for a strip footing on soil with uniform strength with depth. However, a review of 17 case histories of embankments reinforced with woven geotextiles or geogrids showed that using N_c = 5.14 underestimates the bearing capacity of reinforced embankments (Humphrey and Holtz, 1986; Humphrey, 1987). This is illustrated in Fig. 2 which shows the embankment height plotted vs. the average undrained shear strength of the foundation. Significantly, four embankments failed at heights 2 m greater than predicted by Eq. 1. Although two embankments failed at heights less than or equal to that predicted by Eq. 1, one (Case 14) probably had inadequate reinforcement and the other (Case 8) was constructed on peat. The latter requires special design considerations which will be discussed in a subsequent section.

Two reasons that Eq. 1 underpredicts the height of reinforced embankments are: (1) the undrained shear strength of clayey soils generally increases with depth; and (2) in some cases the soft soils are underlain by a stronger layer at a shallow depth relative to the height of the embankment. The influence of these factors have been discussed by Humphrey (1986, 1987), Humphrey and Holtz (1987), Bonaparte, et al. (1987), Rowe and Soderman (1987b), and Jewell (1988).

Bearing capacity factors for foundation soils whose strength increases with depth and limited thickness of foundation soils were developed using plasticity theory by Davis and Booker (1973) and Matar and Salencon (1977). The bearing capacity factors for the combined case of strength increasing with depth and limited thickness have been developed by Rowe and Soderman (1987b). The

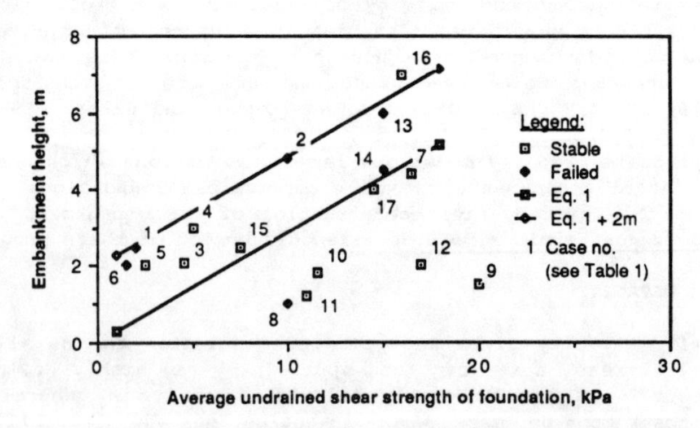

Figure 2. Embankment Height vs. Undrained Shear Strength of Foundation (after Humphrey, 1986)

bearing capacity factors are expressed in terms of the dimensionless quantity $\rho_c b/c_{uo}$ where ρ_c is the rate of increase in c_u with depth, b is the effective width of the footing, and c_{uo} is the undrained shear strength directly beneath the footing as shown in Fig. 3.

The plasticity solutions are for a rigid footing of width b but embankments generally have a trapezoidal shape, so an approximation must be made to determine the equivalent width of the embankment. From plasticity considerations, the pressure at the edge of a rigid footing is $(2 + \pi)c_{uo}$. It is assumed that the effective width of the footing b will extend between the points on either side of the embankment where the applied pressure γh is equal to $(2 + \pi)c_{uo}$ where γ is the unit weight of embankment fill. Thus, the thickness of embankment fill h, where the applied pressure is $(2 + \pi)c_{uo}$ is

$$h = (2 + \pi)c_{uo}/\gamma \qquad (2)$$

and hence from Fig. 4

$$b = B + 2n(H - h) \qquad (3)$$

where B is the embankment crest width, H is the embankment height, and n is the cotangent of the slope angle (Rowe and Soderman 1987b).

The bearing capacity of the rigid footing of width b is given by

$$q_u = c_{uo}N_c + q_s \qquad (4)$$

where q_s is a uniform surcharge pressure applied to the soil surface outside of the footing width and N_c is the bearing capacity factor obtained from Fig. 3. Inspection of Fig. 4 shows that the triangular edge of the embankment is providing a surcharge that would increase stability (Rowe and Soderman, 1987b). The magnitude of the surcharge is estimated as follows. The lateral extent of the plastic region involved in the collapse of a rigid footing extends a distance x from the footing where x is approximately equal to the minimum of d as determined from Fig. 5 and the actual thickness of the deposit D, i.e.

$$x = \min(d,D) \qquad (5)$$

Distributing the applied pressure due to the triangular distribution over a distance x gives

$$q_s = n\gamma h^2/2x \qquad \text{for } x > nh \qquad (6)$$

$$q_s = (2nh - x)\gamma h/2nh \qquad \text{for } x \leq nh \qquad (7)$$

The ultimate bearing capacity computed from Eq. 4 may then be compared with the average applied pressure q_a due to the embankment over the width b, viz.

$$q_a = \gamma[BH + n(H^2 - h^2)]/b \qquad (8)$$

The safety factor is then determined as $SF = q_u/q_a$.

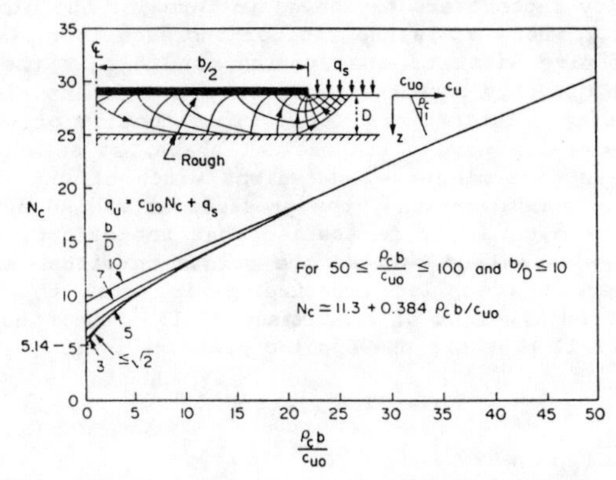

Figure 3. Bearing Capacity Factor for Non-Homogeneous Soil (synthesized from results by Davis and Booker, 1973, and Matar and Salencon, 1977, by Rowe and Soderman, 1987b)

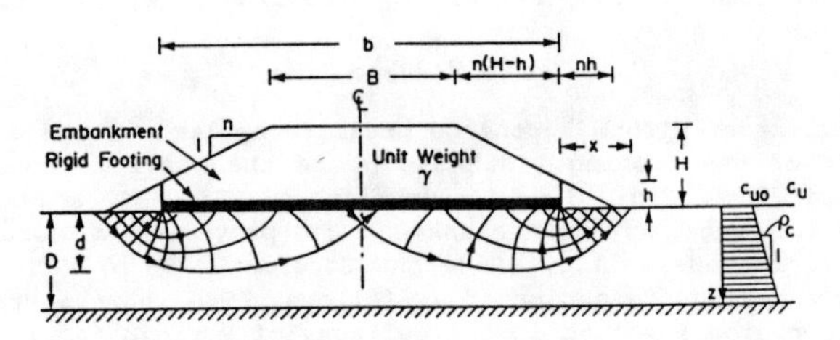

Figure 4. Definition of Variables Used to Estimate Collapse Height for a Perfectly Reinforced Embankment (Rowe and Soderman, 1987b)

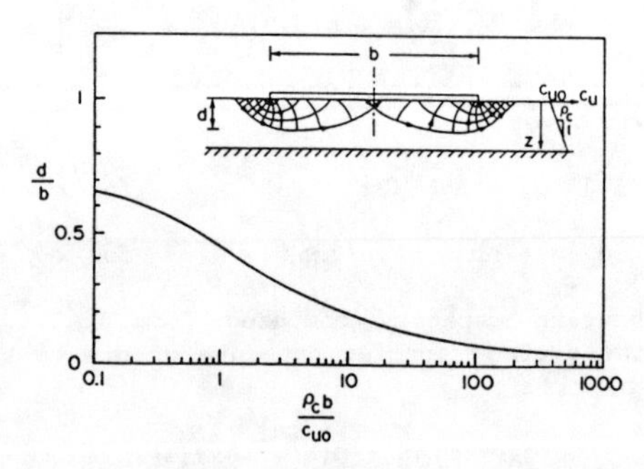

Figure 5. Effect of Nonhomogeneity on Depth of Failure Zone Beneath a Rough Rigid Footing (modified from Matar and Salencon, 1977, by Rowe and Soderman, 1987b)

Jewell (1988) showed the influence of the magnitude and direction of the shear stress at the embankment foundation interface on the bearing capacity factor. Embankments with no reinforcement apply an outward shear stress τ to the surface of the foundation soil as shown in Fig. 6a. This is analogous to a footing subjected to an outward shear stress (Fig. 6b) and the bearing capacity factor decreases as the magnitude of the outward acting shear stress increases as shown in Fig. 6c. In contrast, the shear stress applied to the top of the foundation soil is inward for embankments reinforced with high modulus reinforcement as shown in Fig. 6d. This is analogous to the stresses which develop beneath a rough footing (Fig. 6e). The inward acting shear stress increases the bearing capacity factor as illustrated in Fig. 6f for the case where strength increases with depth. Thus, a major benefit provided by the high modulus reinforcement is that the shear stress applied to the top of the foundation is changed from acting outward to acting inward. Jewell (1988) presents an approximate method to compute the allowable vertical stress taking into account the magnitude of the shear stress at the embankment/reinforcement interface.

GLOBAL STABILITY

Global stability is used to determine the reinforcing force required to produce the desired safety factor. This is generally

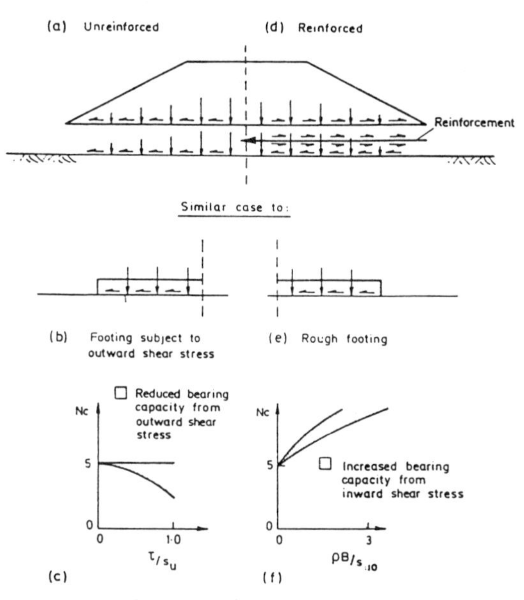

Figure 6. Mechanics of Reinforced Embankment on Soft Soil (After Jewell, 1988; Reprinted by permission of Elsevier Applied Science, Copyright, 1988)

done using a conventional limiting equilibrium stability analysis modified to account for the reinforcing force. An illustration showing simplified Bishop's method of slices including a horizontal reinforcing force is shown in Fig. 7. The reinforcement provides an additional resisting moment equal to $y(F_R)$ where F_R is the reinforcing force and y is the vertical distance to the center of the circle. The equation for the safety factor for simplified Bishop's method with a reinforcing force are given in Humphrey (1986, 1989).

Some investigators have suggested that the reinforcing force be taken tangent to the circle, in which case, the moment arm would be the radius of the circle r. However, it is the opinion of the authors that the deformations required to orient the fabric along the failure surface are so large that for all practical purposes the embankment will have failed before this condition is reached. In some cases the effective moment arm of the reinforcing force may lie between the distance to the horizontal reinforcement and the circle radius (Rowe and Soderman, 1985). However, it is unconservative to use the radius. So it is recommended that the reinforcing force be taken as horizontal which is conservative (Humphrey, 1986).

The resisting force provided by the reinforcement would be expected to alter the normal stress on the failure surface. Where the failure surface passes through a frictional soil, this would alter the shear strength. However, Humphrey (1986) used FE analyses to show that the normal stress on the failure surface increased only slightly where it passes through the fill. This is illustrated in Fig. 8 for an embankment constructed on a 2.3-m thick crust with an undrained shear strength of 6 kPa underlain by soil whose strength increased with depth. It is seen that in the embankment fill the normal stress on the critical failure surface determined by the FE method is increased only slightly from the unreinforced to the reinforced case. The resulting increase in shear strength would cause less than a 1% increase in the safety factor calculated by simplified Bishop's method. The normal stress calculated from the simplified Bishop method is also very similar. In contrast, the normal stress near the toe of the failure surface was decreased

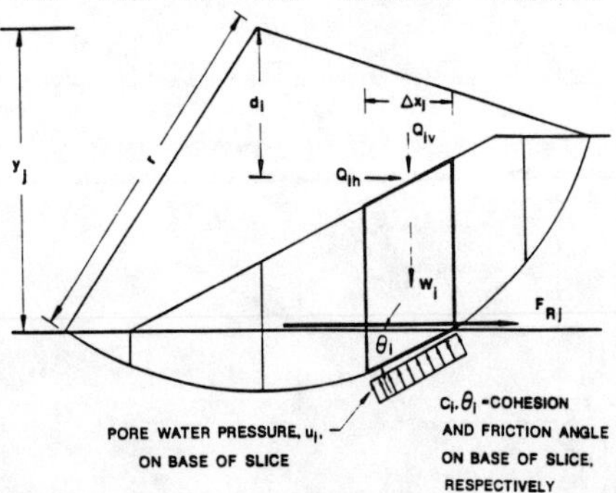

Figure 7. Simplified Bishop's Method of Slices Including Horizontal Reinforcing Force Showing Forces Acting on i^{th} slice (After Humphrey, 1986)

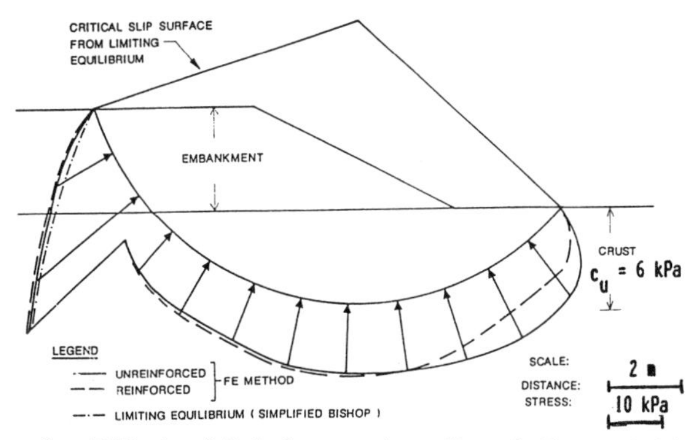

Figure 8. Effect of Reinforcement on Normal Stress Acting on Critical Failure Surface (After Humphrey, 1986).

substantially by the reinforcement but this would not affect the shear strength since the strength in the foundation soil is based on total stresses. Thus, in many cases simplified Bishop's method, modified to account for the reinforcing force, can be used with negligible error in the computed shear strength in the fill.

Rowe and Soderman (1984, 1987b) found that modified limiting equilibrium methods give similar safety factors to those predicted using finite element methods provided the reinforcing force predicted by the finite element method is used in the modified limiting equilibrium analysis. However, in some cases modified limiting equilibrium analyses which use circular slip surfaces can give misleading results for embankments on soils whose strength increases with depth or when an underlying stronger layer is at a shallow depth beneath the embankment (Jewell, 1988; Leshchinsky, 1986). For both these cases the critical failure surface is shallow because the failure surface avoids the deeper stronger soil as shown in Fig. 9a. However, a shallow circle underestimates the driving force in the embankment. To avoid this Jewell (1988) recommends that the driving force in the fill be calculated using an active Coulomb wedge and that a circular surface be used only in the foundation soil as shown in Fig. 9b. The driving force in the embankment is then the active earth force acting at the 1/3 height of the embankment as shown in Fig. 9c. It would be prudent for cases with a shallow failure surface to determine the critical failure surface for both a circular failure surface passing through the embankment and foundation as well as with Coulomb active earth pressure in the fill and a circular failure surface in the foundation. This recommendation is equally valid for reinforced and unreinforced embankments. Alternately, Leshchinsky (1986) proposed that a logarithmic spiral be used for the failure surface which results in a steeper failure surface in the fill. In some cases, a sliding block failure mode may be more realistic. A solution for this mode which includes a reinforcing force has been presented by Milligan and La Rochelle (1984).

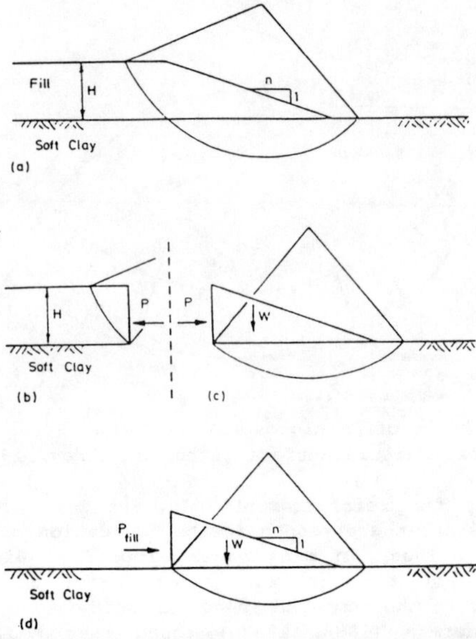

Figure 9. Conventional Slip Circle Analysis Separated to Account for Active Force in Fill (after Jewell, 1988; Reprinted by permission of Elsevier Applied Science, Copyright, 1988)

DEFORMATION

Excessive deformation will cause unwanted settlement and limit serviceability of the embankment. In some cases, deformation may be so large that for practical purposes the embankment has failed even if complete collapse has not occurred. Haliburton, et al. (1978) proposed to limit deformations by placing an upper limit on the allowable strain in the reinforcement. The required reinforcing force determined from global stability $F_{R(req'd)}$ and the allowable strain in the reinforcement ϵ_a are then used to establish the required modulus of the reinforcement $E_{req'd}$ as follows

$$E_{req'd} = F_{R(req'd)}/\epsilon_a \tag{9}$$

The value of ϵ_a that is generally used is 10%, with 5% sometimes used for strain softening soils (Fowler, et al., 1986). However, there is considerable evidence that in many cases a reinforcement strain of 10%, or even 5%, cannot be achieved before "practical failure" of the embankment is reached (Boutrup and Holtz, 1983; Humphrey, 1986; Rowe and Soderman 1987a,b).

"Practical failure" is defined as the fill thickness when the maximum net fill height is obtained and not the fill thickness required to achieve 'true' uncontrolled collapse (Rowe and Soderman,

1987a,b). Practical failure can be illustrated using the results of a FE analysis by plotting the net embankment height above the original ground surface vs. embankment thickness (i.e., thickness of fill placed) as shown in Fig. 10 for three reinforcement moduli J. It can be seen that after the point of maximum net fill height, placement of additional fill causes a net decrease in the embankment height as the embankment sinks into the underlying foundation soil.

The reinforcement strain at failure depends on foundation properties, geotextile modulus, and embankment geometry (Rowe and Soderman, 1985, 1987a,b; Humphrey, 1986). This is illustrated in Fig. 11 which compares failure pressures obtained from FE analysis, with those calculated from limiting equilibrium strength profiles. The horizontal dashed lines indicate the collapse pressure obtained from the FE results while the asterisks refer to the FE "practical" failure pressures. The solid curves give the failure pressures obtained using modified slip circle limiting equilibrium analyses for the three values of allowable geotextile strain indicated. It is evident from Fig. 11 that for each different case, there is a specific allowable strain value which will result in agreement between limit equilibrium and finite element failure pressures. However, this specific allowable strain value varies depending on the foundation properties and geotextile modulus. For example, an allowable strain of 5% seems to work reasonably well for the case in Fig. 11 provided the reinforcement modulus is less than 2000 kN/m. However, if the geotextile modulus is 4000 kN/m, an allowable strain value of 5% yields a failure pressure which is 30% greater than the collapse pressure.

For the case of a reinforced embankment constructed on a soft soil with uniform strength and underlain by a stronger layer Rowe and Soderman (1985) developed a simple procedure to determine the allowable strain in the geotextiles. However, for the common case where the foundation soil increases in strength with depth the failure mechanism is different so the allowable strains predicted by Rowe and Soderman (1985) are not valid. Thus, for many cases there is no simple way to estimate the strain or the force in the reinforcement at failure. The only alternatives are to make a conservative estimate based on experience or to perform a FE analysis. Rowe and Mylleville (1990) have indicated that for many practical cases involving a soil which does not experience strain softening, an allowable strain of 5% produces reasonable results. However, particular caution is required with high strength crusts and/or very high modulus reinforcement.

For foundation soils which experience strain softening, the strains which occur in the foundation must also be considered. It has been shown by Mylleville and Rowe (1988) and Rowe and Mylleville (1989, 1990) that the shear strain in the foundation can be significantly greater than the strain in the reinforcement as illustrated in Fig. 12. It is seen that the maximum shear strain in the foundation is 10% yet the maximum strain in the reinforcement is only 2.5%. It is not adequate to simply limit the strain in the reinforcement and then use peak strength in the foundation. It may be necessary to use: (1) post peak undrained shear strengths for the foundation soils if shear strain in the foundation exceed the strain

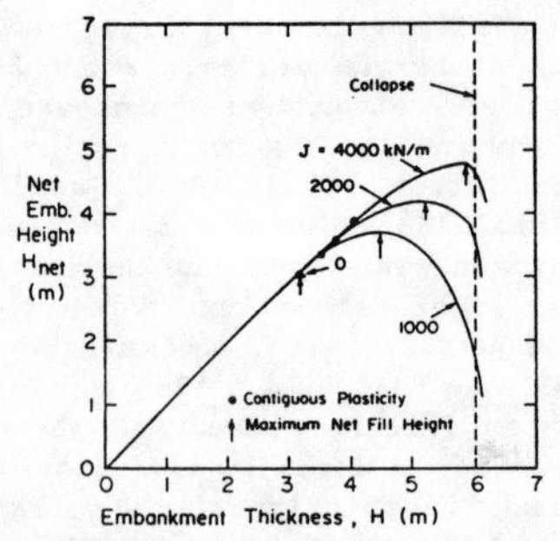

Figure 10. Net Fill Height Versus Fill Thickness for Various Reinforcement Moduli (modified from Rowe and Soderman, 1987a).

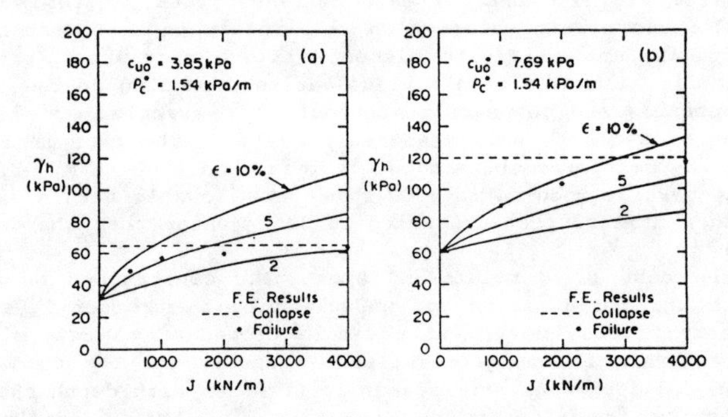

Figure 11. Comparison of Limit Equilibrium and Finite Element Results (after Rowe and Soderman, 1987a).

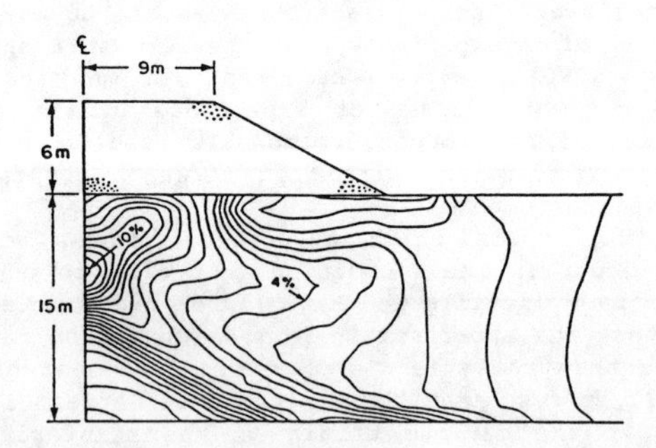

Figure 12. Shear Strain Contours at 6 m Fill Thickness: E_u/c_u = 125 and J = 4000 kN/m; 0.5% Contour Interval (after Rowe and Mylleville, 1989).

at the peak strength; (2) a smaller limiting strain in the reinforcement; and/or (3) use a very high modulus reinforcement. When designing reinforced embankments on soils where strain softening may occur, particular caution is required and a finite element analysis may be justified.

Construction of reinforced embankments on peats presents two problems not associated with embankments on clay foundations: (1) deformations are very large; and (2) significant drainage of the peat will occur during construction so an undrained stability analysis cannot be used (Rowe, et al., 1984). Rowe and Soderman (1985, 1986) have presented guidelines for reinforced embankment design on peat. Their studies showed that the effective use of reinforcement for embankments on peat requires: (1) careful control of the rate of construction; (2) consideration of the thickness of the peat and the presence of any soft underlying layer; and (3) the modulus of the reinforcement. They presented guidelines for the selection of an appropriate geosynthetic modulus for a range of situations involving different thicknesses of peat (3-8 m) and soil conditions beneath the peat. A key part of these guidelines is that the maximum excess pore pressure in the peat should not exceed 34% of the change in total major principal stress at any point. This requirement can be met by controlling the rate of construction (see Rowe, 1984). Where there is doubt, pore pressures should be monitored and construction controlled to ensure that excessive rates of construction are not adopted. One of the authors (Rowe) is aware of unpublished cases where the proposed guidelines have worked well; he is also aware of cases where failure to control the rate of construction and consequent build-up of excess pore pressures as recommended by Rowe (1984) and Rowe and Soderman (1985, 1986) has resulted in failure of reinforced embankments on peat.

As noted by Rowe, et al. (1984) and Rowe (1984) when dealing with embankments on peat, geosynthetic reinforcement and good construction practice may: (1) increase stability of the embankment and hence allow construction of higher embankments; (2) reduce lateral deformations; and (3) allow construction of a surcharge (which might not otherwise be possible) which can be used to minimize long term creep deformation of the final embankment. However, reinforcement will not significantly reduce consolidation settlements of the peat. If settlement is to be reduced then options such as lightweight fill should be considered (e.g., see Rowe, 1984; Rowe and Soderman, 1985, 1986).

It is important to recognize that peat is a very complex and difficult material. The uncertainties associated with any design method should not be underestimated. Nevertheless, the available data would suggest that the guidelines proposed by Rowe and Soderman (1985, 1986) can provide reasonable design for soil conditions similar to that examined in their studies as long as their guidelines on the rate of construction and the maximum excess pore pressure are followed.

ROLE OF FINITE ELEMENT ANALYSIS

The primary roles of finite element analysis are (a) for research into the behavior of reinforced embankments including validation of simplified methods of analysis, as discussed in this paper; and (b) for complementing conventional analyses on large/important projects or where the anticipated conditions are such that the validity of simplified approaches may be questioned (Rowe and Soderman, 1987b). However, finite element analyses require a more thorough determination of soil properties than limiting equilibrium methods and an experienced modeler to perform the analyses.

CONCLUSIONS

1. Haliburton's (1978) design procedure still provides a useful framework for design of reinforced embankments. However, results from plasticity theory, limiting equilibrium, and finite element analyses show that there are several places where improvements in the original procedure are needed.
2. Bearing capacity gives an upper bound to the vertical stress that can be applied by a highly reinforced embankment and hence an upper limit to the height to which a reinforced embankment with this geometry can be constructed. If the foundation soils increase in strength with depth or are of limited thickness this should be accounted for in the bearing capacity analysis as well as the surcharge provided by the toe of the embankment (see Rowe and Soderman, 1987b).
3. Conventional limiting equilibrium analyses modified to account for the reinforcing force can give a reasonable estimate of global stability. However, if the critical circle is shallow, the driving force from the embankment fill will be underestimated. To remedy this the driving force from the fill should be determined from the Coulomb active pressure (e.g., see Jewell, 1988).
4. Using a fixed value of the allowable strain in the reinforcement of 5-10% is unconservative for many cases. The reinforcement strain that will develop before failure occurs is a function of the foundation properties, geotextile modulus, and embankment geometry. Rowe and Soderman (1985) presented a method to estimate the allowable strain in the reinforcement for the case of foundation soils with uniform strength and limited depth. For other foundation conditions the allowable strain must be estimated based on experience or using finite element analysis. Some guidance has been given in this paper and by Rowe and Mylleville (1990).
5. Shear strains in the foundation may be many times greater than the strain in the reinforcement. This must be considered when choosing the design strength of strain softening soils (see Mylleville and Rowe, 1988; Rowe and Mylleville, 1989, 1990).

REFERENCES

1. R Bonaparte, RD Holtz & JP Giroud, Soil Reinforcement Design Using Geotextiles and Geogrids (STP 952), Ed JE Fluet, Jr, ASTM, Philadelphia, 1985, 69-116.

2. E Boutrup & RD Holtz, Analysis of Embankments on Soft Ground Reinforced with Geosynthetics, Proc. VIII European Conf. SMFE, Helsinki, 1983, 469-472.
3. JP Carter & N Balaam, AFENA: Users Manual, Geotechnical Research Centre, The University of Sydney, 1989. [This version was modified by Rowe, Soderman, and Mylleville to allow analysis of reinforced embankments.]
4. BR Christopher & RD Holtz, Geotextile Engineering Manual, Rpt FHWA-TS-86/203, FHWA, Washington, DC, 1985, 1044 pp.
5. EH Davis & JR Booker, The Effect of Increasing Strength With Depth on the Bearing Capacity of Clays, Geotechnique, 23(4), 1973, 551-563.
6. J Fowler & RM Koerner, Stabilization of Very Soft Soils Using Geosynthetics, Proc. Geosynthetics'87, New Orleans, 1987, 289-300.
7. J Fowler, J Peters & L Franks, Influence of Reinforcement Modulus on Design and Construction of Mohicanville Dike No. 2, Proc. 3rd Intl. Conf. on Geotextiles, Vienna, 1986, 267-271.
8. TA Haliburton, CC Anglin & JD Lawmaster, Testing of Geotechnical Fabric for Use as Reinforcement, Geot. Testing J., ASTM, 1, Dec., 1978, 203-212.
9. RD Holtz, Design and Construction of Embankments on Very Soft Soils, 1989 Miles S. Kersten Lecture, Minn.Geot.Eng. Conf., St.Paul, MN, 1989, 35 pp.
10. DN Humphrey, Design of reinforced embankments, Rpt FHWA/IN/JHRP-86/17, Joint Highway Research Project, School of Civil Eng., Purdue Univ., W.Lafayette, IN, 1986, 423 pp.
11. DN Humphrey, Dsc of Current Design Methods by RM Koerner, B-L Hwu & MH Wayne, Geotextiles and Geomembranes, 6(1), 1987, 89-92.
12. DN Humphrey, Geosynthetic Reinforced Embankments, Proc 1989 Maine Section ASCE Technical Seminar, Augusta, 1989, 17 pp.
13. DN Humphrey & RD Holtz, Reinforced Embankments - A Review of Case Histories, Geotextiles and Geomembranes, 4(2), 1986, 129-144.
14. DN Humphrey, WO McCarron, RD Holtz & WF Chen, Finite Element Analysis of Plane Strain Problems with PS-NFAP and the Cap Model -- User's Manual, Rpt FHWA/IN/JHRP-86/18, Joint Highway Research Project, School of Civil Eng., Purdue Univ., W.Lafayette, IN, 1986, 184 pp.
15. RA Jewell, The Mechanics of Reinforced Embankments on Soft Soils, Geotextiles and Geomembranes, 7(4), 1988, 237-273.
16. D Leshchinsky, Short-term Stability of Reinforced Granular Embankments Over Clayey Foundation, Rpt CE-61, Univ. of Delaware, prepared for US Army Corps of Engineers, 1986.
17. M Matar & J Salencon, Capacite Portante a Une Semelle Filante sur Sol Purement Coherent D'epaisseur Limitee et de Cohesion Variable Avec la Profondeur, Annales de l'Institute Technique du Batiment et des Travaux Publics, Supplement No. 352 (Juillet-Aout 1977), Serie: Sols et Fondations, No. 143, 95-107.

18. WO McCarron, Soil Plasticity and Finite Element Applications, Ph.D. Thesis, School of Civil Eng., Purdue Univ., W.Lafayette, IN, 265 pp.
19. V Milligan & P La Rochelle, Design Methods for Embankments Over Weak Soils, Proc. Symp. on Polymer Grid Reinforcement in Civil Engineering, Science and Engineering Research Council, London, 1984, 8 pp.
20. BLJ Mylleville & RK Rowe, Some Considerations in the Design of Geosynthetic Reinforced Embankments on Clayey Foundations, Proc. 3rd Canadian Symp. on Geosynthetics, Kitchener, Ontario, 1988, 29-33.
21. RK Rowe, LEPSSIA - Large Strain Soil Structure Interaction Analysis Program: Users Manual, Geotechnical Research Centre, Faculty of Engineering Science, The University of Western Ontario, London, Canada, 1982.
22. RK Rowe, Recommendations for the Use of Geotextile Reinforcement in the Design of Low Embankments on Very Soft/Weak Soils, Research Report GEOT-1-84, Geotechnical Research Centre, Faculty of Engineering Science, The University of Western Ontario, 1984.
23. RK Rowe, MD MacLean & KL Soderman, Analysis of Geotextile-Reinforced Embankment Constructed on Peat, Canadian Geot.J., 21(3), 1984, 563-576.
24. RK Rowe & BLJ Mylleville, Consideration of Strain in the Design of Reinforced Embankments, Proc. Geosynthetics'89, San Diego, 1989, 124-135.
25. RK Rowe & BLJ Mylleville, Implications of Adopting an Allowable Geosynthetic Strain in Estimating Stability, Proc. 4th International Conference on Geotextiles, Geomembranes, and Related Products, The Hague, 1990, 131-136.
26. RK Rowe & KL Soderman, Comparison of Predicted and Observed Behaviour of Two Test Embankments, Geotextiles and Geomembranes, 1, 1984, 143-160.
27. RK Rowe & KL Soderman, An Approximate Method for Estimating the Stability of Geotextile-Reinforced Embankments, Canadian Geot. J., 22(3), 1985, 392-398.
28. RK Rowe & KL Soderman, Reinforced Embankments on Very Poor Foundations, Geotextiles and Geomembranes, 4(1), 1986, 65-81.
29. RK Rowe & KL Soderman, Geotextile Reinforcement of Embankments on Peat, Geotextiles and Geomembranes, 2(4), 1985, 277-298.
30. RK Rowe & KL Soderman, Reinforcement of Embankments on Soils Whose Strength Increases With Depth, Proc. Geosynthetics'87, New Orleans, 1987a, 266-277.
31. RK Rowe & KL Soderman, Stabilization of Very Soft Soils Using High Strength Geosynthetics: The Role of Finite Element Analyses, Geotextiles and Geomembranes, 6(1), 1987b, 53-80.

SOLID MODELLING FOR SITE REPRESENTATION IN GEOTECHNICAL ENGINEERING

By Norman L. Jones[1], M.ASCE and Stephen G. Wright[2], M.ASCE

ABSTRACT: Three dimensional computer models of construction sites and subsurface stratigraphy are useful for visualization, analysis and design in geotechnical engineering. A useful technique for creating such models is the geometric technique known as "solid modelling." A number of computer-based solid modelling systems are commercially available. However, the available modelling systems are not well-suited for modelling earth masses because of the complex nature of the surfaces involved. To overcome this difficulty a technique employing triangulated irregular networks (TIN's) and set operations has been developed. This technique facilitates the construction of complex three-dimensional models of earth masses from data which are typically available to geotechnical engineers. These data usually consist of surface topographic surveys and borehole logs.

Using the technique described, a geotechnical engineer can create a three-dimensional model of a complex site. Complicated excavations can be modelled and cross sections, or "fence" diagrams, can be constructed anywhere on the model to display the soil stratigraphy at the site. Volumes of excavation and fills are also easily computed.

INTRODUCTION

Geotechnical engineers represent results of subsurface investigations in two-dimensions. Although actual sites are three-dimensional, two-dimensional representations are used because they are simple and can be created easily. Three-dimensional representations are more complex and until recently the tools for creating such representations have not been available.

The advent of faster computers, and especially the development of improved graphics software and hardware, in recent years has provided new opportunities for site representation in three-dimensions. It has become possible to represent almost any object fully in three-dimensions. Objects can be manipulated and viewed from any desired angle and at almost any desired scale. Coloring and shading can be used to make an object appear more realistic and permit the user to more easily visualize its features. These tools also aid greatly in the initial construction of the model of an object as well as in the modification and use of the model in design.

"Solid modelling" is a powerful technique that has developed along with the advances in computer graphics. Solid modelling makes it possible to represent the full

---

[1] Graduate Research Assistant, Dept. of Civil Engrg., The Univ. of Texas, Austin, TX 78712-1076.
[2] Ashley H. Priddy Cent. Prof., Dept. of Civil Engrg., The Univ. of Texas, Austin, TX 78712-1076.

three-dimensional form of objects and to reshape and manipulate objects easily. Various attributes of objects, such as their volume and mass properties can also be determined easily. Solid modelling is also an ideal candidate for modelling earth masses and the results of subsurface investigations.

Solid modelling was originally developed and introduced in the fields of mechanical and aerospace engineering. In these fields, the shape and complexity of objects is relatively simple and well-defined as compared to those of the earth's surface and subsurface. Accordingly, the existing solid modelling techniques require further adaptation to make them suitable for modelling earth masses. This paper discusses the adaptation and application of solid modelling techniques to earth masses and describes the current limitations of solid modelling.

SOLID MODELLING

The term "solid modelling" is used to describe any technique for representing the three-dimensional geometry of an object in the computer such that the volume of the object is unambiguously defined. Numerous representation schemes for solids have been developed (Requicha 1980). The constructive solid geometry (CSG) and boundary representation (B-Rep) schemes are two of the most common schemes. In the CSG scheme solids are formed by combining a series of relatively simple solid shapes (spheres, cubes, pyramids, etc.). Boundary representation schemes are based on the fact that an N-dimensional object can be defined by a set of N-1 dimensional boundaries (Figure 1). For example, a three-dimensional solid object can be defined by a set of two-dimensional surfaces that completely enclose the volume of the object. In practice, the surfaces are typically polygons. The two-dimensional polygons are defined by a set of one-dimensional line segments and the line segments are defined by zero-dimensional vertices. Realistic images of B-Rep solid models can be rendered on the computer by doing hidden surface removal and color shading of the polygons. The B-Rep scheme is used in the modelling process described in this paper.

The volume of a solid object can be treated as a subset of space. Accordingly, it is possible to combine two solids using the set operations: Union, Intersection, and Difference (Figure 2). This allows the user of a solid modelling program to start with relatively primitive shapes and combine them using the set operations to create a more complex solid model. The procedures and algorithms for performing the set operations are complicated and compute intensive. Designing and implementing these algorithms is the most difficult aspect of developing a solid modelling.system

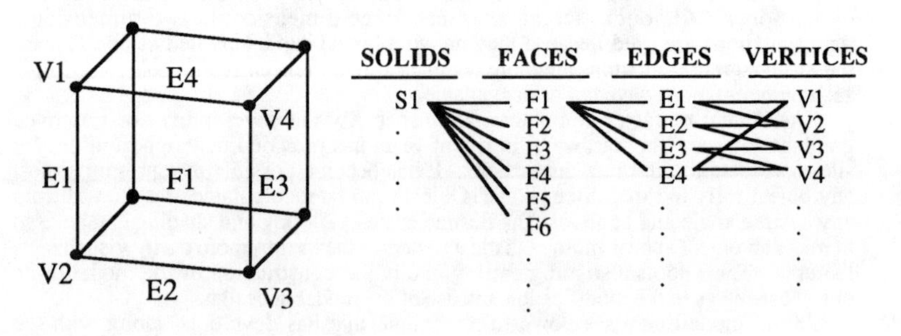

Figure 1. Boundary Representation Scheme for Solid Models.

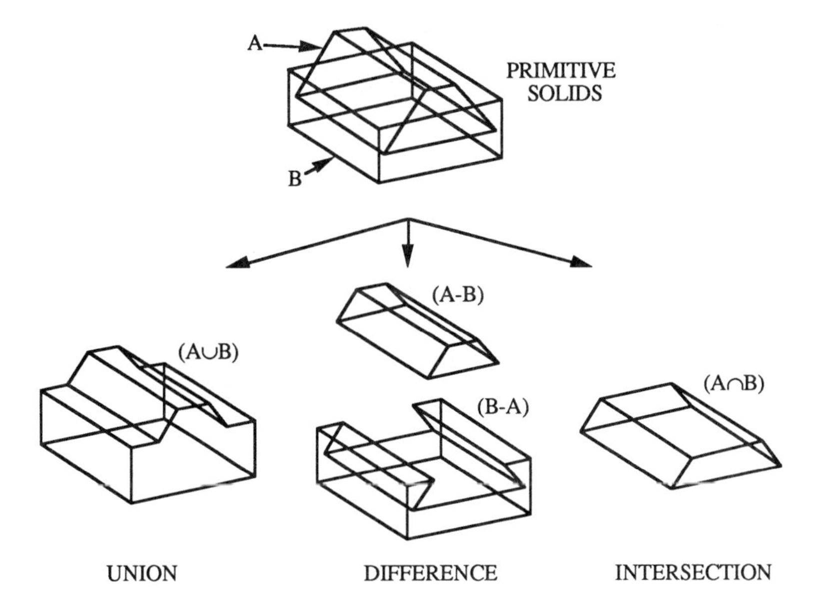

Figure 2. Solid Models Can be Combined with Set Operations.

SOLID MODELLING WITH COMMERCIAL CAD SYSTEMS

A demonstration study was performed by the authors using a commercial solid modelling system called CATIA. The purpose of the study was to investigate the feasibility of modelling construction sites using solid modelling techniques (Jones, 1988). CATIA is a software package developed in France by Dassault Systems and distributed in the United States by IBM Corporation. CATIA is widely used in aerospace and mechanical engineering applications.

Several simple sites were modelled as part of the study. Solids representing various subsurface layers were created. "Solids" representing fill or excavation volumes were then created and used to modify the original site geometry using the set operations described above. Details and illustrations are presented by Jones (1988) and are excluded from this paper due to length limitations.

Experience with CATIA showed that solid modelling has significant potential for applications in geotechnical engineering. However, there are a number of difficulties with commercial solid modelling systems. Such systems have been designed for applications where the surfaces being modelled are smooth and uniform. The surfaces of earth masses are usually more complex and irregular. The techniques employed in commercial solid modelers for converting surfaces into solids are often slow and cumbersome. For applications in geotechnical engineering, surface modelling techniques which are better-suited for modelling earth masses are needed and new techniques are required for converting the surfaces into solid models.

SURFACE MODELLING TECHNIQUES

Many techniques have been developed for surface modelling ranging from simple grid techniques to more sophisticated spline techniques (Franke 1982; Lancaster & Salkauskas 1986). A surface modelling technique that is well-suited for modelling earth masses should possess several features including:

1. The surface modelling technique must allow the creation of irregular surfaces with abrupt discontinuities. Faults, ridges, and valleys must be modelled correctly.
2. The technique must be capable of handling scattered data as input because the main source of input will be from borehole logs and survey data. This requirement makes techniques based on regular grids unsuitable.
3. The final surface should pass through the original data points on which the surface is based rather than approximate the points in a least squares sense.
4. It should be possible to reshape or sculpt the surface interactively. This requirement precludes the type of surfaces referred to as "global" surfaces. The parameters of global surfaces must be recomputed each time any of the data points defining the surface are changed. Examples of global techniques include most inverse distance weighted schemes and kriging (Davis 1986; Journel & Huijbregts 1978; Shepard 1968).
5. The surface must be amenable to computing set operations. This requires that it be possible to compute intersections of surfaces once the surfaces become the boundaries of the solid model. For practical purposes it requires that the surface should either be composed of planar polygons or very simple algebraic surfaces.

A surface modelling technique that meets the requirements listed above is the triangulated irregular network or TIN. The surface shown in Figure 3 is a TIN surface. TIN surfaces are constructed by connecting the xy projection of a set of scattered data points with a series of edges to create a network of triangles.

Typically, a criterion or heuristic is used to guide the construction of the TIN so that long, thin triangles will be avoided. A popular criterion is the Delauney criterion (Lee & Schacter 1980; Watson 1981; Watson & Philip 1984).

If the surface being modelled is assumed to vary linearly across each of the triangles, the TIN defines a piecewise linear surface that passes through all of the data points. Breaklines or feature lines can be incorporated into the TIN to ensure that ridges, valleys, and other distinct features are modelled accurately. The shape of the TIN can be changed simply by changing the coordinates of the data points and adding supplemental data points in areas where extra refinement is desired.

SOLID MODELLING OF EARTH MASSES

In mechanical and aerospace engineering applications solid models of objects such as engine parts are typically constructed by combining several simple primitive shapes such as cubes, spheres, prisms, and cylinders with set operations to achieve the final desired shape. Occasionally, the surfaces making up the boundary of the solid are defined separately and joined together to define the solid. These techniques are not well-suited for constructing solid models of soil and rock masses because the surfaces of earth masses are highly complex and irregular. A more convenient technique for

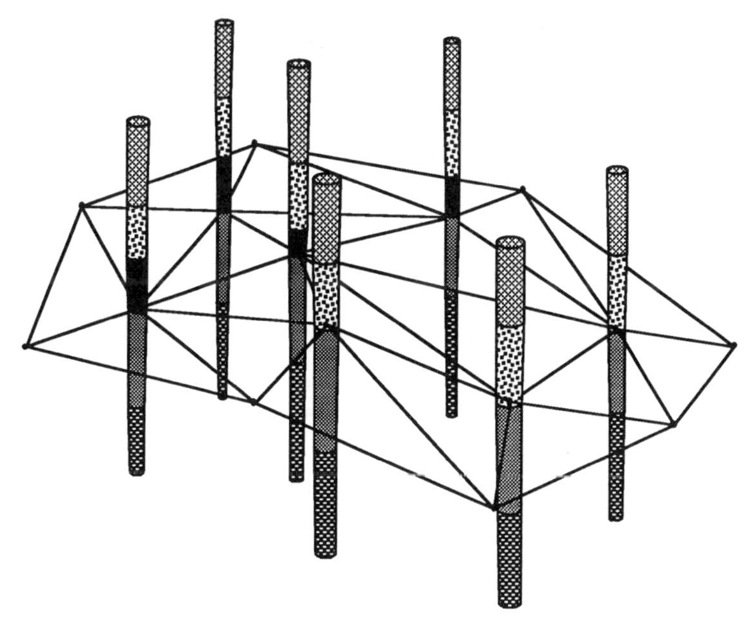

Figure 3. Boreholes and TIN Surface Passing Through Selected Borehole Contacts.

constructing solid models of earth masses has been developed consisting of the following fundamental steps:

1. Construct complex surfaces to represent the interfaces between layers.
2. Convert surfaces into primitive solids.
3. Combine the primitive solids using set operations.

Construction of Surfaces. The first step involves creating a TIN surface from a set of scattered data points and refining or shaping the surface until the surface accurately represents the surface of the soil layer being modelled (Figure 3). Since the subsurface data points typically come from borehole logs, the data points may be very sparse. The addition of several supplemental data points is often required in order to "smooth" the surface sufficiently. The location and elevation of the supplemental points can be specified explicitly as each point is generated.

It is also possible to refine or "smooth" the entire TIN at once by adding a series of supplemental points at special locations dictated by a user-defined smooth mathematical surface that interpolates the original data points. This algorithm for adding supplemental points is described by Jones and Wright (1991).

Conversion of Surfaces Into Primitive Solids. The second step involves converting the surfaces created in step 1 into temporary solid primitives that represent rough approximations of the soil layers. The conversion is accomplished by

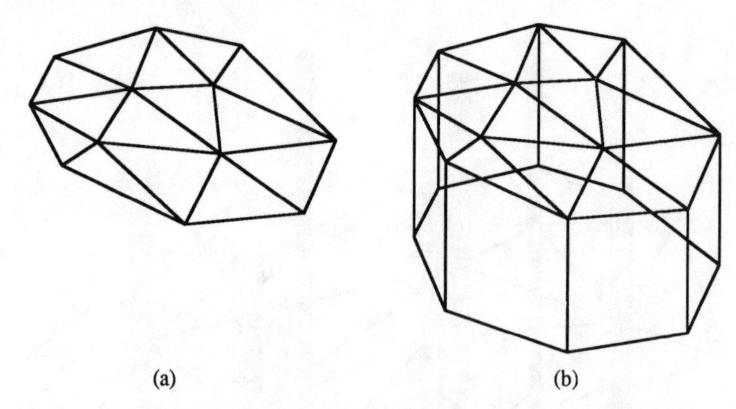

Figure 4. (a) Surface Representing a Soil Interface Defined by a Triangulated Irregular Network. (b) Conversion of the Surface in (a) to a Solid by Extruding the Perimeter of the TIN.

projecting the outer boundary (perimeter) of the TIN down to a horizontal plane as shown in Figure 4. Each of the triangles becomes a face of the new solid. New planar faces are created around the perimeter of the solid and one large face is created at the base of the solid. The faces around the perimeter are quadrilaterals; the base of the solid is a polygon. The elevation of the horizontal plane is typically chosen so that the resulting solid is below the lowest point of interest.

Combination of Primitive Solids Using Set Operations. The final step of the modelling process involves the combination of the primitive solids constructed in step 2 into solid models of the soil layers. This is accomplished using set operations. The primitive solids representing layers adjacent to a given solid are subtracted from the solid to produce the model of the soil layer. In this way, portions of the soil layer that intersect other soil layers are "trimmed" away from the soil layer.

The three step process for constructing a solid model is illustrated in two-dimensions in Figure 5. Three surfaces, labelled p, q, and r, are constructed in step 1 and are shown in Figure 5a. The primitive solids P, Q, and R formed by extruding the surfaces in step 1 are shown in Figure 5b. Step 3 of the modelling process is illustrated in Figure 5c. Primitive Q is subtracted from primitive P to produce the temporary solid P-Q. Primitive R is then subtracted from P-Q to produce layer P. Layer Q is formed by subtracting primitive R from primitive Q. The completed solid models of the soil layers are shown in Figure 5d.

In addition to significantly simplifying the process of constructing solid models of soil layers, the three step modelling process helps to ensure that no artificial voids are introduced between soil layers and that adjacent soil layers match up precisely at their interfaces. By using the simpler intermediate solids to produce the final models of the soil layers through set operations, the soil layers are assured to be well-formed solids, i.e., the faces of each solid completely enclose the volume occupied by the solid and the solid has no dangling faces or edges.

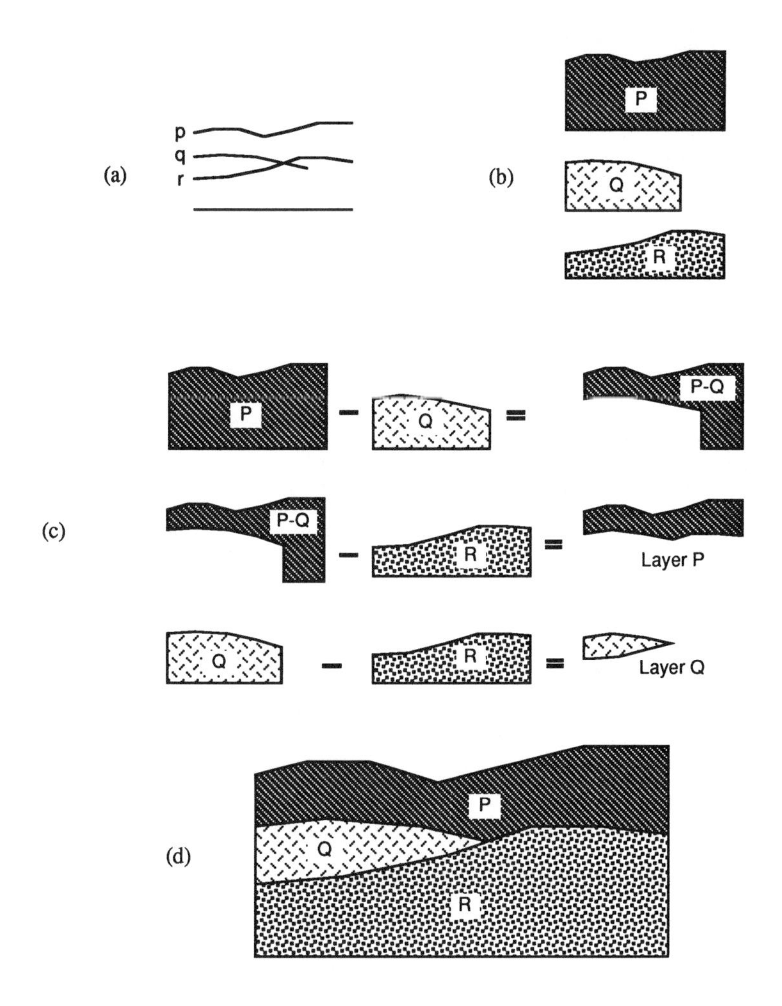

Figure 5. The Three Step Process for Creating Solid Models of Soil Layers. (a) Step 1: Construction of Surfaces. (b) Step 2: Extrusion of Surfaces Into Solids. (c) Step 3: Creation of Layers Through Set Operations. (d) Completed Solid Model of Soil Stratigraphy.

GEOSOLID

A computer program, "GEOSOLID" has been developed to implement the solid modelling process described above. The computer program facilitates the construction of TIN surfaces from scattered data points read in from an ASCII file. These data points typically represent direct survey measurements of the terrain (top) surface. Borehole logs are read from a file and displayed graphically in three dimensions. Interfaces between soil layers are selected interactively on the screen and converted to data points which are then used to define a TIN like the one shown in Figure 3. Extra points are added to the surface and the surface is edited or shaped until the surface corresponds to an acceptable representation of the surface being modelled. The surface is then converted to a solid and manipulated using the set operations as described previously. These operations are all done interactively on an Apple Macintosh II computer.

SAMPLE SITE

The proposed site for a warehouse located in the Eastern United States was modelled using the three step solid modelling process described above. Thirty-three exploratory boreholes were drilled at the site. The borehole data revealed a complex series of alluvial layers underlain by limestone bedrock. The elevation of the surface of the bedrock varied greatly from region to region. It was difficult to visualize the spatial relationship of the soil layers and the bedrock using only the borehole logs on paper and cross sections drawn by hand between sets of adjacent boreholes.

Borehole data were entered into the computer program GEOSOLID. Three-dimensional display of the boreholes on the computer screen were then examined. The boreholes were displayed from several different viewing angles in order to develop a mental image of the spatial relationships of the soil layers. Solid models of the soil layers were then constructed using the three step modelling process. Because the soils were highly heterogeneous, it was necessary to simplify the problem somewhat to make the modelling process feasible. Materials composed primarily of sand (sand, silty sand, clayey sand, etc.) were modelled as a single unit labelled "sand." Likewise, predominantly clayey materials were lumped together as one material labelled "clay." The other materials modelled were the limestone bedrock and a layer of sandstone boulders. A simple solid model of the warehouse was also constructed. The warehouse model was a polyhedron composed of polygons representing the floor, roof, and exterior walls of the warehouse.

A series of cross-sections through the solid models is shown in Figure 6. The cross-sections were created by computing the intersections of the plane of the cross-section with each of the polygons making up the solid. This results in a series of edges forming polygons which represent the cross-section. The solids used to compute the cross-sections in Figure 6 are shown in Figure 7a.

The solids shown in Figure 7b represents the soils directly beneath the warehouse. To create the solids shown in Figure 7b a special temporary solid was constructed: A polygon representing the base of the warehouse was created and extruded downward to create a solid prism whose cross-section corresponded to the perimeter of the warehouse. This solid was then intersected with the solid models of the soil layers shown in Figure 7a to create the solids shown in Figure 7b.

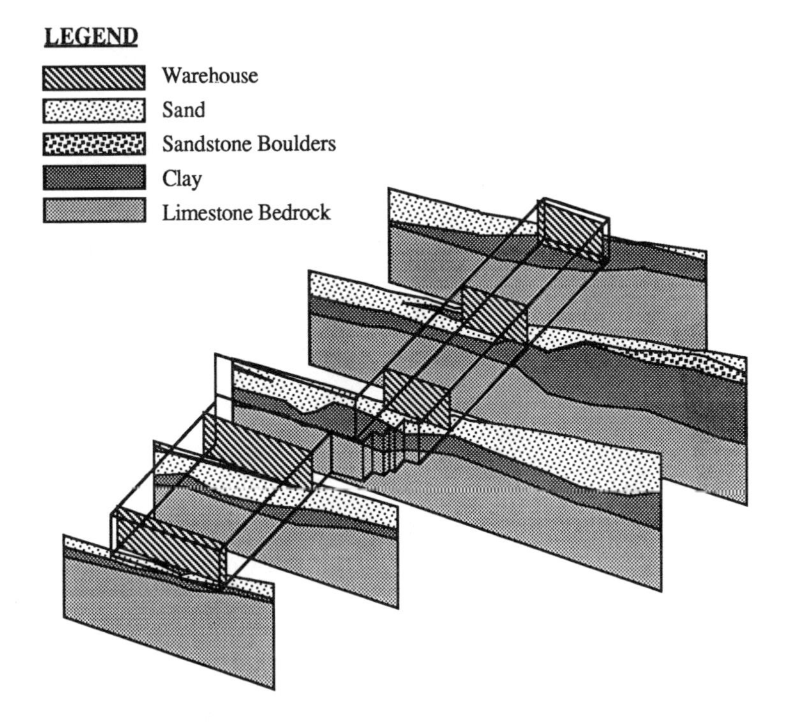

Figure 6. Wireframe Display of Warehouse Solid Model Superimposed on Cross Sections Computed From Solid Models of Underlying Soil Stratigraphy.

CONCLUSIONS

Solid modelling is a powerful technique for creating and manipulating three-dimensional representations of objects using the computer. However, the commercial solid modelling systems have been developed for mechanical and aerospace engineering applications and they must be extended before they become useful for geotechnical engineering applications.

A three-step modelling process employing TIN surfaces has been developed for application of solid modelling to characterize geotechnical sites. This process works well for creating solid models of earth masses. The process is relatively simple and results in well-formed solids, with no erroneous voids or "dangling" edges or faces.

Using the solid modelling techniques described in this paper three-dimensional models can be created for complex sites. Such three-dimensional models greatly help the engineer to understand the spatial relationships of the objects at the sites as well as the subsurface stratigraphy.

Solid models of soil and rock can also be used to support complex three-dimensional analyses. Several algorithms have been developed for automatically

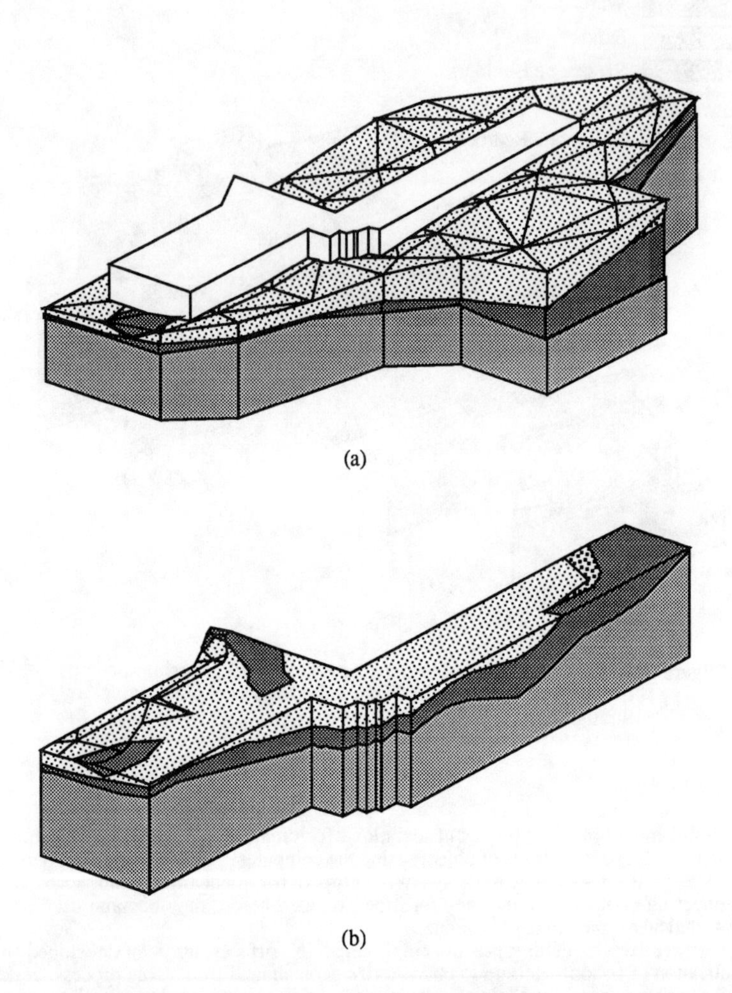

Figure 7. Solid Models of (a) Warehouse and Soils and (b) Soils Directly Beneath Warehouse.

converting complex solid models into three-dimensional finite element meshes (Baker, 1989; Schroeder and Shepard, 1989). These algorithms can be used to create three-dimensional finite element meshes from solid models for both seepage and stress analyses. The three-dimensional solid models should also have potential for three-dimensional slope stability analyses.

It is also relatively easy to compute volumes from the solid model of an earth mass. Computation of earthwork quantities therefore become a significant derivative benefit of the three-dimensional modelling process.

ACKNOWLEDGEMENTS

The work presented in this paper was performed as part of Grant MSM-8717452 from the National Science Foundation.

REFERENCES

1. Baker, T. J., Automatic Mesh Generation for Complex Three-Dimensional Regions Using a Constrained Delauney Triangulation, Engineering With Computers, Springer-Verlag, New York, 5(3/4), Summer/Fall 1989, 161-175.
2. Davis, J. C., Statistics and Data Analysis in Geology, John Wiley & Sons, New York, 1986, 550 p.
3. Franke, Richard, "Scattered Data Interpolation: Tests of Some Methods," Mathematics of Computation, 38(157), Jan., 1982, 181-200.
4. Jones, N. L., Applications of Computer-Aided Design Methods for Site Characterization in Civil Engineering, M.S. Thesis, The University of Texas at Austin, May, 1988.
5. Jones, N. L., & S. G. Wright, An Algorithm for Smoothing Triangulated Surfaces, Journal of Computing in Civil Engineering (ASCE), Jan., 1991.
6. Journel, A. G. & CH. J. Huijbregts, Mining Geostatistics, Academic Press, New York, 1978, 600 p.
7. Lancaster, P. and K. Salkauskas, Curve and Surface Fitting, Academic Press, London, 1986, 280 p.
8. Lee, D. T., & B. J. Schacter, Two Algorithms for Constructing a Delauney Triangulation, International Journal of Computer and Information Sciences, 9(3), June 1980, 219-242.
9. Requicha, A. A. G., Representations for Rigid Solids: Theory, Methods, and Systems, Computing Surveys, 12(4), December 1980, 437-464.
10. Schroeder W. J., & M. S. Shepard, An O(N) Algorithm to Automatically Generate Geometric Triangulations Satisfying the Delauney Criteria, Engineering With Computers, Springer-Verlag, New York, 5(3/4), Summer/Fall 1989, 177-193.
11. Shepard, D., A Two-Dimensional Interpolation Function for Irregularly Spaced Data," Proc. 23rd Nat. Conf. ACM, 1968, 517-523.
12. Watson, D.F., Computing the N-Dimensional Delauney Tesselation with Application to Voronoi Polytopes, The Computer Journal, 8(2), 1981, 167-172.
13. Watson, D. F. & G. M. Philip, Systematic Triangulations, Computer Vision, Graphics, and Image Processing, 26(2), 1984, 217-223.

SMART TUNNEL:
Automated Generation of Structural Models For Cut-And-Cover Tunnels

Brian Brenner (1), Sam S. C. Liao (2), and Cynthia Gagnon (3)

ABSTRACT: This paper reviews the development and application of SMART TUNNEL, a computer program that generates structural models for cut-and-cover tunnel sections from CAD graphics. Issues discussed include preparation of input for the program, program flow, including sorting routines and calculation of model support conditions, and application of the software to analysis and design.

INTRODUCTION

Cut-and-cover tunnels have been traditionally analyzed and designed in section. Engineers would develop a frame model to represent the box sections. Different loadings would be calculated and applied to the model. These loadings include vertical loads from soil, construction and live load surcharge, upward pore pressure on the base slab, and a group of lateral loads including the effects of pore pressure, lateral effective stress of soil, and lateral stress due to surface surcharge. The tunnel also has its own dead load. Analysis would be performed using an indeterminate method like moment distribution, or, more recently, with the assistance of a frame analysis computer program like STAAD-III or STRUDL. Based on the results of the analysis, the components of the cut-and-cover tunnel would be sized. Concrete slabs would be typically designed as bending members. Steel frames and prestressed concrete members likewise would be designed in similar fashion. There would be some amount of iteration to the process. The engineer, following sizing of the structure, may need to revisit the analysis with more accurate proportions for member size.

Even with use of computers, frame models are time-consuming to prepare and run. Input for the analysis includes the geometry of the tunnel section, member properties, material constants, support conditions, and specification of loading. For geometry, the engineer draws a stick model representing the tunnel section. Node points and members connecting the nodes are identified and

----

(1)- Senior Structural Engineer, Parsons Brinckerhoff
(2)- Professional Associate, Parsons Brinckerhoff
(3)- Structural Engineer, Bechtel Civil, Inc.

numbered. The positions of the nodes and resulting members are determined, in part, by not only the shape of the tunnel, but by refinement required by the engineer. For example, several nodes may be placed along the base slab to further define reaction points representing supporting soil. Required data, therefore, is a list of x-y coordinates for the nodes, and a list of member connectivity data defining the members.

Each member must be assigned initial trial analysis properties such as sectional moment of inertia and cross sectional area. For a concrete slab tunnel, typically a one foot longitudinal section is selected for modelling, and the corresponding member properties are calculated based on the assumed depth of the slab. This data is calculated and added to the list of analysis input. Likewise, material properties must be calculated and assigned to the members. For example, each member in the frame analysis must receive a specification for modulus of elasticity and a value for material unit weight.

Composite steel and concrete tunnels require additional calculation input, especially if the analysis is to be run as a "transformed" section in steel or concrete.

Computer frame models require specification of base support. This is typically handled in one of two ways (Figure 1). For the simple tunnel box shown, two imaginary support points are placed in the model. The designer then calculates an upward uniform vertical load to be placed below the base slab such that all vertical loads are balanced. The resulting reactions at the imaginary support points are zero. Using the second method, the engineer specifies spring supports. This will be subsequently referred to as the Beam on Elastic Foundation (BOEF) method. The value of the spring constant is determined from consideration of the underlying strata and its modulus of subgrade reaction [(For example, see Terzaghi (1955), Bowles (1988) and ACI (1988)]. As shown in Figure 1, the BOEF method will result in a non-uniform base slab pressure, in contrast to the conventional design method of balancing the vertical loads. The resulting soil pressure calculated by the BOEF method depends on the width of the tunnel box, and the relative stiffness of the soil. If the soil was infinitely stiff, all load would be transferred to the soil directly beneath the walls of the box. If the soil behaved as a fluid, the load would be uniformly distributed below the base slab.

At this point, all that remains to be input into the model is the loading conditions. Preparation of these calculations is typically the most time consuming for the tunnel section analysis. Consider the summary of tunnel loads shown in Figure 2. The value of each load must be calculated for each tunnel member. Horizontal members are typically specified to receive uniform loads. Vertical members are typically loaded in a trapezoidal pattern. Note that loading groups must also be considered. For example, many tunnel projects require study of unbalanced loading conditions caused by variations in pore pressure on sides of the tunnel, and other groups of loads. Also, the designer must be concerned with construction staging. Figure 2 illustrates typical final loads, but loads induced during the staged construction must be considered.

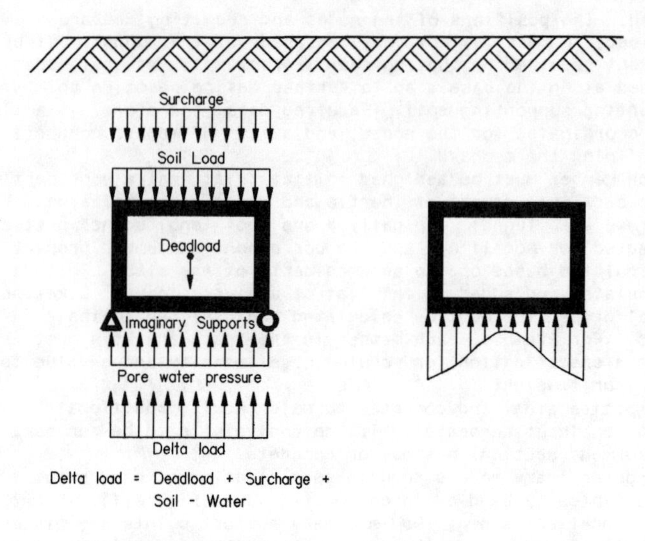

VERTICAL LOAD BALANCING BOEF SUPPORT

FIGURE 1

TUNNEL LOADING DIAGRAM

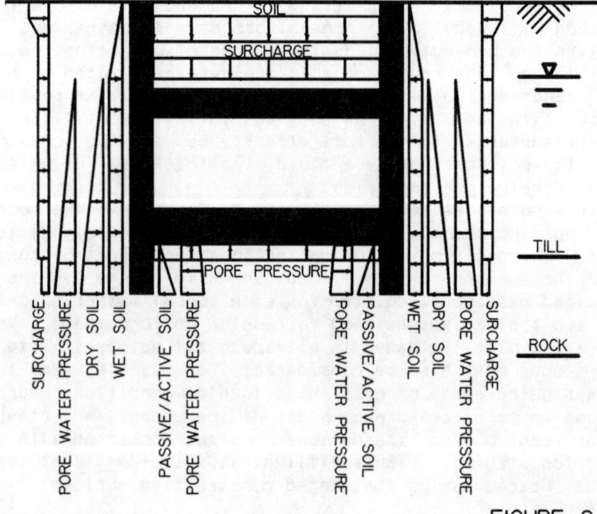

FIGURE 2

When all factors are considered, the cut-and-cover tunnel analysis can become complex and time consuming. To facilitate the work, we have developed software to automatically generate cut-and-cover tunnel models directly from CAD graphics. The program, SMART TUNNEL, reads AUTOCAD DXF-format files and other ASCII files, and it prepares structural analysis input files for the program, STAAD-III. SMART TUNNEL is described below.

PROGRAM FLOW

Figure 3 shows a flow chart for the program. It lists input files used by the program, and output files prepared by the program.

Program Input. The first step in generating the structural frame is the specification of joints and member connectivity. SMART TUNNEL reads in an ASCII geometry file, containing joint numbering and member numbering in STAAD-III format (sample, Figure 4). This file can be created in three ways: (1) From an AUTOCAD line drawing, create a DXF file containing the entities of the frame model; (2) From a pre-formatted Lotus 123 (or compatible) spreadsheet with the joint and member incidence geometry, export an ASCII file; (3) Use an ASCII file editor to create a file with the joint and member incidence geometry.

Next, geotechnical parameters are considered. The current version of SMART TUNNEL [2.0] assumes 3 layers of material: fill/clay, till and rock, matching general soil conditions in Boston (it is envisioned that the next revision will include provision for additional material layers). The following geotechnical parameters are required for calculation of lateral loading and support conditions: coefficient of lateral pressure (active and passive), k, for clay/fill, till, and rock; elevations of grade, till, and rock at left and right edges of the tunnel section; elevation of water table at left and right edges of the section; average blow count in clay/fill layer at left and right edges of section; unit weights of clay/fill, till and rock;

This data is read from an ASCII file that can be created manually, generated from a geotechnical database, or developed from a pre-formatted spreadsheet.

With the completion of the geotechnical parameter read-in, section properties of the structural members of the tunnel are considered. Each member is assigned member properties: cross-sectional area and moment of inertia The data is developed in one of three ways: from a pre-formatted spreadsheet, from AUTOCAD attribute information entered in the CAD drawing, or via a limited capability of SMART TUNNEL to "guess" initial section properties, considering approximate loading conditions and member geometry.

Next, the program processes a default input file, in which special parameters for the analysis can be specified. Input for this section permits generation of models with different framing. For example, all roof slabs can be treated as pinned connections to the tunnel walls. The assignment data is read from a pre-formatted spreadsheet, or from the ASCII default information file.

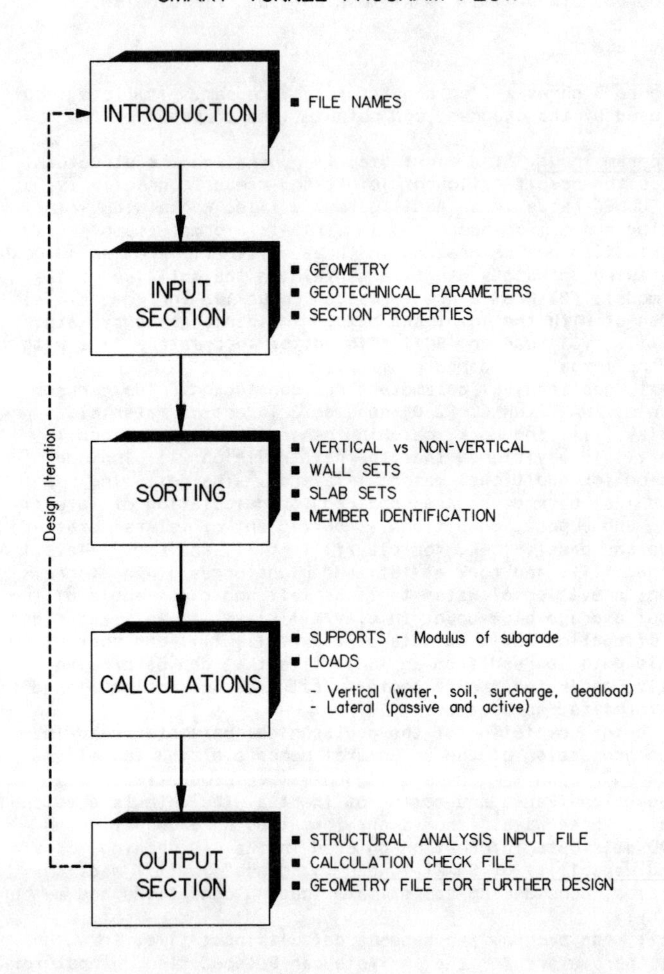
SMART TUNNEL PROGRAM FLOW
INTRODUCTION
FILE NAMES
INPUT SECTION
GEOMETRY
GEOTECHNICAL PARAMETERS
SECTION PROPERTIES
SORTING
VERTICAL vs NON-VERTICAL
WALL SETS
SLAB SETS
MEMBER IDENTIFICATION
CALCULATIONS
SUPPORTS - Modulus of subgrade
LOADS
- Vertical (water, soil, surcharge, deadload)
- Lateral (passive and active)
OUTPUT SECTION
STRUCTURAL ANALYSIS INPUT FILE
CALCULATION CHECK FILE
GEOMETRY FILE FOR FURTHER DESIGN
Design Iteration

FIGURE 3

SAMPLE MODEL GEOMETRY

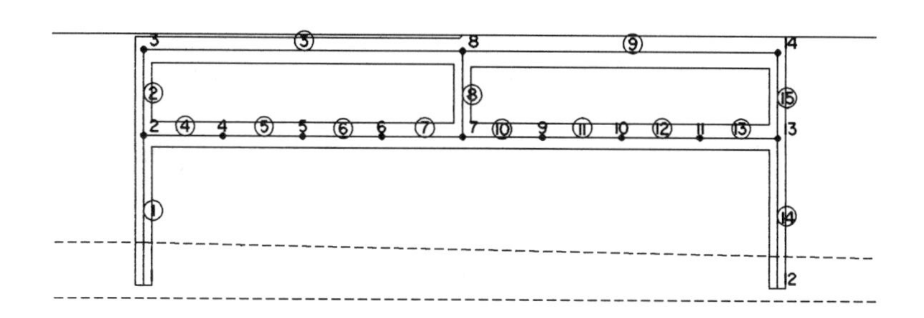

LEGEND

I Joint number

① Member number

```
PLANE SPRING SUPPORTED PLANE FRAME
UNITS KIP FEET
JOINT COORDINATES
I 0 I0
2 0 50
3 0 90
4 I0 50
--- etc ---
MEMBER INCIDENCES
I I 2
2 2 3
3 2 4
--- etc ---
MEMBER PROPERTIES
I 2 3 PRISM YD 4 YZ I
--- etc ---
CONSTANTS
E 3600 ALL
DENSITY 8.68E-5
SUPPORT
I I2 FIXED BUT MZ KFY II25
4 5 6 7 9 I0 II FIXED BUT MZ FX KFY 900
LOADING I
MEMBER LOAD
I TRAP Y 5.0 3.0
2 TRAP Y 3.0 2.0
3 UNI Y -4.0
4 TO 7 UNI Y 2.0
--- etc ---
PRINT MEMBER INFORMATION
```

FIGURE 4

Sorting. Refer to the flow chart shown by Figure 3. With input complete, SMART TUNNEL moves on to its calculation section. As a first step, the geometry of the members is sorted.

All members are stored in different arrays depending on the type of member. The first sort distinguishes between vertical members and non-vertical members. A vertical member is defined as one in which the x-coordinates for nodes I and J are equal.

Vertical members are stored in wall arrays. The non-vertical members are assembled between walls and stored in slab arrays. The code accomplishes this task in the following way (Figure 5):

Starting at the bottom of a wall, the code will search through the horizontal member set and determine that member which shares a node number with the wall member. With this member found, the code then "assembles" the full slab, sequentially seeking the subsequent attached member that makes up the slab and storing it in the slab array. Slab assembly is complete when the next wall to the right is hit. The program then moves up the wall and seeks another slab, continuing until all the slab arrays to the right of that wall are assembled. With all slabs assembled, the next wall to the right is selected. The process is repeated until all slabs are assembled in the slab arrays.

With the slabs assembled, the sorting code determines which slabs are base slabs, which are roof slabs, and which are "interior" slabs. Figure 5 shows a sketch of the three slab types.

With the slabs sorted, assembled, and identified, SMART TUNNEL processes the walls (Figure 5). The sorting routine assigns a position code to each member along a wall. The code is assigned for the location to the right of the wall member and to the left of the wall member, whether the member is inside a box (i.e., no lateral loading), on top of the roof slab, or below the base slab.

A final sort is made for wall sets of elements. SMART TUNNEL will distinguish between cofferdam walls which are integrated with the final structure, and regular tunnel walls. The rigid cofferdam walls, which may be slurry walls, are assumed to end in a spike (note Figure 5). Regular walls do not end in a spike. The sorting routine makes this distinction, and stores the appropriate code.

Supports. With sorting complete, support conditions are calculated. Spring constants are assigned to support points beneath the spiked slurry walls, interior walls, and base slabs. Members and joints for these slabs and walls have previously been identified by the sorting routine (described above). The code determines the position of the support point in clay/fill, till, or rock, and it then calculates the appropriate spring constant. A method used to calculate spring constants is described below:

Modulus of Subgrade Reaction. The current version of SMART TUNNEL includes a simple specification of the modulus of subgrade reaction (k) based on the recommendations by Terzaghi (1955). However, Terzhaghi's recommendations do not account for cases where the compressible layer beneath the foundation is of limited depth. To account for this in SMART TUNNEL, an adaptation of Terzaghi's recommendations was made.

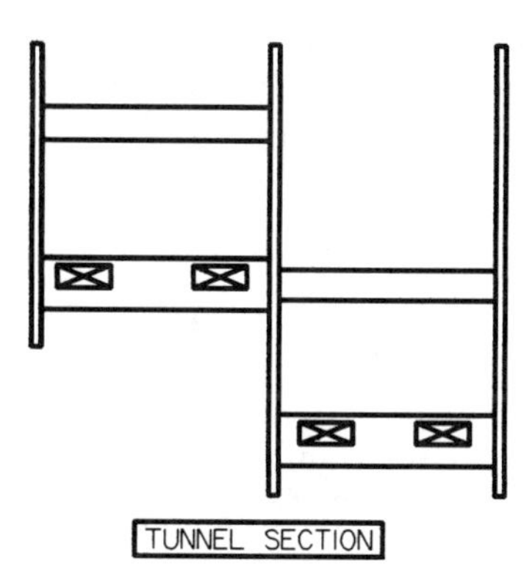

Vertical = Common x-coordinated for nodes i and j
Non-Vertical - All other members.

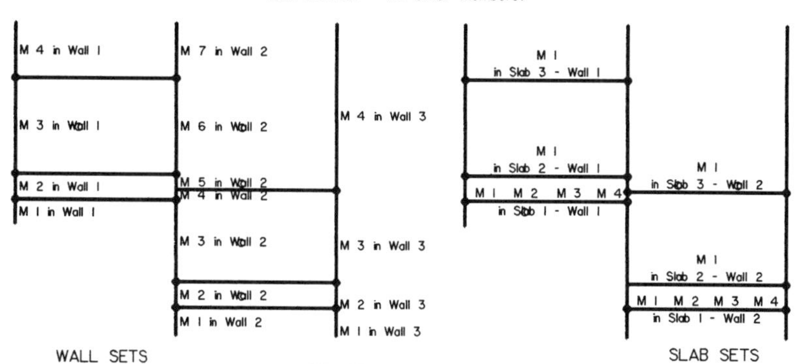

TUNNEL MODELS

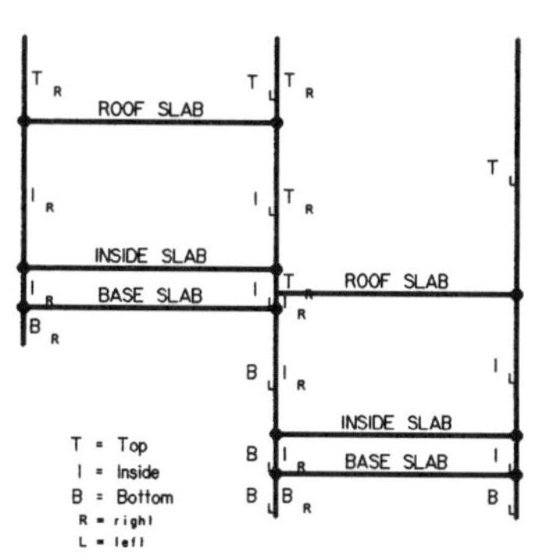

FIGURE 5

For a foundation (e.g. the bottom slab of the tunnel) bearing on a homogeneous soil stratum with unlimited depth, Terzaghi recommends the following formulas for evaluating the modulus of subgrade reaction:

$$\text{For Clay: } k = k1 / B \qquad (1)$$

$$\text{For Sand: } k = k1 / [(B+1)/(2B)]^2 \qquad (2)$$

where k1 denotes the modulus of subgrade reaction measured for a 1 ft square plate, and B is the width of the foundation in units of feet.

Assuming that the depth of influence of a foundation is 1.5 B, the existence of a relatively rigid layer (e.g. till or rock beneath a layer of soft clay) at a depth greater than 1.5 B would have no influence on the specification of the modulus of subgrade reaction k. However, if the depth to the rigid layer is less than 1.5 B, then there would be an increase in k due to the limited soil depth. [For an intuitive comparison, consider a bar with axial stiffness k = AE/L, where A is the cross-sectional area, E is the modulus, and L is the length of the bar. Decreasing L, which corresponds to decreasing the depth of the compressible layer would lead to an increase in k.]

To account for the increase in k due to limited soil depth, Terzhaghi's formulas were modified as follows:

$$\text{For Clay: } k = k1 / [(2/3)d] \qquad (3)$$

$$\text{For Sand: } k = k1 \{[(2/3)d + 1]/[(4/3)d]\}^2 \qquad (4)$$

where d is the depth (in feet) below the foundation to the relatively rigid layer. The assumption of a depth of influence of 1.5 B is based on a recommendation from Whitman (1981). Note that for d = 1.5 B, Eqs. 3 and 4 are identical to Eqs. 1 and 2. For d > 1.5 B, Eqs. 1 and 2 would apply.

Additional research involving the collocation of BOEF analysis models with finite element solutions is currently being performed to further refine the above procedure. The approach is not intended to be the last word on application of the BOEF method for tunnel analysis and design. It does, however, serve as an example of a way this type of calculation can be automated and incorporated in the tunnel section model generator.

At this point, a word of caution is added concerning use of the BOEF method for cut-and-cover tunnel design. The modulus of subgrade reaction for soil, upon which spring constants are calculated, is more a theoretical parameter than one which can be easily measured in the field. Determination of appropriate values is a topic of ongoing research. Also, many structural software packages, including STAAD-III, do not easily deal with the problem of springs in tension. Conceptually, there are states of loading where soil springs will have induced tension. At this point of analysis, the springs should be released, and an iterative analysis should be performed. It is up to the designer to be aware of this problem, and model the structure and supporting soil accordingly.

Vertical Loads. The next calculation the program performs is for vertical loading. Three kinds of vertical loads are calculated. Slabs previously identified as roof slabs by the sorting routine receive roof slab loading. This consists of a live load surcharge, the weight of backfill, and the slab's own weight.

Backfill weight is based on the soil density input in the default file. Slab self-weight is calculated by considering the density of the member and slab dimensions from the CAD drawing. Note that live load calculations vary significantly in underground projects. Later revisions of SMART TUNNEL will include generalized code for specification of live load surcharge. For now, the program has a built in live load calculation based on Central Artery Project Criteria, a modified AASHTO HS-20 loading.

Base slabs are loaded up (positive Y direction) by pore water pressure, calculated by considering the distance from the water table to the bottom surface of the slab in question.

For the tunnel's own weight, the model generator will either read a member load from the CAD drawing, or specify a self-weight command in STAAD-III. The first option is useful for unusual tunnel framing, like composite sections, where it is difficult to use the self-weight command directly.

Lateral Loads. For lateral load calculation, consider Figure 3. The lateral load can be separated into three components: pore water pressure, lateral soil stress, and lateral stress due to an overhead surcharge. The pore water pressure is equal to the depth from the water table multiplied by the unit weight of water. The code calculates a start and ending pressure, and applies a trapezoidal load to the vertical member.

The lateral soil stress and surcharge stress are assumed to be calculated by a lateral stress coefficient multiplied by the corresponding vertical effective soil stress (vertical stress minus pore pressure) or vertical surcharge stress. This calculation is not quite as straight-forward, because the factor to be used can vary greatly. A range can be identified as running from the active pressure coefficient, to the passive pressure coefficient. SMART TUNNEL uses values of the coefficient for each material layer as it is assigned in the geotechnical input file.

Lateral loads are further complicated by soil layer boundaries. Conceptually, each soil layer has a different coefficient of lateral pressure. Therefore, while the vertical stress is a continuous function with depth, the lateral stress will be a discontinuous function, with discontinuities at each layer boundary.

The position of the wall member, obviously, affects the type and magnitude of lateral load. The simplest case is a vertical surface on the inside of a tunnel between a top and bottom slab. These surfaces are identified by the sorting routine, and no loads are applied. Vertical surfaces to be loaded are found above the roof slab, along the outside of exterior walls, and below the base slab. Vertical surfaces above the roof slab and along exterior walls have trapezoidal member loads calculated as previously described.

More complex problems are introduced for vertical surfaces below the base slab and under the tunnel. In this case, SMART TUNNEL assumes that there is no overlying vertical stress due to soil at the start of the loading condition (the point at the base slab). Therefore, the calculated component of vertical soil stress is significantly smaller than it would be if a thickness of overburden was included. However, full pore pressure is calculated below the base slab for lateral loads.

Program Output. SMART TUNNEL prepares three output files: a STAAD-III analysis file, calculation pages prepared for the purpose of checking, and an ASCII file containing information about the geometry sort.

A completed structural frame analysis is generated in STAAD-III syntax (Sample, Figure 4). Briefly, the analysis file consists of these sections: introduction, units; joint coordinates; member connectivity; material property constants; member properties, including moment of inertia and cross sectional area; member framing releases (for pinned connections); member supports; member loading, vertical, lateral, and self-weight; member loading, groups of loads; joint loads; commands related to specification of analysis; and commands related to preparation of results of analysis.

Other STAAD-III commands, which appear in the analysis file, control utility functions. Examples are commands which control the current units settings, and a command which plots a line diagram showing the frame. With completion of this file, the structural analysis is ready to run.

SMART TUNNEL produces a set of pages in which all assumptions and calculations are clearly labelled and presented in a format conducive to manual checking. In this way, it is intended that the program fit comfortably in the traditional process of preparation of calculations, and not be treated as a "black box" that magically prepares analysis and design without verification.

SMART TUNNEL prepares design files, in ASCII code, which contain listings of groups of members and their designations. These files are subsequently read by separate programs for structural member design.

DISCUSSION OF PROGRAM USE

SMART TUNNEL was developed, in part, to assist studies of Central Artery cut-and-cover tunnel sections. For this project, the elevated Central Artery viaduct in downtown Boston (I-93) will be replaced by a depressed expressway. For structural type studies of Central Artery tunnels, sections were drawn every 100' along the 6000' alignment. For these tunnels, slurry walls will be incorporated in the final tunnel structure. Structural models were generated for a variety of conditions and assumptions. Variations were introduced in the way the air vents were framed, the way the slabs connected to the cofferdam walls, and other variations.

The layout for Central Artery cut-and-cover tunnels is complex, with some sections having seven tunnel boxes across in various positions vertically and horizontally. For these sections, SMART TUNNEL helped to reduce the amount of time required for design.

THE CUT-AND-COVER TUNNEL ANALYSIS SYSTEM

SMART TUNNEL is intended to be one link in a group of programs that automate many aspects of cut-and-cover tunnel design. The system includes the following parts: (1) Tools to assist in automated generation of CAD tunnel sections from computerized plan and profile data; (2) Tools to assist in automated generation of the structural model from the CAD section; (3) Design programs for design of the tunnel structural components; (4) A database, and link to the CAD graphics, to enable the designer to regenerate the computer section analysis following design. This will permit easier iteration; (5) Direct links to a computerized database of geotechnical parameters. Therefore, the designer will be able to electronically specify tunnel plan and profile data, and then extract required geotechnical data for the analysis automatically.

SUMMARY

A computer program, SMART TUNNEL, has been developed to generate structural frame models for cut-and-cover tunnel directly from CAD graphics. The program reduces the amount of time and effort required for frame analysis of the tunnel sections. It is intended to fit within a system of programs to automate many aspects of cut-and-cover tunnel design.

ACKNOWLEDGMENTS

This paper describes work supported by the William Barclay Parsons Fellowship Program, a research program conducted by Parsons, Brinckerhoff, Quade and Douglas (PBQD). Results were used to assist in work on the Central Artery/ Tunnel Project, currently under design in Boston. The authors wish to thank PBQD, the Massachusetts Department of Public Works (MDPW) and Bechtel/Parsons Brinckerhoff (B/PB) for their encouragement in preparing the paper, and for granting permission for its publication. Specifically, the authors wish to acknowledge: William Twomey, Robert Albee, and Anthony Ricci of the MDPW; Donald Marshall, Charles Carlson, K.K. See-Tho, and Jeff Brunetti of B/PB; and Paul Gilbert, Chairman of the PBQD Career Development Committee, responsible for administration of the Parsons Fellowship Program.

REFERENCES

ACI (1988). "Suggested Analysis and Design Procedures for Combined Footings and Mats, "ACI Structural Journal, 55 (3), 302-324.

Bowles, J.E. (1982). Foundation Analysis and Design, 4th Edition, McGraw Hill, New York.

Terzhaghi, K. (1955). "Evaluation of Coefficients of Subgrade Reaction," Geotechnique, 5(4), 297-326.

Whitman, R. V. (1981). Soil Dynamics- course lecture notes, Department of Civil Engineering, MIT, Cambridge, MA.

GEOTECHNICAL ENGINEERING WORKBENCH

Gregory B. Baecher[1], M. ASCE and Dwight A. Sangrey[2], M. ASCE

ABSTRACT: The development of software and computer hardware technology provides the basis for a new generation of computer-aided engineering tool which promises to significantly transform the way geotechnical site characteriztion and analysis is performed. This new generation of CAE tool is based on a conceptual model of spatial relationships among objects populating a project site, attribute information associated with those objects, and relationships among the objects. These tools, which we call geotechnical workbenches, provide the engineer with an environment for integrating site data and analyses, visualizing site conditions, and performing "what if" analytical calculations. The SitePlanner™ system is used to illustrate the power of geotechnical workbench technology for practical site characterization and engineering analysis.

BACKGROUND

It is news to no one that computer-aided engineering (CAE) technology [1]continues to develop at an rapid pace. Geotechnical engineers--although early users of modeling applications--have been slow to take advantage of modern computer-aided engineering. Nonetheless, a major change is taking place in geotechnical engineering practice because the hardware and software technology needed to drive CAE systems for the difficult needs of geoscience professionals has at long last become available. This paper describes how CAE technology is evolving in the geosciences and how that evolution will affect geotechnical engineering.

The result of evolving CAE technology for geotechnical engineering practice is what we call a *geotechnical engineer's workbench.* This workbench is a design and analysis environment sitting on an engineer's desk, which

---

1 President, ConSolve Incorporated, 70 Westview Street, Lexington, MA 02172

2 President, Oregon Graduate Institute of Science and Technology, 19600 NW Von Neumann Drive, Beaverton, OR 97006

integrates all the data from a project site and all the tasks which the engineer faces into a common, graphics-based model. The workbench is an highly

interactive electronic replacement for the maps, databases, drafting boards, calculators, and scratchpads that geotechnical engineers have traditionally used to collect and analyze data, design site plans, model engineered and natural structures, and prepare reports and drawings.

A geotechnical engineer's workbench is neither a "computer model" nor an applications program. It is a working environment--much like an electronic spreadsheet is for the financial analyst--in which data can be manipulated, plans can be tested, and color hardcopy products can be generated. A geotechnical engineer's workbench is a place where the engineer can work minute-by-minute to build conceptual or quantitative models of a site, analyze physical systems, and evaluate design decisions (Figure 1).

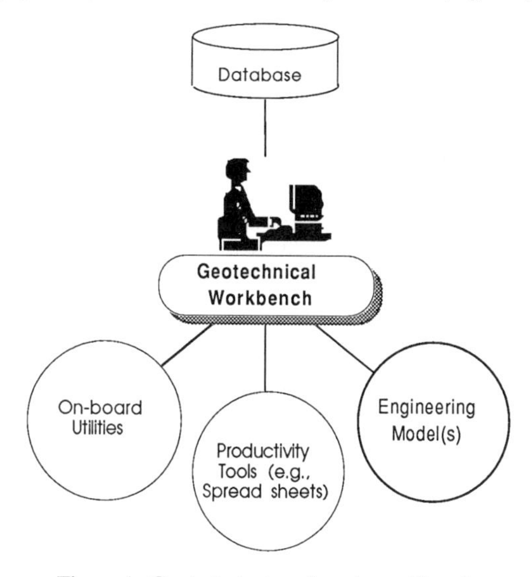

Figure 1--Geotechnical engineer's workbench

EVOLUTION OF COMPUTER-AIDED ENGINEERING TOOLS

The computer-aided design industry (CAD) is roughly divided into three segments: mechanical (MCAD), electrical (ECAD), and architecture, engineering and construction (A/E/C). The A/E/C market is, by most analysts' assessment, about five years less mature than its MCAD and ECAD equivalents. This has disadvantages, in that civil engineers at present do not enjoy the productivity enhancement that other branches of engineering do, but on the other hand, it means that parallels are available from which to forecast developments in the A/E/C market.

These other CAD segments have evolved in technological stages suggested by Figure 2. It appears that A/E/C CAD is evolving along a similar path.

Initially, computer-aided drafting systems were developed. These appeared on mainframes and minicomputers, but quickly migrated to less expensive, distributed platforms. Drafting systems automated the function of producing drawings and maps. They were originally housed in dedicated support groups in an engineering organization, and were operated by draftsmen or technicians.

Soon, engineering users and designers recognized that most of the data they dealt with were defined spatially and referenced to drawings and plans. The need developed to associate attribute data with drawn or mapped features, from which project databases could be developed. In the MCAD and ECAD markets these databases were soon linked to manufacturing and other operations. In A/E/C, and more specifically mapping, these databases led to the development of geographic information systems which associated attributes to geographically mapped features. As with drafting systems, the database design systems were typically allocated to support groups in the engineering organization.

The third generation of technology is model-based engineering systems. These are not housed in support organizations such as a CAD shop but provide automation directly to the engineer at his own desk. For example, in the MCAD market these model-based tools are typified by parametric and constraint-based design systems. In the ECAD market they are typified by concurrent engineering systems. A model-based tool organizes spatial design information within the context of a model of the physical world. Because the information is structured within such a conceptual model, it can be manipulated in intuitive ways by the engineer and seamlessly integrated into application programs, into productivity tools such as spreadsheets, and into report generators. This generation of model-based tools is only now appearing for A/E/C users, and promises a rapid and significant impact both on engineering productivity and on engineering capability.

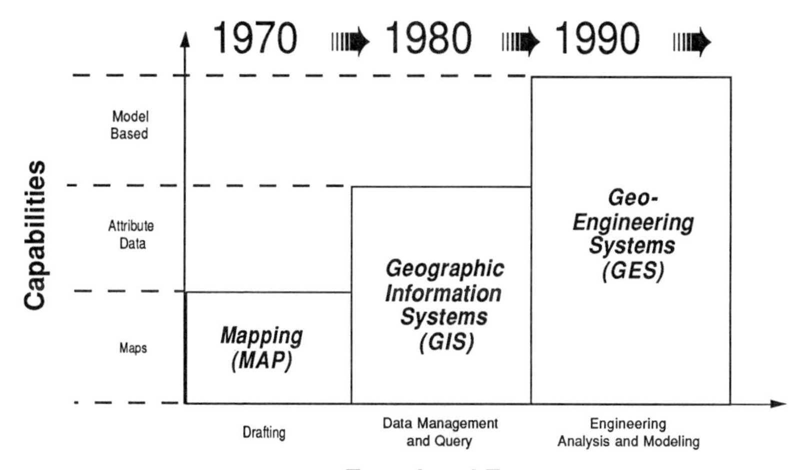

Figure 2--Evolution of CAD/CAE technology.

MODEL-BASED CAE

The new generation of CAE tools for geotechnical engineering--geo-engineering systems--are distinctly different from computer-aided drafting (mapping) and geographical information systems (GIS). Mapping and GIS developed out of user needs for cartographic products and for the management of geographically defined data. This history is reflected in the current use of GIS, shown in Figure 3. The application for which GIS was specifically developed and for which it is ideally suited involves an administrative user managing a large but relatively simply structured database which needs to be queried, maintained, and mapped geographically. Geographic information systems (including mapping) is the fastest growing sector of the US CAD/CAM industry. In 1989, total US shipments of GIS and mapping product hardware and software was almost a billion dollars (Dataquest, 1990) (Figure 4).

The geotechnical engineer or geologist has different needs. Geotechnical applications involve an analytical user manipulating a small but relatively complex database which needs to be integrated with modeling, needs to be dynamically modified as site characterization proceeds, and needs to be visualized in three or higher dimensions. The engineer needs to integrate data and analytical tasks, and needs to satisfy reporting requirements. These capabilities are not provided by GIS (Table 1).

With a model-based system, features and feature attribute data are stored within a structured model of space. This model can be visualized in many ways--as oblique views, profiles, fence diagrams, plan views, etc.--and may be visualized simultaneously in different ways (Figure 5). The system, however, stores information in its spatial model, not in the views. That is, the system does not store a "map" of a site. The map is merely a view of the model. In this way, alterations to the model are broadcast to all views, and the model can be linked to analysis applications and other utilities.

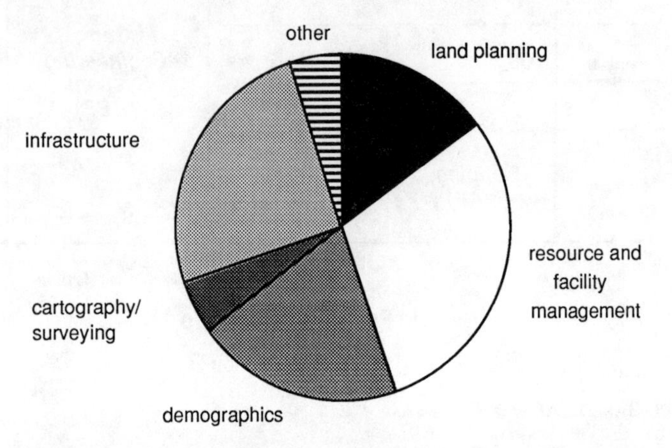

Figure 3--Approximate distribution of GIS applications (vendor survey)

The enabling technology for model-based CAE systems in geotechnical applications is object-oriented programming. The object view of nature provides a powerful basis for model-based systems. People think of the world as empty space filled with objects. These objects have attribute information associated with them, and they have interrelational properties between one another. A volume of space can be filled by any number of objects, or can be empty of objects. One attribute that an object carries is its x-y-z location. An object can move with time as it does its locational attributes change. This property of treating location as yet another attribute is in sharp contrast to drafting and GIS.

The object view of nature contrasts to the GIS view of space as layers of information (Goodchild and Kemp, 1990). In the GIS view, the world is continuous, locations are defined by a coordinate system and attribute information is defined with respect to that coordinate system. Attributes are not defined with respect to objects, but are stored in layers. One layer might store the distribution of surface soil type, another layer might store the average annual rainfall, another might store aquifer thickness, and so on. Each layer captures the variation of one variable over the surface of the earth. The database can be queried to determine the value of any variable at any place. Data models in this layer view of the world typically are of three types: raster (continuous geographic variation is approximated by a grid of discrete cells or pixels), polygon (the world is divided into irregular patches with piecewise linear boundaries within which some attribute has a constant value), and TIN (triangulated irregular network--in which the world is divided into triangles and attribute variation is approximated by a plane within each triangle).

For engineering purposes, an object view of the world has several advantages. Among these are that well-defined objects can be easily moved or modified in time and space, and all the attribute data and relations to other objects correspondingly move or change with the modifications. Objects have identity which persists through processing or modeling. Entire classes of objects can have attribute data or relational properties which are inherited from the class to the instances of objects within the class (e.g., when water samples collected in a boring are sent to the laboratory, because they are instances of the class of objects "measurements in borings" they inherit information on boring location, date, personnel, and other attributes pertinent to borings; and because they are instances of the class "chemical samples," they also inherit information on chemical classes and other attributes pertinent to chemical samples).

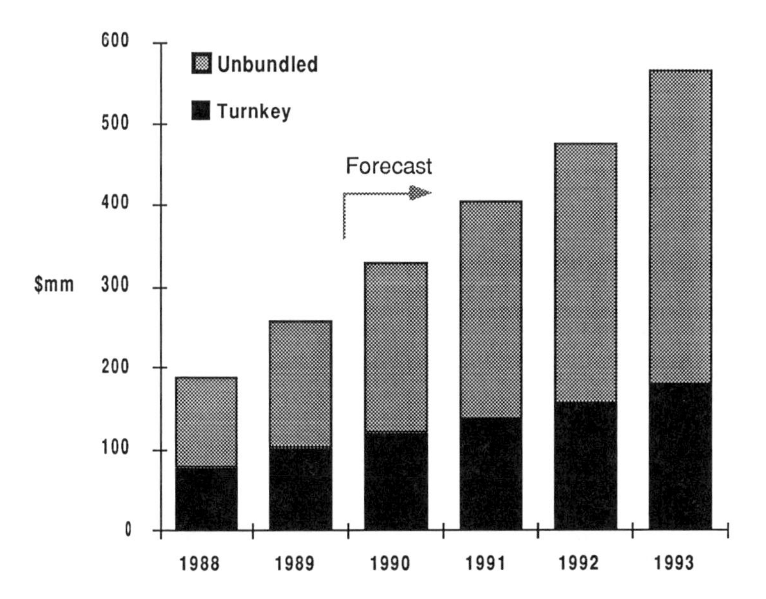

Figure 4--Growth of mapping and GIS market

DIGITAL TERRAIN DATA (DTD)

A second enabling technology for geotechnical engineering workbenches is the increasing availability of digital terrain data. These data are available for topographic elevation and for topographic features (i.e., vegetation cover, drainage features, transportation networks, etc.). The principal sources of digital data are the US Geological Survey (USGS) and the Defense Mapping Agency (DMA).

The USGS and DMA provide digital terrain products of various forms, as summarized in Appendix A. These may be grouped by elevation data (i.e., topographic relief), vector feature data, raster feature data, and other digital terrain data. At present, DTD are limited to geographic information. The availability of digital geological maps remains limited.

	GIS feature strengths:	**CAE feature strengths:**
Concept	Database	Workbench
Scale	Geographic scale	Site scale
Data structure	Static, large databases that exist prior to a project (i.e., archival databases)	Dynamic, smaller databases that are generated fresh with each new project
Uses	Relational queries, administrative reporting	3-D simulation, modeling, analysis, and intelligent reasoning
Visuals	Mapping and cartography	Interactive 2D and 3D on-screen visualization

Table 1 -- Distinction between model based CAE systems and GIS

OBJECT-ORIENTED PROGRAMMING AND DATABASES

The rapid increase of interest in object-oriented programming in the last few years has led to confusion in definition and interpretations (Kim and Lochovsky, 1989). In essence all object-oriented programming approaches share a common focus on encapsulating information in entities (i.e., objects). Object-oriented programming languages formalize encapsulation and encourage programming in terms of objects rather than in terms of programs and data. This leads to highly modularized code which can be mixed and matched for particular applications, and with which the user can interact in intuitive ways.

Objects. Object models in object-oriented programming languages usually encapsulate objects, data and operations (methods) in a single interface. They hide data structures and the implementation of operations from the user or programmer. Objects communicate with one another by message passing. One object sends a message to another object and the second object itself selects a method by which it reacts.

Instantiation is perhaps the most basic mechanism of object-oriented programming. Programming languages in general provide some built-in data types, for example integers or floating point numbers, which can be instantiated as needed. Object-oriented techniques provide a way for programmers to define and instantiate their own objects. An object class defines a set of operations, a set of hidden instance variables and a set of hidden methods that implement operations. When a new instance of an object is created it has its own instance variables and shares methods with other instances in the same class. Each new instance of an object is generated from its object class and brings worth it a template data structure.

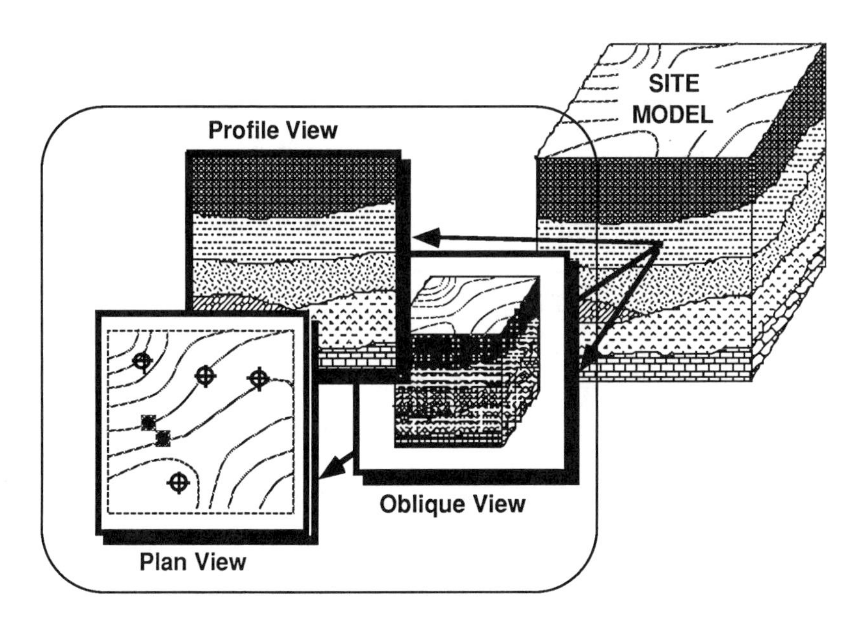

Figure 5--CAE system operates with a spatial model of the site, allowing the user different, simultaneous views of the data.

Object-oriented vs. relational databases. An object-oriented database is a system that provides database-like support for persistence, transactions, queries and the like for objects (Maier, 1989). Objects in this context means encapsulated data and operations. The development of object-oriented databases is an intensely watched area of work in software engineering and also in GIS.

Relational databases consist of a collection of relations or tables and a specification of domains within the tables. Individual entries within those tables are atomic values. Each table has a fixed number of columns and a variable number of rows, the order of which is inconsequential. For example, column headings might be attribute names for the relation called borings and the rows might be the tuples of borings. These relations specify the basic components of borings.

Data in tables are related by matching column values rather than through explicit links. The major advance created by the relational data structure model was that it provided a formal mathematical theory for database development. The significant advantages of relational databases are first, that the basic tabular form of the model is conceptually simple; second, the language of data manipulation is declarative and can be optimized to yield reasonable performance; and third a great degree of data independence is provided.

On the other hand while the tabular form of relational databases is conceptually simple and has advantages, it also has disadvantages. The principal disadvantage is that the very simplicity of the tabular form makes it difficult to express the semantics of complex objects. Its limitations have not been a problem for many traditional applications as financial accounting, however, it has been a serious limitation in CAE applications.

CAE databases. Traditionally, database systems have been uncommon components of computer-aided engineering (CAE) or of real-time, analytical decision support systems. The reason is that hierarchical and relational data structures lack the performance and modeling necessary for CAE applications. Most CAE systems perform their own data management on top of the file system used by the computer operating system. Since the OS file system has no capability to protect integrity or provide specialized storage mappings, in effect most CAE systems operate without an organized database.

Those CAE systems which are built on top of a relational database typically use the database in a quasi-batch mode. The CAE system reads a piece of the database required at start up and builds a record structure in virtual memory. Operations analysis and design checking are done in virtual memory, then at the end of the session virtual memory structures are converted back into tuples and written to the relational database. This means that a typical session starts with downloading information into the CAE application, operating it on the application and then reloading it back to the database. Because the virtual memory data structure contains little or no information about the object structure about the design it provides poor paging performance. The overall result is a batch-like operation with less than real time speed. CAE systems performing data management in virtual memory also forgo other database features of relational database management systems, such as concurrency control, integrity checking, buffer management, recovery and so forth.

Relational database management systems are highly optimized for the tasks they are normally asked to perform, typically associative retrieval. Associative access to disk however, is not the limiting factor in using RDBMSs for CAE, the limit is that conventional database systems are too slow at fetching and storing individual fields. A typical RDBMS transactions involves getting a tuple or small number of tuples from a relation and updating them, or selecting large groups of tuples and performing common operations on all of them. On the other hand a CAE session starts by retrieving pieces of information that are of interest to the designer and analysis at hand, but then continues on with a large dissimilar fetch and store operations. The access paths to the data of these fetch and store operations follow other logical structures of the database itself but not the activity or associativity of the entities being modeled in the real world.

The CAE or decision support system uses the record structures from the programming language that the application is written in to obtain the performance needed on fetching and storing single values of data. The reasons that RDBMS Systems are slow at such operations has to do in large measure with the bias of commercially available database management systems toward data processing applications.

SITEPLANNER

We believe that the SitePlanner CAE system is the first of the new generation of engineering workbenches aimed directly at geotechnical and environmental engineering users. Using object-oriented programming and object database technology, the most important features such a system brings to the geotechnical engineer are:

- Seamless integration of data and tasks,
- High quality data visualization, and
- "What if" analysis and modeling.

Seamless Integration of data and tasks. Seamless integration of data and tasks means that the user can sit at the workbench and address all the data on a site, whether from topographic surveys, sampling programs, borings, geophysics, or whatever source. He can manipulate those data, perform statistical analysis, and develop a conceptual model of the site. He can also perform engineering analyses using numerical applications or other routines running off the same database. This means that there is no data transfer, no translation of data sets, and no moving form one platform to another in analyzing data and building models (ConSolve, 1989).

High quality visualization. Visualization means that the user can use data on a site to draw high quality pictures of maps, profiles, oblique views, or other representations of a site in order to explore data, establish site characterization models, or perform other interpretations. Visualization also allows the user to communicate to others--for example, engineering colleagues, permit authorities, or owners--what the situation at a site is, how a design will affect the site, and how design alternatives might provide different solutions to a problem.

"What if" analysis. "What if" analysis allows the user to perform sensitivity studies and evaluate alternative scenarios on a site database or design situation in a highly interactive, real-time manner. This provides the user with the capability of quickly comparing a broad spectrum of decision alternatives by automatically linking site data to analytical routines and data visualization capabilities. In essence, the geotechnical workbench provides the geotechnical engineer with a capability similar to what the electronic spreadsheet provides the financial planner.

Future of CAE in Geotechnical Engineering

The advent of model-based CAE technology for the earth sciences heralds a new stage in the professional practice of geotechnical engineers. These systems will bring powerful analysis capabilities to the engineer's desk, where site data and design decisions can be brought together in flexible ways to assess site characterization information and forecast the implications of design alternatives. They will also create an even greater need than now exists for electronically captured site data, and thus presage the development of more powerful sensor and data acquisition systems. These developments will fundamentally change the way geotechnical engineers work.

REFERENCES

1. ConSolve Incorporated, *ASP/Advisor: An intelligent spatial decision support system for the layout of field ammunition supply points,* Final Report to the U.S. Army Human Engineering Laboratory, Aberdeen Proving Ground, MD, 1989.
2. Dataquest Incorporated, *Survey of the AEC and GIS Market*, Dataquest Incorporated, San Jose, CA, 1990.
3. Goodchild, M.F. and K.K. Kemp, *NCGIA Core Curriculum*, National Center for Geographic Information and Analysis, University of California at Santa Barbara, 1990.
4. Kim, W. and F. H. Lochovsky, *Object-oriented Concepts, Databases, and Applications*, ACM Press, 1989.
5. Maier, D., "Making Database Systems Fast Enough for CAD Applications," in Kim, W. and F. H. Lochovsky, *Object-oriented Concepts, Databases, and Applications*, ACM Press, 1989.
6. Samet, H., *Applications of Spatial Data Structures*, Addison Wesley Publishing Co., 1990.
7. Samet, H., *The Design and Analysis of Spatial Data Structures*, Addison Wesley Publishing Co., 1990.
8. Department of Defense, Defense Mapping Agency (1989), *Draft Supplemental Guidance for Users of 1:50,000 or 1:250,000 Scale Interim Terrain Data Products*, August 1989.

APPENDIX A--SOURCES OF DIGITAL TERRAIN DATA

U.G. GEOLOGICAL SURVEY

Elevation. Digital Elevation Models (DEM) are digital records of terrain elevations for ground position at regularly spaced intervals. DEM are available from the USGS for 7.5 minute topographic quadrangles and 1:250,000 maps. The 7.5 minute data approximate a at grid spacing of 30m. The 1:250,000 data are on 3 arcsec grid. DEM are only available for select USGS quadrangles, but are expanding in availability. The USGS DEM correspond to DMA DTED Level I and DTED Level II data, respectively (see below).

Vector Planimetric Data. The USGS Digital Line Graphs (DLG's) provide vector representation of planimetric information from USGS quadrangles (i.e., line map data). These data include boundaries, transportation, hydrography and US Public Land Survey System. DLG's are available for some 7.5 and 15 minute maps, and for some 1:100,000 scale maps. DLG's for 1:2,000,000 scale maps are available for the entire U.S.

Vector Thematic Data. Land Use and Land Cover (LULC) data are derived from thematic overlays of 1:250,000 and 1:100,000 scale maps. LULC data are registered to the base maps and include information on urban or buildup land, agricultural land, rangeland, forest, water, wetland, barren land tundra, and perennial snow or ice. Other LULC data include political units, hydrologic units, Federal land ownership, and census country subdivisions.

DEFENSE MAPPING AGENCY

Elevation. Digital Terrain Elevation Data (DTED) Level I data correspond to 1:250,000 military planning maps, providing 3 arcsec x 3 arcsec data (approximately 100m spacing). DTED Level I has been available as a standard DMA product since early 1970's, and is approximately 99% complete world-wide. Digital Terrain Elevation Data (DTED) Level II data correspond to 1:50,000 tactical maps, providing 1 arcsec x 1 arcsec data (approximately 30m spacing). DTED Level II is available only for limited geographical areas, primarily in West Germany and South Korea.

Vector Feature Data. Tactical Terrain Data (TTD) is DMA's standard specification for future world-wide digital terrain information. TTD correspond to the Terrain Analysis Vector Database with the addition of "urban and special features." TTD will be accompanied by DTED Level II, and are planned also to include offshore bathometric data from Marine Corps' charts of coastal landing areas. TTD will be similar to 1:50,000 scale products and will include T-Line Maps (TLM) for annotation. At present, TTD is a prototype phase product with coverage only of a 15 min. x 15 min. cell of Fort Hood, TX.

Interim terrain data (ITD) correspond to DMA's Terrain Analysis Vector Database and contain only a part of the eventual TTD. ITD contain data on f ive terrain overlays: soils, slope, vegetation, drainage, transportation, and obstacles. They are accompanied by DTED Level I. ITD have been available as standard DMA product since 1989. ITD is currently provided on approximately 1000 map sheets at 1:50,000 and 1:250,000 scales. Digital Feature Attribute Data (DFAD) was originally developed for radar and simulation uses and is not precise enough for most civil engineering uses.

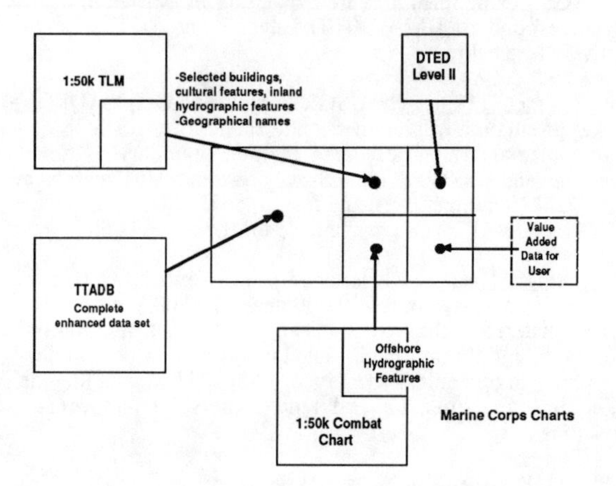

Figure A1--Tactical Terrain Data Content

Raster Data. Arc Digitized Raster Graphics (ADRG) is 24 bit raster color representation of terrain data. ADRG contains three types of information: map image data, graphics data (elevation tint, boundary diagrams), and text information (classification, publication data, etc.). ADRG is available on CD-ROM, with four 1:50,000 maps per disk. This is to be a standard DMA product in FY90, available at 1:50,000, 1:250,000, 1:500,000, and 1:1,000,000 scales. Electronic Map Data (EMD) is the Army's replacement for ADRG, which is planned to be available in FY91. EMD will provide raster map background combined with raster scanned color separates to provide data layers (in contrast to ADRG's single layer).

Other DMA Products. Digital Chart of the World (DCW) is now being developed by DMA to be a topological vector database of the world based on the Operation Navigation Chart (ONC) at 1:1,000,000 scale. The DCW is a prototype in anticipation of the larger effort of producing world-wide TTD in the mid 1990's. World Vector Shoreline (WVS) is a topological vector database of shorelines and political boundaries at 1:250,000 scale.

State-of-the-art On Geotechnical Laboratory Testing

by Clarence K. Chan,[1] Member and Jorge B. Sousa,[2] Associate Member

Abstract: This paper attempts to present some of the new avenues open to the geotechnical testing engineer through the development of microcomputers. The use of personal computers has become an indispensable part of any material testing laboratory. Advances in microcomputer technology provides a relatively inexpensive way to develop apparatus where the accuracy and power of feedback closed-loop control systems are combined with the flexibility, automation, and speed capabilities of digital equipment.

The combination of such features has had significant impact in laboratory testing. Examples are presented of such impacts specifically describing how:

- Previously impossible tests have been made feasible,
- Study of some variables which could not have been done without the microcomputer,
- Testing speed has brought about significant advantages,
- Complex tasks have been made easy by the automation that can be accomplished by microcomputers.

This paper presents a review of papers and other publications specifically focusing on the use made of the microcomputer and related technology to improve geotechnical materials testing. It also presents the authors' views on the future trends of geotechnical laboratory testing.

---

[1]Research Engineer, Department of Civil Engineering University of California Berkeley, Davis Hall, Berkeley, CA 94720

[2] Assistant Research Engineer, Institute of Transportation Studies, University of California Berkeley, 480 Richmond Field Station, 1301 South 46th Street, Richmond, CA 94804

Scope

Laboratory testing usually involves a number of steps including: i) Specimen Preparation (sampling, fabrication, molding, conditioning of specimens), ii) Testing (equipment and environmental control and data acquisition) and iii) Data Analysis (interpretation and presentation of data and report preparation) (see Figure 1). This paper focuses on the use of the computer in applications that are related to the actual testing of the specimens and data analysis. Although a number of automated data acquisition systems have existed in geotechnical testing laboratories for a number of years this paper will not concentrate on those but rather on the control aspects of the testing using microcomputer based systems. The capability of using microcomputers to control the testing systems is a cornerstone in the development of new tests and test methodologies.

Components of a Testing System

Advances in microcomputer technology provide a relatively inexpensive way to develop apparatus wherein the accuracy and power of feedback closed-loop control systems are combined with the flexibility, automation, and speed capabilities of digital equipment. Figure 2 shows a block diagram of the major components of any modern laboratory testing system utilizing the computer. The three major components are the Testing Apparatus, the Interface Hardware and the Software.

The **Testing Apparatus** is usually composed of input transducers (such as actuators, servo-valves, solenoid valves, electro-pneumatic valves, stepping motors, pressure regulators, thermal regulators, etc), a reaction frame, a specimen, an environmental chamber and output transducers (such as LVDTs, load cells, pressure transducers, thermistors, etc).

The **Interface Hardware** is composed of the microcomputer, controllers for the input transducers, signal conditioners for the output transducers and analogue to digital (A/D) and digital to analogue (D/A) converters. These converters are responsible for all the communications between the testing processes and the controlling and data acquisition algorithms of the software.

The **Software** is a program or series of programs designed to acquire data and/or control the testing apparatus. The sophistication of the programs depends on the testing needs. A program can range from extremely simple when just data acquisition from a testing apparatus is required to very complex when simultaneous control of different testing apparatus is essential.

The characteristics and requirements of some of these components have been described by Kuerdis & Vaid (1991).

Control Algorithms

Conventional test equipment largely operates on the basis of an open loop principle. An improvement to this system was brought about by the development of analogue feedback closed loop equipment. With the advent of the microcomputer such capabilities can be easily implemented and expanded. Figure 3 illustrates the essential elements of an open loop testing system using a microcomputer based system. The software sends a signal through the D/A (digital to analogue) converter to a controller to actuate an input transducer, which in turn, activates an actuator causing some excitation to a specimen. In this approach it is

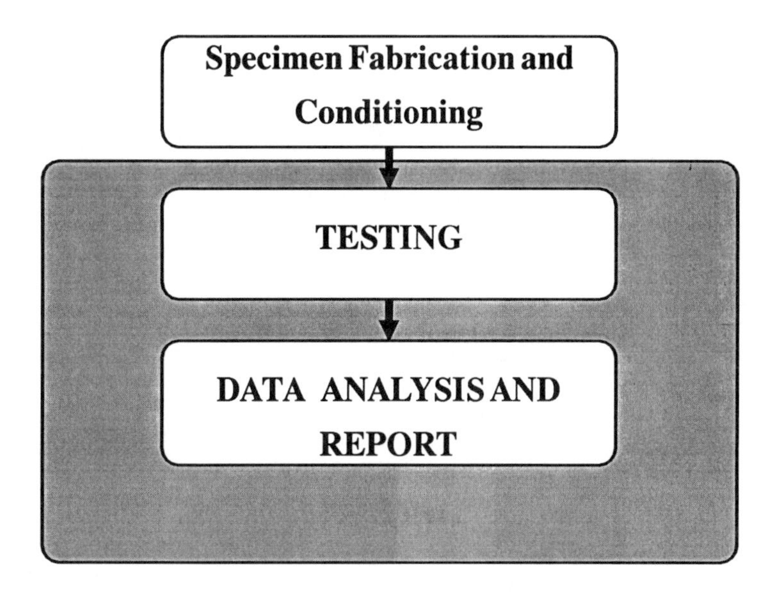

FIG. 1 - Diagram of steps followed on a laboratory test program

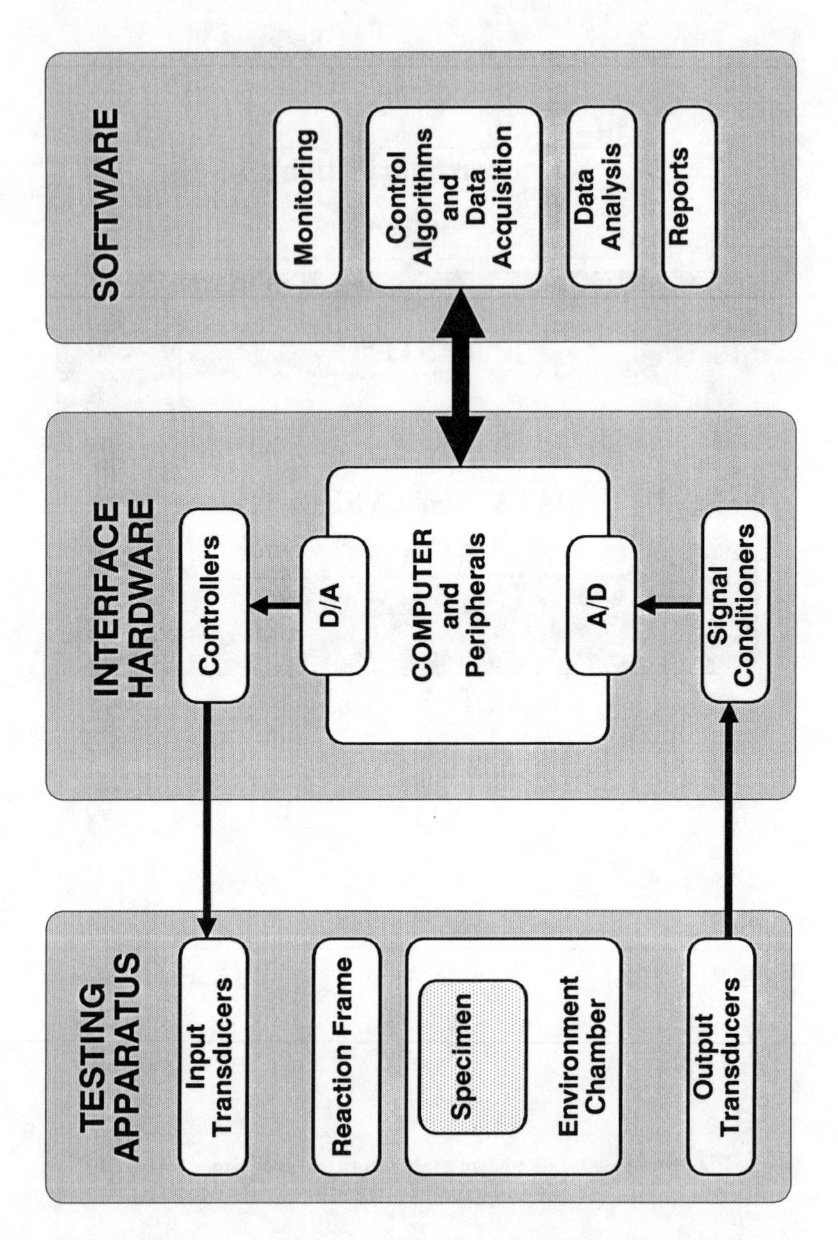

FIG. 2 - Block diagram of a modern design for a laboratory testing system

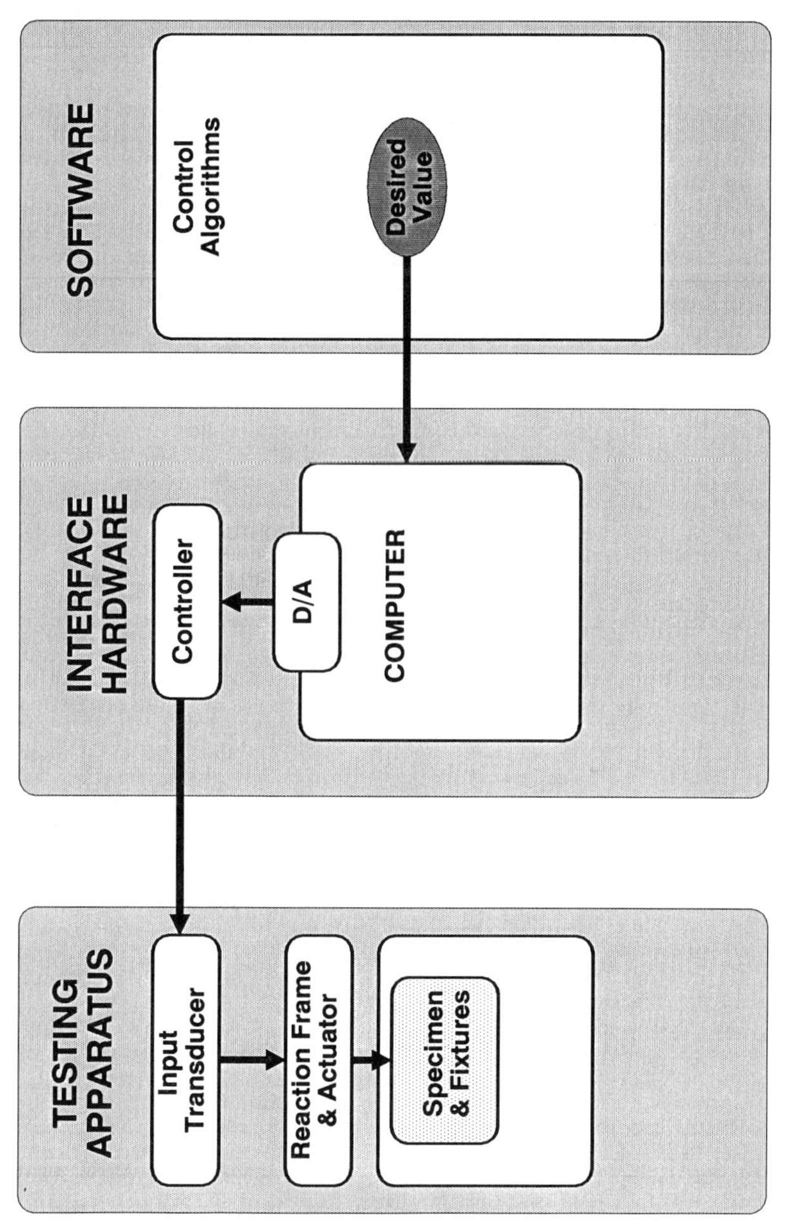

FIG. 3 - Block diagram of a open loop control algorithm

assumed that the response caused by the excitation is the desired one. No checking mechanism is provided.

Figure 4 illustrates the capabilities of a feedback closed loop control system. In this type of algorithm the software has the capability of setting a desired value for any given variable and also the capability to monitor if that value was obtained. As with the open loop algorithm the software sends a signal through the controller to the input transducer to cause an excitation on the specimen. However what makes this algorithm so powerful is that the excitation sent to the specimen is at each instant only an adjustment, programmed to bring the obtained value as close as possible to the desired value. The response of the specimen caused by the adjustment signal is monitored through a signal produced by an output transducer routed through a signal conditioner and a analogue to digital (A/D) converter back into the control algorithm.

Microcomputer based single channel open loop and closed loop systems present some relative advantages over analogue systems mainly because they are generally part of a software package which provides data acquisition and analysis capabilities thus minimizing data manipulation. Where the microcomputer based feedback closed loop systems go beyond the capabilities of analogue systems are in the generalization of the concept of feedback closed loop to several variables (i.e controlling, not only a desired variable, but a complete state of stress or strain in a specimen). With these multi-variable closed loop systems it is possible, today, to control several closed loops each interdependent on the other by complex algebraic algorithms.

Figure 5 presents a diagram illustrating these capabilities. The control algorithms of the software create a desired state that should be present in a specimen. The state currently present in the specimen is compared with the desired state and adjustment signals are computed. The adjustment signals are sent by several analogue to digital converters to the input transducers through the controllers. The input transducers cause the actuators to excite the specimen. The specimen's response is monitored through the output transducers and their readings are routed through the signal conditioners and A/D converters into the control algorithm. These obtained values are the variables in the algebraic equations on the algorithms to compute the current (obtained) state in the specimen. The extreme flexibility and power of these digital loops resides in the capability to implement almost any equation (function) in software. Although these algorithms can be quite complex their implementation and use can be made relatively accessible and user-friendly if the software is built within a window oriented graphical mouse driven interface with menus and interactive screen prompting to help the operator run the test.

Recent Developments

Sousa (1986) and Sousa and Monismith (1986 and 1987) presented the first fully digital feedback closed loop control software for micro-computers capable of controlling two hydraulic actuators with testing rates of up to 30Hz (see Figure 6). This system was used to determined dynamic properties of pavement materials under stress control. This is an example of a system with two digital closed loops (similar to that presented in Figure 5). The adjustment values were digitally computed and sent directly to the servovalves through D/A converters. The software was capable of executing 5000 loops/sec/actuator and was developed to

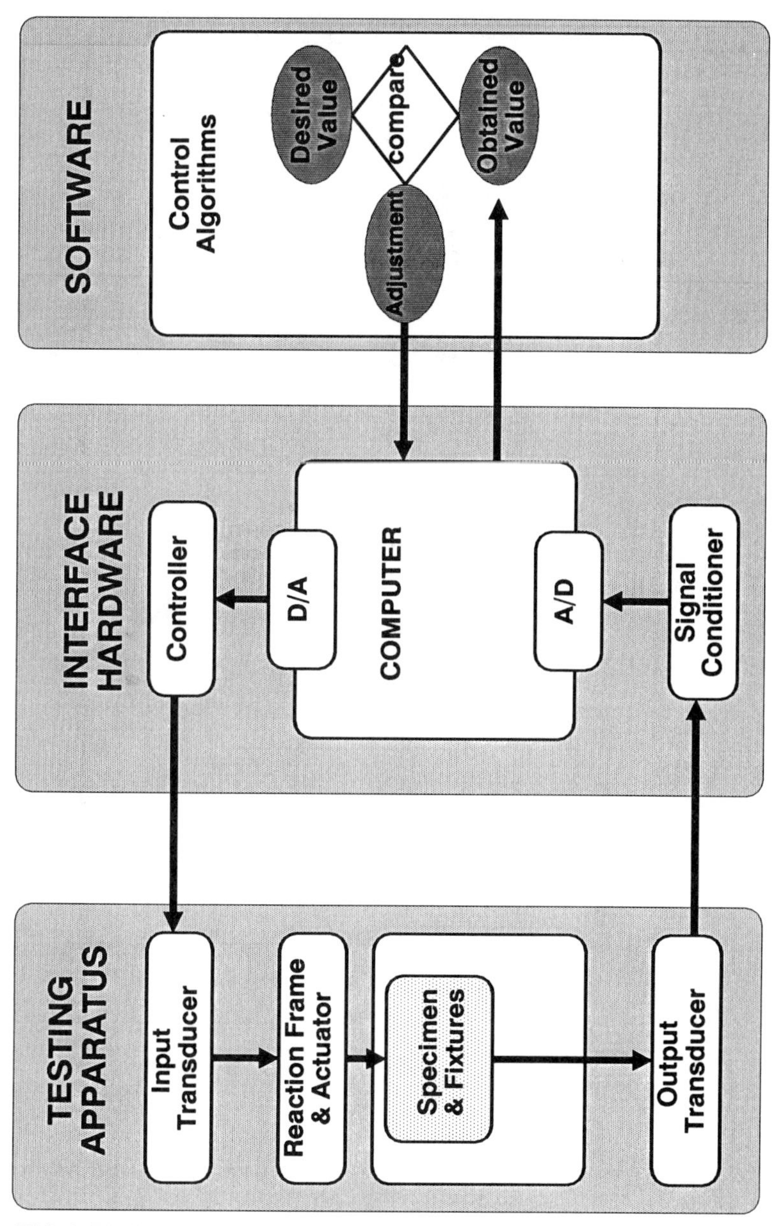

FIG. 4 - Block diagram of a closed-loop algorithm

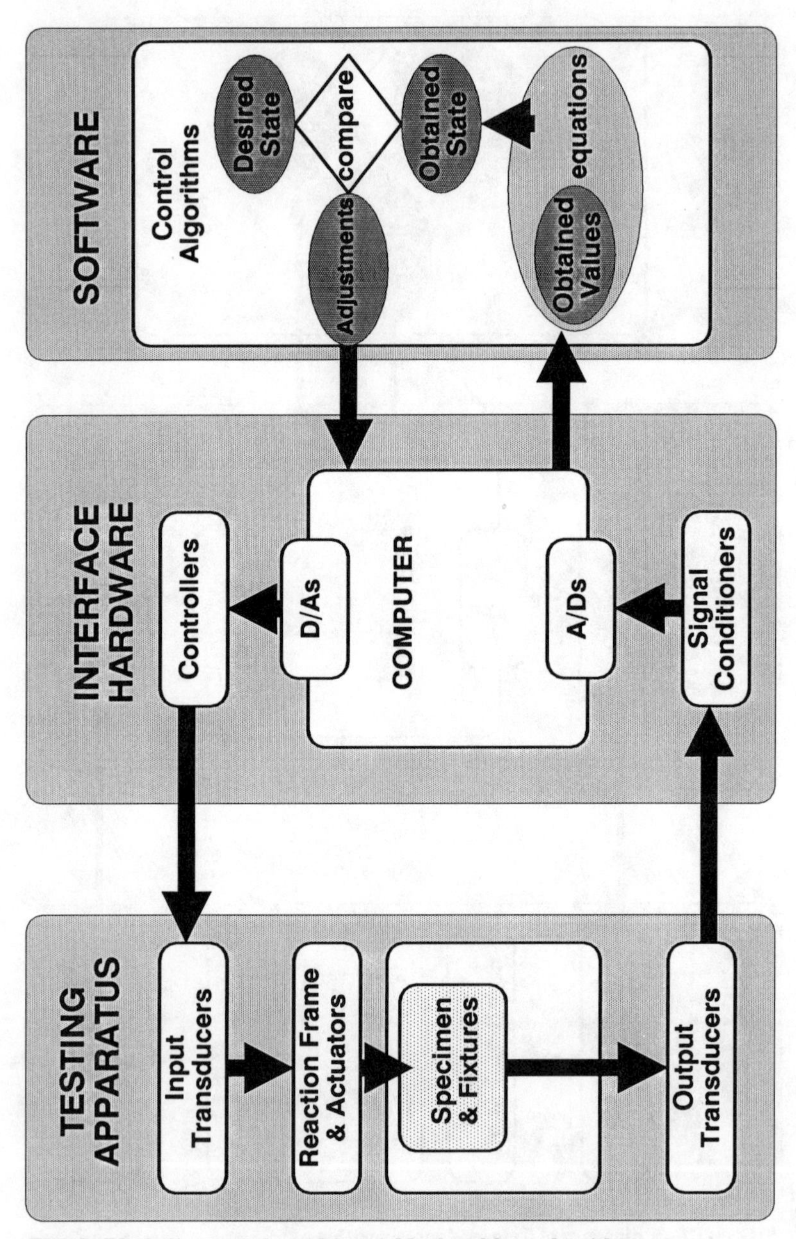

FIG. 5 - Block diagram of a multi-variable closed-loop algorithm

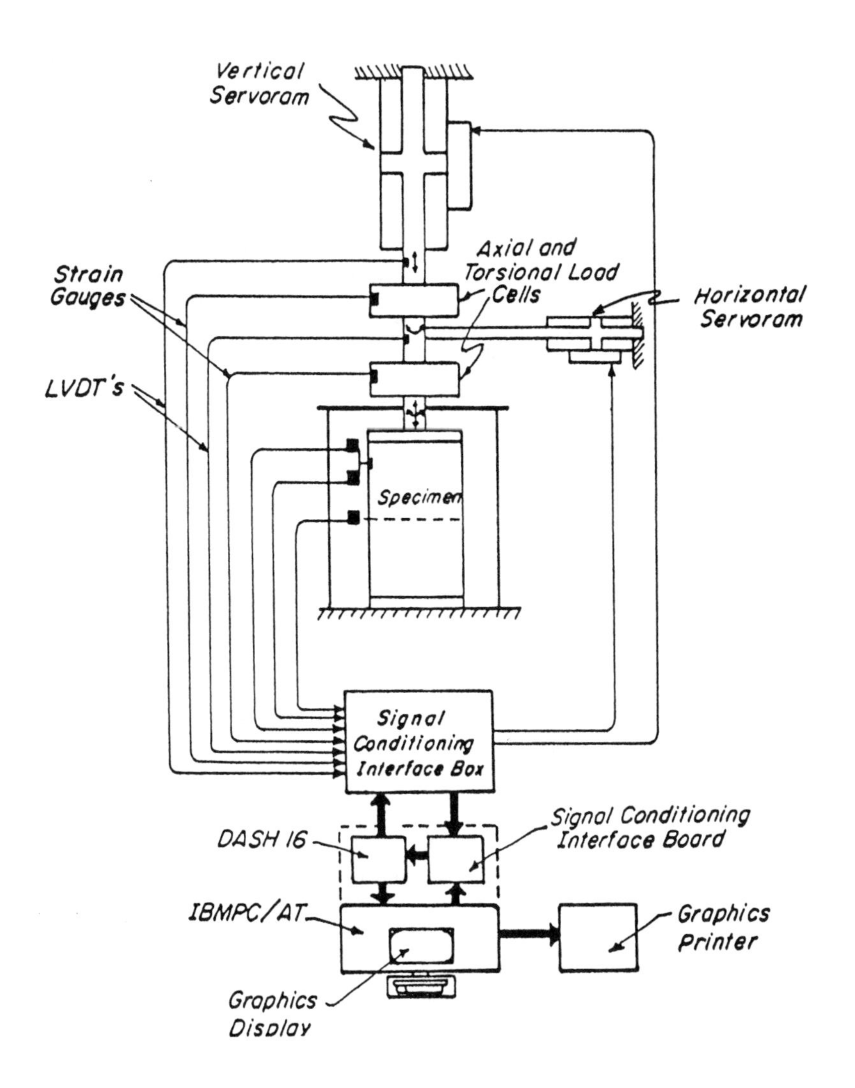

FIG. 6 - Data Acquisition and Control Schematic (Sousa and Monismith 1987).

control the excitation of hollow cylindrical specimens on the dynamic loading system (see Figure 7). The main advantage of this system over analogue controllers was the cost and also the capability added by the software to provide adaptative control in cyclic testing. The adaptive control adjusted the gains on the feedback closed loops to either prevent resonance or adjust for changes in the modulus of the specimen. Schewbridge (1987) used this apparatus to test properties of reinforced sand under plain strain conditions and Schewbrige and Sousa (1991) investigated the dynamic properties of unreinforced, uniaxial and biaxially reinforced sands. These tests were conducted under axial stress control and torsional shear cyclic strain control.

ATTS-Automated Triaxial Testing System - is a two channel system controlling axial load and chamber pressure and recording data on up to eleven channels. This is a dedicated system used for advanced class work and also as a research tool. The learning time for this system is about 1 hour. The current software is IBM compatible and is an upgrade of the system described by Li, Chan, and Shen (1988). The software uses two feedback closed loop algorithms to control the axial load and chamber pressure using electro-pneumatic transducers. More detailed description of these system is presented by Sousa and Chan (1991).

With a microcomputer-based servo-system Shibuya (1988) was able to use a hollow cylinder apparatus to apply precisely specified sequences of stresses to the specimen. The axial, torsional, internal and external pressures are applied through four identical stepper motors linked to air regulators. The microcomputer directly controlled the shear stress, the mean principal stress, a parameter that describes the relative magnitude of intermediate principal stress and the orientation of the major and minor principal stress to the x and y axis. This is an example of multi-variable feedback closed loop (see Figure 5) where the software was used to control, at each time interval, the desired state in the specimen.

A cuboidal shear device was interfaced to a microcomputer by Sivakugan et al. (1988) (see Figure 8). A pair of solenoid valves (one normally closed and the other normally open) were used to activate or deactivate the pressure that controls each of the three principal stresses. Stresses are controlled by a feedback closed loop algorithm that regulates the opening and closing of the valves to achieve the target values.

A computer controlled hydraulic triaxial testing system utilizing three sets of stepping motors to screw drive hydraulic cylinders which in turn applied pressures or displacements on the specimen was reported by Menzies (1988).

Sheahan, Germaine and Ladd (1990) made major improvements on the commercial system of Menzies (1988), by introducing transducers for direct measurements of the axial load, axial deformation, and pore pressures. These independent measurements by-passes the indirect measurements that relied on the ideal performance of a complex mechanical system used in the closed loop feedback configuration. The authors also pointed out other important items, such as: system compliance in the mechanical and hydraulic system, minimizing sample disturbance during setup, modification for extension testing, and software changes to allow gradual backpressure saturation and secondary compression after primary consolidation.

Pradhan, Tatsuoka, and Horii (1988) described a torsion shear apparatus and the computer-based servo-control system using electro-pneumatic transducers. Teachavorasinskun, Shibuya and Tatsuoka (1991) used the torsion shear apparatus

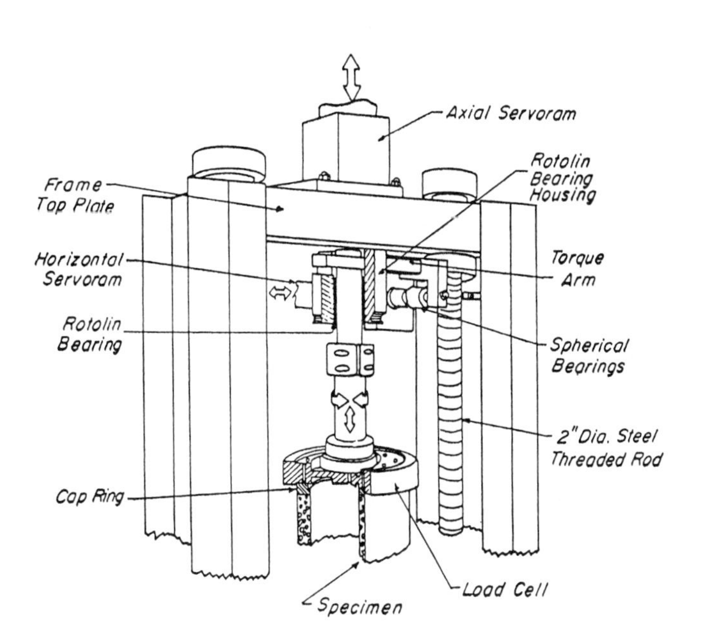

FIG. 7 - Loading System (after Sousa and Monismith 1987).

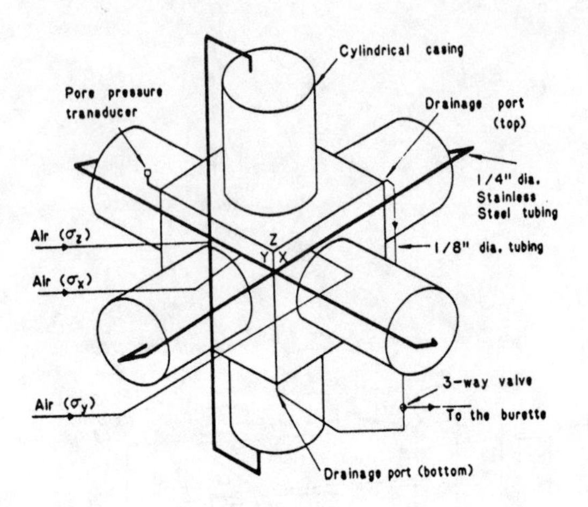

Isometric view of the cuboidal shear device.

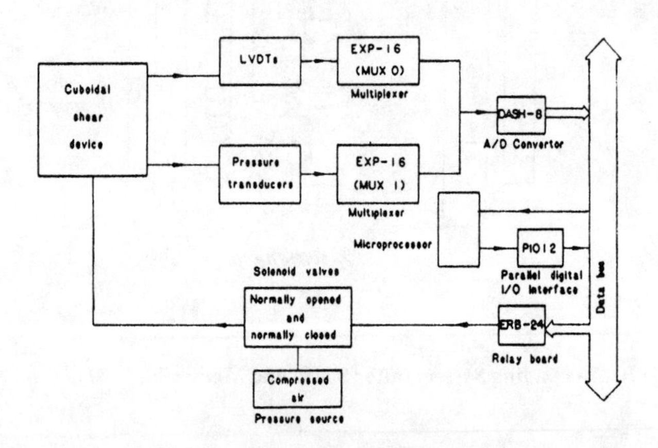

Schematic diagram of hardware interfacing.

FIG. 8 - Cuboidal Shear Device and Control Scheme (Sivakugan et. al 1988).

on two sands with improved measurements on the shear strains from 10-6 to those at the peak strength. Fig. 9 shows the outline of the current servo-system, while Fig. 10 shows the open loop system that was used for cyclic loading.

Brandon (1990) developed an automated back-pressure saturation device that minimizes operator attention (see Figure 11). The possibility of overconsolidation during back pressure is eliminated by monitoring the maximum effective stress in the specimen and keeping this below the desired consolidation pressure. By means of a feedback closed loop, the back pressure is increased at the fastest rate possible without overconsolidating the specimen.

An interesting example of computer control capabilities is that implemented by Nicholson (1990). He used the Automated Testing System (ATS) software from Digital Control Systems to simultaneously control his axial actuator (under stress control) and his servo-controlled injection piston (see Figure 12) to inject water into the specimens. This software is an evolution of the program used by Sousa and Monismith (1986) running under the Microsoft Windows environment; Sousa and Chan (1991) presented a description of the capabilities of this software. The compliance mitigation procedure used in Nicholson's studies involved predetermining the magnitude of volumetric compliance as a function of effective stress and soil parameters, and the use of computer-controlled injection or removal of water to continuously eliminate membrane compliance effects. At the initiation of the test, readings are recorded for the current effective confining stress and the position of the injection system cylinder pistons. After a short specified time interval, additional readings of effective stress are taken. The new readings of effective stress are then compared to the previously recorded readings. For any changes in effective confining stress, the program calculates the volumetric error induced by membrane compliance based on pre-determined values for the material being tested and the sample geometry, and commands the servo-controlled injection cylinder pistons to move accordingly. The loop continues until termination of the test. The program has options to overcome possible irregularities that might be encountered. These include: (a) the capability to change the rate of injection (time interval between the runs of the injection control loop), and (b) injection at given time intervals of a specified percentage of the calculated volumetric error. It is also possible to set a minimum effective stress below which no further injection-corrections will be implemented. This last parameter is necessary to adjust for the non-linearity of the semi-log plot of unit compliance as a function of effective confining stress at very low levels of effective stress. This system represents an example of a multi variable closed loop control (see Figure 5) using a complex algorithm in the control of the injection actuator.

Boulenger (1990) has utilized the bi-directional simple shear apparatus designed at the University of California at Berkeley (see Figure 13) to model the multi-directional seismic loads imposed by earthquakes on earth embankments or slopping ground surfaces. The main feature of this equipment is the capability to apply bi-directional monotonic and cyclic simple shear loads. The ability to apply bi-directional simple shear loads was accomplished by mounting two horizontal tables on orthogonal track bearings, with one of the tables riding on the top of the other. Test specimens are circular in shape and may be confined in either a wire wrapped or plain rubber membrane. Chamber pressure can be applied in order to back-pressure saturate samples or to provide an isotropic confining pressure. The automated control system of the device was provided by ATS with the software

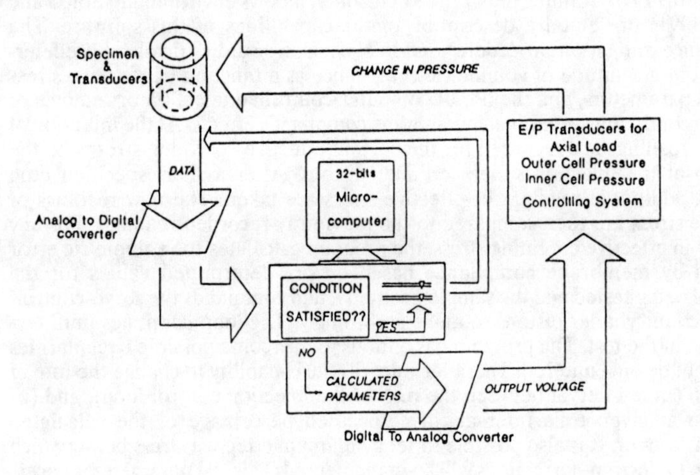

FIG. 9 - A servo-control to simulate simple shear (Teachavorasinskun et. al 1991).

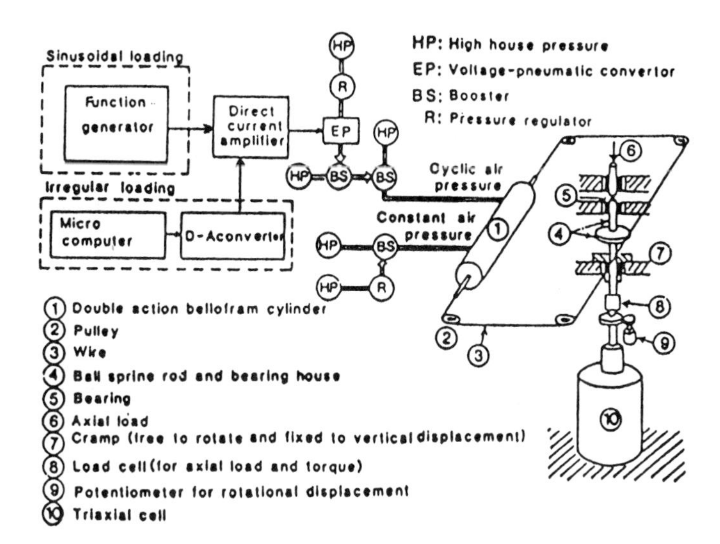

FIG. 10 - Cyclic loading system on Hamaoka sand (Teachavorasinskun et. al 1991).

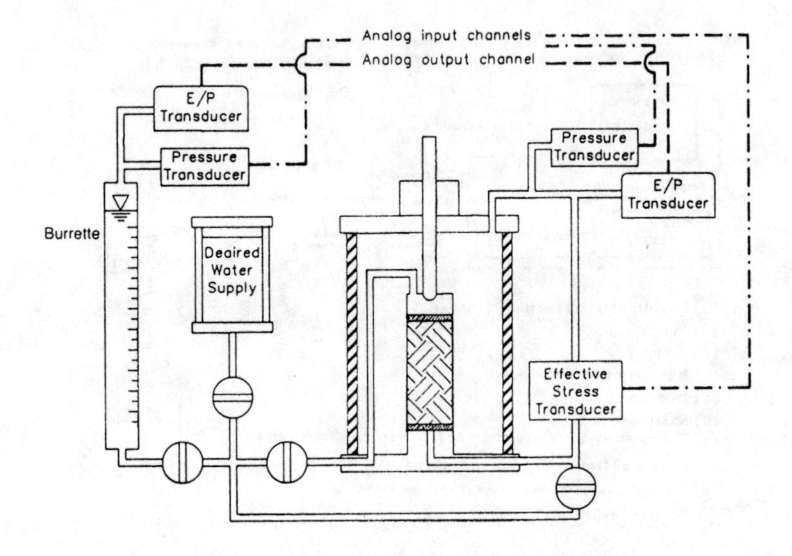

FIG. 11 - The Virginia Tech Automatic Saturation Device (Brandon et al. 1990).

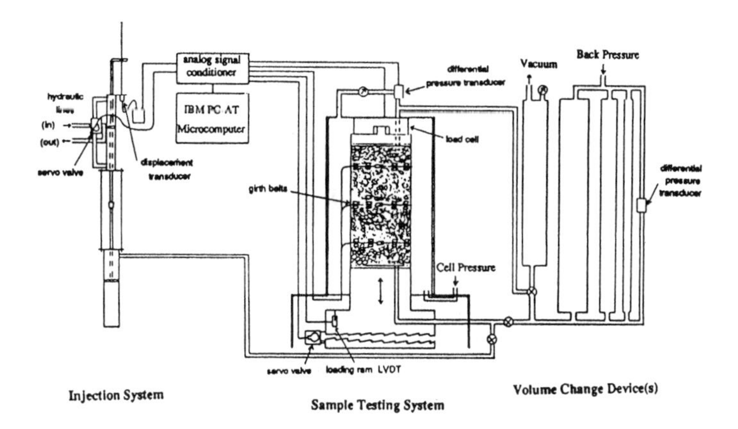

FIG. 12 - Schematic Illustration of Large-Scale Testing Set up for Testing 12-Inch Diameter Specimens (Nicholson 1990).

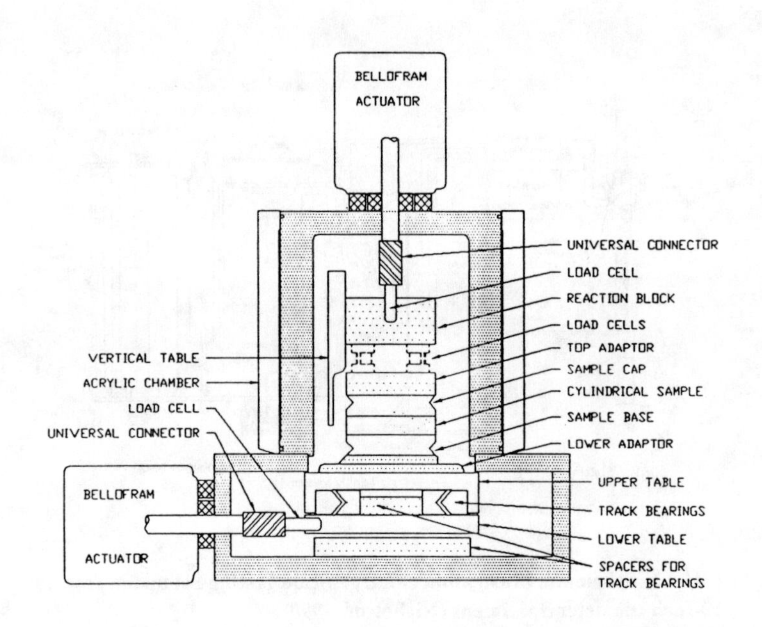

FIG. 13 - Diagram of the Bi-directional Shear Device (Boulenger 1990).

updated to control four feedback closed loop systems. This capability was used to provide simultaneous control to the two shear loads (stress control) and the confining pressure while holding the axial displacement constant (strain control) after consolidating under stress control. The software controls four air servovalves under feedback closed loop control at a rate of up to 2000 loops/sec/valve.

Trends

The philosophy of software development will be directed towards systems that are flexible and powerful, and that have short learning curves. Recognizing that computer literacy among potential users will vary widely, these requirements can only be met with advanced software design. Systems powerful enough to automate and control complex testing processes, easily programmable by engineers, flexible in a research environment, and rapidly mastered by technicians for routine use, will have to be built on a powerful graphics environment. Additionally, realizing the value of accurate performance and time efficient data analysis and report generation, the software packages will have to provide report ready output options.

The existence of these capabilities will permit the development and implementation of new and more complex routine standard tests. Although these tests might be more complex from an engineering point of view, providing more accurate material properties, they will, probably, be performed in less time and with greater ease, through the use of automated software control.

New research oriented tests will be developed based on the expanded resources made available by the possibility of directly controlling complex states of stress/strain which might be able to better simulate in-situ conditions heretofore difficult to simulate in laboratory. As new theoretical developments are advanced the need will exist for their validation through laboratory tests under ever complex conditions that can only be obtained by multi-variable control algorithms.

Summary

This paper presents a brief discussion of new avenues open to the geotechnical testing engineer through the development of microcomputers. The use of personal computers has become an indispensable part of any material testing laboratory. Advances in microcomputer technology provides a relatively inexpensive way to develop apparatus where the accuracy and power of feedback closed-loop control systems are combined with the flexibility, automation, and speed capabilities of digital equipment.

Also presented has been a review of papers and other publications specifically focusing in the use made of the microcomputer and related technology to improve geotechnical materials testing.

It is apparent that one of the greatest advantage brought about by the introduction of the microcomputer on a testing environment is the capability of executing tests under multi-variable feedback closed loop control where complex states of strain and stress can be created following predetermined theoretical formulations.

REFERENCES

Boulanger, R.W. (1990) "Pore Water Pressure Generation and Liquefaction Behavior of Saturated Cohesionless Soils Subjected to Unidirectional and Bidirectional Simple Shear Stresses," Dissertation to be submitted in partial fulfillment of the requirements of a Ph.D. degree at the University of California, Berkeley.

Brandon, T., Duncan, J., and Cadden, A., (1990), "Automatic Back-Pressure Saturation Device for Triaxial Testing", ASTM Geotechnical Testing Journal, Volume 13, Number 2, June 1990.

Li, X.S., Chan, C. K. and Shen, C. K., (1988), " An Automated Triaxial Testing System", Advanced Triaxial Testing of Soil and Rock, ASTM STP 977, pp. 95-106.

Kuerbis, R.H. and Vaid, Y.P., (1991), "Data Acquisition and Test Control in the Laboratory", Geotechnical Engineering Division of ASCE-Geotechnical Engineering Congress June 1991, Boulder, Colorado,

Nicholson, P.G., (1990), "Measurement and Elimination of Membrane Compliance Effects in Undrained Testing of Coarse Granular Soils" Ph. D. Thesis, Stanford University, California

Menzies, B. K.,(1988)"A Computer Controlled Hydraulic Triaxial Testing System.", Advanced Triaxial Testing of Soil and Rock, ASTM STP 977, pp. 88-94.

Sheahan, T. C., Germaine, J. T., and Ladd, C. C.,(1990)"Automated Triaxial Testing of Soft Clays: An Upgraded Commercial System," Geotechnical Testing Journal, GTJODJ, Vol. 13, No. 3, September 1990, pp. 153-163.

Shibuya, S., (1988), "A Servo System for Hollow Cylinder Testing of Soils" ASTM Geotechnical Testing Journal, Volume 11, Number 2, June 1988.

Sivakugan, N., Chameau, J., Holtz, R. and Altschaeffl, A., (1988), "Servo-Controlled Cuboidal Shear Device" ASTM Geotechnical Testing Journal Volume 11, Number 2, June 1988.

Sousa, J.B., (1986), " Dynamic Properties of Materials for Pavement Design". Ph. D. Thesis , University of California, Berkeley, pp.400.

Sousa, J.B. and Chan, C.K., (1991) "Computer Applications in the Geotechnical Laboratories of the University of California, at Berkeley", Geotechnical Engineering Division of ASCE Geotechnical Engineering Congress, June 1991, Boulder, Colorado.

Teachavorasinskun, S., Shibuya, S., and Tatsuoka F., (1991) "Stiffness of Sands in Monotonic and Cyclic Loading Simple Shear." Geotechnical Engineering Division of ASCE-Geotechnical Engineering Congress, June 1991, Boulder, Colorado.

UNDRAINED STRENGTH OF NC CLAY UNDER 3-D CONDITIONS

Poul V. Lade[1], M. ASCE

ABSTRACT: The undrained shear strength of normally consolidated clays observed in laboratory experiments has been found to vary in a systematic manner with the highest strength in compression, the lowest in extension, and the simple shear strength in between. This variation has often been attributed to effects of anisotropy. A study of the three-dimensional, undrained behavior of laboratory prepared, remolded, isotropic normally consolidated Edgar Plastic Kaolinite has produced results with similar variation in undrained shear strength as found in natural clays. The results of this experimental study have been compared with predictions from a recently developed isotropic constitutive model for frictional materials such as clay. The predictions compared favorably with all aspects of the experimental results. Notably the variation in undrained shear strength obtained from the isotropic clay was satisfactorily accounted for by the isotropic constitutive model. The primary factor in the variation of the undrained shear strength is the presence of the third stress invariant, I_3, in the model. Natural clay deposits exhibit cross-anisotropic behavior, and this also affects the variation of the undrained shear strength. The relative influence of these and other factors are evaluated on the basis of experimental results from the literature.

INTRODUCTION

The undrained shear strength is the most important material property used for evaluation of short term stability of slopes, retaining structures, and footings in normally consolidated clay. Triaxial compression tests on natural clays have shown that the undrained shear strength primarily varies with consolidation pressure and overconsolidation ratio (5). When normalized on the major principal consolidation pressure, the resulting normalized undrained shear strength ratio, s_u/σ'_{vc} (in which σ'_{vc} = vertical consolidation

[1]Prof., Dept. of Civ. Engrg., Univ. of California, Los Angeles, CA 90024.

pressure = σ_{1c}'), has been found to be constant for normally consolidated clays. Results of triaxial compression and extension tests as well as simple shear tests show that this ratio is constant for each type of test, but the value of s_u/σ_{vc}' varies in a systematic manner with the highest strength in compression, the lowest strength in extension, and the simple shear strength in between. This variation has often been attributed to effects of anisotropy.

An evaluation of the undrained shear strength variation under three-dimensional stress conditions is presented here. This evaluation is based on predictions from an isotropic constitutive model which is capable of capturing the experimentally observed stress-strain, pore pressure, and strength behavior of a remolded, isotropic normally consolidated clay under three-dimensional loading conditions. Natural clay deposits exhibit cross-anisotropic behavior, and the effects of this and other factors on the undrained shear strength variation are discussed on the basis of experimental results in the literature.

PREDICTION OF UNDRAINED SHEAR STRENGTH

Constitutive Model. A constitutive model has been developed on the basis of a thorough review and evaluation of data from experiments on frictional materials, such as clay, sand, concrete, and rock (3,7,8). The framework for the evaluation and subsequent development consisted of concepts contained in elasticity and work-hardening plasticity theories. In addition to Hooke's law for the elastic behavior, the framework for the plastic behavior consists of a failure criterion, a nonassociated flow rule, a single yield criterion that describes contours of equal plastic work, and a work-hardening/softening law. The functions that describe these components are all expressed in terms of stress invariants.

The new model employs a single, isotropic yield surface shaped as an asymmetric teardrop with the pointed apex at the origin of the principal stress space. The yield surface, expressed in terms of stress invariants, describes the locus at which the total plastic work is constant. The total plastic work (due to shear strains as well as volumetric strains) serves as the hardening parameter and is used to define the location and shape of the yield surfaces. The use of contours of constant plastic work as yield surfaces results in mathematical consistency in the model, because the measure of yielding and the measure of hardening are uniquely related through one monotonic function. In addition, application of a single yield surface produces computational efficiency when used in large computer programs. The nonassociated flow rule is derived from a potential function that describes a three-dimensional surface shaped like a cigar with an asymmetric cross section.

The model is devised so that transition from hardening to softening occurs abruptly at the point of peak failure.

Thus, transition does not involve any points at which the hardening modulus is zero, but the pointed peak is hardly noticeable in actual comparisons with experimental data.

Characterization of a soil, such as normally consolidated clay, involves 11 constant parameter values. These may be determined from results of isotropic compression (including unloading-reloading) and three consolidated-undrained triaxial compression tests.

3-D Experiments. The single-hardening model has been employed for prediction of the three-dimensional behavior of remolded, isotropic normally consolidated, undrained clay. A series of consolidated-undrained cubical triaxial tests with different but constant values of $b = (\sigma_2-\sigma_3)/(\sigma_1-\sigma_3)$ was performed on Edgar Plastic Kaolinite. The measured stress-strain relations, pore pressure responses, and strengths served as a basis for evaluating the capabilities of the model. Overall acceptable and accurate predictions were produced for the normally consolidated clay (6).

Undrained Shear Strength. One of the interesting aspects of the model predictions is the variation of the normalized undrained shear strength, s_u/σ_c' (in which σ_c' = isotropic consolidation pressure = σ_{1c}'), under three-dimensional conditions. A constant value of this ratio implies that Tresca's failure criterion applies to the soil under undrained conditions. Fig. 1 shows the variation of s_u/σ_c' with b. Clearly, the experimental results for the remolded, normally consolidated EPK clay show that Tresca's failure criterion is not applicable and that, in fact, s_u/σ_c' is not constant but decreases with increasing b-value. The decrease in this ratio from $s_u/\sigma_c' = 0.54$ in triaxial compression (b=0) to

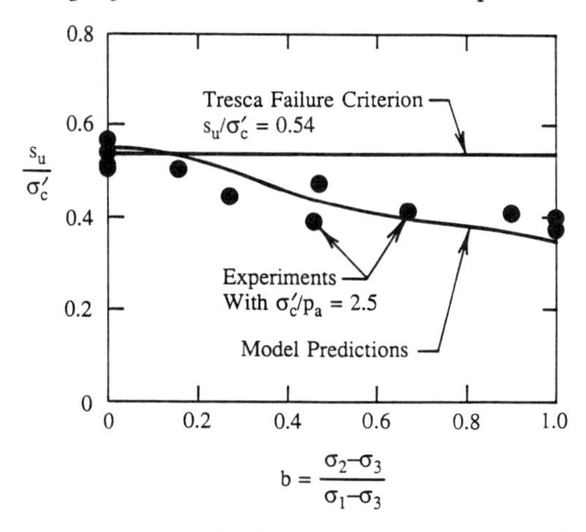

Figure 1. Comparison of Measured and Predicted Normalized Undrained Shear Strength for Cubical Triaxial Tests on Edgar Plastic Kaolinite.

$s_u/\sigma_c' = 0.38$ in triaxial extension (b=1) corresponds to a change of 30%. This substantial drop in normalized undrained shear strength is not accounted for by the Tresca failure criterion, which is most often assumed to hold for total stress stability analyses of soil structures. The decrease in s_u/σ_c' from triaxial compression to plane strain conditions (b = 0.40 to 0.45), for which most analyses procedures have been developed, is in the order of 20% for the remolded EPK clay.

The predictions of maximum shear strengths from the single hardening model (see ref. 6) are also shown in Fig. 1. The model predictions generally follow the experimental observations. For triaxial compression, the value of $s_u/\sigma_c' = 0.55$, and for triaxial extension $s_u/\sigma_c' = 0.34$, which corresponds to a 38% decrease. This decrease is a little larger than observed, but the predictions for plane strain conditions (b = 0.40 to 0.45) correspond very well with the experimental observations. The overall comparison of predicted and measured normalized undrained shear strength indicates that the single hardening constitutive model is capable of predicting the behavior under three-dimensional conditions with good accuracy.

The variation in undrained shear strength with b has often been attributed to effects of anisotropy (e.g.5). However, the experiments performed for this study were carried out on laboratory prepared, remolded, isotropically consolidated EPK clay, and the predictions of undrained behavior were made by an isotropic model. The differences in undrained shear strength observed and predicted for different test conditions could therefore not be due to anisotropy.

The variation in strength as a function of b is accounted for by the presence of the third stress invariant, I_3 ($= \sigma_1' \cdot \sigma_2' \cdot \sigma_3'$), in the model. Fig. 2 shows the predicted effective stress paths and undrained shear strengths for triaxial compression and extension tests on EPK clay in the triaxial plane. The teardrop shaped yield surface corresponding to the isotropic compression stress state is also indicated on this figure. This yield surface is asymmetric with regard to the hydrostatic axis due to the presence of I_3 in the formulation. Thus, for any given mean normal stress, the stress difference, $(\sigma_1 - \sigma_3)$, is highest in compression and lowest in extension. For three-dimensional stress states in between compression and extension, the stress difference is in between these extreme values (see ref. 6).

When undrained stress paths are calculated, the required balance between elastic volumetric strains, ε_v^e, and plastic volumetric strains, ε_v^p, causes the pore pressures to develop in such a manner as to maintain undrained conditions with no volumetric strains ($\varepsilon_v^e + \varepsilon_v^p = 0$). The effective stresses therefore follow a path for which this condition is fulfilled. In the initial stages of an undrained test on normally consolidated clay, the plastic volumetric strains are compressive and this requires the elastic volumetric strains

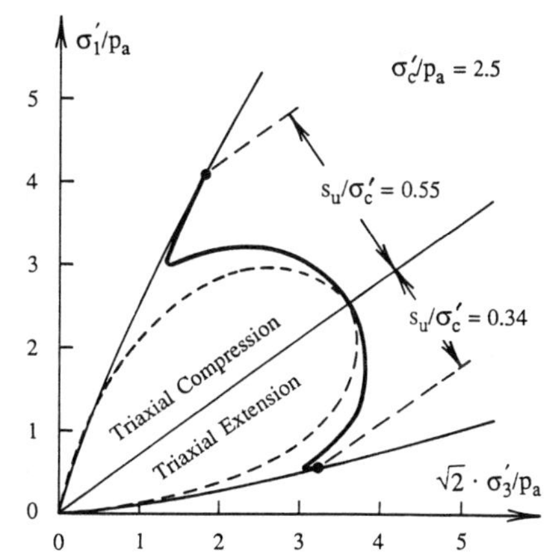

Figure 2. Predicted Stress Paths and Undrained Shear Strengths for Triaxial Compression and Extension Tests on Edgar Plastic Kaolinite Shown in Triaxial Plane.

to be expansive and of equal magnitude. Elastic strains are small in nature, and elastic volumetric expansion can only occur for decreasing mean normal stresses. The yield surface is therefore pushed out very little thereby producing only small amounts of plastic volumetric compression. Consequently, the effective stress path initially follows and to some degree reflects the shape of the yield surface. Since the effective stress point is always located on the current yield surface during loading, the effective stress path in Fig. 2 indicates how much the yield surface has been pushed out from its initial position.

It follows from this behavior pattern that the effective stress path is controlled by the shape of the yield surface and by the plastic compressibility of the clay. In the tests on EPK clay shown in Fig. 2, the effective stress paths reach transition zones and move into regions where the plastic volumetric strains become expansive. The effective stresses therefore change direction and follow paths with increasing mean normal stresses along which the elastic volumetric strains are compressive. The effective stresses finally reach failure conditions at the effective stress failure surface, which is also expressed in terms of I_3. This change in direction of the effective stress path may not always occur. It depends on the plastic compressibility of the clay at high stress levels near the failure surface. Clays with high plasticity indices and high compressibilities exhibit continuously decreasing effective means normal stresses during undrained shear. The effective stress paths continue to reflect the shape of the yield surface, and they may exhibit decreasing stress differences similar to those indicated by the teardrop shaped yield surface near the origin of the stress space (see Fig. 2). The maximum stress difference and therefore the undrained shear strength for such clays are reached inside the effective stress failure surface.

EXPERIMENTAL OBSERVATIONS OF UNDRAINED SHEAR STRENGTH RELATIONS

To investigate whether the behavior pattern observed from the tests on EPK clay is generally obtained, results of undrained tests on normally consolidated clays in triaxial compression and extension as well as in simple shear (plane strain) tests were collected from the literature (1,4,10). Fig. 3 shows the variation of normalized undrained shear strength with plasticity index for these three test conditions. For each clay, the highest value of s_u/σ'_{vc} was obtained in triaxial compression, the lowest value was obtained in triaxial extension, and the value from simple shear tests was in between. To demonstrate this pattern more clearly, the data have been plotted in Fig. 4 as ratios of undrained shear strengths in simple shear (SS) and in triaxial extension (TE) divided by the undrained shear strength in triaxial compression (TC). For each clay, the highest ratio (= 1) is obtained in compression, the lowest in extension, and the ratio for simple shear is between these two extremes. This sequence of undrained shear strength ratios is obtained for each clay, but two clays show the same strength in triaxial compression and simple shear. Thus, all normally consolidated clays exhibit the same undrained strength pattern as observed for the EPK clay.

It is interesting to note that the largest differences in strengths occur for clays with low plasticity indices, and these differences decrease with increasing plasticity index.

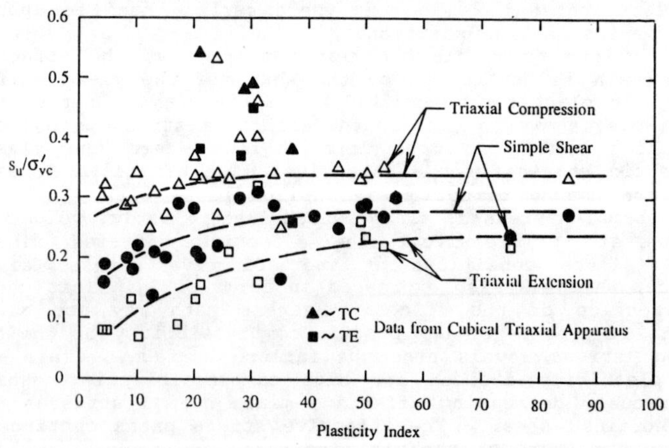

Figure 3. Variation of Normalized Undrained Shear Strength with Plasticity Index and 3-D Stress State.

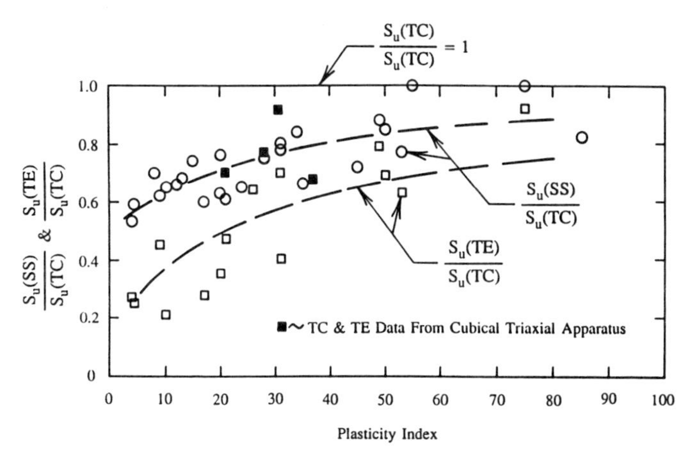

Figure 4. Variation of Undrained Shear Strength in Simple Shear (SS) and Triaxial Extension (TE) Normalized on Undrained Shear Strength in Triaxial Compression (TC).

This variation in undrained shear strength will be discussed further below.

The results of triaxial compression and extension tests performed in a cubical triaxial apparatus on four normally consolidated clays (including the EPK clay reported above) produced lower variation in undrained strengths than indicated by the general pattern in Figs. 3 and 4. Three of these test series were performed on remolded, isotropic clays. The fourth test series was conducted on isotropically consolidated, cross-anisotropic San Francisco Bay Mud. The possible reasons why they fall outside the general pattern is addressed below.

FACTORS AFFECTING THE UNDRAINED SHEAR STRENGTH

Several factors influence the undrained shear strength variation under three-dimensional stress conditions. These are discussed in detail below.

Effects of I_3 and Compressibility. The effects of the shapes of the yield surface and the failure surface, which both include I_3 in their formulations, have been explained in general above (Fig. 2). The locations and shapes of these surfaces interact with the compressibility of the clay to produce effective stress paths that lead to undrained shear strengths whose variation with b is largely reflected by the shape of I_3 = constant in an octahedral plane. Fig. 5 shows the cross-sectional shape of the three-dimensional effective stress failure surface for a clay in an octahedral plane. This surface has a smoothly rounded triangular shape with the largest stress difference in triaxial compression (b=0), the

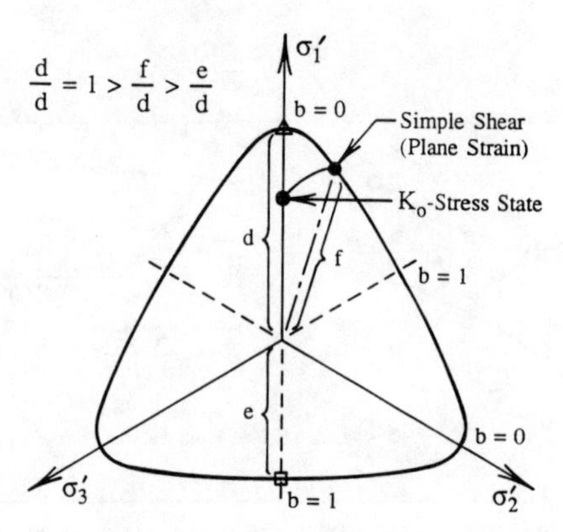

Figure 5. Shear Strength Ratios in Octahedral Plane.

lowest in triaxial extension (b=1), and with values in between for 0 < b < 1. The simple shear (i.e. plane strain) stress state at failure corresponds to b-values typically in the range from 0.3 - 0.5 for normally consolidated clays. The stress differences, $(\sigma_1-\sigma_3)$, which are proportional to d, e, and f in Fig. 5, form ratios whose relative magnitudes are similar to the undrained shear strength ratios shown in Fig. 4.

For isotropic clays, the pore pressures that develop under the three different stress conditions are similar in magnitude, but the pore pressure at failure in extension is typically slightly higher than that in compression, as indicated in Fig. 2. The failure stress points therefore correspond to slightly different mean normal stresses and octahedral planes. However, the trend acts to separate the undrained shear strength ratios more than indicated on Fig. 5.

Effective friction angles for clays in triaxial compression decrease with increasing plasticity index as shown in Fig. 6. This results in a change in the shape of the three-dimensional effective stress failure surface, as shown in Fig. 7. The cross-sectional shape, expressed through I_3, becomes more rounded, and the ratio of stress differences in extension and compression, (e/d), therefore increases with decreasing effective friction angle and increasing plasticity index. This increase in (e/d) is similar to the increase in undrained shear strength ratio with increasing plasticity index shown in Fig. 4. Thus, for similar pore pressures at failure, the presence of I_3 in the yield and failure criteria appears to account for the variation in undrained shear strength ratio with plasticity index.

Effect of Anisotropy. Natural, normally consolidated clays are deposited by sedimentation in the vertical direction and consolidated under K_o-conditions. These clays are cross-anisotropic with the vertical direction as an axis of

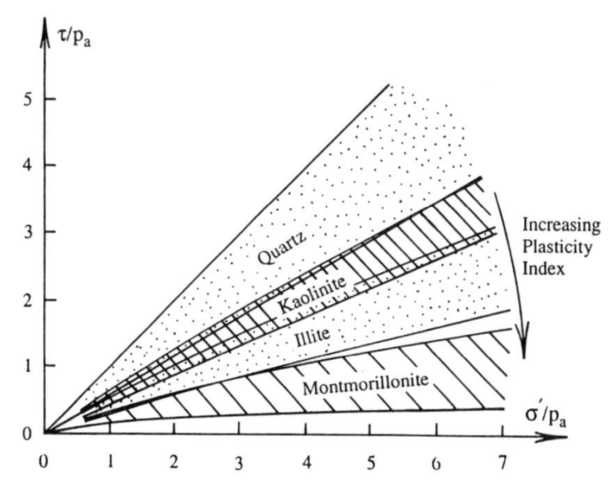

Figure 6. Ranges in Effective Failure Envelopes for Soils Composed of Pure Clay Minerals or Quartz (Adapted from Olson (11)).

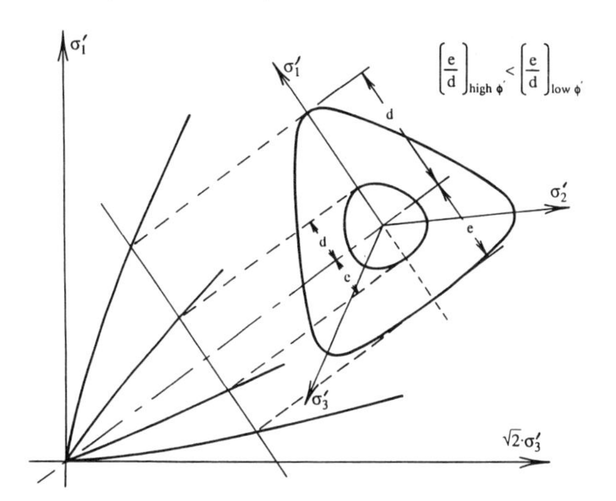

Figure 7. Variation of Shear Strength Ratio in Octahedral Plane with Friction Angle in Triaxial Compression.

material symmetry. The effect of cross-anisotropic behavior caused by K_o-consolidation and further enhanced by aging is demonstrated in Fig. 8. Effective stress paths for undrained triaxial compression and extension tests on resedimented Boston Blue clay show effects of initial isotropic and initial K_o-consolidation. The undrained shear strength ratio for the tests with initial isotropic compression is 0.83, whereas

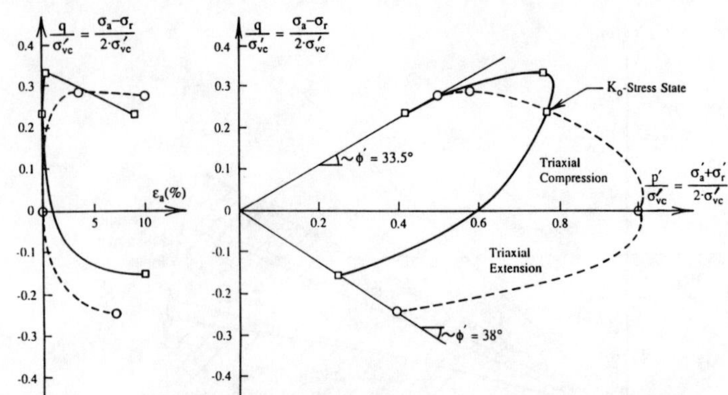

Figure 8. Undrained Stress-Strain Behavior and Effective Stress Paths for Isotropic and K_o-Consolidated, Resedimented Boston Blue Clay in Triaxial Compression and Extension (Adapted from Germaine and Ladd (2)).

that for the tests with initial K_o-consolidation is 0.46. Thus, the anisotropic behavior causes additional changes in the undrained shear strengths measured in the two types of tests.

The effective stress paths in Fig. 8 all end up on the same effective stress failure surface. This failure surface exhibits higher friction angle in extension than in compression, and this is accounted for by the presence of I_3 in the failure criterion. Whereas the pore pressures at failure are similar in magnitude for the isotropically consolidated specimens, the pore pressures for the K_o-consolidated specimens are quite different at effective stress failure. More importantly, the maximum stress difference is obtained well before the effective stress failure envelope is reached in compression, whereas the maximum stress difference occurs at the time the effective stress envelope is reached in extension. In the compression test with initial K_o-consolidation, creep or aging effects have pushed the yield surface out such that the initial response to loading is essentially elastic, as indicated by the steep stress-strain curve. The maximum stress difference occurs immediately as the yield surface is reached. The subsequent large plastic strains correspond to decreasing stress differences, and the effective stress path reflect the shape of the yield surface until the effective strength envelope is reached. Similar effects of aging are not observed in extension, because the large changes in stresses and strains required to reach extension erase the preconsolidation effects produced by aging. Effects of aging are more difficult to ascertain for isotropically consolidated specimens, because the initial stress path directions tend to be perpendicular, or almost perpendicular, to the hydrostatic axis for specimens with and without aging effects. When the effective stresses reach higher stress levels, any effects of aging have been erased by the strains produced on the way up to failure.

Most of the experiments shown in Figs. 3 and 4 were performed on K_o-consolidated, cross-anisotropic clays, and they show greater differences in undrained strength ratios than observed for the cubical triaxial tests with initial isotropic consolidation. The differences in behavior are explained, in part, by the effects of anisotropy as produced by initial sedimentation and subsequent K_o-consolidation and aging effects.

Effects of Localization. Localization of plastic strains to form shear planes in a test specimen occurs past peak failure in triaxial compression on normally consolidated clays. Shear planes do therefore not have any effect on failure, whether drained or undrained, in triaxial compression. In triaxial extension tests, the shear planes cut across the specimen at a relatively shallow angle, and their development may therefore be hastened by the presence of horizontal bedding planes in natural clays. In the conventional triaxial extension test, the soft membrane allows the shear planes, which are planes of weakness, to develop fully. This involves only a small part of the specimen, and generally the part with the lowest density. In comparison, the cubical triaxial apparatus used at UCLA is designed to impede the gross development of shear planes in any direction. The entire specimen is deformed uniformly, and this provides a better measure of the clay properties at the average specimen density. The two modes of failure in extension tests result in considerably different effective friction angles (9), but the influence on the undrained shear strength may be less pronounced and is not well-established. Nevertheless, natural clays deposited with horizontal bedding planes may exhibit effects of localization in extension tests with soft membranes, whereas development of shear planes is impeded in the cubical triaxial apparatus. This may have contributed to the higher undrained strength ratios obtained from the cubical triaxial tests.

CONCLUSION

The undrained shear strength of normally consolidated clays has been observed to vary in a systematic manner with the highest strength in compression, the lowest strength in extension, and the simple shear strength in between. The causes of this variation under three-dimensional stress conditions have been evaluated. The predictions of undrained shear strengths from an isotropic constitutive model compare favorably with measured values for a remolded, isotropically consolidated clay. This variation for the isotropic clay can be accounted for by the presence of the third stress invariant, I_3, in the model. Additional effects are observed in natural clays in which cross-anisotropic behavior is produced by initial sedimentation and subsequent K_o-consolidation and aging. Finally, the effects of shear planes which may develop prematurely in extension tests on natural clays are considered, but this effect has not yet been established conclusively.

REFERENCES

1. T. Berre & L. Bjerrum, Shear Strength of Normally Consolidated Clays, Proc. 8th Intl. Conf. SMFE (1.1), Moscow, 1973, 39-49.
2. JT Germaine & CC Ladd, Triaxial Testing of Saturated Cohesive Soils, Adv. Triax. Test. Soil Rock, ASTM STP 977, Ed RT Donaghe, RC Chaney & ML Silver, Am. Soc. Test. Mat., Philadelphia, 1988, 421-459.
3. MK Kim & PV Lade, Single Hardening Constitutive Model for Frictional Materials, I. Plastic Potential Function, Computers & Geotechnics, 5(4), 1988, 307-324.
4. CC Ladd & L Edgers, Consolidated-Undrained Direct-Simple Shear Tests on Saturated Clays, Res Rep R72-82, MIT, Cambridge, 1972.
5. CC Ladd & R Foott, New Design Procedure for Stability of Soft Clays, J. Geot. Engrg. Div. (ASCE), 100(GT7), 1974, 763-786.
6. PV Lade, Single-Hardening Model with Application to NC Clay, J. Geot. Engrg. (ASCE), 116(3), 1990, 394-414.
7. PV Lade & MK Kim, Single Hardening Constitutive Model for Frictional Materials, II. Yield Criterion and Plastic Work Contours, Computers & Geotechnics, 6(1), 1988, 13-29.
8. PV Lade & MK Kim, Single Hardening Constitutive Model for Frictional Materials, III. Comparisons with Experimental Data, Computers & Geotechnics, 6(1), 1988, 30-47.
9. PV Lade & J Tsai, Effects of Localization in Triaxial Tests on Clay, Proc. 11th Intl. Conf. SMFE (2), San Francisco, 1985, 549-552.
10. R. Larsson, Undrained Shear Strength in Stability Calculation of Embankments and Foundations on Soft Clays, Can. Geot. J., 17(4), 1980, 591-602.
11. RE Olson, Shearing Strength of Kaolinite, Illite, and Montmorillonite, J. Geot. Engrg. Div. (ASCE), 100 (GT11), 1974, 1215-1229.

LABORATORY TESTING AND PARAMETER ESTIMATION FOR TWO-PHASE FLOW PROBLEMS

Dobroslav Znidarčić[1], AM. ASCE, Tissa Illangasekare[2], AM. ASCE and Marilena Manna[3]

ABSTRACT: A new testing technique for the determination of suction-saturation relationships for various soils is presented. This relationship is basic to the modeling of unsaturated flow and multiphase flow in soils with applications in soil physics, geotechnical engineering and groundwater engineering. The technique utilizes a flow pump for a precise control of water flow in and out of the sample, enabling accurate determination of the degree of saturation. Draining and wetting cycles and hysteresis loops are readily obtained from a continuous experiment on a single specimen. The complete automation reduces the operator involvement and the possibility of errors.

INTRODUCTION

Transport of chemicals and waste products through soil-water systems due to accidental spills and disposal of hazardous wastes often result in simultaneous flow of two or more liquid phases through soils. Typical examples are spills of non-aqueous phase liquids (NAPL) or their release from leaking underground tanks. Models of multi-phase flow in soils are used to simulate and predict the movement of the contaminants through soils and to evaluate the impact and effectiveness of proposed remediation schemes. These models require the knowledge of flow characterization parameters that include relative permeability and partial pressure-saturation relationships for various phases. The parameters are determined from laboratory experiments that are often time consuming, or are estimated from published data for similar soils (Luckner et al. 1989). This paper

---

1 Assistant Professor, 2 Professor and 3 Graduate Research Assistant, Department of Civil, Environmental and Architectural Engineering, University of Colorado, Boulder, CO 80309-0428

presents a novel technique for determining suction-saturation curves of a soil sample for both draining and wetting cycles with an arbitrary number of hysteresis cycles in between. The technique utilizes a flow pump for precise control of flow rates and a precision differential transducer for measuring induced head differences. The experiment is five to ten times faster than the conventional pressure plate method (Klute 1986). The continuous nature of the experiment facilitates its automation and reduces the danger of operator induced errors.

EQUIPMENT AND TESTING PROCEDURE

A conventional triaxial cell is used to house the soil sample. The bottom platen is equipped with a high air entry porous stone which prevents the air from the unsaturated sample to enter into the water saturated system. A rubber o-ring is placed between the base and the ceramic porous stone to prevent air entrance around it. This design facilitates the change of ceramic plates with different air entry values in the same equipment. The top platen has a conventional coarse porous stone. Both top and bottom platens have two ports for easy flushing and saturation of the system. The sample is enclosed into a latex membrane which is secured to the platens with o-rings. A schematic drawing of the sample assembly with the porous stones is shown in Figure 1. The sample is 71 mm in diameter and typically 30 mm high.

The triaxial cell is connected to three water reservoirs with independently controlled pressures (Figure 2). One reservoir controls the confining cell pressure, while the other two are used for controlling pressures at the top and bottom platens. The flow pump is connected to one of the ports in the bottom platen. The flow pump consists of a custom made stainless steel syringe and a commercially available driving mechanism. It is capable of precisely controlling low flow rates with a resolution of 10^{-6} ml/s. Its detailed description and main characteristics are given by Aiban and Znidarčić (1989). A precision differential pressure transducer is connected between the top and bottom of the sample. As will be explained later, it measures the capillary suction in the sample.

The test starts by saturating the sample and all the lines in the system. This is achieved by first applying vacuum and then by flushing deaired water from lower to higher parts of the system. Finally, a back pressure of at least 200 kPa is applied to the

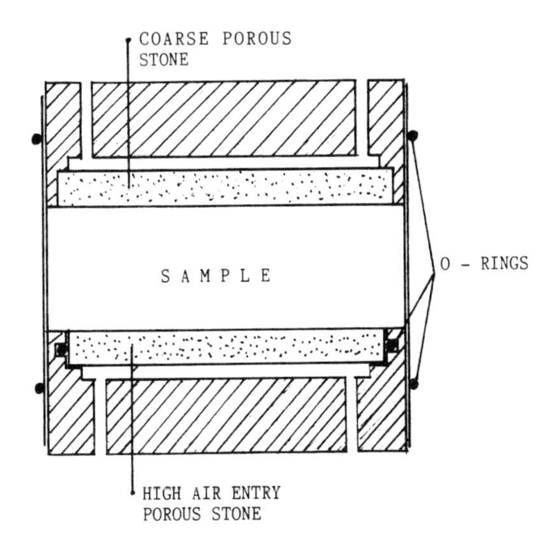

Figure 1 - Sample assembly

pore water system to further facilitate complete saturation. The full saturation of the high air entry porous stone is the main challenge in this configuration. The problem is solved by oven drying the fine porous stone prior to its installation and by percolating the deaired water through the system and the sample under the back pressure. It is usually sufficient to percolate the water overnight.

Once the sample and the whole system is saturated the water is drained from the part of the system connected to the top platen without reducing the back pressure. Thus, the lines leading to the top cap and the space behind the coarse porous stone are filled with air under pressure equal to the applied back pressure, while the sample and the rest of the system remains saturated with water. The water is then withdrawn from the bottom of the sample at a constant rate with the flow pump and the corresponding pressure difference is measured with the differential pressure transducer. A PC based data acquisition system is used for continuous recording of the pressure change. The transducer records the difference between the air pressure (back pressure)

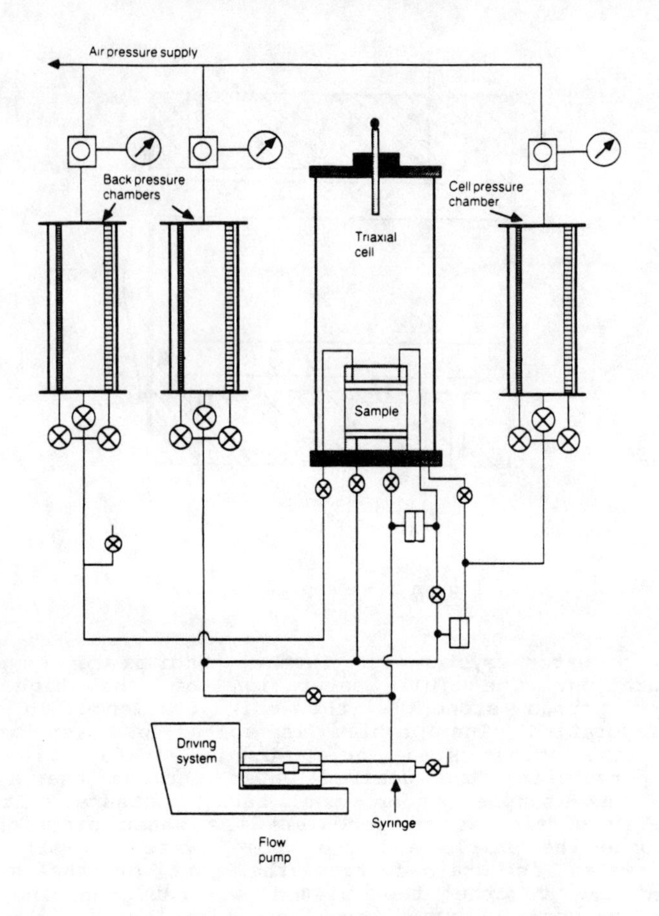

Figure 2 - Testing system

at the top of the sample and the water pressure below the high air entry porous stone. This pressure difference is thus equal to capillary suction present in the sample. Note that these measurements are made at an elevated back pressure so that even when the water pressure is substantially lower than air pressure, it is still above atmospheric pressure and there is no risk of water cavitation. This is the same concept as in the conventional axis translation technique for measuring negative water pressures in unsaturated soils (Fredlund 1979).

An example record of transducer output versus time is shown in Figure 3. For some time at the

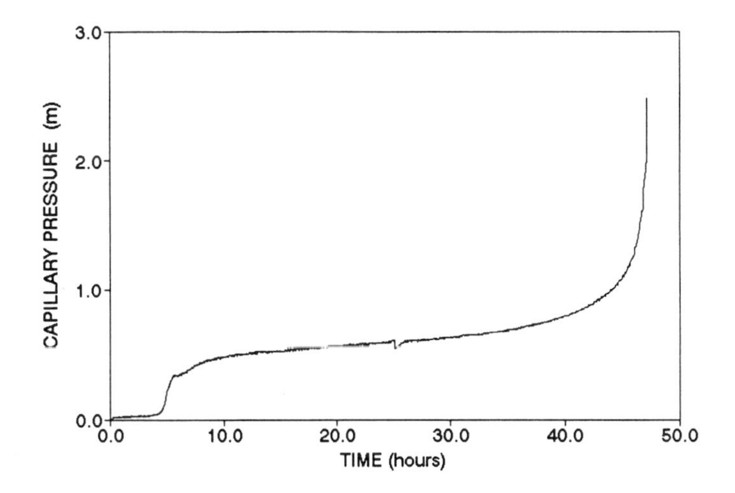

Figure 3 - Suction - time record

beginning of the experiment no significant pressure change is recorded. During this period water is being removed from the coarse porous stone at the top of the sample, while the sample remains fully saturated. The sudden increase in pressure difference corresponds to the instant at which all the water is removed from the corse porous stone in the top cap and the sample starts to desaturate. From this point on the amount of water removed from the sample is calculated by multiplying the known flow rate, as precisely controlled by the flow pump, with the elapsed time. From this data and from the sample porosity the degree of saturation at any time is calculated. As mentioned earlier, the differential transducer measures the pressure difference between water pressure below the high air entry porous stone and the air pressure above the sample. Since both water and air are continuous phases through the sample the measured pressure difference is equal to the capillary suction within the sample. Thus, the data needed to construct the capillary suction-saturation curve is obtained from the experiment. For the pressure time record shown in Figure 3, suction-saturation curve is presented in Figure 4.

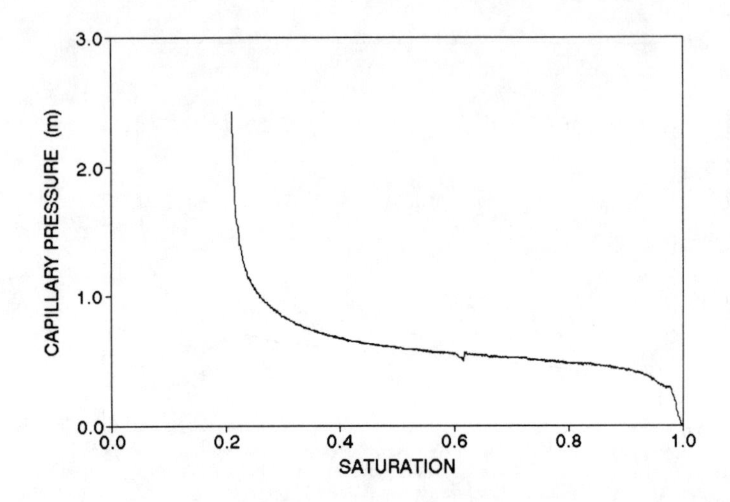

Figure 4 - Suction - saturation relation for #125 sand

The critical test parameter is the applied flow rate for the withdrawal of the water from the sample. If the rate is slow the test will be long, but a too high rate will cause a nonuniform degree of saturation within the sample. In the later case, the assumption made in the analysis, that at any moment the degree of saturation is uniform and can be calculated from the amount of water withdrawn from the sample, will be invalid and the resulting saturation-suction relation will be in error. The optimal test velocity has to be determined experimentally for different materials.

The test described above will produce the suction-saturation curve for drainage from fully saturated state to the residual saturation. However, at any time during this process the test can be terminated by stopping the flow pump and then reversing the flow direction to obtain the wetting branch of a hysteresis loop. After achieving the desired degree of saturation on the wetting curve, a new drainage process can be started. These cycles can be repeated as many times as desired. Note that the process is continuous and automated. The operator intervention is needed only to reverse the flow direction between drying and wetting cycles. A continuous pressure record for a test with two hysteresis loops is presented in Figure 5, and the

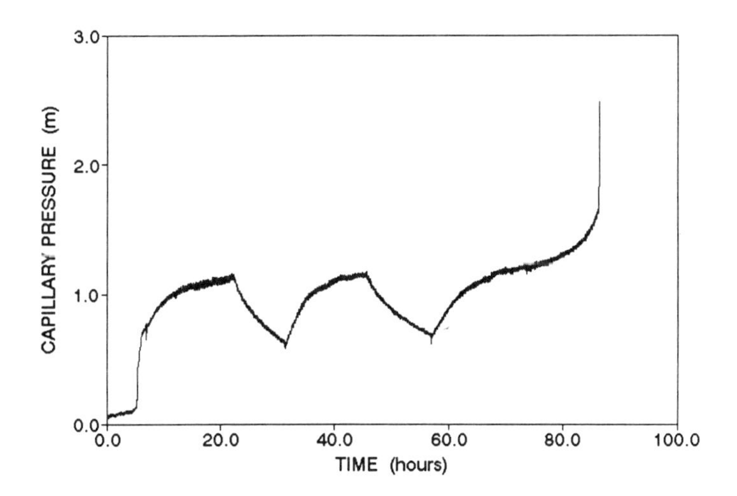

Figure 5 - Suction - time record for drying and wetting cycles

corresponding suction-saturation relation is shown in Figure 6.

RESULTS AND DISCUSSION

The described testing procedure has been applied for obtaining suction-saturation curves for several sand samples. Drying curves for three uniform sands having different grain sizes are presented in Figure 7. The numbers associated with each curve correspond to the sieve number through which the sand had been sieved. It is clear that for finer sand, suction is

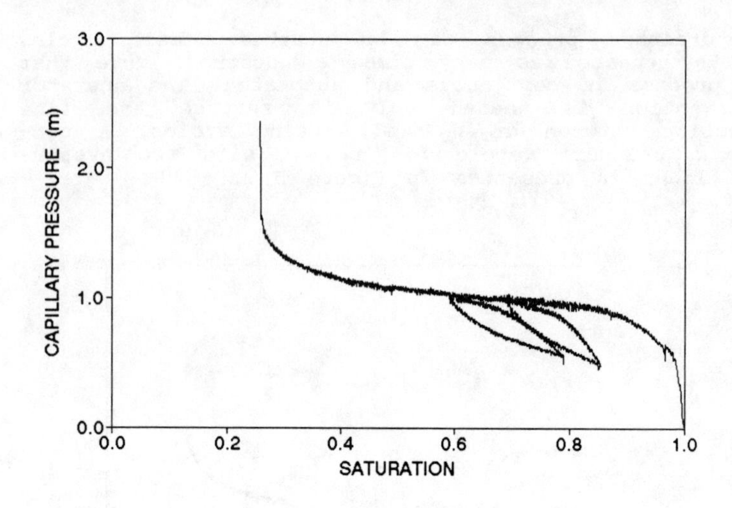

Figure 6 - Suction - saturation relation with hysteretic loops

higher for the same degree of saturation. The experimental results are used to calibrate the parameters of the Van Genuchten (1980) equation for the suction-saturation relation in the form:

$$S = \frac{1 - S_r}{[1 + (\alpha h)^{\beta}]^{\left(1 - \frac{1}{\beta}\right)}} + S_r$$

The model parameters α and β for the three sands as well as porosities n and residual saturations S_r are given in Table 1.

Table 1: Model parameters and soil properties

Sand	α	β	n	S_r
# 30	13.50	4.5	0.49	0.26
# 70	2.70	5.5	0.51	0.30
#125	1.78	6.5	0.41	0.22

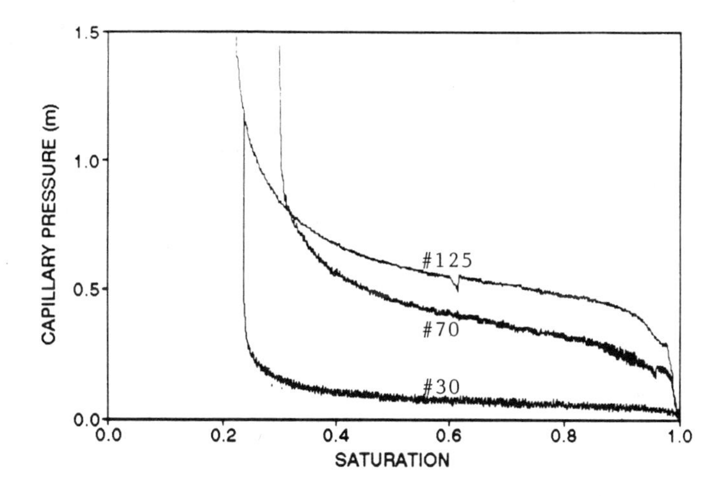

Figure 7 - Suction - saturation relations for three sands

The results presented above are consistent and there is no reason to question their validity. However, the question of their accuracy has to be discussed. The suction-saturation relation is obtained from the measured pressure difference in the experiments and from calculated degree of saturation. An accurate measurement of pressure difference poses no problem with the use of the precision differential transducer with the proper pressure range. An accuracy of less than 1 centimeter head is easily obtained. The degree of saturation is calculated from the withdrawn water whose quantity is precisely controlled by the flow pump. However, this volume is subtracted from the initial volume of water which is calculated from the sample volume and porosity. Thus, the accuracy of the saturation measurement depends on how accurate these measurements are. Their relative error can be reduced by increasing the size of the specimen. However, the height of the specimen should not be increased since this would increase the test duration and saturation variability within the specimen. Namely, for higher samples the elevation head difference between the top and bottom may become significant especially for coarse grained materials for which the capillary fringe is relatively small.

The analysis of the described test is based on the assumption that during the suction change the overall sample volume remains constant. That means the soil skeleton is rigid. While this assumption is certainly an acceptable approximation for sands tested in the described project, for fine grained materials it would lead to gross error. An independent measurement of soil skeleton volume change will be needed for these materials and a more elaborate analysis procedure should be developed. We leave the description of such an analysis to a later paper.

CONCLUSIONS

The presented results demonstrate the usefulness of the newly developed testing technique for the determination of suction-saturation relationships for sandy materials. The test is significantly shorter than the conventional pressure plate technique and due to its continuous nature it is readily automated. Both draining and wetting cycles and multiple hysteresis loops can be obtained on a single sample with the minimal operator intervention. Presently the technique is limited to sandy materials but its extension to silts and clays is feasible with slight modifications in the equipment and the analysis procedure.

ACKNOWLEDGEMENTS

The authors would like to gratefully acknowledge the support of EPA through the Hazardous Waste Research Institute at Louisiana State University and US EPA Robert Kerr Environmental Research Laboratory at Ada, Oklahoma for this research.

REFERENCES

Aiban, S.A., and Znidarčić, D. 1989. Evaluation of the flow pump and constant head techniques for permeability measurements, Geotechnique, 39(4), 655-666.

Fredlund, D.G. 1979. Second Canadian Geotechnical Colloquium: Appropriate concepts and technology for unsaturated soils, Can. Geotech. J., 16, 121-139.

Klute, A. (Ed.) 1986. Methods of Soil Analysis, 1, Physical and Mineralogical Methods, Agron. Monogr. 9, 2nd ed., American Society of Agronomy, Madison, Wisc.

Luckner, L., van Genuchten, M.Th., and Nielsen, D.R. 1989. A Consistent Set of Parametric Models for the Two-Phase Flow of Immiscible Fluids in the Subsurface, Water Resources Research, 25(10), 2187-2193.

van Genuchten, M.Th. 1980. A closed-form equation for predicting the hydraulic conductivity of unsaturated soils, Soil Sci. Soc. Am. J., 44(5), 892-898.

VOLUME-CONTROLLED APPROACH FOR DIRECT MEASUREMENTS OF COMPRESSIBILITY AND HYDRAULIC CONDUCTIVITY

James D. Gill[1], Harold W. Olsen[2], M.ASCE,
Karl R. Nelson[3], M.ASCE

ABSTRACT: A volume-controlled approach has been developed for simultaneous and direct laboratory measurements of compressibility and hydraulic conductivity on clay specimens. Flow pumps are used to provide volume control. One flow pump consolidates the sample at a constant rate of volume change while another flow pump periodically superimposes constant flow rates through the specimen. The void ratio-effective stress relationship is continuously monitored and the hydraulic conductivity-effective stress relationship is evaluated by the steady-state component of the pore pressure difference across the specimen induced by the superimposed constant flow rates. In comparison with the traditional combined approach where falling-head or constant-head permeability tests are performed between increments of a conventional step-loading consolidation test, the new volume-controlled approach requires far less time and much smaller gradients, and it provides much more detailed information on the interrelationships among void ratio, effective stress, and hydraulic conductivity. In comparison with indirect permeability determinations from the conventional constant rate of deformation (CRD) test, the new approach provides closely similar results and is not constrained by the minimum deformation rate required for obtaining permeability data from a CRD test.

INTRODUCTION

Detailed information on the variation of compressibility and hydraulic conductivity with void

---

1- Staff Engr., Michael W. West & Associates, 8906 W. Bowles, suite 290 Littleton, CO 80123.
2- Res. Engr., U.S. Geological Survey, Box 25046, MS 966, Denver, CO 80225.
3- Assoc. Prof, Dept. of Engineering, Colorado School of Mines, Golden, CO 80401.

ratio is necessary for state-of-the-art analyses of clay consolidation behavior (Znidarcic et al. 1984, Kabbaj et al. 1986). Without this information, it is generally necessary to assume a constant value for the coefficient of consolidation, C_V, a lumped parameter which varies with both the compressibility and hydraulic conductivity. Existing laboratory approaches for providing this information include (1) traditional incremental-loading consolidation tests with constant-head, falling-head, or constant-flow (flow-pump) permeability tests between loading increments; and (2) continuous-loading (constant rate of deformation-CRD, controlled gradient-CG, and continuous loading-CL) consolidation tests where hydraulic conductivity can be indirectly determined from the pore pressures that develop during the test.

One of the fundamental limitations of traditional incremental-loading consolidation tests is that strain-rate effects are obscured by the large variation in strain rate that occurs during and between loading increments. In consequence, no rational basis is provided to account for strain-rate effects in estimating field behavior from this data. One example of the importance of this limitation is the research by Leroueil et al. (1985) which defines the strain-rate effect on the preconsolidation stress of natural clays. This limitation also is shared by the controlled gradient (CG) consolidation test developed by Lowe et al. (1969), in which the pore pressure gradient is held constant, and the continuous loading (CL) test developed by Janbu et al. (1981), in which the pore pressure-load ratio is held constant.

Other limitations of the incremental loading approach include the time required to perform a series of load increments and constant-head or falling-head permeability tests between increments, and the effects of high pore-pressure gradients that occur during both the loading increments and the permeability tests. For example, several weeks may be needed to perform a series of incremental loads and falling-head tests on a low permeability clay, and gradients typically exceed 100 (Pane et al. 1983). In recent years it has been demonstrated that the need for high gradients and/or long measurement times with the constant-head or falling head methods can be readily minimized with the constant-flow (flow-pump) method (Olsen 1966, Olsen et al. 1985, Aiban and Znidarcic 1989). However, the previously mentioned limitations of the incremental loading approach still occur even when constant-flow permeability tests are performed between the load increments.

Limitations due to variations in strain-rate are avoided with the constant rate of deformation (CRD) consolidation test pioneered by Crawford (1964) and further developed by numerous researchers, including Smith and Wahls (1969), Wissa et al. (1971) and Znidarcic et al. (1986). With this method, continuous stress-strain relationships are obtained in substantially less time than is needed with incremental loading tests and at lower pore-pressure gradients. Moreover, strain-rate effects on the stress-strain relationship can be measured. Also, continuous hydraulic conductivity-void ratio relationships can be inferred from consolidation induced pore pressure differences across test specimens. However, to obtain pore pressure differences of sufficent magnitude for indirect permeability determinations, a minimum strain rate on the order of 5×10^{-7}/s is required for some clays (Leroueil et al. 1985), and this strain rate can be orders of magnitude higher than those in the field which usually do not exceed 10^{-9}/s (Crawford 1988).

This paper describes a recently developed approach for conducting constant-flow (flow-pump) permeability tests during the progress of a CRD consolidation test. This combined volume-controlled approach provides direct hydraulic-conductivity measurements and is not constrained by the minimum strain rate required for obtaining permeability data from a CRD test. Experimental data are presented that illustrate the capabilities of this approach and its technical and practical advantages over the traditional incremental loading, falling head, and continuous loading methods.

COMBINED VOLUME-CONTROLLED APPROACH

The method presented here uses a volume-controlled approach for both the compressibility and the permeability aspects of the test. Flow pumps, devices that deliver a range of small constant flow rates, are used to provide volume control. One flow pump is used to control the sample volume and consolidate the sample at a constant rate of volume change (equivalent to the CRD test). Another flow pump is used periodically during the test to superimpose constant flow rates through the sample. The effective stress at the top of the specimen and the pore pressure difference across the specimen are monitored. These data are used directly to give the void ratio-effective stress relationship. The hydraulic conductivity-effective stress relationship is evaluated from the steady-state component of pore-pressure difference across the sample induced by the superimposed flow rates.

EXPERIMENTAL SYSTEM

A diagram of the system is presented in Figure 1.

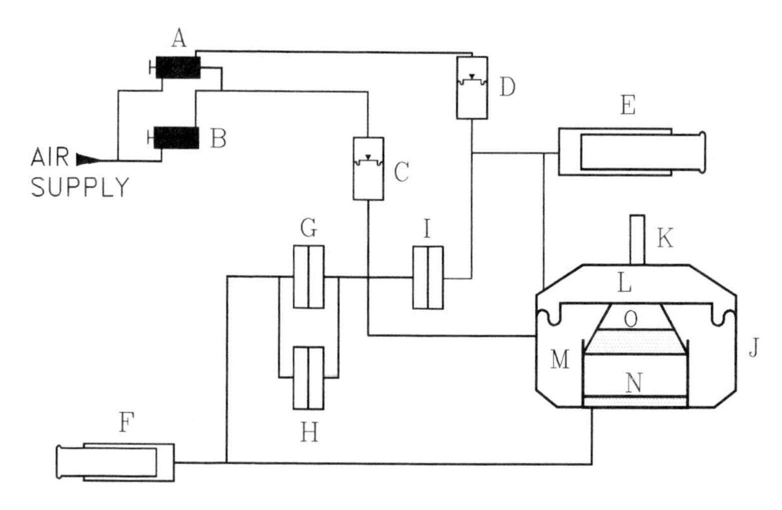

Figure 1. Schematic Diagram of Experimental System

An Anteus back-pressure consolidometer (J) was adapted to perform the test. The sample (N) is surrounded by a Teflon-coated cylinder (fixed ring) and has porous stones in contact with its top and bottom surfaces. Single drainage occurs from the top of the sample during consolidation. The back pressure is controlled by the absolute pressure regulator (B). The back pressure is converted from air pressure to fluid pressure using a bellofram system (C), which prevents air from going into solution in the pore water. This pressure is applied to the back-pressure chamber (M) of the consolidometer and is communicated to the sample via the top porous stone. Deformation is induced by infusing a constant flow rate from the deformation flow pump (E) into the load pressure chamber (L) causing the piston (O) to move down at a constant rate. Flow-pump permeability tests are performed by infusing a constant flow rate from the permeability flow pump (F) into the base porous stone and measuring the gradient induced thereby using the pore- pressure differential transducers (G&H). The vertical effective stress at the top of the sample is measured using another differential transducer (I). Deformation is measured using a linear variable differential transducer (K). A differential pressure regulator (A) and load pressure bellofram (D) are provided to keep the specimen under stress control for the initial portion of the test (seating load and initial permeability testing). A high-range pore-pressure transducer (H) can be used if pore pressures become high enough to damage the sensitive low-range transducer (G). The data aquisition system (not shown) includes a panel meter and a three channel analog strip chart recorder, which provides a continuous record of effective stress, base pore pressure, and deformation.

CONDUCT OF TEST

The data and test results described below were obtained on an undisturbed natural clay specimen of medium consistency with 25 percent finer than 2μm, a plasticity index of 45 and a Unified Soil Classification of CH. The specimen was carefully trimmed and placed in the oedometer and saturated for 12 hours at a back pressure of 275 kPa under a very small seating load.

The specimen was consolidated from an intial void ratio of 1.48 to a void ratio of 0.698 at a strain rate of 2 x 10^{-6}/s. Figure 2 illustrates the complete set of data including time-history plots of: total stress, base pore pressure, and void ratio, obtained during the test. An initial flow-pump permeabilty test (A) was performed resulting in a hydraulic conductivity of 1.9 x 10^{-7} cm/s. Using the deformation flow pump, the specimen was then consolidated at a constant rate of volume change. The result is a linear decrease in void ratio as indicated on the void ratio time-history plot. The total stress increases nonlinearly until the deformation flow pump is turned off. The pore pressure time-history plot shows how base pore pressure responds during consolidation and also during flow-pump permeability tests. Once the deformation pump is turned off the sample is held at a constant volume for approximately 24 hours, over which time eight flow-pump permeability tests were performed. The total stress noticably decreased during this period (constant volume relaxation) in response to the change in strain-rate. During rebound, three flow-pump permeabilty tests were performed.

EVALUATION OF HYDRAULIC CONDUCTIVITY AND COMPRESSIBILITY

Eighteen flow-pump permeability tests were performed during the course of consolidation. Flow rates ranging from 7.55 x 10^{-6} cm^3/s to 3.78 x 10^{-5} cm^3/s were superimposed through the consolidating specimen resulting in changes in pore pressure-gradient ranging

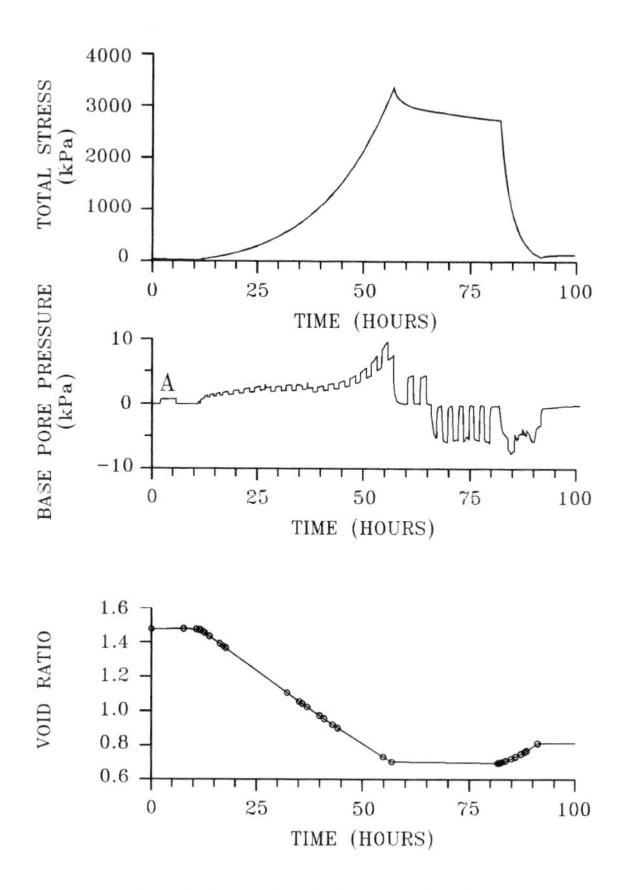

Figure 2. Complete set of Data Obtained During Test

from 2 to 10 (cm/cm). Figure 3 illustrates a close-up view of the base pore-pressure data used to evaluate hydraulic conductivity. After CRD consolidation was initiated using the deformation flow pump, the base pore pressure went through an initial transient response. Once the base pore pressure had stabilized into what is labeled as the CRD compression baseline, flow-pump permeability tests were initiated. Each permeability test was run until steady state flow was clearly established. Figure 3 shows that a few minutes of response time was required to reach steady-state, and after 1 to 2 hours each test was stopped. A few minutes later, the base pore pressure had returned to the value associated with the consolidation portion of the test (CRD compression baseline).

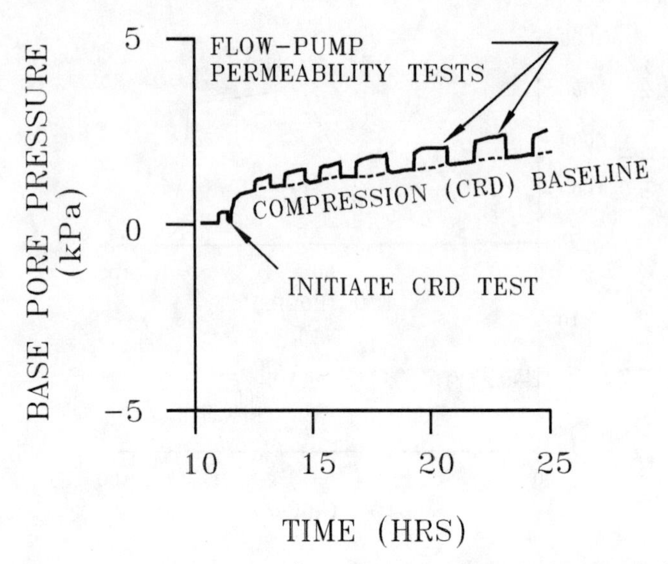

Figure 3. Close-up view of Base Pore Pressure Data

Hydraulic conducitivity, k, can be directly evaluated using Darcy's law:

$$k = Q/iA \tag{1}$$

where Q is the applied flow rate superimposed by the permeability flow pump; i is the hydraulic gradient, which is determined from the steady state component of excess base pore pressure (relative to the CRD compression baseline) induced by the flow pump test; and A is the cross-sectional area.

Hydraulic conductivity can also be evaluated indirectly (Wissa et al. 1971) using the equation developed for the CRD test:

$$k = \frac{(dh/dt)}{2u_b} (\gamma_f) (H) \tag{2}$$

in which (dh/dt) = deformation rate, γ_f = specific weight of the fluid, H = sample height and u_b is the base pore pressure due to consolidation (pressure from CRD compression baseline). This equation is derived from applying Darcy's law at the drainage boundary assuming the pore-pressure distribution is parabolic during consolidation. This equation can be applied when the flow-pump permeability tests are not being performed, which thus provides a convenient comparison between directly measured and back-calculated hydraulic conductivity values. An example of this comparison is presented in the results.

Compressibility can be directly evaluated from the test data based on equations developed for continuous-loading consolidation tests (Lowe et al. 1969, Smith and Wahls 1969, Wissa et al. 1971, and Janbu et al. 1981). At low pore pressure gradients, the equations are all of the same form:

$$\sigma' = \sigma_{tot} - 2/3\, u_b \qquad (3)$$

where σ' = effective stress at the midplane of the sample, σ_{tot} = total stress measured at the top of the sample and u_b = pore pressure measured at the base of the sample. The constant, 2/3, is based on the parabolic pore pressure distribution assumed to occur during a low-gradient continuous consolidation test. The effective stress is reported against average void ratio. The first derivative of this relationship ($\partial e/\partial \sigma'$) is the coefficient of compressibility, a_v. The coefficient of compressibility and the hydraulic conductivity can be used together to calculate the coefficient of consolidation:

$$C_v = \frac{k\,(1+e)}{a_v\, \gamma_f} \qquad (4)$$

where k and a_v are based on direct measurements taken during the test and e is the void ratio.

TEST RESULTS

The basic relationships that can be derived from the test are the void ratio-effective stress and the hydraulic conductivity-effective stress relationships. These relationships are illustrated in Figure 4 using semi-log and log-log formats, respectively. The hydraulic conductivity data is obtained from the direct flow-pump measurements. The data from these two basic relationships can be combined to derive the void ratio-hydraulic conductivity and the coefficient of consolidation-effective stress relationships. Figure 5 presents these relationships in semi-log format. The results indicate that the material was slightly overconsolidated. Enough hydraulic conductivity values were obtained to give a detailed look at the permeability behavior. Hydraulic conductivity decreased in an irregular manner during consolidation. It also decreased somewhat during the constant-volume-relaxation

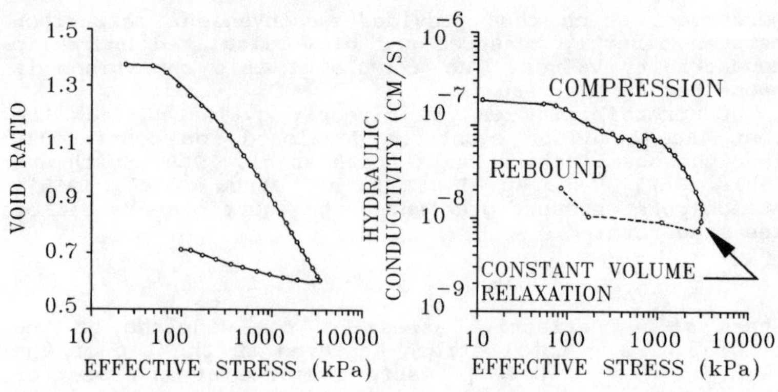

Figure 4. e-log σ′ and log k-log σ′ Curves

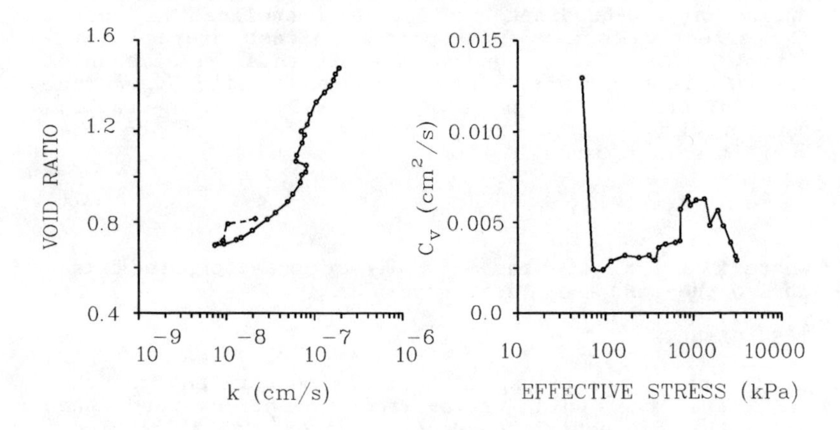

Figure 5. e-log k and C_v-log σ′ curves.

portion of the test. The void ratio-log hydraulic conductivity relationship is relatively linear at strains less than 20 percent as suggested by Tavenas et al. (1983). The coefficient of consolidation, C_v, is clearly not constant during consolidation.

COMPARISON WITH EXISTING METHODS

These data provide a convenient way to compare directly versus indirectly obtained values of hydraulic conductivity. Values directly obtained from the flow-pump tests are compared to values indirectly obtained using the equation for the CRD test. This comparison is presented in Figure 6 where the void ratio-hydraulic

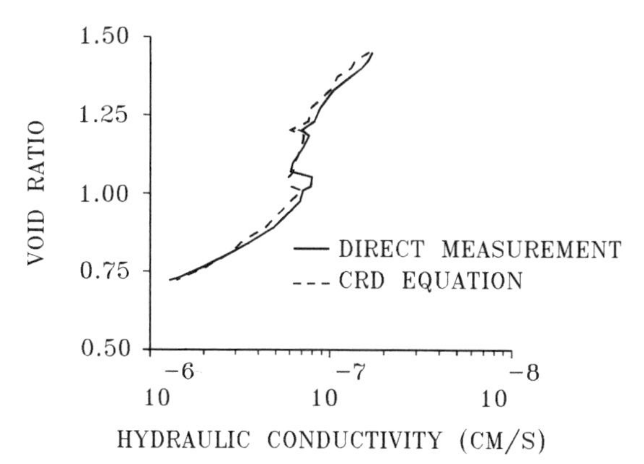

Figure 6. Comparison of Directly-Measured versus Back-Calculated Hydraulic Conductivity versus Void Ratio Relationships

conductivity relationships from each method are plotted. The results agree quite well except for local, minor variations. This agreement is expected since the critical assumption for both methods, constant void ratio across the sample, is nearly met during the test. The practical advantage of the direct method over the indirect method is that a minimum deformation rate is not required in order to develop sufficiently measureable pore pressures needed to indirectly determine hydraulic conductivity.

When compared with the more traditional step-loading/falling-head approach, the new method has several advantages: (1) lower hydraulic gradients are used and flow-rate measurements are eliminated, (2) strain-rate effects can be readily measured, (3) the testing time can be much shorter, and (4) a more detailed hydraulic conductivity-void ratio relationship is obtained.

CONCLUSIONS

The volume-controlled approach for directly measuring compressibility and hydraulic conductivity has several advantages over the traditional methods currently being used in practice. The method employs the technical advantages of both the CRD consolidation test and the flow-pump permeability test. These advantages include (1) consolidation at a constant rate of deformation, which enables the study of strain-rate effects on compressibility; (2) many direct measurements of hydraulic conductivity can be performed using the flow-pump method resulting in a detailed picture of permeability behavior, (3) direct flow rate measurements during the permeability tests are eliminated, which aviods their associated error, (4) the permeability tests can be performed at low gradients (typically less than 10), which minimizes the effect of seepage-induced consolidation (Pane et al. 1983); and (5) the minimum deformation rate required to develop measurable pore pressures and then to back calculate hydraulic conductivity is eliminated.

REFERENCES

1. Aiban, S. A., and Znidarcic, D., Evaluation of the Flow Pump and Constant Head Techniques for Permeability Measurements, Geotechnique, V.39, No. 4, 1989.p.655-666.

2. Crawford, C. B., Interpretation of the Consolidation Test, Journal of the Soil Mechanics and Foundations Division, ASCE, 90, SM 5, 1964, p. 87-102.

3. Crawford, C. B., On the Importance of Rate of Strain in the Consolidation Test." Geotechnical Testing Journal, GTJODJ, Vol 11, No. 1, March 1988, p. 60-62.

4. Janbu, N., Tokheim, O., Senneset K., Consolidation Tests with Continuous Loading, Proceedings of the Tenth International Conference on Soil Mechanics and Foundation Engineering, Stockholm, 1981, v. 35.

5. Kabbaj, M., Oka, F., Leroueil, S., Tavenas, F., Consolidation of Natural Clays and Laboratory Testing, Consolidation of Soils: Testing and Evaluation ASTM STP 892, R. N. Yong and F. C. Townsend, Eds., American Society for Testing and Materials, Philidelphia, 1986, p. 378-404.

6. Leroueil, S., Kabbaj, M., Tavenas, F. and Bouchard, R., Stress-Strain-Strain Rate Relation for the Compressibility of Sensitive Natural Clays, Geotechnique 35, no. 2, 1985, p. 159-180.

7. Lowe, J.,III, Jonas,E., and Obrician, V., Controlled Gradient Consolidation Test, Journal of the Soil Mechanics and Foundations Division, ASCE, 95, SM1, Proceedings Paper 6327, 1969, p. 77-97.

8. Olsen, H.W., Darcy's Law in Saturated Kaolinite, Water Resources Research, v.2,no.6,1966, p.287-295.

9. Olsen, H. W., Nichols, R. W., and Rice, T. L., Low Gradient Permeability Measurements in a Triaxial System. Geotechnique, v. 35, no. 2,1985 p. 145-147.

10. Pane, V., Croce, P., Znidarcic, D., Ko, H.Y., Olsen H.W., and Schiffman, R.L., Effects of consolidation on Permeability Measurements for soft clay, Geotechnique, v. 33, No. 1., 1983, p. 67-72.

11. Smith R. E., and Wahls, H. E., Consolidation Under Constant Rates of Strain, Journal of the Soil Mechanics and Foundations Division, ASCE 95, SM2, Proceedings Paper 6452, 1969 p. 519-539.

12. Tavenas, F., Jean, P., Leblond, P., and Leroueil, S., The permeability of natural soft clays. Part II: Permeability Characteristics, Can. Geotech. J., 20, 1983, p. 645-660.

13. Wissa, A. E. Z., Christian, J. T., Davis, E. H., and Heiberg, S., Analysis of Consolidation at Constant Strain Rate. Journal of the Soil Mechanics and Foundations Division, ASCE, 97, SM10, Proceedings Paper 8447, 1971, p. 1393-1413.

14. Znidarcic, D., Croce, P., Pane, V., Ko, H.-Y., Olsen, H.W., Schiffman, R. L., The Theory of One-Dimensional Consolidation of Saturated Clays: III. Existing Testing Procedures and Analyses Geotechnical Testing Journal, gtjodj, Vol. 7, No.3 Sept. 1984, p. 123-133.

15. Znidarcic, D., Schiffman, R.L., Pane, V., Croce, P., Ko H.-Y. & Olsen, H.W., The theory of one-dimensional consolidation of saturated clays: Part V, constant rate of deformation testing and analyses, Geotechnique 36, 1986, No. 2, p.227-237.

TRANSIENT DRAINAGE FROM UMTRA TAILINGS

Ned B. Larson[1] and Timothy J. Goering[2]

ABSTRACT: The Uranium Mill Tailings Remedial Action (UMTRA) Project is charged with the task of isolating radioactive tailings from the environment. This involves preventing both the surface and subsurface migration of contaminants to groundwater. The tailings and other contaminated materials are consolidated in covered disposal cells. A significant amount of effort has been made to design effective covers that will prevent surface migration of contaminants and to keep infiltration a minimum. These covers are fairly impermeable, and will generally have an annual infiltration flux averaging between 1E-7 and 1E-9 cm/sec. However, the initial transient drainage from the tailings may greatly exceed the moisture flux through the cover due to the wet nature of the tailings. Consequently, to methods to quantify the water draining from the tailings and to evaluate its impacts on groundwater quality are needed to assess compliance with the EPA's groundwater protection standards for the UMTRA Project. This paper discusses the UMTRA Project's approach to evaluate the transient drainage from tailings.

INTRODUCTION

The uranium mills at many of the UMTRA Project sites have been inactive for more than 20 years. During this time, the uranium mill tailings have been uncovered and exposed to precipitation. The shape and surface of the tailings piles at the end of milling operations generally has limited surface runoff of precipitation. Therefore, precipitation on the tailings piles either infiltrates into the tailings or evaporates. Because vegetation is almost nonexistent on the tailings piles, little or no transpiration occurs, and much of the water enters the tailings. This water often perches on lower permeability layers of slimes and fine-grained materials, resulting in a condition of near saturation in the tailings. When the tailings are nearly saturated, the moisture flux from the tailings approaches the saturated hydraulic conductivity of the tailings. However, as the tailings dry, their hydraulic conductivity decreases with time.

1 - Assistant Project Manager, UMTRA Project, Jacobs Engineering Group Inc., 5301 Central Ave. NE, Albuquerque, NM 87108

2 - Assistant Manager of Hydrology, UMTRA Project, Jacobs Engineering Group Inc., 5301 Central Ave. NE, Albuquerque, NM 87108

Therefore, the moisture flux from the tailings can be represented by the following decay function:

$$q = dW/dt = at^{-b}$$

where q is the flux, W is the water content, t the time, and a and b are empirical coefficients related to the boundary conditions and conductance properties of the soil (Hillel, 1980).

However, because empirical coefficients for different types of soil vary or have not been determined, analytical solutions to the transient drainage equation are not feasible. Therefore, a numerical approach is used to solve for moisture flux as a function of water content and time.

SOLUTION

Since this problem involves vertical drainage of the tailings seepage, it is often treated as a one-dimensional problem. However, a two-dimensional analysis may be required if material discontinuities, such as interbedded layers of sands and slimes, affect the flow direction. A number of computer programs exist for modeling unsaturated flow through a soil. Existing variably-saturated flow models were evaluated to assess which models were best suited for the problem, and the finite element model UNSAT2 (Neuman and Davis, 1983) was selected. This well-documented public domain computer program is readily available at a modest price.

Material Properties. The material properties required for the UNSAT2 program include porosity, residual moisture content, saturated hydraulic conductivity, and the relationship for the unsaturated hydraulic conductivity and the moisture content. This relationship was determined using empirical correlations from soil characteristic curves determined by laboratory testing. The relationship between the hydraulic conductivity and soil moisture suction can be estimated (Maulem 1976) from the relationship between soil moisture suction and moisture content.

The soil moisture characteristic curve is measured in the laboratory. The shape of the curve depends on whether the soil is wetting or drying, a phenomenon known as hysteresis. Since this analysis is only concerned with the drying of the tailings, the effects of the wetting portion of hysteresis can be ignored. The drying soil moisture curve is typically obtained using the pressure plate test (ASTM D2325 and ASTM D3152). This data is reported as the resulting moisture content of the soil at various pressures (Table 1). Because this moisture content data is reported at various pressures, it is discontinuous and does not produce a smooth and continuous curve. To produce such a curve, Van Genuchten (1985) proposed the use of curve fitting techniques. The curve fitting yields two parameters called alpha and N, which describe the curve. Table 2 shows typical soil and moisture parameters for different tailings types.

Figure 1 shows soil tension versus volumetric moisture content for tailings sands, sand-slimes, and slimes plotted and then regressed using the Van Genuchten (1985) techniques. Figure 2 shows the same data for each of the materials but the percent saturation is plotted instead of the moisture content. The regressed curve produces a smooth, continuous representation of the soil moisture characteristic model. By transforming this line with a numerical function, the hydraulic conductivity can also be calculated with a smooth function. Figure 3 shows the unsaturated hydraulic conductivity versus moisture content for each type of material, and Figure 4 shows the hydraulic conductivity versus percent saturation. This data was analyzed using a computer program named RETC, written by Van Genuchten (1985).

Table 1. Typical laboratory data for the relationship between soil tension and volumetric moisture content.

OBSERVATION NO.	Soil Tension (cm)	Volumetric moisture content
1	7807.600	0.0830
2	3218.300	0.0860
3	1058.500	0.0900
4	515.700	0.0950
5	317.600	0.1090
6	136.500	0.1490
7	62.200	0.1960
8	30.800	0.2550
9	15.000	0.3320
10	7.800	0.3560
11	2.000	0.3840
12	1.000	0.3870

Table 2. Example Van Genuchten curve-fitting parameters

Tailings Material Type	Saturated Moisture Content*	Residual Moisture Content*	Hydraulic Conductivity (cm/sec)	Alpha	N
Sands	0.3857	0.0779	9.80 E-4	0.0517	1.7874
Sand-Slimes	0.4032	0.0355	3.35 E-4	0.0244	1.7058
Slimes	0.6522	0.3066	1.84 E-4	0.0153	1.3119

*All moisture contents presented are volumetric

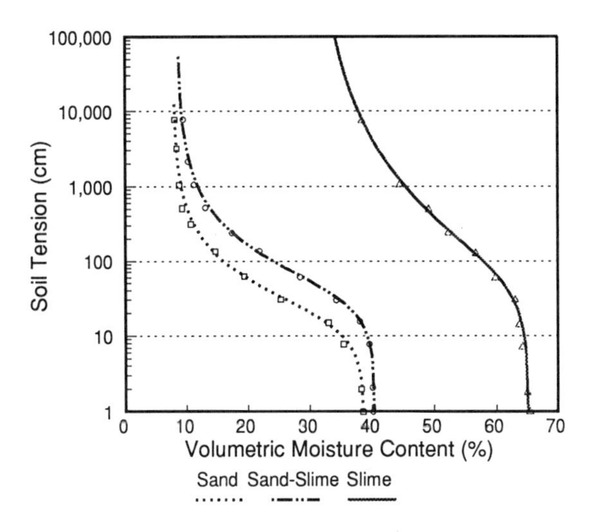

Figure 1. Soil tension versus volumetric moisture content

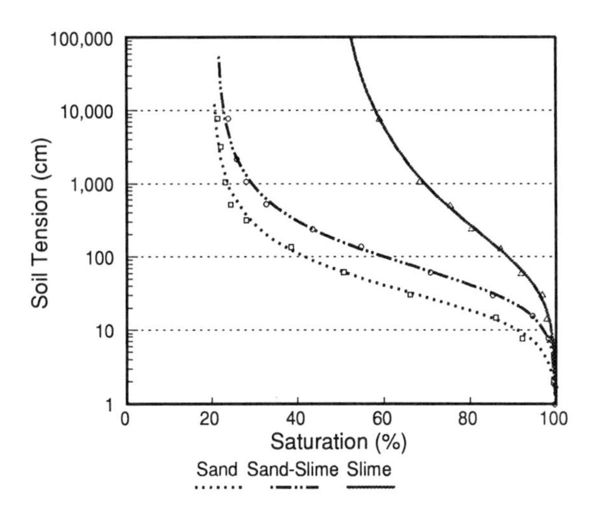

Figure 2. Soil tension versus percent saturation

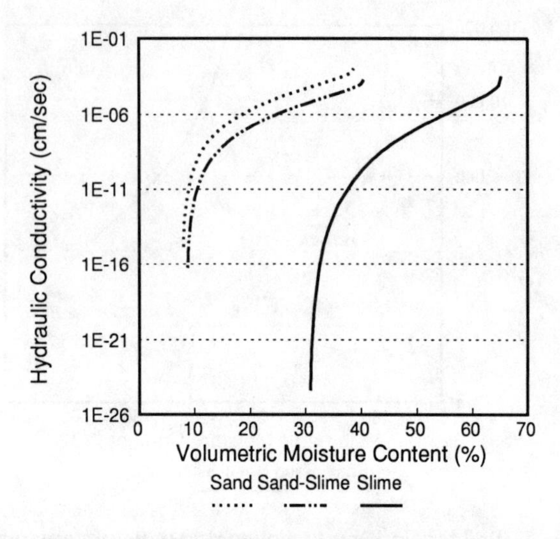

Figure 3. Unsaturated hydraulic conductivity versus moisture content

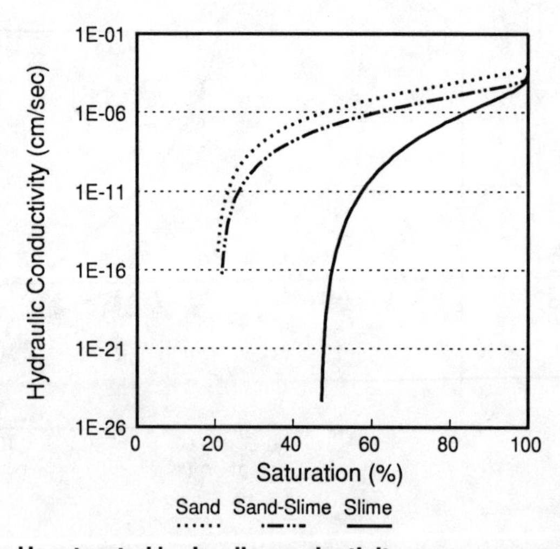

Figure 4. Unsaturated hydraulic conductivity versus percent saturation

ANALYSIS

Because the UNSAT2 computer program uses the finite element method to solve the governing equations, a finite element grid was set up that represented the soil conditions and pile geometry being analyzed. An example of a two-dimensional grid that was used for the Durango, Colorado, tailings site is shown on Figure 5.

Boundary Conditions. The boundary conditions for the analyses require estimating the flux through the constructed cover. Depending on the layers and materials used to construct the cover, this flux, which serves as the upper boundary condition, generally ranges from 1E-7 to 1E-9 cm/sec. The lower flux value was used when bentonite was incorporated into the cover design.

The initial soil tension in the tailings is another parameter required for the analysis. This is calculated by determining the moisture content of the tailings following construction, and then selecting the corresponding soil tension from the soil moisture characteristic curve. By knowing the initial moisture content, the initial tension in the tailings can be determined, as shown on Figure 6. If the tailings were transported and then compacted, the initial moisture content is assumed to be the moisture content at compaction. If the tailings were left in place and a cover constructed over them, the initial moisture content is assumed to be the in-situ moisture content of the tailings.

The lower boundary condition of the finite element grid was treated as gravity drainage that allows the water to freely exit the system being modeled. The water leaving the system is reported both in terms of flux and cumulative drainage.

RESULTS

The computer analyses were performed on an 80386 personal computer workstation. Even though the simulation period was 500 years, the analysis required a relatively short period of time due to the speed of the workstation. The model output was plotted to simplify interpretation.

Figure 7 shows moisture flux from the tailings with time for the Durango tailings pile. In general, the flux and time of transient drainage are direct functions of the relative percent of fine-grained materials in the tailings. The sandier the tailings, the faster the water will drain and the clayier the tailings, the slower the drainage process will be. As can be seen from Figure 7, most of the excess water in the tailings drains within 150 years. Once a steady state is achieved, the water leaving the bottom of the pile is equal to the water entering the pile through the cover.

Figure 8 shows the cumulative outflow of water leaving the pile with time. Since the flux of water leaving the pile decreases with time, the cumulative flux leaving the pile reflects this reduction of flux with time.

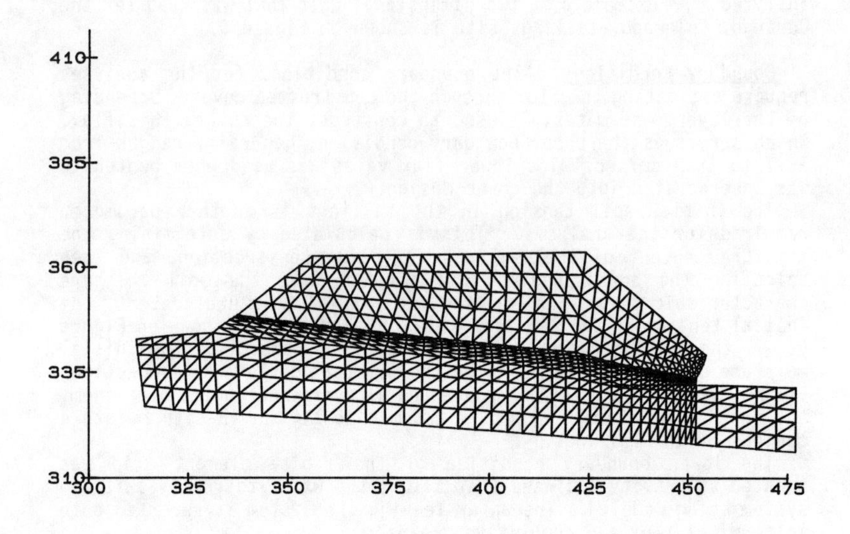

FIGURE 5. Finite elements grid for the Durango tailings pile

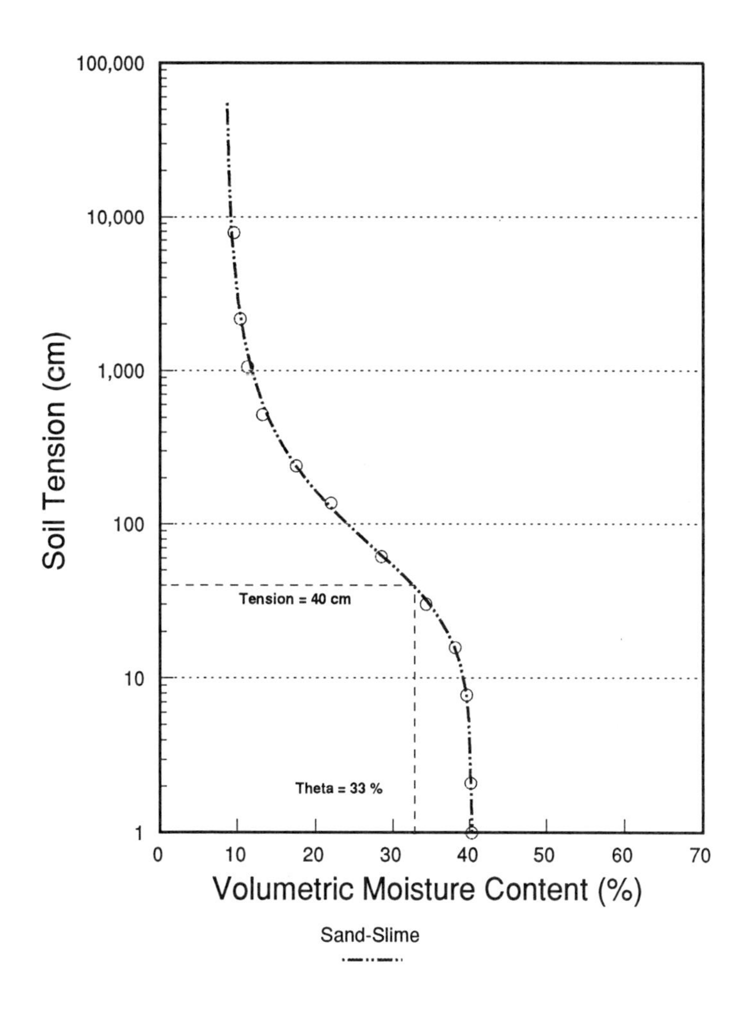

Figure 6. Determining initial suction from the moisture content

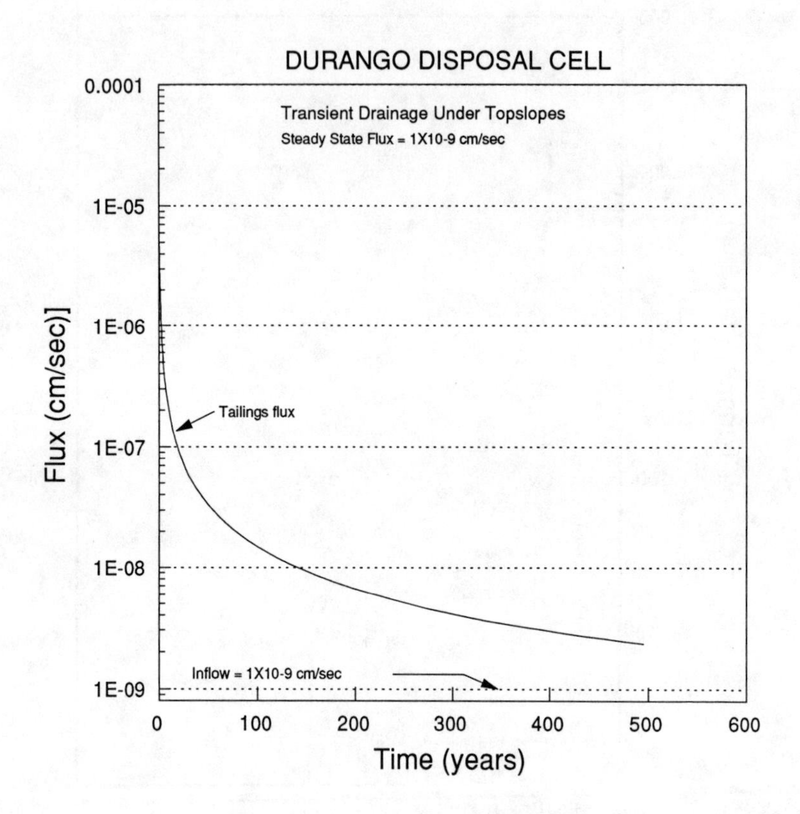

Figure 7. Flux versus time from the Durango tailings

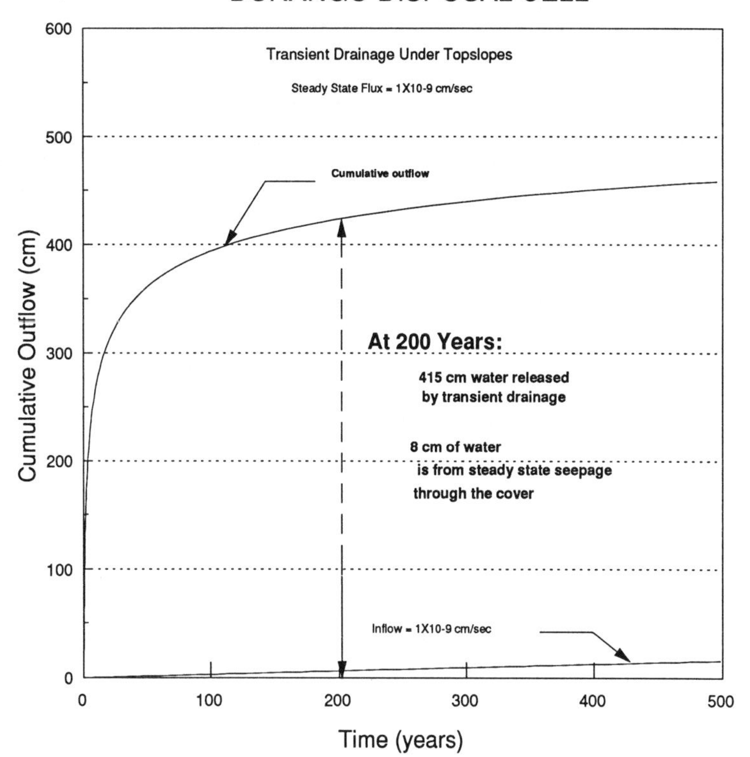

Figure 8. Cumulative outflow of water from the Durango tailings pile with time

CONCLUSIONS

As can be seen in the previous figures, the flux leaving the tailings piles can greatly exceed the long-term steady state flux through the disposal cell cover. Consequently, this seepage can have a consequently greater impact on the groundwater quality than the steady state seepage from the tailings. Transient drainage must be considered when analyzing environmental impacts and demonstrating compliance with the EPA groundwater protection standards.

In the cases where the tailings are relocated to another disposal site, the tailings can be dried to reduce the amount of water that is available to drain out of the pile. The additional effort of drying the tailings after they have been excavated may be significant, and drying can be a costly process. However, the costs are often justified, as controlling or eliminating the transient drainage may be necessary to protect groundwater quality.

At the sites where the tailings are left in place, the alternatives are much more limited. Not only will the tailings drain water over an extended period of time, but once the cover is placed, additional water will be released due to consolidation caused by the weight of the cover. However, the additional cost of excavating, drying, stockpiling, and then replacing the tailings in the original footprint is generally far too great to be considered as a remedial action. In these conditions, the magnitude and duration of the transient drainage are still evaluated to determine the effects of the drainage on the existing groundwater contamination from the tailings pile.

REFERENCES

1. L.A. Davis and S.P. Neuman, Documentation and User's Guide: UNSAT2-Variably Saturated Flow Model, Prepared for Division of Waste Management Office of Nuclear Material Safety and Safeguards, USNRC, Washington D.C. NUREG/CR-3390 WWL/TM-1791, 1983.
2. D. Hillel, Applications of Soil Physics, Academic Press, New York, 1980, 385 p.
3. Y. Mualem, A new model for predicting the hydraulic conductivity of unsaturated porous media, Water Resources Research, 12 (3), 1976, 513-522.
4. M. Van Genuchten, On describing and predicting the hydraulic properties of unsaturated soils, Annals Geophysicae, 3(5), 1985, 615-628.

HYDROCARBON WASTE STABILIZED WITH A CEMENTED CLAY MATRIX

Joseph P. Martin[1], M. ASCE, John S. Browning III[2], A.M. ASCE and Mary Ann Susavidge[3], S.M. ASCE

ABSTRACT: A hydrocarbon sludge is to be stabilized by microencapsulation in a clay matrix cemented with a pozzolanic lime-fly ash mixture. Goals for the landfill monolith include dimensional stability, low leachate production rates, and contaminant immobilization. Volumetric constraints require that the stabilized mixture have high porosity to contain the sludge.

The basic mixture determined in other studies is sludge, clay, fly ash, and lime in a 1.0/0.75/0.75/0.25 weight proportioning, respectively. The focus herein is to optimize lime replacement with portland cement to better handle a more liquid sludge. Small amounts of cement improved strength, but high lime replacements adversely affected immobilization. Low hydraulic conductivity and durability could not be achieved, such that an insulative and impermeable landfill cap is necessary. The effects of compactive effort and leaching time are also addressed.

INTRODUCTION

Acidic sludge has been stored in lagoons at a site underlain by residual silty clay and gneiss. While minimal offsite impact has been detected from the decades of storage, continued exposure is unacceptable. Stabilization and redeposition in onsite landfill monoliths is one solution to the problem of providing waste disposal space while minimizing the potential for uncontrolled contaminant release.

Sludge solidification often employs hydraulic cements or industrial byproducts such as fly ash or kiln dust in

---

1 - Assoc. Prof.of Civil Engineering, Drexel University, Philadelphia, PA 19104.
2 - Sr. Engr. NTN Consultants, Exton, PA 19341
3 - Grad. Stnt., Civil Engineering, Drexel University, Philadelphia, PA 19104.

pozzolanic mixtures (10,12,14). The result is a uniform and dimensionally stable subgrade for the final cap (11). Solidification also provides some degree of contaminant immobilization, supplementing the external containment. The mechanism could be simply physical isolation in an impermeable matrix (microencapsulation), but chemical fixation may also occur, such as by neutralization, precipitation, partitioning and sorption (6,13).

This sludge is an emulsion with a wide range of low volatility compounds. Solidification with portland cements and pozzolans only gave poor results. However, encapsulation in a clay matrix bound with a lime-fly ash admixture produced a medium stiff consistency, low permeability, and low carbon solubility.

The site has limited capacity or "air space" over the lagoons, such that the sludge-filled porosity of the stabilization product had to be in the range of 40-45%. The clay structure is expanded to accomodate sludge, with the cement compensating for the disturbance. The properties of the sludge-clay-pozzolan mixtures depend upon component proportioning, moisture content, age, sludge consistency and compactive effort (1,7). These factors were varied in an empirical study to yield soil-like mixtures that produced basic effluent with low carbon content when permeated with distilled water or dilute acidic solvents (8). A field study demonstrated the practicality of the method (15).

DESIGN CRITERIA

Characterization of a stabilization technique often centers on a few tests: unconfined compression, leaching, and hydraulic conductivity or permeability. These indicate mechanical, chemical fixation, and contaminant transport restriction, respectively (16). Durability tests also indicate resistance to matrix deterioration.

As this sludge is not hazardous, the design can be site-specific, with some semi-quantitative criteria:

- Fit in the available space
- Provide uniform support for the cap
- Support heavy equipment placing overburden layers
- Limit air emissions during stabilization
- Restrict permeation through the stabilized mass
- Produce neutral leachate with low organic carbon

Analysis of in-situ conditions (Figure 1) provides other insights. The maximum overburden stresses will be below 5,000 psf, but a stiff product is still desirable. While low permeability is necessary, it will tend to retard consolidation. If the mixture is saturated, only restricting construction rates will minimize post-closure settlement. It is thus desirable that the mixture be unsaturated. Providing an unsaturated condition at the end of primary consolidation will minimize long-term fluid expulsion by secondary consolidation.

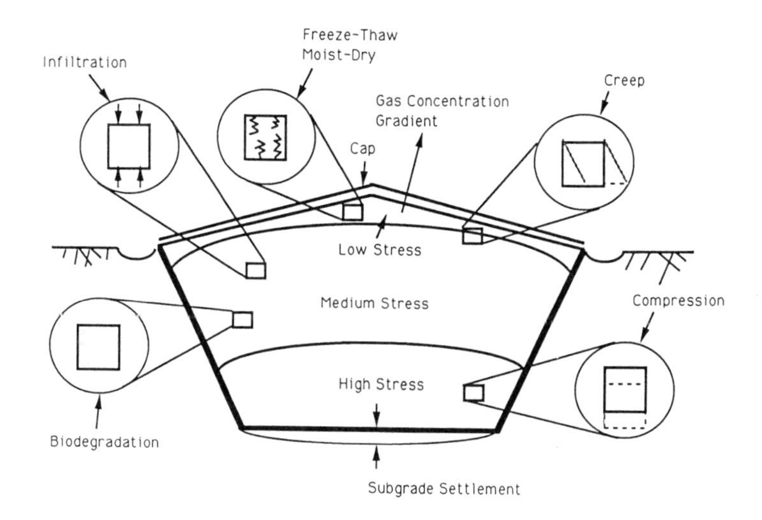

Figure 1. In-situ Environment of a Stabilized Landfill

The stiffness of this clay-like material is more critical than its strength. The unconfined compression test is useful for optimizing proportioning and mixing time (10,14). High strength might also indicate development of an entrapping structure. However, a strength of 50 psi (340 kN/M^2) that is often specified for stabilized hazardous wastes isn't required for stability of the Figure 1 scenario (1). A more realistic criteria for this project was an as-compacted strength of 5-10 psi,and rapid strength gain to support overburden.

There is no reason to expect that a cemented clay will be durable over the life of the waste fill (e.g. "forever"). Thus, it was accepted that an insulative as well as an impermeable cap would be necessary.

The lack of impact on local groundwater after decades of sludge storage in unlined lagoons implied low permeability and high sorptive capacity in the natural clay subgrade (13). This clay is also a workable construction material. It was arbitrarily decided to attempt to match or exceed the geotechnical properties (strength. stiffness, permeability) of the compacted onsite clay. This approach worked well for most of the viscous sludge batches, using a clay/fly ash/lime mixture(2,9), but did not provide sufficient strength for the some of the more liquid sludges. A portland cement "booster" appeared to be necessary.

WASTE MATERIALS

Sludge

The sludge consists primarily of long-chain aliphatics and polycyclic aromatics. It passes both EP Toxicity and TCLP tests in terms of hazardous material concentrations. Hazardous volatiles (Boiling point < 100°C) were not detected above the micrograms/liter (ppb) range. Sulfur and semivolatile and nonvolatile hazardous constituents are present up to several hundred ppm. Only traces of chlorinated compounds or heavy metals were detected.

The sludge from the most liquid lagoon used in all work reported herein has an ash content of 4.5% and an organic carbon content (TOC) of 35.3%. Most of it displays a consistency similar to that of a slow-curing roadway asphalt. To indicate a relative mobility, the sludge was split into "solid" and "liquid" portions, based upon the fraction volatilized from a stabilized mixture at 105°C.

Native Clay

The onsite soil is a clayey silt, derived from weathering of the gneiss bedrock. X-ray diffraction tests indicate that the predominant clay-size mineral is kaolinite. Index properties are shown on Table 1. The Standard Proctor maximum dry unit weight is 110 lb/ft^3, with an optimum moisture content of 16%.

TABLE 1

Clay Index Properties

Property	Value
Cation exchange capacity (EPA method 9081)	49 meq/gm
Specific gravity	2.72
Liquid limit	31%
Plastic limit	22%
Shrinkage limit	12%
% finer than #200 mesh	86%
% finer than 0.002 mm	6%

The natural clay exists as readily dried and pulverized loamy aggregations. An important issue is the mixing and compactive effort applied to the stabilized material. Permeability and compressibility studies of the clay were thus done at the two extremes of the structure in which the sludge could be encapsulated (9). Undisturbed samples showed very high compressibility, and a permeability of 4.5×10^{-5} cm/sec in a flexwall permeameter at low confining stresses (5-10 psi). Compaction to about 90% standard proctor unit weight at a moisture content of 20% produced a stiffer material, with a compression index (C_c) of 0.20 and a recompression index (C_r) of 0.04. The coefficient of consolidation (C_v) was about 0.011 in^2/min in the 1000-5000 psf stress range. Permeability was reduced by compaction, and fixed-wall tests at 20% moisture showed 3×10^{-8} cm/sec.

Unconfined compresion tests showed some thixotrophy. Remolded and compacted samples had strengths in the range of 12 psi (80 kN/M^2) in the 15-20% moisture range, increasing 50% to 75% with 30 day's curing.

Additives

Lime neutralizes, reduces clay plasticity, and is a component in the pozzolanic reaction (17). It also appears to condition the clay to immobilize the sludge. Hydrated dolomitic lime was used to minimize heat and volatilization, as air quality will probably govern the project permitting.

Type F fly ash (not self-cementing) was obtained from a nearby powerplant. It serves as a moisture absorbent to improve the blending of sludge and clay, and then participates in the pozzolanic cementing. Fly ash also appears to serve as a continuing source of alkalinity during permeation with acidic solvents, thus improving the longevity of the stabilization process.

The fly ash used has a specific gravity of 2.54, a median grain size (d_{50}) of 0.03mm, and 14.6% unburnt carbon content (Loss on Ignition test). The Standard Proctor dry unit weight was 83 lb/ft^3.

To establish the best-case scenario, the lime and the fly ash were blended in several ratios over a range of water contents from 10% to 50%. Unconfined compressive strengths of 75 psi (500 kN/m^2) and higher at moisture contents of 25% to 35% were obtained in 30 days. A lime-fly ash blend in the ratio of 1:3 was chosen for the stabilization, but addition of small amounts of portland cement to this blend substantially increased strength.

BASIC MIXTURE

In the empirical studies, mixtures were described by relative weight proportions in the order:

acidic sludge/clay/fly ash/hydrated lime

The clay proportion is on a dry weight basis. Other variables are the consistency of the raw sludge, the clay water content, and the moisture content of the complete mixture measured with 105°C oven drying. The moisture content includes volatile ("liquid") organic material as well as water. Jar tests indicated that organic fixation was not very sensitive to the clay moisture content. Consequently, the focus of clay moisture content control was to obtain workable fresh mixtures.

The optimum mixture for most of the sludge is:

1.0 sludge/0.75 native clay/0.75 fly ash/0.25 lime

The resulting fresh mixture had 11.1% moisture content (24 hours oven drying), with the source clay at 17% moisture. Compaction in a Proctor mold at both 100% and 50% Standard Proctor effort yielded moist unit weights of 98 lb/ft^3 (15.4 kN/M^3) and 91 lb/ft^3 (14.8 kN/M^3) respectively. Strength samples were compacted at 100% Proctor effort in teflon molds, extruded, and cured in sealed containers. Lower compaction effort did not yield consistent results. A typical sludge batch showed initial strength about 10 psi, increasing to 16 psi (75 kN/M^3) at 30 days, and 22 psi (115 kN/M^3) at 60 days. Samples cured in the molds showed 40% higher strength (10).

Other properties such as compressibility were similar to those of the compacted native clay. A permeability of 2 x 10^{-5} cm/sec was obtained in the plexiglass permeameters with light compaction, decreasing to 3.5 x 10^{-7} cm/sec with 100% proctor effort. All tests were run after 14 days of curing with a thin water film on the surface of the sample to avoid dessication. Permeability results were insensitive to gradients up to 60 cm/cm.

SLUDGE CONSISTENCY EFFECTS

Lower strengths were obtained with more liquid sludge, even when the clay moisture content was decreased to 11%. Therefore, some lime was replaced with type II portland cement, still maintaining the 0.25 "parts" proportion. Figure 2 shows results of varying the lime/cement ratio, with 100/0 representing no cement substitution and 70/30 representing 30% replacement of the lime.

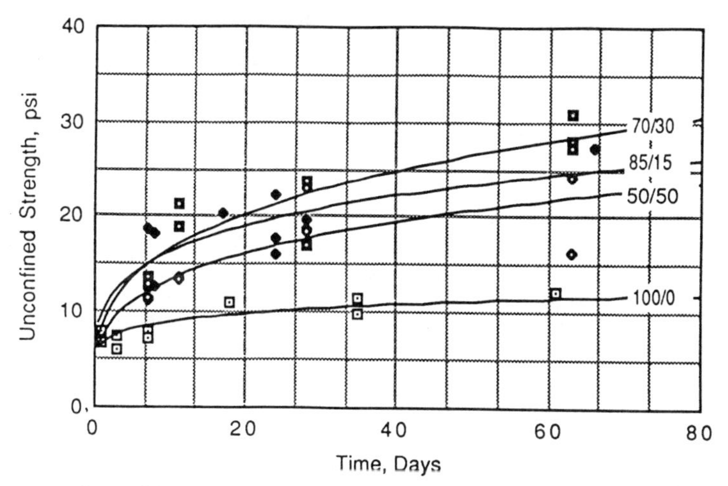

Figure 2. Effect of Cement Addition on Strength Gain

The 85/15 lime:cement (4% cement overall) was the optimum mixture. Apparently, this is as far as the cement can affect the development of structure without running into a shortage of water for hydration. However, it also appears that the early strength gain provided by the portland cement has a symbiotic effect in allowing the slower pozzolanic reaction to proceed.

Using drier clay and adding cement improved compaction. A moist unit weight of 100 lb/ft^3 was obtained with 100% Proctor effort. The fresh mixture moisture content (water and volatiles) water was 22%, indicating that the sludge was much more liquid than that in earlier investigations. The specific gravity of the solids calculated from the original solid components was 2.3.

TABLE 2
Mixture Components by Volume (Unreacted)

Solids	
Silty clay	14.4%
Fly ash	16.9%
Lime & Cement	4.8%
Voids	
Sludge "Solid"	21.1%
Sludge "Liquid"	28.6%
Water	6.5%
Air	7.7%

The computed porosity is 63%, but some sludge acts like a solid in the hardened material. Calculations show 21% of the volume is occupied by "solid" sludge, not volatilized with 24 hours of 105°C drying. Total sludge filled porosity is about 49%, yielding a twofold swell over its present volume. The fresh mixture includes only about 4% water from the clay and fly ash, but more was contributed by sludge, or produced by the neutralization reaction between hydrated lime and acids.

The original 0.25 lime proportion had provided the best carbon fixation. Jar tests indicated that replacing more than about 20% of the lime with cement increased mobilization, implying little contribution from the cement. Hence, the mixture for the low viscosity sludge was set at the 85/15 lime/cement ratio, compatible with strength optimization (Fig.2). The net proportioning is:

1.0/0.75/0.75/0.21/0.04

Sludge/Native Clay/Fly Ash/Hydrated Lime/Type II Cement

IMMOBILIZATION RESULTS

Permeability/Hydraulic Conductivity

200 g of stabilized mixture was compacted in 2.5" diameter falling-head permeameters and cured for two weeks. Trial results were similar with dilute sulfuric acid, dilute acetic acid, and distilled water permeants, with pH values of 2.5, 4, and 6, respectively (13). Thereafter, only distilled water was used.

One set of permeameters were packed with the no cement mix, each sample receiving 20 blows/lift with a Harvard miniature hammer. Gradients were controlled to limit velocity and detention time variations. One sample was subjected to a gradient varying between 26 and 37 cm/cm, and the other had a range from 13 to 24. The former displayed a permeability of 2.1×10^{-5} cm/sec, while the latter was 3.2×10^{-5} cm/sec. This is well within experimental scatter.

Another set of tests were run on the optimized cement content mixture, 1.0/0.75/0.75/0.21/0.04, with the higher gradient range. One sample received 20 blows/lift and the other 10 blows/lift. The lightly compacted sample had a permeability of 6.0×10^{-5} cm/sec, while the slightly denser sample showed 1.0×10^{-5} cm/sec. While very low permeabilities (10^{-7}) were not obtained, infiltration will be better restricted with compactive remolding.

Permeation Effluent Quality

Figures 3 and 4 show the effluent quality results. The vertical axis represents the percent of the carbon in the sample mobilized, as indexed by Total Organic Carbon (TOC) of the effluent or leachate.

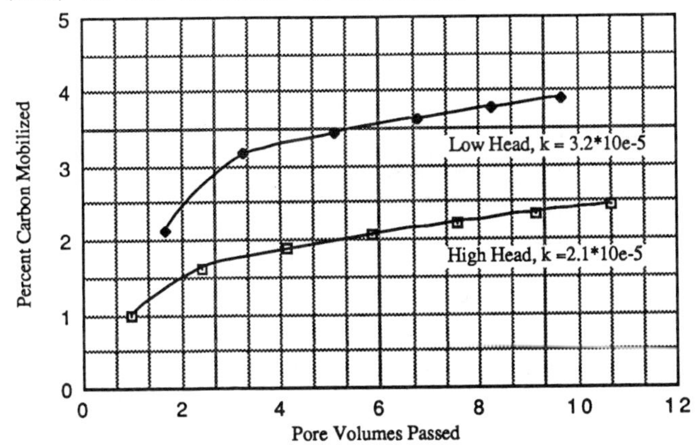

Figure 3. Accumulated Effluent Quality of Sludge Stabilized Without Cement; Gradient Effects

It can be seen that the immobilization is reasonably successful. Less than 4% of the sludge was mobilized at throughputs that could represent many decades of infiltration. The variation in results is most readily explained by permeant hydraulic detention times. The implication is that the rate-determining step in contaminant release is carbon dissolution from where it is entrapped or absorbed in the pore structure, not its transport by advection and diffusion in main channels. The cement addition showed no effect on immobilization.

The highly compacted cemented sample was permeated to almost 40 pore volumes. Effluent quality improved after a "first flush" to about 200 mg/l after 6 pore volumes.

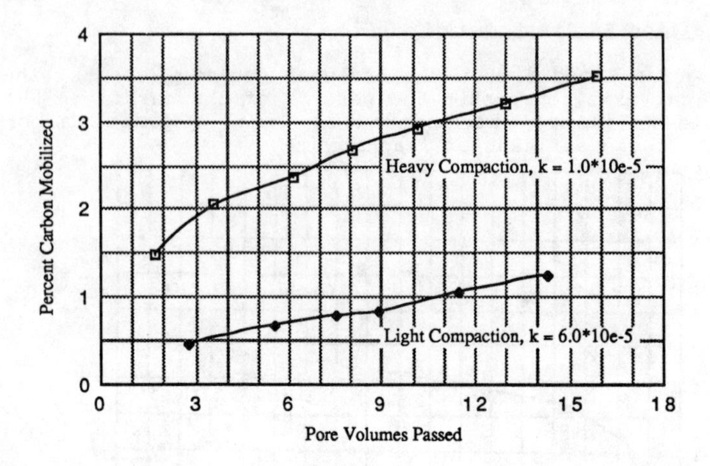

Figure 4. Accumulated Effluent Quality of Sludge Stabilized With Cement: Compaction Effects

CONSOLIDATION/COMPRESSION

Cutter rings were pressed into material compacted in Proctor molds. The 1" high, 2.5" diameter samples were cured two weeks before loading. Tests on earlier mixtures showed minimal difference between load doubling and fixed-increment loading (9). Each increment was retained for 48 hours to examine secondary consolidation.

Figure 5 shows compressive and unloading stress vs. strains for the two mixtures, with and without the 4% portland cement (2). There is no noticeable difference in compressibility, in contrast to cement effects on the shear strength shown on Figure 2.

Primary consolidation can generate leachate, but this is unlikely if the as-compacted saturation is in the range of 85% or less. Strains of 5% to 6% are expected in the lower layers under overburden stresses of about 5000 psf which will only expel air in the pores. Consolidation samples were saturated, and the pozzolanic hydration was ongoing. The test results are thus probably conservative.

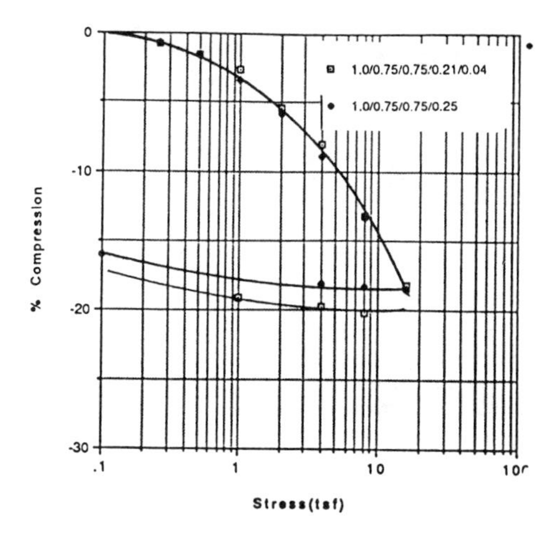

Figure 5. One Dimensional Compression Results

Analysis of consolidation behavior to separate primary and secondary consolidation was hampered by the type III time curves (5) that do not inflect at 100% primary consolidation. Creep effects are manifest early in the deformtion. Using Su's method as described in (4) to determine completion of primary settlement, coefficients of consolidiation of about 4 to 5 x 10^{-5} cm^2/sec were computed for both mixtures in the 2000-6000 psf range. The secondary consolidation index (in strain terms) was computed to be in the range of 0.01 to 0.015, in the medium to high range. Post-construction subsidence is to be expected, but if the cap sideslopes are fairly flat, subsidence will probably be fairly uniform and thus not seriously affect cap stresses or alignment.

SUMMARY

This project involves stabilization and redeposition of a non-hazardous hydrocarbon sludge in permanent, capped landfill monoliths. Site constraints require high sludge-filled porosity. A procedure to encapsulate the sludge in a pozzolan-cemented natural clay matrix was developed. Empirical work established optimum component proportionings, which worked well for most onsite sludges in terms of producing a solid with mechanical properties similar to a stiff clay, low permeability, and high capacity to immobilize the sludge.

However, the goal of uniformity was complicated by the variable sludge consistency. An attempt to compensate by replacing some of the lime with portland cement and using a drier clay was partly successful. With only 4% cement unconfined compressive strength was greatly increased, and the unit weight was marginally improved. However, compressibility, permeability or ability to chemically fixate the hydrocarbons were not significantly affected. After a "first flush" immobilization was still achieved.

CONCLUSIONS

For this type of stabilization, the unconfined compressive strength is not a good indicatior of other properties, such as permeability and compressibility, where a correlation might be expected to exist. Cement addition has the most value in improving construction conditions, but will probably not influence post-closure behavior. These limitations can be considered in the design, primarily with cap design to insulate, restrict infiltration, and accept some post-closure subsidence.

Key influences were the structure of the clay matrix and the interference on cement reactions by the sludge. The cemented clays encapsulate and also fixate the waste by complex and interrelated mechanisms, including lime conditioning of clay. The empirical investigations show that it isn't necessary to fully understand internal mechanisms to be able to predict mechanical, seepage and contaminant migration behavior by customary engineering principles and methods.

ACKNOWLEDGEMENTS

The work was sponsored by Sun Refining & Marketing Co., Dr.Arthur J.Raymond, Project Director. Most of the data in the latter half of the paper is from Reference (1).

REFERENCES

(1) JS Browning III,"Stabilization and Solidification of a Hydrocarbon Refining Sludge: Engineering Optimization and Performance Analysis" M.S.C.E. Thesis, Drexel Univ, Phila PA, 1990

(2) JS Browning III and FJ Biehl "Evaluation and Analysis for Subsidence of Stabilized Sludge" Proc. 22nd Mid-Atlantic Industrial Waste Conf., Philadelphia. PA. July 1990, pp 594-609

(3) MJ Cullinane, "An Assessment of Materials that Interfere with Solidification/Stabilization Processes" Final Report for U.S. EPA IAG No.SW-219306080-01-0 (1988)

(4) BM Das, Advanced Soil Mechanics McGraw-Hill, New York, 1983

(5) GA Leonards and AG Altschaeffl "Compressibility of Clay" Jour. of Soil Mech. & Found. Div., ASCE 90 (SM5), 1964

(6) PG Malone and RJ Larson, "Scientific Basis for Hazardous Waste Immobilization", Hazardous and Industrial Solid Waste Testing: Second Symp., ASTM STP 805, R.A.Conway and W.P.Gulledge, eds., 1983.

(7) JP Martin, AJ Felser and EL Van Keuren, "Hydrocarbon Waste Stabilization for Landfills", Proc. ASCE Specialty Conf. for Waste Disposal, Ann Arbor MI, pp June 1987.

(8) JP Martin, JS Browning III, K Adams, and WT Robinson "Modeling Mobilization from Stabilized Refinery Waste Deposits by Sequential Leaching" Proc Petroleum Hydrocarbons and Organic Chemicals in Ground Water, NWWA-API, Houston TX, Nov, 1989

(9) JP Martin, FJ Biehl, JS Browning III, and EL Van Keuren,"Constitutive Behavior of Clay and Pozzolan-Stabilized Hydrocarbon Refining Waste" Proc. Geotechnics of Waste Fills, ASTM STP 1070,A. Landva and G.D. Knowles, Eds. San Francisco, CA, June 1990

(10) DS Morgan, JI Novoa and AH Halff, "Oil Sludge Solidification using Cement Kiln Dust", J. Env. Eng. Div., ASCE 110 (EE5), pp 935-949, 1984.

(11) WL Murphy and PA Gilbert "Estimation of Maximum Cover Subsidence Expected in Hazardous Waste Landfills". Proc., 10th Ann. Research Symp., U. S. EPA, April 1984

(12) CL Smith and DJ Frost, "Secure Landfilling with Pozzolanic Cementing", Proc. 1st Ann. Conf. on Hazardous Waste Mgmt., Phila, Pa., pp 153-160, 1983.

(13) WT Robinson, "Characterizing the Leaching Potential of Hydrocarbon Wastes from a Stabilized Mixture" M.S. Environ. Engineering Thesis, Drexel Univ, Phila PA, 1987

(14) U.S.EPA "Handbook for Stabilization/Solidification of Hazardous Wastes" EPA/540/2-86/001, 1986

(15) EL Van Keuren, J Martino, J Martin and A.Defalco, "Pilot Field Study of Refinery Waste Stabilization" Proc. 19th Mid-Atlantic Industrial Waste Conf., Bucknell Univ., Lewisburg, PA, June 1981.

(16) WC Webster, "Role of Fixation Practices in the Disposal of Wastes" ASTM Standardization News, 1984.

(17) HT Winterkorn, "Soil Stabilization", Foundation Eng. Handbook, H.F. Winterkorn and H.Y. Fang, Eds., Van Nostrand Reinhold 1975.

FLOW OF SURFACTANT FLUID IN NONAQUEOUS PHASE LIQUID-SATURATED SOILS DURING REMEDIAL MEASURES

Raymond N. Yong, [1], M.ASCE, Abdel-Mohsen O. Mohamed[2] and Diaa S. El Monayeri[3]

ABSTRACT: The flow of surfactant fluid in nonaqueous phase liquid (NAPL) - saturated soil has been studied experimentally. It is seen that as the temperature increased, the percentage of NAPL removed from the sample increased. The experimental investigation has shown that the clay ratio in the NAPL/soil mixture affects the surfactant flow rate. This can be attributed to the change in sample porosity with time and also to increasing clay/NAPL ratio. The porosity profile along the pathlines changed due to migration of NAPL and clay located during the surfactant injection process. Furthermore,the rate of NAPL and clay migration increased with increasing surfactant inlet pressure head. This was true, however, only to a certain inlet pressure head after which these changes leveled off.

INTRODUCTION

Nonaqueous phase liquids (NAPL), such as organic solvents and petroleum hydrocarbons, are presently some of the most frequent groundwater contaminants. Spills, leaking underground tanks, and fuel pipelines are nearly everyday occurrences.

The current remedial measures for NAPL can be divided into soil remediation and aquifer remediation. As explained by Nyer and Sklandany (1989), soil remediation techniques are: (1) soil excavation and off-site disposal; (2) in-situ soil venting; (3) in-situ biodegradation; and (4) above ground or in-situ chemical oxidation. During soil excavation large amounts of contaminants can volatize which can lead to significant air pollution. Experience has shown that excavation and disposal of large amounts of soil is very costly and becomes more and more economically limited. Soil venting on the other hand

[1]William Scott Professor of Civil Engineering and Applied Mechanics, Director, Geotechincal Research Centre, McGill University, Montreal, Canada, H3A 2K6

[2]Research Associate, Geotechnical Research Centre, Adjunt Professor, Civil Engineering Dept., McGill University, Montreal, Canada, H3A 2K6

[3]Professor, Dept. of Civil Engineering, El Monofia University, El Zagazek, Egypt.

relies on moving air past the contaminated to induce volatization. Efficiency is related to the vapor pressure and boiling point of the contaminant. Not all compounds can be dealt with in this way.

However, all of the compounds found in NAPL can be degraded by bacteria. This method of bioremediation involves adding nutrients and oxygen to the soil to induce bacterial proliferation. As shown by Wilson and Brown (1989), the technique can target both the adsorbed and dissolved phases. Bacteria have the capability to attach themselves directly onto droplets of contaminants held within the porous matrix. This can occur because many of these NAPLs use a source of oxygen (usually hydrogen peroxide and catalyst) to destroy the hydrocarbons present.

Aquifer remediation can be achieved by: (1) pumping and treating the water on-site; and (2) in-situ bioremediation. Dissolved petroleum compounds can be pumped and treated on-sites using a wide range of technologies such as air-stripping, carbon adsorption, and biological treatment.

Further research is needed to address the problems encountered in in-situ pumping and extraction - if efficient and economic long-term solutions to groundwater contamination problems from NAPL chemicals is to be achieved. This experimental study examines the flow of surfactant fluid in NAPL- saturated soil under different injection pressures.

MATERIAL AND METHODS

Materials

The samples of bunker fuel oil (6C) used in the study was supplied by Gulf Oil Canada Limited, with the following properties (1) flush (P) = 88°C; and (2) sulphur weight = 2.6%. The gas chromatogram is shown in Fig. 1. The surfactant used is a combination of emulgin 05 and emulgin 010 at a ratio of 9:1 (El Monayeri, 1983). A triple mixture, consisting of carbon disulphide, acetone and methanol in the ratio of 70:15:15 by volume, was used as a solvent to extract the oil from the substrate. This mixture has been known to remove four to five times more oil than petroleum ether (Yong and Sethi, 1975). The soils used in this study were montmorillonite and sand, and their x-ray diffraction patterns are shown in Fig. 2. The chemical analysis of the montmorillonite is given in Table 1.

Sample Preparation

A known weight of bunker oil was added to a dry mixture of sand and clay (montmorillonite) and mixed thoroughly with a spatula. The bunker oil was preheated to obtain better mixing conditions. The ratios of clay to sand (W_c/W_s) were 0,0.05,0.15, and 0.25, while the ratio of the oil to sand (W_o/W_s) was fixed at 0.15.

Test Procedures

The oil/soil mixture was packed into a lucite cylinder with dimensions of 50 mm I.D. and 100 mm long. Equal amounts of the mixture were compacted to a length of $\approx$ 30 mm. The porosity of the mixture was controlled by the clay/sand ratio. The cylinder was then placed in a hot-water bath adjusted to the desired testing temperature. The samples were ready for testing when temperature equilibrium was attained. The surfactant solution was heated

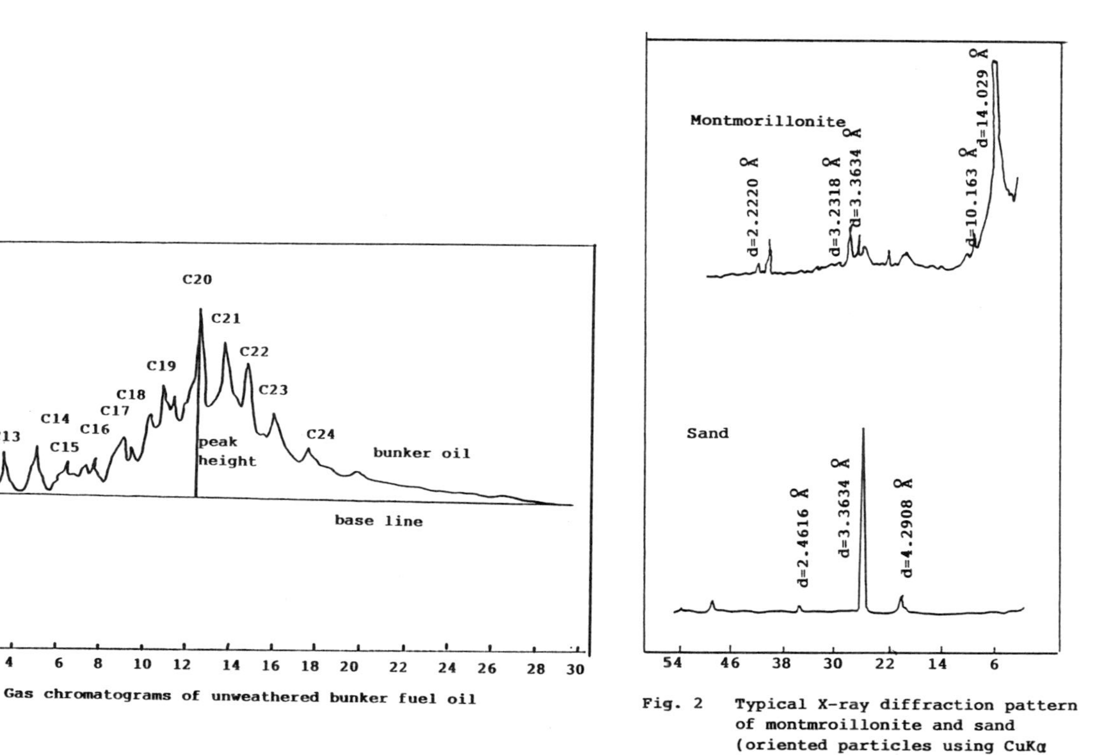

Fig. 1 Gas chromatograms of unweathered bunker fuel oil

Fig. 2 Typical X-ray diffraction pattern of montmroillonite and sand (oriented particles using CuKα radiation)

to the same temperature as that of the soil sample. It was then injected with the surfactant, using a constant head device at one end of the cylinder. The surfactant supply system consisted of two separate units: the surfactant supply tank, and the constant temperature tank. This arrangement allowed for a continuous supply of heated surfactant. After injecting the surfactant through the oil/soil mixture for a specified period of time, samples of the oil/soil mixture were taken from the inlet and outlet positions. The samples were oven dried, and the surfactant content was determined as the ratio of the difference between the sample weight before and after the drying process, to the dry weight of solids. The oven dried soil samples were crushed and then placed in thimbles, where they were washed with the triple mixture solvent. The soil samples were washed with the solvent until no further color developed in the solvent. The oil content, then, was determined as the ratio of the difference between the weight of the sample before and after washing to the weight of the sand. After washing out the oil, the samples were air dried. They were then mixed in a sodium bicarbonate solution to disperse the montmorillonite. The supernatant liquid was removed, and the procedure was repeated until, practically, no more clay was observed in the supernatant. The ratio of the clay to sand by weight was then determined.

APPROACH

The experimental program consisted of studying the flow of surfactant fluid in oil-saturated soils under different injection pressures. In the tests, the surfactant was allowed to flow into a semi-infinite cylindrical mixture of oil and soil, under a constant injection head, at the plane of contact between the porous plate and the oil/soil sample. Fig. 3 is a layout of the apparatus used in the injection tests. The effect of temperature, injection pressure head, porosity, and oil/clay ratio on the surfactant flow rate was studied.

The variables considered were as follows: (1) one type of oil (bunker fuel oil)and one type of soil (montmorillonite mixed with #70 silica sand) for the oil/soil mixtures; (2) one temperature for both the injected fluid and the oil/soil samples, namely 70^oC; (3) five different inlet injection pressure heads (20,50,100,150 and 200 mm); (4) eight different injection periods (10, 20, 40, 80, 120, 160, 200 and 250 min), and (5) four different clay/sand ratios (0, 0.05, 0.15 and .25 W_c/W_s).

The experiments consisted of injecting surfactant for different periods of time and measuring at the end of the injection time, the following: (1) surfactant content; (2) clay content; (3) oil content, and (4) resultant porosity.

EXPERIMENTAL RESULTS

Surfactant Profile during the Injection Process

Since the primary mechanism controlling the uptake of oil compounds by surfactant injection is essentially one of interfacial tension, it would be expected that the viscous forces present in the injected phase will play a role of great importance (Dullien, 1979). The degree of saturation of both fluids, i.e., the relative saturation of oil and surfactant, also determines the amount of the displacement (McCaffery and Bennion, 1974), and hence the rate at which the surfactant content changes. For these reasons it was considered important to investigate the surfactant content along the flow paths.

Two samples were taken for measurement from the experimental cylinders,

Table 1 Chemical Analyses of Montmorillonite

	%		%
SiO_2	49.4	Ca_O	1.80
Al_2O_3	20.10	Na_2O	0.43
TiO_2	0.4	K_2O	1.12
Fi_2O_3	3.7	H_2O moisture	5.00
FeO	n.d.	H_2O loss at 900°C	15.15
MgO	2.55		

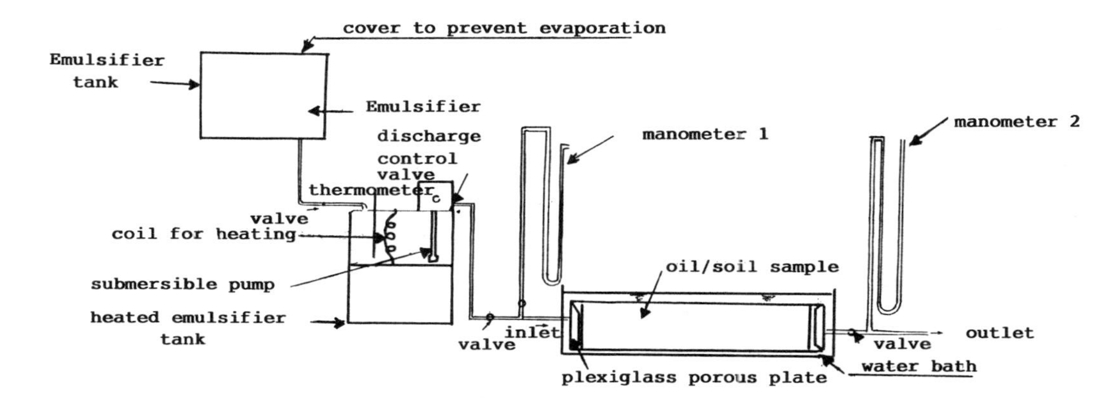

Fig. 3 Layout of apparatus for injection series tests

one from the inlet (x = 0.0 mm) and the other from the outlet (x = 100 mm) of the injection tube; This was done after a certain injection periods (e.g., 40 80, 120, 160 , 200, and 250 min). In each sample, the surfactant content was determined by weight loss after oven drying (this procedure was then repeated for different injection periods).

The distribution profiles of the surfactant content are shown in Figs. 4 and 5. The graphs show that, generally, the surfactant content was higher at the inlet position, and also in samples with a lower clay content, indicating a higher proportion of replacement of the oil and clay by the surfactant. The results also show that the rate of change of surfactant content, along the the flow direction, increased as the clay content decreased. This may be attributed to the fact that decreasing the clay content increased the porosity, and hence, the flow velocity. Increasing the injected fluid velocity increases the momentum transferred to the stationary fluid (i.e., oil) surrounding the stationary sand particles (Bird, et al. 1960). Thus, an external stress is exerted on the stationary fluid by the moving fluid. Consequently, a shear force will arise in the direction of the moving fluid (surfactant) at the oil/surfactant interface. This force must be large enough to overcome the resistance to flow of oil as well as to initiate the emulsification process. When this shear stress exceeds a certain threshold, oil and fine particles, due to the initiation of the emulsification process, are brought into the moving fluid phase. Hence, more oil and clay will be removed.

It should be mentioned here that the migration of oil and clay is dependent, not only on the surfactant flow rate, but also on the pore diameter, or porosity, of the medium.

Clay Profile during the Injection Process

The clay content measurements were made for samples taken from the inlet and outlet positions of the injected surfactant. The oil material was extracted from the soil sample and the clay content was determined by dispersion in sodium bicarbonate solution. These results were used as an indication of the rate of clay migration.

Figures 6 and 7 show the amount of clay extracted from samples with originally different clay-to-sand ratios, W_c/W_s (namely, 0.05, and 0.25). The figures show that the overall clay migration, over the tested period (250 min), decreased as the ratio W_c/W_s increased. At low clay-to-sand ratios (0.05) the change in clay content were measurable after a period of time ranging from 10 to 40 min, depending on the injection pressure head. Additionally, the rate of clay removal, and hence the rate of clay migration, was significantly lower at lower injection pressure heads. This may be attributed to the fact that the rate of momentum transfer at higher injection pressures was enough to supply the necessary seepage force for clay migration.

The amount of clay removed depended on the injection pressure heads; at low pressure (20 and 50 mm) it reached a maximum after $\sim$ 80 min, whereas at higher pressures (100, 150 and 200 mm) it reached a maximum after $\sim$ 160 min, and then leveled off (Fig. 6). This leveling off is probably due to the low clay-content paths created by the injected surfactant, and the resulting lower clay content.

At higher clay/sand ratio (0.25) the overall clay particle migration was lower as shown in Fig. 7. These results can be compared with those of Bodman and Harradine (1938) who worked on six different unsaturated soils, using water as a permeant. Their results indicate the possibility of particle migration at low clay contents, but almost no movement of particles at clay content of

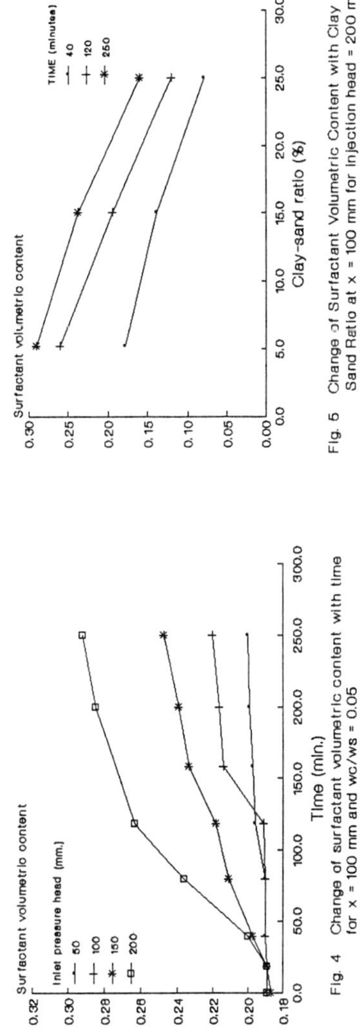

Fig. 4 Change of surfactant volumetric content with time for x = 100 mm and wc/ws = 0.05

Fig. 5 Change of Surfactant Volumetric Content with Clay Sand Ratio at x = 100 mm for injection head = 200 mm

Fig.6 Percentage clay removal with time under different injection pressures (montmorillonite .05 wc/ws)

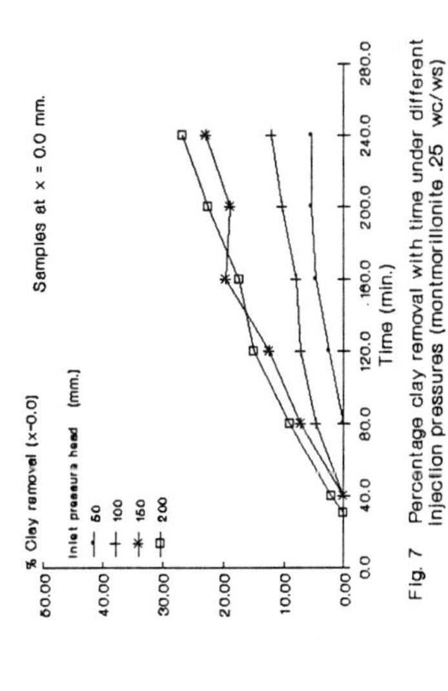

Fig. 7 Percentage clay removal with time under different injection pressures (montmorillonite .25 wc/ws)

60% or higher. It is observed that Frenkel et al. (1978) found that the migration of montmorillonite depended on the type and concentration of the electrolytes in the permeant. These moving fine particles were found to be responsible for plugging or unplugging the flow channels in the skeleton pores (Hansbo, 1973).

Oil Profile during the Injection Process

To study the effect of surfactant flow rate on the oil concentration, the soil samples tested had initial oil contents of 15% (i.e., $W_o/W_s = 15\%$), and clay contents of 0,5,15 and 25% respectively. The tests were carried out at one temperature, 70^oC, and under different surfactant injection heads (20,50,100,150 and 200 mm). The oil content after different injection periods was determined at the surfactant inlet and outlet positions.

The data obtained for the percentage oil removal are shown in Figs. 8 and 9 for W_c/W_s equals 0 to and 25%, respectively; these values correspond to clay/oil ratios (W_c/W_o) of 0, and 1.67, respectively. The results show a rapid increase in the percentage oil removal after an elapsed time ranging from 10 to 80 minutes, depending on the injection pressure and the initial percentage of clay materials. In addition, the type of substrate and hence its adsorption capability, have a considerable influence on the soil uptake. Thus, when the surfactant fluid was injected into a mixture free of clay (Fig. 8), the overall oil removal was higher than in samples containing clay (Figs. 9). This may be attributed to the fact that the clay minerals interact with organic compounds in their vicinity (Yong and Sethi, 1975). The nature of the interaction, however, is not well understood. This interaction may, in part, contribute to the lower percentage of oil removal. At the same time, increasing the clay content in the sample decreased the sample porosity. Thus, under the same injection pressure, the seepage velocity and hence the momentum transferred to the oil film surrounding the soil, will be lower; this lower momentum transfer may also contribute to the lower efficiency of oil- detachment, (and hence the percentage oil removal) process.

On increasing the injection pressure, it was observed that more oil was removed, i.e., more oil migrated. This was true up to a certain value of injection head, after which the increase in oil migration rate leveled off. For instance, on increasing the injection head from 20 to 150 mm for the clay-fines sample, the percentage of oil removed, after an injection period of 250 minutes, increased from $\sim$ 10% to $\sim$ 54%; on the other hand, a further increase of the head up to 200 mm only increased the percentage oil removal to 56% (Fig. 8).

The ratios of surfactant to oil and clay along the flow pathlines will vary with time due to the washing out of oil and clay materials. It is expected that the changes in these ratios will strongly influence the whole flow process. For example, Larson et al. (1981) showed experimentally that the permeabilities of the oil and water phases, in two-phase flow systems, depended only weakly on the viscosity ratio when the ratio is of the order of unity. On the other hand, when the viscosities are widely different (two or more orders of magnitude), the permeability becomes strongly dependent on the viscosity ratio. In addition, other reports in the literature show that the permeability depends on (a) the fluid phase configuration, which is controlled by wettability (Graham and Richardson, 1959), and (b) fluid saturation and saturation history (Naar et al., 1962).

Porosity Distribution and Pore volume Change

Soil porosity, grain size and pore coordinations are among the most important factors governing the flow process through the soil. Marczal (1968) pointed out that if a mass contains more than 5% of fine particles (e.g., silt and clay), the behavior of the grain assembly is determined by the fine particles, due to their high specific surface. The flow of immiscible fluids in a porous medium may be conveniently classified into two categories:

i. steady state i.e., all macroscopic properties of the system are time invariant at all points,
ii. unsteady sate, i.e., properties change with time.

The existence of an unsteady state is attributed to the migration of both oil and clay particles, under the influence of the injection pressure, through the oil/soil mixture samples. Once the emulsification process occurs, the migration of both oil and clay particles takes place through the carrier fluid, i.e., the injected surfactant. The injected fluid will carry these materials from one part of the sample to another. The migration rate is highly influenced by the flow rate of the carrier fluid, and by the coalescence and retention of the moving particles, which may occur during the flow (Mason and Mackay, 1963). It would be expected that the leaching out of oil and clay from the sample would continue indefinitely, under the influence of the injection of the surfactant. Accordingly, a continuous increase in porosity should be observed at any position of the sample. This would be true in the absence of screening, or filtration (Le Goff et al,. 1970).

The changes of porosity as a function of time for various clay/sand ratios and injection heads are shown in Figs. 10 and 11. It is seen that the pore volume changes with time, indicating the unsteady state condition of the problem. Referring to Fig. 10, it is observed in the 5% clay samples that for an injection pressure head of 200 mm, the porosity did not change ($\sim$ 0.43) after 80 minutes, but changed to 0.44 after an injection period of 250 minutes. The change in porosity along the flow direction, at the same time intervals, was observed for all tested samples; the changes, however, are not identical. The differences in the rate of change of porosity in terms of time and flow direction may be attributed to: (a) the difference in total input energy, due to the different times of injection. This energy is used for emulsifying the oil and /or clay, and thus less clay is leached out at the shorter time interval, and (b) the possible retention of the suspended clay particles.

The retention mechanisms have been attributed to the existence of the following retention sites, (LeGoff et al., 1970): (1) surface sites: the particle is retained on the surface of a porous bed grain; (2) crevice sites: the particle becomes wedged between the two convex surfaces of two grains; (3) construction sites: the particle cannot penetrate into a pore of a smaller size than its own, and (4) cavern sites: the particle is retained in a sheltered area, a small pocket formed by several grains.

The rate of porosity change along the for pathlines, due to the change in the clay content, varies with time. For example, in the case of $5\%W_c/W_s$ montmorillonite content, as the injected pressure increased, the rate of porosity change increased. This was true only up to 160 min of surfactant injection into these samples. On the other hand, as the injection time increased the rate of change of porosity decreased along the flow pathlines. This can be attributed to the fact that clay migration leveled off as shown previously. In other words, clay particles were removed from some of the pores, creating properly connected pathlines. These pathlines can act as effective porosity and tortuosity flow path (Douglas and Yong, 1981) through which the carrier fluid, i.e., the surfactant, will flow. As the clay content increased, i.e., at .15

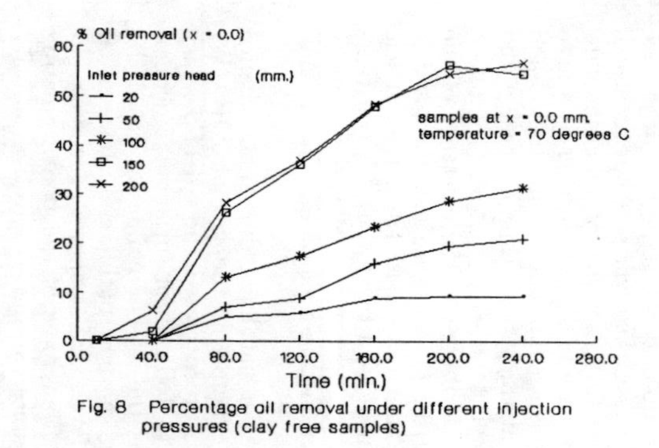

Fig. 8 Percentage oil removal under different injection pressures (clay free samples)

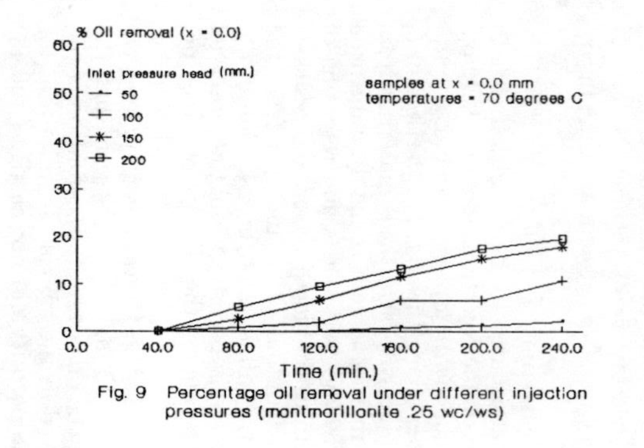

Fig. 9 Percentage oil removal under different injection pressures (montmorillonite .25 wc/ws)

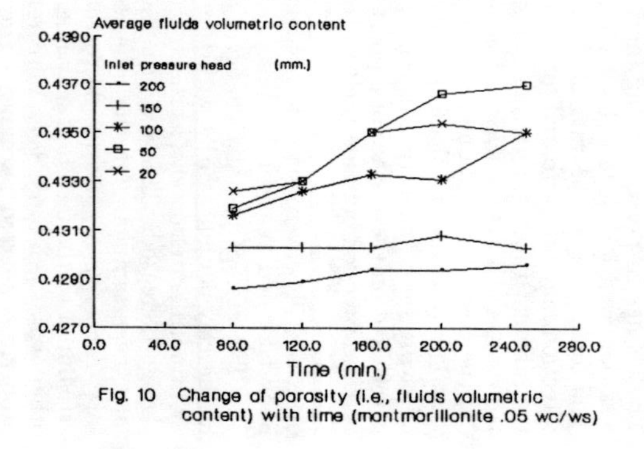

Fig. 10 Change of porosity (i.e., fluids volumetric content) with time (montmorillonite .05 wc/ws)

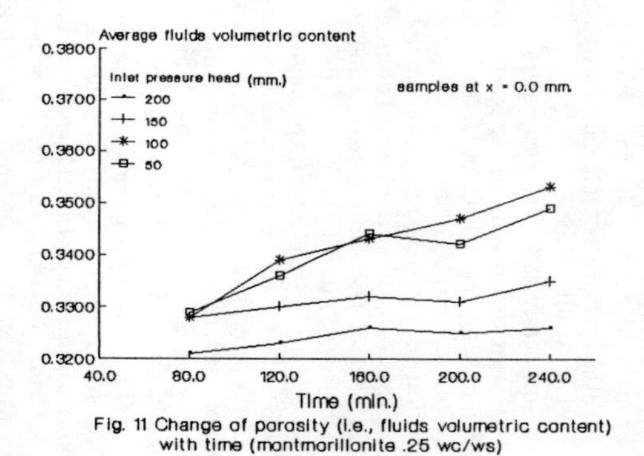

Fig. 11 Change of porosity (i.e., fluids volumetric content) with time (montmorillonite .25 wc/ws)

and .25 W_c/W_s clay content, similar behavior was observed.

At a higher clay content, i.e., 25%W_c/W_s, the percentage removal of clay particles was lower than in the case of lower clay contents, i.e., 5 and 15%. Thus, the larger the number of loose clay particles, i.e., not rigidly attached to any position, the greater the opportunity for them to move under the application of a seepage force. Further increase of seepage force, by increasing the injection pressure, could cause further movement of some of these loose particles. This was observed in all the tested samples, at all clay loads (i.e., 5, 15, and 25% W_c/W_s). As the clay content increased, the overall amount of clay removed, decreased. This observation is similar to that of Frenkel et al. (1978).

CONCLUSIONS

From the experimental study, the following statements can be made:

(1) for the same injection head, the surfactant content was higher at the inlet position, and also in samples with a lower clay content, indicating a higher proportion of replacement of the oil and clay by the surfactant. The results also show that the rate of change of surfactant content, along the flow direction, increased as the clay content decreased. This may be attributed to the fact that decreasing the clay content increased the porosity and hence the average velocity.

(2) the overall clay migration, over the tested period, decreased as the clay-to-sand (W_c/W_s) ratio increased. At low clay-to-sand ratios, the change in clay content was measurable after a period of time ranging from 10 to 40 min., depending on the injection pressure head. Also, the rate of clay removal and hence the rate of clay migration, was significantly lower at lower injection pressure heads. This may be attributed to the fact that the rate of momentum transfer at higher injection pressures was enough to supply the necessary seepage force for clay migration.

(3) there was a rapid increase in the percentage of oil removal after an elapsed time ranging from 10 to 80 minutes, depending on the injection pressure and the initial percentage of clay materials. In addition, the type of substrate and hence its adsorption capability,have a considerable influence on oil uptake. It was shown that, in the case of samples free from clay, the overall oil removal was higher than in samples containing clay which attributed to the interaction of clay. This can be attributed to the interaction of clay with organic compounds in their vicinity.

(4) the change of porosity along the flow direction, at the same time intervals, was observed in all tested samples. The differences in the rate of change of porosity in terms of time and flow direction can be attributed to the difference in total input energy and the possible retention of the suspended clay particles.

Acknowledgement

This study was supported under a Grant from the Natural Sciences and Engineering Research Council (NSERC) of Canada, Grant No. A- 882.

REFERENCES

1. RB Bird, WE Stewart & GN Lightfoot, Transport Phenomena, John Wiley and Sons, New York, 1960,
2. GB Bodman & EF Harradine, Mean Effective Pore Size and Clay Migration During Water Percolation in Soils, Soil Sci. Soc. Amer. Proc., Vol. 3, 1938, pp. 44-5l.
3. E Douglas & RN Yong, Validity of Darcy's Law for One-Dimensional, Unsaturated Flow in Kaolinite-Water-NaCL Systems, Amer. Soc. of Agriculture Eng., Vol. 24, 1981, pp. 657-662.,
4. FAL Dullien, Porous Media Fluid Transport and Pore Structure, 1979, Academic Press.
5. H Frenkel, JO Goertzen & SD Rhoades, Effects of Type and Content, Exchangeable Sodium Percentage, and Electrolyte Concentration on Clay Dispersion and Soil Hydraulic Conductivity, Soil Sci. Soc. Amer. Proc. 42, 1978, pp. 32-39.
6. JW Graham & JG Richardson, Theory and Application of Imbibition phenomena in recovery of soil, Tech. Note, Feb., J. Pet. Tech. 11, 1959, pp. 65-69.
7. RG Larson, LA Scriven & HT Davis, Displacement of Residual Non-wetting Fluid from Porous Media, Chem. Eng. Sci., 36, 1981, pp. 57-62.
8. EJ Lefebvre du Prey, Factors Affecting Liquid- Liquid Relative Permeabilities of a Consolidated Porous Medium, Soc. Pet. Eng. J., 13, 1973, pp. 39-46.
9. P LeGoff, DM Leclerc & JP Herzig, Flow of Suspensions Through Porous Media - Application to Deep Filtration, Symp. on Flow Through Porous Media, Washington, D.C., Chemical Soc. Publ. 1970, pp. 129-158.
10. L Marczal, New Formulation of the Grain Size Determining the Behavior of Grain Assemblies, Proc. 3rd Conf. Soil Mech. Found. Eng., Budapest, 1968, pp. 175-189.
11. SG Mason & GDM Mackay, Some Effects of Interfacial Diffusion on the Gravity Coal Escence of Liquid Drops, J. of Coll. Sci. 18, 1963, pp. 674-683.
12. FG McCaffery & DW Bennion, The Effect of Wettability on Two-phase Relative Permeability, J. Can. Pet. Technol. 13, 1974, pp. 42-53.
13. J Naar, RJ Wygal & JH Henderson, Three- phase Imbibition Relative Permeability, Soc. Pet. Eng. J., 1962, pp. 254-258.
14. EK Nyer & GJ Sklandany, Relating the Physical and Chemical Properties of Petroluem Hydrocarbons to Soil Aquifer Remediation, GWMR, Winter, 1989.
15. SB Wilson & RA Brown, In-Situ Bioreclammation: A Cost Effective Technology to Remediate Surface Organic Contamination, GWMR, Winter, 1989.
16. RN Yong & AJ Sethi Compositional Changes of a Fuel Oil from an Oil Spill Due to Natural Weathering, Water, Air and Soil Pollution 5, 1975, pp. 195-205.
17. RN Yong, AMO Mohamed & DS EL Monayeri, Nonaqueous Phase Liquid NAPL Treansport in Porous Media and Its Cleanup, Paper Presented at IASTED Int. Conf. on Modelling, Simulation and Optimization, Montreal, Canada May 22-24, 1990.

HAZARDOUS WASTE STABILIZATION USING ORGANICALLY MODIFIED CLAYS

by
Jeffrey C. Evans[1], M., ASCE
George Alther[2]

ABSTRACT: Recent studies have demonstrated that organically modified clays can be effective for the stabilization/solidification of hazardous materials including waste sludges and contaminated soils. Organic modification is accomplished through the replacement of naturally occurring inorganic cations within the clay by long-chain organic cations. The resulting modified clay is organophilic in nature and organic contaminants are adsorbed by the organically modified clay surfaces.

The effectiveness of organophilic clays in the adsorption of organic contaminants is quantified by laboratory methods. For these studies, the sorptive capacity of the organically modified clay was evaluated using a modified sedimentation test (a free swell test). Organically modified clays were mixed with different inorganic and organic fluids to study the impact of fluid properties upon swelling as a comparison of the relative capacity of organophilic clays to adsorb organic fluids.

Organically modified clays coupled with cementitious additives are used effectively in the stabilization/solidification process. For these studies, organically modified clays were evaluated as part of stabilization/solidification research involving acidic petroleum sludges. The chemical effectiveness of organically modified clay in the stabilization of the organic compounds was determined using the Toxicity Characteristic Leaching Procedure (TCLP). The extract was then analyzed using gas chromotography/mass spectrometry (GC/MS). Total organic carbon was also used as a means for comparison as a bulk indicator of stabilization effectiveness. Inorganic species were quantified using atomic adsorption techniques.

These studies show that organically modified clays can be effective for the stabilization/solidification of hazardous wastes and contaminated soils. The large number of sites with organic contaminants create the need for new technologies such as the use of organically modified clays. With the utilization of organically modified clays, stabilization/solidification is an effective alternative for the remediation of organic hazardous waste sites.

---

[1]Associate Professor, Dept. of Civil Engineering, Bucknell University, Lewisburg, PA 17837
[2]President, Bentec, Inc., Ferndale, MI 48220

INTRODUCTION

The use of stabilization/solidification technologies for the remediation of hazardous waste sites is on the increase. Earlier approaches to site remediation focused on containment and pump and treat options. This approach is costly and ground water may remain in place for many, many years. As a result, alternative technologies which alter the nature of the contaminants have shown increasing favor.[1] These technologies include incineration, biological treatment and stabilization/solidification.

Historically, stabilization/solidification technologies have shown considerable promise and modest utilization for the fixation of inorganic wastes. These wastes include metal hydroxide sludges for the plating industry, etc.[2]. The success with inorganics relates to (1) the lack of alternatives (e.g. metals are not biodegradable and do not change atomic structure when incinerated) and (2) readily available and understood mechanisms (e.g. precipitation, adsorption, chelation). However, the use of stabilization/solidification for organic wastes has been limited for the same reasons. First, many organics are biodegradable into less toxic forms and second, the mechanisms for binding organics into and inorganic solidified structure are less well-understood.

The application of organically modified clays to waste stabilization has resulted in the development of materials and processes which can be effectively utilized for stabilizing organic wastes.[3] Further, studies into the mechanisms of stabilization have revealed the nature of the process. This paper presents (1) the nature of organically modified clays, (2) results of free swell tests which demonstrate difference in organically modified clay performance, and (3) results of studies on the stabilization of acidic petroleum sludges with organically modified clays.

ORGANICALLY MODIFIED CLAYS

Naturally occurring clays have aluminosilicate mineral structures with a net charge imbalance within the crystalline layers due to isomorphous substitutions. This charge imbalance is neutralized by exchangeable inorganic cations at the surface. Organically modified clays are made by replacing the inorganic cations with organic ions. For example, the structures of montmorillonite, a two-to-one clay, and the organically modified montmorillonite are shown on Fig. 1.[4]

The organic cations used in clay modifications are typically quaternary ammonium ions although a wide range of organics have been used to design the clay for specific performance. A detailed discussion of the numerous factors which influence the organic modification of clays is described elsewhere.[5]

During organic modification of a clay, intercalation (a process in which a molecule enters the interlayer space and forces apart the silicate layers.[6]) is accomplished employing ion exchange, in which a cationic surfactant, (typically a quaternary amine), replaces the exchangeable sodium, calcium, and magnesium ions on the surface of the clay. As a result of this exchange, the clay changes from a hydrophilic to an organophilic medium. After replacement of the inorganic cations with organic cations, the clay swells and disperses in a wide range of organic fluids. During the replacement of the exchangeable inorganic cation, the surfactant forces the clay platelets apart

Table 1 Free Swell Test Results

CLAY	Deionized Water (ml)	Acetic Acid (ml)	Acetone (ml)	Aniline (ml)	Carbon Tetrachloride (ml)	Hexane (ml)	Kerosene (ml)	Xylene (ml)	Average (ml)
Attapulgite	15.5	7.0	6.5	8.0	6.0	7.0	7.0	7.0	6.93
Attatone T (Bentec)	8.5	13.0	20.0	7.0	11.0	8.0	8.5	12.0	11.36
Attatone J (Bentec)	16.0	8.0	19.0	14.5	23.0	15.0	17.5	26.0	17.57
BC 90 (Bentec)	8.5	11.0	11.5	7.5	7.0	18.0	18.0	20.0	13.29
Bentonite	34.0	4.5	6.0	4.5	5.0	5.0	5.0	5.0	5.00
Bondtone (NL Baroid)	7.0	8.0	13.0	7.5	0.0	40.0	20.0	36.0	17.79
Claytone APA (Southern Clay Products)	14.0	19.0	17.5	23.0	0.0	9.0	21.0	50.0	19.93
Claytone 40 (Southern Clay Products)	0.0	13.0	14.0	12.0	0.0	12.0	22.0	48.0	17.29
2HT (NL Baroid)	6.0	8.0	11.0	8.0	15.0	47.0	27.0	43.0	22.71
P-1 (Bentec)	9.0	12.0	13.5	11.0	0.0	13.0	18.0	27.0	13.50
P-11 (Bentec)	8.0	9.0	10.0	9.0	25.0	25.0	22.0	25.0	17.86
P-40 (Silicate Technology)	7.0	9.0	13.0	8.0	33.0	13.0	26.0	20.0	17.43
PC-1 (Bentec)	8.0	8.0	26.0	8.0	20.0	45.0	33.0	24.0	23.43
PT-1 (Bentec)	7.0	9.0	22.0	8.0	10.0	15.0	32.0	22.0	16.86
Suspentone (NL Baroid)	10.0	11.0	15.0	9.0	6.0	28.0	35.0	33.0	19.57
TS-55 (NL Baroid)	6.0	7.0	8.0	7.0	40.0	34.0	24.0	28.0	21.14

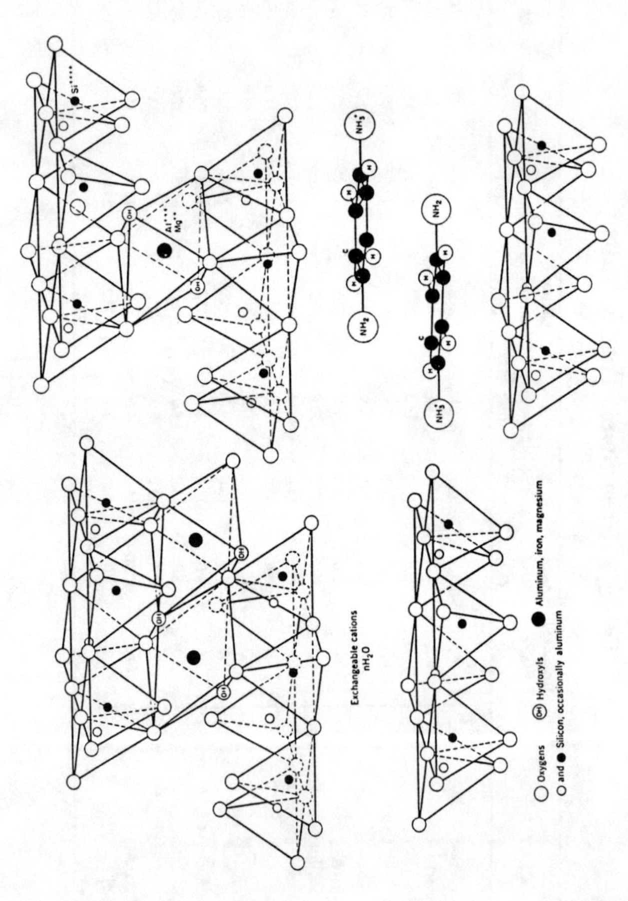

Figure 1 Organically Modified Clay Structure

imparting additional porosity to the organoclay. This process is known as pillaring of the clay. Early studies have shown that the organophilic property of a clay is most fully developed when a carbon chain length of at least ten is utilized and maximum swelling capacity is obtained with a carbon chain length of twelve.[7] An amine with twelve carbon atoms results in coverage of just over half of the available clay surface. The unmodified portion remains hydrophillic in nature.

Several types of bonds are found between the organoclay and an organic compound.[8] The primary bonds are electrostatic between the organic cations and the charged clay surface but physical, non-coulombic forces, also contribute to adsorption.[9] Van der Waals' forces dominate the adsorption process when large cations are involved since the principal interactions occur between the adsorbed organics and not between the organics and the clay surface. If the organic cations contain radicals, such as NH_3^+ groups, which interact with the surface oxygen molecules of the silicates, hydrogen bonding may be extensive. The pillaring effect described above results in space for the organic molecules to be physically bound, or absorbed, between the clay layers. Chemical adsorption of the organic molecule to the organically modified clay is however the most important mechanism.

MECHANISMS OF ORGANIC WASTE STABILIZATION

It is necessary to examine the physical and chemical interactions between the stabilizing agents and the organic contaminants to better assess the nature of the stabilization process. Successful stabilization of organic contaminants with organically modified clays employs the mechanisms of macroencapsulation, microencapsulation, and adsorption.

Macroencapsulation is the mechanism by which waste constituents are physically captured in voids of the stabilized material.[10] The contaminants are held in discontinuous pores with the stabilized materials.

Contaminants may also be confined in the crystalline structure of the solidified matrix at a microscopic level (microencapsulation). As with macroencapsulation, degradation of the stabilized mass into smaller particles, may result in release of the organic contaminants.

Adsorption is the mechanism by which contaminants are electrochemically bound to stabilizing agents. Adsorption is a surface phenomena and the bonding may be through van der Waal's or hydrogen bonding. Adsorbed contaminants within the stabilized matrix require greater environmental stress to cause release into the environment compared with those bound by microencapsulation and macroencapsulation. Organically modified clays employ adsorption for the stabilization of organic wastes. Organic contaminants are adsorbed to the clay.[11]

Certain stabilization processes precipitate inorganic contaminants from the waste. Precipitates include hydroxides, sulfides, silicates, carbonates and phosphates. This mechanism is applicable to the stabilization of inorganic wastes such as metal hydroxide sludges but not for organic wastes. Chemical reactions occurring during stabilization process may result in a waste with reduced toxicity. Detoxification is any process by which a chemical constituent is changed into another constituent or form of the same constituent that is either less or non-toxic, such as the reduction of chromium in the $^{+6}$ valence state to chromium in the $^{+3}$ valence state during stabilization with cement based materials.[12] The trivalent chromium has low solubility and toxicity in contrast to the high toxicity of hexavalent chromium. Reduction in valence state can result from a combination of ferrous sulfate and sodium sulfate.[13] Leaching of the reduced chromium poses a lesser threat to the environment than the leaching of the original hexavalent chromium.

FREE SWELL TEST DATA

The sorption capacity of organically modified clays can be evaluated in a relative sense by a sedimentation or free swell test. In this test, 50 ml of the test fluid is contained in a 100 ml graduated cylinder. Clay (2.5 grams) is sprinkled into the fluid and allowed to settle to the bottom. The mixture is periodically swirled as necessary to assist contact of the clay with the test fluid. The free swell volume is the volume of the gelled clay in the graduated cylinder, after twenty-four hours. This procedure was adopted from laboratory procedures already developed.[14] A similar set of investigations with organically modified clays was published by Jordan.[15]

The clays tested, the name of the supplier, and the free swell test results are presented on Tab. 1 along with the test fluids (acetic acid, acetone, aniline, carbon tetrachloride, deionized water, hexane, kerosene, and xylene). The test fluids were concentrated.

The test fluids are grouped based on the physical and chemical properties that govern their interactions with clay minerals.[16] The first grouping, organic acids, have acidic functional groups such as acetic acid, phenols and carboxylic acids. These fluids are capable of donating protons to react with and dissolve clays. In the free swell tests with acetic acid, the largest swell volume, 19 ml, was measured for Claytone APA although the other organically modified clays had similar swell volumes. The bentonite had the smallest volume with 4.5 ml, indicating little interaction with this unmodified clay.

Organic bases are positively charged and have a strong affinity for the negatively charged clay surface. They may also be strong enough as proton acceptors to dissolve components of the clay. Aniline is an organic base (and an aromatic amine). For this test fluid, Claytone APA had a swell volume of 23 ml, the largest. Claytone 40 and PT-1 had considerably lower swell volumes, 12 ml and 11 ml respectively, while the other organoclays had values of approximately 8 ml. Bentonite was the lowest of all the clays tested, with a swell volume of 4.5 ml.

Neutral polar organic compounds do not have a net charge but an asymmetrical distribution of electron density creating a dipole moment. Because of this dipole moment, the fluid competes for adsorption sites on the negatively charged clay particles. Acetone, a ketone, was employed as a neutral polar organic fluid. PC-1 had the highest swell volume (26 ml) followed by PT-1 (22 ml) and Claytone APA (17.5 ml). The lowest swell volumes were found for the unmodified clays, bentonite and attapulgite.

Neutral nonpolar organic compounds do not exhibit a net charge or a significant dipole moment. These fluids have little affinity for the negatively charged clays surfaces. Carbon tetrachloride, hexane, kerosene, and xylene are neutral nonpolar organic fluids used in these studies. Carbon tetrachloride produced several cloudy solutions which were recorded as having swell volumes of zero. In other samples, portions of the clay settled to the bottom and other portions floated on the top. For these situations, the two volumes were summed to compute the free swell volume. TS-55 had the largest swell volume, 40 ml, and P-40 was next with 33 ml. P-11 and PC-1 also had significant swell volumes, with values of 25 ml and 20 ml respectively. The other clays had comparatively small swell volumes or were completely cloudy with a reported a swell volume of 0 ml.

Hexane, kerosene, and xylene produced significantly greater free swell in the organophilic clays than the other test fluids. As expected, the unmodified clays, bentonite and attapulgite, had the lowest free swell (between 5 and 7) for the three test fluids. For hexane, 2HT had a swell volume of 47 ml, which was the largest value, and PC-1 was next with a volume of 45 ml. Claytone APA, which swelled the most in some of the other fluids, had the smallest swell volume of the organically modified clays with a volume of only 9 ml. This illustrates the importance of the type of clay modification in the clay adsorption capacity. Kerosene produced a maximum swell volume of 35 ml with Suspentone and somewhat lower values of 33 ml and 32 ml were reported for PC-1 and PT-1 respectively. With xylene, Claytone APA sorbed the entire 50 ml of test fluid and Claytone 40 sorbed to 48 ml. 2HT had the next highest swell volume of 43 ml.

Deionized water was also used in the investigations. Water is polar and is attracted to the negative surface of unmodified clays. Bentonite swelled to 34 ml while attapulgite had a swell volume of 15.5 ml. The organically modified clays are generally hydrophobic and this characteristic is reflected in their small swell volumes in deionized water. One product, Claytone 40, was so hydrophobic that it would not be wetted by the water and floated on top of the water.

The analysis of free swell volume data enables a comparison of the organophilic clays for their average sorption capacity. All of the organically modified clays except one, have average free swell volumes between 15 ml and 24 ml. The two unmodified clays, bentonite and attapulgite, have average free swell volumes below 7 ml, showing they did not swell appreciably in the concentrated organic fluids.

The sorption capacity of organically modified clays as indicated by free swell tests have several limitations for quantitative use in evaluating stabilization suitability. The clay quantity was held constant on a gravimetric basis, but not on a volumetric basis. The density of the clays is in part determined by the manufacturing process. There are two processes, wet and dry, to organically modify clays. The wet process is more expensive and produces a less dense clay without the inert materials which remain in the dry process. Organophilic clays manufactured by a dry process contain about 20 to 30 percent inert minerals resulting in a denser and less reactive product.[17] Organically modified clays differ in the base clay and organic compound reacted with the clay. The base clays vary with regard to their reactivity and the amount of impurities which are present. The organic compound is usually a quaternary or aliphatic ammonium salt and affects the affinity for sorbing other organics. These factors influence the performance of the clay with regard to its sorption capacity as measured in the free swell and stabilization tests utilized in these investigations.

STABILIZATION/SOLIDIFICATION DATA

Testing and Evaluation Program. As part of on-going investigations into the stabilization/solidification of petroleum sludges, over 250 mixes were prepared using various stabilization formulations. The sludge is described as a black tar-like material having a pH varying from 1 to 5 with an average loss on ignition of over 90%. The raw sludge and the stabilized sludge were analyzed using the Toxicity Characteristic Leaching Procedure (TCLP).[18] In this test, the material is leached in acid and the leachate is analyzed for organic constituents using a gas chromatograph and mass spectrometer (GC/MS). The inorganic constituents in the extract are analyzed using ion chromatography. For the samples leached with sulfuric acid, TOC was also measured as a bulk indicator of the concentration of organic constituents. It has been previously demonstrated that using sulfuric acid yields essentially the same TCLP result as acetic acid.[19] TOC is a useful parameter to evaluate the effectiveness of stabilization as it indicates gross organic content of the extract from the TCLP test.

The concentration of phenol in the extract of the TCLP was also examined to evaluate the effectiveness of stabilization. Phenol is typically found in the raw sludge and is relatively mobile in the environment, making this a useful parameter to gauge stabilization effectiveness.

Unconfined compression strength tests were also conducted on the stabilized sludge samples. Since the organics interfere with the hydration reactions of cement and pozzolan,[20] it was anticipated that the strength of the stabilized mass would correlate to the effectiveness of stabilization. That is, all else being equal, increased strength indicates reduced interference of the organics with the cementitious reactions although previous studies have shown that strength alone does not correlate to stabilization effectiveness.[21] Beyond that, project strength requirements are minimal; only that necessary to support the final cover.

Examining the data as a whole revealed an average TOC of 93 mg/l for materials stabilized using organically modified clays as a sorbent versus 204 mg/l for the other mixes. Likewise the strength of mixes using organically modified clays was 39 psi versus 23 psi for the others. These data indicate, exclusive of the other variables, the improved stabilization effectiveness of organically modified clays.

Sludge Variability. The variability of the raw sludge results in variability in test results on the stabilized sludge. The lower the pH of the sludge, higher TOC concentration of the stabilized mass.

Selected Test Results. The effect of the nature of the sorbent upon the effectiveness of stabilization is illustrated on Fig. 2. In these mixes, finely ground cement (Microfine MC-500 was used as a binder with unmodified and modified clays used as sorbents. The ratio of sludge to sorbent to binder was 1.0 to 0.4 to 0.25 for this series. As shown, the TOC in the TCLP extract for bentonite and attapulgite binders was about three times that for the organically modified clay. The unconfined compressive strength for samples mixed with unmodified clays was an order of magnitude weaker than the organically modified clay mixes.

Increasing the sorbent quantity had the expected result of increasing the stabilization effectiveness. Shown on Fig. 3 are two sets of mixes where the variable was sludge to sorbent ratio. As shown, increasing the sorbent decreased the phenol concentration in the TCLP extract and increased the strength. Thus, the more organically modified clay used as a sorbent, the greater the stabilization effectiveness.

As expected, the ratio of sludge/binder/sorbent influences the effectiveness of any individual stabilization mix. Despite the variability, it is clear that mixes with organically modified clay more effectively stabilize the sludge than mixes with other binders. Shown on Fig. 4 is a bar graph of the test results for mixes 1-3, 5-8 and 10. Each of these mixes used the same sludge in a ratio of approximately 1.0 to 0.4 to 0.3 to 0.1 as sludge to sorbent to binders (flyash and lime). Thus only the nature of the sorbent varied. Sorbents used included attapulgite, bentonite, and four different organically modified clays (Bondtone, Claytone APA, Claytone 40 and Suspendtone). From this graph, the TOC is about an order of magnitude greater for the bentonite and attapulgite sorbents than for the organically modified clay sorbents. These results on a heterogeneous sludge support the previous findings that organically modified clays sorb organics better than unmodified clays. Further, they are consistent with adsorption data that show differences in adsorption between clays manufactured with different organic modifiers.

The sludge used in this study was highly organic; the loss on ignition averaged over 90%. Thus the amount of sorbent required to stabilize this difficult waste was quite high (40% by weight of sludge). Studies of variability of test results coupled with reduced sorbent content were made using six replicates and a sludge to sorbent to binder ratio of 1.0 to 0.2 to 1.0 where the binder was Type I cement and the sorbent was an organically modified clay (PT-1 for Bentec, Inc.). The phenol content averaged 2084 ppb with a standard deviation of 306 ppb. The unconfined compressive strength averaged 3.7 psi with a standard deviation of 0.4 psi. These data show that for the same sludge the stabilization and analysis process yields reasonably reproducible results. The data also show that with a highly organic sludge as used for these studies, the sorbent addition must be substantial (40% or more). Without sufficient sorbent the contaminants are more readily extracted for the stabilized matrix and the strength is low. This need for substantial additive for highly organic sludges reduces the cost effectiveness of the technology. The premise that organically modified clays out perform conventional clays as sorbents, however, is still supported.

The results of a series of stabilization mixes using organically modified clays showed differences in performance between materials. In this series, the mix with the lowest TOC and phenol concentration had the highest unconfined strength. As the TOC and phenol concentrations progressively increase and the unconfined compressive strength progressively decreases. The data show the internal consistency expected but not always found between the parameters used to measure effective stabilization and the individual mix performance.

The influence of sorbent type upon stabilization effectiveness was also studied where the binder is Type I portland cement. The superior performance of the organically modified clays in organic waste stabilization is demonstrated by the resulting lower concentrations of TOC and phenol and higher unconfined compressive strength as compared with other sorbents.

SUMMARY AND CONCLUSIONS

Stabilization/solidification is becoming increasingly important as a method of waste treatment and site remediation. Studies have demonstrated the adsorption capacity of organically modified clays through free swell testing and adsorption isotherms. The studies described in this paper have demonstrated that organically modified clay enhances stabilization of organic wastes, even those with a high organic content (90% loss on ignition). However these studies have also demonstrated that, where the waste has a high organic content, significant quantities of organically modified clays are required. Given the relatively high cost of organically modified clays ($0.50 to $2.00 per pound), stabilization may not be a cost effective alternative for highly organic waste.

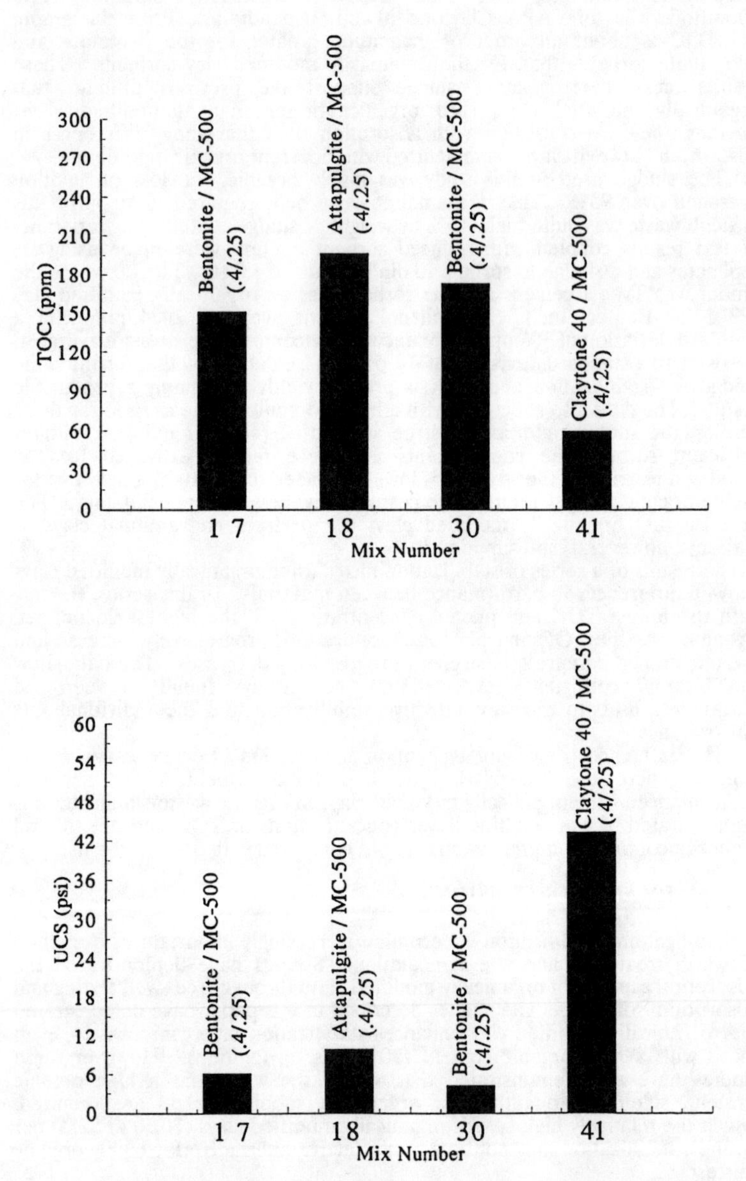

Figure 2 Influence of Sorbent Type upon Waste Stabilization

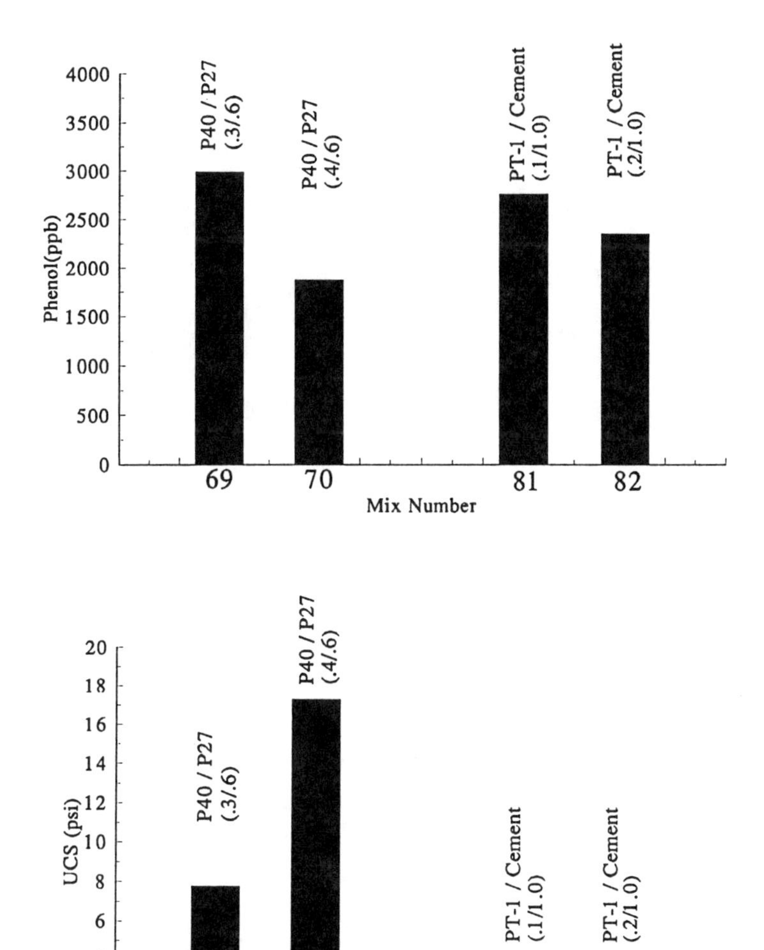

Figure 3 Influence of Sorbent Quantity upon Waste Stabilization

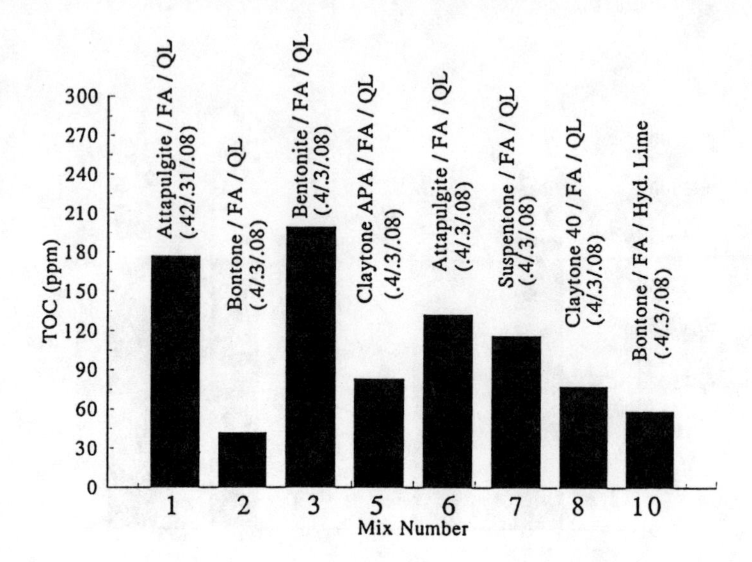

Figure 4 Influence of Sorbent Quantity upon Total Organic Carbon

ACKNOWLEDGEMENTS

These investigations are part of a larger research project funded by the Sun Refining and Marketing Company, the Ben Franklin Partnership of Pennsylvania and the earth Technology Corporation.

The authors appreciated the cooperation of the clay suppliers, Bentec, NL Baroid, Silicate Technology Corporation, and Southern Clay Products, who provided the clays for the laboratory investigations and technical assistance regarding their manufacture and the Hercules Cement Co. for providing cement for our research. Appreciation is also extended to Chris Bailey and Holly Borcherdt who provided laboratory assistance and to Lewis Albee and Jason Strayer for assistance with the figures, extractions, and laboratory analysis.

REFERENCES

[1]SAIC, "Remedial Action at Waste Disposal Sites," EPA Handbook (Revised), Prepared for EPA's Office of Research and Development, Hazardous Waste Engineering Research Laboratory, Office of Emergency and Remedial Response, EPA/625/6-85/006, October 1985.
[2]Conner, J. R., Chemical Fixation and Solidification of Hazardous Wastes, Van Nostrand Reinhold, NY, NY, 692 pp., 1990.
[3]Beall, Gary W., "Method of Immobilizing Organic Contaminants to Form Non-Flowable Matrix Therefrom," U.S. Patent Number 4,650,590, March 17, 1987.
[4]Grim, Ralph E., Clay Mineralogy, Second Edition, McGraw-Hill Book Company, 1968.
[5]Mortland, M.M., "Clay-Organic Complexes and Interactions," Advances in Agronomy, Volume 22, Academic Press, Inc., pp. 75-117, 1970.
[6]Lagaly, Gerhard, "Clay-Organic Reactions," Interactions. Phil. Trans. R. Soc. London, A-311, pp. 315-332, 1984.
[7]Jordan, John W.; "Organophilic Bentonites. I Swelling in Organic Liquids," The Journal of Physical & Colloidal Chemistry, Volume 53, Number 2, pp. 294-306, February 1949.
[8]Gibbons, J.J.; and Soundararajan, R., "The Nature of Chemical Bonding Between Modified Clay Minerals and Organic Waste Materials," American Laboratory, Volume 20, Number 7, pp. 38-46, July 1988.
[9]Raussell-Colom, J.A.; and Serratosa, J.M., "Reactions of Clays with Organic Substances," in Chemistry of Clays and Clay Minerals, ed. A.C.D. Newman, pp. 371-422, John Wiley & Sons, New York, 1987
[10]Environmental Laboratory, "Guide to the Disposal of Chemically Stabilized and Solidified Wastes". EPA-IAG-D4-0569, September 1982.
[11]Cadena, F. and Jeffers, S. W., "Use of Tailored Clays for Selective Adsorption of Hazardous Pollutants", *Proc. of the 42nd Ind. Waste Conf.*, Lewes Publishers, Chelsea, MI, 113, 1988.
[12]Conner, J. R., Chemical Fixation and Solidification of Hazardous Wastes, Van Nostrand Reinhold, NY, NY, 692 pp., 1990.
[13]Carpenter, C. J.; "Ferrous sulfate/sodium sulfide chromium reduction metals precipitation," **Proceedings of the 5th National Conference on Hazardous Waste and Hazardous Materials**, Las Vegas, NV, pp. 52-56, 1988.
[14]Hettiaratchi, J.P.A.; and Hrudey, S.E., "Influence of Contaminant Organic-Water Mixtures on Shrinkage of Impermeable Clay Soils with Regard to Hazardous Waste Landfill Liners," Hazardous Waste & Hazardous Materials, Volume 4, Number 4, pp. 377-388, 1987.
[15]Jordan, John W.; "Organophilic Bentonites. I Swelling in Organic Liquids," The Journal of Physical & Colloidal Chemistry, Volume 53, Number 2, pp. 294-306, February 1949.
[16]Brown, K.W.; and Anderson, D.C., "Effects of Organic Solvents on the Permeability of Clay Soils," Project Officer Robert E. Landreth, Municipal Environmental Research Laboratory, EPA-600/2-83-016, March 1983.
[17]Beall, Gary W. Personal Communication, 1988.

[18]Federal Register, "Appendix I to Part 268 - Toxicity Characteristic Leaching Procedure (TCLP)," Environmental Protection Agency, 40 CFR Part 268, Volume 51, Number 216, pp. 40643-40653, Friday, November 7, 1986.
[19]Toner, K. and Pancoski,
[20]Lea, F.M., The Chemistry of Cement and Concrete, Chemical Publishing Company, NY, NY, 1971, 727 pp.
[21]Evans, J.C.; Pancoski, S.E.. "Stabilization of Petroleum Sludges" **Superfund '89 - Proceedings of the 10th National Conference**, November 27-29, 1989.

INSTALLATION OF A GROUT CURTAIN AT A HAZARDOUS WASTE LANDFILL

Herff N. Gazaway[1], Aff.ASCE; Richard M. Coad[2], M.ASCE;
and Kenneth B. Andromalos[3], M.ASCE

ABSTRACT: An existing inactive hazardous waste landfill located in Niagara Falls, New York was exhibiting migration of contaminants due to lateral groundwater flow in the underlying dolomitic limestone bedrock. It was determined that the installation of a vertical barrier to redirect the groundwater flow around the landfill would be the key element in the abatement design. Further review indicated that a grout curtain would be the most cost-effective method of installing this vertical barrier. The design, which included extensive testing of numerous grout materials for compatibility with known contaminants at the site, required the installation of a single-row grout curtain some eighty feet (26 meters) deep by 2,500 feet (820 meters) long.

Slurry grout materials selected and used on this project included fly ash, Types I and V Portland cement, and the largest quantity of ultrafine cement used to date in the United States. All grout injection and verification testing was performed utilizing automated monitoring and recording equipment, facilitating computerized correlation and evaluation of field data.

This innovative combination of conventional grout curtain design with the more recent advances in grouting materials and monitoring technology yielded a vertical barrier providing significantly reduced permeability of the underlying bedrock in accordance with the original design objectives. Further, in what may be the largest application of this type to date, this technique appears to be a viable method for control of subsurface contaminant migration in the ever growing hazardous waste management field.

INTRODUCTION

For a period of some four decades during the mid-1900s, this site served as a repository for industrial wastes containing

1 - Proj. Mgr., Geo-Con, Inc., P.O. Box 17380, Pittsburgh, PA 15235
2 - Sr. Proj. Engr., Woodward-Clyde Consultants, 5120 Butler Pike, Plymouth Meeting, PA 19462
3 - Group Mgr., Geo-Con, Inc., P.O. Box 17380, Pittsburgh, PA 15235

various hazardous compounds. Upon discovering similar contaminants in the groundwater beneath and downgradient of the fill area, the Owner closed and capped the site and installed numerous monitoring wells, together with extraction wells on the downgradient side of the site in order to remove and treat the contaminated water as a means of mitigating contaminant migration offsite.

The review and analysis of data obtained from this work as well as other investigative work at the site indicated that the primary conduits for contaminant migration in the area were bedding fractures in the underlying bedrock. Further review yielded the concept of a vertical barrier installed on the upgradient sides of the site in order to divert natural groundwater flow around the fill area, thereby improving the containment of offsite contaminant migration and minimizing future groundwater pumping and treatment rates.

SOIL/ROCK PROFILE

The soil and rock profile at the site consisted of approximately 10 to 20 feet (3 to 6 meters) of fill and natural glacial till overlying a nearly horizontally bedded dolomite rock formation. The fill along the grout curtain alignment generally consisted of a silty clay with varying amounts of sand and gravel. Bedrock consisted of a thinly to massively bedded dolomite with core recovery typically in the 90 to 100 percent range. Depending on the degree of fracturing, the Rock Quality Designation (RQD) of the bedrock cored varied from 30 to 100 percent. Within the top 60 to 70 feet (20 to 23 meters) of bedrock, six distinct bedding plane fracture zones were identified.

SELECTION OF TREATMENT METHOD AND PROCEDURES

Based on the soil/rock profile and previous geotechnical applications, the use of grout curtains was chosen as the method of installing an upgradient vertical barrier on this site. The installation of grout curtains by the injection of cement-based grouts, under pressure, into pervious rock formations is a process which has been used for decades throughout the world to reduce the permeability of these formations. Grout curtains have been utilized primarily as a means of protecting dams by limiting water flow through foundation materials beneath the dam embankment. Where high water head differentials exist on either side of the grout curtain, as in the case of a dam, the grout curtain sometimes may consist of the installation of multiple parallel rows of grout injection holes.

The design of the grout curtain on this project was based on current international standards-of-practice for the design of grout curtains, which often involve a single line grout curtain, relatively thick grout mixes, fine grained grouting materials, and relatively higher injection pressures than are commonly used in the U.S. In addition, the design was based on the general practice of extending the grout curtain into an underlying, relatively impervious stratum.

GROUT COMPATIBILITY TESTING

As reported by Weaver, et al[1], extensive laboratory testing was performed prior to the start of the work in order to assure the compatibility of the proposed grout formulations with the various industrial wastes known to exist at the site. This testing included the use of contaminated water from the site, not only in the preparation of grout mixes, but also as the curing medium for some of the grout samples.

This testing yielded the selection of three basic grout formulations: a Type I cement and Class F flyash formulation to be used in relatively open formation conditions, a neat Type V cement grout for use in median conditions, and a neat MC-500 microfine cement grout for relatively tight zones. All of these mixes exhibited final permeabilities ranging from 1×10^{-9} to 3×10^{-10} cm/sec, thus providing highly acceptable physical properties for the intended application.

CONSTRUCTION PROCEDURES

The grout curtain was constructed using the single-line, split spacing method. This was accomplished by drilling and grouting vertical holes to form a curtain approximately 2,545 feet (835 meters) long and 80 feet (26 meters) deep. Primary holes were placed on 40 foot (13 meter) centers, with spacings becoming progressively smaller through quaternary holes on 5 foot (1.5 meter) centers. Grout was mixed at mobile batch plants and pumped at controlled pressures through a single pneumatic packer set at various depths in each hole. Each hole was pressure tested and grouted in 10 foot (3 meter) intervals. Figure 1 shows a typical section of the grout curtain.

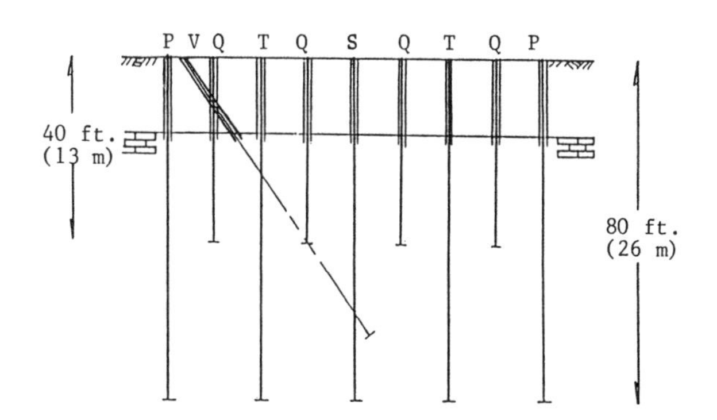

P-Primary, S-Secondary, T-Tertiary, Q-Quaternary, V-Verification

Figure 1. Typical Section of Grout Curtain

Prior to the initiation of production work, a test section was installed to determine whether descending (down-stage) or ascending (up-stage) grouting methods would be used for production grouting. Results of the test section indicated that the ascending stage grouting method was technically acceptable and more cost effective, thus allowing its selection as the method to be used for the production work.

Since the work involved the possible exposure of workers to chemical hazards during the drilling and grouting operations, a site specific health and safety plan was implemented, which included the setup of a three zone work site. These zones consisted of exclusion, contamination reduction, and support zones, as are schematically represented in Figure 2.

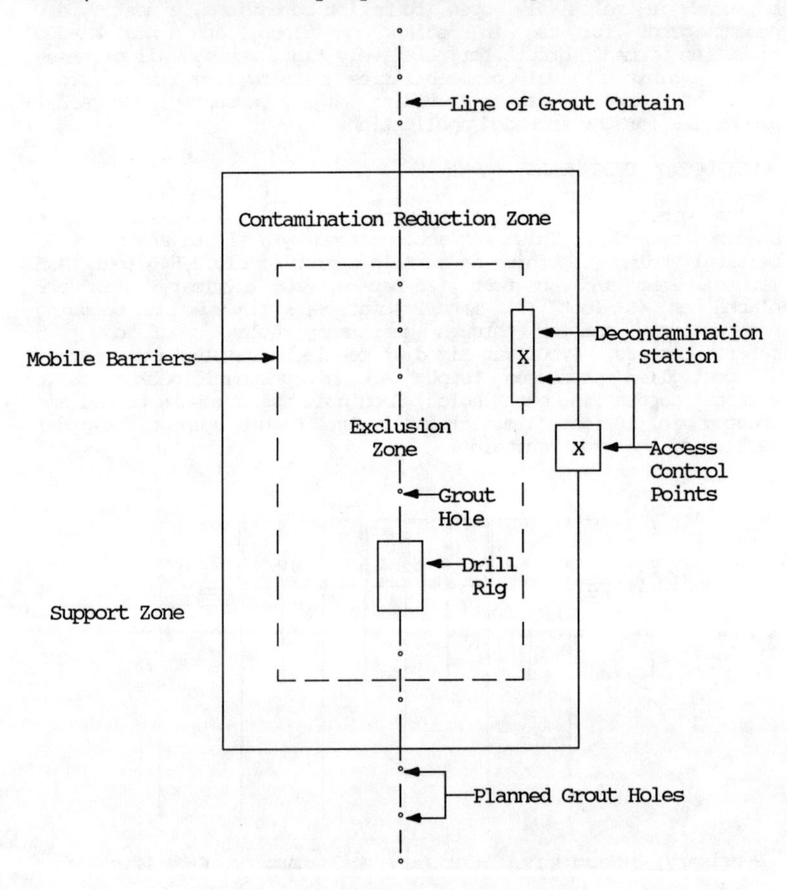

Figure 2. Three Zone Work Site

Actual zone configurations during the course of the work were significantly elongated parallel to the grout curtain to allow access to numerous grout hole locations without the need for frequent relocation of work zone barriers, as shown in Figure 3.

Figure 3. Elongated Three Zone Work Area

Work within the exclusion zone was generally performed under modified Level C personal protection. This level of protection required workers to wear chemical resistant overalls, outer and inner chemical resistant gloves, chemical resistant boots with steel toe and shank, hard hat, and chemical splash goggles or safety glasses. An air purifying respirator was also carried with each worker at all times, and worn when air monitoring indicated that airborne contaminants were present. The personal protection gear requirements were ultimately raised to a Level B for work in one section of the site. Airborne contaminant levels were such that full face respirators and supplied breathing air were required for the protection of the workers in this area.

Production drilling on this project was performed using two different drilling rigs and techniques. A truck mounted rotary drill rig using hollow stem augers was used to set a 3 inch (76 mm) I.D. casing through the overburden and into bedrock. All casings were then grouted into bedrock. This same drill rig was also utilized later for verification core drilling. Drilling of the bedrock was performed using a self contained hydraulic crawler drill. Rotary percussion drilling techniques were utilized to advance the grout holes, with drill foam and water as a circulating medium to remove the drill cuttings from the grout holes. All cuttings and drilling fluids were considered as contaminated, and were contained and stored in appropriate collection tanks for proper treatment offsite. The complexity of drilling in a hazardous environment is illustrated in Figure 4.

Figure 4. Drilling in Hazardous Environment

After the completion of drilling, each grout hole was pressure tested in approximately 10 foot (3 meter) intervals throughout the length of the drill hole in rock. Water was injected into each interval at a constant pressure for a minimum of 3 minutes. The results of each water test was expressed in Lugeon units, where one Lugeon approximately equates to a permeability of 1.3×10^{-5} cm/sec.

The grout on this project was mixed using portable trailer mounted grout mixing plants. Each plant consisted of a high-speed colloidal mixer, two agitated holding tanks, and two progressive cavity pumps. Up to three grout plants were utilized simultaneously during the production work.

Grout from each mixing plant was then pumped through hoses and in-line automated monitoring equipment into each interval of a grout hole. The grout monitors (Figure 5) were capable of automatically maintaining the grout pressure at preprogrammed limits, as entered by the grout technician. The grout monitors were also equipped with an integral strip chart recorder which recorded both pressure and flow rate over time. The information contained on these charts was used in review and evaluation of the work.

The range of grout mixes that were utilized on this project are summarized in Table 1.

Figure 5. Grout Monitor with Strip Chart Recorder

Mixture Type	Mix Designation	Water:Cement Ratio by Weight
Flyash Cement		
Type I Cement and	FAC3	0.6:1
Class F Flyash	FAC4	0.5:1
Neat Cement		
Type V Cement	NEAT1	0.5:1
	NEAT2	1.0:1
	NEAT1A	0.75:1
	NEAT1B	0.66:1
Microfine Cement		
Microfine 500 Cement	MC500	1.0:1
	MC501	1.5:1
	MC502	2.0:1

Table 1. Grout Mixes Utilized

During the early stages of the work, grout mix flow charts were developed and later used as a guideline for determining which mix should be injected in a given interval. An example of this type of chart is given in Figure 6. The use of these charts proved to be an invaluable aid to production efforts.

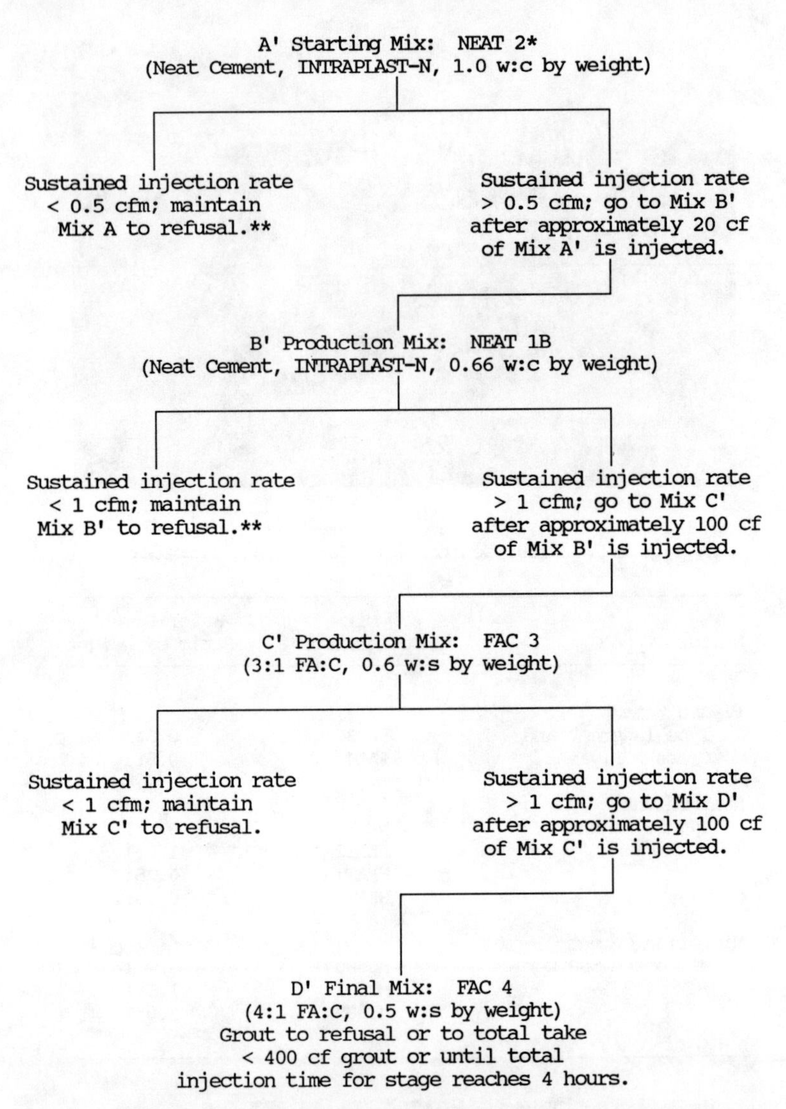

* For use in primary and secondary hole stages with permeabilities > 50 L, after injecting any leftover MC-500 A.

**Do not continue grouting beyond 4 hours or approximately 400 cf.

Figure 6. Sample Grout Mix Flow Chart

Throughout production work, field testing and analysis was performed. Field testing included density, viscosity, sedimentation and temperature measurement of the various grout mixes. Analysis included the development of onsite grout profiles to summarize and evaluate drilling, water test results, grouting results and grouting sequence. Upon completion of all grout holes within a given 80 foot (26 meter) long section, a 30 degree (from vertical) angled NX size core verification hole was drilled to evaluate the effectiveness of the grouting operations. Upon completion, these verification holes were also pressure tested and grouted. The rock core samples were inspected for the presence of grout in horizontal and vertical fractures.

RESULTS

The project required some 9 months of work spanning the summer construction seasons of 1988 and 1989. Also, a labor force often in excess of 20 persons was required exclusive of supervision, engineering and inspection staff, and health and safety personnel. With a total completed curtain length of 2,545 feet (835 meters) requiring a total of 565 holes, the average completion rates for the project were on the order of three holes completed per ten hour work day.

A total of some 2,250,000 pounds (1,020,600 kilograms) of grout solids were injected into a total of 39,264 feet (12,882 meters) of grout hole. A summary of average grout takes by hole type is presented in Table 2, which shows the progressive reduction in grout takes through the various stages of the work. The distribution of the various grout materials, expressed as a percentage of total grout solids, is shown in Table 3.

Hole Type	Average Grout Take lb/ft	Average Grout Take kg/m
Primary	182	252
Secondary	104	144
Tertiary	47	65
Quaternary	24	33
Angle Verification	17	23

Table 2. Grout Takes by Hole Type

Grout Material	Percent of Total Solids
Portland Cement - Type I	6
Portland Cement - Type V	63
Flyash - Type F	12
Ultrafine Cement	19

Table 3. Grout Solids Used by Grout Type

Table 4 summarizes the results of the water tests performed at various sections and depths of the grout curtain for those intervals encountering bedding fracture zones. A review of this table shows a significant reduction in the Lugeon (permeability) value of the rock after grouting, particularly in the highly permeable zones.

Grout Curtain Section	Depth Feet	Interval Meters	Highest Median Lugeon Value	Verification Hole Median Lugeon Value	Lugeon Value Reduction Ratio
West	20-30	7-10	74	1 (6)[(a)]	74
	40-50	13-17	47	2 (5)	23
	50-60	17-20	43	7 (4)	6
N.W.	20-30	7-10	8.5	.1 (9)	85
	40-50	13-17	23	.1 (10)	230
	50-60	17-20	76	4 (10)	19
	60-70	20-23	26	.5 (3)	52
N.E.	20-30	7-10	28	.1 (5)	280
	40-50	13-17	4	.1 (11)	40
	50-60	17-20	64	5 (11)	13
	60-70	20-23	26	4 (6)	6
	70-80	23-26	2.2	11.5 (4)[(b)]	0.2[(c)]
East	20-30	7-10	2.5	0 (1)	0.0[(c)]
	30-40	10-13	3	.1 (3)	30
	50-60	17-20	2	3 (4)	0.7[(c)]
	60-70	20-23	89	2.5 (4)	34
	70-80	23-26	2.6	1.5 (4)	1.2[(c)]

(a) Number of test values.
(b) Higher frequency of vertical fractures resulted in anomalous Lugeon values for this interval.
(c) Intervals encountering only minor bedding plane fractures.

Table 4. Summary of Water Test Results

CONCLUSION

The results of the grouting program show that the more permeable zones of the bedrock have been substantially grouted along the grout curtain alignment. It can be drawn from these results that this grout curtain will substantially improve the hydraulic control of the existing groundwater recovery system and reduce the amount of pumping required to effect a hydraulic control of the landfill area.

In addition, the improvements and innovations in grout curtain design, equipment, materials, and quality control techniques as used on this project have greatly enhanced the technology of cement grouting in the U.S., such that grout curtains can be relied upon to provide effective control and containment of both nonhazardous and hazardous underground fluids. The use of grout curtains for environmental applications such as this will increase as this vast market continues to develop and technological advancements continue to be made.

REFERENCES

1. KD Weaver, JC Evans & SE Pancoski, Grout Testing for a Hazardous Waste Application, Concrete International (ACI), July 1990, 45-47.

SEISMICALLY INDUCED PERMANENT DEFORMATIONS OF RETAINING WALLS

Sreenivas Alampalli[1] and Ahmed–W. Elgamal[2]

ABSTRACT: A two dimensional wall–soil model which accounts for wall/soil resonant dynamic response is proposed. The retaining wall is represented by a bending beam with a base yielding rotational spring and a translational slide element. Backfill soil is represented by a 2D longitudinal–shear beam. The wall and supported soil interact through a Winkler type nonlinear no–tension spring system. Model formulation is presented. Numerical simulation of centrifuge tests are presented. The seismic response of a 15 m high cantilever retaining wall is also studied.

INTRODUCTION

A number of analytical models which investigate the dynamic behavior of retaining wall systems have been proposed. Tajimi [16] used an elastic two dimensional wave propagation theory to investigate earth pressures on a basement wall assumed to undergo periodic vibrations of horizontal translation and rotation. Scott [11] proposed a one dimensional elastic shear beam to model backfill soil connected to a supporting wall by a system of Winkler springs. Wood [17] using a plane strain elastic analysis, studied idealized representations of wall–soil systems. Arias et al. [2] developed a 2D longitudinal–shear model to analyze fixed rigid wall response. In addition to modeling the vibration of the wall–soil system, some nonlinear soil–structure interaction aspects may have a major influence on potential plastic deformations. Wall–backfill soil as well as wall–base interaction control the magnitude of dynamic earth pressure and resulting wall deformation. Prakash [9] outlines a method for calculating permanent sliding displacement of retaining walls. In this method, wall and soil are modeled as a single degree of freedom mass supported by a nonlinear yielding base spring. Prakash et al. [10], Nadim and Whitman [7], and Siddarthan et al. [13] show that the rotational deformation of the wall structure may be very significant in some cases and should be accounted for in analysis procedures. Nadim and Whitman [6], and Siddarthan et al. [12] investigated retaining wall response using a nonlinear plane strain finite element analysis incorporating

---

1 – Engineering Research Specialist I, Engineering R & D Bureau, NYSDOT, Bldg. 7A, Room. 600, State Campus, Albany, NY 12232.

2 – Assistant Professor, Department Of Civil Engineering, Rensselaer Polytechnic Institute, Troy, New York 12180.

slip elements along the active failure wedge boundaries and the wall base–soil interface. Importance of amplification of backfill motion in dynamic analysis of wall–soil systems is emphasized. Nazarian and Hadjian [8] emphasize the importance of a numerical model which includes a no–tension wall–soil interface, simultaneous rotation and translation of the wall base and radiation damping effects.

In this paper a model which accounts for wall/soil dynamic response is proposed. Free vibration response of the wall and soil system is presented. Forced vibration response in which the free vibration solution is employed is also detailed. In the absence of quantitative full–scale data, the developed model is used to numerically simulate the results obtained from two gravity retaining wall centrifuge tests [3,14]. Seismically induced permanent displacements of a 15 m high retaining wall system are also studied.

FORMULATION OF GENERAL MODEL

A wall model is assumed to interact under dynamic loading with a soil model through a system of Winkler–type springs. Two dimensional in–plane vibration conditions are assumed. Response features incorporated in this model are discussed below followed by a presentation of the governing equations.

Features Of Proposed Dynamic Model: Some of the features incorporated in the proposed dynamic wall–soil model (see Fig. 1) are included in currently available models and some are unique to this model. The salient features are:

a) Simultaneous wall translation and rotation: A translation slide element is available at the retaining wall base. Under dynamic loading conditions, the sliding yield force of this element is (Fig. 2a):

$$F_y = \text{Active Earth Pressure Force x (F.S. against sliding} - 1.0)$$

where F.S. is factor of safety.

F_y may be further adjusted to include effects of earthquake vertical component, and increased static earth pressure and reduced base friction due to pore pressure buildup.

A rotational elastic perfectly plastic spring is also available at the base (Fig. 2b). Spring yield moment is chosen to account for wall rotational failure under dynamic loading;

$$M_y = \text{Overturning Moment x (F.S. against overturning} - 1.0)$$

Alternate expressions for M_y may be defined if wall rotation occurs about points other than the wall toe. M_y may be also chosen to represent the resistance to bending of a plastic hinge which may develop at the wall stem–base juncture. Modification to account for effects of pore–pressure buildup are also possible.

b) No–tension wall–soil interface: A nonlinear Winkler type spring system is chosen as a wall–soil interface. No tension properties are intended to represent the dynamic component of wall–soil interaction. These nonlinear

Winkler springs may be adapted to fill any gaps (partially or fully) created during transient seismic interaction. Such gaps may occur due to wall sliding, wall rotation or due to difference in inertia and stiffness of wall and soil. The nonlinear Winkler springs may also be used to exert additional stresses on the wall which may arise from backfill densification.
c) Flexible wall model: An Euler bending beam represents the wall. In many practical cases one or two modes of this beam (1 or 2 degrees of freedom) will provide sufficient accuracy in defining the dynamic wall response. This wall model provides a realistic boundary for wall–backfill dynamic interaction. Gravity as well as flexible walls may be modeled. In addition, a base mass may be included to simulate the wall foundation.
d) Soil model with ground motion amplification: A simple soil model is proposed. Only lateral vibration is included (2D longitudinal–shear beam). In most cases, few degrees of freedom (10 or less) will represent soil response with sufficient accuracy. As mentioned earlier, no–tension Winkler springs allow wall–backfill soil interaction. Other backfill boundaries (base and far end) are modeled by appropriate spring–dashpot mechanisms. Note that the soil base boundary may be aligned with the available geological profile (Fig. 1). Boundary springs and dashpots can be chosen so as to account for far–field compliance and radiation damping effects. Elasto–plastic nonlinear hysteretic soil properties may be included in this model [1].

In brief, the following aspects of wall–soil response are accounted for:
- Wall flexibility
- Simultaneous wall translation and rotation
- No–tension wall–soil interface
- 2D soil geometry
- Amplification of dynamic/earthquake input motions
- Radiation damping effects

A small matrix equation (15x15 or less) will in general incorporate the above features and provide solutions of sufficient accuracy.

Free Vibration Response Of Wall Model. The retaining wall is represented by a one dimensional Euler bending beam with a fixed base. Dynamic soil pressure on the wall is represented by continuous Winkler type springs alongside the wall as shown in Fig. 1.

The equation of motion for the wall system (Fig. 1) can be written as:

$$m\, u_{b,tt} + E'I\, u_{b,zzzz} + k\, u_b = 0 \tag{1}$$

along with the boundary conditions:

$$u_b = 0, \qquad u_{b,z} = 0 \qquad \text{at } z = 0$$

$$u_{b,zz} = 0, \qquad u_{b,zzz} = 0 \qquad \text{at } z = H_w$$

where u_b is the wall displacement relative to its base, E' is the modulus of elasticity of the wall material, I is the moment of inertia, m is the mass per unit length, H_w is the height of the wall, k is the wall–backfill interaction spring constant, t is time, and $(_{,t})$ denotes time differentiation. Natural frequencies and mode shapes $\psi_i(z)$, $i = 1,2,...\infty$ of this system can be easily obtained [1,4].

The bending beam described above will be supported on a stick–slip frictional base of mass m_{bs} [15] and a torsional spring (Fig. 1). This base support system (degrees of freedom u_{bs} in translation and θ in rotation) will be used later to allow for wall sliding and overturning.

Free Vibration Response Of Soil Model. Lateral vibration of the backfill soil is represented by a two dimensional model (Fig. 1). The soil model interacts on one side with the wall through Winkler type springs. At the far–end of the soil a fixed, free, or spring type boundary condition may be assumed. The soil base can also be fixed or spring supported. In the following, the spring supported formulation is presented. By assuming appropriate spring constants, fixed or free conditions can be easily obtained. In this section, soil properties will be assumed constant with depth for simplicity.

The equation of motion representing the soil mass (Fig. 1) can be written as (e.g., Arias et al. [2]):

$$G\, u_{,zz} + E\, u_{,xx} = \rho\, u_{,tt} \tag{2}$$

along with the boundary conditions,

$$G\, u_{,z} = K_b\, u \quad \text{at } z = 0, \qquad G\, u_{,z} = 0 \quad \text{at } z = H_s$$

$$E\, u_{,x} = K\, u \quad \text{at } x = 0, \qquad E\, u_{,x} = -K_f\, u \quad \text{at } x = L$$

where G is the shear modulus of soil, E is Youngs modulus of soil, u is the relative displacement of soil at any point (x,z), ρ is the mass density of soil, L and H_s are the length and height of the soil domain respectively, and K, K_f and K_b are various boundary spring constants (Fig. 1). Natural frequencies and mode shapes $\phi_i(x,z)$, $i=1,2,3,...\infty$ can be derived in closed form for this soil system [1,4].

Forced Response Of Combined Wall–Soil Model. The above discussed wall and soil models are used to represent the vibrational response of combined wall–soil systems. The wall base, as mentioned earlier, will be allowed one translational and one rotational degree of freedom. In translation, the wall will be allowed to slide away from the backfill once the dynamic lateral forces exceed the base frictional force as dictated by the static factor of safety against sliding. An additional elastic–perfectly–plastic base rotational spring will also allow rotation (away from the backfill) if base moment exceeds the available resistance as dictated by the static factor of safety against overturning.

a) Wall Base Translation: When the base sticks to the foundation soil, i.e., no relative motion between base and foundation soil, the non–sliding condition will be:

$$u_{bs,t} = 0 \tag{3}$$

in which, u_{bs} is the wall base displacement relative to the ground. Eq. 3 will hold as long as (see Figs. 1 and 2a):

Wall inertial forces + wall and backfill interaction spring forces $- F_y \leq 0$ (4)

As soon as the stick condition given by Eq. 4 fails, slip occurs and wall base translation in a sliding mode [15] is expressed by (Figs. 1 and 2a):

$$m_{bs}(u_{g,tt} + u_{bs,tt}) + \int_{H_d}^{H_s} \Big[m \big\{ u_{g,tt} + u_{bs,tt} + u_{b,tt} + (z - H_d)\theta_{,tt} \big\} + F_{ws} \Big] dz - F_y = 0 \tag{5}$$

where m_{bs} is the base mass (Fig. 1), u_g is ground displacement, θ is wall base rotation, $H_d = H_s - H_w$, F_{ws} is the nonlinear no–tension wall–backfill interaction pressure (Fig. 3), and F_y is a user specified base yield force (Fig. 2a).

b) <u>Wall Base Rotation</u>: Retaining wall base rotation may be expressed by (Figs. 1 and 2b):

$$M + \int_{H_d}^{H_s} \Big[m \{ u_{g,tt} + u_{b,tt} + u_{bs,tt} + (z - H_d)\, \theta_{,tt} \} + F_{ws} \Big] (z - H_d) dz = 0 \tag{6}$$

where $M = K_t\ \theta$, and K_t = rotational spring stiffness. Dynamic base moment will not be allowed to exceed a prescribed value M_y ($M \geq - M_y$), after which, permanent base rotation occurs (see Fig. 2b).

c) <u>Wall–Soil Model</u>: The wall will interact with backfill soil through a no–tension Winkler type springs (k) as shown in Fig. 1. Under dynamic excitation, these springs will sustain compressive forces only (Fig. 3). It may be noted from Fig. 3 that the force–displacement relation of the springs (k) will always fill any tensile gaps and maintain continued interaction between wall and backfill.

Considering interaction forces and ground motion excitation, the equations of motion of the wall–soil system can be written as [1,4]:

$$\rho\, u_{,tt} - G\, u_{,zz} - E\, u_{,xx} = - \rho\, u_{g,tt} \qquad \text{(Soil System)} \tag{7}$$

$$m\, u_{b,tt} + E'I\, u_{b,zzzz} + F_{ws} = -m\,\{u_{g,tt} + u_{bs,tt} + (z - H_d)\,\theta_{,tt}\}$$

(Retaining Wall) (8)

Employing the mode shapes defined earlier for the wall system and for the soil system to represent the solution in the respective domains, a matrix equation is obtained [4]. Newmark's predictor–multi–corrector implicit scheme with or without user specified numerical damping is employed [1,4] to obtain a step–by–step solution of this matrix equation. Note that due to the nonlinear force–displacement relations (Figs. 2 and 3), iterations will in general be performed at each time step so as to achieve a specified convergence tolerance.

Proportional viscous damping may be introduced as a weighted combination of the inertial and stiffness terms. Radiation damping may also be conveniently included through viscous boundary dampers. Radiation damping in proportion to the stiffness terms derived from the various soil boundary springs (K, K_f and K_b) will achieve this goal.

FORMULATION OF SIMPLIFIED SOIL MODEL

In the present analysis a simplified version of the previously described general wall–soil model is employed [1]. A model suitable for purposes of comparison with available centrifuge test data is derived. In a typical centrifuge test, soil boundaries are essentially rigid with about equal wall and soil height. In this simplified model (Fig. 4), the following assumptions are made; 1) soil is fixed at the far end, 2) soil base is fixed, and 3) wall height is equal to soil domain height ($H_s = H_w = H$, and, $K = k$).

SIMULATION OF CENTRIFUGE MODEL DYNAMIC RESPONSE

In view of the virtual absence of actual recorded retaining wall strong–motion, no quantitative comparisons with actual case histories are possible. The proposed simplified numerical model is used to simulate two centrifuge test results conducted on a gravity retaining wall model [3,14].

Bolton and Steedman [3], and, Steedman [14] subjected a rectangular gravity wall of stiff plywood construction to base shaking. The tested model along with the locations of different accelerometers and LVDT's is shown in Fig. 5. This model, at a simulated gravity of 80 g, corresponds to a 10.8 m high, 7.2 m wide prototype, with an equivalent mass of 21800 Kg/m and resting on a dry sand backfill of 2.4 m depth. Sand particles were glued to the model base prior to placement to ensure rough surface contact. The face of the retaining wall was left smooth. Accelerations in soil behind the wall and on the wall as well as wall movements (at wall base, center and top) were recorded. The wall was subjected to two different harmonic (primarily) base lateral excitations imparted by "the Bumpy Road" mechanism; Case (i): Maximum lateral base acceleration amplitude of 0.211 g, and Case (ii): Maximum lateral base acceleration amplitude of 0.334 g. Input and recorded motions (in prototype scale) are shown in Fig. 6 for these two cases.

The proposed numerical model is used to simulate the above centrifuge results. In the computational model, a nonlinear base rotational spring (K_t $= 4.5 \times 10^9$ Kg m/radian) is employed. Other modeling parameters are: Youngs Modulus of beam = 6.9×10^9 Kg/m^2, mass density of beam = 240 Kg sec^2/m^4 wall–backfill interaction spring k = 5.0×10^5 Kg/m, Youngs modulus of soil = 3.8×10^7 kg/m^2, Shear Modulus of soil = 1.4×10^7 Kg/m^2, mass density of soil ρ = 174 Kg sec^2/m^4, and proportional viscous damping co–efficient = 3%. The yield values for translation and rotation (maximum base force = 17180 Kg/m and maximum base moment = 102,000 Kg m/m) are obtained so as to match the results of case (i) and these values were then used to simulate case (ii). The computed results are shown in Fig. 7. It is noted that computed accumulated displacements (Fig. 7) in case (ii) stop at 8 secs. as opposed to 12 secs. in the actual test (Fig. 6). Also in the centrifuge test, an additional permanent displacement of 0.32 m is accumulated in the time frame 8–12 secs. In this time frame, input excitation (Fig. 6) is of fairly low amplitude. This may indicate that other factors not included in the present numerical model (e.g. partial loss of contact between wall base and soil) may have played a significant role in the centrifuge test results after 8 secs. of shaking.

SEISMIC RESPONSE OF CANTILEVER RETAINING WALL

The proposed simplified model is used in this section to study the seismic response of a retaining wall–backfill system with possible wall translation and rotation. The chosen concrete cantilever retaining wall is of 15 m height and 1.2 m thickness supporting a 150 m long dense sand backfill. Properties of the system are: Elastic Modulus of beam = 3.2×10^9 Kg/m^2, mass density of beam = 240 Kg sec^2/m^4, wall–backfill interaction spring constant k = 5×10^5 kg/m, Youngs Modulus of soil = 3.12×10^7 Kg/m^2, Shear Modulus of soil = 1.2×10^7 Kg/m^2, mass density of soil = 216 Kg sec^2/m^4, proportional viscous damping parameter = 3%, soil internal friction angle = 36° and nonlinear wall base rotational spring K_t = 4.5×10^9 Kg m/radian. A base translational yield force of F_y = 31590 Kg/m and a base yield moment of M_y = 157950 Kg m/m are used (corresponding to a F.S. of 1.5 against translation and rotation). A scaled El Centro 1940 S00E earthquake acceleration record is used to represent ground (base) excitation in the present analysis. The maximum acceleration of this scaled record is 0.60 g. Numerical results obtained are shown in Fig. 8. It is found that all the accumulated displacement is due to rotation with no base translation. Failure due to rotation is apparently more probable given an equal factor of safety in translation and rotation. A residual pressure is maintained on the retaining wall at the end of dynamic shaking (Fig. 8).

SUMMARY AND CONCLUSIONS

A computational model for the dynamic analysis of retaining wall–soil systems is developed, and its formulation is presented. The retaining wall is represented by a bending beam with a base yielding rotational spring and a translational slide element. Backfill soil is represented by a 2D shear beam with radiation damping effects. The wall and supported soil interact through a Winkler type nonlinear no–tension spring system. This model is used to simulate numerically the response of a gravity retaining wall model [3,14]. A seismic analysis of a 15 m cantilever retaining wall is also conducted. Assuming an equal factor of safety in translation and rotation, this cantilever wall is found to be more vulnerable to a seismically induced rotational failure.

ACKNOWLEDGEMENTS

This research is supported by NCEER Grant No. 873011.

REFERENCES

1. S Alampalli, Earthquake Response Of Retaining Walls; Full Scale Testing And Computational Modeling, Ph.D. Thesis, Department Of Civil Engineering, Rensselaer Polytechnic Institute, Troy, NY, 1990.
2. A Arias, FJ Sanchez–Sesma & E Ovando–Shelley, A Simplified Elastic Model For Seismic Analysis Of Earth–Retaining Structures With Limited Displacements, Proc. International Conference On Recent Advances In Geotechnical Engineering, Rolla, Missouri, May 1981.
3. MD Bolton & RS Steedman, Modelling The Seismic Resistance Of Retaining Structures, Proc. XI International Conference On Soil Mechanics And Foundation Engineering, San Francisco, August 1985.
4. A–W Elgamal & S Alampalli, Dynamic Response Of Retaining Walls Including Supported Soil Backfill, submitted for journal publication, 1990.
5. J Lysmer & RL Kuhlemeyer, Finite Dynamic Model For Infinite Media, J. Of Engineering Mechanics Division, ASCE, Vol. 95, 1969.
6. F Nadim & RV Whitman, Seismically Induced Movement Of Retaining Walls, J. of Geotechnical Engineering, ASCE, Vol. 109, July 1983.
7. F Nadim & RV Whitman, Coupled Sliding And Tilting Of Gravity Retaining Walls During Earthquakes, Proc. Eighth World Conference On Earthquake Engineering, San Francisco, CA, July 1984.
8. HN Nazarian & AH Hadjian, Earthquake Induced Lateral Soil Pressures On Structures, J. of Geotechnical Engineering, ASCE, Vol. 105, 1979.
9. S Prakash, Soil Dynamics, McGraw–Hill book Company, NY 1981.
10. S Prakash, VK Puri & JU Khandoker, Displacement Analysis Of Rigid Retaining Walls In Rocking, Proc. International Conference On Recent Advances In Geotechnical Engineering, Missouri, April 1981.
11. RF Scott, Earthquake Induced Earth Pressure On Retaining Walls, Proc. Fifth World Conference On Earthquake Engineering, Rome, Vol. 1, 1973, pp. 103–145.
12. R Siddarthan, GM Norris & E Maragakis, Deformation Response Of Rigid Retaining Walls To Seismic Excitation, Proc. Fourth International Conference On Soil Dynamics And Earthquake Engineering, Mexico City, Mexico, October 1989.
13. R Siddarthan, S Ara & JG Anderson, Seismic Displacements Of Rigid

Retaining Walls, Proc. Fourth US National Conference On Earthquake Engineering, Palm Springs, California, Vol. 3, May 1990
14. RS Steedman, Modelling The Behavior Of Retaining Walls In Earthquakes, Ph.D. Thesis, presented to Cambridge University, 1984.
15. L Su, G Ahmadi & IG Tadjbakhsh, A Comparative Study OF Performances Of Various Base Isolation Systems, Part 1: Shear Beam Structures, Report MIE–153, Clarkson University, October 1987.
16. HB Tajimi, Dynamic Earth Pressures On Basement Wall, Proc. Fifth World Conference On Earthquake Engineering, Rome, June 1973.
17. JH Wood, Earthquake Induced Soil Pressures On Structures, Ph.D. Thesis presented to the California Institute Of Technology, Pasadena, California, May 1973.

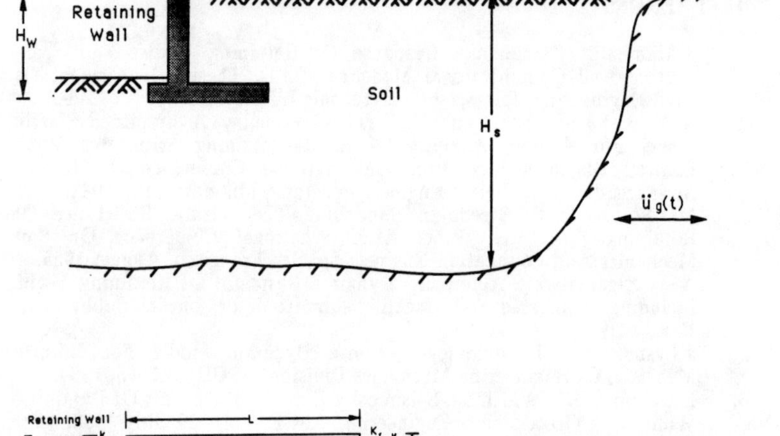

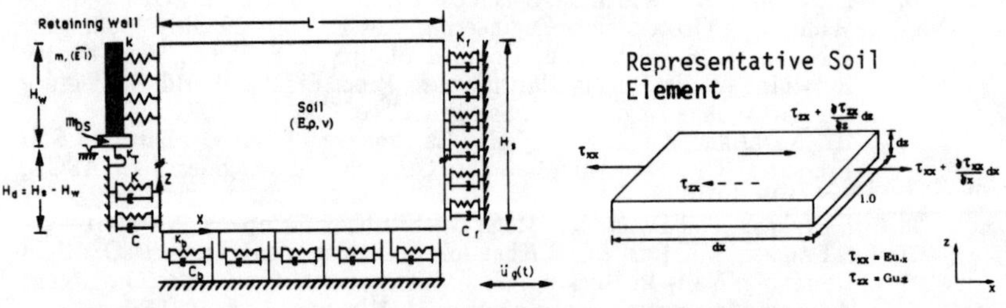

Figure 1. Wall-Soil Configuration And Corresponding Computational Model.

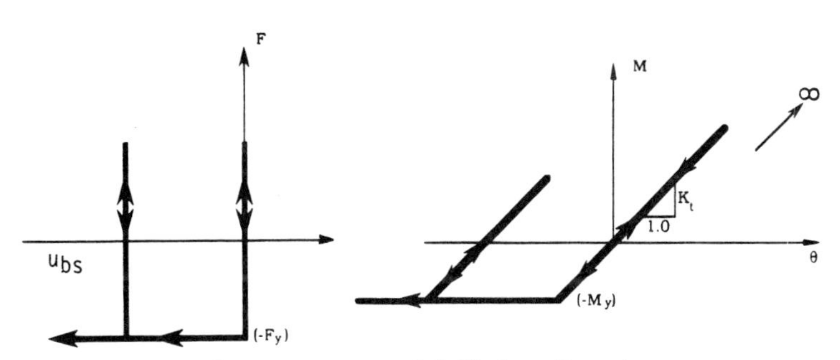

u_{bs} =Wall Base Displacement.
F =Wall Base Force.

θ=Wall Base Rotation.
M=Wall Base Moment.

(a) Wall Base Translational Sliding Element.

(b) Wall Base Rotational Spring.

Figure 2. Nonlinear Wall Base Translational And Rotational Formulation.

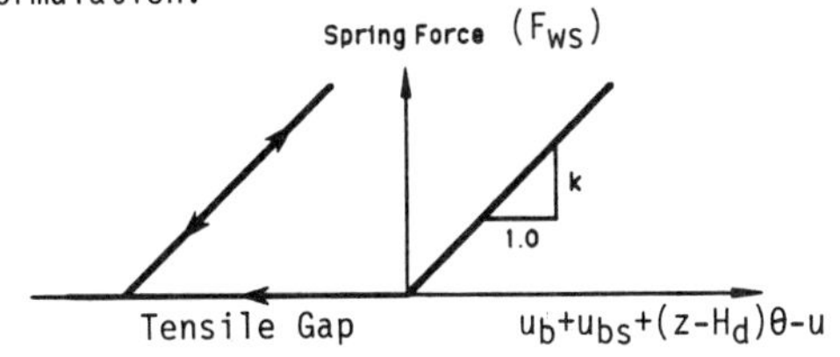

Figure 3. Wall-Backfill Soil Interaction Nonlinear Spring Formulation.

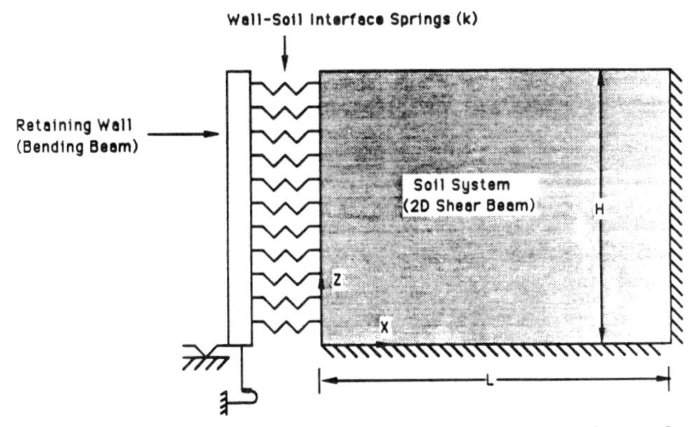

Figure 4. Simplified Wall-Soil Computational Model.

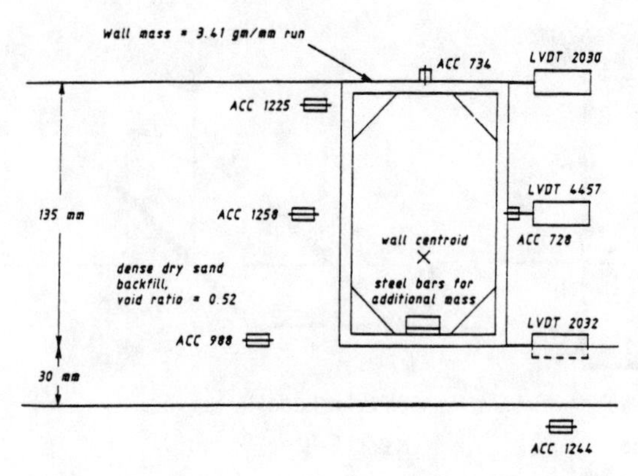

Figure 5. Gravity Retaining Wall Centrifuge Model [14].

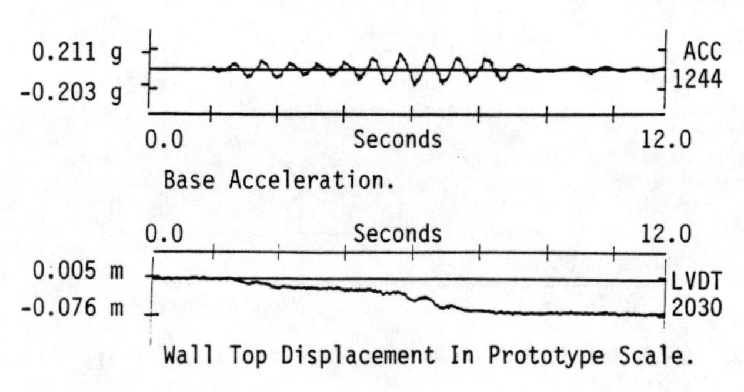

(a) Recorded Test Data Of Case (i).

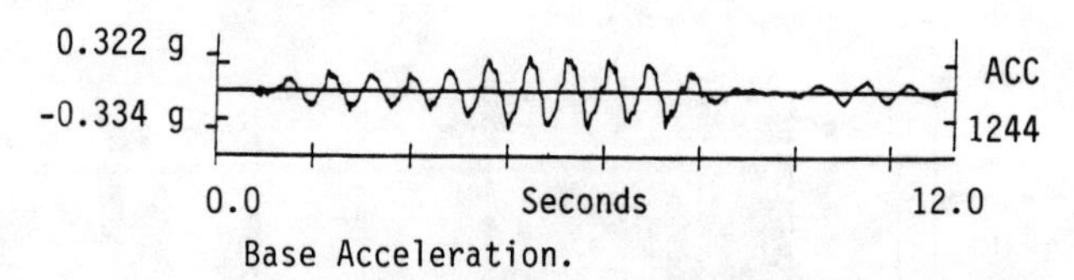

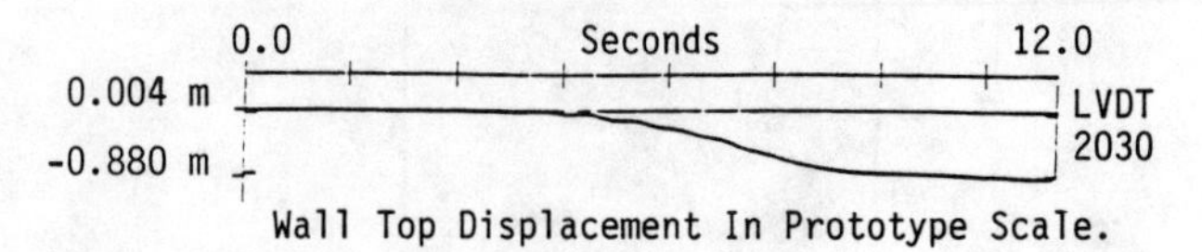

(b) Recorded Test Data Of Case (ii).

Figure 6. Gravity Retaining Wall Centrifuge Test Results [14].

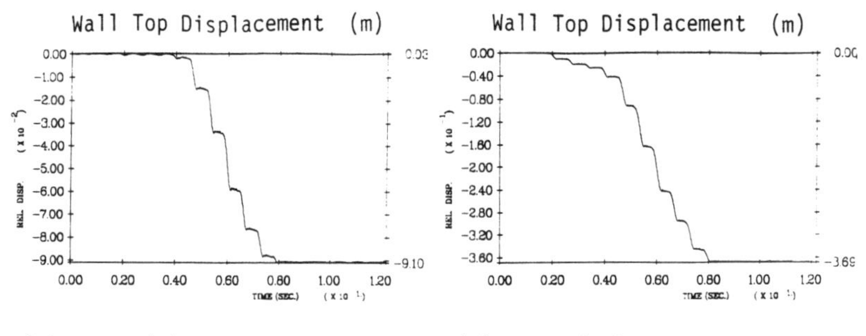

(a) Case (i). (b) Case (ii).

Figure 7. Numerical Simulation Results Of Centrifuge Wall Model Tests.

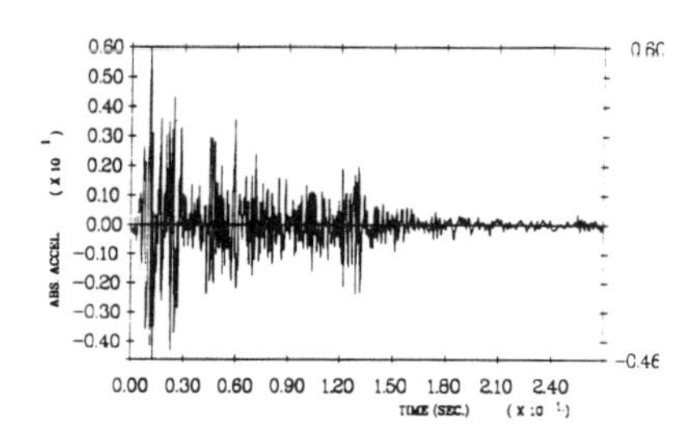

(a) Base Excitation (m/sec/sec).

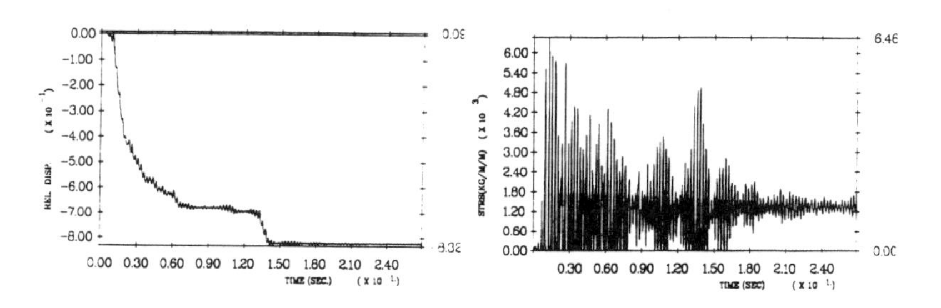

(b) Rel. Displacement At Wall Top (m). (c) Stress History At Wall Top (Kg/m/m).

Figure 8. Numerical Simulation Of 15 m High Cantilever Wall-Soil System.

THREE DIMENSIONAL MODELING OF GROUND WATER FLOW AND TEMPERATURES AT BONNEVILLE DAM, OREGON

Dirk Baron[1], David H. Scofield[1], M.ASCE, Ansel G. Johnson[2], Richard S. Malin[2], and James D. Graham[3]

ABSTRACT: Results of a study for the Bonneville Fish Hatchery including field monitoring of ground water levels, pumping rates, and ground water temperatures, as well as hydrogeologic analysis and numerical modeling, are presented. This investigation is part of the overall design and construction of a new navigation lock at Bonneville Lock and Dam. The Bonneville Fish Hatchery well field pumps up to 1.1 m^3/s from a relatively small alluvial aquifer located adjacent to the Columbia River. The approach channel for the new navigation lock will pass through the existing hatchery well field, so a new well field has to be developed. Concerns for this well field include possible water shortages during deep dewatering to permit lock excavation, potential thermal degradation of the water supply, and potential biological and chemical contamination during and after construction of the lock and lock approach channel. As the final product of the project a numerical model of the ground water system at the site integrating results of field monitoring, subsurface exploration, and hydrogeologic analysis was developed. The model will be used for aquifer management during construction and for operation of the new well field, with the goal of maximizing the yield of water without deterioration of the resource.

INTRODUCTION

This paper presents the results of a multi-year detailed hydrogeologic investigation of the Bonneville Fish Hatchery well field. The work was initiated at the Portland District Corps of Engineers by the authors beginning in 1986. The study was a part of the geotechnical investigation for the design of the new navigation lock at Bonneville Lock and Dam. The construction of the new navigation lock necessitates the relocation of the existing Bonneville Fish Hatchery well field.

The Bonneville Fish Hatchery and associated well field is part of the Bonneville Lock and Dam project. The Oregon shore of the Bonneville project is shown on Figure 1, Site Map. The site map shows the location of the hatchery, the existing well field, the new well field, and the new navigation lock under construction. Included on the site map are the locations of the various ground water monitoring instruments.

The hatchery raises approximately 30 million salmon a year and is an important component in helping to maintain the salmon runs on the Columbia River. A critical

1 - Hydrogeologists, Squier Associates, Inc., P.O. Box 1317, Lake Oswego, Oregon 97035.
2 - Department of Geology, Portland State University, Portland, Oregon 97207-0751.
3 - Geologist, NEA, Inc., 10950 S.W. 5th Street, Beaverton, Oregon 97005.

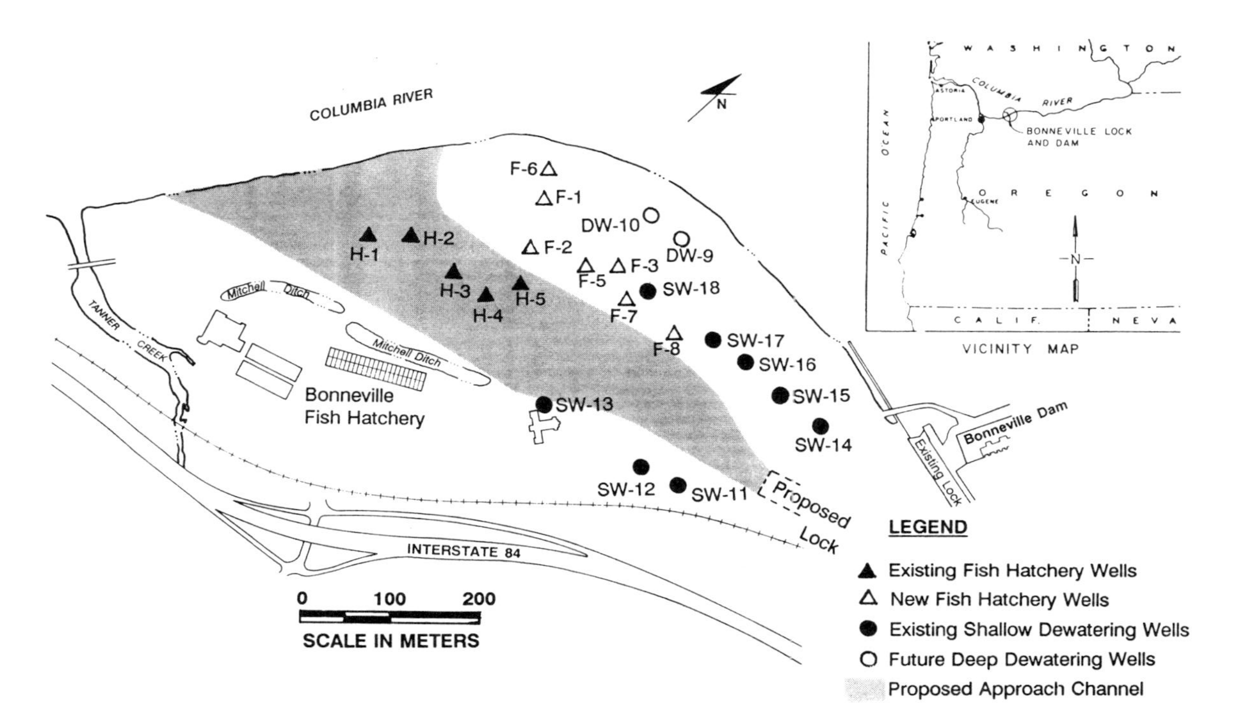

Figure 1. Vicinity Map of Bonneville Dam and Site Plan of the Downstream Terrace Area.

element for the success of the hatchery is a continuous supply of large quantities of clean and thermally moderate water. Highest survival rates of fish are achieved with water in a temperature range between 9.5 and 11.0^{o}C, and free from chemical contamination and biologic pathogens. The existing well field with five large production wells provides up to 1.1 m^3/s of water that is free of fish pathogens, chemically pure, and within a narrow temperature range. The wells pump from a confined alluvial aquifer located adjacent to the Columbia River. For the existing well field the aquifer system effectively filters out pathogens and moderates the seasonally fluctuating river temperatures. It was not known, however, how relocation of the well field and the excavation of the approach channel would affect water quality and temperatures.

To assess concerns about meeting the hatchery requirements for water quantity and quality during and after construction, the Corps of Engineers initiated a detailed hydrogeologic investigation. The results of this investigation were incorporated into the development of a three dimensional ground water model for the site.

WELL FIELD MONITORING AND DATA COLLECTION

The Corps of Engineers initiated studies to evaluate the hydrogeologic characteristics of the aquifer for the purpose of understanding the existing aquifer conditions, establish baseline conditions, design a new well field, and design the dewatering system. The hydrogeologic characteristics of the aquifer were determined through an extensive instrumentation and monitoring program. All permanent instrumentations for measurements of pumping rate and temperature for individual new hatchery wells will be automated as part of a long term instrumentation and monitoring program.

Water Level Monitoring. Water level monitoring was accomplished by an automated data acquisition system (ADAS) consisting of 28 pressure transducers installed in 14 borings. Most of the borings monitored two aquifer zones. The ADAS provided real time monitoring to allow for automated changes in dewatering pumping rates to accommodate variations in fish hatchery well pumping rates. The purpose of water level monitoring was to provide data for characterization of the aquifer system, to provide information on the piezometric head in the various layers, and to provide information relating to the dewatering progress.

Ground Water Temperature Monitoring. Ground water temperature profiles were measured in selected piezometers at two week intervals. Ground water temperatures were measured by lowering a thermistor and recording the temperature at predetermined depths. The data collected showed substantial noise due to construction techniques employed in the installation of piezometers. The ground water temperature profiles did show the locations of rapid flow of ground water of contrasting temperatures.

River Stage and Temperature Monitoring. River stage and temperature monitoring was initially done by the Bonneville Powerhouse as part of normal project operations. River stage is collected on an hourly basis below the powerhouse. However, a dedicated pressure transducer was later installed in the river to monitor river level directly on a continuous basis. River level fluctuates up to 0.6 meters per hour and two meters per day in response to changes in river flow due to power generation demands. Water temperature is collected on a daily basis within the generating units. This temperature is a representative temperature of the river as the water is homogenized as it passes through the turbine units.

Hatchery Water Demand Monitoring. Hatchery water pumping rates for the existing wells were recorded manually by hatchery personnel. The data were estimates with approximate times of starting and stopping different pumps. Pumping rates for each individual well were measured approximately every two weeks or whenever a well was changed. Temperatures at each well were measured every two weeks.

SITE GEOLOGY AND HYDROGEOLOGY

Based on subsurface explorations, analysis of pumping tests, water level data from manual and the automated data acquisition system, and ground water temperature data, the extent and properties of different hydrogeologic units were determined and a conceptual understanding of the ground water system was developed (1,2).

A sequence of alluvial sediments up to 100 m thick, which contains the major hydrogeologic units, has accumulated in the Columbia River Gorge at Bonneville. The bedrock that underlies the alluvial sediments and the rocks that form the walls of the Columbia River gorge are of relatively low permeability and do not contribute to the hatchery water supply. Figure 2 shows a schematic cross-section for the site.

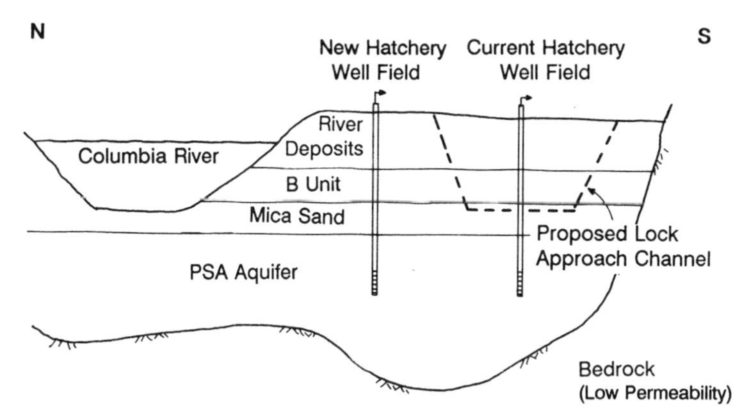

Figure 2. Schematic Cross-Section showing Geologic Units, Locations of the Current and the New Fish Hatchery Well Field, Columbia River, and the Proposed Lock Approach Channel.

The deepest alluvial unit, the pre-slide alluvium (PSA), consists of sand and gravel of variable permeability. Discontinuous lenses of gravels within the PSA are the water source for both the old and the new hatchery well field. This unit is overlain by two units of fine-grained deposits, the mica sand and the b-unit. The uppermost unit in the alluvial sequence are recent river deposits (RD) which are again very permeable. Ground water in this unit is unconfined and in direct connection to the Columbia River.

The PSA aquifer is separated from the upper unconfined aquifer (RD) by the two aquitard layers, mica sand and b-unit. The PSA aquifer also appears to be effectively separated from the Columbia River, probably by these layers or possibly by silty sediments on the bottom of the river channel. Direct recharge from the river into the PSA aquifer does not appear to occur to a large extent (2). Instead recharge occurs over a large area by vertical leakage through the aquitard intervening layers. Pumping in the PSA aquifer generates a large cone of depression which sets up a hydraulic gradient between the overlying deposits and the aquifer. This gradient causes water to leak downward in a distributed manner into the aquifer. Although the vertical hydraulic conductivity of the aquitard layers is low, the large area over which flow is induced supplies sufficient recharge to yield high quantities of water to wells constantly. Water that leaks downward is replaced by water from the Columbia River that flows into the system through the aquitard layers and the highly transmissive river deposits.

GROUND WATER TEMPERATURES

Ground water temperatures are influenced by the seasonal river temperature fluctuation and by the flow of ground water, i.e. mass of flowing water per unit mass of aquifer. Under present conditions the aquifer system moderates the river temperature fluctuations considerably. Figure 3 shows the temperature of the water pumped from wells H-2 through H-5 and the Columbia River temperature for the year 1988. While the river temperature varies between three and 22°C, water temperatures in the existing fish hatchery wells fluctuate only between seven and 14°C. Temperatures in the fish hatchery wells also lag behind the annual river cycle by about three to four months. The amount of temperature fluctuation and the lag time for individual wells depends on their proximity to the Columbia River. Generally, wells closer to the river have higher temperature fluctuations and temperatures and follow the seasonal river temperature cycle with a shorter lag time. Wells further away from the river generally have smaller temperature fluctuations and lag behind the seasonal river temperature cycle by up to four months. However, another important factor that influences water temperatures at individual wells is the rate at which they are pumped. Pumping at high rates tends to pull in river water faster, thus increasing temperature fluctuations and shortening the lag time. The operation of the well field is manipulated to optimize the yield of water in the desired temperature range between eight and 11°C throughout the year by using these effects.

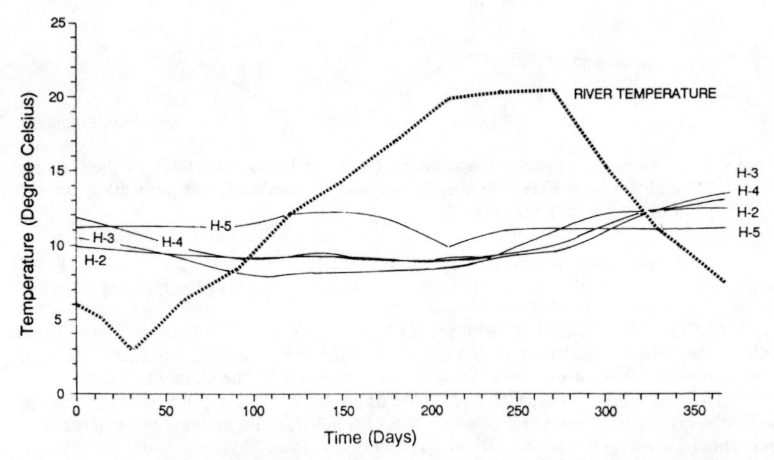

Figure 3. Annual Temperature Cycle of the Columbia River and the Current Hatchery Wells. The Temperature Data are from 1988.

In addition to the hatchery wells, ground water temperatures profiles were measured in piezometers distributed throughout the study area. Generally, the annual ground water temperature fluctuation is the highest in the vicinity of the river and in the shallow unconfined aquifer. Away from the river and in the deep PSA aquifer the annual fluctuation is smaller. In the deep aquifer temperature fluctuation appears related to pumping which drives the ground water flow in the system. The fluctuation is the highest above the depth where the hatchery wells are screened. In the deeper parts of the PSA aquifer, where no active pumping occurs, temperatures are nearly stable.

NUMERICAL GROUND WATER MODELING

Model Selection. Criteria for the selection of a ground water modeling program were:

1. capability to simulate a three dimensional ground water system;
2. capability to simulate heat transport;
3. capability to simulate solute transport;
4. capability to include heterogeneous and anisotropic porous medium properties; and
5. availability of documentation and support.

The ground water modeling program HST3D (3) was selected as the software most suitable for the study. The program allows for a fully three dimensional simulation of ground water flow with associated heat and solute transport using finite-difference techniques. This program is the simplest available to solve the problem as defined.

Initially, the program was run on an Amdahl mainframe computer. However, because of high costs for time on the mainframe and the recent availability of powerful microcomputers, runs were switched to an Intergraph Interserve 200 computer. The run times for individual simulations were slower on the microcomputer. However, data handling and trouble shooting on the microcomputer was significantly easier and faster resulting in a shorter turnaround time.

Discretization. A total area of 1,000 by 1,500 m, including the downstream terrace area and extending beyond the Columbia River to the northwest was included in the model. The model represents a thickness of 69 m, corresponding approximately to the saturated thickness of the aquifer system in the downstream terrace area (2). The modeled area was discretized into a grid of 23 by 17 by 9 nodes. Horizontal grid spacing was varied from 50 m to 200 m according to the detail required in different areas. Vertical grid spacings correspond to the average thicknesses of individual stratigraphic units.

Six hydrogeologic units with different properties were included. The extent of individual units was assigned according to subsurface data from borings in the downstream terrace area. The bedrock was represented by a zone of low permeability, thus essentially excluding it from the simulation. A total of 24 water wells representing existing and future hatchery wells, existing shallow dewatering wells, and future deep dewatering wells were included in the model.

The Columbia River was represented by a set of specified pressure and temperature boundary condition nodes. Pressure, which corresponds to hydraulic head, and temperature at the boundary nodes can be set according to river stages and temperatures that are to be simulated. Pressure and temperature at these nodes can be changed during a simulation, thus making it possible to simulate long periods (annual cycle) of changing river stage and temperature.

Ground Water Flow Calibration. Two independent sets of conditions were used to calibrate the model for ground water flow. The first set of conditions was a pumping test performed in one of the fish hatchery wells (2). Calculated water levels were compared to water levels measured in piezometers during the pumping test. Permeabilities of hydrogeologic units were varied until a satisfactory match of observed and calculated water levels was achieved.

A second set of conditions was used to verify the calibration. Figure 4 shows the comparison of observed and calculated water levels in the deep and shallow aquifers for the final verification run. Water levels in shallow and deep piezometers, under stress from pumping from both shallow and deep wells, were used for the ground water flow verification. Observed water levels in both the deep and the shallow unconfined aquifer were matched to water levels calculated by the model. This made it possible to evaluate the interaction between the shallow and the deep aquifer, an important relationship for the evaluation of the effect of the future approach channel on the PSA Aquifer. The procedure for the verification was to run both sets of conditions with the same hydrogeologic parameters for the different stratigraphic units. Then permeabilities of hydrogeologic units were varied and the simulations of the two different sets of conditions were run again. Changes in water levels as a result of changed permeabilities were noted and the permeabilities that gave the best match of observed and calculated water levels for both sets of conditions were determined. It was found that the horizontal hydraulic conductivity of the PSA aquifer and the vertical hydraulic conductivity of the mica sand unit were the parameters that largely control the water levels in the aquifer system. Values for hydrogeologic parameters determined from prior analysis and the values that give the best results in the model calibration and verification are very similar. The fact that no unrealistically large changes in the hydrogeologic properties were necessary to achieve a satisfactory match of observed and calculated water levels, increases the confidence in the model calibration. The successful calibration and verification of the ground water flow model supports the belief that the conceptual model on which it was based adequately describes the ground water system at the site.

GROUND WATER TEMPERATURE CALIBRATION

Selection of Calibration Conditions. Measured river water and ground water temperatures from the four-month period including July, August, September, and October of 1988, were used to calibrate the model for heat transport. This period was chosen for the following reasons:

1. The temperature difference between river water and ground water at the beginning of this period is large. The effect of warm river water entering the cold aquifer is expected to be large and easier to observe.
2. River temperature during July, August, and September is almost constant at approximately 20°C. A front of warm river water is expected to enter the aquifer system at a constant temperature over a prolonged period of time.
3. Warm river water entering the ground water system during the summer and excessively warming up the PSA aquifer above the optimum temperature is a primary concern for the new well field.

Observed Temperatures. River water temperatures and measured average ground water temperatures in the fish hatchery wells were used to calibrate the model for heat transport. Measured water temperatures in the hatchery wells rose between 1.5°C and 2°C during the period from July to October, 1988. The magnitude of temperature change appears to depend on proximity to the Columbia River and on the pumping rate. Wells close to the river and wells that are pumped continuously at high rates show greater changes. This behavior has been observed in the past and the hatchery personnel uses this knowledge to operate selected pumps to avoid undesirable temperature changes.

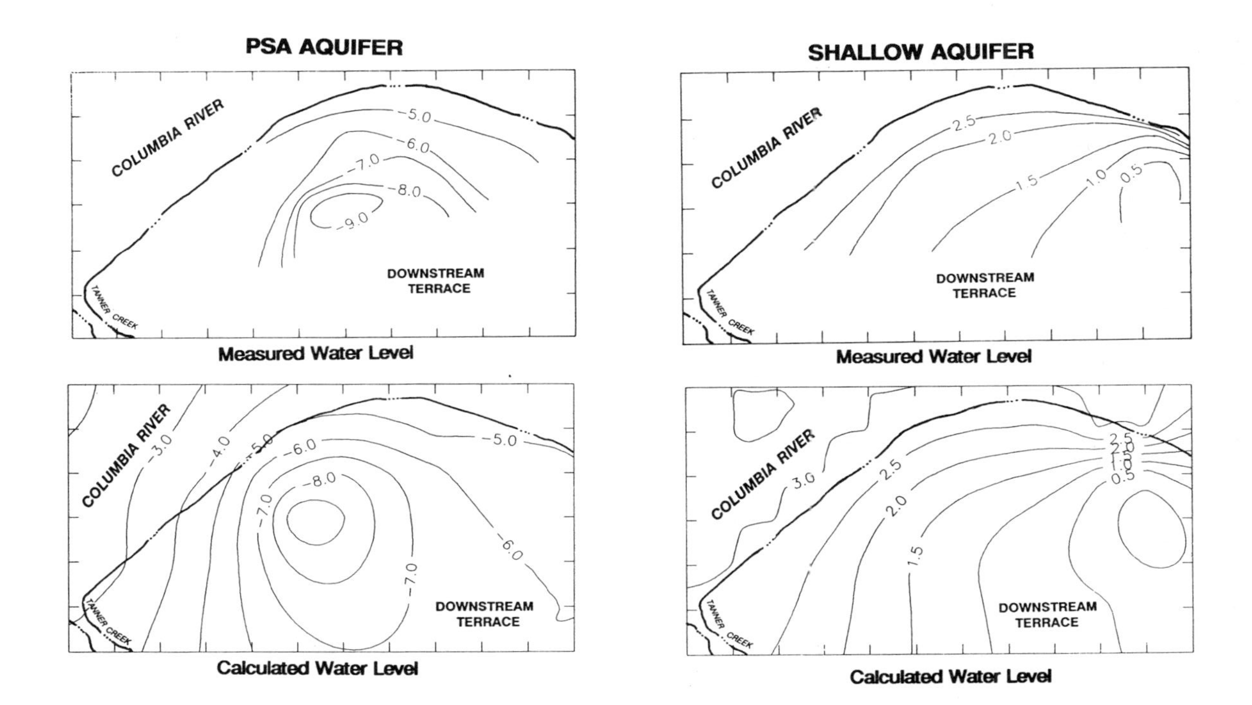

Figure 4. Ground Water Flow Verification. Measured and Calculated Water Levels in the Shallow Unconfined Aquifer and in the Deep PSA Aquifer are shown.

During the summer, wells close to the river (H-1 and H-2) are shut off and wells away from the river are pumped. This makes it possible to limit the average temperature rise to about 2°C.

Temperature Calibration. To simulate the four-month period from July to October, 1988, ground water temperature was initially set to 9°C, representing the average ground water temperature at the beginning of July. River temperature was set to 20°C and left at that temperature for 90 days representing July, August, and September, with relatively constant river temperatures. After 90 days the river temperature was changed to 17°C representing the beginning of the cooling of river water in the fall. Pumping rate from the well field was set representing the average pumping rate during the four-month period.

Thermal conductivity and heat capacity of the aquifer materials were initially set to appropriate values reported in the literature. During the calibration these parameters were varied within a range that, according to the literature, can be expected for alluvial deposits. In addition, aquifer dispersivity was varied. The effect of changes in these parameters on ground water temperatures was evaluated in order to achieve the best possible match between measured and calculated temperatures.

Discussion of Temperature Calibration Runs. Evaluation of the four temperature calibration model runs suggest that ground water temperatures are not sensitive to changes in the thermal properties of the aquifer material within the ranges of values tested. Varying heat capacity and thermal conductivity showed only negligible effects. The porous medium dispersivity appears to control the heat distribution in the aquifer system. High dispersivities result in fast warming of the aquifer system. Lowering the dispersivity resulted in slower spreading of warm river water.

A dispersivity of 50 m (longitudinal) and 5 m (transverse) gave a reasonable match between measured and calculated water temperatures in the hatchery wells. Figure 5 shows the calculated temperatures in the hatchery wells. Temperatures rise between 4.5°C in well H-1 and 1°C in well H-4. Average water temperature in the well field in the model rises about 2°C, corresponding to the observed average rise in water temperature in the well field. It is important to note that in the simulation the pumping rates in the wells were constant and corresponding to the average pumping rate from the well field. In the real well field pumping rates are continuously adjusted to avoid undesirable warming up of individual wells, thus achieving relatively constant temperature. Therefore, temperature differences between individual wells, are higher in the simulation than in the real well field. The similar average temperature rise in the simulation and in the real well field indicates that the thermal properties and the dispersivity of the porous media in the model are reasonable.

SIMULATION OF ANNUAL TEMPERATURE CYCLES

Simulation of the Annual Pumping and Temperature Cycle the of Existing Well Field. The calibrated and verified model was run using the actual pumping rates of the fish hatchery wells modified every two weeks for 1988. Figure 6 shows the river temperature, and the temperature of the water in the pumping wells simulated for pumping rates during the year 1988. The initial river temperature was 6°C, and the aquifer was started at 10.5°C. Wells H-4 and H-5 show the slowest change, and the least change with the river temperature. The river temperature varies from a low of 3.5 °C in February to a high of 22 °C in September. Wells H-1 and H-2 have the highest variations, and are often shut off when their temperature gets too far outside the usable temperature range.

A comparison with Figure 3 shows that the amount of temperature variation in the wells closely approximates the temperatures reported to the fish hatchery. The model is able to simulate the temperature of the water being pumped.

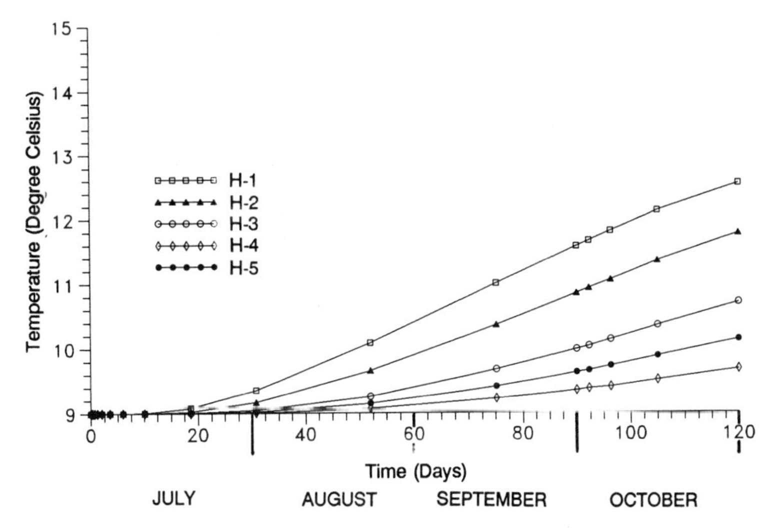

Figure 5. Calculated Ground Water Temperatures from a Simulation of the Four Month Period from July through October, 1988. For this Simulation the Initial Ground Water Temperature was set to 9°C; after 90 Days the River Temperature was Lowered to 17°C, corresponding to the cooling of the River in the Fall.

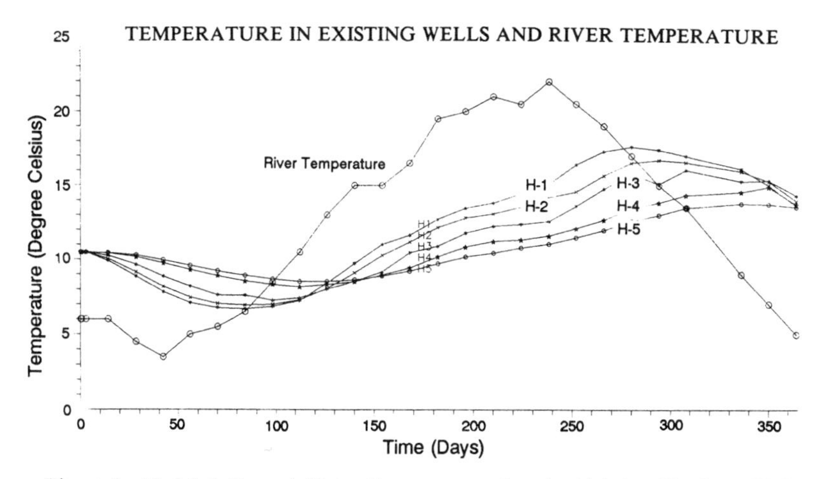

Figure 6. Modeled Ground Water Temperatures for the Existing Hatchery Wells. Simulation for January 1, 1988 through December 31, 1988.

Simulation of the Annual Pumping and Temperature Cycle of the Future Well Field. A primary concern for the fish hatchery operation is the temperature of the water to be produced by the new wells. To evaluate this, a simulation was run using wells at the new locations. The pumping rates and the river temperature were taken from the 1988 pumping simulation, but assigned to five new wells labeled F-1 through F-3, and F-5 and F-6. The proposed well site F-4 was deleted before construction began. Figure 7 shows the results of this simulation. Wells F-2, F-3 and F-5 show the least variation with the river temperature, while F-6 and F-1 show the highest temperature changes. This confirms earlier observations that temperature changes are largest close to the river. A comparison with the simulation of the existing well field (Figure 6) shows that the three wells with the smallest temperature changes behave similarly to the old hatchery wells, with perhaps 1°C more variation. In the model, moving of the wells causes only a slight increase in the temperature of the water in the wells farther away from the river (F-2, F-3, F-5) under the same pumping conditions. The temperature of the wells close to the river (F-1 and F-6) increases an additional 3°C in the fall, according to the model.

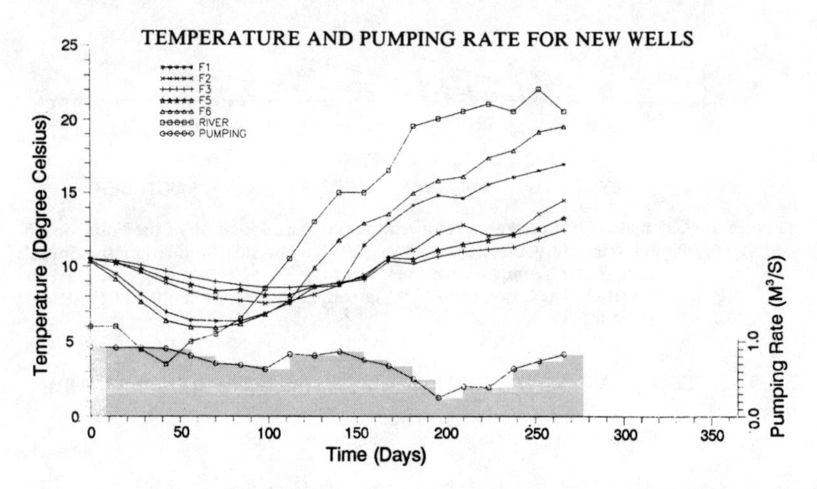

Figure 7. Simulation of the Groundwater Temperatures Using the New Hatchery Wells. Using 1988 River Temperature and Pumping Rates.

SUMMARY AND CONCLUSIONS

A three dimensional ground water model has been constructed which reproduces the temperature of the Bonneville fish hatchery well system. The model was tested using pumping tests, and comparing the temperature in the pumping wells. The aquifer buffers the water temperature which travels from the Columbia River into the wells. The temperatures in the wells track the river temperature changes, with several months lag time after the temperature changes in the river. The magnitude of the temperature variation in the wells is considerably less than that in the river. A considerable difference in the thermal behavior of the wells can be seen, depending on distance from the river and pumping rate. Generally, wells close to the river, and wells that are pumped at high rates show larger temperature fluctuations. According to modeling results, it appears that the new fish hatchery wells will behave quite similarly to the old hatchery wells, but may have slightly higher temperature variations than the original wells. Variations may be about $1^{o}C$ higher in the new wells farther away from the river, and up to $3^{o}C$ higher in the new wells close to the river.

The model will be used to examine dewatering effects on the temperature of the water for the fish hatchery, simulation of other configurations during construction of the locks. Introduction of the navigation lock channel into the model will be attempted. The model should be useful in helping to operate the system of wells, once the construction has been completed.

ACKNOWLEDGEMENTS

The authors wish to acknowledge the assistance provided by the U.S. Army Corps of Engineers, Portland District. They provided time, and resources to allow each of the authors to work on this project over the past few years.

REFERENCES

1. S.R. Sagstad, S. Farooqui, and D. Scofield, Dewatering and Water Supply Considerations: New Bonneville Navigation Lock, Proceedings of the Conference on Northwestern Ground Water Issues by the Association of Ground Water Scientists and Engineers, Portland, Oregon, 1987.

2. D. Baron, Analysis and Numerical Simulation of the Ground Water System at the Bonneville Navigation Lock Site, Oregon, M.S. Thesis, Portland State University, Portland, Oregon, 1990, 117 p.

3. K.L. Kipp, HST3D: A Computer Code for Simulation of Heat and Solute Transport in Three Dimensional Ground-Water Flow Systems, U.S. Geological Survey Water-Resources Investigations Report 86-4095, 1987, 517 p.

FEM ANALYSIS OF STAGED CONSTRUCTION FOR A REINFORCED EARTH WALL

Nelson N.S. Chou[1], M.ASCE , Sao-Jeng Chao[2], S.M.ASCE,

Ching S. Chang[3], M.ASCE, and Jimmy Ni[4], M.ASCE

ABSTRACT: A Reinforced Earth (RE) wall, approximately 1000 feet (304.8 m) long and up to 40 feet (12.2 m) in height, was constructed over a ten feet (3.0 m) thick layer of very soft clay east of Julesburg, Colorado. The external stability of the wall was questionable with regard to bearing capacity and slope stability. The analyses indicated imminent failure if specific construction precautions were not adopted.

Staged construction was recommended to increase the safety factor for external stability, especially for bearing capacity. Horizontal deformations and settlements were measured in each stage at several critical locations to detect symptoms of potential failure during and after construction. Up to 10 inches of settlement was measured with no noticeable distress in the RE wall.

Finite element analyses were performed using the program DACSAR (Ohta & Iizuka, 1989). The program is capable of simulating the sequential construction operation, nonlinear elasto-visco-plastic behavior of soil, and time-dependent deformation. The anisotropic Cam-Clay model was adopted to simulate the foundation clay behavior whereby the consolidation-induced anisotropy is taken into account. The foundation sand and backfill were simulated by using the Duncan-Chang (Hyperbolic) model.

The results of the finite element analyses compared rather well with the monitored deformations. The effectiveness of using staged construction is analyzed, and the results discussed.

INTRODUCTION

The Colorado Department of Highways (CDOH) constructed a grade separation structure over the Union Pacific Railroad tracks on state highway 385 near Julesburg, Colorado. To allow for future widening of the railroad roadbed and to provide sufficient clearance from the tracks, a Mechanically Stabilized Earth (MSE) wall was proposed with a Reinforced Earth (RE) wall design chosen. The wall was approximately

1 - Senior Geotechnical Engineer, Colorado Department of Highway, 4340 East Louisiana Ave. CO 80222.

2 - Research Assistant, Dept. of Civil Engineering, Univ. of Mass., Amherst, MA 01003.

3 - Professor, Dept. of Civil Engineering, Univ. of Mass., Amherst, MA 01003.

4 - Associate Professor, Dept. of Civil Engineering, Tamkang Univ., Taipei, Taiwan.

1000 feet (304.8 m) long, with a height ranging from 5 feet (1.5 m) at both ends to 40 feet (12.2 m) near the center, as shown in Figure 1. The CDOH Geotechnical Section had conducted a study on the external stability of the wall [1]. In this paper, finite element analysis was performed to 'recover' the wall movement during each stage of construction. In addition, the stress path of the soft foundation soil during construction was investigated.

SITE CONDITION AND LABORATORY TEST

Seven test holes were drilled by CDOH along the proposed RE wall alignment. Figure 1 shows the generalized subsurface strata and wall profile. The cross section of the RE wall is shown in Figure 2. Generally, the subsurface at the proposed location consists of soft to very soft silty clay overlying loose to dense sand and gravel. Bedrock was encountered at depths between 28 to 70 feet (8.5 to 21.3 m). The water table was encountered approximately 4 feet (1.2 m) below the ground surface, and "pumping" of the ground surface occurred when construction vehicles passed. The subsoils in the area were divided into the strata shown in Table 1.

Table 1: Subsoil Properties of RE Wall Site

Depth (ft)	Soil Description	Blow Counts[1] (per ft)
0 - 4	CLAY, silty, medium stiff(crust)	3-6
4 - 14	CLAY, silty, very soft	1-4
14 - 21	SAND, loose	5-9
21 - 31	sand, medium dense to dense	20-64

(1 ft = 0.3048 m)
1 - Standard Penetration Test (per AASHTO 206-74)

Laboratory tests were conducted primarily to study the engineering characteristics of the soft clay stratum. The tests included classification, vane shear, direct shear, unconfined compression, triaxial UU, CIU, CAU, and consolidation. The silty clay was classified as CL by the Unified Soil Classification and as A-7-6 by the AASHTO Classification system. The following soil properties were obtained from the laboratory tests, as shown in Table 2.

Table 2: Laboratory Test Results

Depth (ft)	Soil Description	W(%)	LL	PI	S_u(psf)	ϕ(deg.)
0 - 4	Silty Clay (Crust)	21.0	26	10	480-1000 822 (Avg.)	0
4 - 14	Soft Silty Clay	39.6	45	23	40-790 415 (Avg.)	0
14 - 21	Loose Sand	12.5	-	-	-	35
21 - 31	Dense Sand	10.5	-	-	-	40

(1 foot = 0.3048 m, 1 psf = 47.88 Pa = 0.04788 KPa)

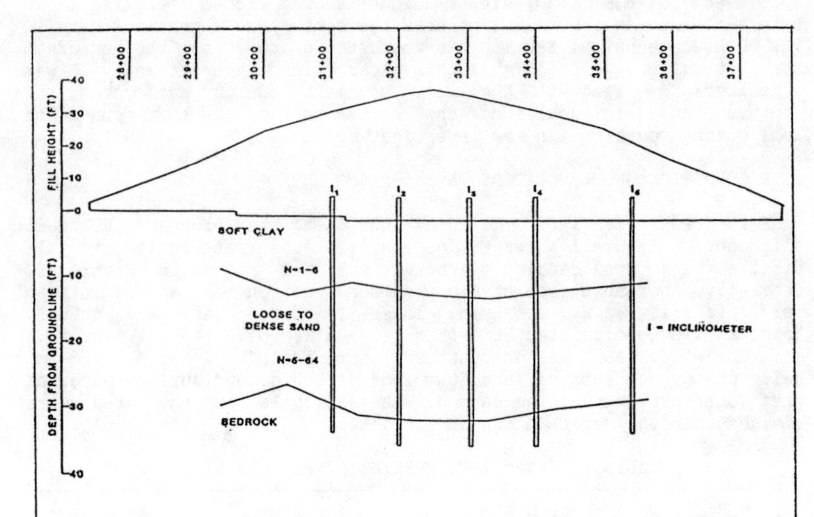

Figure 1: Wall Profile and Generalized Subsurface Strata (1 ft = 0.3048 m)

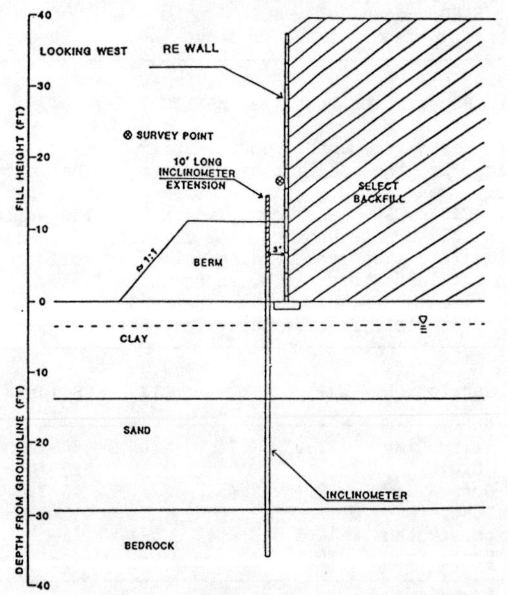

Figure 2: Typical Cross-Section of Reinforced Earth Wall (1 ft = 0.3048 m)

DESIGN CONSIDERATION

The RE wall system must satisfy the requirements of both external and internal stability. After review of the design plan, the internal stability of the wall (designed by the manufacturer) was judged adequate, but its external stability was questionable. The bearing capacity and overall slope stability of the soft clay layer were the major concerns.

A low safety factor of 0.86 for bearing capacity was obtained, based on the undrained shear strength of 415 psf for the soft clay. Staged construction was used to improve the bearing capacity. It was recommended to build the wall in 4 phases, each 10 feet (3.0 m) in height, with at least a 1 month waiting period between each phase.

During construction of the RE wall, the undrained shear strength (S_u) will increase by consolidation. The gain of undrained shear strength, dS_u, can be approximated from the following simple equation [2]:

$$dS_u = m \times I \times d\sigma_c \times U\% \tag{1}$$

Where m = Slope of S_u vs. σ_c, I = Influence factor of loading, $d\sigma_c$ = Increase in mean value of consolidation pressure = $d(\sigma_1+\sigma_2+\sigma_3)/3$, and U (%) = Percentage of consolidation.

Total shear strength gain with resulting bearing capacity safety factor improvements are presented in Table 3. The table shows the undrained shear strength of the clay between the 4 - 14 feet (1.2 - 4.3 m) interval (S_u=415 psf) along with the shear strength improvement (dS_u) at each 10 feet (3.0 m) increment of staged construction. From the table, it can be seen that the safety factors of bearing capacity, during each stage of construction, increased significantly when the increased strength was taken into account.

Table 3: Comparison of RE Wall Bearing Capacity With and Without Staged Construction Considerations

Fill Height (feet)	dS_u (psf)	S_u+dS_u (psf)	Safety Factor: Without Staged construction	Safety Factor: With Staged construction
10	0	415	1.78	1.78
20	180	595	1.05	1.27
30	205	800	0.89	1.14
40	230	1030	0.86	1.10
Long term	409	1439	--	1.54

(1 foot = 0.3048 m, 1 psf = 47.88 Pa = 0.04788 KPa)

An instrumented monitoring system including inclinometers, liquid settlement transducers, and survey points (installed on the wall panels) was proposed to measure the performance of the wall during and after construction. The uses of the instrumented monitoring system were designed to detect early stages of potential wall failure and also provided field data to validate the FEM analysis.

FEM ANALYSIS

Constitutive Models

The effectiveness of using staged construction for strengthening wall stability was investigated by the finite element method (FEM) of analysis. The FEM program used in this investigation, DACSAR (Deformation Analysis Considering Stress Anisotropy and Reorientation), was developed by Iizuka and Ohta [3], and modified by the authors. The constitutive models adopted in this program include:

(1) Anisotropic Cam-Clay model with viscosity
(2) Hyperbolic model
(3) Linear elastic model

(1) Anisotropic Cam-Clay model

The behavior of the soft clay foundation was simulated by the anisotropic Cam-Clay model developed by Sekiguchi and Ohta [4]. This model incorporates simultaneously the effect of time and stress-induced anisotropy in formulating a constitutive law for normally consolidated clay.

Rather comprehensive experimental studies on volumetric-deformation characteristics of clay have been carried out at Kyoto University, Japan. On this basis, Ohta and his co-worker have proposed an anisotropic Cam-Clay model for anisotropically consolidated clay, whereby the stress-induced anisotropy is taken into account. As implied in its name, this model can be regarded as an extension of the Cam-Clay model proposed by Roscoe et al. [5].

This model also considers the effect of the reversal in direction of shearing on the dilatancy. Here, dilatancy is defined as volume changes which occur under loading with p' (i.e., $(\sigma'_1+\sigma'_2+\sigma'_3)/3$) being held constant.

A rheological model considering creep, strain rate, and stress relaxation is also included in this program. By incorporating all the features mentioned above, i.e., elasto-plasticity, dilatancy and rheology, Sekiguchi and Ohta [4] suggested the following equation for calculation of volumetric strain:

$$v = \frac{\lambda}{1+e_o} \ln\left(\frac{p'}{p'_o}\right) + D\,\eta^* - \alpha \ln\left(\frac{\dot{v}}{\dot{v}_o}\right) \qquad (2)$$

where v = Volumetric strain
λ = 0.434 × compression index C_c
p = mean effective stress = $(\sigma'_1+\sigma'_2+\sigma'_3)/3$
e_o, p_o = e & p at the end of K_o consolidation, respectively.
D = Coefficient of dilatancy (Shibata, 1963) [6]

$\eta^* = \sqrt{\frac{3}{2}(\eta_{ij}-\eta_{ijo})(\eta_{ij}-\eta_{ijo})}$, in which η_{ij} is the (i,j) component of 'pressure-normalized deviatoric stress tensor' $(= \frac{s_{ij}}{p})$, and η_{ijo} is the value of η_{ij} at the end of anisotropic consolidation $(= \frac{s_{ijo}}{p_o})$.

α = Secondary compression index

$\dot{v}$ = Volumetric strain rate

$\dot{v}_o$ = Initial value of $\dot{v}$, i.e., volumetric strain rate at the time immediately before the change of loading.

The model assumes that the soil element in an initial state of volumetric strain rate ($\dot{v}_o$) is just undergoing creep under constant effective stress before the change of loading.

(2) Hyperbolic model
The foundation sand, crust, embankment fill and berm were modeled by using the hyperbolic stress strain relationships developed by Duncan, et al. [7]. This model was developed for modeling three important characteristics of the stress-strain behavior of soils, namely, nonlinearity, stress-dependency, and inelasticity, by varying the values of Young's modulus and bulk modulus appropriately as the state of stress varies within the soil.

In this model, five parameters (i.e., k, n, R_f, c, and ϕ) are required to define the stiffness and strength characteristics, and three parameters (i.e., k_b, m, k_{ur}) are required to define the volume-change characteristics.

(3) Linear elastic model
The composite of the RE wall (i.e., granular backfill, reinforcement and facing) was modeled by the simple linearly elastic model. Since the rigidity of the RE wall composite may be several orders higher than that of the foundation soil, it is believed that this assumption is justified. Therefore the discrete method, i.e., representing the steel reinforcement by bar elements and the soil-reinforcement friction by interface elements, was not adopted in this analysis.

Soil Parameters The soil parameters of the elasto-viscoplastic and hyperbolic models used in the finite element analyses are shown in Tables 4 and 5, respectively. The parameters for the soft foundation clay were determined directly from the triaxial and 1-D consolidation tests. The RE wall mass was assumed to have a relatively large stiffness value.

Table 4: Soil Parameters Used in the Finite Element Analysis – Elasto-viscoplastic Model (Foundation Soft Clay)

D	Coefficient of dilatancy	0.022
Λ	Irreversibility Ratio	0.643
M	Critical state parameter	1.370
ν'	Effective Poisson ratio	0.3
k_x	Coefficient of permeability of x direction (ft/day)	0.003
k_y	Coefficient of permeability of y direction (ft/day)	0.003
σ'_{vo}	Preconsolidation pressure (psf)	1720
K_o	Coefficient of earth pressure at rest	0.428
α	Coefficient of secondary compression	0.00124
$\dot{v}_o$	Initial volumetric strain rate (day^{-1})	0.0000036
λ	Compression index in the e and $\ln(p'/p'_o)$ relationship	0.0826
e_o	void ratio corresponding with σ'_{vo}	0.765
λ_k	Gradient of e and ln(k) relationship	0.0826

Table 5: Soil Parameters used in the Finite Element Analysis – Hyperbolic Model (Foundation Sand, Crust, Embankment Fill and Berm)

Material property description	* 1	* 2	* 3	* 4	* 5
Unit weight (pcf)	120	115	110	120	100
Young's modulus number, k	1100	410	200	326	250
Young's modulus exponent, n	0.36	0.69	0.00	0.25	0.00
Failure ratio, R_f	0.85	0.90	0.80	0.90	0.80
Bulk modulus number, K_b	900	260	100	150	130
Bulk modulus exponent, m	0.00	0.15	0.00	0.30	0.00
Cohesion, C (psf)	0	0	400	1900	500
Angle of internal friction (deg.)	40	35	14	7	14
At-rest lateral earth pressure coeff., K_o	0.50	0.50	0.50	0.80	0.50

* - 1 - Sand, medium dense to dense
- 2 - Sand, loose
- 3 - Clay, Silty, medium stiff (crust)
- 4 - Embankment material (clay)
- 5 - Berm material (random)

FEM RESULTS

Comparison of FEM Analyses and Field Measurements

During the period of construction, the behavior of the wall was monitored by the instrumentation. The fill height reached a maximum of 40 feet (12.2 m) on December 3, 1987. The settlement and lateral deformation measurements were taken between 6-02-87 and 12-03-87, during the staged construction.

The FEM program was employed to simulate the exact construction procedure, i.e., sequential loading plus staged construction, with a waiting period of one month between each stage. Figures 3 and 4 show, respectively, the original finite element mesh and the deformed mesh at the end of construction.

The results of the lateral movements of the foundation soil versus depth at Station 32+00 (i.e., where the wall is highest) and the corresponding FEM results are shown in Figure 5. The inclinometers were installed 3 feet in front of the wall, and the field data indicated that there was up to 2.5 inches (63.5 mm) of lateral movement at this location. The predictions from the FEM match rather well with the field data which was collected on four different dates. The maximum value for lateral movement was 5.82 inches (147.8 mm) for the FEM analysis.

The settlements at Station 32+00 and the corresponding FEM results are shown in Figure 6. The field data indicated that the settlement of the wall footings at this location was 9.5 inches (241.3 mm) at the end of construction, which matches well with the FEM analysis of 8.88 inches (225.6 mm). The results from FEM analysis on the four different measuring dates again show close agreement with the field data.

Strength Increase Due to Staged Construction

Staged construction can significantly improve the bearing capacity due to the increase of undrained shear strength (S_u) by consolidation. The benefits of staged construction can be investigated by the FEM analysis, which simulates the construction procedure. Figure 7 shows the stress path during the RE wall construction for a typical element which represents the soft clay underneath the footing at Station 32+00. The location of this element is shown in Figure 3 (shaded element). From Figure 7-(a), it is clear that without staged construction, the Effective Stress Path (ESP) would reach the failure (K_f) line. Due to the effect of staged construction, with the contribution of consolidation (dashed lines Figure 7-(b)), the effective stress path for the element would be below the failure line. Since there is no element in the entire mesh reaching failure, it was concluded that the overall stability of the RE wall is adequate.

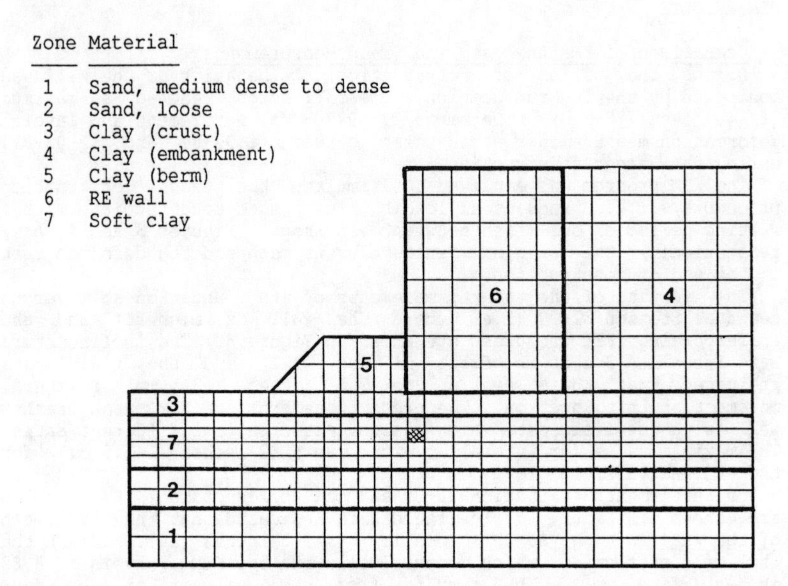

Figure 3: Finite Element Mesh

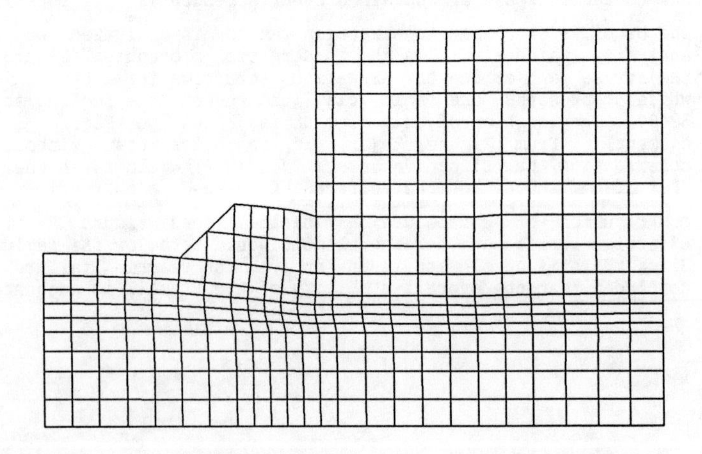

Figure 4: Deformation of the Mesh at the End of Construction
(Displacement enlarged by 5 times for plotting)

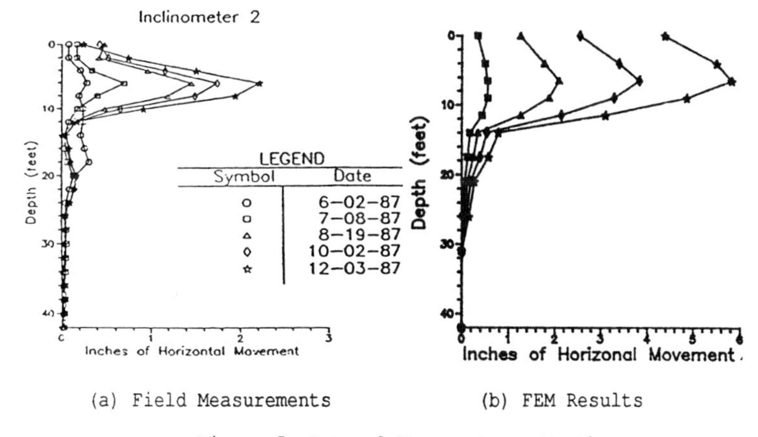

(a) Field Measurements (b) FEM Results

Figure 5: Lateral Movements vs Depth

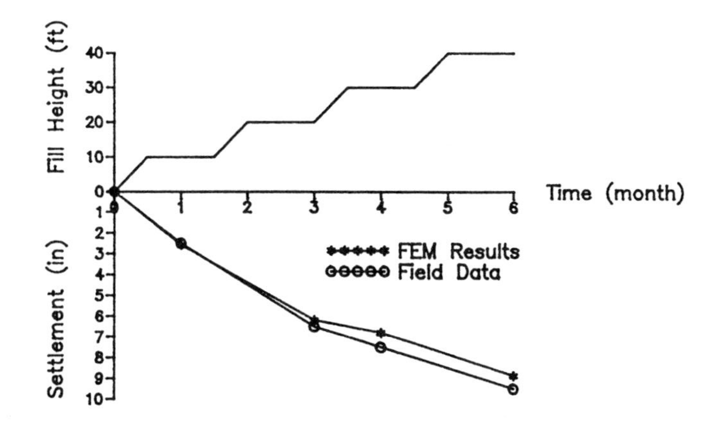

Figure 6: Settlements vs Time

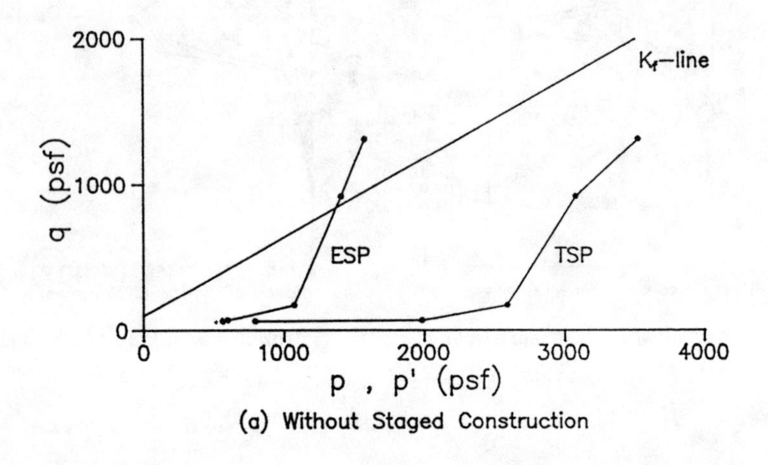

(a) Without Staged Construction

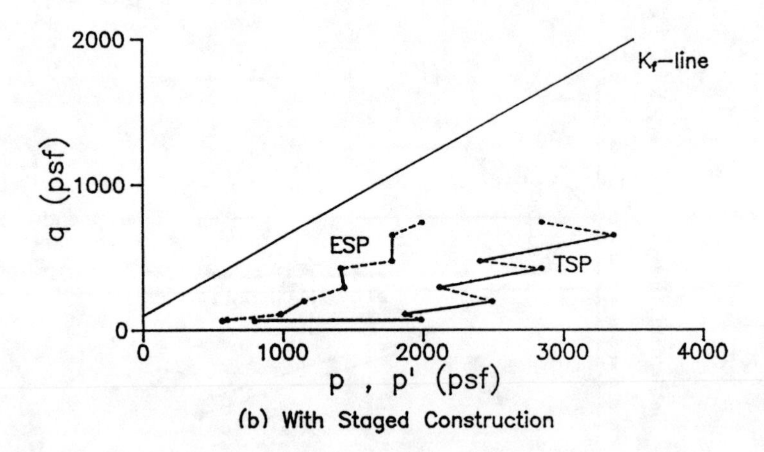

(b) With Staged Construction

Figure 7: Stress Path for a Typical Element

DISCUSSION AND CONCLUSION

The incremental construction of the RE wall was successfully simulated by the finite element analysis using the program named DACSAR. The nonlinear elasto-viscoplastic behavior of the foundation soil and time-dependent deformation can be incorporated in the FEM analysis. The field instruments provided valuable data for the comparison with the FEM results, which demonstrated that FEM can rather well recover the field performance under various stages of construction. It also indicates clearly from the stress path study that, due to the consolidation effect, staged construction can significantly improve the stability of the wall.

ACKNOWLEDGEMENTS

The authors are indebted to Dr. Atsushi Iizuka and Prof. Hideki Ohta of Kanazawa University for their assistances in the appropriate use of the DACSAR program.

REFERENCES

1. N Chou and CK Su, External Stability of a Reinforced Earth Wall Constructed Over Soft Clay, Design and Performance of Earth Retaining Structures, ASCE, Geotechnical Special Publication No. 25, 394-408, 1990.
2. N Chou, KT Chou, CC Lee, and KW Tsai, Preloading by Water Testing Eliminated Sand Drains for a 65,000 Ton Raw Water Tank in Taiwan. Proceedings of the 6th Southeast Asian Conference on soil Engineering, 1980.
3. A Iizuka and H Ohta, A Determination Procedure of Input Parameters in Elasto-Viscoplastic Finite Element Analysis, Soils and Foundations, Vol. 27, No. 3, 71-87, Sept. 1987.
4. H Sekiguchi and H Ohta, Induced Anisotropy and Time Dependent in Clays, Proc. 9th ICSMFE, Specialty Session 9, Tokyo, Japan, 229-237, 1977.
5. KH Roscoe, AN Schofield, and A Thurairajah, Yielding Of Clays in States Wetter than Critical, Geotechnique, Vol. 13, No. 3, 12-38, 1963.
6. T Shibata, On the Volume Changes of Normally-Consolidated Clays, Annuals, Disaster Prevention Research Institute, Kyoto University, No. 6, 128-134, 1963 (in Japanese).
7. JM Duncan, P Byrne, KS Wong, and P Mabry, Strength, Stress-Strain and Bulk Modulus Parameters for Finite Element Analysis of Stresses and Movements in Soil Masses, Geotechnical Engineering Research Report, Department of Civil Engineering, University of California, Berkeley, 1980.

ROTATIONAL FAILURE MECHANISMS USING MULTIPLE FRICTION CIRCLES

By Douglas A. Crum, A. M. ASCE[1]

ABSTRACT: Taylor's friction circle is a classical method for slope stability analysis. A method for utilizing composite failure mechanisms constructed of three friction circles is presented. The multiple friction circle method is ideally suited for bearing capacity calculations of eccentrically loaded foundations and rotational failures of retaining walls and similar structures. The multiple circle method has been extended to include friction in the material strength and utilize computational schemes such that it can be programmed with optimization capabilities rather than being used as a graphical or semi-graphical method. Some examples are given demonstrating ultimate bearing capacity calculations of foundations with eccentric loads and nearby slopes.

INTRODUCTION

The Bureau of Mines conducted this research as part of it's investigation of dump point accidents. Through full scale model tests and analysis by conventional methods, the dump point problem (of a haul truck near the crest of an embankment) was found to be influenced by three dimensional effects (1) and eccentric loads (2). The friction circle method addresses the latter.

TAYLOR'S FRICTION CIRCLE

Taylor's Friction Circle Method derives its name from D. W. Taylor who first used the method to produce slope stability charts (8), and the circles associated with the graphical method. The friction circle method has been used primarily to determine the stability of slopes. The method considers a circular failure surface in the two dimensional plane, which can be viewed as a cross section of an infinitely long cylinder in three dimensions.

The advantage of the friction circle method is that it is sensitive to the location of forces. This makes it ideal for investigation of rotational failure mechanisms influenced by eccentrically loaded footings or retaining walls.

Collapse mechanisms composed of a single circular failure surface are limited in their ability to closely approximate actual rupture surfaces generated in the collapse of a failure mass including structural components (such as footings,

---

[1]Project Engineer, STS Consultants LTD, Plymouth, MN, 55447.
Formerly with Twin Cities Research Center, U.S. Bureau of Mines, Minneapolis, MN, 55417.

retaining walls, and dams). Gudehus (5) showed that composite failure mechanisms can be constructed of more than one friction circle if the centers of each three intersecting circles lie on a line. In this study, a more general approach is presented that is an expansion of Gudehus' method. Since evaluation of the equations is time consuming, the method has been incorporated in a computer code.

GENERAL METHOD AND DERIVATION OF EQUATIONS

The friction circle method relies on the ability to express the stresses in the soil mass as two resultant forces (a cohesive and intergranular force) resisting motion for each circular arc. If these forces can be calculated by static equilibrium, then the stability can be assessed for any soil strength parameters c and ϕ, where c is the cohesion in the soil and ϕ is the internal angle of friction.

Resultant Cohesive Force. Assume that the cohesive shear stress can be described as a function of position along the slip circle by $t(\theta)$, where θ is an angle measured counter-clockwise from the center of the slip circle (see Fig. 1). The orientation and location of the resultant cohesive force can be determined by integrating the cohesive shear stress $t(\theta)$ along the circular arc. But, if the cohesive shear stress is not constant, the integral will have to be performed for every trial failure surface. Thus, the friction circle method assumes a homogeneous soil such that the cohesion is constant along the entire length of the slip surface. This creates a constant shearing stress tangent to the slip surface.

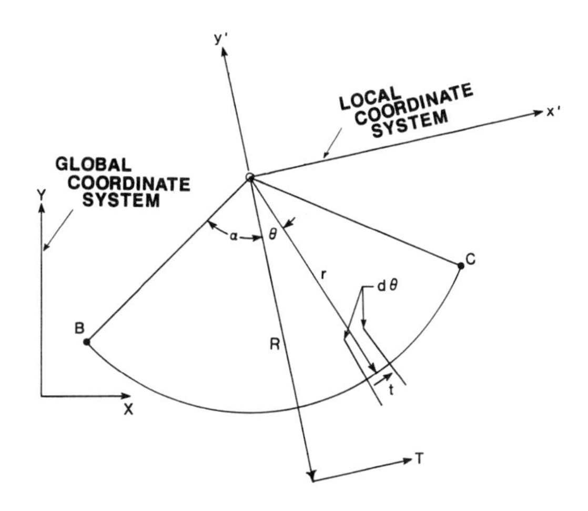

FIGURE 1. Resultant Cohesive Force along a Circular Arc

An expression for the orientation and location of the resultant cohesive force T, where T is the resultant of the resisting cohesive shear stress $t(\theta)$ along the circular arc BC (see Fig. 1), is obtained by considering equilibrium of forces and moments. By considering a local coordinate system (x', y') with its origin at the circle's center, equilibrium of forces and moments about any point, A, requires that the resultant cohesive force must be moved out from the circular arc

$$R/r = \alpha/\sin\alpha\,, \tag{1}$$

valid for $-90^\circ < \alpha < 90^\circ$, in which R and r are radii and 2α is the subtended angle. Further explanation of the derivation is given by Crum (2). The result simplifies the cohesional resistance to a force that can be used in the static resolution of the forces and is valid for equilibrium about any point in the plane. The friction circle method is most easily used by determining the force required for stability T from resolution of forces, and determining the required cohesion at equilibrium from T.

Resultant Intergranular Force. Similar to the cohesive force, the intergranular force must be expressed as a resultant. The intergranular force is taken to be the resultant of the normal force and the resistance due to friction. For cohesive materials ($\phi=0$) the intergranular force degenerates to a normal to the sliding surface. Assuming a mobilized friction angle along a moving rupture surface, the "intergranular force" is inclined to the base at the internal angle of friction of the soil so that the line of action is everywhere tangent to a smaller circle with radius $r^{\phi} = r\sin\phi$. This smaller circle is known as the "friction circle".

As with the cohesive force, the intergranular force will be some distribution $f(\theta)$ along the arc. There are several methods to express the force, F, as a resultant of the surface tractions (minus the shearing resistance due to cohesion) along the arc (2, 9). Here, we have adopted the conventional approach and assumed the force distribution $f(\theta)$ can be approximated by the resultant (a single force F located at θ^*, where $-\alpha \leq \theta^* \leq \alpha$). This can be written mathematically via the Dirac delta function, $\delta(x)$,

$$F = f(\theta) r \delta(\theta^*)\,. \tag{3}$$

This assumption for the force distribution results in a lower bound for the factor of safety to other circular methods of analysis that satisfy statics (11). Since a closed-form solution can not be obtained for most stability problems, this makes the friction circle method a valuable tool for determining lower bound solutions. By use of the Dirac delta function, the resultant forces can be expressed

$$\begin{aligned} F_{x'} &= F\sin\eta\,, \\ F_{y'} &= F\cos\eta\,, \\ M &= Fr\sin\phi\,, \end{aligned} \tag{4}$$

where the angle, η, is the orientation of the resultant, F, in the global coordinate system and M is the moment about the circle center. The angle η is a function of the orientation of the arc in the two dimensional plane, the point of application of the force on the arc and the inclination of the force on the arc at the angle of friction. Further explanation of the derivation is given by Crum (2).

Resolution of Forces. In the case of a single friction circle, the system of forces consists of only three: the intergranular force F, cohesive force T, and resultant actuating force (Q+W), where Q is a load applied on the footing and W is the soil weight. In the case of only three force vectors, it is required that they must pass through a point to satisfy moments. In this case, the problem is simplified since the point of application of the intergranular force is fixed. In a two wedge mechanism defined by three circular arcs, there are a minimum of four forces acting on each wedge. The significance of this is that the points of application of the intergranular forces are not specified. Thus, points Q_i, which define the locations of the intergranular forces are said to be located off the center of each arc by an angle $\Delta\alpha_i$. The general schematic showing the failure mechanism and all the resultant forces with strategic points is shown in Fig. 2. In Fig. 2, F_i and T_i are equilibrium forces, Q, W_L, and W_R are actuating forces, A_3, B_1, D_2, M, K_i, and Q_i are strategic points where the subscripts refer to the circular arc they lie on, η_i, θ_i, α_i, and $\Delta\alpha_i$ are angles, and r_i and r_i^ϕ are radii of the slip circle and friction circle respectively, where i = 1, 2, 3.

The values of $\Delta\alpha_i$ are unknown and will affect the orientations of the forces so that the force polygon (fig. 3) is not unique for a given configuration of the circle centers. Thus, it is necessary to optimize the angles $\Delta\alpha_i$ to determine the optimum orientations of the forces.

Satisfying equilibrium of forces and moments for each soil wedge requires six equilibrium equations for six unknowns (F_1, T_1, F_2, T_2, F_3, T_3). The matrix of Eq. 5 was formulated to solve the system of forces for the case of only vertical applied loads and body forces. Notation is shown in Fig. 2, where B_i is the solution vector, a_{ij} are the matrix coefficients, e is the load eccentricity, and η_i and θ_i are the orientations of the intergranular forces F_i and the cohesive forces T_i respectively. By using equilibrium of forces of the global system and right wedge, and taking moments about point Q_3 and the second circle center, we have

$$\begin{bmatrix} \sin\eta_3 & \sin\theta_3 & \sin\eta_2 & \sin\theta_2 & 0 & 0 \\ \cos\eta_3 & \cos\theta_3 & \cos\eta_2 & \cos\theta_2 & 0 & 0 \\ 0 & 0 & \sin\eta_2 & \sin\theta_2 & \sin\eta_1 & \sin\theta_1 \\ 0 & 0 & \cos\eta_2 & \cos\theta_2 & \cos\eta_1 & \cos\theta_1 \\ 0 & 0 & 0 & 0 & a_{55} & a_{65} \\ a_{16} & a_{26} & a_{36} & a_{46} & 0 & 0 \end{bmatrix} \begin{Bmatrix} F_3 \\ T_3 \\ F_2 \\ T_2 \\ F_1 \\ T_1 \end{Bmatrix} = \begin{Bmatrix} Q+W_R+W_L \\ 0 \\ Q+W_R \\ 0 \\ B_5 \\ B_6 \end{Bmatrix}$$

where
$$B_5 = W_L(Q_{3x}-G_{Lx}) \; ,$$
$$B_6 = W_R(C_{2x}-G_{Rx}) + Q(C_{2x}-b/2-e) \; , \qquad (5)$$
$$a_{36} = -R_2\sin\phi \; ,$$
$$a_{46} = -R_2\alpha_2/\sin\alpha_2 \; ,$$

and the moment arms a_{55}, a_{65}, a_{16}, a_{26} are the orthogonal distances from point C_2

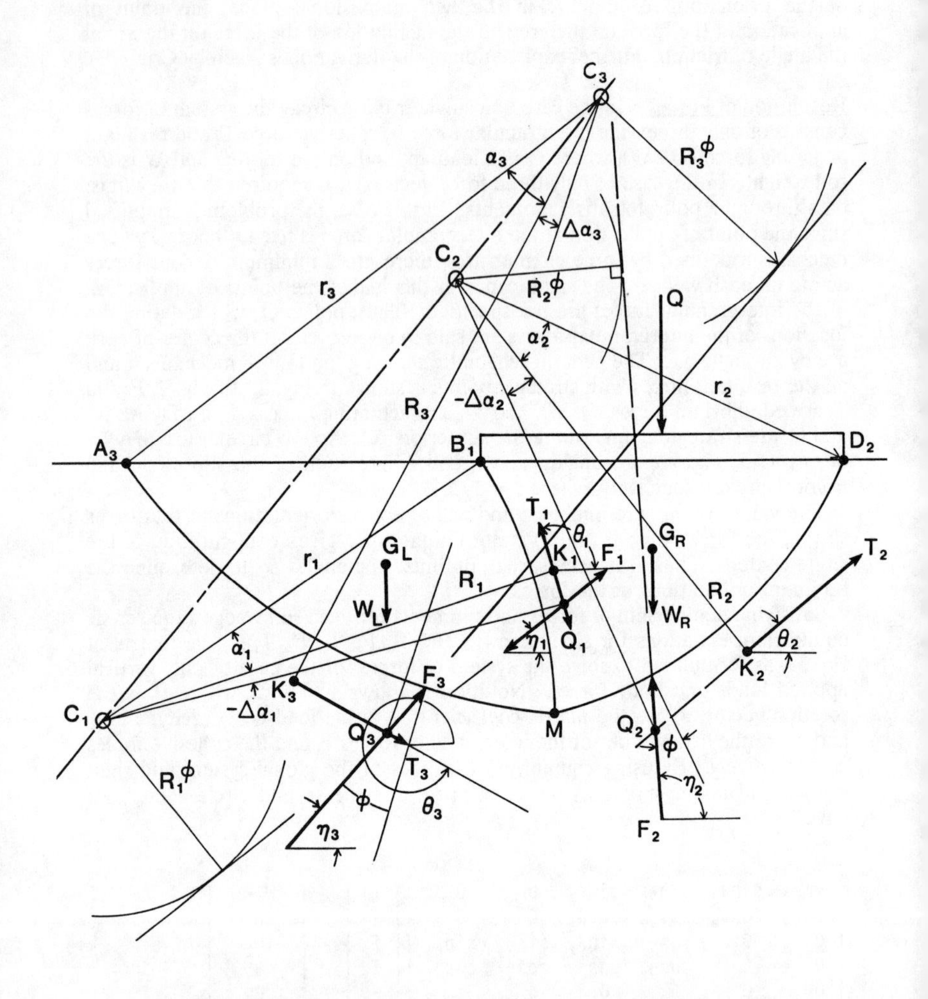

FIGURE 2. General System of Forces on Two-Wedge Mechanism

and Q_3 to the lines of action of the forces. The moment arms are positive for forces acting clockwise about the points and negative for forces acting counter clockwise. If the second circular arc is reversed, so it is convex instead of concave, then the parameters a_{36} and a_{46} are positive for forces acting counter clockwise and negative for forces acting clockwise. Points G_L and G_R are the centers of gravity of the left and right soil wedges.

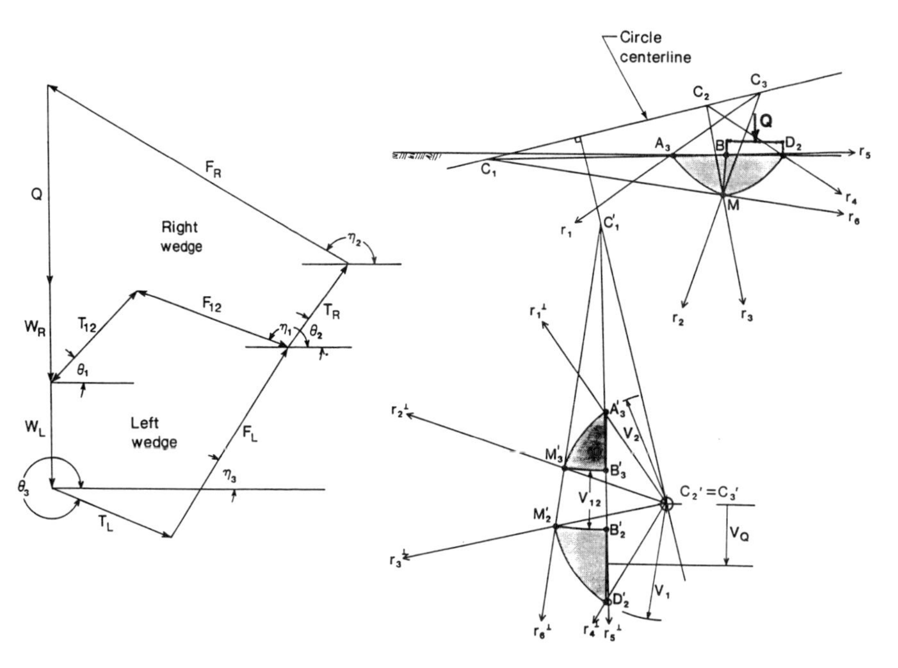

FIGURE 3. Force Polygon for Two-Wedge Mechanism

FIGURE 4. Schematic Showing a Failure Mechanism and Hodograph Construction

Kinematics. For an admissible collapse mechanism, it must be possible to describe the motion that takes place upon collapse. This can be presented in a hodograph, which is a graphical display of velocities of all components of the mechanism.

Construction of the hodograph is based on drawing the velocities so that they are tangent to the sliding surfaces (fig. 4). For the friction circle, this can be accomplished most easily by drawing the velocities perpendicular to the circle radii. When the hodograph is completed, each solid wedge of soil is represented by an identically shaped wedge of arbitrary size rotated 90°. Note that the hodograph does not define the actual magnitude of velocities, but only ratios of velocities.

Once the hodograph is completed, the relative velocities of the forces can be measured by determining the point of application of the force on the body in the hodograph. The direction and relative magnitude of the velocity of any point in the moving body can be measured by drawing a vector from the stationary point to the point of interest.

Factors of Safety. The friction circle method has adopted the factor of safety (FS) based on energies:

$$FS = \delta D/\delta A \,, \tag{6}$$

where δD is the dissipated energy of resisting forces and δA is the actuating energy of existing loads. Since the critical angle of friction (ϕ^*) differs from the measured angle of friction (ϕ) and ϕ^* is required to orient the forces, it is necessary to separate factors of safety for friction and cohesion. By taking moments of the resisting forces and actuating forces about the circle center, an expression for the factor of safety consistent with Eq. 6 is:

$$FS = \frac{F\sin\phi + T\alpha/\sin\alpha}{Q(L/r)} \,. \tag{7}$$

in which L is the radial distance from the circle center to the line of action of the actuating force Q. Eq. 7 must hold as cohesion becomes small ($T \to 0$), for which

$$FS_\phi = \sin\phi/\sin\phi^* \,. \tag{8}$$

Thus, the critical angle of friction ϕ^* to be used in the analysis can be calculated from the measured friction angle ϕ and a desired factor of safety. Since the factor of safety for friction FS_ϕ must be incorporated into the geometry of the problem, it is only necessary to solve for the factor of safety for cohesion. Again, by taking moments about the circle center, the factor of safety for cohesion may be expressed

$$FS_c = 2cr\,\sin\alpha/T \,, \tag{9}$$

where $2cr\sin\alpha$ is the shearing resistance. This expression is valid for a single circular arc or can be taken as a summation of the numerator and denominator for an average factor of safety for the mechanism.

An alternative to Eq. 9 is to express the actuating energy δA in terms of the actuating force Q instead of the shearing resistance. Since the velocities of the forces T and Q are not equal, they are retained in the expressions for the energies

$$\delta D = (2\alpha rc)\cdot V_c$$

$$\delta A = Q\cdot V_Q \tag{10}$$

in which V_c is the velocity of the cohesive force T, and V_Q is the velocity of the load Q. Substituting Eq. 10 into the factor of safety (Eq. 6) leads to

$$FS_c = [2\alpha rc\cdot V_c]/[Q\cdot V_Q] \,. \tag{11}$$

The advantages of Eq. 11 are that the system of forces does not need to be solved and that it yields an upper-bound kinematic solution (4). For composite failure mechanisms, the numerator is summed over the circular arcs and the denominator summed over each actuating force, which may also include the weights of the soil wedges in addition to the load Q. The disadvantage of Eq. 11 is that it is not affected by friction and thus always gives a solution for purely cohesive ($\phi=0$) material.

During optimization of a mechanism, the configuration is adjusted such that the factor of safety is minimized. Selection of a particular expression for the factor of safety as an optimization parameter can have a drastic influence on the optimization path. It is noted that the factor of safety for cohesion of Eq. 9 can be written for any individual surface generated by the i'th circle as

$$FS_c^i = 2cr_i\sin\alpha_i/T_i \tag{12}$$

or as a summation for the mechanisms as a whole,

$$FS_c^{ave} = \frac{2c\ \Sigma(r_i\sin\alpha_i)}{\Sigma\ T_i} \quad . \tag{13}$$

For frictional materials ($\phi>0$), the following assumptions were postulated:
Assumption 1: ϕ and c are fully mobilized on all surfaces where movement occurs.
Assumption 2: The factor of safety is taken as a summation over surfaces where movement occurs.
Assumption 3: The minimum factor of safety is found where the local factors of safety, FS_i, are also minimum and equal on each surface where movement occurs.

Assumption 3 was administered by periodically minimizing the largest local factor of safety FS_i and then converging the factor of safeties such that they were equal. This assumption guarantees that the maximum amount of shearing resistance is fully utilized on each surface. Convergence was done by choosing the maximum of $|FS_1\text{-}FS_2|$, $|FS_1\text{-}FS_3|$, and $|FS_2\text{-}FS_3|$ as the optimization parameter to be minimized. Thus, the optimization procedure adjusts the mechanism with respect to two separate parameters.

DISCUSSION

Optimization. The optimization scheme that was utilized in code OPTIC uses a modified term based on the direct search method proposed by Hooke and Jeeves (6). The convergence/minimization scheme is necessary since optimization of FS_c^{ave} (Eq. 13) resulted in erroneous results characterized by local factors of safety for individual arcs that varied widely and sometimes a wide variation in velocities of forces. However, Eq. 13 is useful since it expresses the factor of safety once the optimum configuration is obtained. Optimization of FS_c via kinetics (Eq. 11) did not experience such problems. The mechanism consistently converged to an optimum configuration. The stability of the optimization with Eq. 11 as an optimization parameter is related to the influence of the hodograph.

In some cases, optimization with respect to two separate parameters leads to convergence toward two distinct configurations. This results in an oscillation between two convergence points and results in a solution that is not optimum and may be inadmissible if some local factor of safety is less than one. Generally, the optimization was found to converge toward one distinct configuration. However, convergence was found to be slower and in some cases not possible as the friction angle was increased. Also, a large load Q with respect to soil weight was usually more likely to result in all positive forces F_i, and T_i for a trial mechanism, so that convergence was faster and more likely to be completed.

Validity of the Method. Although the method generally utilizes equilibrium of forces to calculate the factor of safety, a factor of safety based on kinematics can be utilized for $\phi=0$. In this case, the method yields a kinematic upper bound solution (4). The solution that includes resolution of the forces for frictional material is marked by two fundamental assumptions that characterize the method as approximate. The first is the approximation of the intergranular force as a resultant via the Dirac delta function. The lower bound solution determined by the singular intergranular force assumption will only be accurate if optimization is acheived. The second is that the failure surface for frictional material would occur along circular surfaces rather than a logarithmic spiral. Thus, the hodograph of the friction circle method is rigorously correct for purely cohesive materials ($\phi=0$) but not for frictional materials ($\phi>0$). In the case of material with friction ($\phi>0$), we make the assumption that the mechanism will slide along circular surfaces; so kinematic admissibility is checked via the hodograph regardless of the value of friction.

Limitations of the Method. Due to the assumptions of a homogeneous soil mass in approximating the resultant forces F and T, the friction circle method is not capable of solutions in multiple layered media. Also, since the method calculates the value of cohesion at limiting equilibrium, it can only be used when a value of cohesion greater than zero is required for stability.

Other limitations of minor importance are related to the calculation procedure. Due to constraints of the method, it is not possible to optimize the mechanism based on an explicit factor of safety. This is because incremental changes in the input parameters during optimization may violate constraints such that adjustment of the optimization parameter (factor of safety) is not possible. Thus, the value of cohesion at limiting equilibrium is determined as an alternative. This is possible since the value of cohesion does not affect the configuration of the optimum failure mechanism. Formulation of the method to solve for the factor of safety or limit load is possible, but would require a trial and error search, or an additional loop in a computer code.

Applications of the Method. The strengths of the multiple friction circle method are in rotational collapse modes. Applications of the method may include horizontal loads on the footing; an option of horizontal gravitational loads on the footing would enable the code to be used for analyses of retaining walls; and inclusion of the weight of concrete in the foundation would allow analysis of gravity walls and dams. It is also possible to form collapse mechanisms composed of 5 or more circles (5). However, the additional programming to formulate a 5 circle method without simplifications would likely be tedious.

RESULTS

The multiple friction circle method has been incorporated in a computer program OPTIC, which is written in compiled BASIC (2). Run time of the program varies widely dependent on the input parameters. However, most solutions are obtained in 2-3 minutes, running on a 80286 chip and 80287 math co-processor, both having a clock speed of 16 MHz. The program assumes a surface load (footing) near a slope as formulated in Fig. 2 and Eq. 5. The program may be used for either kinematic or resolution of forces. It should be noted that the factor of safety based on kinetics is always less than or equal to the factor of safety based on equilibrium of forces. Two demonstrations of results are given below.

Fig. 5 shows the non-dimensional bearing capacity (q/c) varying with respect to the non-dimensional eccentricity (e/b), where q is the surface load on the footing [force/area], c is the cohesion, e is the eccentricity and b is the width of the footing. The solution for $(e/b)=0$ is the well known Prandtl solution. Code OPTIC determined a value for $(q/c)=5.30$ for the Prandtl solution which compares well with the exact solution of 5.14. The conventional assumption for treatment of eccentric loads is to reduce the capacity of the footing by the ratio $1-2(e/b)$ (10). This is shown as a straight line on Fig. 5. Code OPTIC converged to this trend at eccentricities $0.09<(e/b)<0.25$.

Fig. 6 shows a nomograph of the problem of a footing on the crest of a slope. The limit load q is a function of the soil properties ϕ, c, γ, where γ is the unit weight of the soil, and b, & β, where b is the footing width and β is the slope angle. For constant values of the slope angle and the internal angle of friction, the dimensionless term $q/\gamma b^2$ varies almost linearly with the dimensionless term $c/\gamma b$. The results for code OPTIC are compared with the results obtained by a rigid block mechanism composed of 2 wedges as shown in the inset on Fig. 6. Results for the rigid block mechanism are given by Papanastiou (7). The rigid block mechanism shows results lower bound to the friction circle method for steep slopes and low cohesion. The friction circle method shows results lower bound to the rigid block mechanism for shallow slopes and high cohesion. This is typical, since the frictional method works best for large values of cohesion and the rigid block mechanism does not conform to a smooth failure surface for the Prandtl mechanism.

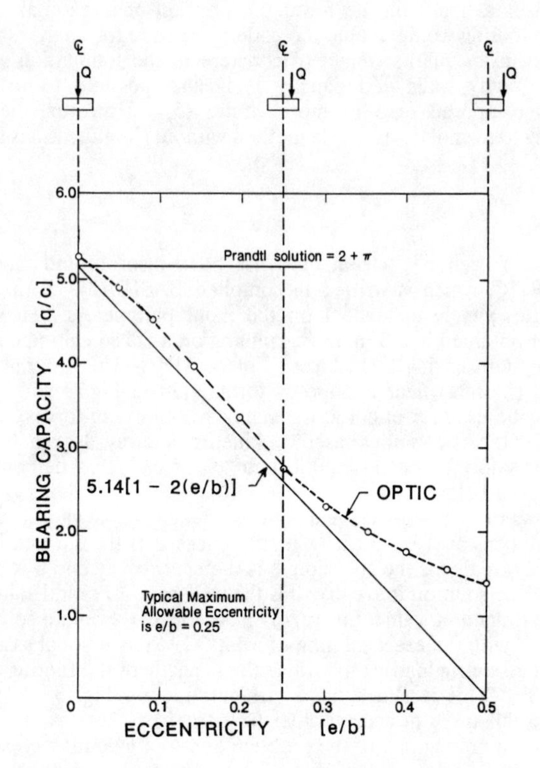

Figure 5. Effect of Eccentric Loads on a Continuous Footing

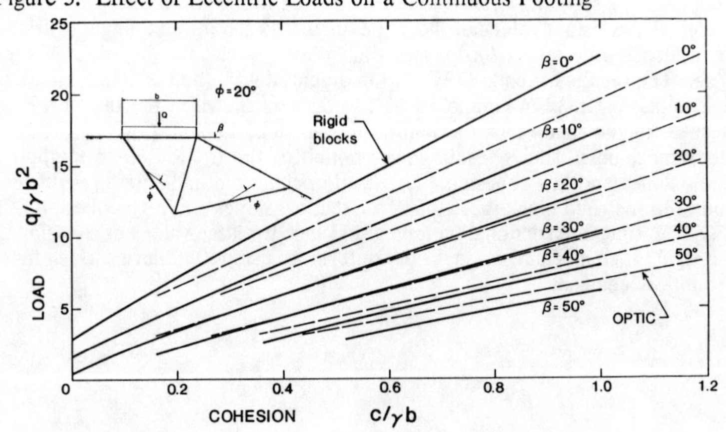

Figure 6. Effect of a Slope at One Edge of a Continuous Footing

CONCLUSIONS

A method for utilizing composite failure mechanisms constructed of three friction circles is presented. The method can be used for determining stability of some simple structures on homogeneous soils. The method is an extension of the well known Taylor friction circle method and follows the work of Gudehus (5). Comparison of bearing capacities with other methods shows the method yields results that are somewhat conservative for small values of cohesion and converge closer to closed form solutions for larger values of cohesion with respect to soil weight and friction angle. Notable Limitations to the method include local minima obtained in optimization, requirement of a homogeneous soil mass, and the task of obtaining an initial mechanism as a starting point prior to optimization.

REFERENCES

1. DA Crum, Application of a Full Scale Dump Point Stability Experiment to Engineering Analysis, Thesis submitted to the University of Minnesota, June 1988, 77p.
2. DA Crum, Rotational Failure Mechanisms Using Multiple Friction Circles, US Bureau of Mines RI-___, 1991, in print.
3. RL Degroot, Dump Point Stability Analysis, Masters Thesis submitted to the University of Minnesota, Dec. 1986, 66p.
4. DC Drucker, HJ Greenberg, & W Prager, Extended Limit Design Theorems for Continuous Media, Quart. Appl. Math., 9(4), 1952, 381-389.
5. G Gudehus, Engineering Approximations for Some Stability Problems in Geomechanics, Advances in Analysis of Geotechnical Instabilities, Univ. of Waterloo Press, 1978, 1-23.
6. JL Kuester & JH Mize, Optimization Techniques with FORTRAN, McGraw Hill, 1973.
7. PC Papanastiou, Passive Earth Pressure and Bearing Capacity Computations with Microcomputers, Masters Thesis submitted to U. of Minnesota, 1986, 76p.
8. DW Taylor, Fundamentals of Soil Mechanics, Wiley, NY, 1948, 700p.
9. I Vardoulakis, Lecture Notes On Foundation Engineering, University of Minnesota.
10. AS Vesic, Chap. 3 in Foundation Engineering Handbook, Ed. HF Winterkorn & HY Fang, Van Nostrand Reinhold, NY, 1975.
11. RV Whitman, Thoughts Concerning the Mechanics of Slope Stability Analysis, Proceedings of the Second Panamerican Conference on Soil Mechanics and Foundation Engineering, Brazil, 1963, 391-411.

ANALYSIS OF LOW EFFECTIVE STRESS CHARACTERISTICS OF GRANULAR MATERIALS IN REDUCED GRAVITY

Emir J. Macari-Pasqualino[1] A.M. ASCE,
Stein Sture[2] M. ASCE, and
Kenneth Runesson[3]

Abstract

Laboratory experiments have shown that the effects of self-weight have significant influence on the stress-strain-strength response of cohesionless granular materials when the confining stress is less than 1 psi (6.9 kPa). In order to simulate the highly nonlinear constitutive properties of soils at low stress levels and under reduced gravity conditions a third stress invariant dependent, cone-cap elasto-plastic constitutive model is employed in a finite element code in order to study nonlinear boundary value problems associated with homogeneous as well as non-homogeneous deformations at low effective stresses. The effect of gravity (self-weight) on the stress-strain response is evaluated.

Introduction

The study of the behavior of granular materials at low effective stresses and in a reduced gravity environment became a subject of great interest in the 1960's, when NASA's Surveyor missions to the Moon initiated the first extraterrestrial investigation of the properties of the Lunar surface and it was found that the regolith exhibited properties quiet unlike those on earth. These and other Apollo missions established that the Lunar surface was composed of a very fine-grained material possessing a small amount of cohesion (Costes, et al., 1969).

In recent years NASA's plans for future exploration and colonization of the Moon and Mars have sparked a renewed interest in this area. A better under-

[1] - Asst. Prof., Dept. Civil Engineering, Univ. of Puerto Rico at Mayaguez, P.O. Box 5000, Mayaguez, P.R. 00709-5000

[2] - Prof., Dept. of Civil, Environmental, and Architectural Eng., Univ. of Colorado at Boulder, Boulder, CO 80309-0428

[3] - Assoc. Prof., Dept. of Structural Mechanics, Chalmers Univ. of Technology, S-41296 Gothenburg, Sweden

standing of the mechanical behavior of granular materials under reduced gravity and at low effective stress levels is of paramount importance for the design and construction of surface foundations, embankments, and buried structures to house Lunar and Martian bases. Gravity induced stresses should not play a part of the material description but should instead be treated as external forces of a boundary value problem similar to imposed tractions. However, gravitational forces are always present in a terrestrial environment and can therefore, not be disregarded in physical experiments. In fact, in laboratory soil specimens the gravity induced stresses can be of the same order of magnitude or even larger than the externally applied tractions or loads, especially when the mean effective confinement stress is lower than approximately 1 psi (6.9 kPa). This can be considered as a threshold value, below which the influence of self-weight stresses should not be neglected.

In order to simulate the highly non-linear characteristics, a three-invariant elasto-plastic constitutive model that simulates the essential response of granular materials was recently proposed (Sture, et al., 1989; Macari-Pasqualino, 1989). The so-called Closest-Point-Projection-Method (CPPM) was used for the integration of the constitutive relations (Runesson, et al., 1990). This algorithm has been implemented in a constitutive driver computer routine that can be used for predicting material behavior under arbitrary combinations of stress and strain components (mixed control) in order to simulate a variety of test conditions in the laboratory.

The model was also implemented in a finite element code in order to study geotechnical problems including the situation where non-homogeneous deformations take place in conventional cylindrical triaxial specimens at low effective stresses. The results are compared to those obtained when the self-weight effects are disregarded.

Three-Invariant Model For Granular Material

The three-invariant model, can be viewed as a generalized version of the model previously devised by Lade (1972, 1975, 1979). The model has a yield surface that consists of two smooth parts comprising the cone and cap surfaces, each of which is defined in three-invariant stress space. One main modification compared to Lade's model consists of replacing the spherical cap yield surface by a drop-shaped cap, which is activated at relatively high confining stresses depending on the initial density of the material. The cone surface was also modified in order to avoid computational problems that are associated with the (plastic corrector) return to multivalued surfaces, which is the case with Lade's model.

The trace of the yield surface in the deviatoric subspace is essentially elliptical as suggested by Willam and Warnke (1975) in their model for concrete.

The ratio of strength in extension and compression is then used as a material parameter, which is clearly a generalization when compared to the model by Lade. The yield surfaces intersect along a plane curve in a given deviatoric (sub)space. The principal shapes in the (p, q) - plane and in the deviatoric plane are shown in Figure 1. The conical yield surface has the shape of an asymmetric cone, whose apex is located at or to the left of the origin in stress space depending on the cohesion of the material.

Since only the cone surface is activated in the low-confinement stress region we give explicit expressions only for this yield surface

$$F(p,q,\theta,\kappa) = f(q,\theta) - \eta(\kappa)(p-p_c) = 0 \qquad (1)$$

where f is a function of the stress state in terms of the effective pressure p, the deviator invariant q and the Lode angle θ, whereas η represents essentially the angle of internal friction. The hardening variable κ is a function of the dissipated plastic work. The parameter p_c is a measure of cohesion.

The function f is chosen as

$$f = q(1+\frac{q}{q_a})^m \, g(\theta) \qquad (2)$$

where q_a is a reference deviator stress, m is a material parameter ($O \leq m \leq 1$), and $g(\theta)$ defines the elliptical shape in the deviatoric plane.

A non-associated flow rule is adopted which has a potential function of the form

$$G(p,q,\theta) = f(q,\theta) - n\eta p \qquad (3)$$

where n is a material constant that dictates the plastic flow direction.

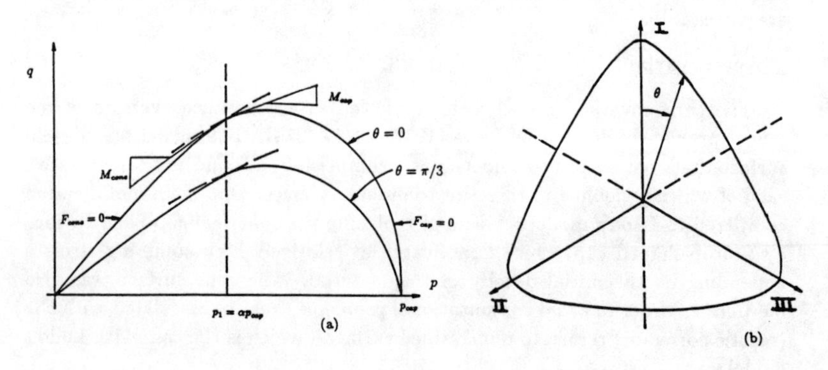

Figure 1: Shape of the cone and cap surfaces;
(a) invariant (p-q) space, (b) deviatoric space

Model Calibration

The calibration of the model was performed in two steps. First, a simple parameter identification or curve fitting was performed by utilizing results from conventional triaxial tests. The second step involves the use of an optimization scheme that is initially performed at the constitutive level, where the effects of self-weight are disregarded (Klisinski et al. 1987).

For each given load history, the experimental and predicted values are compared by measuring the distance between them in a certain norm. The individual norms for a number of load histories comprise the objective function, and the chosen optimization algorithm searches for a local minimum. More specifically, assume that the experimental values of the stress and strain states are known in terms of the principal components for a given point defined as

$$\sigma^{exp} = [\sigma_1^{exp}, \sigma_2^{exp}, \sigma_3^{exp}]^T \tag{4a}$$

$$\epsilon^{exp} = [\epsilon_1^{exp}, \epsilon_2^{exp}, \epsilon_3^{exp}]^T \tag{4b}$$

It is assumed that either ϵ_i^{exp} or σ_i^{exp}, $i = 1,2,3$, have been prescribed in the experiment and the other values are recorded. Thus it is possible to utilize data from experiments conducted under either strain, stress, or mixed control. The response functions

$$\sigma^{pre}(\mathbf{z}, t) = [\sigma_1^{pre}(\mathbf{z}, t), \sigma_2^{pre}(\mathbf{z}, t), \sigma_3^{pre}(\mathbf{z}, t)]^T \tag{5a}$$

$$\epsilon^{pre}(\mathbf{z}, t) = [\epsilon_1^{pre}(\mathbf{z}, t), \epsilon_2^{pre}(\mathbf{z}, t), \epsilon_3^{pre}(\mathbf{z}, t)]^T \tag{5b}$$

represent the predicted behavior for the same choice of control variables as those employed in the experiment, when a certain choice of model parameters $\mathbf{z}$ is made. These functions have a continuous character, but the results are obtained at discrete points as a result of numerical integration of the constitutive relations. It is noted that the calculated points may not necessarily match those of the experiments. Typically, a larger number of theoretical points than experimental points are available for the calibration process.

For each experimental point, the closest distance to the predicted response curve is calculated. In practice this requires interpolation between the predicted points. Upon the introduction of suitable scaling factors μ_σ and μ_ϵ to normalize the stress and strain values to a common dimension, one may define

$$\bar{\sigma}_i = \mu_\sigma \sigma_i, \quad \bar{\epsilon}_i = \mu_\epsilon \epsilon_i \tag{6}$$

One may now compute the Euclidean norm

$$d(\mathbf{z}) = \left(\sum_{i=1}^{3}(\bar{\sigma}_i^{exp} - \bar{\sigma}_i^{pre}(\mathbf{z}))^2 + \sum_{i=1}^{3}(\bar{\epsilon}_i^{exp} - \bar{\epsilon}_i^{pre}(\mathbf{z}))^2\right)^{\frac{1}{2}} \tag{7}$$

where d denotes the distance between the two response curves. Summing all distances d_j for all experimental points provides the objective function for a particular experimental curve

$$E(\mathbf{z}) = \frac{w}{k}\sum_{j=1}^{k} d_j(\mathbf{z}) \tag{8}$$

where k is the number of experimental points and w is a weight number that characterizes the importance given to the experimental response curve under consideration.

A second optimized calibration is performed at the finite element level, where the boundary value problems associated with the triaxial compression tests at 1-g are analyzed, and self-weight effects, frictional boundary effects, and membrane stiffness effects are taken into account (Macari-Pasqualino, 1989). This can be quite expensive in a computational sense, since the entire finite element analysis must be performed many times in order to optimize the material parameters and thus provide better correspondence between the experimental and analytical load-displacement responses. Figure 2 presents the comparison between the predictions performed at 0-g and 1-g employing the same material parameters in both cases. Figure 2 also presents the comparison of the same two situations but after the parameters for the 1-g case had been optimized. The resulting set of parameters can be considered to be the true material parameters since their determination should be unaffected by self-weight and boundary conditions.

Prediction Of Boundary Value Problems

The parameters obtained from the optimization analysis are used to model a classical problem in geotechnical engineering to examine better the performance of the revised Lade model and its integration scheme. The problem selected was that of a strip footing located on the crest of a slope under plane strain conditions. Analyses involving 0-g, 1/6-g and 1-g conditions were performed, while the material parameters remained the same, in order to study how failure and load-settlement characteristics of the loaded slope are influenced by gravity. Analyses of the same slope were also performed where the isotropic tensile strength was given a very low value ($p_c = -1\ kPa$), which gives a cohesion of $c = 0.8\ kPa$. In this case the focus of the analysis effort is to assess the influence that a very small amount of cohesion would have on the response of a slope, where even very small values of self-weight stresses could result in a loss

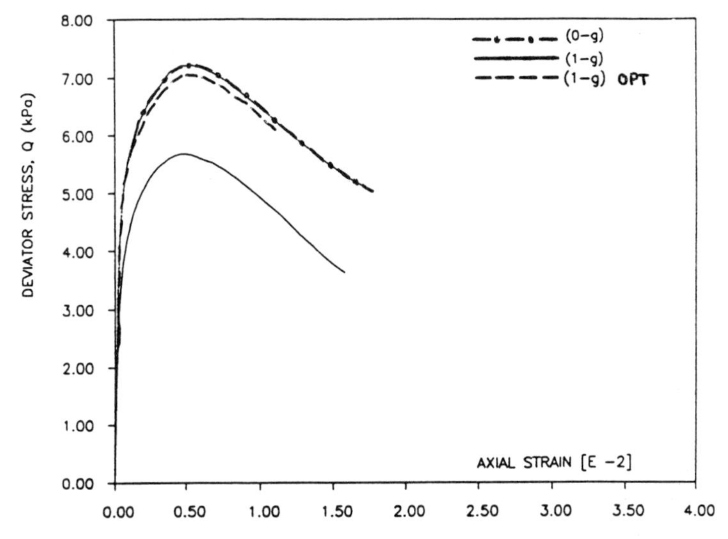

Figure 2: Comparison of finite element predictions for 0-g and 1-g conditions, with and without optimized parameters

of stability. It should also be noted that these low strength values (c, p_c) are also typical of those found in terrestrial moist unsaturated fine sands.

Triaxial test results from a medium-dense sand presented by Gemperline (1983) were used to calibrate the model. The calibration of the parameters was initially performed by physical identification and subsequently by optimization at the constitutive level, as described in the previous section. Figures 3 and 4 present the comparisons between the triaxial test results and the model predictions at the constitutive level for two confinement levels. It is observed that the comparison of the responses over a relatively wide stress range is quite good. In Figure 4 it should be noted that the leveling-off in the experimental volumetric strain curve is probably caused by the formation of a shear band in the triaxial specimen soon after peak strength was reached. Localization effects and formation of shear bands was not modeled in this case, since the focus is on assessment of material parameters.

Figure 5 shows the FEM mesh (9-node isoparametric elements) of a slope and strip footing located at the crest of the slope. The footing was pushed vertically downward onto the slope under displacement control. The analysis was performed for the cases of 0-g, 1/6-g, and 1-g conditions. Figure 6 presents

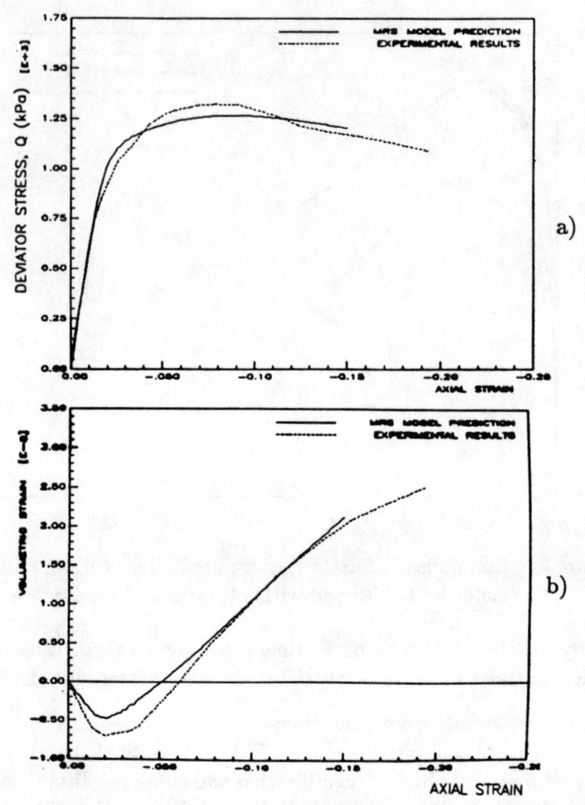

Figure 3: Comparison of laboratory and optimized model parameter responses for 345 kPa confinement; (a) deviator stress-axial strain, (b) volumetric strain-axial strain

the results of the analyses in terms of response curves relating the average vertical contact stress and vertical displacement of the footing. One can see that the effect of cohesion is quite pronounced in all three cases. However, if one compares the results of the three gravity levels individually for each of the three values of the cohesion or tensile strength, it is evident that for the case of lower cohesion ($p_c = -1.0\ kPa$) the influence of gravity is more pronounced than for that of a high cohesion value ($p_c = -100.0\ kPa$). Figure 7 presents such comparisons for the cases when $p_c = -1\ kPa, -10\ kPa$, and $-100\ kPa$, respectively. It is of particular interest to note how global stiffness and strength are affected by small changes in cohesion.

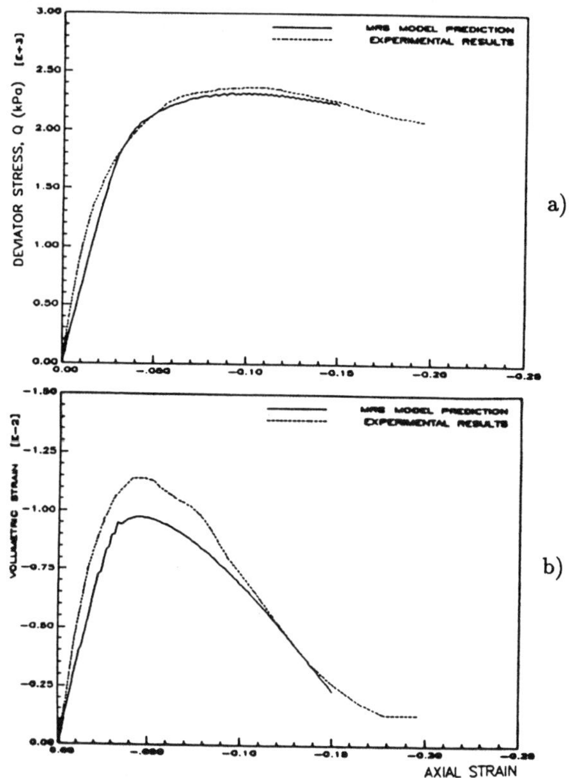

Figure 4: Comparison of laboratory and optimized model parameter responses for 640 kPa confinement; (a) deviator stress-axial strain, (b) volumetric strain-axial strain

Conclusions

The revised Lade (MRS-Lade) model simulates very well the highly non-linear strength and stress-strain behavior of soils at very low as well as high effective stresses. An important feature of the proposed model is that dilatancy is also pressure sensitive. In contrast to earlier models, this version does not explicitly involve the third stress invariant, but a composite ellipse description, which makes the analysis of the model and formulation of the various constitutive equations far simpler to perform. The present model is also different from the earlier version in that it has a work hardening/softening formulation that allows for strain-softening to a realistic residual strength level.

The use of optimization techniques for finding model parameters have been shown to produce better results over wide ranges of stress states and loading paths. Estimates of the model parameters were initially obtained by physical identification of experimental data, and they were further improved by the optimization process. A procedure for obtaining material parameters from the solution of a boundary value problem, such as triaxial test, was devised, whereby the effects of self-weight of the laboratory specimens' response were accounted for.

The comparison of the predictions of a strip footing located on the crest of the slope under 0-g, 1/6-g, and 1-g conditions show that the effect of gravity significantly increases the bearing capacity of the foundation. It was also shown that even very small amounts of cohesion have a significant influence on the bearing capacity of a soil and the overall foundation stiffness.

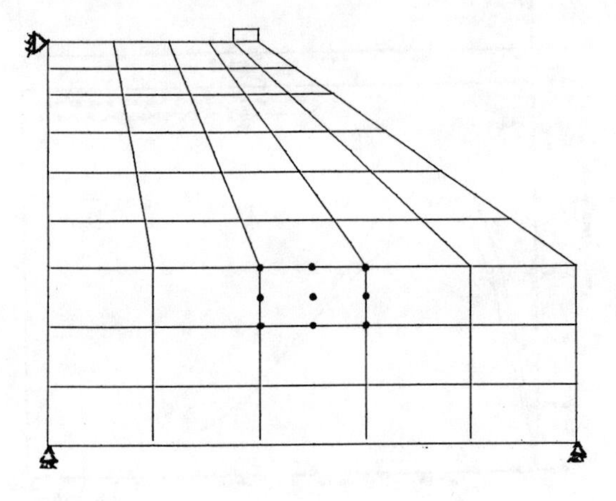

Figure 5: Undeformed FEM mesh showing the slope and footing

Acknowledgements

The authors gratefully acknowledge financial support provided from NASA Grant NAGW-1388 to the Center for Space Construction, University of Colorado at Boulder.

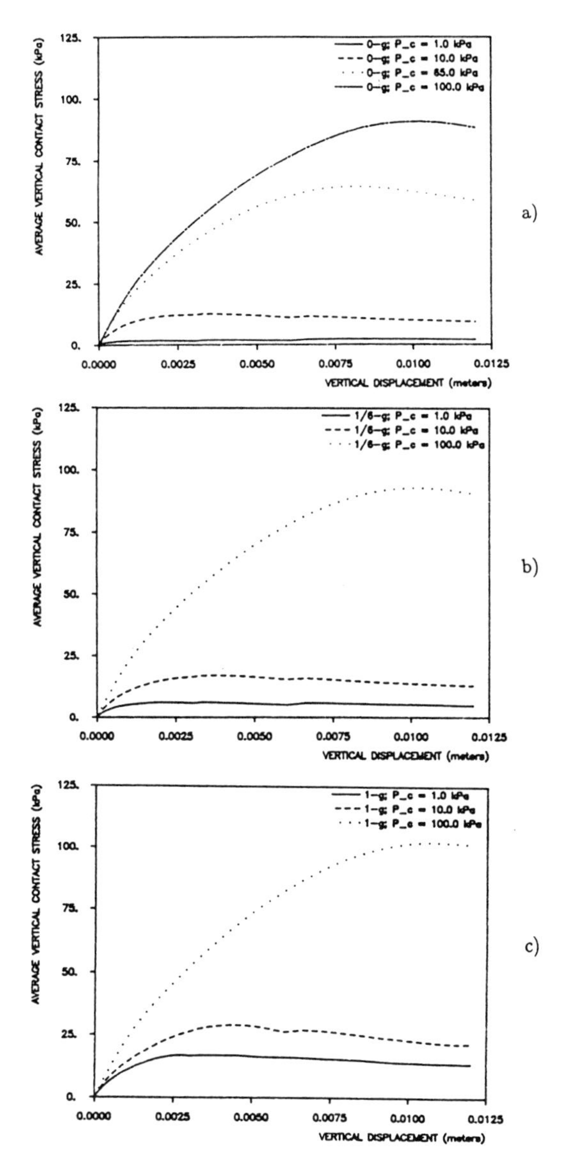

Figure 6: Load-settlement responses of a footing on the crest of a slope for p_c = -1, -10, and -100 kPa under (a) 0-g, (b) 1/6-g, and (c) 1-g

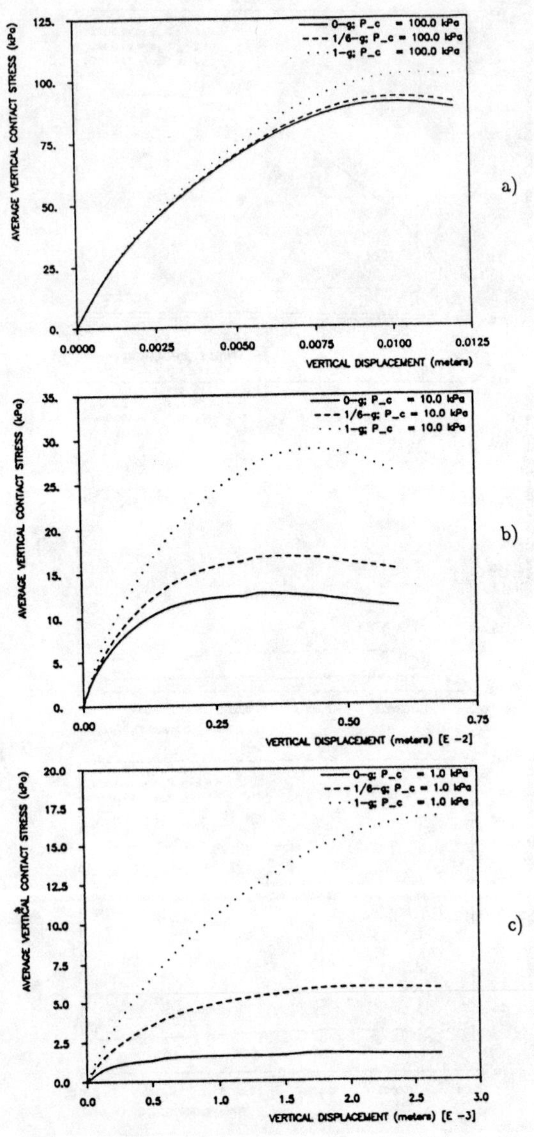

Figure 7: Load-settlement responses of a footing on the crest of a slope under 0-g, 1/6-g, and 1-g for p_c = (a) -100 kPa, (b) -10 kPa, and (c) -1 kPa

References

1. W.D. Carrier, The Lunar Source Book, Lunar and Planetary Institute, Chap. 9, Cambridge Press, 1990, in press.

2. N.C. Costes, W.D. Carrier, J.K. Mitchell, and R.F. Scott, Apollo 11: Soil Mechanics Results, J. Geot. Eng. (ASCE), 7704, 1969, p. 2045-2082.

3. M.C. Gemperline, Centrifugal Model Tests for Ultimate Bearing Capacity of Footings on Steep Slopes in Cohesionless Soils, M.S. Thesis, University of Colorado at Boulder, 1983, 124p.

4. M. Klisinski, M.M. Alawi, S. Sture, H-Y. Ko, and D.M. Wood, Elasto-Plastic Model for Sand Based on Fuzzy Sets, Proceedings of Constitutive Equations for Granular Non-Cohesive Soils, Case Western University, Saada and Bianchini, Eds., Balkema, Rotterdam, 1988, p. 325-343.

5. P.V. Lade, The Stress-Strain and Strength Characteristics of Cohesionless Soils, Ph.D. Dissertation, University of California at Berkeley, 1972.

6. P.V. Lade, and M. Duncan, Elastoplastic Stress-Strain Theory of Cohesionless Soil, J. Soil Mech. Fnd. Div. (ASCE), Vol. 101, GT-10, 1975 pp. 1037-1053.

7. P.V. Lade, Stress-Strain Theory for Normally Consolidated Clay, Proceedings of the Third International Conference on Numerical Methods in Geomechanics, Ed. W. Wittke, Vol. IV, Publ. A.A. Balkema, Rotterdam, 1979 p. 1325-1337.

8. E.J. Macari-Pasqualino, Mechanical Behavior of Granular Material at Low Confining Stress Levels and Under a Reduced Gravity Environment, Ph.D. Dissertation, University of Colorado, August 1989, p. 180.

9. K. Runesson, E.J. Macari-Pasqualino, and S. Sture, An Implicit Integration Algorithm for a Cone-Cap Plasticity Model for Granular Materials, in preparation.

10. S. Sture, K. Runesson, and E.J. Macari-Pasqualino, Analysis and Calibration of a Three-Invariant Plasticity Model for Granular Materials, Ingenieur-Archiv, January, 1989.

11. S. Weihe, Implicit Integration Schemes for Non-Smooth Yield Criteria Subjected to Hardening/Softening Behavior, M.S. Thesis, University of Colorado at Boulder, Boulder, Colorado, 1989.

12. K.J. Willam and E.P. Warnke, Constitutive Model for Triaxial Behaviour of Concrete, Colloquium on Concrete Structures Subjected to Triaxial Stresses, ISMES, Bergamo, IABSE Report No. 19, Proc. Vol. 3 (1974), Zurich, pp. 1-30.

USE OF STABILIZED FLY ASH FOR SEEPAGE CONTROL

Jason Y. Wu[1], M.ASCE

ABSTRACT: A study of using kiln dust stabilized fly ash (KDSFA) for seepage control has been done. Many studies have indicated that the stabilized fly ash may be used as a hydraulic barrier material. Most of the reagents used for stabilization are lime, cement, and bentonite. None of them incorporates the use of kiln dust. This paper evaluates the possibility of this innovated approach. The KDSFA material collected from a power station in New Jersey was tested for physical properties, moisture-density, strength, compressibility, and permeability. Additional cement and bentonite were used in an attempt to lower the permeability.

The kiln dust stabilization reduced the moisture content and increased the workability. The compacted KDSFA showed a high strength and a low compressibility. It exhibited a base permeability of 10^{-6} cm/s. Cement added up to 6% did not reduce the permeability. Bentonite may lower the permeability to 10^{-8} cm/s. The results of this study present that the kiln dust stabilized fly ash can be a satisfactory barrier material for seepage control. It also shows a great potential for use as a low permeability liner material at a hazardous waste site.

INTRODUCTION

Fly ash is a waste by-product resulting from the coal combustion, primarily produced at electric power generating plants. It is a silt-sized material similar in many ways to natural silty soils (Edil, et al., 1987). Fly ash is an artificial pozzolan. It is not cementitious itself, but when mixed with lime or other calcium compounds in the presence of moisture it becomes cementitious (GAI, 1988).

In the United States, fly ash has been increasingly used as a construction material. However, compared to the enormous quantities produced, it is still underutilized (GAI, 1988). Most of the fly ash is directly disposed of in landfills, placed in dry storage, or sluiced to impoundment (Usmen, 1988). All of these

---

1 - Senior Geotechnical Engineer, Public Service Electric and Gas Company, Newark, NJ 07101.

methods lead to management problems on the limited land-use. The handling and disposal of large volume of sluiced fly ash are particularly troublesome.

Kiln dust is a by-product of the Portland cement manufacturing process. It contains about 45% calcium oxide and thus can improve the cementitious properties of the fly ash. The kiln dust stabilized fly ash (KDSFA) acts as a cohesive silt. It presents a significant potential as a cost-effective seepage barrier material to repair a leaking earth structure such as a fly ash pond.

Many studies have indicated that stabilized fly ash and its modifiers appear to be promising for use as a hydraulic barrier material (Bowders, et al., 1987). Most of the reagents used for stabilization are lime, cement, and bentonite. None of them incorporates the use of kiln dust. This paper evaluates the possibility of this innovated approach.

MATERIAL

The fly ash material tested was taken from a coal-fired generating station in New Jersey. It was sluiced to an ash pond from the power units for interim storage. When the pond was full, kiln dust was added to stabilize the fly ash. Based on the site experience, a 15% (by volume) kiln dust was used to indurate the sluiced fly ash. Representative samples of stabilized and unstabilized fly ash were taken from the pond for testing.

The fly ash used is class F with little pozzolanic behavior. Table 1 presents the typical physical and chemical properties of this material. Kiln dust used was supplied by a local cement manufacturing plant. Table 2 shows the chemical analyses of this material.

Table 1. Physical and Chemical Properties of the Fly Ash

Property	Value
Specific Gravity	2.54
Coefficient of Uniformity	2.50
Liquid Limit	16.8
Plastic Limit	N.P
SiO_2	47.0%
Al_2O_3	28.0%
Fe_2O_3	16.1%
TiO_2	1.5%
CaO	2.3%
MgO	1.0%
SO_3	1.4%
P_2O_5	1.0%
Na_2O and K_2O	1.0%
C	0.7%

Table 2. Chemical Analyses of Kiln Dust Used

SiO_2	16.0%
Al_2O_3	3.9%
Fe_2O_3	2.1%
CaO	45.8%
MgO	3.1%
SO_3	5.4%
Na_2O	0.5%
K_2O	2.4%
Cl	0.2%
Loss on Ignition	20.3%

PHYSICAL PROPERTIES

Table 3 presents the physical properties of the KDSFA materials tested. The moisture content dropped about 60 to 100% after stabilization. The reaction of kiln dust and fly ash apparently absorbed most of the free water in the mixture. The stabilization also caused the mixture to have coarser particles and a higher specific gravity.

Table 3. Summary of Physical Properties Test Results

Sample No.	In-situ Moisture (%)	Specific Gravity	Sieve Analysis % Passing Mesh No. 50	100	200
P_u-1[a]	37.0	2.42	99.2	96.8	84.7
P_u-2	41.0	2.44	-	-	-
P_s-1[b]	20.5	2.53	79.2	74.3	68.5
P_s-2	21.9	2.52	78.5	66.9	62.6
P_s-3	21.6	2.52	82.6	78.7	74.5
P_s-4	23.5	2.68	88.0	84.0	78.6
P_s-5	21.8	2.57	89.4	84.5	78.2
P_s-6	22.6	2.44	85.1	79.6	72.7

a - unstabilized fly ash samples

b - kiln dust stabilized fly ash samples

COMPACTIBILITY TEST

The compactibility of the KDSFA material was studied by performing modified Proctor tests (ASTM D-1557) on samples collected at various locations on site. Table 4 presents the results of these tests.

Table 4. Summary of Compaction Test Results

Sample No.	Maximum Dry Density (pcf)	Optimum Moisture Content (%)
Fly Ash	98.0	18.5
C-1	94.3	22.6
C-2	100.7	17.3
C-3	98.9	20.0
C-4	95.1	21.5
C-5	100.8	19.6
C-6	101.0	20.6
C-7	96.8	21.7
Mean Value	98.2	20.4

1 pcf = 0.157 kN/m^3

The test results indicated that the stabilization yielded a similar maximum dry density and a slightly higher optimum moisture content (OMC). The mixing of kiln dust with fly ash tends to absorb the free water in the mixture. Therefore, it needs more water to achieve the peak density as compared to the unstabilized fly ash. Bowders, et al. (1987) reported similar results on compacted cement-treated fly ash samples. The optimum moisture content of the fly ash is very close to the saturation moisture content (GAI, 1979). Fly ash compacted on the wet side of the optimum tends to be saturated and thus difficult for compaction. The stabilization is helpful to increase the workability of the mixture. The KDSFA material showed a compaction character similar to those of silty clay or clayey silt.

STRENGTH AND COMPRESSIBILITY

A hydraulic barrier should be stable and without excessive deformation during service. Strength and compressibility, therefore, must be evaluated for KDSFA materials. Samples were compacted to various densities using materials collected on site. They were cured for 3 days before testing. To simulate the field condition of a hydraulic barrier, samples were tested at saturated condition. To observe the strength behavior at different stress conditions, two types of tests were performed. One is direct shear, consolidated drained test. The other is triaxial compression, unconsolidated undrained test. Table 5 presents the test results.

Table 5. Summary of Shear Strength Tests

Sample No.	Angle of Internal Friction, ϕ (degree)	Cohesion c (psf)[a]
Fly Ash[b]	34.0	209
DS-1[b]	39.0	330
DS-2[b]	35.9	240
DS-3[c]	30.0	130
DS-4[c]	33.6	160
TS-1[d]		2006
TS-2[d]		2030

a - 1 psf = 44.78 kPa.
b - direct shear consolidated drained test, sample compacted to 95% of modified Proctor density.
c - direct shear consolidated drained test, samples compacted to 90% modified Proctor density.
d - triaxial unconsolidated undrained test, samples compacted to 90% of modified Proctor density.

The strength behaviors of compacted KDSFA materials are similar to those of compacted clayey fill. At the same degree of compaction, the stabilized fly ash showed a higher strength than that of unstabilized. The shear strength increased with increasing compaction effort. The shear strength of compacted KDSFA material at either total stress or effective stress condition was high. It can be considered as stiff to very stiff soil according to geotechnical classification (Das, 1984). The compressibility values of compacted KDSFA materials are given in Table 6. Test results are comparable to those of natural fine-grained soils. The typical compression index of fly ash ranges from 0.1 to 0.25. The recompression index was reported ranging from 0.02 to 0.04 (GAI, 1979). The stabilization appeared to have insignificant effect on the compressibility.

The values of overconsolidated pressure generally increased with increasing compaction efforts. The minimum value observed was 10 tsf (957.6 kPa). This is equivalent to the weight of about 150 to 200 feet (45.72 to 60.96 m) of overburden soils. This height is not likely to occur for a hydraulic barrier at normal use. Therefore, the compacted KDSFA material can be considered as overconsolidated soil with a negligible compressibility at most circumstances.

Table 6. Summary of Consolidation Test Results

Sample No.	Compression Index	Recompression Index	Overconsolidated Pressure (tsf)[a]
T-1[b]	0.20	0.01	35
T-2[b]	0.21	0.01	25
T-3[c]	0.17	0.02	10
T-4[c]	0.20	0.02	25

a - 1 tsf = 95.76 kPa
b - samples compacted to 95% of modified Proctor density.
c - samples compacted to 90% of modified Proctor density.

PERMEABILITY

Permeability has been considered as the primary performance criteria for a hydraulic barrier (Johnson, et al., 1985). Fourteen permeability tests were conducted to investigate the effects of kiln dust stabilization on the permeability of fly ash. Tests were conducted on each sample using a flexible wall permeameter. Table 7 presents the test results.

The permeability of unstabilized fly ash was 9.3 x 10^{-6} cm/s. It decreased by about 4-fold to 2.2 x 10^{-6} cm/s after stabilized with kiln dust. The increase of compaction only reduced the permeability slightly. The material with such permeability can be used as seepage barriers for non-hazardous waste sites (Usmen, et al., 1988). To allow for more applications, the material should have a permeability as low as possible. Current design criterion for an earthern liner at a hazardous waste site is to achieve a permeability of less than 1 x 10^{-7} cm/s (EPA, 1985). Cement and bentonite have been widely used for hydraulic barrier constructions for their low permeability characteristics (Bowders, et al., 1987). Additional cement or bentonite was mixed with the KDSFA material in an attempt to further lower the permeability. Cement was not mixed with bentonite in the same sample. This is because that the cement is adverse to the expansion of bentonite and thus limited the reduction of permeability (Usmen, et al., 1988). Samples were prepared to have a dry density of about 95 pcf (14.915 kN/m^3) and an initial moisture content of about 25%.

The addition of cement up to 6% reduced the permeability only to 1.1 x 10^{-6} cm/s. It appeared to be insignificant on the reduction of permeability. Bowders, et al. (1987) reported that the permeability presented no reduction for cement contents at 9% to 15%. Therefore, further increase the amount of cement was not performed.

Table 7. Summary of Permeability Test Results

Group No.	Sample No.	Sample Description	Permeability (cm/s)
1	P_1-1	Fly Ash - 100% compaction	9.3×10^{-6}
	P_1-2	KDSFA - 90% compaction	6.5×10^{-6}
	P_1-3	KDSFA - 90% compaction	5.4×10^{-6}
	P_1-4	KDSFA - 95% compaction	5.3×10^{-6}
	P_1-5	KDSFA - 100% compaction	2.2×10^{-6}
2	P_2-1	KDSFA + 2% cement[a]	2.5×10^{-6}
	P_2-2	KDSFA + 4% cement	1.3×10^{-6}
	P_2-3	KDSFA + 6% cement	1.1×10^{-6}
3	P_3-1	KDSFA + 2% bentonite-S[b]	1.9×10^{-6}
	P_3-2	KDSFA + 4% bentonite-S	1.0×10^{-6}
	P_3-3	KDSFA + 6% bentonite-S	5.9×10^{-7}
	P_3-4	KDSFA + 4% bentonite-U[c]	3.7×10^{-7}
	P_3-5	KDSFA + 4% bentonite-U[d]	7.4×10^{-8}
	P_3-6	KDSFA + 4% bentonite-C[e]	8.1×10^{-7}

a - type I Portland cement, % by weight.
b - sodium bentonite, saline seal- 100, % by weight.
c - sodium bentonite, ultra-seal N, % by weight.
d - bentonite hydrated before mixing, % by weight.
e - calcium bentonite, % by weight.

The addition of bentonite to the KDSFA presented a more pronounced effect on permeability compared to cement. The permeability reduced with the increase of bentonite. Using a low purity bentonite (SS-100) of 6%, the permeability dropped to 5.9×10^{-7} cm/s. It was only about 4-fold below that of without the bentonite. When a high purity bentonite (Ultra Seal-N) of 4% was used, the permeability dropped about one order of magnitude lower than that of without the bentonite. Bentonite of higher purity presents more expansion, viscosity, and colloidal activity. Permeability is in inverse proportion to these properties. The test results were in agreement with this explanation.

Sample P_3-4 exhibited a permeability of about 5-fold higher than that of sample P_3-5. Both samples were made of 4% bentonite (Ultra Seal-N). However, during mixing with KDSFA, dry bentonite was used for P_3-4, and pre-hydrated bentonite was used for P_3-5.

Bentonite normally required at least 48 hours to achieve a full hydration. A fully hydrated bentonite particle may expand 10-to-15- fold of its dry bulk (Wu, 1990). Bentonite hydration thus exhibits a strong effect on the permeability. In this study, samples were prepared by (1) mixing all solid particles in dry, (2) adding water, and (3) compacting to the specified density. The hydration and the expansion of bentonite tended to be restrained due to the limited mixing time and the particle confinements. Therefore, larger void spaces could develop in the sample and cause a higher permeability. Test results supports this hypothesis.

The test results demonstrated that the permeability of KDSFA material could be lowered to satisfy the requirement of EPA. It appears to have a great potential for use as a low permeability liner material at a hazardous waste site. However, the mix of hydrated bentonite with KDSFA material in the field will be troublesome. High purity bentonite also is expensive. It appears to be not commercially effective for permeability reduced to 10^{-8} cm/s due to construction problems and material cost. The bentonite modified KDSFA materials showed a higher permeability compared to those of soil-bentonite or fly ash-sand mixtures as reported by others (Garlanger, et al., 1987; and Vesperman, et al., 1985). Bentonite appeared to be not in favor of fly ash or kiln dust stabilized fly ash for permeability reduction.

Garlanger, et al. (1987) reported a permeability of 1×10^{-8} cm/s for a sand-bentonite liner with 5% bentonite. Vesperman, et al. (1985) studied the permeability of fly ash-sand mixtures. They found the permeabilities were in the range of 10^{-8} and 10^{-9} cm/s for samples containing 100% and 40% fly ash. Edil, et al. (1987) conducted permeability tests on compacted mixes of fly ash, sand, and bentonite. They reported that the addition of bentonite up to 10% did not reduce the permeability.

Wu (1989) conducted sedimentation tests to study the effect of fly ash on bentonite expansion. He found that the bentonite expansion was inhibited significantly by calcium and other high valence ions released from the fly ash. These ions replace the sodium in a bentonite crystal and reduces the thickness of double layer. Based on Gouy-Chapman theory, decreasing the thickness of double layer tends to reduce the expansion, and thus has higher permeability. The higher permeability obtained in this study can be attributed to the calcium- riched environment in the mixture due to fly ash and kiln dust. Sample, P_3-6, tested with calcium bentonite showed a lower permeability than that with sodium bentonite verified this assumption. The bentonite modified KDSFA materials present a significant potential as a successful hydraulic barrier. However, it appeared to be not commercially effective to lower the permeability to the order of 10^{-8} cm/s due to construction problems and material cost. Further research is required to develop a better procedure for permeability reduction.

SUMMARY AND CONCLUSIONS

The addition of kiln dust to fly ash reduced the moisture content of the mixture and increased the workability. At the same compaction, the stabilized fly ash exhibited a higher strength than that of unstabilized. The compacted stabilized material showed a characteristic similar to those of a silty clay or a clayey silt. It has a high strength and a very low compressibility.

The compacted KDSFA material showed a base permeability of the order of 10^{-6} cm/s. The kiln dust stabilization caused a reduction in permeability by about 4-fold. The addition of cement provided insignificant effect on permeability. Use of bentonite up to 6% decreased the permeability to the order of 10^{-7} cm/s. The expansion of the bentonite tended to be limited in the mixture due to the effect of ion replacements. The permeability of a bentonite mixture is highly dependent upon the purity and the hydration of the bentonite. Sample mixed with a high purity and fully hydrated bentonite reduced the permeability to the order of 10^{-8} cm/s and meet the criteria of a hazardous waste liner. However, it appeared to be not commercially effective due to construction problems and material cost. Further research is required to develop a better procedure for permeability reduction.

The results of this study present that the kiln dust stabilized fly ash can be a satisfactory barrier material for seepage control. It also shows a great potential for use as a low permeability liner material at a hazardous waste site.

REFERENCES

1. JJ Bowders, Jr., MA Usmen & JS Gidley, Stabilized Fly Ash for Use as Low-Permeability Barriers, Geotechnical Practice for Waste Disposal '87 GSP 13), Ed RD Woods, ASCE, New York, NY, 1987, 320-333.
2. BM Das, Principles of Foundation Engineering, Brooks/Cole, Monterey, CA, 1984, 73 p.
3. TB Edil, PM Berthouex & KD VEsperman, Fly Ash as a Potential Waste Liner, Geotechnical Practice for Waste Disposal '87 (GSP 13), Ed RD Woods, ASCE, New York, NY, 1987, 447-461.
4. EPA, Draft: Minimum Technology Guidance on Double Liner Systems for Landfills and Surface Impoundments - Design, Construction, and Operation, Rpt EPA 530-SW/85-014, U. S. Environmental Protection Agency, Washington, D.C., 1985, 71 p.
5. GAI Consultants, Inc., Fly Ash Structural Fill Handbook, Rpt EPRI EA-1281, Elec. Power Res. Inst., Palo Alto, CA, 1979, 3-10 p.
6. GAI Consultants, Inc., Fly Ash Construction Manual for Road and Site Applications (1), Rpt EPRI CS-5981, Elec. Power Res. Inst., Palo Alto, CA, 1988, S-2 p.
7. JE Garlanger, FK Cheung & BS Tannous, Quality Control Testing for a Sand-Bentonite Liner, Geotechnical Practice for Waste Disposal '87 (GSP 13), Ed RD Woods, ASCE, New York, NY, 1987, 488-499.
8. AI Johnson, RK Forbel, NJ Cavalli & CB Pettersson, Overview, Hydraulic Barriers in Soil and Rock (STP 874), ED AI Johnson, RK Forbel, NJ Cavalli & CB Pettersson, ASTM, Philadelphia, PA, 1985, 1-6.
9. MA Usmen, Preface, Proc. Disposal and Utilization of Electric Utility Wastes, ED MA Usmen, ASCE, Nashville, TN, 1988, iii-iv.
10. MA Usmen, JJ Bowders, Jr. & JS Gidley, Low Permeability Liners

Incorporating Fly Ash, Proc. Disposal and Utilization of Electric Utility Wastes, ED MA Usmen, ASCE, Nashville, TN, 1988, 50-65

11. KD Vesperman, TB Edil & PB Berthouex, Permeability of Fly Ash and Fly Ash-Sand Mixtures, Hydraulic Barriers in Soil and Rock (STP 874), ED AI Johnson, RK Forbel, NJ Cavalli & CB Pettersson, ASTM, Philadelphia, PA, 1985, 289-298.
12. JY Wu, Permeability and Volume Change Characteristics of Bentonite-Sand Mixes in a Contaminant Environment, Doctor of Engineering Dissertation, New Jersey Institute of Technology, Newark, NJ, 1989, 153-160 p.
13. JY Wu, Properties of a Treated-Bentonite/Sand Mix in Contaminant Environment, Physico-Chemical Aspects of Soil and Related Materials (STP 1095), Ed KB Hoddinott & RO Lamb, ASTM, Philadelphia, PA, 1990, 47-59.

YIELD STRESS AND FLOW OF BENTONITE IN MIXTURE SEALANTS

Shoung Ouyang[1] and Jaak J.K. Daemen[2]

ABSTRACT: The use of bentonite-based mixtures to form hydraulic barriers has greatly increased in recent years. Mixtures of bentonite and crushed rock or sand are also being considered as prominent candidate materials for backfilling nuclear waste repositories. Of particular interest is the long term performance of such sealants. This paper addresses the susceptibility of such structures to piping as a result of bentonite flow. The yield stress of Wyoming bentonite is established as a function of water content. The relation allows prediction of the critical hydraulic gradient at which flow of bentonite takes place. The critical gradient may be treated as the maximum allowable gradient in the design of waste or water containment facilities. The information presented can also be applied to bentonite grouting and filter design.

INTRODUCTION

Sodium bentonite, due to its very low permeability, high swelling potential, chemical stability, and ion adsorption ability, is considered a prominent candidate material for backfilling nuclear waste repositories (14,20). The addition of sand or crushed rock to bentonite is also under investigation (2,7,11).

The sealing performance of bentonite-based sealants under diverse conditions needs to be studied (4,24). The seal components may be required to retain adequate sealing performance for a long time (8). Such requirements pose a great challenge to the long term assessment of sealing performance. For example, piping, seepage erosion, and channelling have been identified to be responsible for the failures of earth dams, embankments, and natural slopes (22,24). Such effects can also be detrimental to the performance of sealants installed against fractured rocks (1,9). The discontinuities, if open, can serve as conduits for the transport of not only water but also bentonite. Piping has been observed as a result of bentonite flow (19).

Unlike flow of water, flow of bentonite will not occur until the shear stress induced by a driving pressure exceeds the yield stress of the bentonite (21). The yield stress of bentonite depends primarily upon its water content (13). If the

---

[1] - Res. Asst., Dept. of Mining & Geological Engineering, University of Arizona, Tucson, 85721.

[2] - Prof., Mackay School of Mines, University of Nevada, Reno, 89557-0139.

yield stress can be established as a function of water content, the critical hydraulic gradient at which flow of bentonite takes place can be predicted.

In this study, yield stresses of sodium bentonite at various water contents (68 to 510%) are determined by forcing bentonite through circular glass tubes. The determination of the yield stress is based on the theory of plastic flow through capillaries. The relationship between the yield stress and water content provides the basis for a proposed bentonite flow model. The model has been examined against nine flow test results. The predicted critical hydraulic gradients are about 1.1 to 2.1 times higher than the experimental ones. Factors that may contribute to the deviation are discussed.

PLASTIC FLOW THROUGH CAPILLARIES

A clay slurry or paste is not a true liquid. Its flow properties lie between those applicable to liquids and those applicable to solids (10,21). Bingham (3) introduced the concept of an ideal material which does not flow until a certain shear stress, the yield stress τ_f is reached and thereafter flows at a rate proportional to the excess shear stress, $\tau-\tau_f$. The rate of shear strain, $d\epsilon_s/dt$, at any point in the material, is given by:

$$\frac{d\epsilon_s}{dt} = \frac{1}{\eta_{pl}}(\tau-\tau_f) \qquad (1)$$

where η_{pl} = plastic viscosity of the material.

The application of Eq. (1) to flow in a circular capillary has been worked out by Buckingham (5), who gave, without detailed derivation, the result in the form:

$$\frac{V}{t} = \frac{\pi R^4 \mu}{8l}\left(P_d - \frac{4}{3}p + \frac{p^4}{3p_d^{\,3}}\right) + \pi R^2 V_R \qquad (2)$$

where V = volume of discharge,
t = elapsed time for discharge V,
R = radius of capillary,
μ = mobility of material, equal to $1/\eta_{pl}$,
P_d = pressure difference over a length l of the capillary,
p = $2l\tau_f/R$
V_R = slip velocity at the wall of the capillary.

The derivation of Eq. (2) can be found in Ouyang (18).

Assuming that the slip effect at the wall is negligible, and expressed in terms of the shear stress and the rate of shear strain, $(d\epsilon_s/dt)_R$, Marsland and Loudon (13) obtain the following equation:

$$\left(\frac{d\epsilon_s}{dt}\right)_R = \frac{4Q}{\pi R^3} = \frac{1}{\eta_{pl}}\left[\tau_R - \frac{4}{3}\tau_f + \frac{1}{3}\frac{\tau^4_f}{\tau^3_R}\right] \quad (3)$$

where Q = volume rate of discharge, the same as V/t in Eq. (2),
τ_R = averaged shear stress at capillary wall, (R/2)(P_d/l).

Eq. (3) gives a flow curve of the type shown in Fig. 1(a). No flow occurs in region I. The material flows according to Eq. (3) in region II. When $\tau_R >> \tau_f$, the slope of the straight portion of the flow curve becomes $1/\eta_{pl}$. Actual clay slurries or pastes do not follow this idealized law exactly but have flow curves of the form given in Fig. 1(b) (13). When the shear stress, τ_R , reaches τ_f , shear failure occurs near the wall of the capillary and the slurry moves forward as a plug (stage II). As the pressure gradient increases, the diameter of the solid plug becomes smaller (stage III) until all the material flows as a viscous liquid. The rate of flow then increases linearly with the pressure gradient (stage IV).

The flow of clay within a matrix of sand or crushed rock particles is analogous to the flow in capillaries. For a hydraulic barrier constructed of bentonite and crushed rock, its sealing ability should be maintained if the clay remains in place. The pressure gradient at which bentonite in the mixture starts to flow is deemed critical and depends upon the yield stress of the clay. When the clay structure becomes flocculated, the shear strength increases (12,15). For bentonite dispersed in distilled water, the yield stress depends primarily upon its water content (13), and is expected to assume a minimum value. In this study, yield stresses of bentonite pastes mixed with distilled water are used to evaluate the critical hydraulic gradient required to initiate the flow of bentonite.

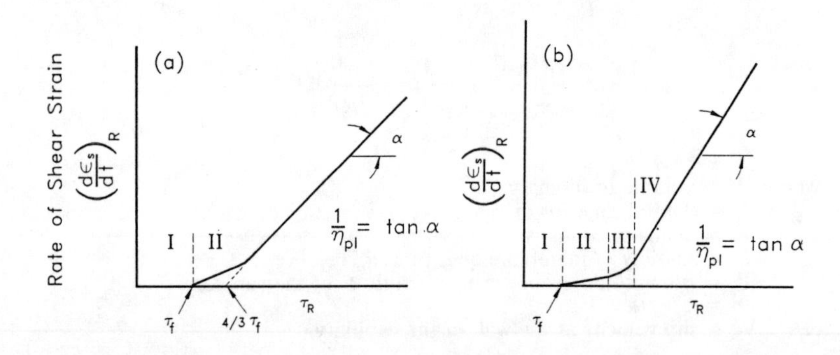

Figure 1. Flow curves of (a) the ideal Bingham material and (b) clay paste, as a function of rate of shear strain and shear stress at the wall of a circular capillary tube. (Adapted from Marsland and Loudon, 1963, p. 16)

YIELD STRESS OF BENTONITE MIXED WITH DISTILLED WATER

The American Colloid C/S granular bentonite used in this study has a liquid limit of 433% and a plastic limit of 50% (18). Bentonite pastes having nominal water contents of 75, 100, 200, 500%, have been prepared and allowed to cure for 72 hours in air-tight containers prior to testing. The pastes are driven through glass capillaries by compressed gas(helium). The experimental design is based on the following logic: (1) Once a clay paste occupies a capillary, a fixed shearing surface is established; (2) By varying the driving pressure and measuring the discharge of the clay paste, a flow curve similar to the one shown in Fig. 1(b) can be obtained. Sample preparation and experimental procedures are described by Ouyang (18).

Experience from several pre-trials has indicated that a constant shearing surface condition is difficult to achieve due to the slow flow rates of the thick bentonite pastes. To prolong the test duration solves the problem but brings about another complication, i.e., a significant change in water content of the clay paste, resulting from migration of water. Shortening the length of the capillaries is found undesirable because the clay slurry near the outlet dries out due to evaporation and therefore jeopardizes the advance of the paste. Since the yield stress of bentonite as a function of water content is the ultimate interest in determining the critical hydraulic gradient for bentonite flow, the experiments are aimed at obtaining such a relationship.

The yield stress of a bentonite paste is determined by narrowing the driving pressures down to a small range within which a slight change of driving pressure results in either a flow or a no flow condition. The yield stress is computed based on the no flow condition. The accuracy of the yield stresses thus determined is limited by the resolution of the measuring devices. Fig. 2 shows the yield stress of bentonite pastes vs. bentonite weight percent. The open circles indicate the shear stress at which no advance of the clay can be detected with a measuring tape of 0.5 mm resolution. Time elapsed before the no flow conclusion is made varies from 30 minutes to more than 24 hours, depending upon water content of the samples. The selection of the time spans is arbitrary. The rate of shear strain calculated accordingly is lower than $2.1 \times 10^{-4} sec^{-1}$. The shear stress computed for the "no flow" condition is then assumed to be the yield stress of bentonite. Also included in Fig. 2 are yield stresses of Wyoming bentonite grouts (triangles) reported by Marsland and Loudon (13).

It is not always easy to find the no flow condition quickly. When it takes a long time to identify such a condition, the water content of the bentonite in the capillary can be quite different from its initial value due to the migration of water. What water content the computed yield stress actually corresponds to requires a careful examination of experimental records. The averaged initial water content is related to the yield stress based on the early no flow condition. Water content of bentonite paste inside the capillary, determined immediately after testing, is used for the no flow condition established in the later part of a test.

RELATION BETWEEN WATER CONTENT AND YIELD STRESS OF BENTONITE

Statistical analysis has been performed on the bentonite flow test results. The equations obtained from curve fitting are given below with the coefficient of correlation (R^2):

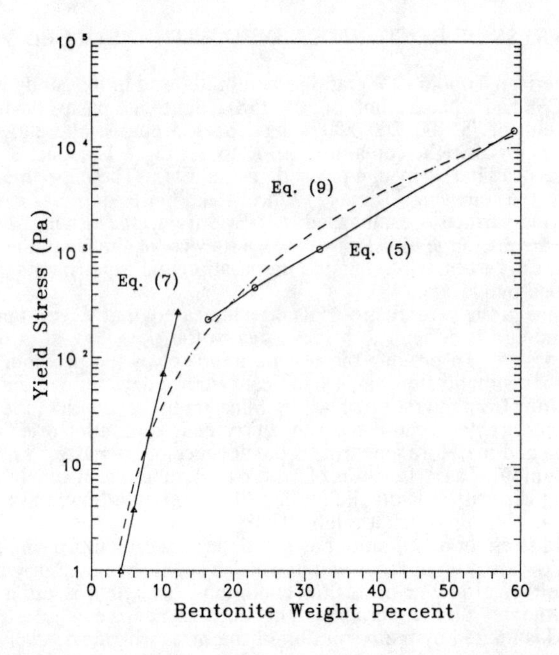

Figure 2. Yield stress of sodium bentonite pastes vs. bentonite weight percent. Triangular points are obtained from Marsland and Loudon (1963).

A. Thick bentonite pastes (for 70% < w < 510%)

$$\text{Log}\ (\tau_f) = 13.728 \times w^{-0.2818} \qquad R^2 = 0.9999 \tag{4}$$

where τ_f = yield stress of bentonite (Pa),
w = water content in percent.

$$\text{Log}\ (\tau_f) = 1.6964 + 0.0417x \qquad R^2 = 0.9998 \tag{5}$$

where x = bentonite weight percent.

B. Thin bentonite pastes (for 720% < w < 2300%, based on the data from Marsland and Loudon (13))

$$\text{Log}\ (\tau_f) = 8.552 \times \text{EXP}(-0.001715w) \qquad R^2 = 0.9979 \tag{6}$$

and

$$\text{Log}\ (\tau_f) = -1.209 + 0.3017x \qquad R^2 = 0.9984 \tag{7}$$

Regression has also been performed on the pooled results of bentonite flow tests (both thick and thin clay pastes). Excluding the anomalous yield stress point for x = 12.5% (w = 723%) shown in Fig. 2, the equations of the best fit are:

$$\text{Log}\ (\tau_f) = 8.81 - 1.07 \times \text{Ln}(w) \qquad R^2 = 0.938 \tag{8}$$

and

$$\text{Log}\ (\tau_f) = -1.76 + 1.44 \times \text{Ln}(x) \qquad R^2 = 0.976 \tag{9}$$

These equations can be used to predict the yield stress of bentonite when mixed with distilled water.

RELIABILITY OF THE YIELD STRESS-WATER CONTENT RELATION

The yield stress of bentonite at a given water content can be considered as the shear strength to resist flow. In this context, the yield stress of bentonite obtained from the flow testing may be different from the shear strength determined using conventional soil test techniques. Some relation between these strength measures should exist. Casagrande (6) has first suggested a possible unique value of the shear strength (2.65 kN/m^2) for most natural fine-grained soils at the liquid limit. Several researchers also report a narrow range of shear strengths of such soils, when measured at the liquid limit: 0.7-1.75 kN/m^2 (23), 0.8-1.6 and 1.1-2.3 kN/m^2 (16), 1.3-2.4 kN/m^2 (26). Although the ranges are small, the shear strength tends to decrease with increasing liquid limit (26). It is speculated that, at the liquid limit of 433%, the C/S granular bentonite should have a shear strength either within or lower than the ranges indicated above. For the given water content, the shear stress computed from Eq. (4) is 0.3 kN/m^2, which is as expected.

The reliability of Eq. (4) has also been examined against published shear strength results obtained from axial shear testing of annulus bentonite seals (17). These annulus seals are installed between an outer plexiglass pipe and an inner steel casing. Table 1 shows the comparisons. Recognizing the differences in the aggregate size and supplier, deviations between the predicted shear strengths from the equation and the experimental results may be considered acceptable.

PREDICTION OF CRITICAL PRESSURE GRADIENT FOR BENTONITE FLOW

The relation between yield stress and water content has been incorporated in a model for predicting the critical pressure gradient for bentonite/crushed rock plugs. The model requires two additional parameters, the water content of bentonite and the representative pore size of the crushed rock skeleton. For the saturated condition, the water content of bentonite in a mixture can be estimated using the following equation:

Table 1. Predicted Yield Stresses and Experimental Axial Shear Strengths for Different Bentonite Products

Clay Type	Water Content	Experimental* Shear Strength (kN/m^2)	Predicted** Yield Stress (kN/m^2)
American Colloid Volclay chip	80.47	15.4	9.70
American Colloid Volclay 3/8" tablets	77.05	18.5	10.86
Wyo-Ben Enviroplug medium (1/4-3/8")	95.33	6.03	6.32
NL Baroid Holeplug (3/4" chip)	83.33	26.2	8.9

*Ogden and Ruff (1989, Table 5-3). Shear strengths measured after the samples set for 72 hours.
**Based on Eq. (4)

$$w_{bs} = \left(\frac{100 V_v \gamma_w}{W_{rb}(100 - w_{rb})} - \frac{1}{G_{sb}}\right) \times 100 \tag{10}$$

where w_{bs} = water content of bentonite at saturation,
V_v = volume of voids between crushed rock particles,
w_{rb} = water content of air-dried bentonite,
r_w = unit weight of water,
W_{rb} = air-dried weight of bentonite,
G_{sb} = specific gravity of bentonite.

The water content thus determined can be used to calculate the yield stress (τ_f), with one of the Eqs. (4) through (9), depending on the water content. The critical pressure gradient ($i_{c,p}$) is then computed as follows:

$$i_{c,p} = \frac{2\tau_f}{R_m} \tag{11}$$

where R_m = representative pore radius of the crushed rock skeleton. The predicted critical gradient ($i_{c,p}$) can be expressed in terms of hydraulic gradient, using the relation of 10.2 m of water = 100 kPa.

Flow tests have been performed on nine mixtures of bentonite and crushed rock. The mixtures are prepared by mixing crushed Apache Leap tuff (Superior, Arizona) with either 25 or 35% bentonite by weight. Three gradation types (A, B, and C) of crushed tuff are shown in Table 2. Hydraulic conductivity of the mixtures appears to be constant for the low range of

Table 2. Particle Size Distribution of Crushed Tuff Gradations

Gradation Type	Sieve Size (mm) 9.42	6.68	4.75	2.0	0.84	0.42	0.25	0.15	.074	C_u
	Weight Percent Retained:									
A		27.0	17.0	23.7	12.7	7.09	4.22	3.8	4.57	16.5
B			26.0	39.5	19.5	7.5	6.0	5.5	6.0	15
C				45	23	10	8.0	6.0	8.0	13

C_u: coefficient of uniformity, d_{60}/d_{10}

hydraulic gradients. As the hydraulic gradient increases, hydraulic conductivity decreases, causing the breakdown of the linear relationship between flow rate and hydraulic gradient. This behavior has been observed for all nine flow tests. An example is shown in Fig. 3. This abrupt departure can not be explained by small-strain deformation and is believed to indicate the onset of bentonite flow (19). The hydraulic gradient corresponding to the deflection point of each flow curve is assumed to represent the critical pressure gradient at which bentonite starts to flow.

In this study, d_{50} (sieve aperture at 50% passing) of crushed rock constituents is assumed to represent R_m. This selection is based on the consideration that, according to Eq. (11), the bentonite will first start flowing in the large

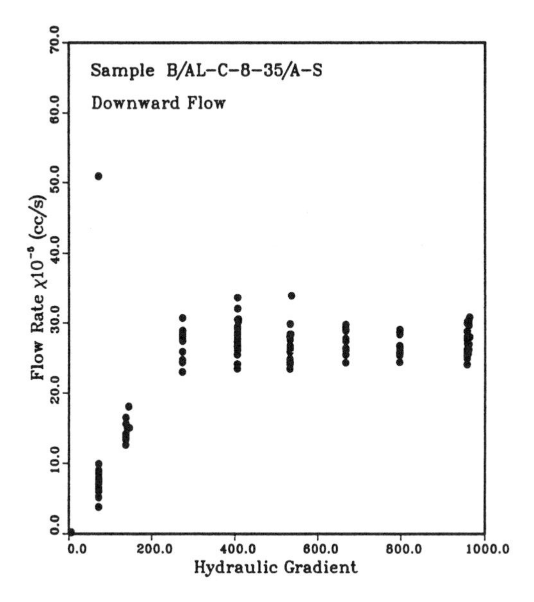

Figure 3. Typical example of flow rate vs. hydraulic gradient. Mixture of 35% bentonite and 65% crushed tuff of gradation A.

Table 3. Predicted and Experimental Critical Pressure Gradients for Bentonite Flow

Sample Number	Saturated Water Content of Bentonite (%)	Critical Gradient Pre-dicted	Critical Gradient Experi-mental	Gradient Ratio Pred./ Exper.	Pore Radius Rm (mm)
B/AL-C-4-25/A-P-B	108.07	244	115	2.12	3.9
B/AL-C-8-25/A-S	115.93	207	120	1.72	3.9
B/AL-C-8-35/A-S	83.42	492	280	1.76	3.9
B/AL-C-4-25/A	120.61	189	170	1.11	3.9
B/AL-C-4-25/B	122.38	290	250	1.16	2.45
B/AL-C-4-25/C	133.33	339	273	1.24	1.72
B/AL-C-4-35/A	79.1	530	424	1.25	3.9
B/AL-C-4-35/B	92.94	560	352	1.59	2.45
B/AL-C-4-35/C	106.43	572	360	1.59	1.72

pores between crushed rock particles for a given yield stress. For a crushed rock mixture of gradation type A, d_{50} is 3.9 mm. This implies a pore diameter of 7.8 mm, close to the maximum particle size of 9.42 mm for crushed rock constituents of type A. Such an estimation for the representative size of large pores appears reasonable, considering the separation of rock particles due to the bentonite filler.

The predicted critical gradients are compared with the experimental ones for the nine mixture samples in Table 3.

DISCUSSION

As shown in Table 3, the proposed model overestimates the critical hydraulic gradient by a factor ranging from 1.11 to 2.12. The discrepancy may be due to any or all of the following: (1) Neglecting slip at the wall of the capillary results in overestimating τ_f . (2) In driving the clay paste inside a capillary, part of the energy is consumed by the accompanying migration of moisture; the actual force effective for the advance of clay paste is less than the product of the driving pressure multiplied by the whole crosssectional area of the capillary. (3) The yield stress of bentonite for a given water content is computed for the no flow condition. Such a condition is established on the basis of observations and hence is limited by the resolution of the measuring device. If the condition identified actually resides in a flow region [e.g. stage II in Fig. 1(b)], the yield stress is overestimated.

Moreover, the saturated water content of bentonite obtained from Eq. (10) is an average one. The water content of bentonite changes as the pore pressure varies during flow testing. The variation of pore pressure leads to consolidation near the outflow end and swelling near the inflow end. This time dependent process can not be eliminated and will necessarily create a non-uniform distribution of water content in the sample. Because of the swelling, the water content of bentonite near the inflow end is expected to be higher than the water content calculated from Eq. (10). The flow of bentonite therefore should first occur in the upper part of the sample.

It is postulated that the critical gradients obtained from the curves of flow rate vs. hydraulic gradient probably correspond to the critical gradient for bentonite flow at a higher water content. For Sample B/AL-C-8-25/A-S, the experimental critical hydraulic gradient of 120 has been used to back calculate the water content according to Eq. (4). The calculation gives a water content of 158, about 36% higher than the 116 obtained from Eq. (10).

For practical purposes, the model can be used to give an approximation of the critical hydraulic gradient at which bentonite starts to flow in the crushed rock matrix. A conservative estimation can be made if the factor 2 in Eq. (11) is removed.

CONCLUSIONS

Yield stresses of bentonite at various water contents (68 to 510%) have been determined by forcing bentonite through circular tubes of different diameters. The logarithm of yield stress vs. water content is best described by a power function. Expressing water content of bentonite in terms of concentration weight percent, the relationship becomes linear, with correlation coefficient, $R^2 = 0.9998$. The critical hydraulic gradients predicted from the proposed bentonite flow model agree reasonably well with those obtained from the flow test results.

For mixture sealants of bentonite and crushed rock or sand, the saturated water content of the clay component can be computed using the phase diagram relation (Eq. 10). Following the established relation between yield stress and water content and the flow model, the critical hydraulic gradient can be estimated without experiments. This method greatly assists in the assessment of the long term sealing performance of bentonite-based mixtures. The critical hydraulic gradient thus defined may be treated as the maximum allowable gradient in the design of waste or water containment facilities. For a known maximum hydraulic gradient, the model and the relation can be used in an inverse way to provide information such as the saturated water content of bentonite and to specify the gradation of the crushed rock or sand allowable for the design of mixture plugs and seals.

The information presented in this paper can also be applied to bentonite grouting and filter design.

Several topics that have been discussed deserve further investigation. Improving the measuring technique of the yield stress, e.g. by more sensitive detection of the onset of flow, is recommended. A model that accounts for the influence of variable and changing water content, and possibly for drying and swelling, would be desirable. The applicability of results obtained from straight constant section flowpaths to highly tortuous flowpaths, e.g. in between crushed rock particles, needs to be evaluated. Studies on the influence of water chemistry on the yield stress of bentonite are recommended.

ACKNOWLEDGEMENT

This work is supported by the Division of Engineering, Office of Nuclear Regulatory Research, U.S. Nuclear Regulatory Commission. The American Colloid company supplied the bentonite. Valuable discussions with Mr. J. Philip, Contract Monitor, are gratefully acknowledged.

REFERENCES

1. B Aisenstein, E Diamant & I Saidoff, Fat Clay as a Blanketing Material for Leaky Reservoirs, Proc. 5th Intl. Conf. on Soil Mechanics and Foundation Engineering, Paris, Vol. II, Division 3B-7, 1961, 523-529.
2. Atomic Energy of Canada Limited, Management of Radioactive Fuel Wastes: the Canada Disposal Program, AECL-6314, Ed J Boulton, Atomic Energy of Canada Research Co., Whiteshell Nuclear Research Establishment, Pinawa, Manitoba, Canada, Oct, 1978.
3. EC Bingham, An Investigation of the Laws of Plastic Flow, Scientific Paper No. 278, U.S. Bureau of Standards, 1916.
4. EP Binnall, SM Benson, L Tsao, HA Wollenberg, TK Tokunaga, & EM Didwall, Critical parameters for High-Level Waste Repository, Vol. 2: Tuff, NUREG/CR-4161, prepared for U.S. Nuclear Regulatory Commission, by Lawrence Livermore National Laboratory, Livermore, California, 1987.
5. E Buckingham, On Plastic Flow Through Capillary Tubes, Proc. of The American Society for Testing Materials, Vol. 24, 24th Annual Meeting, 1921, 1154-1161.
6. A Casagrande, Notes on the Design of the Liquid Limit Device, Geotechnique, Vol. 8, No. 2, 1958, 84-91.
7. DA Dixon, MN Gray, and AW Thomas, A Study of the Compaction Properties of Potential Clay-Sand Buffer Mixtures for Use in Nuclear Fuel Waste Disposal, Engineering Geology, Vol. 21, 1985, 247-255.
8. JA Fernandez, PC Kelsall, JB Case, and D Meyer, Technical Basis for Performance Goals, Design Requirements, and Material Recommendations for the NNWSI Repository Sealing Program, SAND 84-1895, prepared for the U.S. Department of Energy, by Sandia National Laboratories, Albuquerque, New Mexico, and Livermore, California, 1987.
9. RE Goodman, and PN Sundaram, Permeability and Piping in Fractured Rocks, ASCE Journal of Geotechnical Engineering, Vol. 106, No. GT5, 1980, 485-498.
10. BA Keen, and GW Scott Blair, Plastometric Studies of Soil and Clay Pastes, Journal of Agriculture Science, Vol. 19, part IV, 1929, 684-700.
11. CM Koplik, DL Pentz, and R Talbot, Borehole and Shaft Sealing, Vol. 1 of Information Base for Waste Repository Design, NUREG/CR-0495, TR-1210-1, prepared for U.S. Nuclear Regulatory Commission by the Analytic Sciences Corp., Reading, MA, 1979.
12. TW Lambe, The Structure of Compacted Clay, ASCE Journal of the Soil Mechanics and Foundation Division, Vol. 84, No. SM2, 1958, pp. 1654-1 to 1654-33.
13. A. Marsland, and AG Loudon, The Flow Properties and Yield Gradients of Bentonite Grouts in Sands and Capillaries, in Grouts and Drilling Muds in Engineering Practice, Butterworth & Co., London, 1963, 236 p.
14. D Meyer, and JJ Howard, Eds, Evaluation of Clays and Clay Minerals for Application to Repository Sealing, ONWI-486, prepared by D'Appolonia Consulting Engineers, Inc., and Material Research Laboratory, the Pennsylvania State University, for Office of Nuclear Waste Isolation, Battelle Memorial Institute, Columbus, Ohio, 1983.
15. JK Mitchell, Fundamentals of Soil Behavior, John Wiley & Sons, Inc., New York, 1976, 422 p.

16. LEJ Norman, A Comparison of Values of Liquid Limit Determined with Apparatus Having Bases of Different Hardness, Geotechnique, Vol. 8, No. 2, 1958, pp. 79-83.
17. FL Ogden & JF Ruff, Axial Shear Strength Testing of Bentonite Water Well Annulus Seals, Dept of Civil Engineering, Colorado State University, Ft Collins, 1989, 98 p.
18. S Ouyang, Sealing Performance Assessments of Bentonite and Bentonite/Crushed Rock Plugs, Ph.D. Dissertation, Dept of Mining & Geological Engineering, University of Arizona, Tucson.
19. S Ouyang & JJK Daemen, Performance of Bentonite/Crushed Tuff Seals for Nuclear Waste Repositories, Waste Management '90, Vol. II, Tucson, Arizona, Feb 25-March 1, 1990, 605-611.
20. R Pusch, Borehole Sealing for Underground Waste Storage, Journal of Geotechnical Engineering, Vol. 109, no. 1, January, 1983, 113-119.
21. GW Scott Blair, & EM Crowther, The Flow of Clay Pastes Through Narrow Tubes, Journal of Physical Chemistry, Vol. 33, 1929, 321-330.
22. JL Sherard, RS Decker, & NL Ryker, Piping in Earth Dams of Dispersive Clay, Proc. of ASCE Special Conf. on Performance of Earth and Earth-Supported Structures, Vol. 1, 1972, 589-626.
23. AW Skempton & RD Northey, The Sensitivity of Clays, Geotechnique, Vol. 3, No. 1, 1953, 30-53.
24. HP Thompson, Review and Comment on the U.S. Department of Energy Site Characterization Plan Conceptual Design Report, NWPO--TR-009-88, prepared by Engineering Co., Inc., in conjunction with Sea Inc., Dunn Geoscience Corp., and W.F. Guyton Associated, Inc., for Nuclear Waste Project Office, Agency for Nuclear Projects, Nevada, 1988.
25. W Wolski, Model Tests on the Seepage Erosion in the Silty Clay Core of an Earth Dam, Proc. of the 6th Intl. Conf. on Soil Mechanics and Foundation Engineering, Montreal, Vol. 2, 1965, 583-587.
26. MS Youssef, AH El Ramli, & M El Demery, Relationships between Shear Strength, Consolidation, Liquid Limit, and Plastic Limit for Remoulded Clays, Proc. of the 6th Intl. Conf. on Soil Mechanics and Foundation Engineering, Montreal, Vol. 1, 1965, 126-129.

CONTAMINANT MIGRATION EVALUATION AT A HAZARDOUS WASTE MANAGEMENT FACILITY

H. D. Sharma[1], M.ASCE, D. M. Olsen[2], M.ASCE, and L. K. Sinderson[3], M.ASCE.

ABSTRACT: A contaminant migration evaluation was conducted as part of the closure requirements for a Hazardous Waste Management Facility (HWMF). This evaluation was performed for the purpose of determining the capability of the clayey silt and silty clay Bay Mud sediments that underlie the site to serve as a barrier to vertical migration of contaminants. The evaluation involved: subsurface stratigraphic characterization, leachate characterization, leachate-soil compatibility laboratory testing, soil-column-leachate laboratory testing, and advection-dispersion contaminant transport modeling of the present field condition and after installation and operation of a proposed Leachate Collection and Removal System (LCRS). The evaluation results indicate that the Bay Mud sediments in conjunction with operation of a LCRS will serve as an acceptable barrier to vertical migration of contaminants at the HWMF.

INTRODUCTION

A Hazardous Waste Management Facility (HWMF) is located at the West County Landfill, Inc. Contra Costa County, California. The HWMF comprises 28-acres and is part of an approximately 200-acre landfill, which was developed on reclaimed marshlands situated on the southeastern shores of San Pablo Bay. Figure 1 shows the configuration of the HWMF and near vicinity. The HWMF is surrounded by a soil-attapulgite slurry cut-off wall that is founded in the underlying, low permeability, sediments locally referred to as the San Francisco Bay Mud (Bay Mud). Closure of the site will include installation and operation of a Leachate Collection and Removal System (LCRS) that will produce an inward and upward hydraulic gradient to the HWMF (EMCON, 1988). The purpose of this evaluation was to determine the effectiveness of the underlying low permeability Bay Mud sediments in conjunction with the LCRS to prevent migration of contaminants to deeper water bearing sand zones of the Bay Mud.

The contaminant migration, considered in this paper, is driven by both hydraulic and concentration gradients. This process can be simulated by an advection-dispersion analytical model that can be derived from the conservation of mass law (Ogata, 1970; Freeze and Cherry, 1979; Rowe et. al., 1988; and Acar and Haider, 1990). According to this model the movement of contaminants and their

1 - Chief Geotechnical Engineer, EMCON Associates, 1921 Ringwood Ave., San Jose, CA 95131.

2 - Project Geotechnical Engineer, EMCON Associates, 1433 West Market Blvd., Sacramento, CA 95834.

3 - Staff Geotechnical Engineer, EMCON Associates, 1433 West Market Blvd., Sacramento, CA 95834.

concentrations can be expressed by the following one-dimensional form of the dispersion-advection equation:

$$\partial c/\partial t = D\,[\partial^2 c/\partial z^2] - v\,[\partial c/\partial z] - [\rho K/n][\partial c/\partial t] \quad (1)$$

Where c = concentration of contaminant, D = dispersion coefficient, K = adsorption coefficient, v = seepage velocity (average linear pore velocity), n = porosity of soil, ρ = bulk density of soil, t = time, and z = distance from relative datum position.

This paper presents the site stratigraphy, leachate chemistry, laboratory determination of advection-dispersion parameters D and K, laboratory determination of hydraulic conductivity, k, and simulation analyses of present and proposed field conditions.

SUBSURFACE STRATIGRAPHIC CHARACTERIZATION

Numerous subsurface investigations have been performed at the site for the purpose of characterizing the vertical and horizontal limits of the solid waste fill, and liquid waste impoundment, the geologic and hydrogeologic conditions, and the geotechnical engineering properties of the underlying Bay Mud soils. The locations of selected borings utilized for this analysis are shown on Figure 1. Each boring was logged for subsurface conditions in accordance with the Unified Soils Classification System (American Society for Testing and Materials, 1989) by an EMCON geologist or engineer.

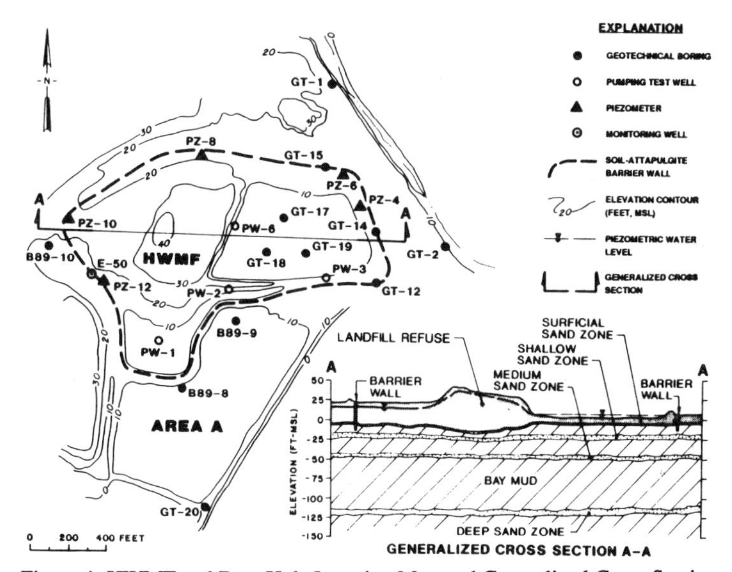

Figure 1, HWMF and Bore Hole Location Map and Generalized Cross-Section

Solid and Liquid Waste. The thicknesses of the solid wastes are estimated to range from 10 to 60 feet (3.0 to 18.3 meters) from historic topographic maps, aerial photographs, and waste manifest records. Borings GT-17, GT-18, and GT-19 were drilled in the liquid waste impoundment and encountered a sludge thickness of 16.5 to 18.5 feet (5.0 to 5.6 meters). The cross-section of Figure 1 shows a generalized interpretation of both the solid and liquid waste thicknesses.

Bay Mud Sediments. The Bay Mud sediments were characterized by drilling and sampling borings GT-1, GT-2, GT-12, GT-14, GT-15, GT-17, GT-18, GT-19, GT-20, B89-8, B89-9, and B89-10. The Bay Mud sediments encountered in these borings generally consisted of saturated, high plasticity clayey silts (MH) and silty clays (CH) that contain four water bearing zones consisting of poorly graded sands (SP), silty sands (SM), and clayey sands (SC): surficial, shallow, medium, and deep. The uppermost sand zone is generally encountered between elevations 0 to minus 5 feet (0 to -1.5 meters) mean sea level (MSL) and is usually in contact with the overlying refuse. The shallow sand zone is usually encountered between elevations minus 10 to 20 feet (-3.3 to -6.6 meters) MSL. The medium sand zone is usually encountered between elevation minus 40 to 50 feet (-12.2 to -15.2 meters) MSL. The deep sand zone is usually encountered between elevations minus 115 to 120 feet (-35.1 to -36.6 meters) MSL.

LEACHATE CHARACTERIZATION AND COMPATIBILITY TESTING

Characterization of the leachate was performed for the purpose of determining the types and concentrations of chemical compounds present in the HWMF leachate. The compatibility tests were performed for the purpose of determining whether the leachate would alter the soil properties thereby, adversely affecting the Bay Mud's capability to serve as an effective contaminant migration barrier.

Leachate Characterization. The leachate was characterized from samples taken form piezometers PZ-4, PZ-6, PZ-8, PZ-10, and PZ-12 and from pumping wells PW-1, PW-2, PW-3, and PW-6, which are all located inside the HWMF, as shown on Figure 1, and screened in the contaminated surficial sand zone. The compounds analyzed include US EPA Priority Pollutant volatile and semi-volatile organic compounds (EPA Test Methods 624 and 625, 1982), metals listed in the California Code of Regulations, Title 22, Section 66699, pesticides (EPA Test Method 608, 1982), gasoline, and pH. The chemicals detected in the HWMF leachate can be grouped as hydrocarbons, ketones, and phenols.

Leachate-Soil Compatibility. Waste fluid conductivity tests were performed to determine the compatibility of the HWMF leachate with the fine grained soils of the underlying Bay Mud obtained from borings GT-17 and GT-20. The test apparatus consists of a flexible wall permeameter equipped with bladder accumulators as shown in Figure 2. The samples were back pressure saturated and then subjected to a differential pressure head from bottom to top of 5 pounds per square inch (34.5 kiloPascals). Both samples were permeated with one pore volume of water taken from well E-50, followed by more than two pore volumes of leachate taken from well PW-1. Well E-50 is screened in the uncontaminated medium sand zone and is considered to have water quality concentrations representative of background conditions.

The effects of the leachate on the Bay Mud were determined by measuring the hydraulic conductivity variations with time as the background ground water and then the leachate permeated the samples while maintaining a constant hydraulic gradient. Plots of hydraulic conductivity versus the number of pore volumes displaced are presented for each sample in Figure 3. The test results indicate that when the Bay Mud is exposed to HWMF leachate the hydraulic conductivity either remains relatively constant (sample GT-20) or decreases slightly (sample GT-17). These observations are consistent with the results presented by Mitchell and Madsen (1987), and Storey and Pierce (1989), which indicate that the hydraulic conductivity does not change when permeated by low concentrations of organic chemicals. Based on these test results the HWMF leachate is considered to be compatible with the underlying fine grained Bay Mud sediments.

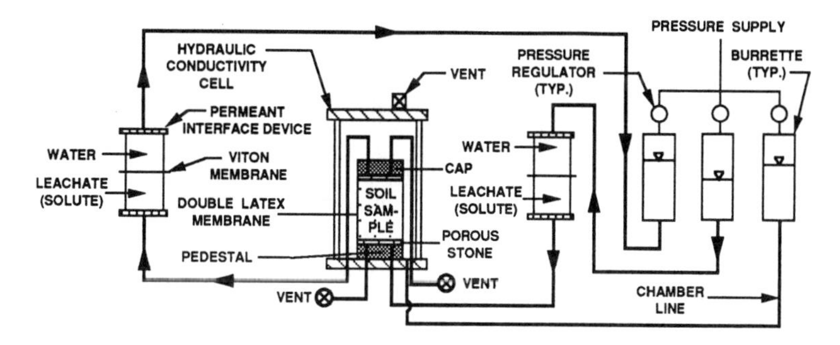

Figure 2, Permeameter and Column-Leachate Test Apparatus

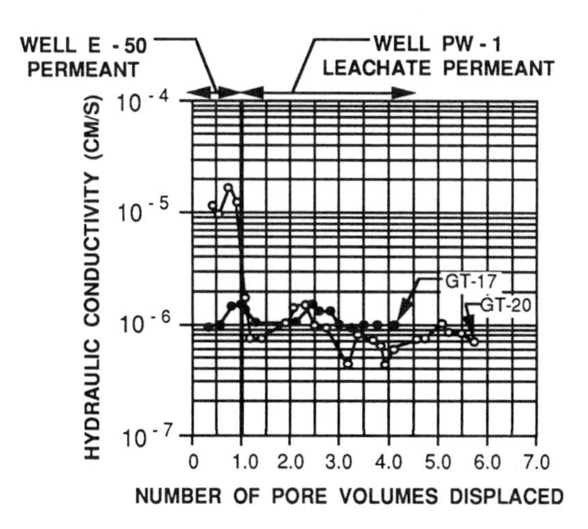

Figure 3, Leachate-Soil Compatibility Test Results

COLUMN-LEACHATE TESTING

The purpose of the column-leachate test is to measure the advection-dispersion coefficients of a soil with specific chemical compounds. These permeant-soil properties are then used to calibrate a solute transport numerical model. The model is used to simulate field conditions so that both travel times and associated chemical concentrations profiles can be predicted. Finally, the field simulation results are used to evaluate and design the site facilities.

Adsorption And Dispersion Coefficients. Transport of solutes in soil is dependent upon the specific advection and dispersion properties of the chemicals and soil. Advection transport is the component driven by the hydraulic gradient. Dispersion transport has two components: mechanical dispersion and molecular diffusion. The dispersion coefficient (D) quantifies the dispersion properties of the soil. The absorption coefficient (K) quantifies the soil property restricting movement of contaminants by absorbing chemical molecules onto the soil particle surfaces.

Laboratory Test Apparatus. The column-leachate test apparatus was modified to accommodate testing of volatile and semi-volatile organic compounds which are the predominant constituents of the HWMF leachate. Each column-leachate test was performed with a conventional triaxial cell, flexible wall, falling head permeameter (permeameter). A closed system was created by attaching bladder accumulators to the inlet and outlet of the permeameter. The bladders accumulators are devoid of air space, thus preventing volatilization of the leachate compounds. A burette pressure board is used to apply a pressure gradient across the permeameter, which causes flow to occur through the soil specimen. The test Apparatus is shown in Figure 2.

Test Procedures. The test is performed by (1) back-pressure saturating a soil specimen with water, (2) measuring its permeability (hydraulic conductivity) with water as the permeant while subjected to a constant hydraulic gradient (representative of expected site conditions), (3) attaching bladder accumulators containing a finite mass, synthetic spike solution (permeant) of key chemical compounds at concentrations that simulate the HWMF leachate, (4) permeating the soil specimen with the spike solution while subjected to the same constant hydraulic gradient, (5) removing the specimen after displacing between 1/3 and 2/3 pore volumes, (6) segmenting the specimen into slices orientated perpendicular to the flow direction, halving each slice, placing each half in a separate jar with an extraction solution (methanol for volatiles and hexane for semi-volatiles), capping with septa lids, and refrigerating, and (7) determining the concentrations of the spike solution compounds in each slice.

Soil Test Specimens. Two relatively undisturbed representative Bay Mud soil specimens taken from boring GT-17 were tested in the column-leachate apparatus. Both soil specimens were classified as high plasticity clayey silts (MH). Table 1 presents the soil properties and test conditions determined for specimen GT-17(A) and GT-17(B) taken from depths below the surface of 30 (9.1 meters) and 40 feet (12.2 meters), respectively.

Spike Test Permeant Solution. A synthetic leachate (spike solution) was fabricated to approximate the HWMF leachate. Use of a spike solution rather than actual HWMF leachate was based on 1) the need to be able to perform conformation testing with the same permeant, 2) for ease of detection, and 3) the chemical compounds present in the HWMF leachate and their concentrations vary to some degree from location to location. Table 2 presents the concentrations of the key chemical compounds used to make up the spike permeant solution.

Analysis of Column-Leachate Test Data. POLLUTE, a computer program developed by Rowe, et al. (1984), was used to evaluate the column-leachate test data. POLLUTE is capable of modeling one-dimensional, contaminant transport through, porous media (soils) while taking into account adsorption of contaminants, changes in leachate concentrations with time, ground water flow in an aquifer lying beneath a clay layer, and both upward and downward hydraulic gradients.

Table 1, Test Conditions and Soil Specimen Properties

Parameter	GT-17(A)	GT-17(B)
permeability (cm/s)	2.2×10^{-7}	4.1×10^{-7}
porosity (%)	50.6	51.5
density (kg/m^3)	1775	1759
length (cm)	8.25	8.20
diameter (cm)	7.05	7.05
hydraulic gradient (cm/cm)	8.52	8.58
pore volume displaced (%)	31.0	60.4

During the column-leachate test the only unknown parameters of equation (1) are the D and K coefficients. These coefficients are estimated with the POLLUTE model by inputting the known parameters given in Table 1 and then adjusting the unknown D and K parameters, through a trial and error process. The most representative D and K coefficients are those that produce a reasonable match between the theoretical concentration profile generated by the POLLUTE model and the observed concentration profile data. Each chemical compound must be evaluated separately to determine its specific D and K coefficients for the tested soil.

In order to obtain a reasonable match between the measured and theoretical concentration profiles it was necessary to model a very thin layer of altered soil along the soil-leachate interface. According to Rowe (1989) the properties of the soil along the soil-leachate interface are altered such that a partitioning effect occurs, whereby the movement of contaminants into the soil is somewhat restricted. However, after entering the soil the contaminant moves according to the advection-dispersion equation presented in the preceding. The cause of this partitioning effect is still unknown. Since the partitioning phenomenon is real, however, it must be taken into account as a thin layer with its own D and K coefficients.

Figures 4(a) and 4(b) show that the measured and theoretical acetone concentration profiles are in excellent agreement for both GT-17(A) and GT-17(B) specimens. Similarly, the measured and theoretical concentration profiles for the other spike solution chemical compounds are in excellent agreement for both test specimens. Table 2 summarizes the chemical compound D and K coefficients for each specimen.

Figure 4(a) also shows two additional theoretical concentration profiles predicted by POLLUTE while using the same D and K coefficients, but for different flow conditions. The theoretical concentration profile for a purely advection condition (no diffusion) does not match at all. However, the theoretical concentration profile predicted for a purely diffusion condition (no advection equals no flow) is fairly close to that of the measured profile. Thus, it can be concluded that contaminant migration in fine-grained soils such as clays and silts is dominated by a diffusion process. Similar results were obtained for the other spike solution chemical compounds.

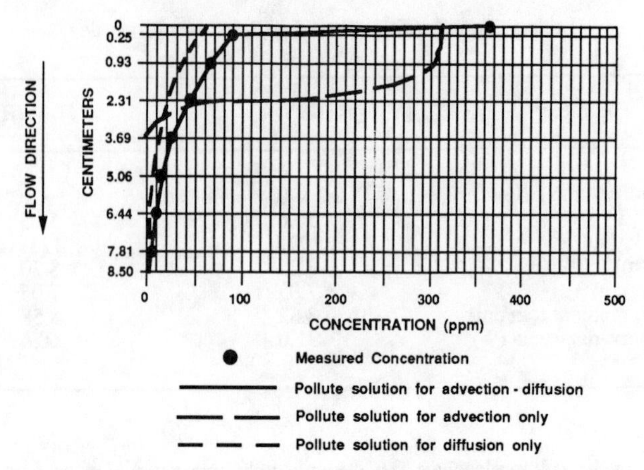

Figure 4(a), Observed vs. Predicted GT-17(A) Concentration Profiles

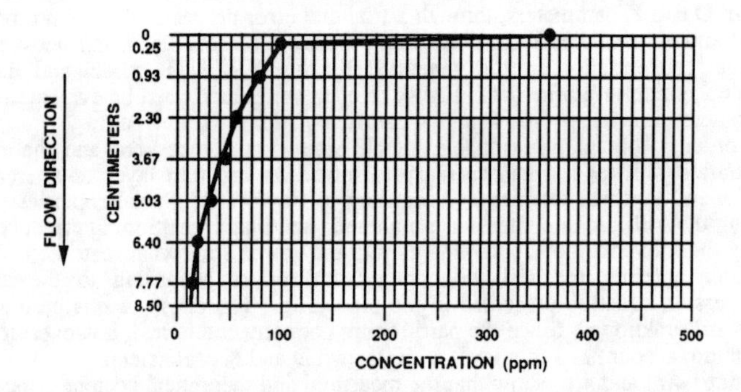

Figure 4(b), Observed vs. Predicted GT-17(B) Concentration Profiles

POLLUTE FIELD SIMULATIONS

Field simulations were performed with POLLUTE to evaluate the migration of contaminants to the shallow, medium, and deep sand zones underlaying the HWMF. The POLLUTE model was separately run with soil properties and D and K coefficient obtained from column-leachate test specimens GT-17(A) and (B). All field simulations were performed with actual leachate concentrations determined from samples taken from monitoring wells at the HWMF. The concentrations used were the highest values reported from samples taken on October 21 and December 6, 1988 (EMCON Associates, 1988). Table 3 presents the relatively constant concentrations of the chemical compounds assumed to be present at the base of the HWMF.

Table 2, Summary of D and K Coefficients

Chemical Compound	Concentration (ppm)[1]	Specimen No.	Layer[2] No.	D (cm^2/day)	K (cm^3/gm)
Acetone	363	GT-17(A)	L1	0.1	11.5
			L2	3.5	0.2
		GT-17(B)	L1	0.1	12.0
			L2	6.0	0.2
2-Butanone	392	GT-17(A)	L1	0.1	12.0
			L2	4.0	0.3
		GT-17(B)	L1	0.1	13.5
			L2	5.0	0.1
Benzene	394	GT-17(A)	L1	0.1	28.0
			L2	4.0	0.3
		GT-17(B)	L1	0.1	23.4
			L2	5.4	0.2
4-Methyl-2 -pentanone	388	GT-17(A)	L1	0.4	18.0
			L2	4.0	0.3
		GT-17(B)	L1	0.1	17.0
			L2	5.0	0.1

1 - ppm = concentration in units of parts per million.

2 - Layer No. 1 represents a thin partitioning layer at the soil-leachate interface and Layer No. 2 represents the Bay Mud sample specimen.

Present Field Conditions (Positive Gradient = downward flow). Presently, a positive hydraulic gradient exists causing downward flow to occur from the surficial to the shallow sand zones. Figure 5 presents a generalized cross-section showing a scenario that was used to evaluate contaminant migration to the shallow sand zone. The hydraulic gradient was estimated from a review of piezometric water levels measured in monitoring wells. Simulations for contaminant migration to the medium and deep Bay Mud sand zones were also run, however, the hydraulic gradient for flow to the medium and deep sand zones are both nearly zero.

Proposed Field Conditions (Negative Gradient = upward flow). The proposed installation and operation of the LCRS, will create a negative hydraulic gradient and will cause upward flow to occur. Figure 6 presents a generalized cross-section showing a the scenario used to evaluate contaminant migration to the shallow sand zone. The hydraulic gradient used was based on design criteria that requires the LCRS to lower the piezometric head to mean sea level. Simulations for contaminant migration to the medium and deep Bay Mud sand zones were also run as in the preceding analyses.

Table 3, POLLUTE Simulations Results Of Contaminant Migration To The Shallow Sand Zone.

Chemical Compound	C_o[1] (ppm)	Sample No.	Concentration Ratios C_t/C_o[2] for Simulation Periods and Gradients							
			25 Years[3]		50 Years		75 Years		100 Years	
			(+)[4]	(-)	(+)	(-)	(+)	(-)	(+)	(-)
Acetone	560	GT-17(A)	1.3	0.2	11.1	1.6	24.4	3.6	37.1	5.9
		GT-17(B)	7.7	0.9	30.6	4.0	50.4	7.2	64.8	10.8
2-Butanone	1200	GT-17(A)	1.9	0.3	13.6	2.5	27.9	5.5	40.9	8.6
		GT-17(B)	5.4	0.4	25.9	2.1	45.4	4.1	60.4	6.2
Benzene	6	GT-17(A)	1.9	0.3	13.6	2.5	28.3	5.5	41.7	8.7
		GT-17(B)	6.2	0.6	27.6	2.8	47.3	5.4	62.0	7.9
4-Methyl-2-Pentanone	87	GT-17(A)	2.1	0.4	14.1	2.6	28.6	5.8	41.7	9.0
		GT-17(B)	5.4	0.4	25.9	2.1	45.4	4.1	60.4	6.2

1 - C_o is the constant leachate concentration at the landfill base.

2 - C_t/C_o a percentage computed from C_t, the concentration at some point within the barrier layer at the end of the specified simulation period, and from C_o.

3 - Duration of simulation period.

4 - The positive (downward) hydraulic gradient simulation results are indicated by (+), and the negative (upward) hydraulic gradient simulation results are indicated by (-).

Field Simulation Results. Based on the field conditions evaluated, chemical concentration profiles were estimated for periods of 10, 25, 50, 75, and 100 years of contaminant migration to the shallow, medium and deep Bay Mud sand zones. Table 3 presents a summary of the simulation results predicted by the POLLUTE model. These results indicate that under present field conditions (downward gradient) it would take about 25 years for the leachate to reach the shallow sand zone at concentrations of between 1 and 8 percent, and 50 years at concentrations of between 11 and 31 percent of the concentration at the landfill base. These simulations also indicated that it would take greater than 100 years for the selected leachate contaminants to reach the medium sand zone and an even greater length of time to reach the deep sand zone.

The simulation results for the proposed field conditions that involve operating an LCRS (upward gradient) indicate that it will take more than 100 years for leachate contaminants to reach the shallow sand zone. The simulation results also indicate that operating the LCRS will significantly inhibit the movement of contaminants to both the medium sand and deep sand zones.

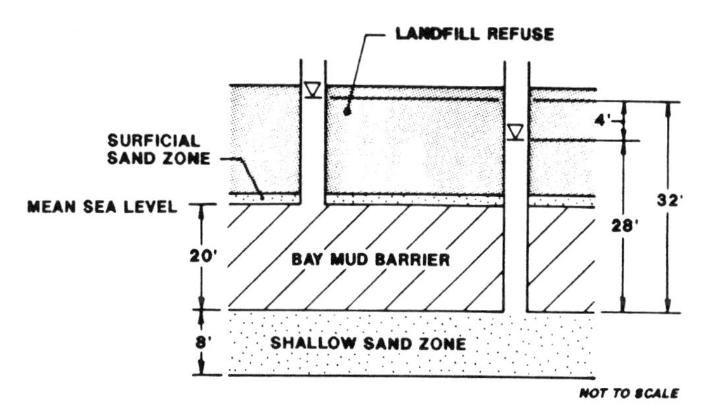

Figure 5, POLLUTE Field Simulation of Contaminant Migration to the Shallow Bay Mud Sand Zone Under Present HWMF Conditions (Positive Hydraulic Gradient = Downward Flow).

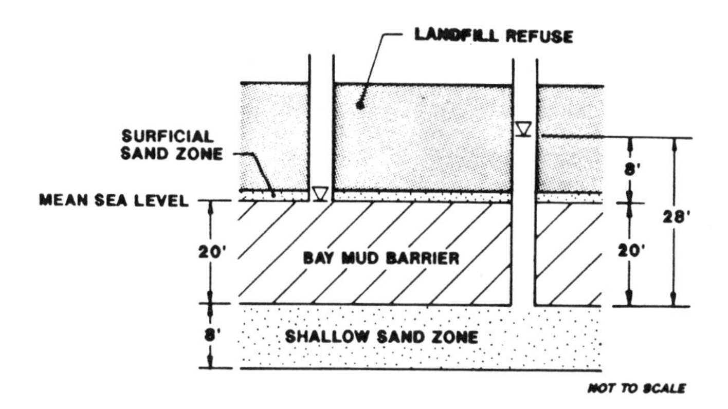

Figure 6, POLLUTE Field Simulation of Contaminant Migration to the Shallow Bay Mud Sand Zone Following Installation and Operation of an LCRS at the HWMF (Negative Hydraulic Gradient = Upward Flow).

Comparison of Simulation Results with Actual Field Data. At the time of this publication submittal, only limited data for the existing field condition were available for comparison with the concentration profiles predicted by the POLLUTE model. These data consist of chemical test results of water samples taken from monitoring wells screened in the shallow sand zone and fine grained Bay Mud soil samples taken from borings at depths about 3 feet below the surficial sand zone, and just above the shallow sand zone. Both soil and water sample test results indicate that the shallow sand zone and the fine grained Bay Mud soil lying just above it do not contain detectable levels of HWMF leachate, therefore both can be considered to be clean.

The POLLUTE simulations indicate that under the present field conditions (downward flow) it will take about 25 years before a small percentage of the HWMF contaminants will reach the shallow sand zone. Therefore, the POLLUTE simulation results are in reasonable agreement with the existing chemical concentration profiles measured at the HWMF.

CONCLUSIONS

Based on the preceding evaluation the following conclusions can be made with regard to the vertical containment at the HWMF:

- The hydraulic conductivity of the clayey silts and silty clays of the Bay Mud will remain relatively unchanged when subjected to HWMF leachate (which are organic chemicals), indicating that these soil are compatible with the leachate.

- There is a partitioning layer at the soil-leachate interface. This layer significantly restricts the movement of contaminants into the underlying Bay Mud soils. The cause of the partitioning phenomenon is still not known.

- Field simulations for the site using measured hydraulic conductivities (k), diffusion coefficient (D) and adsorption coefficient (K) determined from the column-leachate test indicate that changing the present hydraulic gradient from positive (downward flow) to negative (upward flow) will significantly reduce migration of contaminants into the shallow sand zone underlying the site. This conclusion is based on up to 100-year simulation periods following installation and operation of the leachate collection and removal system.

- Field simulations for the site using measured soil properties k, D and K coefficients determined from the column-leachate test also indicate that contaminants will not migrate into the medium and deep Bay Mud sand zones underlying the site. This conclusion is based on up to 100-year simulation periods following installation and operation of the leachate collection and removal system.

- Based on the preceding analysis it can be concluded that the tested fine grained Bay Mud soils are an effective barrier to the vertical migration of contaminants from the HWMF. This situation will be significantly improved when the LCRS is installed and operated.

REFERENCES

Acar, Y.B. and Haider, L. (1990). "Transport of Low-Concentration Contaminants in Saturated Earthen Barriers." *Journal of Geotechnical Engineering*, American Society of Civil Engineers, Geotechnical Engineering Division, Vol. 116, No. 7, July 1990.

American Society for Testing and Materials, Section 4, Volume 04.08, (1989). ASTM D2487-85 "Test Method for Classification of Soils for Engineering Purposes", and ASTM D2488-84 "Practice for Description and Identification of Soils (Visual-Manual Procedure).

EMCON Associates (1988). "Hazardous Waste Management Facility Closure and Post Closure Plan, West Contra Costa County Sanitary Landfill, Richmond, California", prepared for Richmond Sanitary Services.

Freeze, R.A., and Cherry, J.A. (1979). *Groundwater.* Prentice Hall, Inc., Englewood Cliffs, New Jersey.

Mitchell, J.K., and Madsen, F.T. (1987). "Chemical effects on clay hydraulic conductivity." *Geotechnical practice for waste disposal,* ., Woods, ed., ASCE, New York, N.Y.

Ogata, A. (1970). "Theory of dispersion in granular medium." *Professional Paper No. 411-I,* U.S. Geological Survey, 36.

Rowe, R.K., Booker, J.R., and Caers, C.J. (1984). "POLLUTE-1D pollutant migration through a non-homogeneous soil: Users Manual." Systems Analysis Control and Design Activity, Faculty of Engineering Science, University of Western Ontario, London, Ont., Canada, Report No. SACDA 84-13.

Rowe, R.K., Caers, C.J., Barone, F. (1988). "Laboratory determination of diffusion and distribution coefficients of contaminants using undisturbed soil." *Canadian Grotechnical Journal,* Vol. 25, No. 1, pp. 108-118.

Rowe, R.K. (1989). personal communications, regarding interpretation of column-leachate test data.

Storey, J.M., and Pierce, J.J. (1989). "Influence of changes in methanol concentration on clay particle interactions." *Canadian Geotechnical Journal,* Vol. 26, pp. 57-63.

EPA Test Methods 608, 624 and 625 (1982). "Methods for Organic Chemical Analysis of Municipal and Industrial Wastewater." EPA-600/4-82-057, James E. Longbottom and James J. Lichtenberg, Editors, U.S. Environmental Protection Agency, Environmental Monitoring and Support Laboratory, Cincinnati OH

California Code of Regulations, Title 22, Section 66699, (22CFR66699).

IMPACT OF LONG TERM LANDFILL DEFORMATION

Stephen J. Druschel [1], M.ASCE and Richard E. Wardwell [2], M.ASCE

ABSTRACT: Cover systems for the closure of refuse landfills must be capable of withstanding stresses caused by differential movements and settlements, to ensure a sound environmental barrier. A key component of the design process is estimating the magnitude of the settlements and evaluating the impact of these movements on the integrity of the cover system. Surface elevations of a 28 ha municipal landfill were monitored by optical survey over a 6-month period, and by aerial survey over a 3-year period. Time since the last acceptance of wastes was between 5 and 12 years. Rates of settlement are calculated and compared to values previously reported for other sanitary landfills. The impacts of the differential movements on the composite synthetic landfill cap are evaluated.

It is determined that much of this landfill is settling at a rate that equals or exceeds any previously suggested rate, probably due to organic decomposition of the wastes. The effects of microbial activity on organic decomposition, promoting both the rate and magnitude of compressive strain, are discussed using data from laboratory experiments simulating conditions found in sanitary landfills. It is concluded that rates and magnitudes of settlement in a sanitary landfill higher than previously suggested may occur well after the end of waste placement, if conditions within the wastes are conducive to organic decomposition, and new design values are suggested. The composite strength of the synthetic cap materials is estimated to be sufficient to maintain its integrity.

---

[1] Geotechnical Engineer, ABB Environmental Services, Inc., 261 Commercial Street, Portland, ME 04112-7050

[2] Geotechnical and Groundwater Engineer, REW, Inc., 190 Bennoch Road, Orono, ME 04473

INTRODUCTION

Preliminary design of a closure cover system for a 28 ha municipal landfill located in Central New England included a 1.5 mm HDPE geomembrane as the primary barrier to infiltrated precipitation. Other elements of the cover system design were an HDPE drainage net below the geomembrane; a second HDPE drainage net directly above the geomembrane; a woven geotextile (Geolon 600), designed for soil separation, filtration, and protection of the geomembrane during construction; and 0.5 m of soil cover. The last acceptance of waste at any location within this facility was between 5 and 12 years prior to the cover system construction. The design engineer assumed that most of the settlement would occur during the construction of the cover. However, during and subsequent to the placement of an interim soil cover 0.3 m thick, the landfill surface was observed to have subsided 0.38 m and 0.46 m in a nine-month period, at two widely spaced monitoring wells which extended through 18 m and 28 m of waste materials to glacial till foundation soils. Numerous cracks about 0.1 m wide and two large cracks about 0.30 m wide were observed in the existing soil cover, suggesting differential movements were taking place. Gas bubbling in puddles and a rotten-egg odor were also noticed, indicating some form of decomposition was occurring in the wastes. These behaviors were observed at numerous locations over the landfill surface, regardless of time since last acceptance of wastes.

Because the observed subsidence exceeded the levels anticipated by the preliminary design, an evaluation was performed by ABB Environmental Services, Inc. to assess the potential settlements of the landfill and the ability of the geomembrane to withstand the stresses that could be caused by associated angular distortion. The first stage of this program involved monitoring the settlements. Comparison of aerial surveys made over a three-year period showed large areas of 0.9 m to 1.2 m average subsidence occurring on the top of the landfill, while some of the landfill flanks displayed up to 0.6 m of surface heave. Although useful to show surficial movement, it is difficult to assess the accuracy of these observations because of the error that may exist between the two surveys. While vertical precision of the photogammetry is on the order of 0.3 m, the horizontal error could be as high as 3.0 m, creating a situation where the results for the relatively flat landfill top are much more credible than those for the flanks.

To alleviate the inaccuracies of the aerial survey, four optical ground surveys were made of the landfill over a six-month period. The results generally confirmed the trends noted by the aerial survey, with large areas of the top and upper flanks settling 0.3 m or more, while some lower flanks of the landfill experienced up to 0.1 m of heave. The greatest movement measured over the duration of the optical surveys was 0.5 m, and angular distortions up to 1.7 percent were observed.

ANALYSIS OF SETTLEMENT DATA

Sowers (1973) suggested five mechanisms responsible for settlement of waste fills:

1. Mechanical: Distortion, bending, crushing, and reorientation of the materials, similar to the consolidation of organic soils.
2. Ravelling: The erosion or silting of fine materials into the voids between larger particles.
3. Physico-Chemical Change: Corrosion, oxidation, and combustion.
4. Bio-Chemical Decay: Fermentation and decay, both aerobic and anaerobic.
5. Interaction: Methane from bio-chemical decay may support combustion, ignited spontaneously from the heat of decay. Organic acids from decay may produce corrosion; volume changes from consolidation may trigger ravelling.

Of these five, Sowers (1973) stated that only mechanical settlement may be analyzed using a primary or load-related settlement model.

In deposits where internal pore pressures can freely dissipate, such as in most municipal solid waste landfills, the bulk of primary settlement may occur so quickly as to be concurrent with the construction operations which are creating the load. Therefore, during initial waste placement, primary settlements caused by self weight of the waste will occur as the load is applied and are believed to be substantially complete upon cessation of waste placement activities (U.S. Navy, 1983).

Settlement mechanisms at the landfill were analyzed using a secondary or time-based settlement model, in which the rate of secondary compression is calculated as the ratio of settlement strain per log cycle of time since waste placement, given by the following equation:

$$C_{\alpha\epsilon} = H_s/[H_o \log (t_2/t_1)] \quad (1)$$

in which $C_{\alpha\epsilon}$ = rate of secondary compression (units of strain/log time), H_s = secondary settlement, H_o = initial height of deposit, t_1 = time since waste placement at beginning of monitoring period, and t_2 = time since waste placement at end of monitoring period. $C_{\alpha\epsilon}$ may be compared between sites to allow characterization of settlement rates.

The rate of secondary compression ($C_{\alpha\epsilon}$) was calculated for nine points on the landfill surface that experienced high subsidence during the monitoring period. Values of $C_{\alpha\epsilon}$ between 0.14 and 0.95 strain per log time were calculated, as shown in Table 1. Note that the values of $C_{\alpha\epsilon}$ are similar for both the younger and older ends of the landfill, which indicates the pervasiveness of the anticipated settlements. A value of t_1 = 5 years was uniformly assumed for the time of the first optical survey, representing a minimum value for the entire landfill. Although waste had been placed over a 30-year period, the use of t_1 = 5 years since the last placement of waste proved a conservative assumption because larger time values would give higher values of $C_{\alpha\epsilon}$.

Table 1. Evaluation of Secondary Settlement Rate

Location	t_1[2] (years)	t_2 (years)	ΔH_s (m)	H_o (m)	$C_{\alpha\epsilon}$[1] (strain/ log time)
Cross Section 1	5.00	5.5	0.36	18	0.48
Cross Section 2	5.00	5.5	0.49	20	0.59
Cross Section 3	3.42	5.5	0.84	28	0.14
Cross Section 3	5.00	5.5	0.30	21	0.34
Cross Section 4	5.00	5.5	0.22	18	0.29
Cross Section 5	5.00	5.5	0.19	26	0.18
Cross Section 6	5.00	5.5	0.23	29	0.19
Cross Section 6	5.00	5.5	1.36	9	0.95
Cross Section 7	5.00	5.4	0.20	18	0.33

NOTES: 1. $C_{\alpha\epsilon} = \Delta H_s/[H_o \log (t_2/t_1)]$
2. t_1 = 5.0 years at date of first optical survey

Values of $C_{\alpha\epsilon}$ calculated for this landfill were compared to values taken from several literature sources regarding settlement of waste deposits under both internal and external loadings (Table 2), and found to be much higher. The suggested rates of the U.S. Navy (1983) for internal loading were the only values which approximated those found at this landfill. Such extreme rates of settlement, although larger than typically observed in field measurements, are not without precedent from laboratory experiments on biochemical decay.

Table 2. Literature Values Regarding the Rate of Secondary Settlement

Location	Loading	Approximate Waste Height	Monitoring Period	$C_{\alpha\epsilon}$ Maximum	Reference
Not listed	external	7.6 m	1.2-5.7yrs	0.015	Sowers (1968)
Conditions unfavorable to decay"	external	suggested rates	N/A	0.02[1]	Sowers (1973)
"Conditions favorable to decay"	external	suggested rates	N/A	0.07[1]	Sowers (1973)
Laboratory small scale	external	0.23 m	10-1000 min	0.037	Rao, et al. (1977)
Laboratory large scale	external	1.5 m	300-900 days	0.13	Rao, et al. (1977)
Morgantown, W. VA	external	3.1 m	100-300 days	0.14	Rao, et al. (1977)

NOTES: 1: assumes $e_o = 4$.
2: assumes time between surveys was 12 months.

Table 2 (Continued)

Location	Loading	Approximate Waste Height	Monitoring Period	$C_{\alpha\epsilon}$ Maximum	Reference
Los Angeles County	internal	18 m	60-72 mos [2]	0.07	Yen & Scanlon (1975)
Los Angeles County	internal	28 m	62-74 mos [2]	0.07	Yen & Scanlon (1975)
Los Angeles County	internal	30 m	74-86 mos [2]	0.06	Yen & Scanlon (1975)
I-85, NJ	external	2-9 m	0-3 yrs	0.04	Burlingame (1985)
Elizabeth, NJ	external	6 m	6-18 mos	0.03	York, et al. (1977)
I-84, CT	internal	13 m	0-5 yrs	0.034	Keene (1977)
N/A	external	suggested rates	N/A	0.02-0.07	U.S. Navy (1983)
N/A	internal	suggested rates	N/A	0.1-0.4	U.S. Navy (1983)

NOTES: 1: assumes $e_o = 4$.
2: assumes time between surveys was 12 months.

Wardwell and Nelson (1981) presented results of a series of experiments designed to determine the effects of organic decomposition on compression strain. Two sets of cellulose fiber, kaolin clay, and water mixtures, one seeded with sufficient nutrients to promote microbial activity, the other deficient in nutrients, were placed under a constant load in a consolidometer, and compressive strains were monitored for a period of 190 days. A representative graph of their results (Figure 1) shows that settlement strains for seeded mixtures were up to double the amounts observed for nutrient-deficient mixtures. The maximum observed settlement strain increased with increased organic content, from 25 percent strain for an organic content of 40 percent to about 55 percent strain for an organic content of 70 percent.

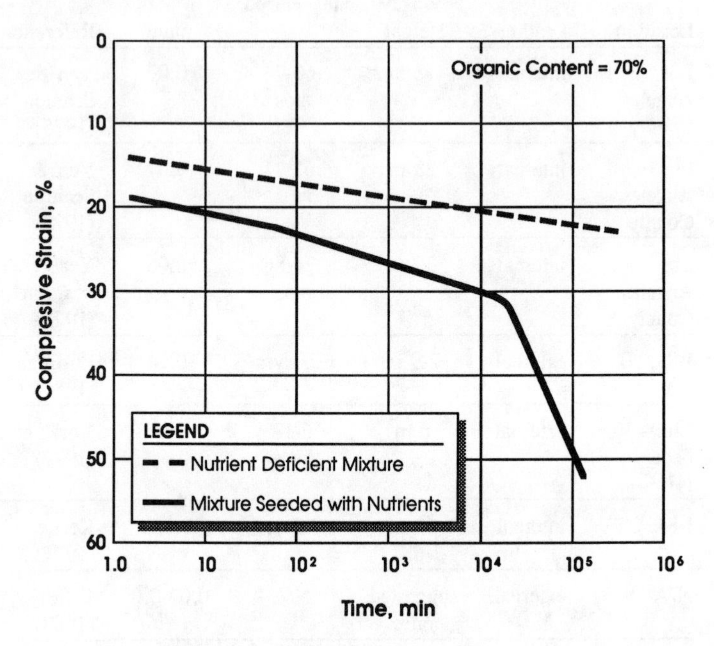

Figure 1. Influence of Organic Decomposition on Settlement Strain (Adapted from Wardwell and Nelson, 1981)

From these experimental data, it may be surmised that landfills partially composed of organic wastes may settle at greater amounts and at accelerated rates due to decomposition of the organic materials. The subject landfill certainly appears to be undergoing organic decomposition, based on the observation of considerable amounts of gas venting from the landfill during the site visits. Further settlements are expected to accompany the decomposition.

The ultimate amount of settlement to occur is believed to depend upon the organic content of the wastes. Al-Khafaji and Andersland (1981) show that for an initial organic fraction of 30 percent, ultimate settlement may reach 32 percent of the initial height, while for an initial organic fraction of 60 percent, ultimate settlement may exceed 58 percent of the initial height. Although the organic content of the wastes within the landfill is not known, an initial organic fraction of between 30 and 60 percent could arguably represent the maximum case. Based on these results and a comparison of the measured rates of settlement with the laboratory data presented in Wardwell and Nelson (1981), it is conceivable that remaining settlement could be on the order of 10 to 20 percent of the landfill height, or 3 to 6 m, assuming decomposition continues to organic depletion.

Construction of the proposed low-permeability cover system will inhibit infiltration of precipitation, and this has the potential to slow the organic decomposition. However, there appears to be enough moisture in the waste materials to support organic decomposition. Since capping the site will also reduce evapotranspiration, it is unlikely that settlement will cease after cover system installation.

EFFECT OF SETTLEMENT ON COVER SYSTEM COMPONENTS

A cover system, specifically the geomembrane layer within, must endure the tensile stresses caused by waste subsidence to fulfill its function of being a passive, low-maintenance barrier to infiltration. Designed to prevent the entrance of water, a cover system becomes an efficient collector and concentrator of that very same water when torn, especially since the hole would probably be located near a low spot created by settlement.

There are three modes of failure from waste subsidence that a cover system needs to withstand: uniaxial strain; supported biaxial strain; and unsupported biaxial strain.

Uniaxial strain occurs when the geomembrane is stretched over a linear feature, such as a long crack, and is analogous to a wide width tensile test. Supported biaxial strain occurs when the cover system deflects into a low spot to be fully resting on the deformed subgrade. Unsupported biaxial strain is caused by the appearance of a void beneath the geomembrane, so that the cover system load is taken up by the tensioning of the geomembrane.

It is important to differentiate between uniaxial and biaxial strain conditions, given the variation in response of HDPE reported by Steffen (1984). The HDPE geomembrane of the cover system design has a specified minimum uniaxial strain of 300 percent or more. However, extrapolation of the results given in Steffen (1984) indicates that biaxial strain failure could occur at about 6.5 percent strain. This value was used as the maximum allowable strain, ϵ_{max}, in the preliminary analysis. Using the beam-deflection model suggested by Gilbert and Murphy (1987) for cover system subsidence, an angular distortion of 34 percent was estimated to cause ϵ_{max}.

A deflection or elongation tolerance depends upon the minimum length over which strain may occur. This length is equal to twice the anchorage distance necessary to resist the ultimate load of the geomembrane, and may be calculated using the following equation:

$$L = 2 \frac{T_{ult}}{\sigma_n (\tan \delta_t + \tan \delta_b)} \tag{2}$$

in which L = minimum length over which strain may occur, T_{ult} = ultimate load of the geomembrane (units of force per width), σ_n = normal stress at geomembrane, δ_t = interface friction angle above geomembrane, and δ_b = interface friction angle below geomembrane.

This length is 13 m for the preliminary cover system design with 0.5 m of soil cover, a specified minimum δ of 16° for the geomembrane-drainage net interface, and a specified T_{ult} of 17.5 kN/m. Applied to the angular distortion of 34 percent previously estimated, a maximum tolerable differential subsidence, Δ_{max}, of 4.3 m is calculated. Note that, in this analysis, Δ_{max} will decrease if σ_n, δ_t, or δ_b increase, or if ϵ_{max} or T_{ult} decrease.

Unsupported biaxial strain of the geomembrane was evaluated through performance of two biaxial stress tests, first on a sample of HDPE alone, then on a sample of HDPE overlying a sample of the woven geotextile. The results of the two tests, plotted in Figure 2 as center deflection versus hydrostatic pressure, show that the geomembrane-geotextile composite sample is better able to support the cover system load over voids and differential settlements than the geomembrane alone.

Tensioned membrane theory (Giroud, et al., 1988), was used to correlate the maximum pressure levels obtained in the laboratory tests with the field conditions of the cover system design, assuming equal levels of strain and tension. Results of the calculations indicate that, for the load imposed by the cover soils, the performance level attained in the laboratory tests will be reached when the composite sample is subjected to a differential settlement greater than 5 m or spans a void of 23 m or wider.

It should be noted that in the comparison of the laboratory results to field conditions, the maximum load considered may not be the failure load, as ultimate strength of the composite sample was not reached during the test. No considerations were made for the effects of polymer creep or for variations in loading rate. No allowance was made for the geomembrane anchoring condition, as a linear development of strength was believed to be unlikely when spanning a void with such a large catenary-shaped deflection. No reduction factors were applied for the influence of material seams, manufacturing or installation defects, or variations in test results. Finally, the effect of a load from ponded water was not considered, because of provisions anticipating ponded water within the maintenance plan.

Subsidence large enough to foster the levels of differential settlement discussed in this evaluation are at the upper bounds of predicted future subsidence for this landfill, although not impossible. However, it was believed that such large differential settlements would be easily discerned during the scheduled post-closure inspections. The suggestion was made that the maintenance plan be reviewed and modified, if necessary, so that appropriate action be specified at a level of differential settlement safe from cover system damage.

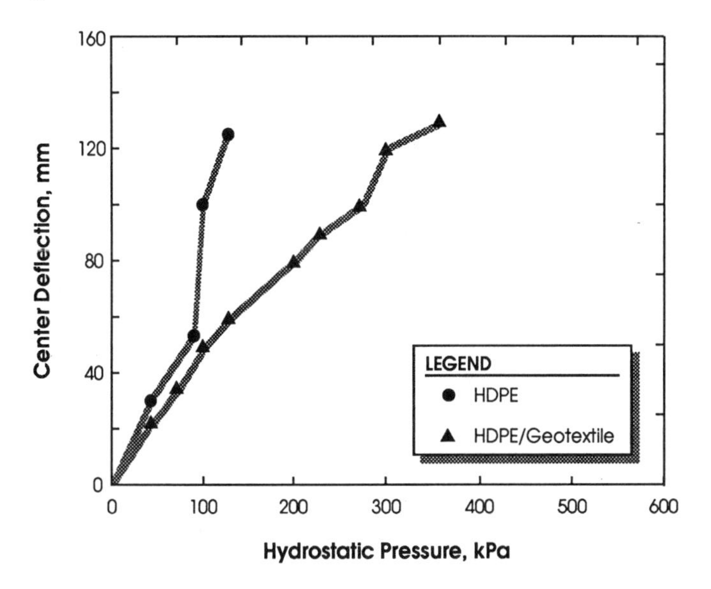

Figure 2. Results of Biaxial Stress Tests

CONCLUSIONS

Settlement at a 28 ha landfill is occurring at a rate that equals or exceeds any rate previously suggested for landfill settlement. Comparison with data reported for laboratory organic decomposition experiments suggests that settlement of landfills may occur at such high rates well after the conclusion of waste placement. The proposed cover system was evaluated for its ability to withstand stresses caused by the potential differential settlements, and was found to be adequate when coupled with an appropriate maintenance effort.

ACKNOWLEDGEMENTS

The authors wish to thank several people for their participation in this project, especially Guy Vaillancourt, Tom Gainer, Don Hunt, and Stan Walker. Ms. Marie Peoples prepared the manuscript and Ms. Dawne Gilpatrick prepared the drawings.

REFERENCES

1. Al-Khafaji, A.W.N., and Andersland, O.B. 1981. "Compressibility and Strength of Decomposing Fiber-Clay Soils," Geotechnique, Vol. 31, No. 4, pp. 297-508.

2. Burlingame, M.J. 1985. "Construction of a Highway on a Sanitary Landfill and Its Long-Term Performance," Transportation Research Records 1031, TRB, Washington, DC, pp. 34-40.

3. Gilbert, P.A. and Murphy, W.L. 1987. "Prediction/Mitigation of Subsidence Damage to Hazardous Waste Landfill Covers," EPA 600/2-87-025, PB87-175386, U.S. EPA, Cincinnati, Ohio.

4. Giroud, J.P., Bonaparte, R., Beech, J.F., and Gross, B.A. "Load-Carrying Capacity of a Soil Layer Supported by a Geosynthetic Overlying a Void," International Symposium on Theory and Practice of Earth Reinforcement, Kyushu, Japan, October 1988, pp. 185-190.

5. Keene, P. 1977. "Sanitary Landfill Treatment, Interstate Highway 84," Proceedings of the Conference on Geotechnical Practice for Disposal of Solid Waste Materials, ASCE Specialty Conference, Ann Arbor, Michigan, pp. 632-645.

6. Rao, S.K., Moulton, L.K., and Seals, R.K. 1977. "Settlement of Refuse Landfills," Proceedings of the Conference on Geotechnical Practice for Disposal of Solid Waste Materials, ASCE Specialty Conference, Ann Arbor, Michigan, pp. 574-598.

7. Sowers, G.F. 1973. "Settlement of Waste Disposal Fills," Proceedings of the Eight International Conference on Soil Mechanics and Foundation Engineering, Moscow, Russia, Volume 2, pp. 207-210.

8. Sowers, G.F. 1968. "Foundation Problems in Sanitary Landfills," Journal of the Sanitary Engineering Division, Proceedings of the American Society of Civil Engineers, Vol. 94, No. SA1, pp. 103-116.

9. Steffen, H. 1984. "Report on Two Dimensional Strain Stress Behavior of Geomembranes With and Without Friction," Proceedings of the International Conference on Geomembranes, IFAI, Denver, Colorado, pp. 181-185.

10. U.S. Navy. 1983. "Soil Dynamics, Deep Stabilization and Special Geotechnical Construction," NAVFAC DM7.3, Department of the Navy, Alexandria, Virginia.

11. Wardwell, R.E., and Nelson, J.D. 1981. "Settlement of Sludge Landfills with Fiber Decomposition," Proceedings, Tenth International Conference for Soil Mechanics and Foundation Engineering, Vol. 2, Stockholm, Sweden, pp. 397-401.

12. Yen, B.C., and Scanlon, B. 1975. "Sanitary Landfill Settlement Rates," Journal of the Geotechnical Engineering Division, Proceedings of the American Society of Civil Engineers, Vol. 101, No. 6T5, pp. 475-486.

13. York, D., Lesser, N., Bellatty, T., Irsai, E., and Patel, A. 1977. "Terminal Development on a Refuse Fill Site," Proceedings of the Conference on Geotechnical Practice for Disposal of Solid Waste Materials, ASCE Specialty Conference, Ann Arbor, Michigan, pp. 810-830.

APPLICATION OF DISCRETE FRACTURE ANALYSIS TO SITE CHARACTERIZATION

W. Dershowitz[1], W. Roberds[1], and J. Black[2]

ABSTRACT: Discrete fracture analysis was originally developed to allow modelling of the influence of fractures on geomechanical and hydrological rock mass behavior. Because of their significant influence on behavior, discrete features often dominate the design and evaluation of site characterization programs. The application of discrete fracture analysis to site characterization is described, utilizing stochastic feature and exploration simulation. Applications are described within the petroleum and hazardous and radioactive waste industries.

INTRODUCTION

Site characterization programs frequently focus on the identification of discrete features, such as faults, fractures, stratigraphic contacts, karsts, and local heterogeneities, which tend to dominate performance. In designing these site characterization programs, it is essential to know the probability that features will be intersected, and the reliability and resolution with which features can be detected and described if intersected. Literature is available on the theoretical problem of intersection between ideally shaped detectors and targets (e.g., Stoyan et al, 1987). Most geological systems, however, are too complex to be described by simple analytical solutions.

Inverse models attempt to identify the single most likely in situ condition, based upon the limited available information. While this is adequate for some applications, it is frequently more desirable to utilize the forward modelling approach, which recognizes the uncertainty in characterization of geologic environments and allows evaluation of alternative interpretations based upon the range of assumptions consistent with available data. This paper describes a forward modelling approach in which basic assumptions are made about the distributions of parameters describing the geology, and these assumptions are then used to generate stochastic realizations

[1] Associates, Golder Associates Inc, 4104 148th Avenue NE, Redmond WA, 98052, (206) 883-0777

[2] Associate, Golder Associates (UK) Ltd, Landmere Lane, Edwalton, Nottingham, UK, +011 (44) 602-456-544

of possible geologies consistent with those distributions. While it is recognized that none of these geologies may exactly reproduce in situ conditions, the forward model realizations are useful in quantifying the range of possibilities within a single geologic concept.

The geologic simulator used in this study is based on the technology of discrete fracture analysis. In contrast to the use of analytical solutions to calculate exploration program effectiveness, this simulator is based upon the use of flexible, realistic three-dimensional geologic conceptual models. These geologic conceptual models incorporate probabilistic descriptions for the structure, heterogeneity, and anisotropy of geologic systems (Geier et al, 1990). Once a realistic prior geologic conceptual model has been developed, the effectiveness of a site characterization program can be evaluated through Monte Carlo simulation by overlaying the exploration program on realizations of the geologic conceptual model.

As summarized in Figure 1, this approach consists of three stages: development of prior geological and engineering conceptual models, description of alternative site characterization programs, and simulation of site characterization programs within the conceptual models.

GEOLOGIC CONCEPTUAL MODELS

Geologic conceptual models describe the geometry, mechanical, and hydrological properties of the features which might be encountered by site characterization. As summarized in Table 1, each of these properties should be expressed probabilistically, to represent both the uncertainty in parameter values and expected spatial variability. Figure 2 shows example realizations of geologic conceptual models for fractures, faults, paleochannels (meander belts) and crevasse splays, as produced by the geologic simulator.

Initially, such quantitative descriptions of a site will be based on little information and will thus entail substantial uncertainties, which manifest themselves as significant differences among realizations. However, it is imperative to identify such uncertainties and to recognize the possible range in the actual properties. In this way, the investigation program can be focused on reducing specific uncertainties.

Often, the information available regarding such properties is qualitative. Techniques for developing quantitative descriptions of uncertainty from various types of information are readily available (e.g., Roberds, 1990).

SITE CHARACTERIZATION PROGRAMS

Site characterization programs are described by combinations of exploration tools. Each tool has a specific geometric coverage, which controls the probability

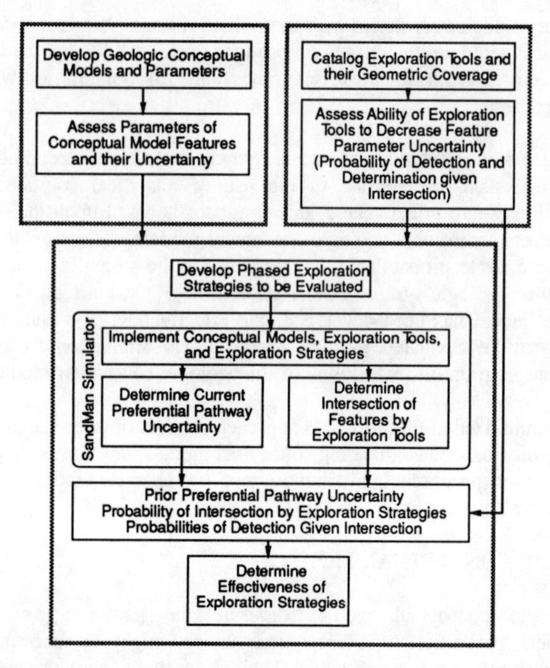

Figure 1. Site Characterization Program Evaluation Approach

of intersection between features and the tool, and a probability of detection and resolution of parameters given intersection. In addition, the exploration program must describe the way the site characterization program will change, depending upon what is found in each stage of exploration. For example, consider a site characterization program as shown in Figure 3 which consists of: (Phase 1) a surface three-dimensional seismic survey in conjunction with one vertical borehole followed by (Phase 2) either two inclined boreholes through each significant potential fault identified by the geophysics or 20 vertical boreholes in areas where faults were not identified by geophysics in order to confirm their absence; followed by (Phase 3) an underground excavation with boreholes at depth. This exploration program is described by the parameters shown in Table 2.

For each element of the exploration program, the engineer must evaluate the range and resolution for detection and quantification of the properties for features of interest. These can be expressed in terms of the following two factors:

- geometry, which controls the probability of intersection
- "power", which controls the probability of true and false detection and determination of parameters given intersection and in situ conditions.

Table 1 - Geological Conceptual Model Parameters and Distributions

Property	Parameter	Typical Distribution Types
Location	Intensity (Area/Volume P_{32}, or Volume/Volume P_{33}) Spatial Structure	beta, gamma, or normal
Hydrology	Hydraulic Conductivity Transmissivity Storativity Porosity	lognormal, normal, uniform, beta
Geometry	Orientation Width Length Thickness Shape/Aspect Ratio Connectivity (intersection/ termination)	Fisher, exponential, normal, lognormal
Mechanics	Strength (c-ϕ, Hoek-Brown) Deformability	normal, lognormal

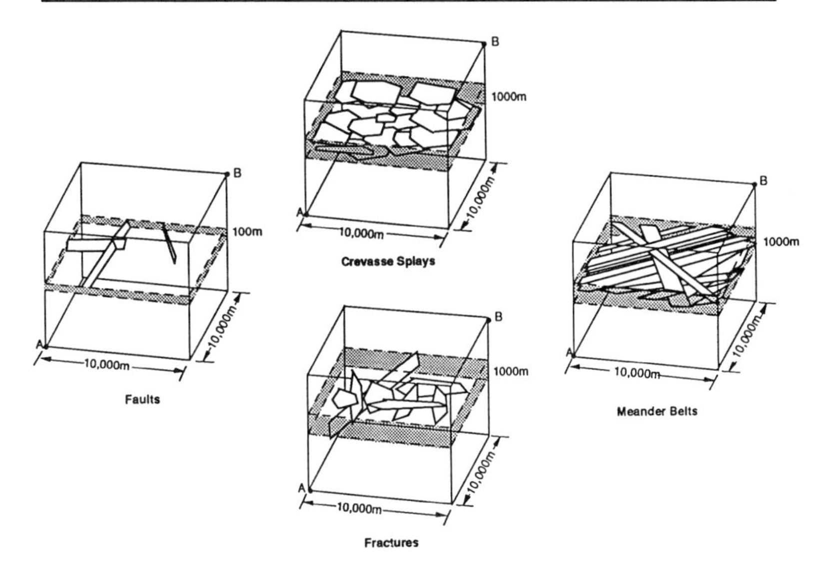

Figure 2. Geological Conceptual Model Simulation

The probability of intersection is a geometric probability (Santalo, 1984). For simple geometries of features and exploration tools, the probability of intersection can be calculated directly. For more complex geometries, it must be determined by simulation, for example, by overlying the exploration program (Figure 3) on the simulated geology (Figure 2).

Once a feature has been intersected, the ability of a site characterization program to reduce uncertainty depends upon the accurate detection of features and determination of their properties. As illustrated in Figure 4, in situ conditions frequently have a strong effect on the probability of accurate or false detection given intersection. In hydrologic testing, for example, accurate detection depends upon both the magnitude of hydraulic conductivities and the contrast in hydraulic conductivity. This can be expressed subjectively in terms of the probability of accurate detection or parameter determination of specific feature F (D_F) by exploration tool A, given intersection of that feature I_F and in situ conditions $\underline{\underline{C}}$, $P[D_F | I_F, \underline{\underline{C}}, A]$. Such probabilities have been assessed for a wide variety of hydrologic, geophysical, and geological detection methods, including core logging, packer testing, cross-hole testing, and three-dimensional geophysics for a variety of geologic conditions (Black, 1990).

Table 2 - Parameters of Exploration Program

Program	Parameter
Surface Geophysics	Coverage (m^3) Probability of detection/parameter determination if intersected Probability of false detection
Boreholes	Borehole geometry and probability of detection/parameter determination if intersected Single Hole Hydrologic Test geometry and probability of detection/parameter determination if intersected Single Hole Geophysical Test geometry and probability of detection/parameter determination if intersected Probability of false detection for each
Drift	Drift geometry and probability of detection/parameter determination if intersected At-Depth Borehole geometry and probability of detection/parameter determination if intersected Probability of false detection for each

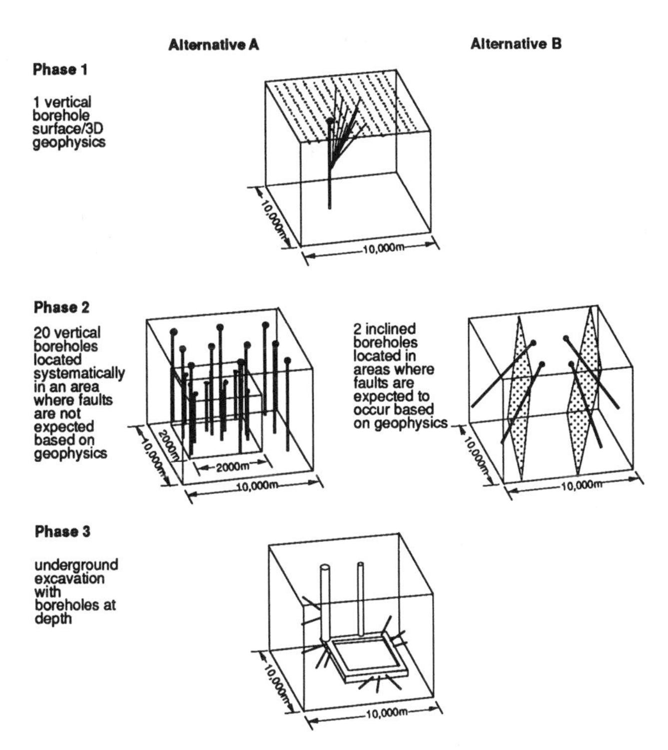

Figure 3. Example of Site Characterization Programs

ANALYSIS

The probability of specific feature detection/parameter determination for a given exploration tool and feature can be derived as follows:

$$P[D_F|I_F,A] = \int_{-\infty}^{\infty} f_c(\underline{\underline{C}})\; P[D_F|I_F,\underline{\underline{C}},A]\; d\underline{\underline{C}} \tag{1}$$

For simple distributions for the uncertainty of in situ conditions $f_C(\underline{\underline{C}})$, and simple relationships between $\underline{\underline{C}}$ and the probability of detection/parameter determination given intersection, equation 1 can be solved analytically. For more complex relationships, numerical integration can be carried out by Monte Carlo methods utilizing, for example, the Lotus add-in @Risk (Palisade, 1990).

Although the probability of feature detection/parameter determination for individual tools within an exploration program is useful in evaluating alternative site characterization strategies, a more comprehensive measure is sometimes required.

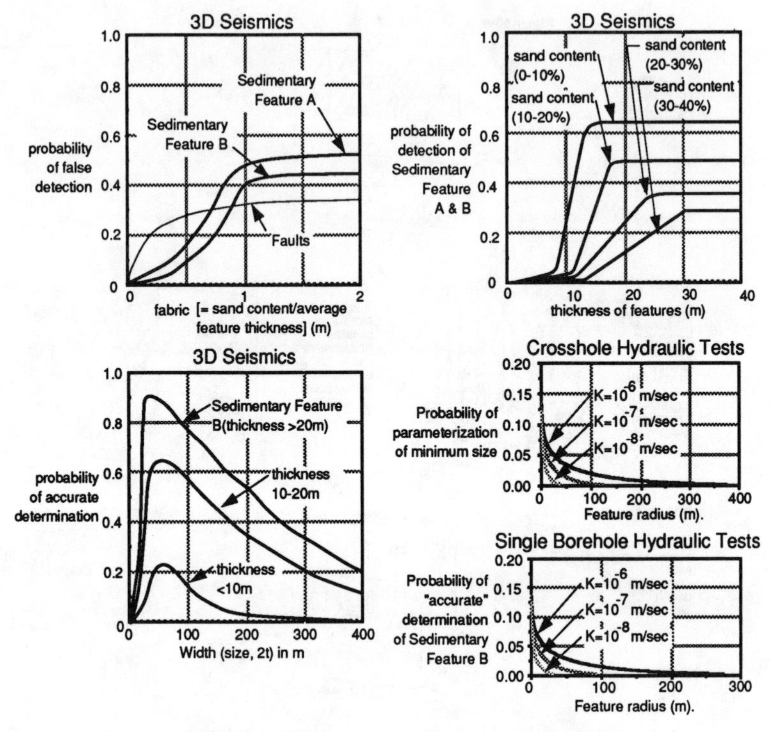

Figure 4. Probability of Detection by Exploration Tools

In this study, that measure is defined by the "probability of success," where success is determined by the goals of site characterization. For example, if the purpose of site characterization is to determine the transmissivity of a specific feature to within one order of magnitude, the probability of success of the exploration program P[S|A] can be derived from the probability of feature intersection by the site characterization program, and the probability of determination of transmissivity within one order of magnitude given intersection, as follows:

$$\begin{aligned} P[S|A] &= P[D_F|A] \\ &= P[D_F|I_F, A]\ P[I_F|A] \end{aligned} \tag{2}$$

$$where\ P[I_f|A] = \int_{-\infty}^{+\infty} P[I_F|\underline{\underline{C}}, A]\ f_c(\underline{\underline{C}})\ d\underline{\underline{C}} \tag{3}$$

by substituting equations 1 and 3 into equation 2

Table 3 - Exploration Program Probability of Success

Site Characterization Tool	Probability of Feature Intersection	Probability of Detection Given Intersection	Mean Feature Rate	Probability of Success
3D Geophysics	95%	70%	3	60%
Vertical Boreholes	2%	99%	3	1%
Angled Boreholes	10%	97%	3	5%
3D Geophysics and Vertical Boreholes	95%	80%	3	65%
3D Geophysics and Angled Boreholes	99%	85%	3	75%
Drift Excavation	99%	99%	3	97%

$$P[S|A] = \int_{-\infty}^{+\infty} P[D_F|I_F,\underline{\underline{C}},A]\ P[I_F|\underline{\underline{C}},A]\ f_c(\underline{\underline{C}})^2\ d\underline{\underline{C}} \tag{4}$$

In many site characterization programs, the goal of the program is to "confirm" (to a specified level of confidence) either the presence or the absence of certain features, such as faults or karsts. Success for this case thus can be expressed in terms of the updated probability that features exist or not after the results of the site characterization program are available:

$$S\ if\ P''[F_O|A] > P_C\ or\ if\ P''[F_{1+}|A] > P_C \tag{5}$$

$$where\ P''[F_o|A] \simeq P[F_o|D_o,A]\ if\ D_o\ occurs \tag{6a}$$

$$\simeq P[F_o|D_{1+},A]\ if\ D_{1+}\ occurs \tag{6b}$$

$$P''[F_{1+}|A] \simeq 1 - P''[F_o|A] \tag{7}$$

The probability of success thus equals the probability of confirming (with a level of confidence greater than Pc) the absence or presence of the features:

$$\begin{aligned} P[S|A] &\simeq P[P''[F_o|A] > P_c] + P[P''[F_{1+}|A] > P_c] \\ &\simeq P[P''[F_o|A] > P_c] + P[P''[F_o|A] < 1 - P_c] \end{aligned} \tag{8}$$

$P[P''[F_o \mid A] > P_c] =$	if $P[F_o \mid D_{1+},A]$	if $P[F_o \mid D_o,A] =$ $<P_c$	$>P_c$
	$<P_c$	0	$P'[D_o \mid A]$
	$>P_c$	$1-P'[D_o \mid A]$	1

$P[P''[F_o \mid A] < 1-P_c] =$	if $P[F_o \mid D_{1+},A]$	if $P[F_o \mid D_o,A]$ $<1-P_c$	$>1-P_c$
	$<1-P_c$	1	$1-P'[D_o \mid A]$
	$>1-P_c$	$P'[D_o \mid A]$	0

$P[S \mid A] =$	if $P''[Fo \mid D1+,A]$:	if $P''[F_o \mid D_o,A] =$ $>P_c$	$<P_c$ and $>1-P_c$	{if $P_c>$ 50%} $<1-P_c$
	$>P_c$	1	$1-P'[D_o \mid A]$	1
	$<P_c$ and $> 1-P_c$	$P'[D_o \mid A]$	0	$P'[D_o \mid A]$
	$<1-P_c$	1	$1-P'[D_o \mid A]$	1

Hence, the probability of successful exploration depends upon the probability that features of interest will not exist if none were detected in exploration $P[F_o \mid D_o,A]$ and on the prior probability of not detecting any features with that exploration program $P'[D_o \mid A]$. The more thorough an exploration program is, the higher the probability that no undetected features will exist following exploration and the higher the probability of detecting features, and thus the higher the probability of success.

The probability of no undetected features can in turn be determined from the prior probability of features and the probability of detecting such features (if they exist) with the exploration program:

$$P[F_o|D_o,A] = \frac{P[F_o]\ P[D_o|F_o,A]}{P[F_{1+}]\ P[D_o|F_{1+},A] + P[F_o]\ P[D_o|F_o,A]} \tag{9}$$

The probability of not detecting non-existing features $P[D_o \mid F_o,A]$ is the complement of false detection. The probability of not detecting any existing features $P[D_o \mid F_{1+},A]$ is a function of the probability distribution for the number of existing features $p[F_i]$ and the probability of not detecting each one independently $P[D_o|F_1,A]$.

$$P[D_o|F_{1+},A]=\sum_{i=1}^{\infty}p[F_i]\ (P[D_O|F_1,A])^i \tag{10}$$

The probability of not detecting an existing feature can in turn be derived from equations 1 and 2.

$$P[D_o|F_1,A] = (1 - P[D_F|I_F,A])\ (P[I_F|A]) \tag{11}$$

EXAMPLE OF SITE CHARACTERIZATION ASSESSMENT

The probability of success determined as presented above is summarized in Table 3 by exploration tool for a simple exploration program. In this example, a binomial distribution has been assumed for $p[F_i]$ with a mean rate of 3 per site, and the confidence threshold P_c for feature detection has been set at 90%. Three-dimensional geophysics, with a high probability of intersecting features, and a high probability of detecting features given intersection, has the highest probability of success. Vertical boreholes, with a low probability of intersecting features, do not have a significant probability of success, or of contributing to success in conjunction with geophysics. Angle boreholes, in conjunction with geophysics, however, have a very high probability of success. The highest probability of success, of course, is for the drift excavation, which has a very high probability of intersection, and a high probability of detection given intersection.

APPLICATION TO PETROLEUM RESERVOIR DESIGN

The detection and characterization of conductive features is very important for the development of fractured reservoirs, and reservoirs with a significant probability of water flooding during production through faults or fractures. In this example, the purpose of site characterization is to locate the exploratory well (which will hopefully become the production well) in a location with a low probability of direct hydraulic connection to water phase saturated zones. In this application, therefore, a successful exploration program is one with the lowest probability of locating a borehole close to (e.g., within 50 m of) a significant fault:

$$P[S|A]=P[R_F>50m|A] \tag{12}$$

The methods described above can be used to determine the probability of success as follows:

- build a conceptual model for faulting based upon current information, providing the prior intensity of faulting

- derive the probability of fault detection (true and false) for each geophysical and geological method included in the proposed site characterization program

- calculate the probability density function for the intensity of faults which would not be detected by the site characterization program

- calculate the probability that an undetected fault will exist within a 50m zone around a borehole

- calculate the probability of success by integrating the probability of undetected faults over the range of possible intensities.

HAZARDOUS WASTE SITE REMEDIATION

Hazardous waste site remediation frequently depends upon implementation of monitoring programs based upon borehole sampling. The effectiveness of these monitoring programs depend upon their ability to detect plumes originating within the controlled area. In general, these programs are evaluated based upon continuum assumptions, such that the density of wells required is controlled directly by the lateral dispersion of solute transport. For example the program MEMO (Einberger, 1990) calculates the "blind zones", in which plume initiation would not lead to detection by the monitoring program. With this approach, the probability of failure of the monitoring program is simply the probability of plume initiation L within the blind zone L_{BZ}:

$$P[S]=1-(P[L]P[L_{BZ}|L]) \tag{13}$$

However, for many sites, discrete pathways, which have very low lateral dispersion, provide another mechanism for undetected leaks, so that the probability of failure is the probability of plume initiation within the blind zone plus the probability of plume initiation coincident with a conductive discrete feature L_{DF} not within the blind zone and not intersected by the monitoring system.

$$P[S]=1-P[L]\left[P[L_{BZ}|L]+(1-P[L_{BZ}|L])(P[L_{DF}|L])(1 - P[D_F])\right] \tag{14}$$

Hence, a monitoring program can be evaluated by simulating the occurrence of discrete fractures based upon a prior geological conceptual model (see Figure 2). The probability of successful leak detection can then be calculated by determining the probability of intersection between the borehole-based monitoring system and the discrete features, as well as between plume initiation (leakage) and the discrete features.

CONCLUSIONS

A geologic conceptual model and exploration simulator has been developed to facilitate forward modelling in evaluation of site characterization programs. Prior evaluation of site characterization programs is appropriate whenever the costs of exploration are high and there is significant uncertainty concerning the possible outcomes of the program. Within the simulator, conceptual models are available which represent combinations of faults, fractures, paleochannels, contacts, and other geologic features. Once a geologic conceptual model has been developed, geologic, hydraulic, and geophysical exploration techniques can be simulated.

ACKNOWLEDGEMENT

The development of the site characterization strategy evaluation methodology described in this paper was supported by Nagra, the Swiss radioactive waste management authority. The valuable contributions of Drs. P. Hufschmied, S. Vomvoris, S. Löw, and R. Andrews to this project were greatly appreciated.

REFERENCES

1. Einberger, C, "MEMO: Monitoring Well Efficiency Optimization Package," Golder Associates Inc., Redmond, WA, 1990.

2. Black, J, "Task 2 Report to Nagra, Assessment of Exploration Methods," Golder Associates (U.K.) Ltd., Nottingham, U.K., 1990.

3. Geier, J, G Lee and W Dershowitz, "FracMan Discrete Fracture Simulation Model: User Documentation," Golder Associates Inc, Redmond, WA, 1990, 89p.

4. Palisade Software, "@Risk: Monte Carlo Simulation for Personal Computers," Palisade Software, Palisade, New York, 1990.

5. Roberds, W, "Methods for Developing Subjective Probability Assessments," Transportation Research Board, National Research Council, Washington, D.C., 1990.

6. Santalo, L, Integral Geometry and Geometric Probability, Addison-Wesley, Reading, MA, 1984, 404p.

7. Stoyan, D, WS Kendall, and J Mecke, Stochastic Geometry and Its Applications, Wiley, New York, 1987, 345p.

MODELLING VERTICAL GROUND MOVEMENTS USING SURFACE CLIMATIC FLUX

Pamela J. Sattler[1] and Delwyn G. Fredlund[2]

ABSTRACT:

An integrated numerical model has been developed which illustrates the relationship between one-dimensional vertical ground movements and meteorological observations. The model consists of four components, namely: 1) a ground movement component, 2) a seepage component, 3) a surface flux boundary condition component, and 4) a Thornthwaite evapotranspiration component. All aspects of the model are presented with an emphasis placed on the surface flux boundary component.

The Thornthwaite potential evapotranspiration computations proved adequate for a first approximation analysis. Based upon several long-term averaging procedures, the actual evapotranspiration was computed to be approximately 70% of the potential evapotranspiration.

The numerical model illustrated the relationship between vertical ground movement and changes in matric suction. The first approximation analysis revealed that it was necessary to consider the infiltration and exfiltration processes separately.

INTRODUCTION

A numerical model has been developed which quantifies the relationship between one-dimensional vertical ground movements (i.e., heave and shrinkage) and meteorological observations. The model is divided into four components as shown in Figure 1. The ground movement model computes the change in the thickness of a soil layer in accordance with a constitutive function relating void ratio and matric suction. The time dependency of ground movements are in accordance with a Darcian type seepage model.

The surface flux boundary condition required for the seepage model is computed from the difference between precipitation and <u>actual</u>

[1] Research Engineer, Dept. of Civil Engineering, Univ. of Saskatchewan, Saskatoon, SK, CANADA S7N 0W0

[2] Prof., Dept. of Civil Engineering, Univ. of Saskatchewan, Saskatoon, SK, CANADA S7N 0W0

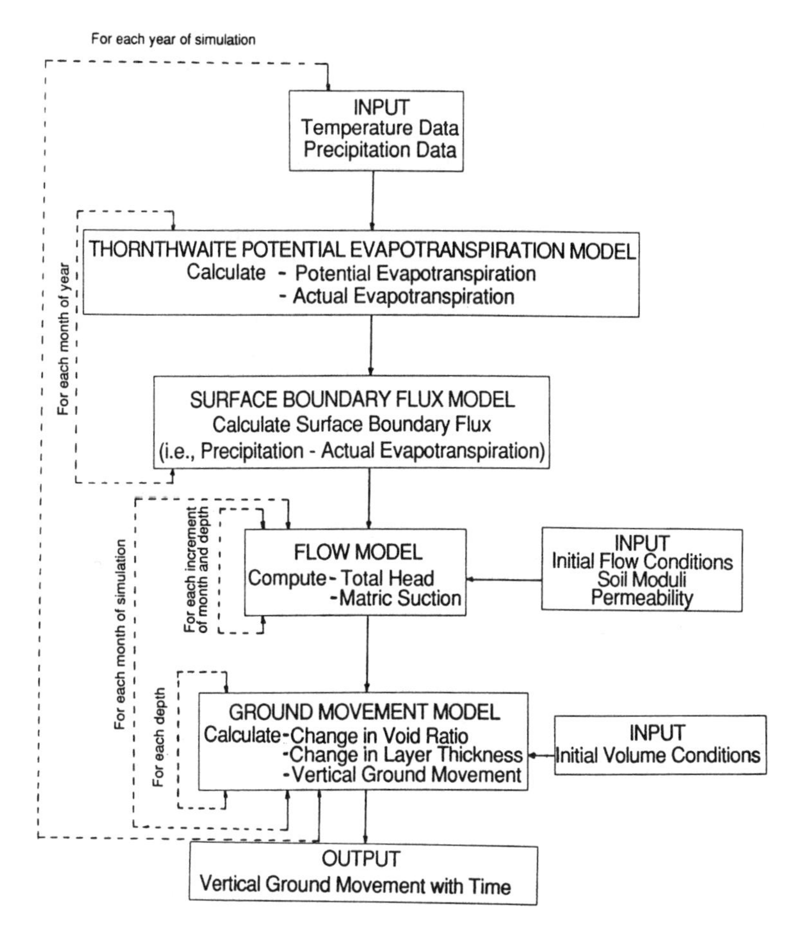

Figure 1. Flowchart for the computer model used to predict vertical ground movements with depth and time

evapotranspiration. Input to the influx portion of the surface flux boundary model is computed from the Thornthwaite (1948) computations for potential evapotranspiration. Based upon several long-term averaging procedures, the actual evapotranspiration was computed to be approximately 70% of the potential evapotranspiration.

This paper describes the one-dimensional numerical model with emphasis placed on the Thornthwaite model and the surface flux boundary model. Further research into quantification of the surface flux boundary condition has recently been initiated (Wilson, 1990).

BACKGROUND

The relationship between meteorological observations and vertical ground movements was illustrated by Hamilton (1965). He summarized the results of two decades of research performed by the Division of Building Research (DBR) of the National Research Council (NRC) of Canada. Vertical ground movements in open fields and beneath building foundations located on expansive soils were measured at various sites. Ground movement measurements were recorded about once a month and water content changes were measured periodically over the span of several years (Hamilton, 1963).

Building slabs were observed to exhibit a gradual heave over long periods of time whereas open-field test plots exhibited more of an upward and downward seasonal movement (Hamilton, 1968). Figure 2 illustrates one typical record of vertical ground movements for an open-field test plot at Regina, Saskatchewan. Increases in water content were observed to accompany movement measurements. During wet periods, both open-field plots and building slabs exhibited upward movement. During dry periods in which vegetation demands become predominant, concrete slab elevations remained unchanged whereas open-field gauges indicated downward movement. It was concluded that there was a relationship between meteorological observations and vertical ground movements.

THE DEVELOPMENT OF A MODEL

The observations recorded by Hamilton (1965) were not placed within the context of a theoretical model. Advances from the study of unsaturated soil behavior during the 1970's have made a mathematical model a logical extension to the field measurements.

Fredlund and Morgenstern (1977) suggested the use of two independent stress state variables to describe the engineering behavior of an unsaturated soil; namely, i) net normal stress $(\sigma - u_a)$ and ii) matric suction, $(u_a - u_w)$ where σ = total stress, u_a = pore-air pressure and u_w = pore-water pressure. In other words, the matric suction is the negative pore-water pressure referenced to pore-air pressure. Changes in matric suction occur due to changes in the microclimate.

The constitutive equation used to describe shrinking and swelling is written as follows:

$$de = a_m \, d(u_a - u_w) \qquad (1)$$

where:
de = change in void ratio
a_m = coefficient of compressibility with respect to matric suction

Vertical ground movements (i.e., one-dimensional volume changes) are computed by summing the changes in layer thicknesses as follows:

$$\Delta H = \sum_{i=1}^{i=n} \Delta H_i \qquad (2)$$

where:

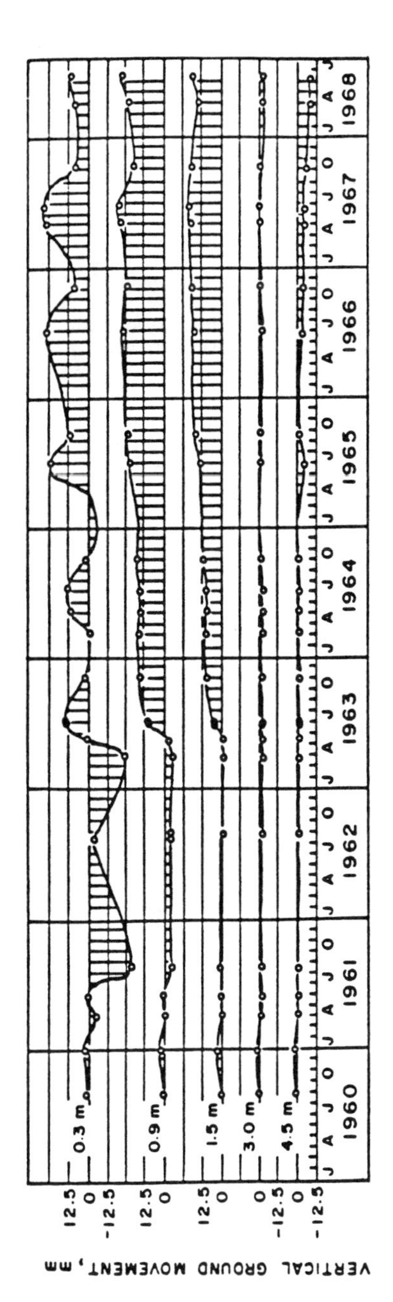

Figure 2. Measured vertical ground movements for an open-field test plot at Regina, Saskatchewan (Hamilton, 1968)

$$\Delta H_i = H_{io} \frac{(\Delta e_i)}{1 + e_{io}} \tag{3}$$

ΔH_i = change in thickness of an individual soil layer
i = number of the soil layer where the soil layers range from 1 to the total number of layers, n
H_{io} = initial layer thickness
Δe_i = change in void ratio for each layer
e_{io} = initial void ratio for each layer

Equations (1), (2), and (3) can be used to compute the vertical ground movement at any depth in the soil. However, the ground movements occur as water becomes available to the soil. The water causes the matric suctions to vary with time as flow occurs from a higher to a lower hydraulic head. The seepage equation is used to describe the flow of water through the soil and thereby solve for the matric suction at any point in space and time. As a first approximation, the transient flow equation can be written:

$$\frac{\partial h}{\partial t} = \frac{k_y}{\rho_w g\, m_2^w} \frac{\partial^2 h}{\partial y^2} \tag{4}$$

where:

h = total hydraulic head, equal to $(Y + \frac{u_w}{\rho_w g})$ where Y is the elevation head above the chosen datum
$\frac{\partial h}{\partial t}$ = the derivative of total hydraulic head with respect to time
k_y = hydraulic conductivity of the soil in the vertical direction
ρ_w = density of water
g = acceleration due to gravity
m_2^w = coefficient of water volume change
$\frac{\partial^2 h}{\partial y^2}$ = the second derivative of total hydraulic head with respect to depth above the datum

The hydraulic conductivity, k_y, for an unsaturated soil is a function of matric suction which in turn is a function of hydraulic head. A more rigorous formulation of the unsaturated soil seepage equation is presented by Lam and Fredlund (1987). However, for the purposes of developing a first approximation model, a linear form of the seepage equation (i.e., Equation (4) above) is proposed. The marching forward finite difference technique was used to solve the transient flow equation.

The volume change modulus, m_2^w, was estimated based upon one-dimensional consolidation data. Values which were used were in the order of 1 x 10^{-2} to 1 x 10^{-5} per kilopascal.

Initial boundary conditions required for the model include the negative pore-water pressure, void ratio, water content, and degree of saturation. Profiles of initial negative pore-water pressures were estimated based upon typical swelling pressure profiles for Regina clay (Yoshida, Fredlund, and Hamilton, 1983) and field matric suction data (van der Raadt, et al., 1987). Profiles of initial void ratio, water content and degree of saturation were established from summaries of statistical properties on Regina clay (Fredlund and Hasan, 1979) and measured water contents at the field locations (Hamilton, 1968).

SURFACE FLUX BOUNDARY CONDITION

For the purpose of first approximation modelling of a flat prairie, it was assumed that all precipitation at the ground surface would contribute to infiltration. Therefore, the surface boundary flux is written as the difference between precipitation and actual evapotranspiration.

Potential Evapotranspiration. Evapotranspiration was defined by Thornthwaite (1948) as "the combined evaporation from the soil surface and transpiration from plants" representing "the transport of water from the earth back to the atmosphere, the reverse of precipitation". Thornthwaite and Mather (1955) defined the term "potential evapotranspiration" as "the amount of water which will be lost from a surface completely covered with vegetation if there is sufficient water in the soil at all times for the use of the vegetation". A great deal of controversy has surrounded the definition of potential evapotranspiration since in practice actual microclimatic data is used to establish a potential evapotranspiration value (Granger, 1989). However, the main purpose in computing potential evapotranspiration is to provide a means of estimating actual evapotranspiration. Potential evapotranspiration establishes an upper limit for actual evapotranspiration since potential evapotranspiration assumes that there is no limit on the available water and actual evapotranspiration is dependent upon available soil water.

Several methods are available for the estimation of potential evapotranspiration (Thornthwaite and Mather, 1955; Penman, 1963; Priestley and Taylor, 1972). The Thornthwaite method was chosen for its simplicity and applicability to sites where meteorological information may be scarce. The relationship is based on mean monthly temperature and average monthly day length:

$$PE = 0.16\,F\left[\frac{10t}{I}\right]^{a} \tag{5}$$

where:

PE = the monthly potential evapotranspiration (mm)
F = sunlight duration correction factor based on Thornthwaite and Mather (1955)
t = mean monthly temperature (degrees Celcius)
I = the sum of the twelve monthly heat indices, i
i = $(t/5)^{1.515}$
a = a variable equal to $0.000000675\,I^3 - 0.0000771\,I^2 + 0.01792\,I + 0.49239$

The object of Thornthwaite's research was the classification of climates for geographical purposes. Sanderson (1948) utilized Thornthwaite's method for computing potential evapotranspiration for large regions across Canada (Figure 3). Thornthwaite's method has been used extensively throughout the world for the classification of climates (Sanderson and Rafique, 1979). Classification is based upon a moisture index describing humidity relative to aridity. Figure 4 presents the classification of Canadian climates based upon Thornthwaite's moisture index.

Estimation of Actual Evapotranspiration. Actual evapotranspiration is less than potential evapotranspiration due to a limited amount of water available in the soil. Several methods were investigated in an attempt to establish a relationship between actual and potential evapotranspiration. These are as follows:

i) By Net Groundwater Recharge Observations. Groundwater recharge rates are based upon average conditions over large areas for long periods of time. Zebarth (1988) conducted an extensive literature review of recharge rates for Southern Saskatchewan and concluded that typical recharge rates were in the order of 5% to 10% of annual precipitation. Assuming a surface boundary flux equal to the lower end

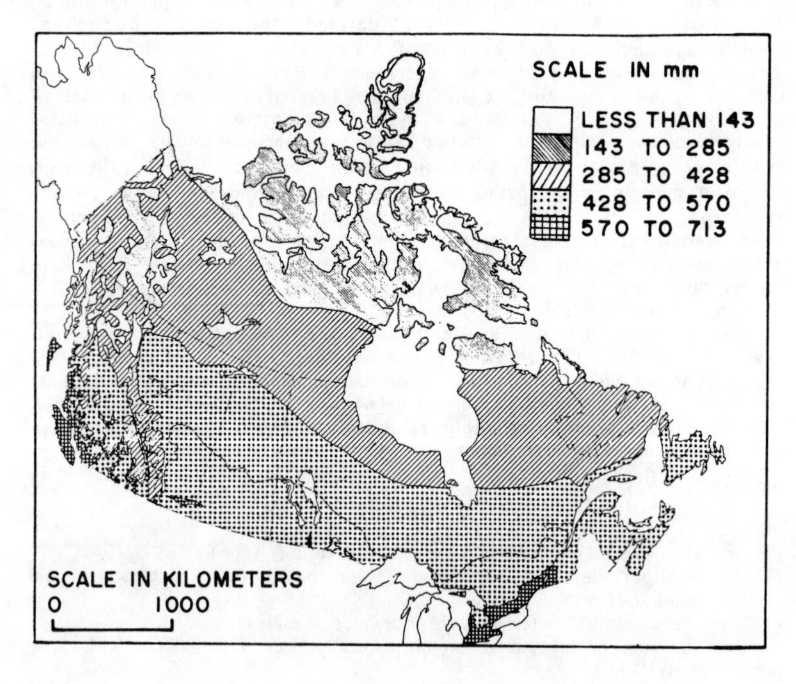

Figure 3. Average annual potential evapotranspiration and climatic types in Canada (Sanderson, 1948)

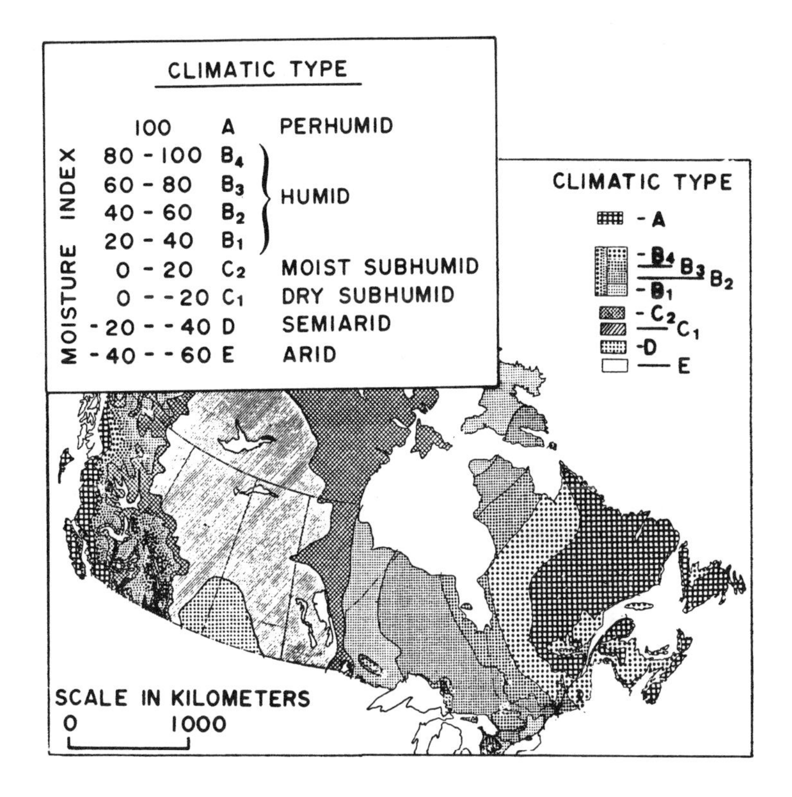

Figure 4. Moisture regions in Canada based on Thornthwaite's classification (Sanderson, 1948)

of the recharge rate (i.e., 5% of precipitation), the surface boundary flux can be written as:

$$F = 0.05\,P = P - AE \qquad (6)$$

where:
F = surface boundary flux
P = precipitation
AE = actual evapotranspiration

By rearrangement, actual evapotranspiration is computed as 95% of precipitation. Precipitation data from Environment Canada, for Regina, Saskatchewan for the period of years from 1930 through 1984 is presented in Figure 5. Mean monthly temperature data for the same location along with the Thornthwaite method (1948) were used to compute potential evapotranspiration for the same period of time. Figure 6 illustrates the ratio between actual and potential evapotranspiration for each year between 1930 and 1984. The average ratio between the estimate of actual evapotranspiration and Thornthwaite potential evapotranspiration is 0.71.

ii) By Equating Long-Term Precipitation and Actual Evapotranspiration. Assuming that over the long term, there is neither a net wetting nor a net drying, long-term precipitation must equal long-term actual evapotranspiration. The total precipitation for the years 1930 through 1984 was assumed to equal the total actual evapotranspiration for the same period. The ratio established between actual and potential evapotranspiration was 0.75.

iii) By Equating Long-Term Mean Monthly Values. The long-term mean monthly precipitation for Regina, Saskatchewan is presented in Figure 7. The mean monthly values are based upon 100 years of record collected by Environment Canada. The assumption is made that the mean monthly precipitation is equal to the mean monthly actual evapotranspiration. The corresponding ratio established between actual and potential evapotranspiration was 0.69.

The above discussion illustrates that a reasonable ratio between actual and potential evapotranspiration is in the order of 0.70. It could also be that 55 years of data may be insufficient to estimate long-term values for the ratio between actual and potential evapotranspiration.

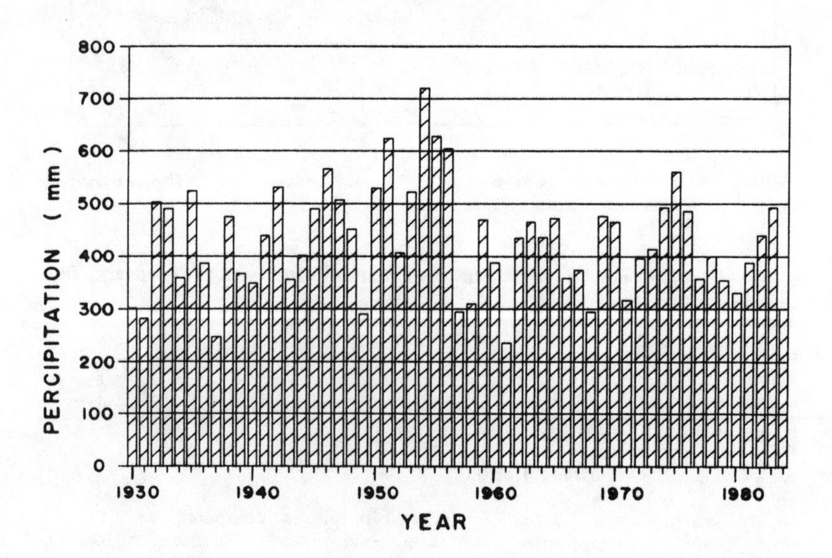

Figure 5. Annual precipitation for 1930 through 1984 at Regina

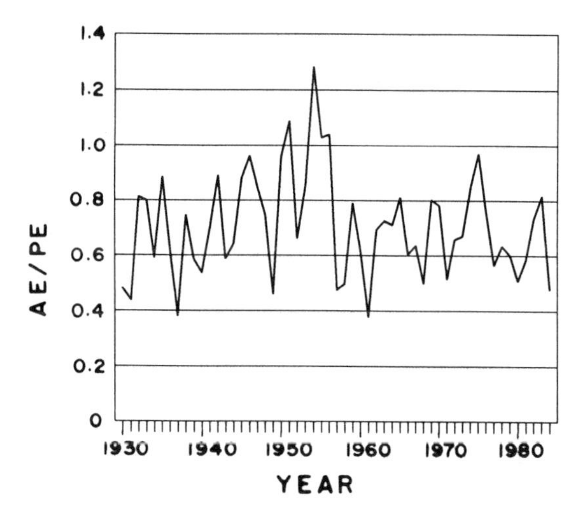

Figure 6. Ratio between actual and potential evapotranspiration based upon the net groundwater recharge method

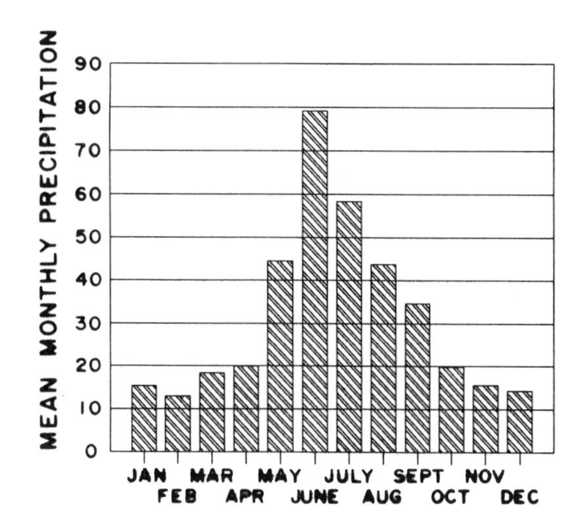

Figure 7. Long-term mean monthly precipitation values for Regina

PRESENTATION AND DISCUSSION OF MODELLING RESULTS

The four components of the model illustrated in Figure 1 illustrate the relationship between matric suction and vertical ground movements. The model shows how increases in matric suction result in shrinkage and decreases in matric suction result in swelling.

Figure 8 illustrates a comparison of simulated vertical ground movements with the recorded ground movements for the Regina location. Movements are in the order of 50 mm at a depth of 0.3 m. Measured surface ground movements were in the order of 75 mm (Hamilton, 1963).

Two independent ranges for the hydraulic conductivity were required: one for the infiltration process and one for the exfiltration process. Values in the order of 10^{-6} to 10^{-8} meters per second were required for modelling infiltration events. These suggest that the soil behaves as though it were relatively permeable during the infiltration process. This can be explained in physical terms by considering the intake of water to the shrinkage cracks evident in the field (Figure 9). The associated volume change behavior is influenced by the macro-structure of the soil mass.

For the exfiltration process, hydraulic conductivities in the order of 10^{-9} to 10^{-11} meters per second were required to obtain reasonable simulations. The exfiltration process appears to occur primarily as a vapor transport process rather than a liquid flow process (Figure 9). The volume change behavior during the exfiltration process appears to be more closely associated with the micro-structure of the soil.

Figure 10 illustrates typical seasonal ground movements for Regina, Saskatchewan. Corresponding typical seasonal suction values vary between 50 kPa and 1400 kPa near the ground surface. The modelling also implies that thermally induced suctions (i.e., winter freezing conditions) could play a significant role in the seasonal vertical ground movements.

CONCLUSIONS AND RECOMMENDATIONS

The numerical model shows the relationship between vertical ground movements and matric suction changes. Reasonable agreement was obtained between measured and simulated movements using soil parameters that vary with depth.

Due to the formation of shrinkage cracks in the field, two ranges of hydraulic conductivity were required for modelling. It was necessary to utilize one for infiltration and one for exfiltration. These ranges reflect the change from water flow during infiltration to predominately vapor flow during exfiltration.

The Thornthwaite potential evapotranspiration computations showed reasonable modelling results with the use of a ratio of 0.70 between actual and potential evapotranspiration. The Thornthwaite potential evapotranspiration computations are suitable for situations in which meteorological data are limited. However, more rigorous physical models would be advantageous. Further research is required to establish better the surface boundary condition and to enlarge the database representing typical soil properties.

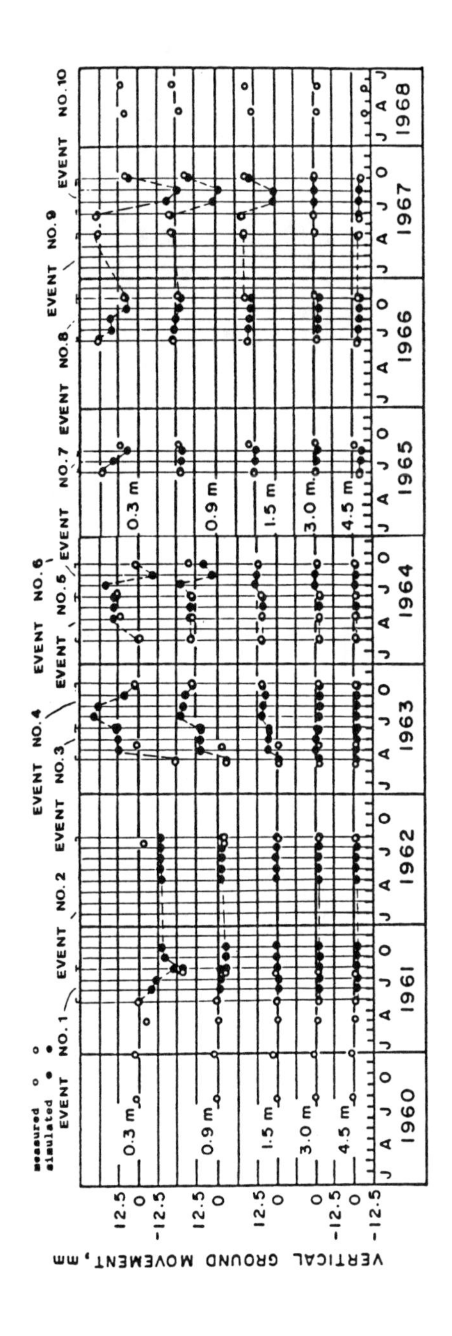

Figure 8. Comparison of predicted and measured vertical ground movements for each of ten events modelled for Regina, Saskatchewan

Figure 9. Schematic diagram illustrating infiltration and exfiltration in a cracked clay soil

REFERENCES

1. DG Fredlund & JU Hasan, Statistical geotechnical properties of Lake Regina sediments, University of Saskatchewan, Department of Civil Engineering, Transportation and Geotechnical Group, Saskatoon, Saskatchewan, Internal Report, IR-9, 1979, 102 p.
2. DG Fredlund & NR Morgenstern, Stress state variables for unsaturated soils, Journal of the Geotechnical Engineering Division, American Society of Civil Engineers, Vol. 103 (GT75), 1977, 447 - 446.
3. RJ Granger, An examination of the concept of potential evapotranspiration, Journal of Hydrology, 1989, 9-19.
4. JJ Hamilton, Volume changes in undisturbed clay profiles in Western Canada, Canadian Geotechnical Journal, Vol. 1, 1963, 27 - 41.
5. JJ Hamilton, Shallow foundations on swelling clays in Western Canada, The International Research and Engineering Conference on Expansive Clay Soils, Texas A & M University, College Station, TX, 1965, 183 - 207.

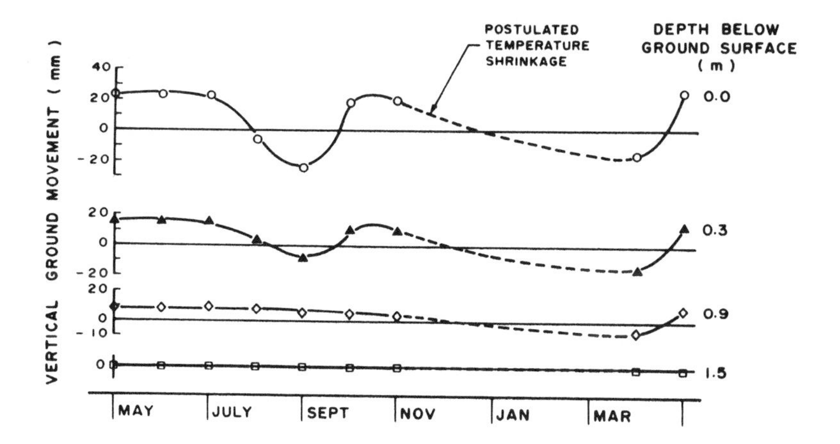

Figure 10. Simulated average vertical ground movements for Regina, Saskatchewan

6. JJ Hamilton, Effects of natural and man-made environments on the performance of shallow foundations, Proceedings of the 21st. Annual Canadian Soil Mechanics Conference, Winnipeg, Manitoba, 1968.
7. L Lam, DG Fredlund & SL Barbour, Transient seepage model for saturated-unsaturated soil systems: A geotechnical engineering approach, Canadian Geotechnical Journal, Vol. 24, 565-580.
8. HL Penman, Vegetation and Hydrology Technical Communication No. 53, Commonwealth Bureau of Soils, Harpenden, Commonwealth Agricultural Bureau, Farnham Royal, Bucks, England, 1963.
9. CHB Priestley, & RJ Taylor, On the assessment of surface heat flux and evaporation using large scale parameters, Monthly Weather Review, Vol. 100, No. 2, 1972, 81 - 92.
10. M Sanderson, The climates of Canada according to the new Thornthwaite classification, Scientific Agriculture, Vol. 28, 1948, 501 - 517.
11. M Sanderson & A Rafique, Potential evapotranspiration and water deficit in Bangladesh using Garnier's modification of the Thornthwaite water balance, Climatological Bulletin No. 25, McGill University, Department of Geography, 1979, 13 - 24.
12. CW Thornthwaite, An approach toward a rational classification of climate, Geographical Review, Vol. 38, 1948, 55 -94.
13. CW Thornthwaite & JR Mather, The Water Balance, (2nd Ed), Publications in Climatology, Vol. 8, No 1, Drexel Institute of Technology, Centerton, New Jersey, 1955.
14. P van der Raadt, DG Fredlund, AW Clifton, MJ Klassen & WE Jubien, Soil suction measurements at several sites in Western Canada, Presented to the Session on Geotechnical Problems in Arid Regions, Transportation Research Board, Committee A2L06TRB, Washington, D.C., 1987.

15. GW Wilson, Soil evaporation fluxes for geotechnical engineering problems", Ph.D. thesis, University of Saskatchewan, Saskatoon, Canada, 1990, 294 p.
16. RT Yoshida, DG Fredlund & JJ Hamilton, The prediction of total heave of a slab-on-grade floor on Regina clay", Canadian Geotechnical Journal, Vol. 20, No. 1, 1983, 69 - 81.
17. BJ Zebarth, Saturated and unsaturated flow in a hummocky landscape in relation to topography and soil morphology, Ph.D. Thesis, Department of Soil Science, University of Saskatchewan, Saskatoon, Saskatchewan, 1988.

AN ANALYTICAL INVESTIGATION OF THE BEHAVIOR OF LATERALLY LOADED PILES

Sujit K. Bhowmik[1], S.M.ASCE and James H. Long[2], A.M.ASCE

ABSTRACT : Behavior of laterally loaded piles is investigated using the non-linear finite element technique. The soil is modeled using a bounding surface plasticity model, and the soil-pile interface is modeled using thin-layer elements. A full-scale lateral load test is analyzed using two- and three-dimensional, linear and non-linear, finite element models, and various aspects of behavior of laterally loaded piles (e.g., p-y response, load-deflection response, bending moments in the pile, effects of slip and gap formation, etc.) are investigated. Results from three-dimensional analysis are compared with those from plane stress and plane strain analyses. The effects of gap formation behind the pile on the distribution of stresses around the pile and on the load-deflection and p-y response are investigated. Results from finite element analyses are compared with the observed pile behavior in a lateral pile load test. Advantages and limitations of various methods of analysis are discussed and important findings are summarized.

INTRODUCTION

The response of a pile to lateral loads is influenced by characteristics of soil, pile, soil-pile interface, and load. The stress-strain-strength behavior of soil and soil-pile interface, geometry and stiffness of the pile, boundary and drainage conditions at the site, and magnitude, nature, and rate of applied loading are among the most important factors that control the behavior of a pile-soil system.

A comprehensive methodology for the analysis of laterally loaded piles, therefore, requires : i) a material model that gives reasonably accurate predictions of soil behavior, ii) a model for the behavior of soil-pile interface, iii) a numerical solution technique, and iv) a realistic idealization of the pile-soil geometry and loading and boundary conditions.

Two- and three-dimensional non-linear finite element programs with selected constitutive models for soil and soil-pile interface have been developed. The ability to simulate various geotechnical loading and boundary conditions (e.g., drained and undrained loading, influence of water table and in-situ and preconsolidation stresses, etc.), and the influence of slip and gap formation at the soil-pile interface have been incorporated.

First, the existing methods of analysis of laterally loaded piles are briefly discussed. Features of the present method of analysis are then described. Results of two- and three-dimensional, linear and non-linear, analyses are presented, and analytical results are compared with the observed behavior of a full-scale lateral pile load test.

EXISTING METHODS OF ANALYSIS

Existing Methods. Existing methods of analysis of laterally loaded piles can be classified broadly into four categories : i) methods based on subgrade reaction approach,

1 - Grad. Res. Asst., Dept. of Civil Engineering, Univ. of Illinois, 2212 Newmark Lab., 205 N. Mathews Avenue, Urbana, IL 61801.
2 - Asst. Prof., Dept. of Civil Engineering, University of Illinois, 2230d Newmark Lab., 205 N. Mathews Avenue, Urbana, IL 61801.

ii) methods based on p–y approach, iii) methods based on theories of elasticity, and iv) methods based on the finite element technique. These methods are discussed briefly in the following paragraphs.

Methods Based on Subgrade Reaction Approach. In the subgrade reaction approach, the pile is idealized as an elastic, transversely loaded beam supported on a series of unconnected springs with stiffnesses that represent the stiffness provided by the soil. This method has been widely used because of its simplicity (e.g., Hetenyi, 1946 ; Matlock and Reese, 1960 ; Davision and Gill, 1963). Consequently, considerable amount of experience has been gathered in applying this method. Theoretically, a disadvantage with this approach is that the soil reaction at any point is taken to be dependent on the pile deflection of that point only. Practically, greater uncertainties are associated with assigning appropriate values to the subgrade reaction modulus to represent adequately the stiffness provided by the soil. Subgrade reaction modulus values are assigned using experience or semi–empirical rules, and they include, implicitly or explicitly, the effects of soil behavior (e.g., stress–strain, strength, non–linearity), soil layering, pile geometry, and the magnitude and duration of lateral load.

Methods Based on p–y Approach. The p–y method can be regarded as a subgrade reaction method that allows the soil resistance (p) at any point along the pile shaft to be a non–linear function of the displacement (y) of that point. A number of methods to obtain p–y curves have been suggested by different investigators (e.g., McClelland and Focht, 1958 ; Reese and Cox, 1969 ; Matlock, 1970 ; Reese and Welch, 1975 ; O'Neill and Gazioglu, 1984; Dunnavant and O'Neill, 1989). These methods are based on results of lateral load tests conducted under specific field conditions, and they exhibit the advantages and shortcomings of the subgrade reaction method, except that soil non–linearity is taken into consideration in these methods.

Methods Based on the Theories of Elasticity. In the methods based on the theories of elasticity, the soil is modeled as an elastic continuum (e.g., Spillers and Stoll, 1964 ; Poulos, 1971) and the pile is modeled as a thin rectangular vertical strip (Poulos, 1971). Methods to extend the elastic analysis to include local yielding and pile–soil separation have also been suggested (Poulos and Davis, 1980). The main advantage of these methods over the subgrade reaction and p–y method is that the continuous nature of soil is taken into account in these methods. The major shortcomings of these methods are : i) soil is treated as a linear elastic material, ii) factors such as soil non–linearity, preconsolidation, pile–soil separation, etc. are difficult to incorporate into the analysis.

Methods Based on Finite Element Technique. The finite element method provides a powerful analytical tool that can be used for the analysis of laterally loaded piles. Many analytical studies of laterally loaded piles based on finite element method have been reported in the literature (e.g., Yegian and Wright , 1973 ; Baguelin et al., 1977 ; Muktadir and Desai, 1986 ; Kooijman, 1989 ; Brown et al., 1989). The main advantages of the finite element method are : i) various geometry and boundary conditions for the soil–pile system can be considered, ii) different forms of constitutive models for soils and soil–pile interfaces can be used, and iii) systematic investigation of various aspects of pile behavior can be conducted since all the parameters can be varied and their influences studied within the same analytical framework.

PRESENT METHOD OF ANALYSIS

Components. The analytical tool used herein employs the finite element method combined with a bounding surface constitutive model for the soil, a thin–layer element for the soil–pile interface, and other features to model geotechnical loading and boundary conditions. These components are discussed briefly in the following paragraphs.

Constitutive Model for Soil. Considerable advancements have been made in modeling of soil behavior in recent years. Many existing models have the capability of simulating soil behavior with reasonable accuracy (e.g., Prevost, 1978 ; Mroz and Zienkiewicz, 1984 ; Banerjee and Yousif, 1986 ; Anandarajah and Dafalias, 1986).

The soil model used herein is the one proposed by Banerjee and Yousif (1986). The model uses the bounding surface plasticity theory and the critical state concept. This model can simulate the behavior of real soils with reasonable accuracy, and the parameters required by the model can be determined easily. It has been shown that the model is capable of simulating the behavior of isotropically and anisotropically consolidated clay soils, having different degrees of overconsolidation, under various stress paths, with a reasonable degree of accuracy (Banerjee et al., 1985 ; Banerjee and Yousif, 1986).

The interpolation rule to compute the plastic modulus at any stress point from the plastic modulus at a stress point on the bounding surface, was modified to improve the predictive capability of the model.

Model for Soil–Pile Interfaces. Large relative displacements may occur between the pile and the surrounding soil during lateral loading of a pile. A gap may form behind the pile near the ground surface which can influence significantly the behavior of the soil-pile system (e.g., Dunnavant, 1986). Therefore, models for laterally loaded piles should include the effects of gap formation by incorporating a model for the behavior of the interface between the pile and the soil.

Many different models to represent interface behavior have been proposed in the literature. Some examples are Goodman et al. (1968), Zienkiewicz et al. (1970), Ghaboussi et al. (1973), Herrmann (1978), and Desai et al. (1984). A thin–layer element adapted from Desai et al. (1984) was used herein to model the interface behavior. These interface elements are used to model the formation of a gap and slip at soil–pile interfaces in a manner described in a latter section.

Numerical Solution Technique. Two– and three–dimensional non–linear finite element programs have been developed to model laterally loaded piles. The soil and the soil–pile interface models have been incorporated. For three–dimensional analysis, 20–node solid isoparametric elements are used whereas 8–node plane isoparametric elements are used for two–dimensional analysis. The non–linear stiffness equations are solved using the frontal solution technique along with the tangent stiffness algorithm.

Gap Formation and Slip at Soil–pile Interfaces. To model gap formation and slip at the soil–pile interface, the stresses in the interface elements are monitored at every iteration step. A gap is assumed to form at an interface location when the stress normal to the pile surface at that location reduces to zero (soil is assumed to have zero tensile strength). When this condition occurs, no stress is transferred from the pile to the soil through the interface element. This phenomenon is modeled by assigning very small incremental normal and shear stiffness to the particular interface element. Slip is assumed to occur at an interface location when the shear stress at that location exceeds the specified interface shear strength. The interface shear strength is specified in terms of the residual friction angle of the soil. When slip occurs, a very small incremental shear stiffness is assigned to the particular element.

DESCRIPTION OF THE LOAD TEST

Load Test. The load test analyzed is a full scale lateral load test on a 1.22 m diameter steel pipe pile in submerged overconsolidated clay at the University of Houston Foundation Test Facility. Detailed descriptions of the site and the test have been given by Dunnavant (1986) and Dunnavant and O'Neill (1989). The geometry and configuration of the pile are shown in Figure 1.

Site Condition. The soil deposit at the site consists of the Beaumont formation (to a depth of about 7.3 m) underlain by the Montgomery formation (to a depth of at least 76 m). Soils in both formations consist of saturated, overconsolidated clays and contain discontinuous fissures and slickensides and isolated sand and silt lenses. The Beaumont soil is of the CH type whereas the Montgomery soil is of the CL type. The lateral load tests were conducted in a 0.6 m deep test pit. The natural ground water table was at a depth of about 1.53 m below the test surface. However, the test pit was flooded with water before the load tests were conducted (Dunnavant and O'Neill, 1989).

Soil Properties. Detailed profiles of various soil properties have been given by Dunnavant (1986). In the analysis, values of these properties were taken from those profiles. The average unit weights of Beaumont and Montgomery soils were 19.8 kN/m^3 and 20.9 kN/m^3 respectively. The undrained shear strength at the site varied between 35 kN/m^2 near the ground surface to 207 kN/m^2 at a depth of 12 m. The elastic modulus varied between 7 MN/m^2 to 70 MN/m^2, the overconsolidation ratio varied between 24 and 4, and, the lateral pressure coefficient varied between 3.5 and 1, near the ground surface and at a depth of 12 m, respectively.

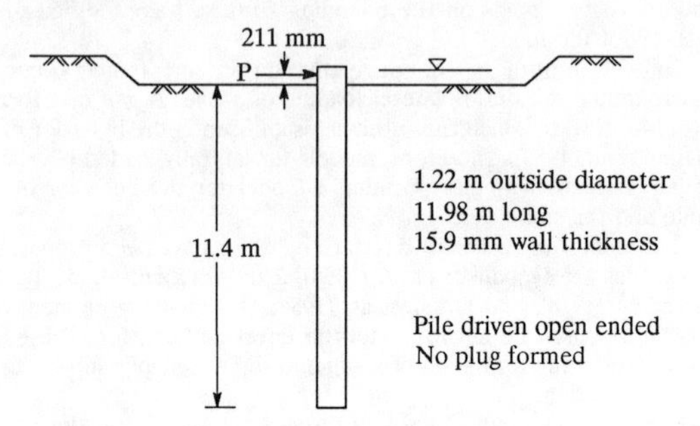

Figure 1. Geometry and Configuration of the Pile

THE FINITE ELEMENT MODEL

Model for 2–Dimensional Analysis. Figure 2 shows the finite element model used for two–dimensional analysis. It consists of 104 8–node plane isoparametric elements forming a total of 349 nodes. The mesh contains 16 pile elements, 8 interface elements and 80 soil elements. The lateral load is applied at the center of the pile along the x–direction. By taking advantage of symmetry, only half of the geometry is used to model the problem.

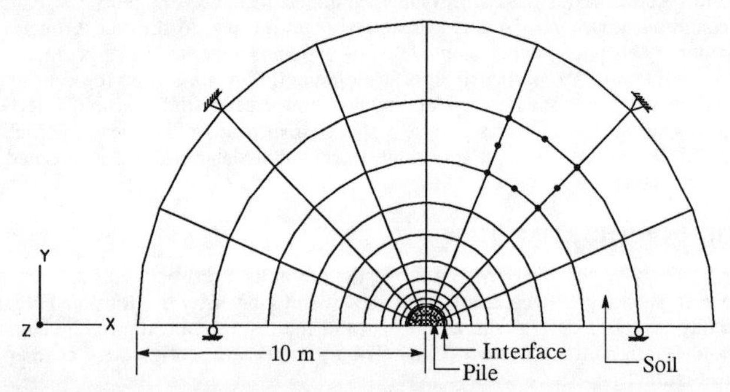

Figure 2. Finite Element Model used for Plane Strain Analysis.

Model for 3–Dimensional Analysis. The finite element model for three–dimensional analysis is shown in Figure 3 . The pipe pile is replaced by an equivalent solid pile of same flexural rigidity (EI). The model consists of 788 20–node solid isoparametric elements forming a total of 3884 nodes. There are 126 pile elements, 598 soil elements and 64 interface elements in the mesh. As before, the lateral load is applied at the center of the pile along the x–direction.

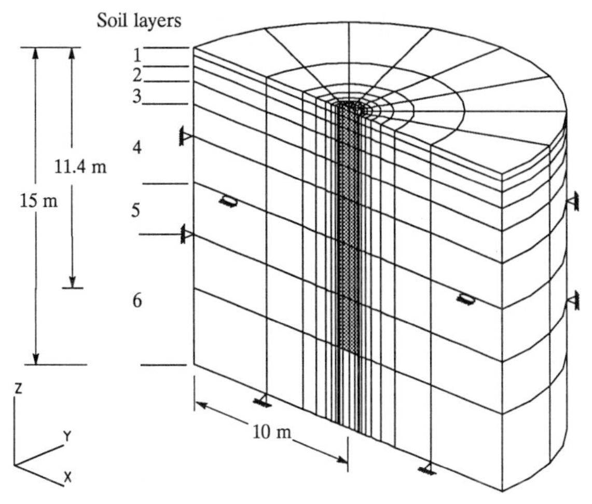

Figure 3. Finite Element Model used for Three-dimensional Analysis.

DETERMINATION OF SOIL PARAMETERS

The soil parameters used in the analysis were obtained from the results of laboratory and field tests reported by Dunnavant (1986). Based on these test results, the soil profile was divided into six layers (Figure 3). Results of triaxial tests on soils from each layer were used to obtain the model parameters for that layer. The soil parameters used for different layers are shown in Table 1, where, ϕ = angle of internal friction, e = void ratio, λ = slope of compression e-$\log_e$ p curve, κ = slope of swelling e-$\log_e$ p curve, K_0 = coefficient of earth pressure at rest, ν = Poissons ratio, OCR = overconsolidation ratio, E = elastic modulus, and β = plastic modulus parameter (used to compute the plastic modulus at any stress point from the plastic modulus at a stress point on the bounding surface).

Table 1. Soil Parameters Used in the Analysis

Layer No.	Thickness (m)	Soil Parameters								
		e	ϕ°	E (MN/m^2)	K_0	OCR	λ	κ	ν	β
1	0.90	0.50	31	9.5	3.0	25.0	0.08	0.03	0.25	0.85
2	0.75	0.67	30	13.0	2.2	16.0	0.08	0.03	0.25	0.78
3	1.10	0.60	30	15.0	1.8	11.0	0.08	0.03	0.25	0.92
4	3.70	0.76	32	21.0	1.5	7.7	0.10	0.03	0.25	0.94
5	2.40	0.50	30	21.0	1.3	7.0	0.10	0.03	0.25	0.80
6	6.15	0.50	30	21.0	1.1	6.0	0.10	0.03	0.25	0.95

The parameters K_0, OCR and E were obtained from the profiles of these parameters reported by Dunnavant (1986). Void ratios (e) were obtained from the water content profile, friction angles (ϕ) were obtained from the results of triaxial tests, and, λ values were obtained from the results of a consolidation test. No test result was available from which κ could be obtained, therefore, a reasonable value of 0.03 was used for all the layers. Poisson's ratio was assumed to be 0.25 for all the layers. The value of the plastic modulus parameter, β, for each layer was obtained by varying β until reasonable agreement was obtained between predicted and measured stress–strain curves for triaxial loading. Some examples of simulated and measured stress–strain curves are shown in Figure 4.

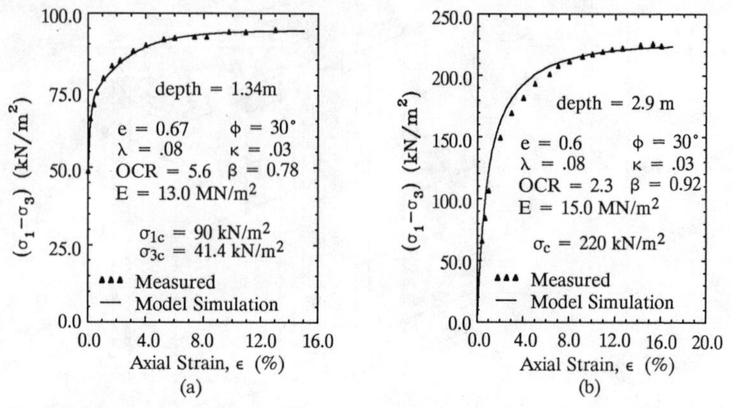

Figure 4. Measured and Simulated Stress–strain Response under Undrained Triaxial Compression Tests : a) K_0–Consolidated, b) Isotropically Consolidated. [Test Results from Dunnavant, 1986].

RESULTS AND DISCUSSION

General. Results of linear elastic and non–linear elasto–plastic, two– and three–dimensional, analyses are presented. In linear analyses, E = 20 MN/m^2 and ν = 0.25 were used for the soil. In non–linear analyses, soil parameters for various layers were taken from Table 1. Results are presented in the form of stress–distribution, p–y response, load–deflection response, and bending moments in the pile.

Stress Distribution. The distribution of mean normal and octahedral shear stresses around a laterally loaded pile obtained from plane strain linear elastic analyses are shown in Figures 5 and 6. In these analyses, the pile surface is assumed to be rough.

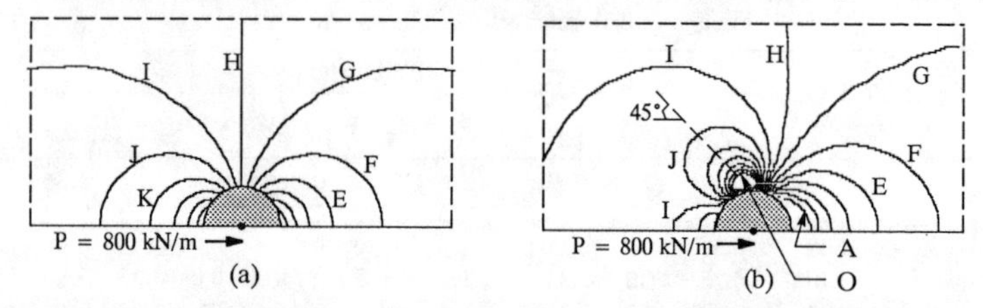

Figure 5. Distribution of Mean Normal Stress Around the Pile (plane strain elastic analysis) : (a) Gap Formation Not Allowed, (b) Gap Formation Allowed. [Contours are at Equal Intervals of 35 kN/m^2 from A=245 kN/m^2 to O=-245 kN/m^2]

Figures 5(a) and 5(b) show the distribution of mean normal stress for cases in which gap formation was prevented, and allowed, respectively. It can be seen that formation of a gap results in a change in shape and position of stress contours. The line of zero mean normal stress (H) shifts in the direction of loading, and higher tensile and compressive stresses are developed (e.g., the maximum tensile stress in Figure 5a is about 245 kN/m^2 whereas it is about 500 kN/m^2 in Figure 5b). Near the pile, tensile stress contours are oriented along a line forming an angle of about 45° with the line of action of the applied load (Figure 5b). Tensile cracks in the soil, when formed, would be expected to form along this line.

Figures 6(a) and 6(b) show the contours of octahedral shear stress for cases with and without gap formation, respectively. When a gap is allowed to form behind the pile, higher shear stresses are developed (e.g., the maximum shear stress in Figure 6a is about 280 kN/m^2 whereas it is about 1000 kN/m^2 in Figure 6b) and the contours behind the pile are oriented at an angle of about 30° with the line of action of the applied load (Figure 6b).

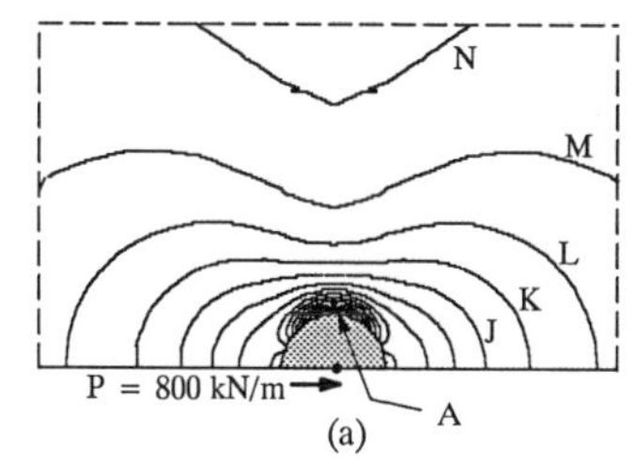

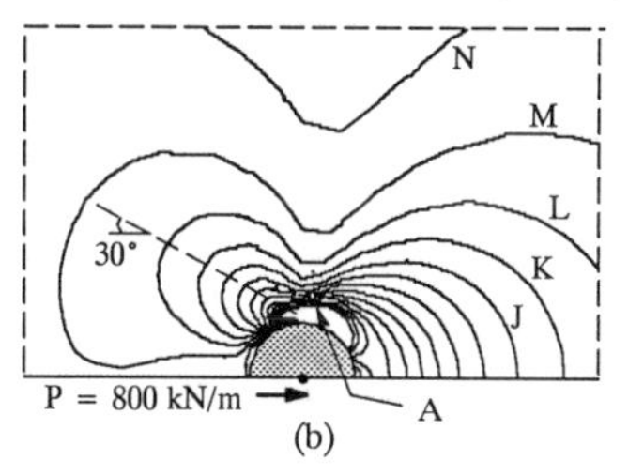

Figure 6. Distribution of Octahedral shear Stress Around the Pile (plane strain elastic analysis) : (a) Gap Formation Not Allowed, (b) Gap Formation Allowed. [Contours are at Equal Intervals of 20 kN/m^2 from A=280 kN/m^2 to N=20 kN/m^2]

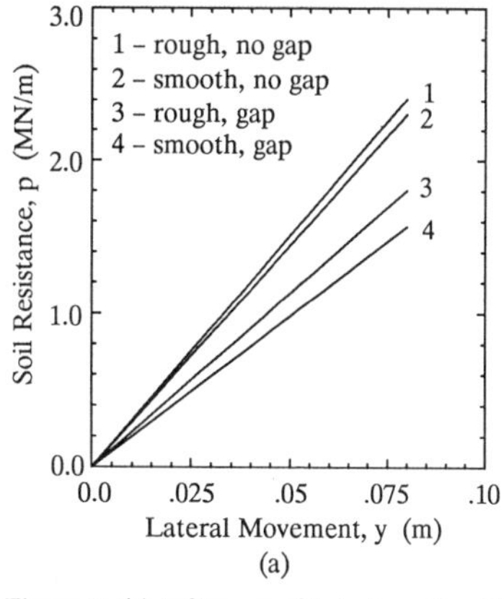

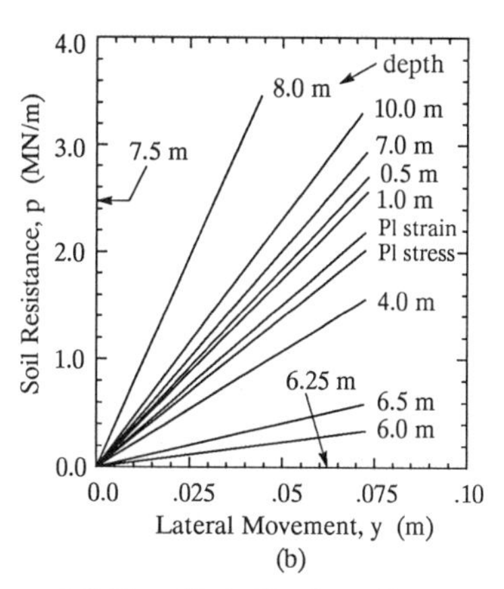

Figure 7. (a) Influence of Interface Conditions on 2-D Plane Strain Elastic p-y Response, (b) Comparison of Plane Stress, Plane Strain and Three-dimensional (at different depths) elastic p-y response.

2-D Plane Strain Linear Elastic p-y Response. The effects of interface conditions on plane strain linear elastic p-y response are shown in Figure 7(a). In this analysis, a rough pile surface is assumed to provide complete adhesion between the pile and the soil, and, therefore, the shear moduli of the interface elements are taken to be the same as those of the adjacent soil elements. For a pile with a smooth surface, it is assumed that no shear stress is transferred from the pile to the soil through the interface elements. This is modeled by assigning very small shear modulus to the interface elements. It can be seen that as the surface of the pile is changed from rough to smooth, the p-y stiffness decreases. However, allowing the formation of a gap behind the pile causes a more significant reduction in the p–y stiffness. These simple results illustrate that the interface deformation mode, particularly the formation of a gap, can have significant influence on the behavior of the pile.

2-D and 3-D Linear Elastic p-y Response. Elastic p–y response at various depths, obtained from three-dimensional analysis, are compared with plane strain and plane stress p–y response, in Figures 7(b) and 8(b). In Figure 7(b), soil resistance is plotted versus lateral deflection, whereas in Figure 8(b), p–y stiffness (soil resistance at a point for unit lateral movement of the pile at that point) is plotted versus depth. In these analyses, pile surface is assumed to be rough and no gap is allowed to form at the soil–pile interface.

In the three-dimensional analysis, the soil resistance at any depth is computed from the states of stress, at that depth, in the interface elements all around the pile. Tractions on the interface elements at every Gauss point elevation along the length of the pile are computed from the states of stress at those points. Soil resistance at any depth is then computed by integrating the components of traction vectors in the direction of applied load along the circumference of the pile. Soil resistance and lateral pile movement along the length of the pile are plotted in Figure 8(a). At any depth, the p–y stiffness is obtained by dividing the soil resistance by the lateral movement of the pile at that depth.

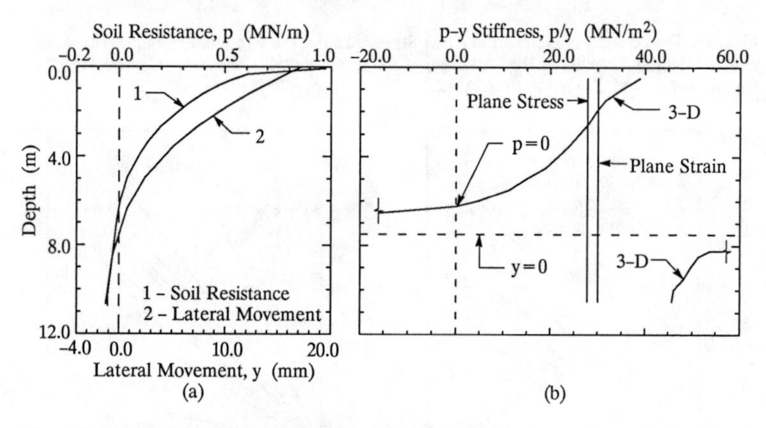

Figure 8. (a) Variation of Soil Resistance and Lateral Pile Movement Along the Length of the Pile (3–D Linear Elastic Analysis), (b) Comparison of p–y Stiffness obtained from Plane Stress, Plane Strain and Three-dimensional Linear Elastic Analyses.

It can be seen from Figures 7(b) and 8(b) that, near the ground surface, the p–y stiffness obtained from 3–D analysis is higher than those obtained from plane stress or plane strain analyses. The p–y stiffness decreases with depth to zero at a depth (6.25 m) where the soil resistance is zero. From this depth, the stiffness becomes negative, further decreases with depth and attains an infinite (theoretically) magnitude at a depth (7.5 m) where the lateral movement of the pile is zero. From this depth, the stiffness again becomes positive and decreases with depth down to the tip of the pile. The reason behind this wide variation of p–y stiffness with depth is that the lateral soil resistance at a given depth is not uniquely related to the lateral pile movement at that depth and, therefore, the depths of zero soil resistance and zero lateral pile movement do not coincide. Whereas, in plane stress and plane strain conditions simulated with 2–D finite element model, the points of zero soil resistance and zero lateral pile movement are assumed to coincide.

Plane Strain and 3-D Non-linear p-y Response. Figures 9 and 10 show the p–y curves at depths of 1.3 m and 5 m, respectively, obtained from plane strain and 3–D non-linear analyses. p–y curves with and without gap formation are compared with those from load test results.

In plane strain p–y curves (Figures 9 and 10), the reduction in ultimate soil resistance due to gap formation is about 26% at a depth of 1.3 m and about 20% at a depth of 5 m. Thus, the effect of a gap on soil resistance is more significant at 1.3 m depth than at 5 m depth. This is because the lateral stresses are lower at 1.3 m depth than at 5 m depth and, therefore, at 1.3 m depth the formation of a gap starts at a lower load than that at 5 m depth.

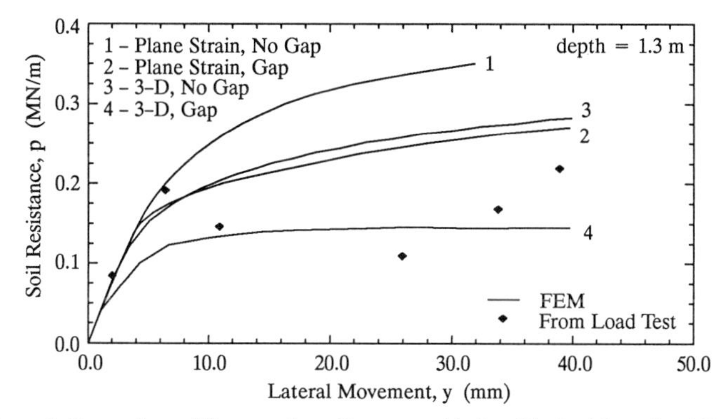

Figure 9. Comparison of Computed p–y Response with that Obtained from Load Test Results (depth = 1.3m). [Test Results from Dunnavant, 1986]

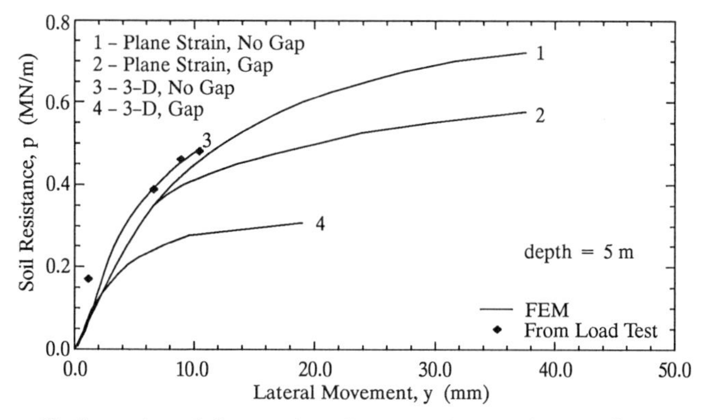

Figure 10. Comparison of Computed p–y Response with that Obtained from Load Test Results (depth = 5 m). [Test Results from Dunnavant, 1986]

The reduction in ultimate soil resistance due to gap formation is more significant in the p–y curves obtained from 3–D analyses than that in the p–y curves obtained from plane strain analyses. At a depth of 1.3 m, the reduction in ultimate soil resistance due to gap formation in the p–y curves obtained from 3–D analysis is about 48%. At 1.3 m depth, the 3–D p–y curve with gap formation allowed agrees reasonably well with the p–y data obtained from load test results. Whereas, at 5 m depth, the 3–D p–y curve without gap formation allowed agrees well with the p–y data obtained from load test results. These results suggest that no gap formed at a depth of 5 m during the load test.

Load–Deflection Response. Pile–head loads are plotted against the pile–head deflections in Figure 11. The load–deflection curve obtained from 3–D non–linear finite element analysis is compared with the measured load–deflection response. When the formation of a gap is allowed in the analysis, the computed load–deflection curve agrees very well with the measured load deflection response. However, if gap formation is not allowed, a higher pile–head load is predicted for a given pile–head deflection.

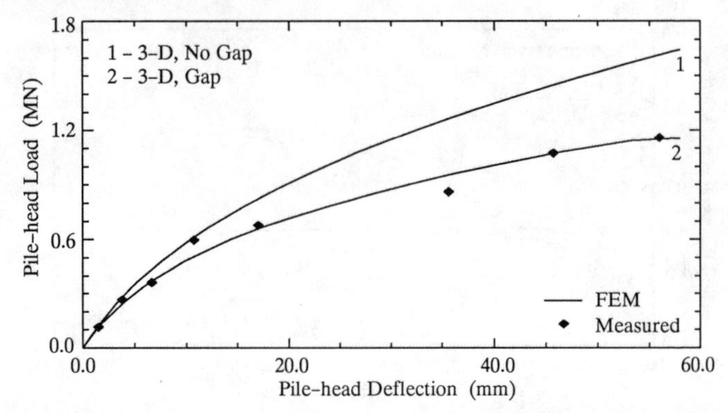

Figure 11. Measured and Computed Load-Deflection response. [Test Results from Dunnavant, 1986]

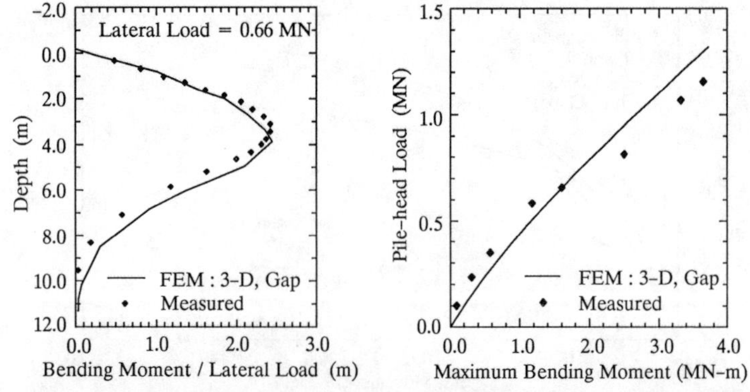

Figure 12. (a) Computed and Measured Normalized Bending Moment along the length of the Pile, (b) Computed and Measured Maximum Bending Moments for various Pile–head Loads. [Test Results from Dunnavant, 1986]

Bending Moment Response. Bending moments in the pile along its length and the maximum bending moments for various pile–head loads, obtained from 3–D non–linear finite element analysis with gap formation allowed, are compared with measured bending moments in Figures 12(a) and 12(b) respectively. The agreement between computed and measured bending moments is reasonably good. The maximum bending moment obtained from the finite element analysis occurs at a depth larger than that of the measured maximum bending moment (Figure 12a). The magnitudes of maximum bending moment for this particular lateral load (660 kN), however, are the same for measured and computed curves. This, however, is not true for other load levels as can be seen in Figure 12(b).

The differences between computed and measured pile behavior (in terms of p-y, load-deflection and bending moment) can be attributed to the following factors :

i) In the load test, cyclic loads were applied at every step. Behavior of the pile under any particular load is influenced to various degrees by cyclic loading under previous loads. In the analysis, all the load increments were applied monotonically.

ii) The pile was driven in place. This installation procedure had an influence on the soil properties. Enough information was not available to quantify this effect and, therefore, it was not taken into account in the analysis.

iii) Most of the soil parameters were obtained from laboratory test results. Soil properties in the field can be different from those obtained from laboratory tests.

SUMMARY AND CONCLUSIONS

An analytical methodology has been developed and used to investigate various aspects of behavior of laterally loaded piles. The soil behavior is modeled using a bounding surface plasticity model and the soil-pile interfaces are modeled using thin-layer isoparametric elements. Two- and three-dimensional non-linear finite element programs incorporating these models have been developed. Pile-soil behavior has been studied in terms of stress distribution around the pile, load-deflection response, p-y response, and bending moments in the pile. Effects of roughness of the pile surface and formation of a gap at the soil-pile interfaces, on the behavior of the pile have been investigated.

Formation of a gap at the soil-pile interface has major influence on the load-deflection response and p-y response of the pile, and on the stress distribution around the pile. From the stress distribution around the pile it has been shown that if tensile cracks form behind the pile due to gap formation, they will form along a line that makes an angle of about 45° with the line of action of the applied lateral load (Figure 5b).

The p-y stiffness at various depths obtained from three-dimensional elastic analysis have been found to be different from those obtained from plane stress and plane strain analyses. Depending on the depth, the p-y stiffness obtained from 3-D analysis can be higher or lower than those obtained from plane stress and plane strain analyses.

Good agreements have been obtained between analytical and measured pile behavior in terms of p-y response, load-deflection response and bending moments in the pile. The methodology presented in this paper has been found to be useful for detailed investigation of the behavior of laterally loaded piles.

REFERENCES

1. A. M. Ananadarajah & Y. F. Dafalias, Bounding Surface Plasticity III : Application to Anisotropic Cohesive Soils, J. Eng. Mech. Div. (ASCE), 112(12), Dec 1986, 1292–1318.
2. F. Baguelin, R. Frank & Y. H. Said, Theoretical Study of Lateral Reaction Mechanism of Piles, Geotechnique, 27(3), Sept. 1977, 405–434.
3. P. K. Banerjee, A. S. A. Stipho & N. B. Yousif, A Theoretical and Experimental Investigation of the Behavior of Anisotropically Consolidated Clay, Chap. 1, Developments in Soil Mechanics and Foundation Engineering, Vol. 2, Ed. P. K. Banerjee & R. Butterfield, Elsevier Appl. Sci. Publishers, England, 1985, 1–41.
4. P. K. Banerjee & N. B. Yousif, A Plasticity Model for the Mechanical Behavior of Anisotropically Consolidated Clay, Int. J. Num. Anal. Meths. in Geomech., 10(5), Sept.-Oct. 1986, 521–541.
5. D. A. Brown, C.-F. Shie and M. Kumar, p-y Curves for Laterally Loaded Piles Derived from Three-dimensional Finite Element Model, Numerical Models in Geomechanics, Ed. S. Pietruszczak & G. N. Pande, Elsevier Appl. Sci. Publishers, 1989, 683–690.

6. M. T. Davison & H. L. Gill, Laterally Loaded Piles in a Layered Soil System, J. Soil Mech. Founds. Div. (ASCE), 89(SM3), 1963, 63–94.
7. C. S. Desai, M. M. Zaman, J. G. Lightner & H. J. Siriwardane, Thin Layer Elements for Interfaces and Joints, Int. J. Num. Anal. Meths. in Geomech., 8(1), Jan.–Feb. 1984, 19–43.
8. T. W. Dunnavant, Experimental and Analytical Investigation of the Behavior of Single Piles in Overconsolidated Clay Subjected to Lateral Loads, Ph.D. thesis, Dept. of Civ. Engng., Univ. of Houston - Univ Park, Aug. 1986, 599pp.
9. T. W. Dunnavant and M. W. O'Neill, Experimental p–y Model for Submerged Stiff Clay, J. Geot. Engng. (ASCE), 115(1), Jan. 1989, 95–114.
10. J. Ghaboussi, E. L. Wilson & J. Isenberg, Finite Element for Rock Joints and Interfaces, J. Soil Mech. Founds. Div. (ASCE), 99(SM10), Oct. 1973, 833–848.
11. R. E. Goodman, R. L. Taylor & T. L. Brekke, A Model for the Mechanics of Jointed Rock, J. Soil Mech. Founds. Div. (ASCE), 94(SM3), May 1968, 637–659.
12. L. R. Herrmann, Finite Element Analysis of Contact Problems, J. Eng. Mech. Div. (ASCE), 104(EM5), Oct. 1978, 1043–1057.
13. M. Hetenyi, Beams on Elastic Foundation, Univ. of Michigan Press, Ann Arbor, Michigan, 1946.
14. A. P. Kooijman, Comparison of an Elastoplastic Quasi Three-dimensional Model for Laterally Loaded Piles with Field Tests, Numerical Models in Geomechanics, Ed. S. Pietruszczak and G. N. Pande, Elsevier Appl. Sci. Publishers, 1989, 675–682.
15. H. Matlock, Correlations for Design of Laterally Loaded Piles in Soft Clay, Proc., Offshore Tech. Conf., 1970, I(1204), 577–593.
16. H. Matlock & L. C. Reese, Generalized Solutions for Laterally Loaded Piles, J. Soil Mech. Founds. Div. (ASCE), 86(SM5), Oct. 1960, 63–91.
17. B. McClelland & J. A. Focht, Soil Modulus for Laterally Loaded Piles, Trans. ASCE, 123, 1958, p. 1049.
18. Z. Mroz & O. C. Zienkiewicz,Uniform Formulation of Constitutive Equations for Clays and Sands, Mechanics of Engineering Materials, Ed. C. S. Desai & R. H. Gallegher, John Wiley and Sons, 1984, 415–449.
19. A. Muqtadir & C. S. Desai, Three Dimensional Analysis of a Pile-group Foundation, Int. J. Num. Anal. Meths. in Geomech.,10(1), Jan.–Feb. 1986, 41–58.
20. M. W. O'Neill & S. M. Gazioglu, Integrated Formulation of p–y Relationships in Clay, A Report to the American Petroleum Institute, Report No. PRAC-82-41-2, Univ. of Houston, April 1984.
21. H. G. Poulos, Behavior of laterally loaded piles : I - Single piles, J. Soil Mech. Founds. Div. (ASCE), 97(SM5), May 1971, 711–731.
22. H. G. Poulos & E. H. Davis, Pile Foundation Analysis and Design, John Wiley and Sons, 1980, 397 pp.
23. J. H. Prevost, Plasticity Theory for Soil Stress-strain Behavior, J. Eng. Mech. Div. (ASCE), 104(EM5), May 1978, 1177–1194.
24. L. C. Reese & W. R. Cox, Soil Behavior from Analysis of Un-instrumented Piles Under Lateral Loading, ASTM, STP 444, 1969, 160–176.
25. L. C. Reese & R. C. Welch, Lateral Loading on Deep Foundations in Stiff Clay, J. Geotech. Eng. Div. (ASCE), 101(GT7), July 1975, 633–649.
26. W. R. Spillers & R. D. Stoll, Lateral Response of Piles, J. Soil Mech. Founds. Div. (ASCE), 90(SM6), June 1964, 1–9.
27. M. Yegian & S. G. Wright, Lateral Soil Resistance - Displacement Relationships for Pile Foundations in Soft Clays, Proc. 5th Offshore Tech. Conf., 1973, 2(OTC1983), 663–676.
28. O. C. Zienkiewicz, B. Best, C. Dullage & K. G. Stagg, Analysis of Non-linear Problems in Rock Mechanics with Particular Reference to Jointed Rock Systems", Proc., 2nd Int. Cong. on Rock Mech., Belgrade, 1970, 3, 501–509.

TIME-DEPENDENT BEHAVIOR OF SHOTCRETE AND CHALK MARL DEVELOPMENT OF A NUMERICAL MODEL

R K Pöttler[1] and T A Rock[2]

ABSTRACT: Numerical methods have proved to be a useful tool for the design of tunnels driven according to the principles of the New Austrian Tunneling Method (NATM).

The usual procedure is to start with a simple conservative model and examine possible tunnel shapes, lining thicknesses, and excavation sequences before construction commences. These numerical models are refined with the help of back analyses based on geotechnical measurements in the course of the excavation works.

The present paper is concerned with the back analysis of the service tunnel excavation on the Channel Tunnel Project. The results were later successfully employed in the design of the further NATM construction works: the running tunnels at Shakespeare Cliff, the tunnels under Castle Hill, and the UK Undersea Crossover Cavern.

INTRODUCTION

The very tight construction program for the UK end of the Channel Tunnel Project required that a rapid versatile method of tunnel construction be employed so as to enable early commencement of Tunnel Boring Machine (TBM) drives and to allow rapid construction of tunnel adits, junctions, and a variety of cross sections. Figure 1 shows the area from which the TBM drives progressed. The NATM was chosen to meet the challenge posed [1, 2] and this choice has been fully justified by the rates of progress achieved.

The art in tunnel lining design is (i) to choose the opening shape so that the ground is not unduly overloaded and (ii) to delay the installation of a stiff lining until ground creep is substantially complete. These requirements are ideally met by the NATM. The use of a

---

1 Dr, ILF Consulting Engineers, Innsbruck, Austria
2 Mott Mac Donald Group, Croydon, England

creeping lining material, e.g. shotcrete, in a creeping but competent ground medium, e.g. chalk marl, has resulted in the design of relatively thin linings especially in comparison with adjacent TBM constructed tunnels where the expanded stiff unyielding lining is installed before the chalk marl has had sufficient time to relax.

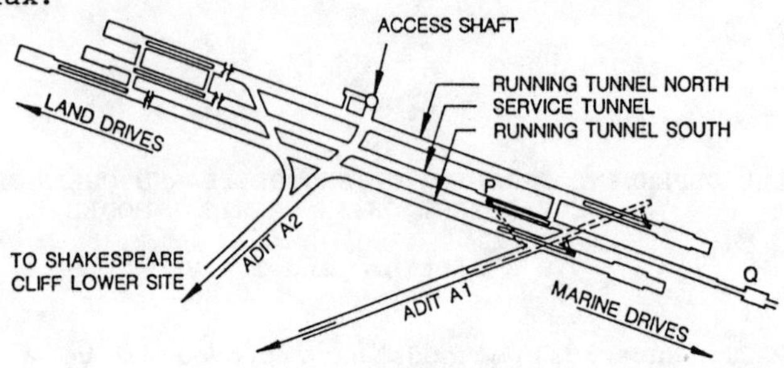

Figure 1 Junction Area: Adit A2 - Main Tunnels [2]

BACKGROUND

In the first four months of 1975, some 250 metres of machine driven segmentally lined service tunnel were constructed as part of the then proposed "Channel Tunnel". This length of tunnel was extensively instrumented and has been continuously monitored since that time. The construction records and instrumented results from this pilot drive were processed as part of the development studies for the present Channel Tunnel Project leading to the production of a lining load versus time curve as shown in Fig.2. From the series of field observations and laboratory creep tests double Kelvin rheological models (DKM) were adopted as best describing the ground-lining interaction for the TBM tunnel drives. Similar models were then adopted as a starting point for the design of the NATM linings.

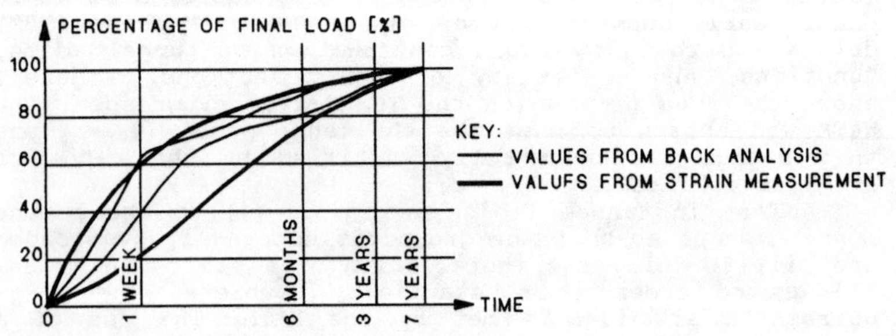

Figure 2 Lining Load vs Time TBM Tunnels

NUMERICAL MODELS

Simple Predictive Models. In the first numerical analyses the ground - structure interaction was modelled in two stages. In the first stage the shotcrete was modelled viscoelastically and the ground was modelled as a time-independent medium with reduced strength capacity. The analysis was then progressed to a second stage in which both the lining and the ground were modelled viscoelastically and the strength capacity of the ground was reduced. This method of prediction though simple proved too conservative when compared with the field observations, and a more sophisticated model was sought.

Model from Back Analysis. Extensive laboratory and field tests combined with numerical investigation had earlier revealed that the behavior of the chalk marl could be best simulated by means of a Double Kelvin Model (Fig.3) wherein the creep strains were driven by deviatoric stresses. The standard FE formulation in 3D is as follows:

$$\dot{\underline{\epsilon}} = \frac{1}{n_1}(\underline{M}^I \underline{\sigma}_1 - E_1 \underline{\epsilon}_1) + \frac{1}{n_2}(\underline{M}^I \underline{\sigma}_2 - E_2 \underline{\epsilon}_2) \qquad (1A)$$

$$\underline{M}^I = \begin{vmatrix} 2/3 & -1/3 & -1/3 & 0 & 0 & 0 \\ & 2/3 & -1/3 & 0 & 0 & 0 \\ & & 2/3 & 0 & 0 & 0 \\ & & & 2 & 0 & 0 \\ & & & & 2 & 0 \\ \text{symm.} & & & & & 2 \end{vmatrix} \qquad (1B)$$

$\underline{\sigma}$ = stress vector, $\underline{\epsilon}$ = strain vector, E_1, n_1, E_2, n_2 = parameters

The relevant parameters derived from 1975 creep tests were determined to be:

Parameter Set 1 (2A)

E_1 = 0.200 GPa E_2 = 0.250 GPa
n_1 = 7 dGPa n_2 = 100 dGPa

Parameter Set 2 (2B)

E_1 = 0.700 GPa E_2 = 0.250 GPa
n_1 = 14 dGPa n_2 = 250 dGPa

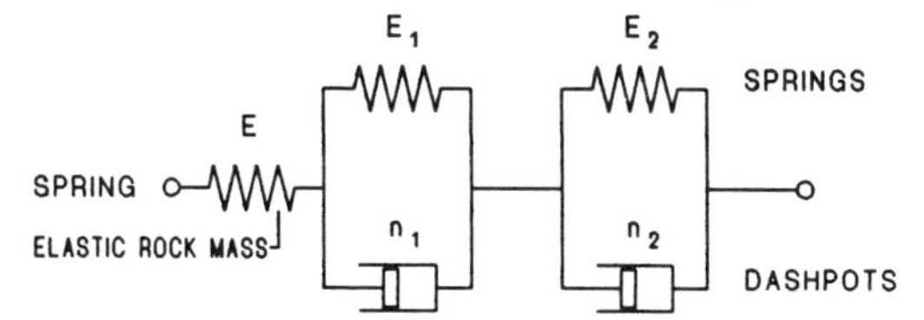

Figure 3 Double Kelvin Model

Reduction of Strength Parameters of Rock Mass. The laboratory tests showed a reduction of the sustainable deviatoric stresses to 70% of their initial value due to long term loading. The following formulae for the decrease in sustainable deviatoric stresses were used, taking account of the fact that approximately 95% of the reduction takes place within 3 months.

$$\sin\phi_2 = \frac{f * \sin \phi_1}{1 - (1 - f) \sin \phi_1} \tag{3A}$$

$$c_2 = \frac{f\, c_1 \cos \phi_1 (1 - \sin \phi_2)}{\cos \phi_2 (1 - \sin \phi_1)} \tag{3B}$$

$$f = 0.7 + 0.3 \exp(- 0.036\, t) \tag{3C}$$

ϕ = angle of friction, c = cohesion, subscript 1 = initial strength parameters, subscript 2 = strength parameters at relevant time, t = time in days

Shotcrete Model. Extensive investigations of the behavior of green shotcrete were described in reference [3].

Creep type behavior is most pronounced during the first hours and days, when the newly installed shotcrete lining is subjected to load resulting from the tunnel driving sequence, particularly 3D face advance effects:

- The shotcrete hardens with age, thus changing its deformation capabilities, especially its ability to creep.
- The creep rate of shotcrete increases hyperlinearly with an increase of deviatoric stresses.

The numerical formulation of the shotcrete model using the Boltzmann theorem reads:

$$\Delta\underline{\epsilon} = \Delta\underline{\epsilon}^{el} + \Delta\underline{\epsilon}^{in} + \Delta\underline{\epsilon}^{th} = \underline{D}^{-1}\Delta\underline{\sigma} + \dot{\underline{\epsilon}}^{in}\Delta t + \dot{\underline{\epsilon}}^{th}\Delta t \tag{4}$$

$\underline{\epsilon}^{th}$ = vector of thermal strain component, $\underline{\epsilon}^{el}$ = vector of elastic strain component, $\underline{\epsilon}^{in}$ = vector of viscoelastic or inelastic strain component, $\underline{D}$ = elasticity matrix, $\underline{\sigma}$ = stress vector

The increase of the modulus of elasticity over time, which is relevant to the elastic deformation, can be expressed using an exponential law [4]:

$$E_t = E_{28} * a \exp(c/t^{0.60}) \tag{5}$$

E_t = modulus of elasticity at time t, E_{28} = modulus of elasticity after 28 days, a,c = parameters, t = time in days

The viscoelastic strain rate for low shotcrete stresses up to 10 MPa is given by Equation(6).

$$\underline{\dot{\epsilon}}^{in}(t) = \frac{1}{3\ B_K} * \exp\left(- \frac{G_K}{B_K} * t\right) * \frac{3}{2} * \underline{M}^I * \underline{\sigma} \quad (6A)$$

$$B_K\ (\sigma_v, t) = B_K^* * t^n * \exp\ (k_2 * \sigma_v) \quad (6B)$$

$$G_K\ (\sigma_v) = G_K^* * \exp\ (k_1 * \sigma_v) \quad (6C)$$

$$\sigma_v^2 = 0.5\ ((\sigma_I - \sigma_{II})^2 + (\sigma_{II} - \sigma_{III})^2 + (\sigma_{III} - \sigma_I)^2) \quad (6D)$$

σ_I, σ_{II}, σ_{III} = principal stresses

G_K^*, k_1, k_2, B_k^* and n are parameters obtained from tests and are applicable to a shotcrete age of 0 to 60 days. Consistent values based on long term creep tests are as follows [3]:

$$G_K^* = 1.312\ \text{GPa} \qquad k_1 = -\ 0.0309 \qquad n = 0.7000$$
$$Bk^* = 17.315\ \text{dMPa} \qquad k_2 = -\ 0.2786 \quad (7)$$

The thermal component influences the results only to a minor extent and can therefore be neglected in engineering practice [4].

NUMERICAL APPROACH OF EXCAVATION PROCESS

The rock-shotcrete interaction near the tunnel face is a function of the non-linear time-dependent excavation process. For a realistic assessment of this problem, time consuming calculations based on the Finite Element Method (FEM) are carried out. The calculation models are 3-dimensional and the time dependence is taken account of by a time-stepping procedure. This approach requires extensive CPU time even on large computer systems and is not very practical in everyday engineering.

Stability analyses are mostly effected by means of 2-dimensional analyses and by simulating the excavation process by three calculation steps. The first step simulates the stress relief ahead of the lining, the second step simulates the stress redistribution after installation of the lining up to the time when the influence of the excavation process on the investigated cross section has become zero. Both calculations are done without a time-stepping procedure. The modulus of elasticity of shotcrete used in the second calculation step is the long term hypothetical modulus. This hypothetical modulus of elasticity (HME) can be determined on the basis of experience and/or back analyses. A more detailed description and derivation of this modulus is given in reference

[6]. A third calculation step takes account of the time-dependent behavior of the shotcrete and, if any, the time-dependent behavior of the rock mass. The calculation is effected by means of a 2-dimensional model using a time-stepping procedure.

The simplified model described is shown in Fig.4: Model A (section I-I) models the stress redistribution due to the excavation process up to the time of installing the shotcrete lining. The unlined opening is modelled and loaded only by part of the total load, which is due to the support of the rock mass ahead of the tunnel face, thus simulating the stress relief ahead of the shotcrete lining.

$$\underline{f}_A = \alpha * \underline{\sigma}_{PRIM} \tag{8A}$$

$$0 < \alpha < 1.0 \tag{8B}$$

$\underline{\sigma}_{PRIM}$ = primary stress vector (in the rock mass)

Model B (section II-II in Fig.4) takes account of the lining and the further stress redistribution. The model is loaded by the remaining part of the total load:

$$\underline{f}_B = (1 - \alpha) * \underline{\sigma}_{PRIM} \tag{8C}$$

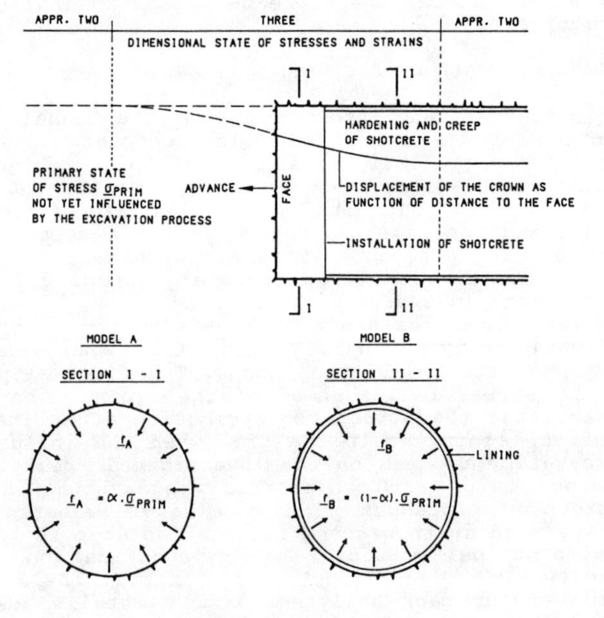

Figure 4 Numerical Model

For calculation step 3 Model B is used again. No further loading takes place.

FIELD MEASUREMENTS

As has already been mentioned, continuous measurements are an essential feature of the NATM. During excavation of the Service Tunnel, the values shown in Fig.5 were measured.

When evaluating the measurements, for the back analysis two periods were differentiated:

Short term (ST): from the moment of excavation up to 14 days after excavation
Long term (LT): from 14 days up to 60 days

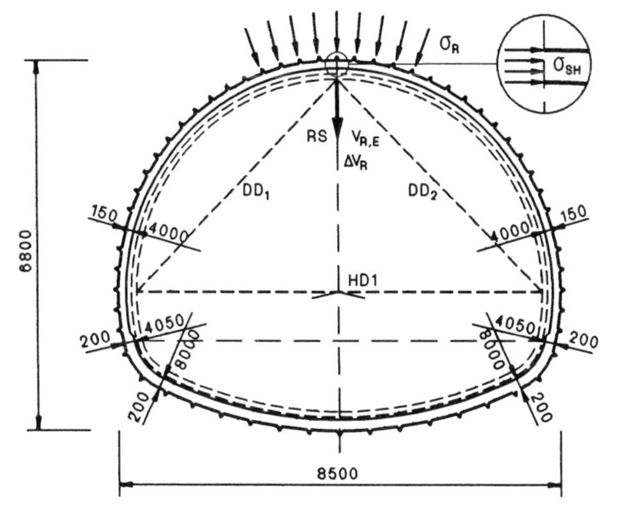

Figure 5 Regular Cross Section of the Service Tunnel

Roof settlements (RS): The roof settlements up to 14 days vary within a range of 2 to 7 mm. The increase in deformations due to long term behavior up to 60 days after excavation amount to approx. 1 to 2 mm. These for roof settlements were derived from the convergence measurements (HD1, DD1 and DD2) as such values are more precise than the direct measurement of roof settlement by levelling.

The short term shotcrete stresses (σ_{SH}) on an average amount to 1 MPa; an increase due to long term behavior has not been observed. The rock mass pressure (σ_R) after 14 days amounted to approximately 0.125 MPa. Here as well, a slight decrease rather than an increase over time is observed. The values mentioned refer to the roof, as the latter is best suited for comparison with the system chosen for the back analysis.

BACK ANALYSIS

Input Parameters and Aims. The input parameters used for the first calculation are best estimate parameters from previous investigations and back analyses.

	Shotcrete:	Rock mass:
Modulus of elasticity	$E_{SH} = 30$ GPa	$E_{RM} = 2.625$ GPa
Poisson's ratio	$v_{SH} = 0.167$	$v_{RM} = 0.30$

The primary stress field was assumed to be isotropic:

$$\sigma_x = \sigma_y = \sigma_z = g * H = 2.36 \text{ MPa}$$

Specific weight $g = 0.0225$ MN/m^3, overburden H = 105 m

The objective was to

- determine the hypothetical modulus of elasticity (HME)
- establish a back analysis of the creep behavior of the rock mass 14 days to 60 days
- predict the creep behavior up to 120 years.

To obtain a better feel for the problem, the influence of creep of shotcrete and that of the chalk marl was dealt with separately.

Influence of HME. For the different values of HME the shotcrete stresses and the additional displacements between 14 and 60 days were computed. For the time after 14 days, the calculation was based on the material law Equations(4), (5) and (6) for shotcrete. A time-dependent behavior of the rock mass was not assumed.

Table 1. Shotcrete stresses and displacements.

	Short term 0 - 14 days			Long term 14 - 60 days		
HME GPa	$V_{R,E}$ mm	σ_{SH} MPa	σ_R MPa	V_R mm	σ_{SH} MPa	σ_R MPa
0.50	4.9	0.6	0.022	0.0	0.2	0.008
1.00	4.9	1.2	0.043	0.1	0.5	0.016
2.50	4.8	2.9	0.106	0.1	1.1	0.039
5.00	4.6	5.6	0.203	0.3	2.0	0.072

$V_{R,E}$ = displacement of shotcrete, σ_{SH} = shotcrete stress, σ_R = rock pressure, V_R = additional displacements from 14 days to 60 days.

The short term shotcrete stresses measured are best reproduced by a model using HME = 1.00 GPa. Due to creep of shotcrete up to 60 days these shotcrete stresses are reduced considerably and the additional displacements are minimal. Using an HME of 5.00 GPa, the short term stresses are by far too high, whereas the stresses computed for the time after 60 days are slightly higher than the actual ones.

Creep of Shotcrete and Rock Mass. In the subsequent calculation set, creep of the rock mass and of the shotcrete was taken into account, giving the following results.

Table 2. Shotcrete stresses and additional displacements after 60 days.

	Parameter set 1 of DKM (Equation 2A)			Parameter set 2 of DKM (Equation 2B)		
HME GPa	V_R mm	σ_{SH} MPa	σ_R MPa	V_R mm	σ_{SH} MPa	σ_R MPa
0.5	10.9	22.2	0.8	9.8	23.5	0.8
1.0	10.6	22.3	0.8	9.8	23.6	0.8
5.0	10.7	22.8	0.8	9.9	24.0	0.9

Conclusion from First Investigations. From the analysis performed, the following conclusion was drawn:

- The behavior up to 14 days is simulated most realistically by an HME of 1.00 GPa.
- There must be creep of the rock mass since the decrease of stresses according to Table 1 was not measured by the pressure cells.
- Creep of the rock mass is certainly not as pronounced as assumed on the basis of the 1975 laboratory test results, as the predicted increase in displacements and stresses overestimate observations (Table 2).
- Due to creep of the shotcrete the stresses are reduced considerably. The higher the original stresses the higher the degree of stabilization.
- As a result of creep of the rock mass the curve of the shotcrete stresses is smoothed. In Table 2 there are only minimal differences between long term shotcrete stresses, whereas these differences were significant in the short term.
- On the basis of the present results of calculations a realistic simulation of the long term behavior of the rock mass can be achieved by varying the parameters E_1, n_1, E_2 and n_2.

Parameter Study on Creep Behavior of Rock Mass. On the basis of the results presented above a parameter study was performed. In a third calculation step, damping parameters n_1 and n_2 were held constant while E_1 and E_2 were varied (Equation(1)). The results are listed in Table 3.

Table 3. Shotcrete stresses and displacements 60 days after excavation. V_R between 14 and 60 days.

E_1 GPa	n_1 dGPa	E_2 GPa	n_2 dGPa	V_R mm	σ_{SH} MPa	σ_R MPa
0.7	7.0	0.25	100	10.6	22.3	0.8
7.0	7.0	2.5	100	3.1	8.9	0.3
10.0	7.0	3.5	100	2.5	7.4	0.3

The additional displacements of approx. 2.5 mm are higher than the measured ones but are within a realistic range. The increase of the rock pressure from 0.04 MPa (Table 1) to 0.3 MPa (Table 3) and that of the shotcrete stresses from 1.2 MPa (Table 1) to 7.4 MPa (Table 3) were not measured on site.

In a further set of calculations the parameters of the DKM were varied by multiplication of the basic parameters.

Table 4. Shotcrete stresses and displacements at 60 days after excavation; V_R between 14 and 60 days

E_1 GPa	n_1 dGPa	E_2 GPa	n_2 dGPa	V_R mm	σ_{SH} MPa
20	7	7.0	100	1.5	4.4
20	14	7.0	200	1.3	4.3
80	56	28.0	800	0.4	1.5

Table 4 shows good agreement between calculation and measurement from the time of excavation up to 60 days. It was generally possible to obtain this coincidence between the numerical model and the measurements, i.e. the same results, with a different parameter combination. In this case the development over time was slightly different, but the final results were not affected.

Based on the above parameters, an extrapolation to 365 days and to 120 years was performed (Table 5). The development of displacements for the last parameter combination up to 120 years is shown on Fig.6. There is only a slight further increase of displacements and stresses after 60 days. Generally, the analyses demonstrate that the increase in stresses and displacements is small in the long term and that the final state will be reached after six months.

Table 5: 365 days and 120 years

				14-365 d		14 d-120 a	
E_1 GPa	n_1 GPa	E_2 GPa	n_2 GPa	V_R mm	σ_{SH} MPa	V_R mm	σ_{SH} MPa
20	7	7.0	100	1.5	4.7	NC	NC
20	14	7.0	200	1.5	5.4	NC	NC
80	56	28.0	800	0.4	1.8	0.4	1.8

NC = not calculated

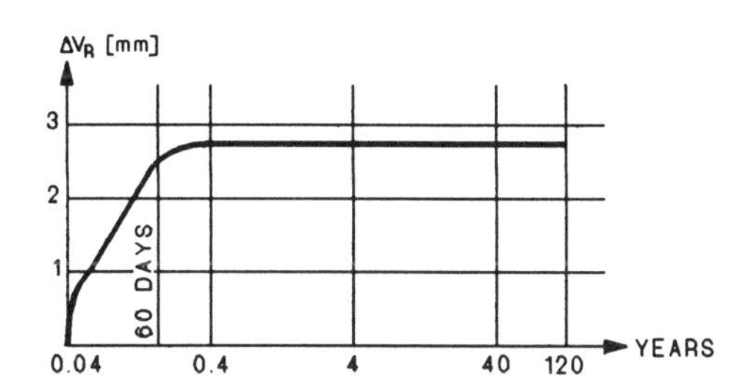

Figure 6 Stresses and Displacements as Functions of Time up to 1 year

CONSEQUENCES FOR THE DESIGN

Originally, a second layer of shotcrete was anticipated in the service tunnel to carry the forces arising from the creep behavior of the rock mass. However, as the calculation revealed, the creep behavior is less pronounced than originally assumed, and the relaxation of the shotcrete matches the additional creep strain imposed by the ground. Therefore the second layer of shotcrete could be omitted.

As creep of the rock mass is of a minor extent, on the basis of this calculation, a simplified viscoelastic model for chalk marl and shotcrete was developed, which allowed further calculations of tunnel sections excavated according to NATM principles in the same geotechnical medium to be performed. This made an economic design for the remaining tunnel sections possible.

The time-consuming calculations carried out were justified because they enabled:

° the existing models to be refined by way of back analyses, which led to substantial cost savings at the construction stage (omission of the 2nd shotcrete

- the design engineer to be provided with information about the actual behavior of the rock mass and the shotcrete, an essential requirement for his understanding of the overall behavior of the tunnel.
- many different possible ground and boundary conditions to be simulated readily.

The consequence of delay or uncertainty in any element of the NATM tunnels on this project were so great as to warrant the most thorough investigation and prediction of ground behavior using the best available tools.

ACKNOWLEDGEMENT

The authors would like to thank TML for their kind permission to publish this paper. They would also like to thank their colleagues who carried out the parameter study and evaluated the geotechnical measurements.

REFERENCES

1. GS Crighton and L Leblond, Tunnel Design, in The Channel Tunnel, Ed T Telford, London/U.K., 1989, pp 95-135.

2. AR Biggart, JRJ King, RD MacKenzie, GA Moore and JA Miles, U.K. Tunnels Construction, in The Channel Tunnel, Ed T Telford, London/U.K., 1989, pp 231- 248.

3. A Petersen, Geostatische Untersuchungen für tiefliegende Regionalbahnen am Beispiel Hannover, Forschungsergebnisse aus dem Tunnel- und Kavernenbau, 12, Ed Technical University of Hannover, 1989.

4. W Weber, Empirische Formeln zur Beschreibung der Festigkeitsentwicklung und der Entwicklung des E-Moduls von Beton, in Betonwerk- und Fertigteiltechnik, 1979, pp 753-759.

5. R Pöttler, Junger Spritzbeton im Tunnelbau: Beanspruchung - Festigkeit - Verformung, Proc. Spritzbetontechnologie, Ed W Lukas & W Kusterle, Innsbruck/Austria, 1990.

6. R Pöttler, Time-Dependent Rock-Shotcrete Interaction. A Numerical Shortcut, accepted for publication in Computers and Geotechnics, 1991, pp 149-169.

ANALYTICAL SOLUTION FOR A COUPLED HYDRAULIC AND SUBSIDENCE MODEL

RAYMOND N. YONG[1], M.ASCE, DA-MING XU[2] AND ABDEL-MOHSEN O. MOHAMED[3]

ABSTRACT: The Trial Function Technique (TFT) is used to obtain the analytical solution for a coupled hydraulic and subsidence model. The solutions for the characteristic hydraulic surface in the vicinity of a pumping well and the average basin subsidence are expressed in the explicit analytical forms. The analytical results are compared with the measured subsidence in the city of Bangkok, Thailand. The analytical results reveal that: a) subsidence is caused mainly by consolidation of the aquitard; b) the hydraulic pressure drop proceeds very slowly in the aquitard compared with that in aquifer; c) the average subsidence in one well system is proportional to $t^{\frac{2}{3}}$.

INTRODUCTION

Subsidence due to underground water pumping (extraction) is related to the phenomenon of consolidation. When water is withdrawn from a well located in the aquifer layer, the hydraulic pressure reduces in the vicinity of the well, appearing as a pear-shaped hydraulic head drop profile. The effective stress increases to compensate the hydraulic pressure drop is a direct result of the deformation-consolidation process, which produces land subsidence (Nutalaya et al., 1986).

[1]William Scott Professor of Civil Engineering And Applied Mechanics, Director, Geotechnical Research Centre, McGill University, Montreal,Canada, H3A 2K6

[2]Research Associate, Geotechnical Research Centre, McGill University, Montreal, Canada H3A 2K6

[3]Research Associate, Geotechnical Research Centre, Adjunct Professor, Civil Engineering Dept. McGill University, Montreal, Canada, H3A 2K6

In view of the requirements in detailed modeling of sub-surface stratification and physical-mechanical properties of a typical multi-layered aquifer-aquitard system, no single mathematical model can be generated to encompass all the complex details. A detailed review is addressed by Sexena (1979). The available analytical methods can be briefly summarized as follows:

a) *Data extension method:* Based on previous data one tries to predict the future. Numerical techniques generally used to extend the data trend are by and large not reliable.
b) *Hydraulic model plus subsidence model:* The hydraulic model is used to obtain piezometric head profiles and the subsidence model is used to calculate the subsidence based on the obtained hydraulic profiles. In this method the coupling effect between the deformation u and hydraulic profiles is neglected.
c) *Aquitard drainage model:* This method regards the subsidence as being mainly due to aquitard drainage. The concept is realistic and is thought to identify the problem. However, the analysis to date (reported in the literature) only addresses subsidence analysis; the piezometric head profile in the aquifer layer is not well treated.
d) *Continuum mechanics method:* This method borrows from principles of solid mechanics. Elastic theory is used to list the deformation equations. It can be used to deal with 2-D or 3-D problems, however, soil properties are not realistically modeled.

This paper presents the analytical solution for characteristic hydraulic surface and average basin subsidence, using a coupled hydraulic and subsidence model based on a two layer geological system. The effects of geometric, physical, and mechanical properties of the soil strata as well as, the external water pumping flux rate are reflected in the analytical solutions. The problem is analyzed for a multi-layer aquifer-acquitard system, with specific attention to the quaternary sediment underlying the city of Bangkok, Thailand. A full description of the urban hydrogeology of the region has been reported by AIT (1982).

MODEL

System Description

Figure 1 shows a schematic representation of double aquitard- aquifer hydrogeologic modeling under well pumping condition. The basic quantities which describe the system are hydraulic excess pressure head, h^*, and deformation, $\bar{u}$. The material properties are permeability, k, and coefficient of volume change, m. The subscript c and s denotes aquitard and aquifer respectively. Due to pumping rate, Q, the pear-shaped hydraulic pressure head is presented in Fig. 2. At any vertical cross section in the aquifer layer it is assumed that h_s^* is constant. Thus, $h_s^* = h_s^*(r,t)$. In the aquitard layer, the excess pressure head is given as $h_c^* = h_c^*(r,z,t)$;where r = radial distance; z = vertical distance, and t = time.

Formulation

1) For aquifer layer:

$$\frac{\partial h_s^*}{\partial t} = C_{vs}\frac{1}{r}\frac{\partial^2 h_s^*}{\partial r^2} \tag{1}$$

where:

$$C_{vs} = \frac{k_s}{m_s} \tag{2}$$

2) For aquitard layer:

$$\frac{\partial h_c^*}{\partial t} = C_{vc}\left[\frac{1}{r}\frac{\partial^2 h_c^*}{\partial r^2} + \frac{\partial^2 h_c^*}{\partial z^2}\right] \tag{3}$$

where:

$$C_{vc} = \frac{k_c}{m_c} \tag{4}$$

The second term of the RHS of Eq. (3) can be neglected due to the fact that the excess pressure head drop in the radial direction is very small in comparison with the head drop in the vertical direction. Thus, Eq. (3) can be reduced to:

$$\frac{\partial h_c^*}{\partial t} = C_{vc}\frac{\partial^2 h_c^*}{\partial z^2} \tag{5}$$

The initial conditions for solution of equations (1) and (3) are given as:

$$h_s^*(r, t = 0) = 0 \tag{6}$$

and

$$h_c^*(r, z, t = 0) = 0 \tag{7}$$

where $t = 0$ denotes the time at which well pumping starts to function. Before $t = 0$, since no water flow is assumed, no excess hydraulic pressure head is developed.

The boundary conditions are given as:

1) on the well's inner surface: using Darcy's Law the following equation is obtained:

$$Q_i = -2\pi r_o H_s k_s \frac{\partial h_s^*}{\partial r} \Big|_{r=r_o} \tag{8}$$

where: Q_i = pumping rate of a ith well, and r_o = radius of the well.

2) at the bottom of the aquifer: it is assumed to be impermeable, i.e.,

$$\frac{\partial h_s^*}{\partial z} \Big|_{z=bottom} = 0 \tag{9}$$

Since h_s^* is a function of z, the above condition is satisfied automatically.

3) at the aquifer-aquitard interface: the hydraulic pressure should be continuous, i.e.,

$$h_c^* = h_s^*(\text{on interface}) \tag{10}$$

When the pressure head in the aquifer drops, i.e., $h_s^* < 0, h_c^*$ will also drop accordingly. At the interface boundary water flow should be continuous, i.e.,

$$k_c \frac{\partial h_c^*}{\partial z} = k_s \frac{\partial h_s^*}{\partial z}(\text{at the interface}) \tag{11}$$

4) at the top of the aquitard: a nonpermeable boundary conditions is assigned,i.e.,

$$\frac{\partial h_c^*}{\partial z} |_{(\text{at top of aquitard})} = 0 \tag{12}$$

5) when r is at infinity, h_c^* and h_s^* should be zero, i.e.

$$h_s^*(r \to \infty, t) = 0$$

and

$$h_c^*(r \to \infty, z, t) = 0 \tag{13}$$

If r extends into a river,the pressure head h_c^* and h_s^*must be assumed to be zero at the r position, since there is no pressure head change.

The subsidence $w(r,t)$, which denotes the vertical downward displacement, is due to consolidation of both the aquitard and aquifer layers. It is given as:

$$w(r,t) = w_s(r,t) + w_c(r,t) \tag{14}$$

where w_s and w_c are the vertical deformations for the aquifer and aquitard layers, respectively. They are given by:

$$w_s(r,t) = \int_0^t \int_0^{H_s} \rho_w g m_s \frac{\partial h_s^*}{\partial t} dz dt \tag{15}$$

and

$$w_c(r,t) = \begin{cases} \int_0^t \int_0^{H_c} \rho_w g m_{cp} \frac{\partial h_c^*}{\partial t} dz dt & (\text{compression/subsidence}) \\ \int_0^t \int_0^{H_c} \rho_w g m_{ce} \frac{\partial h_c^*}{\partial t} dz\ dt & (\text{expansion/rebound}) \end{cases} \tag{16}$$

To evaluate the subsidence (in an engineering sense), it is appropriate to introduce an "average subsidence" concept. Let r_f denote the front of the progressing pressure head. The average compression/subsidence in both aquifer and aquitard are defined as:

$$w_s^*(t) = \frac{\int_{r_o}^{r_f(t)} r w_s(r,t) dr}{\int_{r_o}^{r_f(t)} r dr} \tag{17}$$

and

$$w_c^*(t) = \frac{\int_{r_o}^{r_f(t)} r w_c(r,t) dr}{\int_{r_o}^{r_f(t)} r dr} \tag{18}$$

Note that the $*$ above the w terms indicate average values. The total average subsidence is thus given as:

$$w^*(t) = w_s^*(t) + w_c^*(t) \tag{19}$$

Equations (17) to (19) describe the subsidence model.

TRIAL FUNCTION TECHNIQUE FOR ANALYTICAL SOLUTION

The trial function technique, TFT, is used to determine the following important quantities:

1) Hydraulic pressure head drop in the well - *this varies with time t under a pumping rate Q_i*.
2) the pressure head drop region which progresses with time t.
3) The aquitard layer subsidence process resulting from vertical drainage of the aquitard.

(a) Generalized Coordinates $S_1(t), S_2(t)$ and $S_3(t)$

The problem is initiated at $t = 0$, when well pumping (extraction) begins and Q_i water flux is extracted. The water pressure head drop at the well is denoted by $S_1(t)$, and the disturbed radial distance measured from the wall of the well is denoted by $S_2(t)$ as shown in Fig. 2.

In describing the water pressure head drop $h_c^*(r, z, t)$, a third generalized coordinate $S_3(t)$ is defined. This specifies the water pressure front in the z' direction in the vicinity of the well in the aquitard layer. $S_2(t)$ advances in the horizontal direction in the aquifer while $S_3(t)$ is vertical in the aquitard. It should be noted that $S_1(t), S_2(t)$, and $S_3(t)$ are positive definite. Because k_s is much larger than k_c, $S_2(t)$ progresses much faster than $S_3(t)$. From the excess pressure head profile shown in Fig. 2, there is no drop in water pressure at position G upon arrival of $S_2(t)$. In the region of $r > S_2(t)$, the aquifer in this region is not yet disturbed at time t. In region of $0 < r < S_2(t)$, the larger the value of r, the lesser is the pressure head drop. $S_1(t), S_2(t)$ and $S_3(t)$ have physical unites of length and are identified as characteristic lines.

To determine the characteristic shapes of $h_s^*(r', t)$ and $h_c^*(r', z', t)$ and the nature of their variations, a trial function technique is used. The following trial function is used to describe $h_s^*(r', t)$.

$$h_s^*(r',t) = \begin{cases} -S_1(t)\left[1 - 2\frac{r'}{S_2(t)} + \frac{r'^2}{S_2^2(t)}\right] & \text{for} \quad 0 \le r' \le S_{2(t)} \\ 0 \quad \text{for} \quad r' \ge S_{2(t)} \end{cases} \tag{20}$$

where:$r' = r - r_o$, and r_o is the radius of the well. It should be noted that the negative sign appears in Eq. (20) is attributed to the decresae in pressure head since $S_2(t)$ is defined as a positive quantity.

The above trial function satisfies the following:

1) At $r' = 0$ (well's position), $h_s^*(0,t) = -S_1(t)$. This is the pressure head drop at the well.
2) At $r' > S_2(t), h_s^*(r',t) = 0$. This characterizes the diffusion process associated with wave front $S_2(t)$.
3) At $r' = S_2(t), h_s^*(S_2(t),t)$ and its derivative with r' are continuous.

In a similar manner, the trial function for $h_c^*((r',z',t)$, can be specified as follows:

$$h_c^*(r',z',t) = \begin{cases} -S_{1(t)}\left[1 - 2\frac{r'}{S_2(t)} + \frac{r'^2}{S_2^2(t)}\right]\left[1 - 2\frac{z'}{S_3(t)} +. \frac{z'^2}{S_3^2(t)}\right] & \text{for} \quad 0 \le z' \le S_3(t) \\ 0 \quad \text{for} \quad z' > S_3(t) \end{cases} \tag{21}$$

where, z' is vertical distance measured from the interface boundary between the aquifer and aquitard layer.

This function satisfies:

1) At $z' = 0$, the water pressure head is continuous, i.e.,

$$h_s^*(r,t) = h_c^*(r,o,t)$$

2) Along the z' direction, the pressure profile is smooth.
3) If $z' > S_3(t), h_c^*(r',z',t) = 0$. This refers to the undisturbed aquitard soil region.
4) At $z' = S_3(t)$, $\frac{\partial h_c^*}{\partial z'} = 0$ i.e. slope is continuous from left to right.

b) $S_1(t), S_2(t)$ and $S_3(t)$ Ordinary Equations

In using Eqs. (20) and (21), three equations for $S_1(t), S_2(t)$ and $S_3(t)$ can be obtained by considering: (1) the pumping relation given by Eq. (8); (2) water mass conservation in the aquifer, and (3) water mass conservation in the aquitard. After lengthy mathematical manipulation, the following equations are obtained:

1)**Pumping relation**

$$\frac{S_1(t)}{S_2(t)} = \frac{Q_i}{4\pi r_o H_s k_s} \tag{22}$$

2) **Water mass conservation in aquifer**

$$\frac{1}{6}\pi H_s \rho_w g m_s S_1(t) S_2^2(t) + \eta Q_i + \frac{\pi}{3} k_c \frac{S_1(t) S_2^2(t)}{S_3(t)} = Q_i \tag{23}$$

where η represents a percentage of water flux loss in Q_i due to leakage from the aquitard to the aquifer layer.

3) **Water mass conservation in an aquitard**

$$\frac{d}{dt}\left[S_1(t) S_2^2(t) S_3(t)\right] = 6 \frac{k_c}{\rho_w g m_c} \frac{S_1(t) S_2^2(t)}{S_3(t)} \tag{24}$$

The initial conditions for $S_1(t), S_2(t)$ and $S_3(t)$ are given by:

$$S_1(0) = S_2(0) = S_3(0) = 0 \tag{25}$$

The total amount of water volume pumped out from $t = 0$ to t is $\int_0^t Q_i dt$. Hence,the average subsidence $w^*(t)$ in the pressure drop region due to pumping is given by:

$$w^*(t) = \frac{\int_0^t Q_i(t) dt}{\pi S_2^2(t)} \tag{26}$$

If $Q_i(t)$ is constant, then

$$w^*(t) = \frac{Q_i t}{\pi S_2^2(t)} \tag{27}$$

or,

$$w^*(t) = \frac{1}{18(1-\eta)} \rho_w g S_1(t) \{3 H_s m_s + S_3(t) m_c\} \tag{28}$$

c) **Some Approximate Solutions for $S_1(t), S_2(t)$ and $S_3(t)$**

In considering a typical field situation, $\frac{m_c}{m_s} \approx 10^{-4}$ and assuming the pumping rate $Q_i(t)$ is constant, the approximate solutions for $S_1(t), S_2(t)$ and $S_3(t)$ solved from Eqs. (22)-(25) are obtained as follows:

$$S_1(t) = \frac{Q_{io}}{4\pi r_o H_s k_s} \left[\frac{72(1-\eta) r_o H_s k_s}{(6 k_c m_c \rho_w g)^{\frac{1}{2}}}\right]^{\frac{1}{3}} t^{\frac{1}{6}} \tag{29}$$

$$S_2(t) = \left[\frac{17(1-\eta) r_o H_s k_s}{(6 k_c m_c \rho_w g)^{\frac{1}{2}}}\right]^{\frac{1}{2}} t^{\frac{1}{6}} \tag{30}$$

$$S_3 = \left[6 \frac{k_c}{\rho_w g m_c}\right]^{\frac{1}{2}} t^{\frac{1}{2}} \tag{31}$$

The average subsidence $w^*(t)$ can be calculated from (26) as:

$$w^*(t) = \frac{m_c}{18(1-\eta)}\rho_w g S_1(t)\, 3H_s(\frac{m_s}{m_c}) + \left[6\frac{k_c}{\rho_w g m_c}\right]^{\frac{1}{2}} t^{\frac{1}{2}} \tag{32}$$

The first term on the RHS is the contribution to the subsidence from the aquifer layer, while the second term is the contribution from the aquitard. The second term is the dominant term, since $\frac{m_s}{m_c} \approx 10^{-4}$.

NUMERICAL RESULTS

Using some typical values for the situation reported in Bangkok, Thailand (AIT, 1982), i.e.,:$Q = 1350000 m^3/$ day; well's number = 11000; Q_i = $122.73 m^3$/day; $r_o = 0.1m; H_s = 100m; k_s = 100m$/day; $k_c = 10^{-4}m$/day; $\rho_w g m_c = 10^{-2}/m; \rho_w g m_s = 6.7.10^{-6}/m$, and $\eta = 0$.
We obtain from the developed model:

$$S_1(t) = 8.06\left[\frac{t}{year}\right]^{\frac{1}{6}} m$$

$$S_2(t) = 8.25\left[\frac{t}{year}\right]^{\frac{1}{6}} m$$

$$S_3(t) = 4.68\left[\frac{t}{year}\right]^{\frac{1}{2}} m$$

and

$$w^*(t) = 0.021\left[\frac{t}{year}\right]^{\frac{2}{3}} m$$

Fig. 3 shows the calculated pear-shaped pressure head profiles in the aquifer at different times t. Fig. 4 shows that the calculated subsidence varies with time t. The average yearly subsidence is around 0.0775 m. Actual field data obtained shows the yearly subsidence to be around 0.05-0.10 m. The estimated yearly subsidence of 0.0775 m falls within the actual range.

SUMMARY

The approximate analytical relationship developed to calculate water table drop $S_1(t)$, pressure head front $S_2(t)$ in the horizontal direction in the aquifer and pressure head front $S_3(t)$ in the vertical direction in the aquitard have been applied to the multi-well pumping situation to permit calculation of the global subsidence in Bangkok. From the results obtained, the features are observed:

1) Subsistance is caused mainly by consolidation of the aquitard due to the fact that m_s/m_c is very small.

2) $S_3(t)$ progresses in the aquitard in a typically slow fashion in comparison with $S_2(t)$ in the aquifer.

3) The basic average subsidence in one well system is proportional to $t^{\frac{2}{3}}$

The effects of geometric, physical and mechanical properties of the soil substrate are reflected in the $S_1(t)$, $S_2(t)$ and $S_3(t)$ explicit formulae.

ACKNOWLEDGEMENT

This study was supported under a Grant from the International Development Research Centre (IDRC) of Canada, Grant No. 3-P-1044- 02.

REFERENCES

1. AIT, Final Report on Investigation of Land Subsidence Caused by Deep Well Pumping in the Bangkok Area, Asian Inst. of Tech. Div. of Geotech. and Trans. Eng., Report NEB-1982-005, Bangkok, 1982.
2. P Nutalaya, S Chandra, & AS Balasubramaniam, Subsidence of Bangkok Clay due to Deep Well Pumping and Its Control through Artificial Recharge, pp. 727- 744, Land Subsidence, edited by A.I. Johnson, Laura Caregognin and L. Ubertini.
3. SK Saxena, Evaluation and Prediction of Subsidence, Int. Conf. on Evaluation and Prediction of Subsidence, ASCE, 1979. pp. 594.

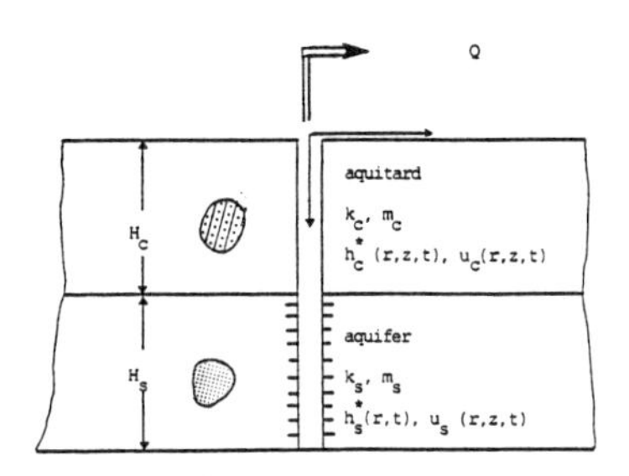

Fig. 1 Double aquitard-aquifer hydrogeologic modelling under well pumping condition

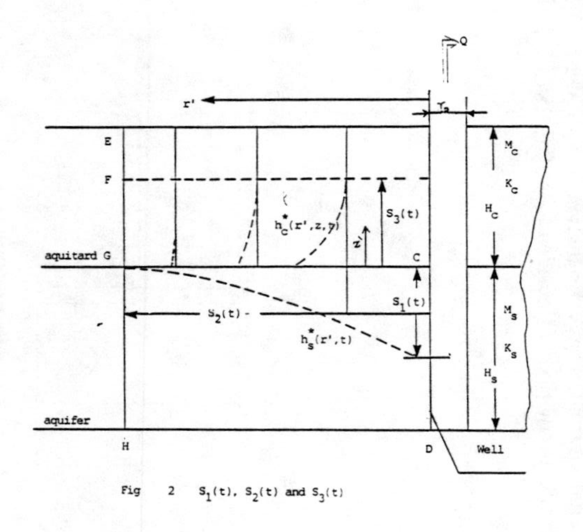

Fig 2 $S_1(t)$, $S_2(t)$ and $S_3(t)$

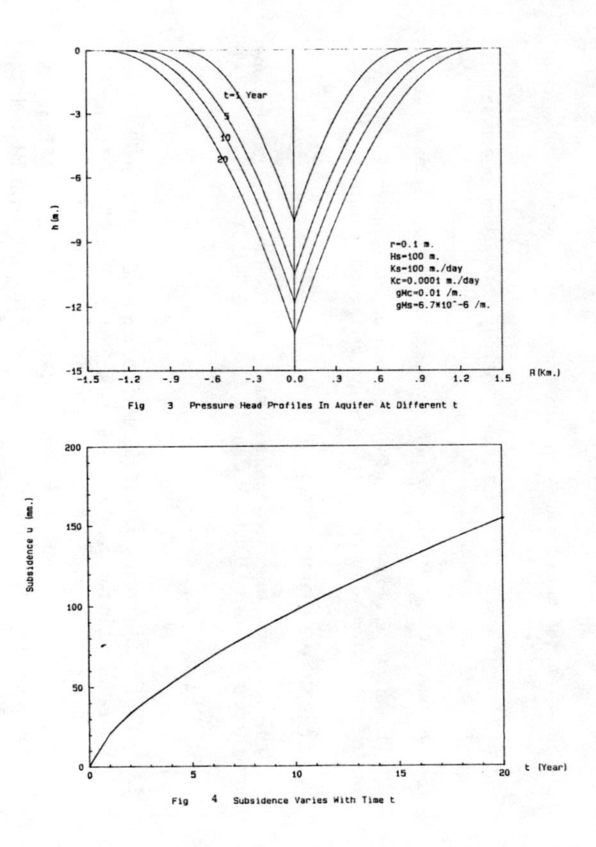

Fig 3 Pressure Head Profiles In Aquifer At Different t

Fig 4 Subsidence Varies With Time t

LIMIT STATES DESIGN - THE EUROPEAN PERSPECTIVE

Niels Krebs Ovesen[1] and Trevor Orr[2]

ABSTRACT: Codes of Practice are described as documents for quality assurance of the design of civil engineering structures. Also described are the basic components of a code of practice and the concept of limit states design. The use of the system of partial factors of safety for safety evaluation is introduced and the use of this system in Danish geotechnical practice for more than forty years is described. A presentation is given of the ongoing effort to introduce in the member countries of the European Communities the so-called Eurocodes which are codes of practice for the design of structures of concrete, steel, timber and masonry. There will also be a Eurocode 7, Geotechnics to deal with the design of foundations and retaining structures. All Eurocodes will be based on the limit states concept and they will make use of partial factors of safety. A comparison between the European and the North American concept of load and resistance factor design is presented.

CODES AS QUALITY ASSURANCE DOCUMENTS

At the onset some definition on the basic concepts of Codes and Standards may be appropriate.

According to the International Standard Organization, ISO (1986), a Code of Practice is a document that recommends practices or procedures for the design, manufacture, installation, maintenance or utilization of equipment, structures or products. It may also be worth noting that the Oxford Advanced Learner's Dictionary of Current English defines a code as "a system of rules and principles that has been accepted by society or a class or group of people".

ISO defines a Standard as a document established by consensus and approved by a recognized body that provides for common and repeated use, rules, guidelines or characteristics for activities or their results aimed at the achievement of the optimum degree of order in a given context.

In the civil engineering context a Code of Practice describes recommended design practice by defining the requirements which are aimed at reaching a reasonable technical level of quality. The code requirements are normally expressed as so-called functional requirements and they are based on scientific/technical principles. The

---

1 - Dr, Man. director, Danish Geotechnical Institute, DK-2800 Lyngby, Denmark.

2 - Dr, Senior lecturer, Trinity College, Dublin, Ireland.

codes of practice will normally avoid standardising certain methods or procedures of design and construction.

Thus it is emphasized that a code of practice in its concept deviates from a standard. A code of practice aims at obtaining a specific technical level of quality while a standard aims at a specific degree of order in a given context.

THE BASIC COMPONENTS OF A CODE

A code of practice for geotechnical design comprises a set of provisions, compliance with which will ensure a reasonable technical quality for common and routine foundations and earth works. Generally speaking, there are two types of code provisions.

One type may be termed "prescriptive measures". They consist of advice on conventional and generally conservative details in the design and specification of control of materials, workmanship, protection and maintenance procedures.

The other type of provisions will normally be formulated as design calculation procedures. There are several components in such calculation procedures, and in Figure 1 an attempt has been made to illustrate these components by means of an example.

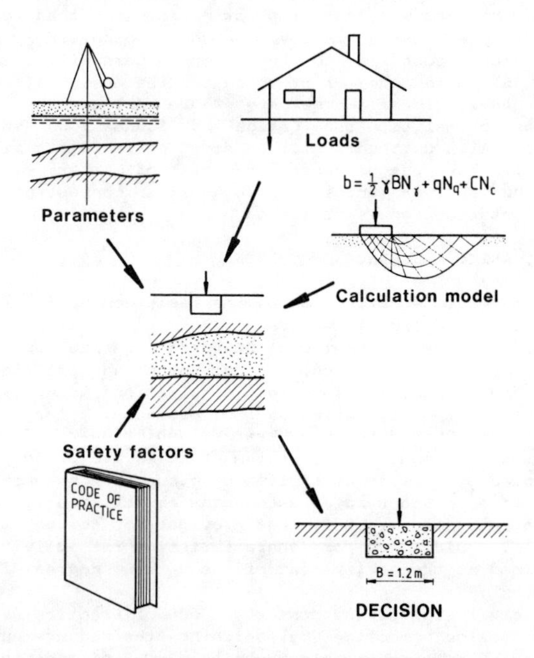

Figure 1. Components of Geotechnical design

Figure 1 illustrates the problems facing the designer of a foundation of a building. In order to design the footings of such a building, the geotechnical engineer may use a code of practice to determine the following four components of the design: Loads, Soil parameters, Calculation procedure and Safety elements.

The loads on the structure consist of the weight of the structure and live loads due to fittings and furnishings, persons, snow, wind, etc. Let us consider as an example the live load on the floors in office buildings. Investigations indicate that this live load in most office buildings will be in the actual range of .2 to .5 kN/m^2. For the design, however, a typical code may specify this live load to be assigned a so-called characteristic value of 2-3 kN/m^2. Therefore it is important to understand that there is more than one answer to the question: Which live load on floors is the correct one to be used in the design of office buildings ? The answer depends on the entirety of the design in which the live load is going to be used.

The shear strength parameters of the soil in question will be determined either from field tests, from element tests in the laboratory, or from empirical relations between the shear strength parameters and the standard classification parameters. Let us consider as an example the use of triaxial tests to determine the angle of shearing resistance for sand. Several questions now arise: Which diameter and which height/diameter ratio should be used for the sample? Should rough or smooth surfaces be used on the top and bottom platens? Which cell pressure should be applied ? All these questions have to be answered in order to determine the test procedure and to evaluate the test results with regard to the angle of shearing resistance.

The calculation procedure used for the design of footings against failure is based normally on the plasticity theory. The bearing capacity for the footing is often determined from the Terzaghi bearing capacity formula. Even though this formula is widely accepted and applied, a large number of questions arise concerning the N_γ value, the shape, depth and inclination factor(s) etc. A comparison by Malcharek et al. (1981) has demonstrated that for $\varphi = 30°$ a range of N_γ values between 8 and 20 can be found in codes of practice from the eight countries: Czechoslovakia, Denmark, Federal Republic of Germany, France, German Democratic Republic, Poland, USA, and USSR.

The safety factors to be applied to the bearing capacity problem may often be specified in a code of practice. Factors of total safety between two and three will normally be considered adequate. However, this also raises a number of questions. Should the safety factor be applied on the load or on the material strength ? Should the same safety factor be used in an effective stress and in a total stress analysis ? etc., etc.

From the above discussion it appears that quite a number of questions have to be answered in order to design a footing. Each code of practice will answer the various questions in different ways.

It is thus important to understand that a code of practice at its very best represents a fine balance between the four components mentioned above. A code of practice cannot be judged on the basis of an isolated evaluation of just one component. A code of practice is not "scientific" by nature. It does not represent "the truth" about the matter in question. It represents a tool by means of which decisions can be made regarding the design of a geotechnical structure, such as a foundation or a retaining wall.

A good code proves its value if it works in practice; and that means if the right decisions are made by means of the code. Right decisions mean designing structures which are sufficiently safe on the one hand and which are economical on the other. In a perhaps too oversimplified manner, it could be said that a good code leads to a situation where only a few structures fail from time to time.

LIMIT STATES DESIGN

In order to ensure an adequate technical quality of a geotechnical structure it is necessary that the structure as a whole and the various parts of the structure fulfil certain fundamental requirements of stability, stiffness etc. during construction and throughout the design life of the structure. In a code of practice the fundamental requirements should be expressed in specific terms as performance criteria.

Whenever a geotechnical structure or a part of a geotechnical structure fails to satisfy one of its performance criteria, it is said to have reached a "limit state". In a code based on "the limit states method" each limit state is considered separately in the design, and its occurrence is either eliminated or shown to be sufficiently improbable.

In civil engineering design it is general practice to distinguish between "ultimate" and "serviceability" limit states. Ultimate limit states involve safety: loss of static equilibrium or rupture of a critical section of the structure. Serviceability limit states involve the inadequacy of the structure to satisfy required standards of: settlement, deformation, utility, appearance, comfort etc.

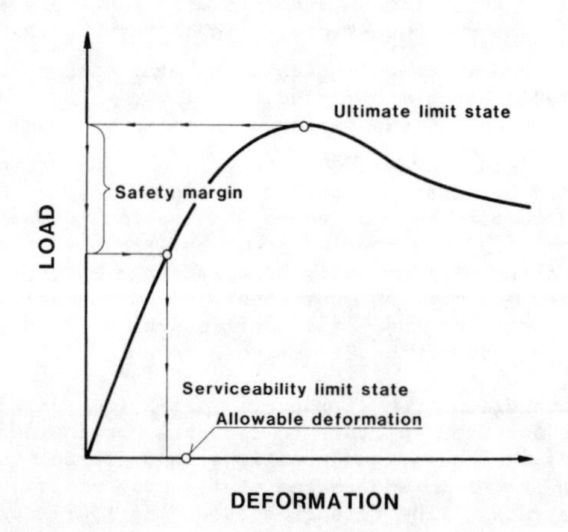

Figure 2. The concept of Limit States Design

The concept of Limit States Design is illustrated in Figure 2 which attempts to emphasize the characteristic difference between the two limit states:

- the structure is prevented from reaching an ultimate limit state by introducing a safety margin expressed by one or more dimensionless numbers and

- the structure is prevented from reaching a serviceability limit state by introducing a constraint (basically in the unit of length) on the movement of some part of the structure.

In geotechnical limit states design two main classes of limit states are considered:

- An Ultimate Limit State at which
  - either a mechanism is formed in the ground
  - or a mechanism is formed in the structure or severe structural damage occurs due to movements in the ground.

- A Serviceability Limit State at which deformation in the ground will cause loss of serviceability in the structure.

Relevant serviceability limit states include settlements which affect the appearance or efficient use of the structure or cause damage to finishes or non-structural elements or vibrations which cause discomfort to people or damage to the content of the building.

In practice it may often be difficult to know which type of limit state will govern the design, and it is therefore often necessary to investigate the ultimate as well as the serviceability limit state.

Whether the ultimate or the serviceability limit state will be decisive in any given case will depend on a number of factors, such as the type of superstructure, the type of soil as well as the dimensions of the foundation and the load acting on it.

THE SYSTEM OF PARTIAL FACTORS OF SAFETY

One important role of a code of practice is to indicate how risks should be dealt with through the introduction of adequate safety margins.

In 1953 Brinch Hansen, in order to introduce safety margins into geotechnical design, proposed the principle of partial factors of safety for the design of foundations and earth retaining structures. Brinch Hansen applied the principle to geotechnical structures in his book on "Earth Pressure Calculation" (1953), and he proposed numerical values for the various partial safety factors. The system was rapidly accepted in foundation engineering practice in Denmark, and from around 1955 virtually all foundations and retaining structures in Denmark were designed according to the new principle.

The numerical values of the factors proposed by Brinch Hansen have undergone minor changes during the past 35 years. In the following a description of the method will be given as it is used in the Danish Code of Practice for Foundation Engineering today, Dansk Ingeniør Forening (1984).

Design values (indices d) of loads are found by multiplying the corresponding characteristic values (indices c) by the respective partial factors:

$$G_d = \gamma_g G_c \qquad Q_d = \gamma_q Q_c \tag{1}$$

where G refers to dead loads, and Q refers to variable loads.

In most foundation engineering problems the dead loads are known with considerable accuracy. Small variations may occur in unit weights and dimensions, and for this reason it might be considered appropriate to use a partial factor of e.g. 1.05. However, not all dead loads will have unfavourable effects, and for those which are favourable, it would be unsafe to multiply by 1,05. On the contrary, these loads should be divided by 1,05. In this way things can get rather complicated, and it is really not worthwhile to accept such complications for the sake of a margin as small as 5%, taking into consideration all the uncertainties involved in soil strength etc. Consequently, according to Danish tradition, all dead loads of structures and of soils are given the partial coefficient $\gamma_g = 1.0$.

Hydrostatic water pressures are known with the same accuracy as dead loads when the water table is given, and for this reason it is logical to use a partial factor of unity for hydrostatic water pressures. Moreover, any other value would lead to similar complications as for dead loads, because the uplift force on a soil mass or a structural element is part of its effective weight.

For a variable load the partial factor $\gamma_q = 1.3$ is normally used. However, in the case of more than one variable load acting, the various loads are normally combined in the following way:

$$S_d = \gamma_g G + \gamma_q Q_1 + \sum_{i>1} \gamma_q \psi Q_i \tag{2}$$

Where

G is the characteristic value of permanent load

Q_1 is the characteristic value of one of the variable loads

Q_i $(i>1)$ are characteristic values of the other variable loads

γ_g and γ_q are the partial load factors

ψ is a load combination factor

For most variable loads a value $\psi = 0.5$ is used. Q_1 is selected to give the most critical combination of loads; if necessary, several alternative loads should be tested to find the most critical. The idea behind the load combination concept is that it is very unlikely that all variable loads will act with their full design value at the same time.

Design values of strength parameters are found by dividing the corresponding characteristic value by the respective partial factor:

$$\tan \varphi_d = \frac{\tan \varphi}{\gamma_\varphi} \qquad c_d = \frac{c}{\gamma_c} \tag{3}$$

According to the Danish tradition a value $\gamma_\varphi = 1.2$ is used. Similarly, a partial factor $\gamma_c = 1.5$ is used in the case of stability or earth pressure problems, while $\gamma_c = 1.8$ is used in the case of bearing capacity problems of footings.

For the bearing capacity of piles or anchors $\gamma_b = 2.0$ is used to obtain the design value of the bearing capacity in the case when the bearing capacity has been found by a geostatic calculation or from a pile driving formula. In the case when load tests have been performed, a value $\gamma_b = 1.4$ is used for the piles tested while $\gamma_b = 1.6$ is used for the other piles. The partial factors given for piles are valid only for the bearing capacity determined by the strength of the soil and not for the pile material.

A special problem exists in geotechnical engineering concerning the selection of characteristic properties for the strength parameters c_c and φ_c of soil and rock. The ground displays a large range of material behaviour, and many different testing techniques are appropriate in order to measure or infer the required material parameters. However, very often it is not possible to obtain a sufficiently large number of test results to derive a characteristic value using formal statistical methods. In geotechnical engineering characteristic values of soil and rock parameters are therefore normally based on careful assessment of the range of values which might govern the field behaviour during the lifetime of the structure. This assessment must take account of geotechnical and other background information, such as relevant data of previous projects and the results of field and laboratory measurements. For parameters for which the relevant values in the field are well established with little uncertainty, the characteristic value may be taken as the best estimate of the value in the field. Where there is greater uncertainty, the characteristic value is somewhat more conservative.

According to the principles of Limit States Design, the design criterion is simply to design for equilibrium in the design limit state of failure. The design criterion could be expressed in the following way:

$$R_d \geq S_d \tag{4}$$

The design resistance effect R_d, which in the case of the design of a footing is the design ultimate bearing capacity of the footing, is calculated on the basis of the design soil parameters defined by Equations (3). S_d is the design load effect calculated on the basis of the principles underlying Equation (2).

The use of partial factors of safety requires special attention with regard to the derivation of design values of earth pressures. The magnitude and direction of earth pressure depends on the material properties of the soil. As a consequence, the Danish tradition has been to calculate design values of earth pressures by introducing partial factors of safety for the material properties and not for the earth pressures as such.

It should finally be stressed that the information given above on the use of partial factors of safety relates exclusively to the ultimate limit states. In the case of serviceability limit states it is normal Danish practice to use unity for the partial factors on the material side, and at the same time to use partial factors equal to unity on the loads, which should then be taken in their frequent combination.

THE EUROCODE SYSTEM

Close economic cooperation was established among the countries in Western Europe during the seventies. As a consequence, among others, a need for harmonised technical specification for civil engineering works arose.

The work on Eurocodes started about 1976. The Eurocodes are intended as common unified design codes for different load bearing structures constructed from different materials. They are intended for common use in all member states and are based on the Limit States Design concept. They are established with the view of promoting the free movement of goods and services within the member states of the European Community - the so-called Internal Market - by removing obstacles arising from differing rules for calculation and design.

The following Eurocodes (EC's) are under preparation:

EUROCODE 1 : Basis of Design and Actions (Loads) on Structures
EUROCODE 2 : Concrete Structures
EUROCODE 3 : Steel Structures
EUROCODE 4 : Mixed Structures of Steel and Concrete
EUROCODE 5 : Timber Structures
EUROCODE 6 : Masonry
EUROCODE 7 : Geotechnics
EUROCODE 8 : Structures in Seismic Zones

The Eurocodes presently being drafted are mainly aimed at the design of buildings. Additional parts will be established with respect to the design of other structures and civil engineering works, such as bridges, masts, chimneys etc.

The responsibility for drafting and issuing the Eurocodes has recently been transferred from the Commission of the European Communities to CEN, the Committee Europeenne de Normalisation, which comprises the 12 community countries as well as the 6 EFTA countries. When the Eurocodes have been issued as European Standards - so-called EN's - they will replace national standards. This replacement will be compulsory. The introduction of Eurocodes as EN's is expected to take 4 to 8 years.

EUROCODE 7: GEOTECHNICS

In 1980 an agreement was reached between the Commission of the European Communities and the International Society for Soil Mechanics and Foundation Engineering (ISSMFE), according to which the society should undertake the drafting of a model code which could be adopted as Eurocode 7. The ISSMFE established an ad hoc committee for this task, which produced a draft model for Eurocode 7 (1987).

In 1988 the Commission of the European Communities established a small Drafting Panel, which was given the task of redrafting the 1987 version of the model code into the Eurocode format.

Eurocode 7 will contain the following chapters:

- Introduction
- Basis of design
- Geotechnical categories
- Geotechnical data
- Fill, dewatering and ground improvement
- Spread foundations
- Pile foundations
- Retaining structures
- Embankments and slopes
- Supervision of construction, monitoring and maintenance

Eurocode 7 is being written in the Limit States Design format, and is based on partial factors of safety. However, since there is only limited geotechnical experience with the partial safety factor format, a good deal of calibration is needed before definitive numerical values of the partial factors of safety can be established.

A number of features of Eurocode 7 deserves further mentioning.

Depending on the character of the individual clauses, distinction is made in all Eurocodes between "principles" and "application rules". The principles comprise general statements and definitions for which there is no alternative as well as requirements and analytical models for which no alternative is permitted unless specifically stated. The application rules are generally recognized rules which follow the principles and satisfy their requirements. It is permissible in the Eurocodes to use alternative design rules different from the application rules given, provided it is shown that the alternative rules accord with the relevant principles and are at least equivalent with regard to the strength, stability, serviceability and durability achieved by the structure.

In Eurocode 7 three "geotechnical categories" are defined in order to establish minimum requirements for the extent and quality of geotechnical investigations, calculations and construction control checks. The following factors shall be taken into consideration when determining which geotechnical category is appropriate to each particular design situation:

- nature and size of the structure,
- special conditions with regard to its surroundings (neighbouring structures, utilities, vegetation, etc.),
- ground conditions,

- groundwater situation,
- regional seismicity,
- influence of the environment (hydrology, surface water, subsidence, seasonal changes of moisture, etc.).

Geotechnical Category 1 includes small and relativly simple structures for which it is possible to ensure that the functional requirements will be satisfied on the basis of experience and qualitative geotechnical investigations and with no risk for property and life.

Geotechnical Category 2 includes structures for which quantitative geotechnical data and analyses are necessary to ensure that the functional requirements will be satisfied but for which conventional procedures of design and construction may be used.

Geotechnical Category 3 includes very large or unusual structures, involving abnormal risks or unusual or exceptionally difficult loading conditions and structures in highly seismic areas. Eurocode 7's specification for Geotechnical Category 2 forms the lower limits for the extent and quality of the necessary investigations and calculations but apart from this no detailed code requirements have been formulated for Geotechnical Category 3.

THE EUROPEAN VERSUS THE NORTH AMERICAN CONCEPT

The concept of Limit States Design with the use of partial factors of safety has developed differently in North America. In comparison with Equation (4), the North American concept may be expressed by the equation

$$\phi R_n \geq \Sigma \gamma_i F_i \quad (5)$$

where R_n is the nominal resistance determined by means of a design calculation model, e.g. a bearing capacity calculation procedure. ϕ is a resistance factor ($\phi < 1$) that accounts for uncertainties in resistance. Its value depends on the variability in strength and the statistical differences between the design models and experimental data.

$\Sigma \gamma_i F_i$ is equal to the summation of the factored load effects for a given load condition. The right-hand side of Equation (5) thus is quite similar in concept to the right-hand side of Equations (2) and (4).

The difference between the European and the North American concept lies in the resistance side as illustrated in Figure 3. Take the design in the ultimate limit state of a footing as an example. According to the North American tradition, the bearing capacity is calculated using soil strength parameters that have not been factored, while the bearing capacity according to the European tradition is calculated as a design value using factored values of shear strength parameters.

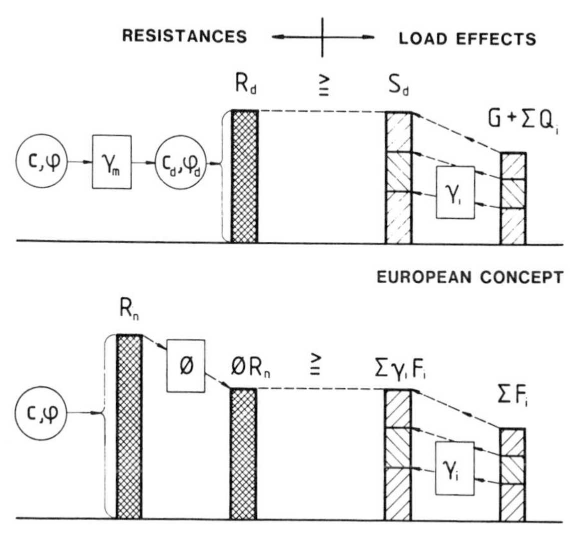

Figure 3. Concepts of partial factors

According to Duncan et al. (1989) different values of load factors are to be used for different loads. The selection of load factors is based on the perceived level of uncertainty in each load. Loads with small uncertainties are assigned load factors slightly greater than unity, while loads with larger uncertainties are assigned larger load factors.

Different resistance factor values are used for different types of resistance and failure mode. Selection of the value of resistance factor Ø is based on the uncertainty involved in the estimation of resistance and depends on the quality of the data and the method of calculation. Resistances with small uncertainties are assigned values of Ø which are only a little less than unity, while resistances with larger uncertainties are assigned somewhat lower values of Ø. For the resistance factor Ø numerical values of between .5 and .85 are normally recommended.

CONCLUDING REMARKS

The following is an attempt to draw some general conclusions on the use of Codes and Limit States Design:

- A Code of Practice represents a fine balance between the four components entering into the design process: Material parameters, loads, calculation methods and safety elements. A Code of Practice is not "scientific" by nature. It represents a tool for making rational design decisions.

- The Limit States Concept represents a logical design principle. It is not in itself a radically new method compared to earlier design practice, but it represents a clear formulation of some widely accepted principles.

- The system of partial factors of safety is today widely used in Codes of Practice for structural design. During the years to come it will find increasing use in geotechnical engineering for the evaluation of the risk of failure and collapse. It may represent a useful tool for the design of traditional and routine geotechnical structures, but it is not a universally applicable system, which can readily be used with fixed numerical values for all geotechnical structures.

- With increased sophistication in design and more stringent requirements from society regarding the performance of structures, it is likely that the estimation of deformations will become more important in the design process. For this reason, the use of sufficiently large factors of safety in stability and bearing capacity calculations to restrict deformations to tolerable values will no longer be acceptable. Instead, geotechnical engineers will have to consider the limit states of collapse and, particularly, excessive deformation separately and carry out separate and increasingly more realistic calculations for these different limit states.

- Closer economic cooperation among countries, particularly in Europe, has shown the need to regulate different activities and the advantages of harmonised technical specifications. In Europe the Eurocodes are being introduced to set the requirements to be met in engineering design. The Limit States approach provides the kind of international design framework needed to ensure that a reasonable technical quality is achieved in geotechnical design.

REFERENCES

1. J Brinch Hansen, Earth Pressure Calculation, The Danish Technical Press, Copenhagen, 1953, 271 p.
2. Dansk Ingeniørforening, DS 415 Code of Practice for Foundation Engineering, (Available in Danish and in English), The Danish Technical Press, Copenhagen, 1984, 64 p.
3. J M Duncan, C K Tan, R M Barker & K B Rojiani, Load and Resistance Factor Design for Bridge Foundations, Proc. Can. Geot. Soc. Symposium on Limit States Design, Toronto, 1989, 47-63.
4. Eurocode 7, Draft Model for Eurocode 7, unpublished document available through the Danish Geotechnical Institute, Lyngby, Denmark, 1987, 216 p.
5. ISO, Guide 2 for General terms and their definitions concerning standardization and related activities, The International Standard Organization, 1986, 47 p.
6. K Malcharek & U Smoltczyk, Vergleich Nationaler Richtlinien fur Berechnung von Fundamenten, Baugrundinstitut, Mitteilung 16, Stuttgart, 1981, 46 p.

CALIBRATION OF LOAD FACTOR DESIGN CODE FOR HIGHWAY BRIDGE FOUNDATIONS

Kamal B. Rojiani[1], Phillip S. K. Ooi[2], and Chia Kiang Tan[3]

ABSTRACT: The probabilistic framework for the new load factor design code for highway bridge foundations being developed for AASHTO is presented and the methodology for calibrating the new code against existing allowable stress design criteria is described. The procedure consists of performing a reliability analysis to determine reliability indices for existing designs, selecting target reliability indices based on the results of the reliability analysis, and then determining performance factors that are consistent with the selected target reliability indices. Results of the code calibration for driven piles and spread footings are presented. The new design criteria are being developed for driven piles, spread footings, drilled shafts, and retaining walls.

INTRODUCTION

Load factor design was adopted by the American Association of State Highway and Transportation Officials (1) for the design of bridge superstructures. However, the AASHTO specification does not include provisions for load factor design of bridge foundations. Allowable stress design is currently the only method available in AASHTO for the design of foundations. The use of two fundamentally and philosophically different design methods - load factor design for superstructures and allowable stress design for foundations - has resulted in duplication of design effort since two sets of loads must be considered, and has hindered the widespread adoption of load factor design. To remove this inconsistency, AASHTO initiated a research program, administered by the National Cooperative Highway Research Program, to develop a rational load factor design method for bridge foundations (NCHRP Project 24-4). This paper presents the probabilistic framework and calibration procedure that forms the basis for the development of load factor design for highway bridge foundations.

PROPOSED CODE FORMAT

The proposed code is based on the limit states concept with load and resistance factors obtained using reliability theory. Limit states are conditions under which a

---

1 - Associate Professor
2 - Graduate Research Assistant
3 - Graduate Research Assistant
Charles E. Via Department of Civil Engineering,Virginia Polytechnic Institute and State University Blacksburg, Virginia 24061

structure or structural element fails to fulfill its design requirements. Two types of limit states are considered in the proposed code: a) ultimate limit states and, b) serviceability limit states. Ultimate limit states are limit states that pertain to structural safety. Examples of ultimate limit states for foundations include overturning, sliding and ultimate bearing capacity failure. Serviceability limit states are the limiting conditions affecting the function of the structure under expected service conditions. They include conditions which may restrict the intended use of the structure. Examples of serviceability limit states are excessive settlement, excessive vibrations, local damage, deterioration, and cracking.

The basic design objective is to keep the probability of a limit state being exceeded below a certain acceptable value. Safety is ensured by verifying that the design resistance is less than the effect of the design loads. The proposed load factor design code is written in a semi-probabilistic format using load and resistance factors. The safety condition is expressed as:

$$\phi R_n > \Sigma \ \gamma_i Q_i \qquad (1)$$

where R_n is the nominal resistance determined through engineering analysis, ϕ is the performance (or resistance) factor that accounts for uncertainties in resistance, ϕR_n is the factored or design resistance, Q_i is the load effect due to nominal load i, γ_i is the load factor corresponding to Q_i and $\gamma_i Q_i$ is the factored load effect. For a satisfactory design the factored nominal resistance should exceed the sum of the factored load effects for a particular limit state. The above design inequality must be satisfied for all applicable load combinations and limit states.

Load factor design is capable of providing more nearly uniform risk levels since it employs several factors instead of one global factor of safety. The load and resistance factors are obtained by a calibration procedure based on reliability theory in which risk levels implied in existing working stress design criteria are determined from statistical analysis of existing data, or where that data is not available, by matching the results of proven design methods.

Load Factors. To make the design of the foundation compatible with the design of the superstructure, the same loads and load factors as given in the current AASHTO load factor design criteria for superstructures are being adopted for the design of foundations. In other words, Section 3 of the current AASHTO code (1) is being adopted with as few changes as possible. This will simplify the design process considerably since the designer will only need to consider one set of loads for the design of both the superstructure and the foundation. In checking serviceability limit states, one or more additional load combinations will be required.

Performance Factors. The safety check requires that the factored resistance should exceed the sum of the factored load effects for each limit state. There are at least two ways in which the factored resistance can be specified. In the first format, an overall "performance factor" ϕ is applied to the resistance side of the design equation. With this approach, the factored resistance for a particular limit state is given by ϕR_n. This is the format followed by ACI (3) for concrete design and by AISC (2) for steel design. The advantages of this approach are its simplicity and familiarity to many designers. With this approach it is possible to include the consequences of failure in the performance factor for that limit state.

The second approach employs partial resistance factors which are applied directly to the individual variables in the resistance equation. In this approach, partial factors are applied to the individual soil strength properties such as cohesion (c) and angle of internal friction (ϕ_f). These factors are typically specified only once and are used for all ultimate limit states. The second approach is more sophisticated, since the partial resistance factors are related directly to the parameters which are the source of variability in strength. An advantage of this approach is that it allows for a more precise calibration over a wide range of soil types and can thus potentially lead to a more uniform reliability. This format has been adopted by the Danish Foundation Code (8) and was also initially adopted by the Ontario Highway Bridge Design Code (12). However, in the Ontario Bridge Design Code different partial factors were used for the same soil parameters for different limit states. A disadvantage of the partial factor approach is that it is not consistent with resistance factors on structural elements which consider overall performance such as bending strength rather than material properties such as concrete compressive strength and yield strength of steel.

The first of the methods described above is being used in the code being developed. This will result in a new code for foundations that is compatible with the existing AASHTO load factor design specifications for bridge superstructures. Values of performance factors for four major foundation types - spread footings, driven piles, drilled shafts, and retaining structures will be specified. Performance factors are being developed for both ultimate and serviceability limit states. For each foundation type, different values of performance factor will be specified depending on the limit state, strength prediction method and the method of determining soil strength. For example, different values of performance factors will be specified for bearing capacity estimated using soil resistance determined using Standard Penetration Tests (SPT), Cone Penetration Tests (CPT), and laboratory tests on undisturbed samples.

CODE CALIBRATION

Code calibration is the process of assigning values to code parameters such as the load and resistance factors. There are several approaches that can be used in calibrating a design code. Codes may be calibrated by judgment, fitting, optimization or a combination of these approaches (10). In this study, a reliability based procedure for code calibration was used. The basic procedure for estimating load and resistance factors consists of the following steps:

1. Estimate the level of reliability implied in the current allowable stress design specifications
2. Based on the above reliability indices, select target reliability indices
3. Determine performance factors that are consistent with the selected target reliability indices.

Several design codes have been calibrated with this basic philosophy. These include the Building Code of Canada (11), the Ontario Highway Bridge Design

Code (12), and the AISC Load and Resistance Factor Design specifications for steel design (3). The process of calibration with the existing allowable stress design criteria ensures a proper design evolution and avoids drastic deviations in designs by the new procedure from existing designs. Instead of specifying a new level of reliability, the new design procedure is based on risk levels implied in the current allowable stress design criteria. Calibration is an important step in development of load and resistance factors. It gives an important overview of the level of reliability underlying current design practice, and ensures that designs obtained with the new specifications will be similar to those obtained using existing procedures.

Reliability Analysis. Risk levels implied in the allowable stress design criteria were determined by computing reliability indices for a range of load ratios and resistance parameters. The reliability analysis is based on the first-order second-moment method. (4,7). For each limit state, a limit state function g() can be written in terms of two variables, R and Q, where R represents the resistance and Q the load effect. For the simple case where R and Q are lognormally distributed random variables, the limit state function can be written as g() = ln(R/Q). Failure occurs when g() < 0. The probability of failure, p_f, is given by

$$p_f = \Pr[\, g(\,) < 0\,] = \Pr[\, \ln(R/Q)\,] < 0 \qquad (2)$$

The probability of failure can be determined from the statistics of R and Q.

Instead of specifying a probability of failure, a common approach is to express reliability in terms of a safety or reliability index, β. The reliability index represents the distance measured in standard deviations between the mean safety margin and the failure limit. For lognormally distributed R and Q, the reliability index can be determined from

$$\beta = \left[\frac{\ln(\overline{R}/\overline{Q}) \quad (1 + V_R^2) / (1 + V_Q^2)}{\sqrt{\ln[\,(1 + V_R^2)\,(1 + V_Q^2)\,]}} \right] \qquad (3)$$

where $\overline{R}$ and $\overline{Q}$ are the mean values and, V_R and V_Q the coefficients of variation (standard deviation divided by mean value) of R and Q. In general, R and Q are functions of other random variables and the limit state function may be non-linear. However, efficient numerical computation procedures have been developed for the computation of the reliability index (6).

To evaluate the level of reliability implied by the current allowable stress design specifications, a database of statistics of load and resistances was assembled. The results of existing work on the statistical analysis of loads, load effects and soil properties was used to assemble this database. The information includes coefficient of variation of the design parameters as well as bias factors that relate nominal values to mean values. Factors considered in the determination of resistance statistics include uncertainties in soil properties, uncertainty due to quality of information and site investigation procedures, and uncertainties in the strength prediction model. The loads considered include dead load, live load, and environmental loads. For the determination of the statistics of the total load effect, it is assumed that the total load effect is the sum of the individual load effects from dead, live and other environmental loads.

Resistance models were developed for each of the four major foundation types: driven piles, drilled shafts, spread footings and retaining walls. For each foundation type, several ultimate and serviceability limit states are being considered. For example, for spread footings the limit states considered include ultimate bearing capacity failure, sliding failure, and excessive settlement. Uncertainties in soil properties and the resistance model were evaluated for each case. The sources of uncertainty considered include model error, inherent spatial variability, and systematic error in soil strength prediction. The analysis of resistance statistics also includes a consideration of the uncertainty in estimating soil strength using different methods such as SPT, CPT and laboratory tests on undisturbed samples. Details of the uncertainty analysis are presented in Ref. (5).

Target Reliability Indices. Risk levels implied by the different procedures for predicting the capacity of a foundation were evaluated. For example, for spread footings the risk levels implied by semi-empirical methods and by rational theory were evaluated. For driven piles, risk levels implied by three rational methods: the α-method (15), the λ-method (16) and the β-method (9) (not related to the safety index β), and two in situ methods, SPT and CPT, were computed. Based on the computed reliability indices, target reliability indices were selected. The selected target reliability indices were as follows: 3.5 for spread footings, 2.0 for driven piles and 2.5 for drilled piers. These target reliability indices form the basis for the determination of performance factors.

Performance Factors. Since it was decided to keep the same load factors as are prescribed in the existing AASHTO load factor design specifications for superstructures, the only code parameters that needed to be evaluated are the performance factors. If the loads consist of dead and live loads, the load factors recommended by AASHTO are 1.3 for dead load and 2.17 for live load. Performance factors were obtained for a range of dead load to live load (D/L) ratios and design parameters. The effect of various design parameters on the computed performance factors was studied and based on this, a single value of performance factor was selected so as to maximize the uniformity of safety index values for a range of conditions. It should be noted that the final values of performance factors will not be based solely on the results of the reliability-based calibration, but will also incorporate the judgment and experience of the research team, consultants and the review panel.

There are several situations in foundation design for which reliable statistical data is not available for computing reliability indices. An example of which includes retaining walls. The lack of full scale tests makes it difficult to obtain accurate statistics regarding prediction and modeling uncertainties. For these cases, performance factors were determined using judgment, and comparison with current AASHTO specifications. The performance factors will be adjusted so that similar designs are obtained as in the current allowable stress design code. The main objective is to transfer the experience with the performance of the old code to the new code. The effectiveness of the new code parameters is being tested against conventional design practice to ensure that the results are reasonable.

DRIVEN PILES

The methods for designing pile foundations selected for calibration include three rational methods for predicting skin friction of piles in clay, the α-method (15), the β-method (9) and the λ-method (16), and two in situ methods which are based on SPT and CPT data. The α-method which is a total stress method, and the λ-method which is a mixed method, both require the knowledge of the undrained shear strength of the clay. The β-method however, is an effective stress method.

Rational Methods. Statistics for the three rational methods were based on the study by Sidi (14) who analyzed the results of numerous pile load tests in cohesive soils. Sidi separated the statistics of the α-and λ-methods into two distinct groups: the first group, called Type I clay, is for clays with undrained shear strengths, S_u, less than 1000 psf and the second group, called Type II clay, is for clays with S_u greater than 1000 psf.

Factors contributing to the uncertainty in pile capacity include the following:

1. Model error where there may be an overall bias in the prediction method.
2. Time and reconsolidation in the case of clays. The capacity will change depending on the time between pile installation and load testing because of the time dependent properties of clay.
3. Inherent spatial variability. Soil properties are known to be correlated between any two points. The uncertainty associated with soil property at a point is larger than if the soil property was evaluated over a certain distance because of the averaging effect. The effect of inherent spatial variability can be included through the use of random field theory.
4. Systematic error in soil strength which accounts for the bias and repeatability of the tests used for measuring soil strength parameters.

The α-and λ-methods are calibrated based on undrained shear strengths measured using laboratory UU tests or field vane tests.

In Situ Methods In situ tests have gained popularity especially as a means of estimating shear strengths of sandy soils. The standard penetration test (SPT) is the most commonly used in situ test in the United States for sandy soils. However, many variations of the test equipment such as type of hammer, length of drill rods, type of sampling spoon, rate of impact and type of drill rods, give rise to different energy levels imparted into the soil (13). Although many of these effects may be minimized by standardizing the test, the variability in the equipment and procedure cannot be completely eliminated .

Reliability Indices for Driven Piles The variation of reliability index versus pile length for the five methods is shown in Figs. 1 and 2. Fig. 1 is for a short span bridge (span = 60 ft) and Fig. 2 is for a long span bridge (span = 250 ft). Based on the results of the reliability analysis, the following observations are made:

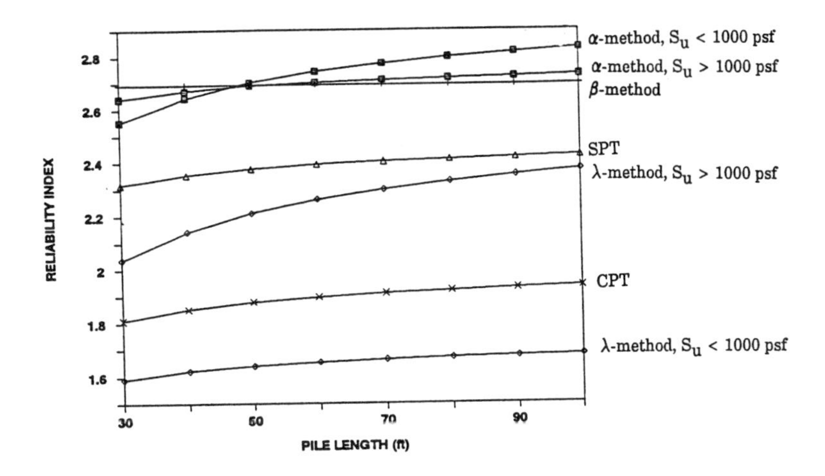

Figure 1 Reliability Indices for Driven Piles, Span = 60 ft (18 m)

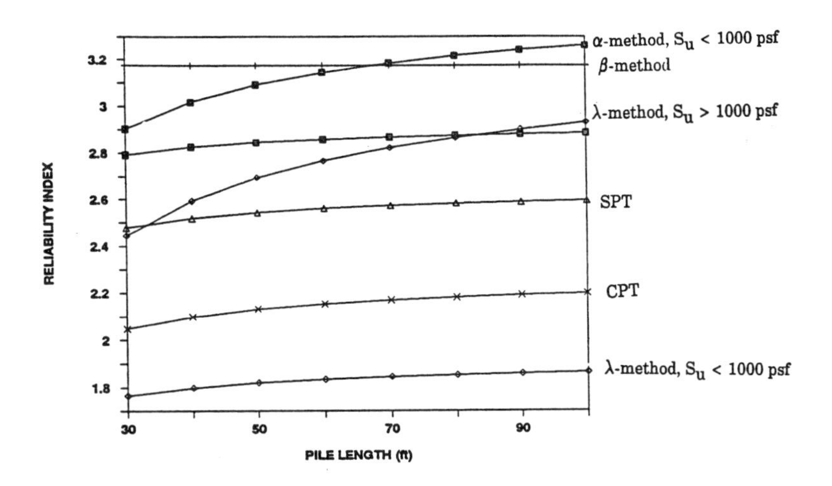

Figure 2 Reliability Indices for Driven Piles, Span = 250 ft (75 m)

1) The reliability index increases with span length. This is due to the fact that the D/L ratio for long span bridges is larger than that for short span bridges. Since the uncertainty in dead load is less than the uncertainty in live load, the total uncertainty in the load decreases as the D/L ratio increases.
2) Reliability indices for driven piles vary between 1.6 and 3.3 for short span bridges. For long span bridges, reliability indices vary between 2 and 3.3

It should be noted that the α, β and λ methods for predicting pile capacities in clay and the CPT method for predicting pile capacities in sand all employ a factor of safety of 2.5. The factor of safety employed with the SPT method is typically 4.

Performance Factors for Driven Piles Performance factors obtained using the calibration process described above with a target reliability index of 2.0 are shown in Table 1. Recommended ϕ values for the five methods are also given in Table 1 These values are based on judgment, experience, as well as the results of the reliability-based calibration procedure. Studies have shown that among the three rational methods for predicting pile capacities in clays, the following is judged to be the order of accuracy of the methods in descending order: (1) the α-method, (2) the λ-method, and (3) the β-method. The values chosen reflect this trend. The calibration results indicate that the performance factor for the β-method is high (0.70). However, the β-method when applied by different engineers can give widely divergent estimates of pile capacity and therefore a lower value of ϕ (0.5) was recommended. The results of the calibration for the in situ methods indicate that the factor of safety of 4 used for the SPT method results in a very conservative estimate of pile capacity. The recommended ϕ factor for the CPT method is 0.55 while for the SPT method a value of 0.45 is recommended. The CPT method is more reliable than the SPT method because of the continuous nature of the measurements made during a CPT test afforded a higher reliability (performance factor) as compared to the SPT. Thus, a higer value of performance factor is recommended for the CPT method than for the SPT method.

Table 1 Performance factors for driven piles

Pile Length	α-method		β-method	λ-method		CPT	SPT
	Type I	Type II		Type I	Type II		
30 ft.	0.70	0.87	0.70	0.49	0.57	0.54	0.46
100 ft.	0.75	0.90	0.70	0.51	0.62	0.57	0.49
Selected	**0.70**		**0.50**	**0.55**		**0.55**	**0.45**

SHALLOW FOUNDATIONS

The ultimate limit states considered for shallow foundations include ultimate bearing capacity failure and sliding failure. Uncertainties associated with the estimation of ultimate bearing capacity and sliding resistance were estimated from a statistical analysis of available data. Bias factors for the strength prediction model were estimated by comparing measured values with predicted values (5). Since different methods may be used to determine a specific soil property, and different models can be used to calculate soil bearing capacity and sliding resistance, the reliability inherent in existing designs is expected to vary with the method used to estimate soil property and the model used to predict soil response.

Ultimate Bearing Capacity Reliability indices associated with existing designs with respect to ultimate bearing capacity of shallow foundations on sands and clays are given in Table 2. The reliability indices range between 3 and 6. These values are compatible with risk levels associated with the design of superstructures.

Results of the reliability analysis show that current design procedures do not provide a uniform level of safety, despite the fact that the safety factors used in conventional design do take into account the different levels of uncertainty associated with the determination of soil properties and strength prediction models. For example, the reliability index for ultimate bearing capacity on sand for semi-empirical procedure based on CPT is smaller than that estimated using SPT results, even though it is well known that the cone penetration test is often a better means for estimating soil property. This suggests that the safety factor of 4 adopted for the procedure based on SPT results may be somewhat conservative when compared with the value of 2.5 used in the procedure based on CPT data.

Table 2 Reliability Indices for Bearing Capacity Failure of Footings

Soil Type	Prediction Model	Safety Factor	Reliability Index	
			D/L = 3	D/L = 10
Sand	Semi empirical procedure based on SPT data	4.0	4.1 - 4.2	4.2
	Semi-empirical procedure based on CPT data	2.5	3.2 - 3.3	3.2 - 3.3
Clay	Rational theory with S_u estimated from lab UU tests	2.5	3.6 - 5.6	3.6 - 5.4
	Rational theory with S_u estimated from field vane tests	2.5	3.6 - 5.3	3.7 - 4.9
	Rational theory with S_u estimated from CPT data	2.5	2.7 - 4.4	2.7 - 4.3

Sliding Failure. Design of footings against sliding failure involves consideration of lateral earth pressures. Prediction of lateral earth pressures is often a formidable task. Its magnitude depends on many factors including soil type and density, stress history, and wall movements. If the wall undergoes sliding failure, the earth pressure acting on it is reduced to the minimum active value. In almost all cases, design earth pressures are higher than the minimum actual value. In the case of sand and gravel backfills, this occurs because conservative values of ϕ_f are used in design. In the case of clayey backfills, this occurs because the values of equivalent fluid pressure used in design are higher than the minimum active earth pressures. There is no precise way to evaluate these biases, and probably no single value can be applied to all conditions. However, for purposes of illustration, the writers believe that it is reasonable to consider a value of bias factor (= minimum active pressure/design pressure) equal to 0.6.

Reliability indices for sliding capacity of footings on sand and clay based on current design practice are summarized in Table 3. As was the case with ultimate bearing capacity, reliability indices are method dependent. With the exception of precast footings on sand, reliability indices are smaller than 3 for a nominal safety factor of 1.5. Reliability indices corresponding to a nominal safety factor of 2.0, range between 3.3 and 4.6. The reliability index values are relatively insensitive to footing width and to the ratio of dead to live load.

Table 3 Reliability Indices for Sliding Failure of Footings

Soil Type	Footing Type	Basis for estimating ϕ_f	Safety factor	Reliability Index
Sand	Cast -in-place	CPT data	1.5	2.4
	Concrete	CPT data	2.0	3.3
		SPT data	1.5	2.4
		SPT data	2.0	3.3
	Precast	CPT data	1.5	3.3
	Concrete	CPT data	2.0	4.2
		SPT data	1.5	3.3
		SPT data	2.0	4.2
Clay	$S_u < 0.5$ *normal	Lab UU Tests	1.5	2.9-3.2
	pressure	Lab UU Tests	2.0	3.8-4.2
		Field Vane	1.5	2.3-2.4
		Field Vane	2.0	3.2-3.4
		CPT Data	1.5	2.6-3.0
		CPT Data	2.0	3.4-4.0
	$S_u > 0.5$ *normal		1.5	3.2
	pressure		2.0	4.2

Performance Factors for Spread Footings Based on the results of the reliability analyses, a target reliability index of 3.5 was selected for ultimate limit states. This target reliability index was then used to determine performance factors. Performance factors for ultimate bearing capacity failure of footings are given in Table 4. Performance factors for sliding failure of footings on sand are given in Table 5. Table 6 shows performance factors for sliding failure of footings on clay.

The recommended values of performance factors are shown in the last column of Tables 4, 5 and 6. These values were selected based mainly on the results of the reliability-based calibration procedure. In those cases where insufficient statistical information was available or where inconsistencies were observed, judgement and experience was applied to arrive at an appropriate value.

Table 4 Performance Factors for Bearing Capacity Failure

Soil Type	Method	Type of Test	Performance Factors		
			Reliability Analysis	Fitting with ASD	Selected
Sand	Semi-empirical	SPT data	0.49 - 0.53	0.37	0.45
	Semi-empirical	CPT data	0.52 - 0.57	0.60	0.55
Clay	Rational theory	CPT data	0.46 - 0.74	0.60	0.50
	Rational theory	Lab UU test	0.60 - 0.85	0.60	0.60
	Rational theory	Field vane	0.60 - 0.97	0.60	0.60

Table 5 Performance Factors for Sliding Failure of Footings on Sand

Footing type	Method for determining ϕ_f	Performance Factors		
		Reliability analysis	Fitting with ASD	Selected
Cast-in-place	CPT data	0.79	0.85	0.80
Cast-in-place	SPT data	0.79	0.85	0.80
Precast	CPT data	0.87	0.85	0.90
Precast	SPT data	0.87	0.85	0.90

Table 6 Performance Factors for Sliding Failure of Footings on Clay

Method for determining S_u	Width (ft)	Performance Factors		
		Reliability analysis	Fitting with ASD	Selected
Lab UU tests	15 to 50	0.93 - 1.03	0.85	0.85
Field Vane tests	15 to 50	0.79 - 0.82	0.85	0.85
CPT data	15 to 50	0.82 - 0.97	0.85	0.80

CONCLUSIONS

The basic concepts and the calibration procedure for the new load factor design code for highway bridge foundations being developed for AASHTO were presented. The development of a load factor design for bridge foundations is expected to: a) create efficiency in design effort, through the use of the same factored loads for design of superstructures and foundations, b) result in more uniform margins of safety in the design of superstructures and foundations, and c) provide a more rational basis for selecting safety margins thus resulting in more economical use of materials. The study also shows that reliability based procedures are useful for dealing with geotechnical problems and illustrates the application of these procedures for the development of a reliability-based design code. However, the results of the reliability based calibration should be combined with engineering judgment and experience so as not to compromise on actual trends in practice.

ACKNOWLEDGEMENT

This work was sponsored by the American Association of State Highway and Transportation Officials, in cooperation with the Federal Highway Administration, and was conducted in the National Cooperative Highway Research Program (NCHRP) which is administered by the Transportation Research Board of the National Research Council.

REFERENCES

1. American Association of State Highway and Transportation Officials, Standard Specifications for Highway Bridges, Fourteenth Edition, AASHTO, Washington, DC, 1989.
2. American Concrete Institute, Building Code Requirements for Reinforced Concrete , ACI, Detroit, 1989.
3. American Institute for Steel Construction, Load and Resistance Factor Design Specification for Structural Steel Buildings, Chicago, 1986.
4. A.H-S. Ang, and C.A. Cornell, Reliability Basis of Structural Safety and Design, J. of Struct. Div., ASCE, Vol. 100, No. ST9, Sept. 1974, pp. 1755-1769.
5. R.M. Barker, J.M. Duncan, and K.B. Rojiani, Load Factor Design Criteria for Highway Structure Foundations, Final Report prepared for the National Cooperative Highway Research Program, Virginia Polytechnic Institute and State University, October, 1990.
6. P. Thoft-Christensen, and M.J. Baker, Structural Reliability Theory and Its Applications, Springer-Verlag, 1982.
7. C.A. Cornell, Structural Safety Specification Based on Second-Moment Reliability, Sym. Int. Assocaition of Bridge and Struct. Engr., London, 1969.
8. Danish Geotechnical Institute, Code of Practice for Foundation Engineering, 3rd Edition, DS 415, Bulletin No. 36, 1985, 53 pp.
9. M.I. Esrig, and R.C. Kirby, Advances in General Effective Stress Method for the Prediction of Axial Capacity for Driven Piles in Clay, Eleventh Annual Offshore Technology Conf., Proc. Paper No. OTC 3406, May 1979, pp. 437-449.
10. H.O. Madsen, S. Krek, and N.C. Lind, Methods of Structural Safety, Prentice Hall Inc., New Jersey, 1986.
11. National Building Code of Canda, National Research Council of Canada, Ottawa, Ontario, 1980.
12. Ontario Highway Department, Ontario Highway Bridge Design Code, 2nd Edition, Ministry of Transportation and Communication, Highway Engineering Division, Toronto, Ontario, 1983, 350 pp.
13. H.B. Seed, and P.De Alba, Use of SPT and CPT Tests for Evaluating the Liquefaction Resistance in Sands, Proc. In Situ '86, Blacksburg, Virginia, June 1986. pp. 281-302.
14. I.D. Sidi, Probabilistic Prediction of Friction Pile Capacities, PhD Thesis, Univ. of Illinois, Urbana-Champaign, Dept. of Civil Eng., 1986, 314 pp.
15. M.I. Tomlinson, Pile Design and Construction Practice, A Viewpoint Publication, London, 1987, 415 pp.
16. V.N. Vijayvergiya, and J.A. Focht, A New Way to Predict Capacity of Piles in Clay, Fourth Annual Offshore Technology Conference, Houston, Texas, Paper No. 1718, pp. 865-874.

THE DEVELOPMENT OF A LRFD CODE FOR ONTARIO BRIDGE FOUNDATIONS

R. Green, M.ASCE[1]

ABSTRACT: A new design procedure (LRFD) for bridge substructure foundations and retaining walls, documented in the Ontario Highway Bridge Design Code (OHBDC), is described. The changes made to factor of safety design to ensure compatibility between structural and geotechnical design procedures in LRFD are highlighted. LRFD aids in clarifying the calculation procedures at the interface between the soil and the structure. No new technical problems result for the geotechnical engineer. However,improved communication between geotechnical and structural engineers is required if the separation of service conditions and ultimate conditions in LRFD is to be successful.

INTRODUCTION

Load and Resistance Factor Design (LRFD) procedures for bridge superstructures and substructures were introduced in Canada in 1979 as part of the first Edition of the Ontario Highway Bridge Design Code (OHBDC1) (16). This new design method came about as a consequence of increases in legal truck loads during the 1970's. There was a need to be assured that the design and evaluation of structures included the changes in legal truck load and length.

The new Code (16) addressed the design of substructures and retaining walls, and the communication and coordination between and structural engineers. These new design procedures were not well accepted by geotechnical engineers nor were they completely successful. The reasons for the problems included the new terminology of LRFD, an incomplete understanding of LRFD concepts, and the questions about the codification of geotechnical design procedures. The perception within much of the geotechnical profession in Canada was that LRFD is based on statistical concepts. This reaction was unexpected as Dansk Standard DS 415 (5) had been in use for some time and is basically a simple rearrangement of factor of safety design (FSD) provisions.

The design process, problems of code writing, earth pressures and distribution, shallow foundations and deep foundations are discussed in this paper.

THE DESIGN PROCESS

Structural design, geotechnical design, and the design of the interface between structure and soil/rock have common requirements to ensure an acceptable level of

[1] - Professor, Department of Civil Engineering, University of Waterloo, Waterloo, Ontario,Canada,N2L3G1

reliability and to minimize loss of function. Uncertainty in design decisions occurs as a consequence of the variability of load effects, the variability in material characteristics and resistance predictions, the imperfections of analysis and an incomplete knowledge of the system being designed. There is little to distinguish structural design or evaluation from geotechnical design or evaluation, though there is a perception that structural design is an "exact" science and that geotechnical design is based on "experience" supported by science. Both design processes involve the recognition of uncertainty, and require sound judgment and the use of experience by the engineer of record.

The majority of design procedures used for foundations or for structures, or both, address one or more limit states. These limit states may be defined formally in a design specification or informally as part of an office procedure. Significant limit states are:

i) An ultimate limit state (ULS) where a failure mechanism forms in the soil/rock, or in a structure, or
ii) A serviceability limit state (SLS) where loss of serviceability occurs in a structure due to deformation of the soil/rock.

An ultimate limit state implies major loss of investment or life, and generally is not immediately nor easily repairable. The probability of an ULS occurring would be of the order of 10^{-3} to 10^{-5} (12). A serviceability limit state implies that the damage or loss of service is repairable with little capital expenditure. In the foundation design for a bridge, there may be one chance in 30 to 50 that differential or total settlement could lead to loss of ride quality. The collapse of a bridge, for example, may cause considerable economic loss and need for replacement. However, a loss of ride quality due to settlement of an approach or footing may be accepted for some time by the user or may be easily repaired with extra fill.

SAFETY CONSIDERATIONS

A system, for example a bridge superstructure or a pile foundation, with resistance, R, subject to a specified load effect, U, will be examined. Different resistance values, R, appropriate to the serviceability limit state, R_s, and ultimate limit state, R_u, will be used in design together with U values incorporating specified loads due to various combinations of vertical and horizontal loads. The provisions for safety will be illustrated for Load and Resistance Factor Design (LRFD), and for the traditional Factor of Safety Design (FSD).

Load Factor Resistance Design. When LRFD concepts are employed, specified loads are factored with a load factor appropriate to the level of uncertainty associated with a given specified load and limit state. Several load combinations are usually employed to determine the maximum destabilizing effect of load so as to maximize probable resistance demands. The design equations for serviceability (SLS) and strength (ULS) become

$$R_s > U \quad \text{(SLS)} \qquad [1]$$

$$I(\phi R_u) > \alpha U \quad \text{(ULS)} \qquad [2]$$

where

ϕ = a resistance factor

α = an average load factor associated with combinations of specified loads, U

R_s = a resistance based on a prescribed deformation

R_u = an ultimate resistance of soil or rock , or a structural component
I = a geotechnical factor applied to $øR_u$ to account for load inclination.

Equations [1] and [2] are equally applicable to structural design or geotechnical design. Different combinations of load may be used for the two processes at the same limit state. Equation [1] appears to be identical to factor of safety design except that R is limited to R_s, a resistance based on a prescribed deformation. Strength considerations are assigned to Equation [2]. A feature which arises with the ULS condition for geotechnical design is that the value of ultimate resistance is a function of angle of inclination of the particular load combination forming U.

In LRFD compatible values of R_s and R_u must be supplied by the geotechnical engineer. For example, for a medium sand and footing width of 4.0 m, R_s may be specified as 220 kPa for a vertical settlement of 25 mm and R_u as 2000 kPa for vertical load. A value of R_u equal to 2000 kPa for a 4 m wide footing may at first sight appear to be large but it is representative for a soil where the angle of internal friction approaches 35 degrees.

The uncertainties covered by the design equations, Equations [2] and [3], include:

i) the specified loads, both structural and geotechnical
ii) the method of analysis, both structural and geotechnical
iii) the geotechnical parameters and hence resistance for a given stratigraphy
iv) the variability in material properties and member resistance (structural).

Factor of Safety Design. For factor of safety design, the following applies:

$$R > U \qquad \text{(FSD)} \qquad [3]$$

where R is the lesser of R_s or $(I.R_u)/F$, and F is a factor of safety. For narrow footings on a granular soil, (R_u/F) will control the choice of R, while R_s will apply for wide footings. In Equation [3]. all uncertainty is assigned to one function, the factor of safety, F. When a single value is used to cover all uncertainties, there is little room for improvement if the uncertainty associated with any variable is reduced through changes in design knowledge or quality assurance.

Factors of safety used in the past vary from 1.3 for earthworks where the problem is almost completely geotechnical to 3.0 for foundations where both geotechnical and structural considerations apply and where loss of life may be a consideration (13). For earthworks and slope stability problems, calculated factors of safety at failure of 14 embankments varied from 1.0 to 1.8 indicating that problems of uncertainty in analysis and the selection of geotechnical parameters can exist (1).

RELIABILITY

OHBDC2 (17) includes specified permanent loads based on as-built observations of Ontario bridges, and specified live loads which are mean maximum loads based on surveys of existing truck traffic projected over a 50 year design life.

In general, predictions of the ultimate resistance of the soil, $øR_u$, may be obtained from one of the following:

a) Empirical Values
b) Assessed Values
c) Geotechnical Equations
d) Partial Coefficients or Soil Strength Factors
e) Reliability Based Resistances

Experience plus calculation are required for resistances predicted from a) and b), while for the calculation of c) to e), the traditional geotechnical parameters of unit weight, cohesion, and angle of internal friction are needed.

Empirical Values. ULS and SLS resistances are based on empirical relationships between resistance and some measure of the geotechnical parameters at a site, say SPT, cone penetration, or pressuremeter data (3). OHBDC3 (18) will permit the use of empirical methods and recommends a resistance factor of 0.5 for empirical bearing resistance values, even though the procedures may be method driven. Empirical values are not identical to presumed values available in many design handbooks. Presumed values appear to apply only to SLS and FSD and cannot easily be enhanced to apply to an ULS situation.

Assessed Values. Sites adjacent to the one of interest frequently have a similar stratigraphy. Data from completed investigations may be of value in an ongoing investigation. It is possible to use ultimate resistances from the completed investigation and an appropriate resistance factor for the new site, taken as 0.5 or 0.6 depending upon the situation, to assess values for the new site (18). An example of such is piles driven through a shallow overburden to a dense till where the resistance of the pile is primarily bearing and where extensive test data from similar sites are available.

Geotechnical Equations. For each design situation, there will be a suite of design equations. A geotechnical engineer will usually favor one or two equations for shallow or deep foundation design based on experience. Each equation will provide a different value of the mean of the ratio of observed to calculated resistance (E) based on an assessment of the geotechnical parameters assessed for the test site. For design:

$$\text{Calculated Factored Resistance} = \text{ø}.E.\ R_u \qquad [4]$$

where R_u is a function of the geotechnical parameters $ø_u$, $ø'$, c_u, c', and γ, and the geometry of the footing or piles. The geotechnical parameters will be assessed from the site investigation by the geotechnical engineer.

Partial Coefficients (Soil Strength Factors). Partial coefficient use for geotechnical design appears to be due to Hansen (8). These coefficients are not based on a reliability assessment of typical soils or rocks but rather are a rearrangement of the FSD values. This rearrangement permitted the separation of uncertainty due to geotechnical resistance (partial coefficients) and uncertainty due to load effects (load factors in LRFD). Dansk Standard DS 415 (5) provides values of partial coefficients for various safety classes. In OHBDC2 (17) partial coefficients are referred to as soil strength factors.

The factored soil strength parameters, c_f and $\tan ø_f$, in OHBDC2 (17) are given by:

a) Cohesion: $c_f = F_c c$ [5]

b) Internal friction: $\tan ø_f = F_ø \tan ø_u$ or $F_ø \tan ø'$ [6]

where F_c and $F_ø$ are soil strength factors for cohesion and friction respectively. The values of F_c are 0.65 and 0.50 for stability and earth pressure respectively, and $F_ø$ has a single value of 0.80 which applies to both earth pressure and resistance calculations. Factored soil strength parameters should be used directly in Equation [4] with the resistance factor (ø) taken as 1.0. If the site investigation is based on SPT or CPT data, these data are used to develop values of geotechnical parameters and hence soil strength parameters for subsequent use in a geotechnical equation. Many Ontario engineers using OHBDC2 (17) believed that this indirect treatment of geotechnical parameter data was a complication and preferred to obtain SLS and ULS bearing resistances directly from SPT values without added approximation.

Reliability Based Values. For the case of a major structure involving a high degree of risk, for example a bridge or an offshore structure, a comprehensive site investigation may be carried out to the extent that the geotechnical parameters for various strata can be described in terms of a mean and standard deviation. It is implied that such a site investigation will reduce uncertainty compared to a more traditional investigation. MacGregor (12) discusses calculation details for factored resistance.

Discussion of Selected Resistances. The selection of factored resistances will be a function of the quality of the site investigation used and the complexity of the site. Refined methods may be inappropriate for a site where the subsurface conditions are extremely variable and uncertain. OHBDC2 (17) recommended that soil strength factors be used for ULS values for shallow and deep foundations. When test data were available say for piles, a global factor of safety was applied in resistance assessments (17). Soil strength factors were not fully appreciated by many users and problems arose. Hence in OHBDC3 (18) these factors will dropped from use.

SHALLOW FOUNDATIONS

Ultimate bearing resistances may be based on SPT data, cone penetration test data, or if the geotechnical parameters are known by using bearing resistance (capacity) theories [14, 20]. Bowles provides data whereby the ratio of observed to calculated bearing resistance for shallow foundations can be calculated (3). A ratio of 1.20 is obtained for Terzaghi's method (20) and 0.86 for Meyerhof's method (14). When well accepted resistance calculation procedures are used, questions still arise as to the conservatism of the calculation procedure and the method to be used to select the 'best' geotechnical parameters.

For a LRFD format the geotechnical engineer should provide a range of ultimate bearing resistance values in terms of both footing width and embedment for a shallow foundation. Figure 1 provides calculated bearing resistance for various widths for an ideal footing founded on the surface of a soil with an average `N' value of about 20 to 25 with a low water table. All resistances increase with footing width, except for the SLS resistance, which is based on an assumed vertical deflection of 25 mm, and is approximately constant for various footing widths. The widths where there is a transition from strength to service are indicated by the points 'a' and 'b' in Figure 1. Ultimate resistances taken clockwise in Figure 1 are a calculated ultimate resistance (R_u), a factored resistance proposed in OHBDC3 ($øR_u$), a FSD resistance ($R_u/3.0$), and a factored resistance ($IøR_u$) including an inclination factor from OHBDC2 for angle 'i' equal to 21.4°. Items of note in Figure 1 include that the factored resistance for a footing width of 4 m is approximately five times the SLS value. Typical foundation reports quote a factored resistance which is frequently only 1.5 times the SLS value for a granular material. The value of 1.5 is believed to be based on a back calculation from an SLS value, assuming a value of F of 3.0 and applying a resistance factor of 0.5. The FSD values (Figure 1) do not include an effect for the inclination of load.

With the inclusion of the inclination factor in the ULS criterion (Equation 2), marked changes in bearing resistances result. For example, if reduction factors are used for an abutment wall where the ratio of a vertical to horizontal force ratio is 0.15 at ULS, a reduction in the calculated ultimate resistance of approximately 40 percent occurs for a granular material. For an earth retaining wall, a typical value of this ratio is 0.40 equivalent to 21.4°, and the ultimate resistance for inclined load is only

20 percent of that for vertical load. The reduction factors quoted above are for footings founded on the surface. An increase in bearing resistance occurs with footing embedment. An additional ULS resistance of about 700 kPa may be added to the values shown in Figure 1 if an embedment of 1.2 m is present - this is the frost cover depth used in much of Southern Ontario. This additional in resistance with embedment is well known and should be considered by the geotechnical engineer. Also the geotechnical report should offer the data which permits the structural engineer a choice between changing footing width or embedment during design.

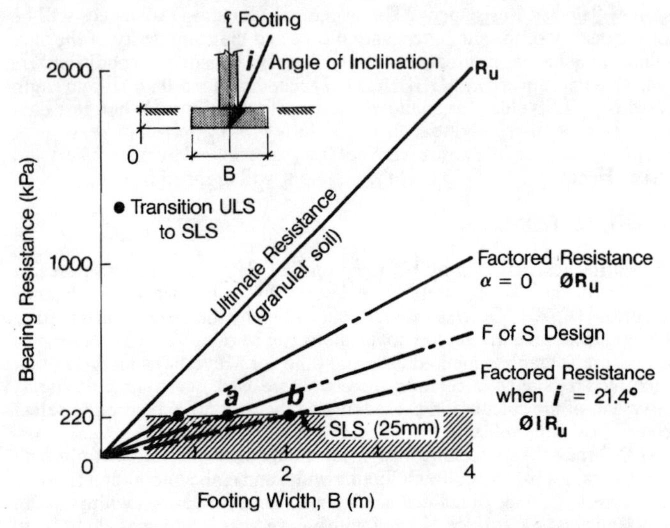

Figure 1 - Various Bearing Resistances and Footing Width

An appropriate reduction factor equation is essential in LRFD design. Few tests of inclined load on footings of some proportion, say 1.0 m by 3.0 m, are available. The data of Muhs and Weiss (16) were compared with the design proposals of Vesic` (21), Hansen (9), and Meyerhof (14). This comparison shows that all three theories provide conservative estimates of reduction factors for granular materials with $ø_u$ equal to about 38°. The ratio of observed to calculated reduction value was 1.4 with a standard deviation of about 0.05. The reduction factor equations of Meyerhof (14) are retained for design use, in graphical form, in OHBDC3 (18). A minimum embedment depth of 1.2 m is assumed for the charts. Where bearing resistances are calculated directly from geotechnical parameters, reduction factor expressions appropriate to the resistance equations should be used (18).

EARTH PRESSURE

It is common in the design of cantilever walls and abutments to use an equivalent fluid pressure representation for the earth pressure of free draining engineered backfill based on either Coulomb's or Rankine's method (2,3). An active pressure condition, K_a, is reached when the shear resistance of the retained material is mobilized for assumed displacements of 0.001 of the wall height, a base rotation of 0.002 or a combination of these. For a retained soil with an angle of internal friction of 30°, the horizontal pressure coefficient, K_a, will be 0.33.

When both the stem and base of the wall are unyielding during installation and compaction of the retained soil, lateral pressures in excess of at-rest pressures (K_o = 0.5, $ø_u$ = 30°) are possible (Figure 2a). The stems of most abutment walls and retaining walls translate or rotate as a consequence of the compaction of each layer of soil from the base to the top of the wall. These movements which reduce the 'locked in' compaction stresses lead to the lateral pressure distribution given in Figure 2b. Procedures for the calculation of the compaction pressures, which are not large for light compaction equipment, are given in (5) and (11). The force effects due to the pressure distributions given in Figure 2b for light compaction can be approximated using an equivalent fluid pressure, K_b, with a value which is midway between active and at-rest pressure.

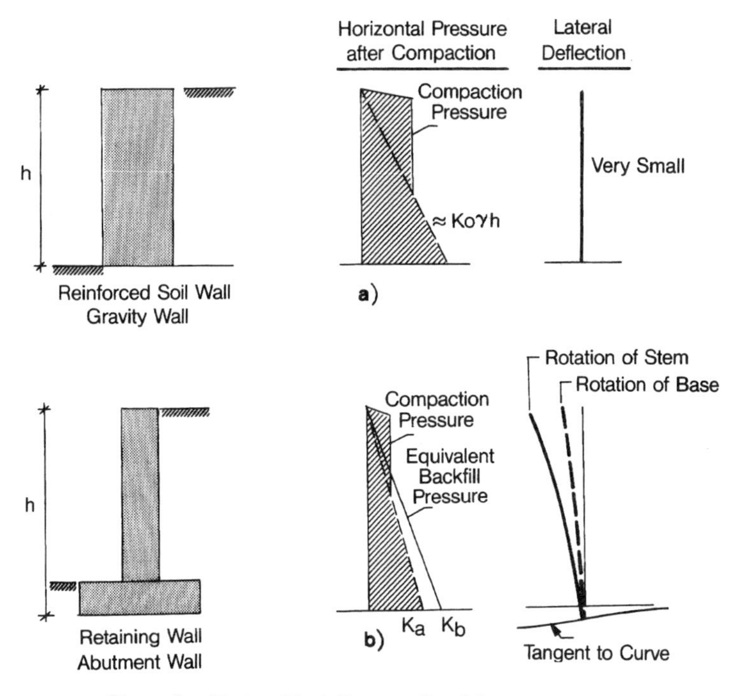

Figure 2 - Various Earth Pressure Conditions

Data from the MTO (6) indicate that $ø_u$ is between 35° and 46° for rock backfill and between 32° and 42° for a granular backfill suitable for highway embankments. Many engineers use a $ø_u$ value of 30° for granular backfill for engineered walls. This is conservative compared to Ontario observations and leads to excessive safety as lateral earth forces for an active condition are overestimated by 30 or more percent.

A number of pressure distributions (Figure 3) may exist following installation of the backfill. In Figure 3a, surcharge and active pressures are shown acting on the wall structure. The destabilizing effects of the earth forces are resisted by the soil beneath the footing base. This soil is assumed to have a factored resistance and to permit movements sufficient to cause active pressure K_a. Figure 3b illustrates the situation when the soil beneath the footing has not reached limiting equilibrium. For this case, bearing resistance is not a design problem and the earth pressures acting on the wall are those of the backfill pressure and any surcharge, K_b. This will occur when the bearing resistance (for soil or piles) present following compaction of the fill is competent, and the foundation does not deform following installation of the retained fill. The wall should be designed to resist the forces from both these pressure conditions, either K_a or K_b plus any surcharge. Thus two separate designs are necessary - one in which the base width is selected (active conditions control) and one in which the structural size of the wall is calculated (backfill pressure conditions control).

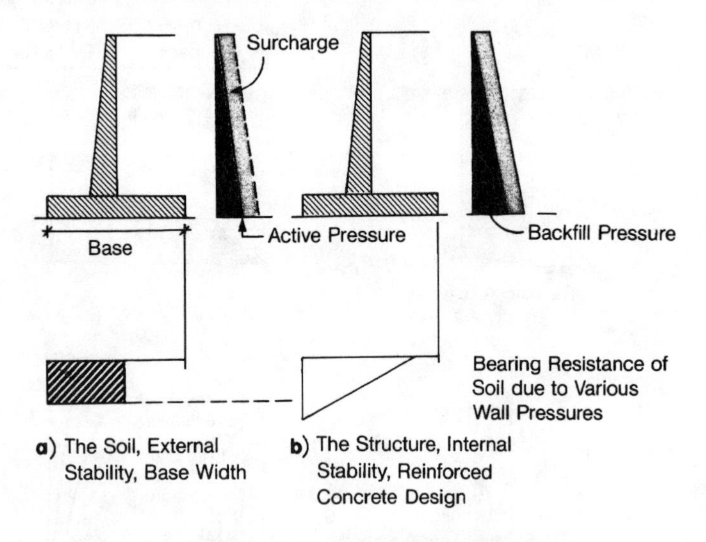

Figure 3 - Earth Pressures and Bearing Resistances

Danish design procedure (5) uses a load factor of 1.0 for all vertical and horizontal earth forces. This procedure was not followed in OHBDC1 (16) and OHBDC2 (17) where a load factor of 1.25 was applied to earth pressures that already included an allowance for uncertainty through the use of factors. Thus, the factored active and factored backfill pressures are numerically identical. This double counting of the safety provisions coupled with reduction factors for inclined load resulted in some footing widths typically 50 percent larger than might be obtained using FSD. These proportions were questioned by design engineers. OHBDC3 (18) attempts to rectify the 'double counting' by only using a load factor to handle uncertainty in active pressure calculation effects. The load factor chosen is 1.25. The uncertainty associated with the horizontal forces and moments due to lateral earth pressure can be handled by using soil strength factors or by load factors but not both.

DEEP FOUNDATIONS

The design of deep foundations requires a knowledge of the axial and lateral resistance of a pile, and an analysis procedure whereby the forces in a pile due to external actions can be calculated. The geotechnical engineer will normally supply values of ultimate resistance for axial load and may provide lateral resistance values.

Figure 4 shows typical load-deflection results for both vertical and horizontal load tests completed in Ontario. The data are for steel piles driven into fine sand (top 3 m) and silty clay. In Figure 4a the ULS and SLS values for vertical load are easily identified. The ULS value is associated with a limiting vertical deflection. The SLS value may be based on a limiting stress value or a limiting deflection value, based on controlling footing deflection or rotation, shown as 10 mm in Figure 4a. Not shown in Figure 4 is a SLS value based on downdrag effects and controlled by the structural resistance of the pile. Figure 4b is perhaps of more value to a designer. The horizontal load - horizontal displacement graph applies to a pile which does not fail as a rigid body when loaded laterally . In this figure, three main features relate to design. The first is an assumed SLS value based on a lateral movement of the pile of 10 mm. The remainder are ULS resistances, one of which is based on the soil passive resistance of the soil and the second is controlled by the structural resistance of the pile including moments due to lateral load and axial stresses. For the majority of simulations completed by the author using Reese's methods (19) the structural ULS resistance of a pile subjected to horizontal load and supported by the soil controls.

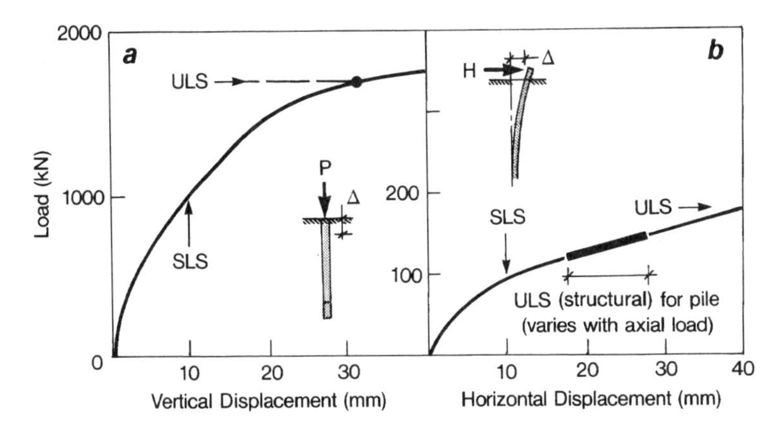

Figure 4 - Load and Deflection Test Data for Both Vertical and Horizontal Loading

There are a number of empirical expressions for the prediction of pile resistance. Many are empirical relationships. Briaud and Tucker (4) developed ratios of observed to calculated values for 98 pile tests using 13 methods of calculation. Several categories were considered including piles driven in sand or clay, and piles in layered soil. Of the thirteen methods, only three yield a ratio of observed to calculated greater than unity. This is not a problem as long as the appropriate value of E and the resistance factor (ø) are applied in Equation 4 for a given analytical method, or if a conservative choice is made of the geotechnical parameters used. For reliability based calculations the distributions of resistance appear to be log-normal but with large values of coefficient of variation frequently in excess of 0.5.

The structural engineer requires both simple and detailed methods for the analysis of pile foundations. Simple methods are required for both preliminary and final analyses, and soil-structure interaction solutions are available whereby final designs can be verified (19). There appears to be no universally accepted method of analysis given in design codes whereby the pile footings with the geometries given in Figure 5 can be analyzed simply so as to include the interaction between vertical and horizontal forces and resistances. OHBDC2 (1) provides a limiting equilibrium solution for the analysis of vertical load on a pile group but is silent as to how the analysis should be modified to include horizontal effects. The forces in individual piles in a pile group is a function of the applied axial load, the eccentricity of load and the horizontal load. It is unreasonable for a method of analysis not to consider all three load effects concurrently

A method of analysis which combines the interaction equation for the calculation of vertical pile forces due to eccentric load on a footing with a simple graphic statics solution to cater for the interaction of vertical and horizontal load is available (10). This procedure is perhaps the simplest of any offered for preliminary design purposes even though compatibility of deformation between the structure and the soil is not considered. An example is given in Figure 5 where one loading case is illustrated.

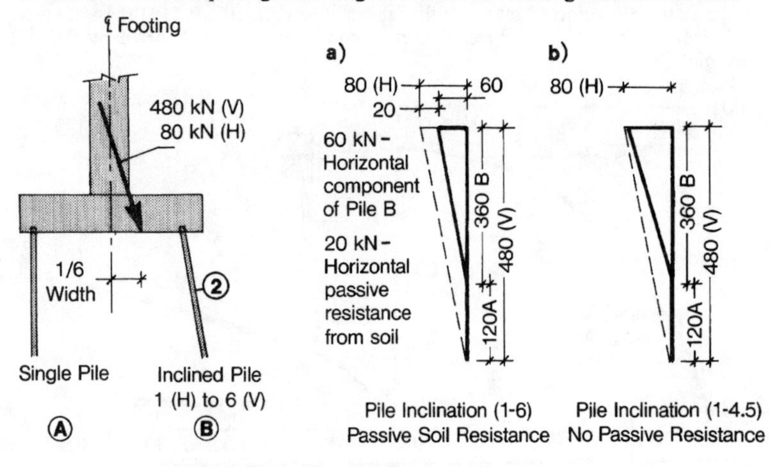

Figure 5 - Pile Force Analysis

The point of application of the vertical load is such as to give equal vertical loads in the single rear pile and each of the two inclined forward piles. For the loading cases and geometry shown, the two inclined (1 to 6) piles only resist 75 percent of the applied horizontal load of 80 kN. Horizontal passive resistance (20 kN) is required from the soil for equilibrium. If all horizontal resistance is assigned to the pile without assistance from the soil, then a pile inclination of 1 to 4.5 is required. This design will only be effective for a single load case. If the ratio of horizontal to vertical load changes then passive resistance is required of the soil. Conservatively chosen factored horizontal passive resistance values are required even if simple hand methods of analysis are used in the absence of the p-y compatibility conditions outlined by Reese (19).

This simple example suggests a simple method based on a LRFD approach for the analysis of pile footings. The method is

a). Select the most common SLS loading condition for the footing, typically dead load plus permanent horizontal load
b). Choose a pile arrangement giving equal axial load in all piles
c). Check this pile arrangement to ensure that all other SLS and ULS load combinations are satisfied.

This last check which might include up to 10 to 15 load combinations would include the passive horizontal resistance at the pile/soil interface specified by the geotechnical engineer.

DISCUSSION

LRFD requires the use of little or no new technology for either the structural or the geotechnical engineer. Site investigation procedures remain the same though it is anticipated that new technology for such investigations will be used in the future to reduce uncertainty regarding soil parameters. For complex structures, the need for a high level of investigation remains the same and the detailed results may provide geotechnical parameters of the quality and quantity necessary for statistically based calibrations and predictions of resistances.

A repackaging of the design information used for FSD design is required for LRFD as this latter procedure addresses SLS and ULS as two separate specific design states. The geotechnical engineer should no longer provide a single bearing value for shallow or deep foundations based on the more conservative of either SLS or ULS resistances. For components where soil-structure interaction is present, serviceability may control aspects involving the soil and ultimate strength may control structural design. Thus, different combinations of load may apply for the proportioning of the width of a footing (geotechnical) than for selecting the footing depth and reinforcing steel for the same footing. Such a design process is not new. It permits consideration of design for extreme loads and combinations of load. Some additional computational effort may be required. This is not a problem if design spread sheets are used.

LRFD demands of a more complete understanding of both the behavior of and the process of design for both soils and structures, than has been associated with FSD. The method leads to more complete designs and permits the use of new data in both design and evaluation.

CONCLUSIONS

LRFD is an appropriate procedure for resolving design problems where interaction between soils and structures is present. Designs evolve where either serviceability or ultimate will control the final design, thus providing a linkage between factor of safety design and ultimate strength design. No new technology is required for LRFD. However, a reassessment of current design processes is required.

ACKNOWLEDGEMENT

The author wishes to thank the Ministry of Transportation Ontario for an opportunity to work on various OHBDC Committees. The views expressed in this paper are those of the author and not of any sponsor.

REFERENCES

1. K Been, The Complexity of soils and its influence on working stress and limit states design. Symposium on Limits States Design in Foundation Engineering, The Can. Geo. Soc., Toronto (Mississauga), Ontario, May, 1989.
2. MD Bolton, Limit states design in geotechnical engineering. Ground Engineering, Vol. 14, No. 6, 1981.
3. JE Bowles, Foundation analysis and design. McGraw Hill, Inc., 1982.
4. J-L Briaud & LM Tucker, Measured and Predicted Axial Response of 98 Piles, J. Geo. Div. Vol. 114, No. 9, Sept, 1988, pp. 984-1001.
5. Dansk Ingenirforening, Code of Practice for Foundation Engineering, Dansk Standard DS 415, English Translation, Danish Geotechnical Institute, Copenhagen, Bulletin No. 36, 1985, 53 p.
6. M Devata, Geotechnical Data on Backfill, MTO, Downsview, Ontario (Personal Communiation), 1990
7. M Duncan et al., Load and Resistance Factor Design for Bridge Foundations. Symposium on Limits States Design in Foundation Engineering, Can. Geo. Soc., Toronto (Mississauga), Ontario, May, 1989.
8. J Brinch Hansen, Limit design and partial safety factors in soil mechanics. Danish Geotechnical Institute, Copenhagen, Bulletin No. 1, 1956, 4 p.
9. J Brinch Hansen, A Revised and Extended Formula for Bearing Capacity. Danish Geotechnical Institute, Copenhagen, Bulletin No. 28, 1970, 21 p.
10. WE Huntington, Earth pressure and retaining walls. J Wiley, 1957.
11. TS Ingold, The Effects of Compaction on Retaining Walls. Geotechnique, Vol. 29, pp. 265-284.
12. JG MacGregor, Safety and Limits States Design for Reinforced Concrete, Can. J. Civ. E., Vol. 3, No. 4, 1976, pp. 484-513.
13. GG Meyerhof, Limit states design in geotechnical engineering. Structural Safety,Vol. 1, No. 1, 1984, pp. 67-71.
14. GG Meyerhof, The ultimate bearing capacity of foundations. Geotechnique, Vol.2, No. 4, 1951.
15. H Muhs & K Weiss. Die Grenztragfähigkeit von flach gegründeten Streinfenfundamenten unter geneigter Belastung nach Theorie und Versuch, Berichte aus der Bauforschung, Heft 101, Mitteilungen der Degebo, Heft 31, Berlin, 1975.
16. OHBDC1 (Ontario Highway Bridge Design Code - 1st Ed). MTC, Downsview, Ontario, Two Volumes, 1979.
17. OHBDC2 (Ontario Highway Bridge Design Code - 2nd Ed). MTC, Downsview, Ontario, Two Volumes, 1983.
18. OHBDC3 (Ontario Highway Bridge Design Code - 3rd Ed). Ministry of Transportation Ontario, Downsview, Ontario, 1991, in preparation.
19. LC Reese, Behavior of piles and pile groups under lateral load. Report No. FHWA/RD-85/106, U.S. DOT, FHWA, Washington, D.C., March, 1986.
20 K Terzaghi, Theoretical Soil Mechanics. Wiley & Sons, Inc. New York, 1943.
21. AS Vesic`, Chap. 3: Bearing Capacity of Shallow Foundations, in Found. Engng.Handbook, Van Norstrand Reinhold Book Co., NY, 751 p.

SUBJECT INDEX

Page number refers to first page of paper.

AUTHOR INDEX

Page number refers to first page of paper.